ENERGY TRANSPORT INFRASTRUCTURE FOR A DECARBONIZED ECONOMY

ENERGY TRANSPORT INFRASTRUCTURE FOR A DECARBONIZED ECONOMY

Edited by

KLAUS BRUN
Ebara Elliott Energy, Jeannette, PA, United States

TIM ALLISON
Southwest Research Institute, San Antonio, TX, United States

RAINER KURZ
Solar Turbines Incorporated, San Diego, CA, United States

KARL WYGANT
Ebara Elliott Energy, Jeannette, PA, United States

Elsevier
Radarweg 29, PO Box 211, 1000 AE Amsterdam, Netherlands
125 London Wall, London EC2Y 5AS, United Kingdom
50 Hampshire Street, 5th Floor, Cambridge, MA 02139, United States

Copyright © 2025 Elsevier Inc. All rights are reserved, including those for text and data mining, AI training, and similar technologies.

Publisher's note: Elsevier takes a neutral position with respect to territorial disputes or jurisdictional claims in its published content, including in maps and institutional affiliations.

No part of this publication may be reproduced or transmitted in any form or by any means, electronic or mechanical, including photocopying, recording, or any information storage and retrieval system, without permission in writing from the publisher. Details on how to seek permission, further information about the Publisher's permissions policies and our arrangements with organizations such as the Copyright Clearance Center and the Copyright Licensing Agency, can be found at our website: www.elsevier.com/permissions.

This book and the individual contributions contained in it are protected under copyright by the Publisher (other than as may be noted herein).

Notices

Knowledge and best practice in this field are constantly changing. As new research and experience broaden our understanding, changes in research methods, professional practices, or medical treatment may become necessary.

Practitioners and researchers must always rely on their own experience and knowledge in evaluating and using any information, methods, compounds, or experiments described herein. In using such information or methods they should be mindful of their own safety and the safety of others, including parties for whom they have a professional responsibility.

To the fullest extent of the law, neither the Publisher nor the authors, contributors, or editors, assume any liability for any injury and/or damage to persons or property as a matter of products liability, negligence or otherwise, or from any use or operation of any methods, products, instructions, or ideas contained in the material herein.

ISBN: 978-0-443-21893-4

For information on all Elsevier publications
visit our website at https://www.elsevier.com/books-and-journals

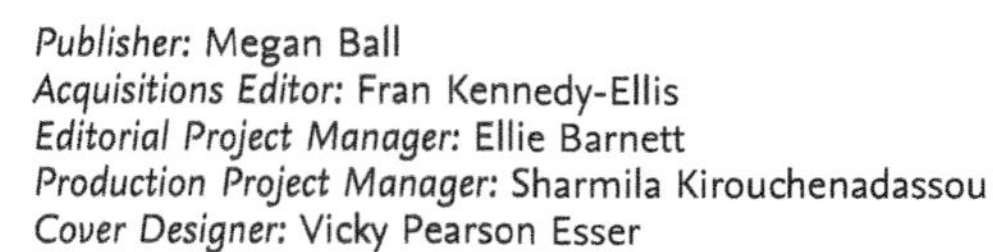
Publisher: Megan Ball
Acquisitions Editor: Fran Kennedy-Ellis
Editorial Project Manager: Ellie Barnett
Production Project Manager: Sharmila Kirouchenadassou
Cover Designer: Vicky Pearson Esser

Typeset by STRAIVE, India

Contents

Contributors

Mahmoud Abdellatif GASCO, Abu Dhabi, United Arab Emirates

Tim Allison Southwest Research Institute, San Antonio, TX, United States

Derrick Bauer Ebara Elliott Energy, Jeannette, PA, United States

Buddy Broerman Southwest Research Institute, San Antonio, TX, United States

Klaus Brun Ebara Elliott Energy, Jeannette, PA, United States

Jon Bygrave Hanwha Power Systems, Houston, TX, United States

Sebastian Freund Energyfreund Consulting, Munich, Germany

Gabe Glynn Atlas Copco, Stockholm, Sweden

Brian Hantz Ebara Elliott Energy, Jeannette, PA, United States

Shane Harvey Ebara Elliott Energy, Jeannette, PA, United States

Matthew Holland Southwest Research Institute, San Antonio, TX, United States

Justin Hollingsworth Southwest Research Institute, San Antonio, TX, United States

Rahul Iyer KCK Group, Cupertino, CA, United States

Enver Karakas Ebara Elliott Energy, Jeannette, PA, United States

Kelsi Katcher Southwest Research Institute, San Antonio, TX, United States

Terry Kreuz National Fuel Gas Company, Williamsville, NY, United States

Rainer Kurz Solar Turbines, San Diego, CA, United States

R.B. Laughlin Department of Physics, Stanford University, Stanford, CA, United States

David T. Sánchez Martínez University of Seville, Seville, Spain

Marybeth McBain Ebara Elliott Energy, Jeannette, PA, United States

Christoph Meyenberg Voith, Crailsheim, Germany

Michael Müller EagleBurgmann, Wolfratshausen, Bayern, Germany

Adam Neil Ebara Elliott Energy, Jeannette, PA, United States

Jordan Nielson Southwest Research Institute, San Antonio, TX, United States

Todd Omatick Ebara Elliott Energy, Jeannette, PA, United States

Kaartik Palaniappan fIndependent Consultant, Kuala Lumpur, Malaysia

Guillermo Paniagua-Perez Purdue University, West Lafayette, IN, United States

Robert Pelton Ebara Elliott Energy, Jeannette, PA, United States

Brian Pettinato Ebara Elliott Energy, Jeannette, PA, United States

Ramees K. Rahman University of Central Florida, Orlando, FL, United States

Peter Renzi Engineering Design Group, LLC, Buffalo, WY, United States

Brandon L. Ridens Southwest Research Institute, San Antonio, TX, United States

Stephen Ross Ebara Elliott Energy, Jeannette, PA, United States

Sterling Scavo-Fulk Ebara Elliott Energy, Jeannette, PA, United States

Avneet Singh Solar Turbines, San Diego, CA, United States

Natalie R. Smith Southwest Research Institute, San Antonio, TX, United States

Subith Vasu Sumathi University of Central Florida, Orlando, FL, United States

Kevin Supak Southwest Research Institute, San Antonio, TX, United States

Matt Taher Bechtel Energy Inc., Houston, TX, United States

James Underwood Solar Turbines, San Diego, CA, United States

Subith Vasu Sumathi University of Central Florida, Orlando, FL, United States

Jason Wilkes Southwest Research Institute, San Antonio, TX, United States

Bernhard Winkelmann S&B Enterprises, La Jolla, CA, United States

Karl Wygant Ebara Elliott Energy, Jeannette, PA, United States

About the editors

Klaus Brun is Global Director of Research and Development at Ebara Elliott Energy, where he leads a group of more than 60 professionals in the development of turbomachinery and related systems for the energy industry. His past experience includes positions in product development, applications engineering, project management, and executive management at Southwest Research Institute, Solar Turbines, General Electric, and Alstom. He holds 15 patents, has authored more than 400 papers, and has published 7 textbooks on energy systems and turbomachinery. Dr. Brun is a fellow of the American Society of Mechanical Engineers (ASME) and won an R&D 100 Award in 2007 for his semi-active valve invention. He also won the ASME Industrial Gas Turbine Award in 2016 and 14 ASME Best Paper/Tutorial Awards. Dr. Brun has chaired several large conferences, including the ASME Turbo Expo and the Supercritical CO_2 Power Cycles Symposium. Dr. Brun is a member of the Global Power Propulsion Society Board of Directors and the past chair of the ASME International Gas Turbine Institute Board of Directors, the ASME Oil & Gas Applications Committee, and the ASME sCO_2 Power Cycle Committee. Dr. Brun founded and chaired the annual sCO_2, TMCES, and GEMS workshops. He is also a member of the API 616 Task Force, the ASME PTC-10 Task Force, the Asia Turbomachinery Symposium Committee, and the Supercritical CO_2 Symposium Advisory Committee. Dr. Brun is currently the executive correspondent of *Turbomachinery International* magazine, a frequent contributor to engineering magazines, and an associate editor of several journal transactions.

Tim Allison is Machinery Department Director at Southwest Research Institute, where he leads an organization that focuses on R&D for the oil and gas, propulsion, and energy industries. His research experience includes analysis, fabrication, and testing of turbomachinery and systems for advanced power generation, industrial, and oil and gas pipeline applications, including high-pressure turbomachinery, centrifugal compressors, expanders, gas turbines, reciprocating compressors, and test rigs for bearings, seals, blade dynamics, and aerodynamic performance. Dr. Allison holds 3 patents, has authored 4 book chapters, edited 2 books, and has published more than 75 papers and articles on various turbomachinery topics. He received the best tutorial/paper awards from the ASME Turbo Expo Oil & Gas and Supercritical CO_2 Power Cycle Committees in 2010, 2014, 2015, and 2018. He has chaired both of those committees at ASME Turbo Expo, the Supercritical CO_2 Power Cycles Symposium, and the TMCES Workshop, and founded and chairs the IPER workshop. He is an advisory board member for the Global Power and Propulsion Society and an associate editor for the *ASME Journal of Engineering for Gas Turbines and Power*.

Rainer Kurz is the Manager, Gas Compressor Engineering, at Solar Turbines Incorporated in San Diego, California. His organization is responsible for the design, research, and development of Solar's centrifugal gas compressors, including aerodynamic, rotordynamic, and mechanical design. Dr. Kurz attended the Universitaet der Bundeswehr in Hamburg, Germany, where he received the degree of a Dr.-Ing. in 1991. He has authored numerous publications as well as several books on turbomachinery-related topics, is an ASME fellow, and won the ASME Industrial Gas Turbine Award in 2013. Dr. Kurz is a member of the Turbomachinery Symposium Advisory Committee, the ASME Oil and Gas Applications Committee, the GMRC Board of Directors, the GTEN Committee, and the GPPS Executive Committee.

Karl Wygant is a professional in the field of advanced turbomachinery technology, where he currently serves as the senior manager of Advanced Technology Programs at Ebara Elliott Energy. In this role, he leads hydrogen and carbon dioxide compressor development initiatives.

He previously held the position of chief operating officer at Hanwha Power Systems Americas, a turbomachinery OEM specializing in integrally geared compressors. During his tenure, he oversaw all aspects of business operations throughout North and South America. In this role, he led new product development within the integrally geared products sector. These products included a newly engineered product line, high-temperature blowers, and DNV/ABS-certified products for LNG carriers. This expertise spans a wide array of turbomachinery applications, including LNG, plant air, air separation, and oil and gas applications. Furthermore, he led the design, development, and commercial launch of a 5-MW supercritical carbon dioxide power system tailored for concentrated power and heat recovery applications.

Dr. Wygant has published more than 30 technical articles and has an educational background, including a PhD from the University of Virginia in mechanical and aerospace engineering. In addition, he holds an MBA from Norwich University, rounding out his comprehensive skill set and demonstrating his commitment to both academic and professional excellence.

Foreword

Energy transport has been fundamental to humanity from our earliest roots and continues to be an issue of monumental importance today. Energy-dense fuels and nutrition were a matter of survival for the early hunter-gatherer cultures of humankind, and challenges with energy transport were a driving factor in the growth of civilization around localized sources of nutrition (agriculture) and other energy/fuel sources (water, heat, wood). In the modern world, global per capita energy consumption is thousands of times that of early civilizations [1,2], and energy drives our economies and civilization. Energy availability remains a matter of critical global importance, and energy transport is therefore a key aspect of trade and energy security.

The steam engine is often recognized as a primary energy enabler of the Industrial Revolution. While this recognition is appropriate, it is also necessary to credit the availability of coal as an easily transportable and storable energy source that fueled the engines and furnaces that advanced civilizations' standard of living [3]. Coal, transported initially in wagons or carts, then in large quantities by canal, and later by rail and ship, brought massive amounts of energy to bear, rapidly accelerating our ability to mechanize manufacturing in the early 1800s. In the late 19th century, large-scale electrification, initially from coal, powered manufacturing development at scale. These improvements enabled mass production, lighting, and many other technological improvements that began to increase the standard and quality of living for civilization for the first time since the development of agriculture and domestication.

The cost-effective production of iron and steel allowed the development and implementation of pipelines, initially for coal-derived oil and gas products in 1843 but then for crude oil following its first production in the 1860s [4]. Oil is more energy dense than coal and can be refined into diesel and gasoline, and liquid pipeline growth paralleled the boom in automotive transportation utilizing these fuels. Continued technological advancements in pipeline manufacturing that improved reliability and reduced leakage enabled the expansion of the natural gas infrastructure, motivated by the low cost and cleaner combustion of natural gas. This expansion was also enabled by the development and improvement of compression technologies, energy demand, and favorable legislation, resulting in a significant buildup in the 1950s and 1960s after World War II.

Energy transport technology was revolutionized again through the development and introduction of liquefied natural gas (LNG) processes and tankers beginning in the 1950s. The boom in hydraulic fracturing paved the way for low-cost oil and gas sources in the early 2000s and was accompanied by a monumental expansion of LNG capabilities. Although LNG transport is more energy intensive than pipeline transport, the favorable economics of LNG shipments vs. pipelines, coupled with the geographic flexibility of a tanker, continue to motivate the expansion of this infrastructure. The world's recent attention to carbon emissions has also buoyed gas demand via pipelines and LNG due to natural gas' lower carbon emissions vs. coal and oil.

In addition to the generation of electric power and to heating applications, energy transport technoeconomics are a driver of technologies for the transportation sector. Air and ground transportation applications drive energy transport requirements to their near extremes due to their needs for high power density, scalability, and compatibility with an extremely broad user base. Automotive and aircraft applications have long relied upon the high energy densities of hydrocarbon fuels, but efforts to reduce carbon emissions are driving the development and commercial adoption of electrochemical batteries, hydrogen, biofuels, compressed natural gas, and other forms of energy transport. Space propulsion is the most extreme application of energy transport due to the high amount of energy required, direct impact of weight on mission success, and availability of significant government funding for space efforts. Although a variety of solid and liquid rocket fuels exist, most rockets have liquid stages due to their better controllability and specific impulse, incorporating onboard fuels such as kerosene, liquid methane, or liquid hydrogen paired with liquid oxygen as an oxidizer.

The history and range of energy transport applications highlight the many important physical properties of a substance that affect its feasibility as an energy carrier. Energy density (both by mass and volume) is of obvious importance, but mass density, transport pressure, material compatibility, propensity for leakage, and parasitic cooling requirements are all important factors that affect the efficiency, cost, and reliability of energy transport. Due to the large scale of many energy transport systems and their interface with large population centers, safety considerations are of importance, including flammability limits in air, need for active cooling, leak dispersion, toxicity, and other factors. Finally, energy demands and the cost of conversion to and from an energy carrier will affect the overall cost and environmental footprint of a system and will be strong determinants of success for various technologies.

With a wide range of energy carriers comes a wide range of technologies that transport energy in the form of chemicals (solid/liquid/gas), heat, or electrons. While all have success in various markets and applications, energy efficiency, security of supply, and cost are of fundamental importance to technoeconomic success. Gas and liquid pipelines have been shown to be 1 and 2 orders of magnitude less expensive (cost per MW-mile) than high-voltage transmission lines while also having 1 and 2 orders of magnitude lower losses, respectively [5–7], resulting in systems that generally transport energy via pipeline for as much distance as possible and convert to electricity fairly close to the point of use. For example, the United States has approximately 3 million miles of oil [8] and gas pipelines compared to approximately 160,000 miles of electric transmission lines [9]. Based on the relative magnitude of pipeline vs. electrical transport and also the expertise of the editor and author teams, electrical transport is not addressed as a specific topic in this book.

Past instances of energy transport disruption underscore the critical significance of energy distribution. The repercussions of these interruptions reverberate through society, revealing the intricate web of dependencies on consistent energy flow. One notable case was in 1965, when a massive power outage swept across the Northeastern United States, affecting 30 million people. While our focus in this book may not center on electrical grids, this historical occurrence clearly demonstrates the far-reaching implications of energy transport challenges. A more recent illustration of such consequences is the Nord Stream pipelines that link Russia to Germany. In 2022, these pipelines suffered substantial damage, disrupting the flow of natural gas—a source constituting 35% of Europe's energy supply [10]. Beyond the ecological repercussions, the aftermath of such a significant energy transport disturbance extends to the global domains of shipping, production, and transportation in order to prevent a humanitarian crisis.

Energy transport cannot be viewed accurately without understanding the storage and conversion details of the end application. Energy conversion imposes losses, so efficient energy systems will generally minimize conversion between chemical, heat, and electrical forms of energy. Most power plants generally produce both electricity and heat, so combined heat and power systems may directly benefit from district heat networks transporting heat directly as hot water or steam. Storage cost and efficiency may drive a choice of energy carrier, and storage can also play a key role in minimizing peak transmission capacity requirements.

The decarbonization and growth of energy demand set the stage for the exploration and potentially dramatic reinvention of energy transport infrastructure. Zero-carbon energy carriers, including hydrogen and ammonia, are being evaluated and funded for economy-wide decarbonization, presenting a host of technical and economic challenges and opportunities for conversion, transport, and storage technologies. The transport of carbon dioxide (although not an energy carrier itself) enables the continued use of existing infrastructure and technology for hydrocarbon-based energy transport. The wide variety of applications that require energy transport is likely to result in a variety of new systems as the world seeks to maintain a reliable, affordable, and sustainable energy system in the future.

We, the editors, are indebted to the contributions from many authors to various chapters throughout this book. We are grateful to all of you for your knowledge and time that resulted in this impactful book! The book is divided into multiple chapters. Chapters 1 and 2 provide an introduction to the topic and an overview of the energy marketplace, respectively. The fundamentals of pipeline transport and compression are described in Chapter 3, and an overview of various energy applications is provided in Chapter 4. Chapters 5–10 are focused on the transport characteristics of various energy carriers and byproducts, including liquefied natural gas, natural gas, hydrogen, carbon dioxide, district heating, and various gas-to-liquid or other decarbonized energy carriers. Finally, Chapters 11 and 12 provide perspectives on future trends, research and development activities, and research facilities for enabling energy transport technology.

Sincerely,
Klaus Brun,
Tim Allison,
Rainer Kurz, and
Karl Wygant

References

[1] V. Smil, World history and energy, in: Encyclopedia of Energy, Vol. 6, Elsevier, 2004.

[2] A. Courtney, Historical perspectives of energy consumption, lecture notes, in: Energy and Resources in Perspective GS 361, Western Oregon University, 2005. https://people.wou.edu/~courtna/GS361/electricity%20generation/HistoricalPerspectives.htm. (Accessed 14 August 2023).

[3] J.A. Tainter, T.G. Taylor, Energy, transport, and consumption in the industrial revolution, in: Behavioral and Brain Sciences, Vol. 42, Cambridge University Press, 2019.
[4] P. Hopkins, Pipelines: past, present, and future, in: Proceedings of the 5th Asian Pacific IIW International Congress, Sydney, Australia, 2007.
[5] K. Brun, The energy infrastructure of the future, in: Proceedings of Global Power and Propulsion Forum, Zurich, Switzerland, 2020.
[6] T.C. Allison, J. Klaerner, S. Cich, R. Kurz, M. McBain, Power and compression analysis of power-to-gas implementations in natural gas pipelines with up to 100% hydrogen concentration, in: Proceedings of ASME Turbo Expo 2021 GT2021-59398, Virtual, Online, 2021.
[7] B.D. James, D.A. DeSantis, J.M. Huya-Kouadio, C. Houchins, G. Saur, Analysis of Advanced H2 Production & Delivery Pathways, U.S. Department of Energy Project ID:P102, 2019. https://www.hydrogen.energy.gov/pdfs/review19/p102_james_2019_p.pdf. (Accessed 14 August 2023).
[8] U.S. Energy Information Administration, Natural Gas Explained: Natural Gas Pipelines, 2022. https://www.eia.gov/energyexplained/natural-gas/natural-gas-pipelines.php. (Accessed 14 August 2023).
[9] U.S. Energy Information Administration, U.S. Electric System is Made up of Interconnections and Balancing Authorities, 2016. https://www.eia.gov/todayinenergy/detail.php?id=27152. (Accessed 14 August 2023).
[10] S. Fleming, International Impacts of the Nord Stream Leaks, 2022. https://globaledge.msu.edu/blog/post/57169/international-impacts-of-the-nord-stream.

Acknowledgments

We would like to thank Rachel Pyle, Brian Hantz, Nicholas Bishop, Leslie Solis-Justice, and Herminia Mares for their tireless efforts and assistance while putting this book together.

Nomenclature

a	speed of sound
A	amplitude
AF	amplification factor
b	impeller exit width
C	flow heat capacity ($\dot{m}c_P$)
c	flow velocity in absolute reference frame
c	specific heat capacity for a solid or incompressible fluid
$C_{\text{effective}}$	effective damping $= C_{xx} - K_{xy}/\omega$
c_v, c_P	specific heat capacity at constant volume and specific heat at constant pressure, respectively
C_{xx}, C_{yy}	direct damping coefficient
C_{xy}, C_{yx}	cross-coupled damping coefficient
D	diameter
D_2	impeller tip diameter
e	energy per unit mass
E	elastic modulus (Young's modulus)
f	frequency
G	shear modulus (modulus of rigidity)
h	enthalpy
$\hat{h}$	convective heat transfer film coefficient
h	height
H	head
i, I	irreversibility per unit mass and total irreversibility (entropy), respectively
I	electrical current
J	polar moment of inertia
k	isentropic exponent
k_{xx}, k_{yy}	direct stiffness coefficient
k_{xy}, k_{yx}	cross-coupled stiffness coefficient
L	length
m	mass
$\dot{m}$	mass flow rate
Ma	Mach number
MW	molecular weight
n	polytropic exponent
n_s	isentropic volume exponent
N	rotational speed
N_c	critical speed
Nu	Nusselt number
P	power
p	pressure
pe	potential energy per unit mass (gz, where z represents elevation)
ρ	density
Pr	Prandtl number of the fluid
q, Q	heat transfer per unit mass, total heat transfer
Q	volume flow rate
r_v	specific volume ratio
R	gas constant for a specific gas
$\overline{R}$	universal gas constant
Re	Reynolds number
SQ	std. flow
s	entropy
T	temperature
u	internal energy per unit mass
$\hat{U}$	overall heat transfer coefficient
U_2	impeller tip speed
V	voltage
v	flow velocity in stationary reference frame
v_d	specific volume at discharge

v_i	specific volume at inlet
w	flow velocity in rotating reference frame
W	weight
W	work
x	fluid quality
X	reactive impedance
Z	compressibility factor
Z	total impedance
ΔP	pressure drop
ε	heat exchanger effectiveness
Φ	exergy
η	efficiency
ρ	density
ψ	head coefficient
μ	absolute (dynamic) viscosity
φ	phase angle
ϕ	flow coefficient
υ	kinematic viscosity
Γ	torque
θ	angular displacement
δ	ratio of specific heats (c_P/c_v), fluid thermal conductivity
α	absolute flow angle
β	relative flow angle
γ	specific heat ratio

Abbreviations

ke	kinetic energy per unit mass ($v^2/2$, where v represents velocity)
CHP	combined heat and power
CSR	critical speed ratio
EGS	enhanced geothermal system
FFT	fast Fourier transfer
HP	horsepower
HVAC	heating, ventilation, and cooling
J-T	Joule Thomson
LNG	liquefied natural gas
MCOS	maximum continuous operating speed
MR	mixed refrigerant
MTPA	million tons per annum
PF	power factor
TEM	thrust equalizing mechanism
TES	thermal energy storage
VFD	variable frequency drive
VSHD	variable speed hydrodynamic drive

Subscripts

1, 2, 3	property at defined point
I, II	first law (or energy) and second law (or exergy) basis, respectively
C, H	heat exchanger cold and hot fluids, respectively
C, T	compressor, turbine, respectively
f	saturated liquid
fg	difference in property for vaporization from liquid to vapor
g	saturated vapor
H	heat source
o	dead state
p	polytropic
r	rejected heat
R, S	heat rejected and supplied, respectively
S	state point that would be reached in an isentropic process
S	isentropic
th	thermal efficiency (refers to energy transformations within the working fluid)

Over dot

$\dot{\square}$	rate or time derivative
$\longrightarrow$	vector
$\sim$	matrix

C H A P T E R

1

One hundred trillion watts

R.B. Laughlin

Department of Physics, Stanford University, Stanford, CA, United States

Human civilization requires an immense amount of power to function. This is reflected, for example, in the light of cities radiated out into space at night, as shown in Fig. 1.1, but it is also familiar from everyday things such as vehicle fuel consumption burdens, heating and cooling bills, lighting, food supplies, water, and so forth [1]. The 2022 world primary energy consumption of 6.0×10^{20} Joules per year (600 EJ y^{-1}) or 1.9×10^{13} Watts amounts to 2400 W, about one electric toaster, for each of the 8 billion people presently living on Earth [2,3]. The consumption is larger, between 5000 W and 20,000 W per person, in wealthy OCED countries and smaller, 500 W or less, in poorer countries. The larger number is the more relevant to future civilization because people in poor countries aspire to become less poor.

In 2021, 82% of the energy of civilization came from the burning of fossil fuels, chiefly coal, oil, and natural gas. As shown in Fig. 1.2, this has remained stably the case for the last two decades, declining only 5% since 2000. All three of these fuels contain carbon and thus put CO_2 into the environment when consumed. The exact number of joules generated per mass of CO_2 emitted varies slightly by fuel, as shown in Table 1.1. Taking the oil value as a representative average for the fuels, we find that civilization dumped 3.7×10^{13} kg y^{-1} of CO_2 into the environment in 2021. The total mass of CO_2 presently in the atmosphere is measured by sea level pressure to be about 3.4×10^{15} kg, so the burning of fossil fuels makes a 1% perturbation to the atmospheric CO_2 each year. About half of this shows up in measured atmospheric concentration increases, with the rest presumably going into the oceans [4].

This alarming increase in environmental CO_2 and its potential to change the world climate for the worse through greenhouse effect heating have led to the ambition of decarbonizing the world economy [5]. Decarbonization is certain to happen on the scale of centuries because the supplies of fossil fuels will eventually exhaust, but the situation in the near term is more problematic. Obtaining energy by burning the cheapest available fuel is built into the working of the world economy, so it is very hard to discourage. Rapid decarbonization would greatly lessen the cumulative damage to the earth and is thus desirable.

An example of what can happen quickly to mitigate the CO_2 burden on the earth is shown in Fig. 1.3. For a number of reasons, including both political pressure and competition from natural gas interests, coal consumption in North America and Europe has declined drastically since 2005. The energy that had been previously provided by this coal must be coming from somewhere else, and in both cases, it is coming from a combination of increased natural gas consumption and renewables. The resulting lightening of the CO_2 burden on the earth has been substantial. Even assuming that the decline is made up entirely of natural gas burning, the CO_2 emission reduction from this effect is 25% of that of all the renewable energy that came into being over the same time, as shown in Fig. 1.1. This is because generating a joule of electric energy by burning coal produces twice the CO_2 as does generate the same joule by burning natural gas.

When the transition away from fossil fuels eventually completes, it will still be necessary to handle physical fuels analogous to coal, oil, and natural gas even if they have to be manufactured. This is because (1) energy can be safely stored over the scale of months only in chemical form, (2) some applications such as aviation require portable energy sources, and (3) transfer of energy by electric transmission line over large distances, such as between continents, is comparatively expensive. Examples of synthetic fuels presently being considered for this purpose include electrolytic hydrogen, ammonia made from this hydrogen, and synthetic methane. All of these are economic proxies for natural gas.

Energy Transport Infrastructure for a Decarbonized Economy
https://doi.org/10.1016/B978-0-443-21893-4.00014-3

Copyright © 2025 Elsevier Inc. All rights are reserved, including those for text and data mining, AI training, and similar technologies.

FIG. 1.1 Lights of earth as seen from space. The fraction of the energy budget of civilization this represents is not known exactly but is typically estimated at about 0.1% of the total, or 2×10^{10} Watts in 2022. The remaining 2×10^{13} Watts consumed is discarded as heat after use and radiated away in the infrared. *Image Credit: Craig Mayhew and Robert Mimmon NASA GSFC from data provided by Marc Imhoff of NASA FSFC and Christopher Elvidge of NOAA NGDC. Data source: U.S. Defense Meteorological Satellite Program (DMSP) Operational Linescan System (OLS). Courtesy of NASA.*

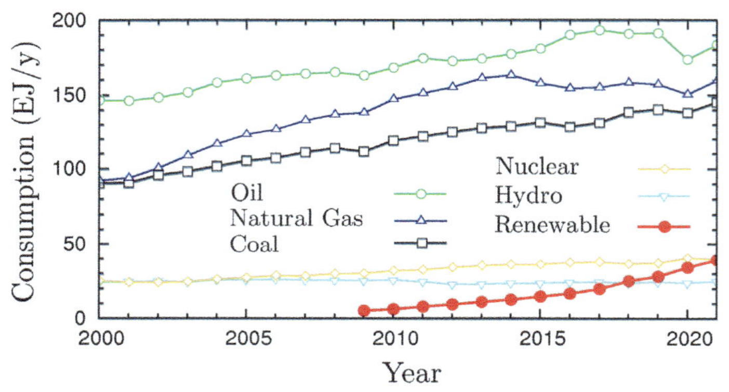

FIG. 1.2 World energy consumption broken down as to source [2]. The masses of fuels burned have been converted to exajoules (10^{18} Joules) using the factors described in Table 1.1. The renewable energy figure is not the total energy delivered but the amount of fossil energy supplanted by the renewables [2]. It is specifically the total electric energy generated from the wind and sun divided by an equivalent power plant thermal efficiency factor of about 0.36, then added to the unadjusted biofuel production (3.9 EJ in 2021) [2]. The nuclear and hydro energy figures have been similarly adjusted for power plant efficiency [2]. *Image Credit: R. B. Laughlin.*

TABLE 1.1 Heat contents of fossil fuels and the energy obtained per kg of CO_2 released into the environment in each case.

Fuel	Energy content	Mass ratio	CO_2 burden
Coal	3.28×10^7 J kg^{-1}	12/44	0.89×10^7 J kg^{-1} CO_2
Oil	4.19×10^7 J kg^{-1}	14/44	1.33×10^7 J kg^{-1} CO_2
Natural gas	4.94×10^7 J kg^{-1}	16/44	1.80×10^7 J kg^{-1} CO_2

Fuels vary in composition, so these values should be viewed as the standards for conversion used in this document. The energy content of coal is picked to be that of anthracite because this gives the proper calibration with the CO_2 produced. The actual heat contents of low-grade coals are 2.0×10^7 J kg^{-1} or less [2]. The oil conversion number is the standard kilogram of oil equivalent 4.1868×10^7 J [2]. The natural gas conversion number is slightly larger than the BP convention of 4.898×10^7 J kg^{-1} [2]. For comparison, the heat of combustion of pure methane is 5.57×10^7 J kg^{-1}.

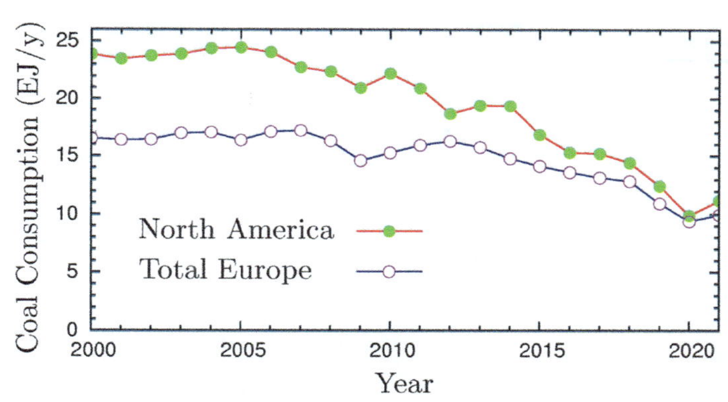

FIG. 1.3 History of coal consumption in North America and Europe [2]. The conversion from tonnes of coal to exajoules (10^{18} Joules) varies by grade, ranging from 1.5×10^7 J kg^{-1} to 2.5×10^7 J kg^{-1} [2]. North America is defined to be Canada, the United States, and Mexico [2]. Total Europe is defined to be all of Western and Eastern Europe including Turkey and Ukraine but not including the Russian Federation or any other former Soviet state. Values for years before 2009 were converted from TOE (tonnes of oil equivalent) at 1 TOE = 1.4868×10^7 Joules. *Image Credit: R. B. Laughlin.*

FIG. 1.4 Mount Vesuvius in Italy, as seen from Naples. At an energy density of 2.0×10^7 J kg^{-1}, the 160 EJ (1.60×0^{20} J) of coal mined and burned in 2021 would make a pile roughly the size of this mountain. Measured as a cryogenic liquid, the volume of natural gas consumed would be about the same, as would the amount of crude oil consumed [2]. All of this fuel presently needs to be transported, much of it intercontinentally. This transportation system is expected to still be needed in the distant future when fossil fuels are no longer used. *Image Credit: Massimo Finizio via Wikimedia Commons.*

The task of transporting all this fuel from its point of manufacture to its point of consumption will remain formidable, even in the future time when the fuels being transported are synthetic. The energy density of synthetic fuels is constrained by chemistry fundamentals to be roughly the same as that of fossil fuels, or perhaps slightly less. The mass of synthetic fuel requiring transport will thus roughly equal the mass of fossil fuel requiring transport today.

Fig. 1.4 illustrates just how large a task this will be. The amount of coal presently mined and burned every year would, if heaped into a pile, be about the size of Mount Vesuvius [2]. The amounts of natural gas (measured in liquefied form) and oil consumed every year are comparable to this. Virtually all of this fuel has to be transported by ship, rail, or pipeline from its source to its point of consumption. In the case of natural gas, approximately 1/3 of the total world production is shipped either by inter-regional pipeline or intercontinentally as liquefied natural gas [2,6]. In the case of crude oil, 2/3 is transported similarly over long distances by ship or pipeline.

The prototype for this energy transport system of the future is today's natural gas handling infrastructure. It is a smoothly functioning system with all the components needed to liquefy, ship, regasify, and distribute electrolytic Hydrogen or simple fuels made from it. There are technical issues to overcome in making this new technology work well for fuels other than natural gas, as is discussed in other chapters of this book. But it is very easy and realistic to imagine a future time in which the oceans of the world are routinely crossed by huge tankers carrying Hydrogen, just as they are crossed today with huge ships carrying natural gas. An example of what one of these would look like is shown in Fig. 1.5.

FIG. 1.5 A typical mid-sized tanker, the *Energy Progress*, taking on an LNG shipment from Wickham Point, UK. The length and width of this ship are 290m and 49m, respectively. The four spherical MOSS tanks have a total capacity of 1.45×10^5 m^3. At an average LNG density of 450 kg m^{-3} this corresponds to 6.5×10^7 kg of LNG with an energy content of 3.21×10^{15}(3.2×10^{-3} EJ). In 2021 world shipping traffic of LNG was about 18 EJ [2,6]. *Image Credit: Ken Hodge via Wikimedia Commons.*

In parallel with the transport of energy as fuel, there will continue to be transported by high-capacity electric power lines, particularly over short distances. The amounts of energy involved are expected to be large. For example, the great bulk of the coal energy shown in Fig. 1.3 is presently used to generate electricity, which then goes out to consumers on transmission lines. In 2021, the world produced 6.29×10^{19} Joules (62.9 EJ) of electric energy from fossil fuel, mostly coal, all of which was transmitted to customers on power lines [2]. Assuming a fleet-average power plant thermal efficiency of 35%, this amounted to 1.80×10^{20} Joules (180 EJ) of primary fossil fuel energy, or about 36% of all the fossil fuel energy consumed in the world [2]. Thus electric infrastructure similar to the one that presently exists could, in principle, transport the power of the world.

However, this is unlikely to happen because (1) there is a need to store a significant fraction of the world's energy for time scales of months and (2) long-haul electric transmission lines are many times more expensive per Watt-kilometer than pipelines [7].

Ultra-high-voltage DC transmission has a slight cost advantage over conventional high-voltage AC transmission for long-distance applications, so it has become the industry standard for this purpose. An example of such a transmission line is shown in Fig. 1.6 [8,9]. There are many similar high-voltage DC lines deployed around the world, particularly in China [10–13].

The cost problem of electric transmission comes from the limitations of dielectric breakdown. For a 2-wire DC line, such as that shown in Fig. 1.6, the maximum power P_{max} in Watts the line can carry is related to the single-wire potential V_0 with respect to ground in volts by

$$P_{max} = 8\pi \sqrt{\frac{\epsilon_0}{\mu_0}} \frac{V_0^2}{ln\,(R/r_0)} \tag{1.1}$$

where $\sqrt{\mu_0/\epsilon_0} = 376.6$ Ohms is the impedance of free space, R is the distance between the wires in meters, and r_0 is the radius of the wires in meters. The origin of this limitation is propagation of signals from load variations backward down the line at the speed of light, which destabilizes the system if the power being transmitted exceeds P_{max}. One would thus like V_0 to be as large as possible. However, its value is limited to about 10^6 Volts by the need to avoid avalanche breakdown and arcing across the medium separating the wires. As a practical matter, this limit is about 1 GW per conductor bundle. The limit applies even more strictly to undersea power cables and underground cables, which are coaxial and thus have conductors in close proximity separated by a solid dielectric [14,15]. For $V_0 = 5.2 \times 10^5$ Volts, $R = 10$m and $r_0 = 0.023$ m Eq. (1.1) gives 3.0×10^9 Watts (3 GW) [9]. Eq. (1.1) is formally four times the power limitation of AC transmission, but this difference is a detail.

Conventional pipelines, such as that shown in Fig. 1.7, are expected to continue outperforming electric transmission lines for long-distance energy transport in the future, although perhaps not with as great margins as they have today [16,17]. A 1.2-m diameter oil pipeline, such as that in Fig. 1.7, transports 50 times more power at peak capacity than the

FIG. 1.6 View of the Pacific DC Intertie in northern Nevada, the first high-voltage DC transmission line built in the United States [8]. It runs 1362 km from the Columbia River to Los Angeles. It has two wires at $\pm 5.2 \times 10^5$. Volts with respect to ground and a design power of 3.2 GW (3.2×10^9 Watts) [9,10]. This is 25% of the 14 GW average demand of the 5-county Los Angeles metropolitan area. If run constantly for a year this line would deliver 1.01×10^{17} Joules (0.1 EJ) or 30 times the amount of energy carried by the ship in Fig. 1.5. *Image Credit: Wikimedia Commons.*

FIG. 1.7 View of the Alaska oil pipeline. It is 1.22 m in diameter and runs 1288 km from Prudhoe Bay on the Arctic Ocean to the port of Valdez. Its maximum throughput in 1988 was reported at 7.44×10^8 bbl y^{-1} [16,17]. This corresponds to 4.2×10^{18} J y^{-1} (4.2 EJ y^{-1}) or 1.35×10^{11} Watts, which is 45 times the capacity of the transmission line in Fig. 1.6 [2]. *Image Credit: Kenneth Gill via Wikimedia Commons.*

electric line in Fig. 1.6 does, and is approximately 100 times cheaper per Watt-kilometer [7]. This factor is reduced to about 15 for natural gas transport because of its lower fluid density, and then further to about 9 for Hydrogen transport because of its lower energy per molecule vis-à-vis natural gas. But even in the case of hydrogen, the cost advantage of a pipeline is enormous.

Natural gas pipelines are capable of transporting a stupendous amount of energy [18]. For example, the sabotaged Nordstream I pipeline under the Baltic Sea, which was about the size of the pipeline in Fig. 1.7, carried natural gas from Russia to Germany with a design capacity of 4.0×10^{10} kg y^{-1} (55 billion m^3 y^{-1}) [19]. The energy content of this gas, 2.0×10^{18} Joules (2 EJ), exceeded 60% of all the natural gas consumed in Germany in 2021, 15% of all the primary energy consumed in Germany in 2021, and the entirety of the nuclear energy produced in Germany in 2001, the peak production year [2]. Had Nordstream 2 come online, it would have doubled this figure to 32% of all primary energy consumed in Germany, a number also exceeding all the nuclear energy produced in France in 2021 [2,20].

Thus the present situation in the energy industry leads us to anticipate that an energy transport system not so different from the one that exists today will be a centerpiece of civilization's coming decarbonized future. The energy being transported will not be mined out of the earth anymore, and there will be a lot more Hydrogen in the energy industry's ships, pipelines, and storage tanks, but everything else is likely to be the same, or at least familiar. Technological improvements are likely to occur, and economic competition among the various technologies will almost certainly rebalance the mix. But overall the energy transport system should transition to the decarbonized future smoothly and without incident.

References

[1] Earth at Night, U.S. National Aeronautics and Space Administration, 2019. NP-2019-07-2739-HQ.
[2] BP statistical review of world energy 2022, Brit. Petrol. (2022).
[3] World Population Prospects 2022, United Nations, 2022. UN DESA/POP/2021/Tr/No.3.
[4] R. Showstack, Carbon dioxide tops 400 ppm at Mauna Loa, Hawaii, Eos 94 (2013) 192.
[5] Intergovernmental Panel on Climate Change, Climate Change 2022: Mitigation of Climate Change, Cambridge University Press, 2023.
[6] 2022 World LNG Report, International Gas Union, 2022.
[7] D. DeSantis, et al., Cost of long-distance energy transmission by different carriers, iScience 24 (2021) 103495.
[8] Pacific Intertie: The California Connection on the Electron Superhighway, Northwest Power Planning Council, 2001. Document 2001-11.
[9] Pacific Direct Current Intertie Upgrade: Draft Environmental Assessment, Bonneville Power Administration, U.S. Department of Energy, 2014. DOE/EA-1937.
[10] Assessing HVDC Transmission for Impacts of Non-Dispatchable Generation, U.S. Energy Information Administration, 2018.
[11] B.R.T. Cotts, J.R. Prigmore II, K.L. Graf, HVDC transmission for renewable energy integration, in: B.W. D'Andrade (Ed.), The Power Grid: Smart, Secure, Green and Reliable, Academic Press, 2017.
[12] H. Zhou, et al., Ultra-High Voltage AC/DC Power Transmission, Springer, 2018.
[13] D. Huang, et al., Ultra high voltage transmission in China: developments, current status and future prospects, Proc. IEEE 97 (2009) 555.
[14] M. Ardelean, P. Minnebo, HVDC Submarine Power Cables in the World, Joint Research Centre, European Commission, 2015. EUR 27527 EN.
[15] Undergrounding High Voltage Electricity Transmission Lines, U.K. National Grid, 2014.
[16] Facts: Trans Alaska Pipeline System, Alyeska Pipelines Service Company, 2021.
[17] Annual Energy Outlook 2012, U.S. Energy Information Administration, 2012, p. 52. DOE/EIA-0383(2012).
[18] G. Molnar, Economics of gas transportation by pipeline and LNG, in: M. Hafner, G. Luciani (Eds.), The Palgrave Handbook of International Energy Economics, Palgrave Macmillan, 2022.
[19] Nord Stream's Twin Pipelines: Part of the Long-Term Solution for Europe's Energy Security, Nord Stream, 2016.
[20] M. Russell, Nord Stream 2 Pipeline, European Parliament Research Service, 2021. PE 690.705.

CHAPTER

2

Energy transport is a cornerstone of the energy supply chain

David T. Sánchez Martínez[a], *Terry Kreuz*[b], *Brandon L. Ridens*[c], *Ramees K. Rahman*[d], *Subith Vasu Sumathi*[d], *Stephen Ross*[e], *James Underwood*[f], *Rahul Iyer*[g], *and Natalie R. Smith*[c]

[a]University of Seville, Seville, Spain [b]National Fuel Gas Company, Williamsville, NY, United States [c]Southwest Research Institute, San Antonio, TX, United States [d]University of Central Florida, Orlando, FL, United States [e]Ebara Elliott Energy, Jeannette, PA, United States [f]Solar Turbines, San Diego, CA, United States [g]KCK Group, Cupertino, CA, United States

Decarbonization and energy transport

Human activities rely on the consumption of energy, presented in different forms: thermal energy from the sun to keep us warm, energy encapsulated in food and beverages supplied to our metabolism, wood for fireplaces, electricity to power heaters and electric/electronic devices at home and the workplace, and gasoline to run our cars. In connection with this, the public debate is today focused on what types of energy must be consumed, which must be phased out, and how much of them humans must make use of anyhow. This discussion has triggered political and legislative actions on both ends of the energy supply chain: primary energy sources and energy consumption.

Today, there is an international, collective effort to substitute decarbonized energy sources for fossil fuels in order to mitigate the effects of greenhouse gas (GHG) emissions on climate change. This initiative started over 25 years ago[a] with the objective of reducing GHG concentrations in the atmosphere to "a level that would prevent dangerous anthropogenic interference with the climate system," set out in Article 2 of the Kyoto Protocol, and accelerated recently in 2015. Indeed, signed in France in 2016, the Paris Agreement widened the scope of work and embraced not only technical and legislative but also financial actions. The generic objective of the Kyoto Protocol is hence translated into more specific tasks: (a) to keep the rise in mean global temperature well below 2°C above preindustrial levels (with an effort to reduce this to 1.5°C) and (b) to accomplish carbon neutrality by 2050. These objectives have shaped the legislative packages implemented across the world recently, with the result of a large penetration of renewable energy sources; between 2000 and 2021, the share of renewable energy sources in power generation increased from almost null to approximately 15%.[b]

The other end of the energy supply chain is also affected by these actions, some of which are very visible, for instance, the phase-out of combustion engines in light commercial and passenger cars (agreements have either been reached or are under discussion to implement these policies across the world: Europe, China, India, and the United States) or the limitations of controlling temperatures in commercial HVAC systems.

[a] December 11, 1997, signature of the Kyoto protocol.

[b] BP Statistical review, 71st edition, 2022.

https://doi.org/10.1016/B978-0-443-21893-4.00009-X
Copyright © 2025 Elsevier Inc. All rights are reserved, including those for text and data mining, AI training, and similar technologies.

Unfortunately, primary energy[c] is most of the time found in areas that are far from where consumption takes place. This poses a need for sustainable energy transportation means from technical (reliably), environmental, social (safely), and economic (cost-effectively) standpoints, which brings a large number of challenges with it. The literature on (primary) energy harvesting is vast, and so is the available work on final energy production. However, energy transport attracts much less attention, and it is typically overlooked in the social and political arenas. There are several reasons for this: It is a highly technical field requiring specific education and training; the impact on energy prices and other aspects such as sustainability (i.e., environmental impact) is not straightforward; the infrastructure is very often hidden from the general public to avoid the *not in my backyard* syndrome. This might also be the reason why the courses on energy transport within most energy and engineering degrees (either Masters or Bachelor) at universities worldwide represent a marginal share of the syllabus..

The situation described in the foregoing paragraphs is bringing about fast, deep changes in the energy landscape that are also strongly affected by the unsteadiness of the sociopolitical arena in recent times: Conventional technologies are being revised to achieve unprecedented efficiency; mature renewable energy technologies are being deployed massively, while new sustainable energy technologies are under development (low-enthalpy geothermal, ocean currents); large-scale and/or long-term energy storage solutions are getting close to the precommercial stage; carbon capture, sequestration, and storage are in the spotlight again; new renewable energy hubs are becoming relevant. Yet, in order for them to be effective and sustainable, these changes must rely on energy transport more than ever, and this work comes to serve as a reference source of information about the technologies involved, their state of the art, and the gaps that must still be bridged, while also paying attention to future trends, infrastructure issues, and next-generation R&D.

Mapping of energy sources

The world is undergoing a very interesting transformation. The telltale signs of climate change, along with geopolitical unrest, have caused much of the world to make use of renewable or zero-carbon energy sources and alternative forms of energy conversion systems. These energy sources and conversion technologies are, however, not always collocated and concurrent with demand. The law of conservation of energy states that energy cannot be created or destroyed but can only be transformed from one form to another or transferred from one object to another. For these reasons, the energy must be stored and transported from the source of generation to the destination of demand. Classically, sources of generation were constructed as needed; as demand increased for a city, a new power plant would be built, and substations added. With the global goal of decarbonization, this old edict no longer holds true, and renewable energy generation requires additional considerations for storage and transportation. This is the core topic of this book, which explores the various forms of energy transportation that have arisen from this new energy landscape. As a first step, the following subsections introduce the distribution of renewable and nonrenewable energy sources and some of the main features of the associated transportation (and trade). This will provide the reader with an idea of the magnitude of the challenge ahead.

Nonrenewable energy sources

Oil and gas are the most abundantly used fuels in most regions across the world, except for Asia-Pacific, where the consumption of coal is still dominant, shown in Fig. 2.1. This means that these energy sources must be produced, extracted, processed (refined), transported to storage hubs, and then distributed to end users, regardless of where these users are.

The extraction of oil was estimated at ~82 million barrels per day in January 2023, with a total energy content of ~140 TWh of energy, extracted from some 1500 oil fields worldwide, 1300 of which located inland and ~200 offshore. Often, these wells produce not only oil but also natural gas, with an estimated global production of over 10 billion cubic meters per day. The production of natural gas regionally is summarized in Table 2.1, showing three types of nations in the world: those that produce and consume vast amounts of natural gas, leveling production and consumption to yield self-sufficiency, such as the United States, Russia, China, Iran, or the countries in the Gulf Cooperation Council. A second category is comprised of nations producing large amounts of natural gas, which is mostly exported, given

[c] Primary energy is, formally, any form of energy available in nature prior to any anthropogenic transformation. In engineering practice, primary energy is that supplied to a specific process or system where downstream energy transformation takes place (i.e., origin of the energy supply chain under analysis).

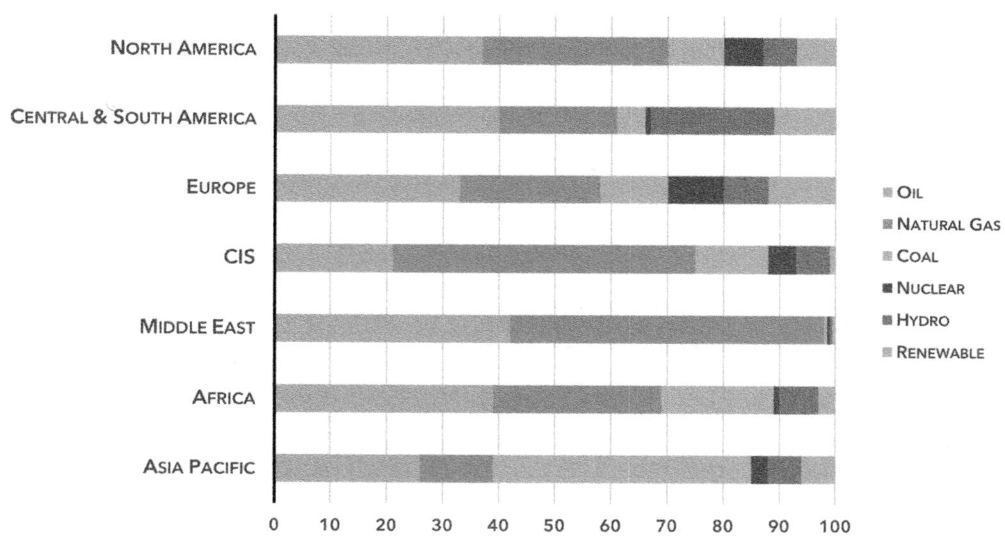

FIG. 2.1 Regional consumption pattern in 2022. *BP Statistical review, 71st Edition, 2022.*

TABLE 2.1 Regional distribution of oil fields [1].

	Gas production		Gas consumption	
Country	**Rank in top 20**	**Billion cubic feet per day**	**Rank in top 20**	**Billion cubic feet per day**
Algeria	9	8.8	–	–
Argentina	–	–	16	4.8
Australia	10	8.8	–	–
Canada	5	14.7	7	9.6
China	6	13.4	3	20.3
Egypt	18	4.0	14	4.9
Germany	–	–	9	7.8
India	–	–	15	4.8
Indonesia	12	6.7	–	–
Iran	3	19.5	4	19.4
Italy	–	–	12	6.2
Japan	–	–	5	10.7
Malaysia	11	7.1	20	4.2
Mexico	16	4.6	8	8.6
Nigeria	17	.3	–	–
Norway	7	11.3	–	–
Pakistan	19	4.0	18	4.4
Qatar	4	17.5	–	–
Russia	2	55.9	2	37.7
Saudi Arabia	8	10.6	6	10.6
South Korea	–	–	19	4.4

Continued

TABLE 2.1 Regional distribution of oil fields [1]—cont'd

	Gas production		Gas consumption	
Country	**Rank in top 20**	**Billion cubic feet per day**	**Rank in top 20**	**Billion cubic feet per day**
Thailand	–	–	17	4.7
Turkmenistan	13	6.4	–	–
UAE	15	6.0	11	7.4
UK	20	4.0	10	7.4
USA	1	72.3	1	75.1
Uzbekistan	14	6.1	13	5.0

the moderate domestic consumption: Iran, Qatar, Indonesia, Australia, Norway, etc. Finally, most countries must import natural gas because they do not (or hardly) produce this fuel, even though they, on the contrary, rely heavily on it to produce power and to serve the industry; this is particularly notable in Japan, South Korea, Argentina, Germany, Italy, etc.

The next step in the oil supply chain is refining, which is typically done in separate facilities from the extraction fields. Fig. 2.2 shows the list of countries with the largest oil refining capacity alongside the largest exporters and importers of oil; the comparison between the former and the latter gives an insight into the need and relevance of energy transport to bridge the distance between oil sources and end users.

	GAS PRODUCTION		GAS CONSUMPTION	
COUNTRY	Rank in Top 20	Billion Cubic Feet per Day	Rank in Top 20	Billion Cubic Feet per Day
ALGERIA	9	8.8	-	-
ARGENTINA	-	-	16	4.8
AUSTRALIA	10	8.8	-	-
CANADA	5	14.7	7	9.6
CHINA	6	13.4	3	20.3
EGYPT	18	4.0	14	4.9
GERMANY	-	-	9	7.8
INDIA	-	-	15	4.8
INDONESIA	12	6.7	-	-
IRAN	3	19.5	4	19.4
ITALY	-	-	12	6.2
JAPAN	-	-	5	10.7
MALAYSIA	11	7.1	20	4.2
MEXICO	16	4.6	8	8.6
NIGERIA	17	4.3	-	-
NORWAY	7	11.3	-	-
PAKISTAN	19	4.0	18	4.4
QATAR	4	17.5	-	-
RUSSIA	2	55.9	2	37.7
SAUDI ARABIA	8	10.6	6	10.6
SOUTH KOREA	-	-	19	4.4
THAILAND	-	-	17	4.7
TURKMENISTAN	13	6.4	-	-
UAE	15	6.0	11	7.4
UK	20	4.0	10	7.4
US	1	72.3	1	75.1
UZBEKISTAN	14	6.1	13	5.0

FIG. 2.2 Oil refining capacity (left) and net exporters (center) and importers (right) of oil [2]. Metrics in million barrels per day (refining capacity) and megatons of oil (export/import).

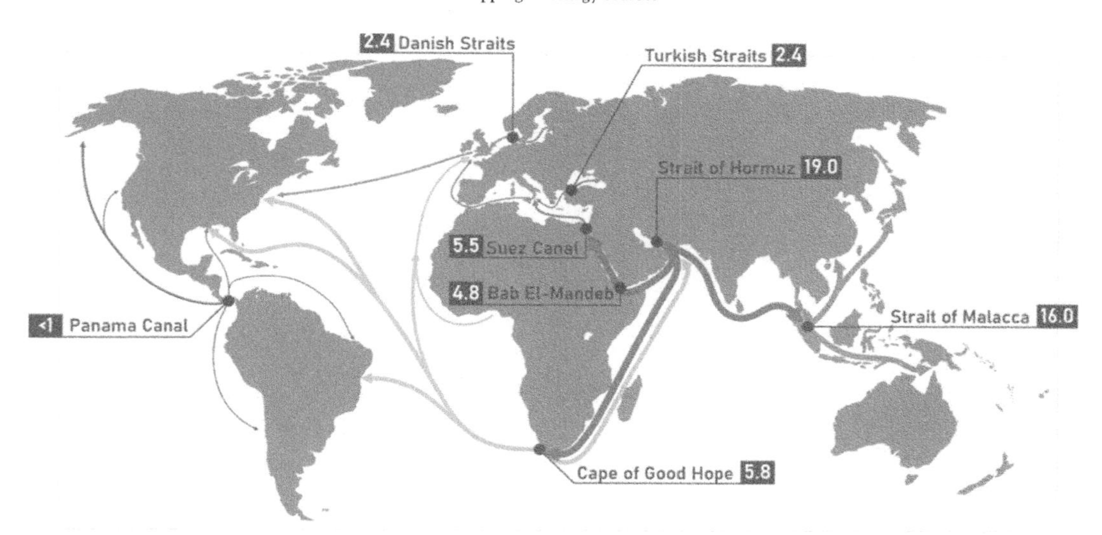

FIG. 2.3 Main oil shipping routes and chokepoints (size of circle and figures report transport volume in million barrels per day). *Adapted from The Economist, US-Iran tensions threaten the world's most important oil-shipping route, June 21, 2019. https://www.economist.com/graphic-detail/2019/06/21/us-iran-tensions-threaten-the-worlds-most-important-oil-shipping-route (Accessed 10 June 2023).*

Transportation routes from oil fields to refineries across the world rely mostly on tankers (~60%) and, to a lesser extent, on oil pipelines, with the main maritime routes traveling from the Middle East to Asia (India, Malaysia, Japan, and China), Europe (Suez Canal), and America (Cape of Good Hope), shown in Fig. 2.3. Natural gas is, on the contrary, transported mostly through gas pipelines or by tankers in the form of liquified natural gas (LNG). This explains the dissimilar topology of the distribution networks of these two energy sources: The cumulative length of large capacity (i.e., distribution or small-scale transportation) pipelines in the world is estimated at ~1.25 million kilometers, 85% of which are gas pipelines and just 15% oil pipelines.

The infrastructure used to supply oil and gas is not driven only by technical considerations; it is strongly affected by sociopolitical conditions as well. This is very visible in Europe, a continent that was strongly reliant on natural gas supplies from Russia (Northern and Eastern Europe) and Algeria (Spain and Northern Italy). Recent stress caused by the war in Ukraine and the strained trilateral relationship between Spain, Morocco, and Algeria over Western Sahara has altered the natural gas supply map, shifting to imported LNG and natural gas from the Caspian Basin through the Southern Gas Corridor, with the aim to ensure security of supply [3]. In particular, the natural gas imported from Russia into Europe (pipeline plus LNG) decreased from 50% in 2019 to less than 25% by the end of 2022, whereas non-Russian LNG imports amounted to this same share (25%). For these and other reasons, the global LNG market is currently booming, and most large consumers of natural gas have plans to increase their regasification capacity in the coming years. In this respect, Fig. 2.4 shows the largest producers and consumers of LNG in the world.

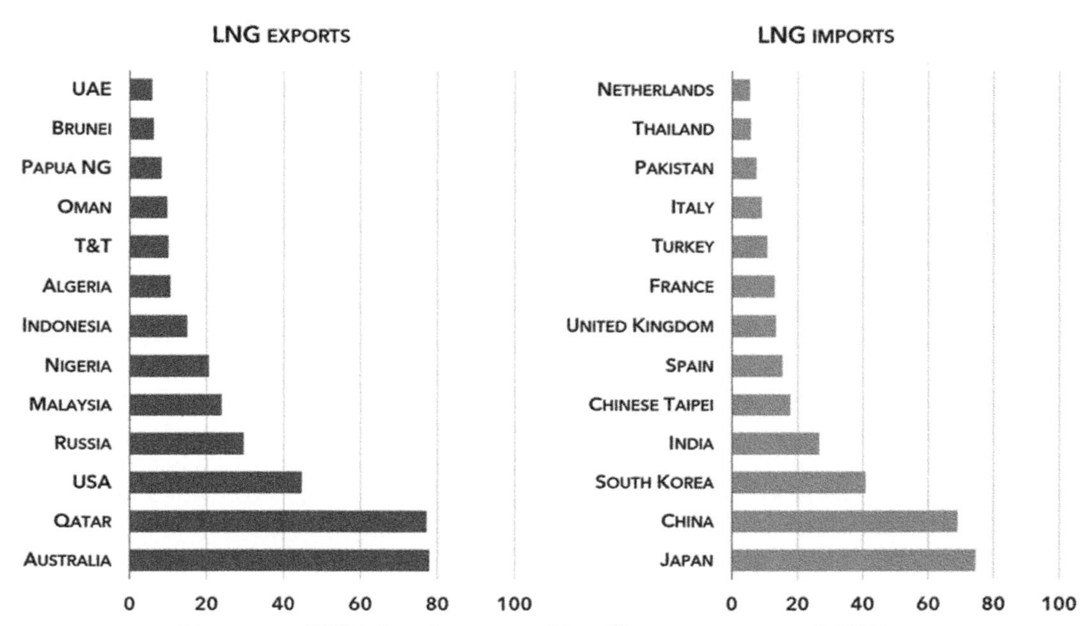

FIG. 2.4 Largest exporters and importers of LNG. Capacity expressed in million tons per annum (MTPA).

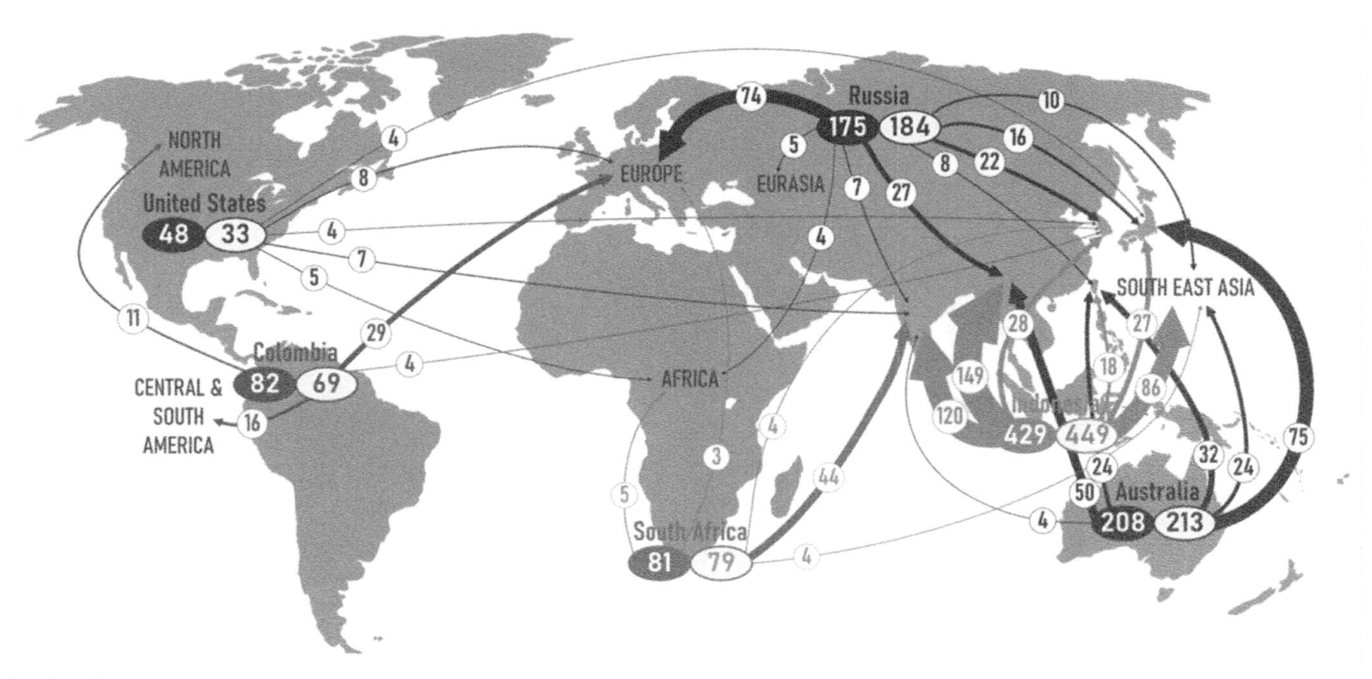

FIG. 2.5 Main trade flows in the thermal coal market, 2019 (MTPA) [4].

Connecting the dots between these charts yields the maritime routes for this supply chain and explains why the number of LNG tanker orders doubled from 2019 to 2020. These aspects of the energy market are discussed in more detail in a later section.

Coal is the second source of primary energy in the world, second to oil and slightly ahead of natural gas. Over one-fourth of the primary energy consumption worldwide comes from coal, but the share of this fuel in the energy mix is marginal in most regions of the world, inasmuch as most of the consumption takes place in Asia as shown in Fig. 2.1. Out of the ~8 billion tons of coal produced worldwide, China produced 4.25 (53%), India 1.1 (13.8%), and Indonesia 0.6 (7.5%), with Australia, Russia, and the United States producing ~0.5 each. The destination of the coal extracted in these countries is nevertheless dissimilar: While China and India use it to satisfy the ever-increasing domestic demand for electricity, Australia and Indonesia dominate the world exports of coal with around 30% of the market each. The situation in the coming years is expected to consolidate this trend since the extraction of coal in China, India, Indonesia, and Australia is on the rise, whereas that in the United States or Europe is declining rapidly. It is deduced from this that the main means of worldwide transportation of coal today is freight rails, whereas the use of tankers covering maritime routes is foreseen to increase in the coming years (Fig. 2.5).

Nuclear energy represents 5% of the world's primary energy consumption. Kazakhstan produces around 45% of the uranium traded internationally (~50 million tons globally), followed by Canada (15%) and Namibia (10%); interestingly, only Canada among these top three producers is found in the list of largest generators of nuclear power. This natural uranium consists mostly of two isotopes, U-235 and U-238, and only the first of these is used to produce electric power in fission nuclear reactors. Unfortunately, the content of U-235 in natural uranium is ~0.27%, and this must be enriched to at least 3%–5% (or more) to be effectively used in civil nuclear reactors; the degree of enrichment depends on the reactor technology used. Due to restrictions set by the need to control the production of highly enriched uranium for military use, the number of facilities to carry out this process is limited; specifically, out of the global enrichment capacity in the world (~61 million tSWU/a), Russia holds almost 50% of it (c.30 million tSWU/a), with France and China being second and third far behind (7500 and <6000 million tSWU/a, respectively). The Netherlands, the United States, the United Kingdom, and Germany are the remaining countries with significant enrichment capacity (4–5 million tSWU/a in each country).

The transportation of radioactive materials is linked to the nuclear fuel cycle shown in Fig. 2.6. Uranium ore is milled and separated from the ore onsite (typically) and then shifted to conversion facilities where it is converted into uranium hexafluoride, UF_6, a gas used for centrifuge separation of U-235 and U-238 isotopes. Lowly enriched uranium (LEU) for civil use is then shipped to fuel fabrication facilities, where the enriched UF_6 is converted into nuclear fuel in the form of ceramic uranium dioxide bars that are eventually shipped to nuclear power stations. The steps along this

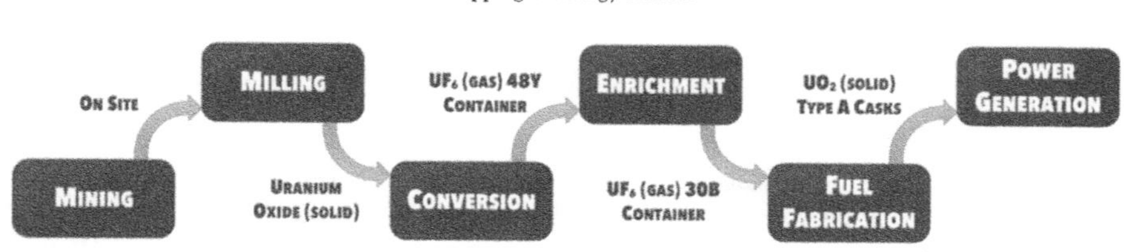

FIG. 2.6 Nuclear fuel cycle (disposal and reprocessing not shown for brevity).

supply chain are regulated by the International Atomic Energy Agency, which provides instructions for the safe transport of radioactive materials, mostly with the aim of preventing the transported mass from reaching criticality [5]. Maritime and rail hauling are the most used means of transportation.

Renewable energy sources

Energy sources classified as renewable are those that are replenished at a higher rate than they are consumed. This means that, in a high-level, nonscientific context, they can be regarded as unlimited and, therefore, sustainable by nature (using this energy today does not compromise the availability of the same energy source for future generations).

In the main, renewable energy sources feature (1) lower energy density, hence needing a larger energy-collection surface (surface of photovoltaic panels, flow area of wind turbines, and basin of a mighty river) than nonrenewable sources, and (2) they cannot be packaged or stored without transformation into another form of energy (except for hydro, and this to a limited extent only). In other words, these energy sources are mostly converted into electric power directly, without an intermediate thermomechanical conversion in a thermodynamic cycle (there are, of course, notable exceptions to this: geothermal and solar thermal energies), and then transported in the form of either electricity or an energy carrier (hydrogen and electrofuels). They also have a strong regional dependency, but this is not different from what has been described for nonrenewable energy sources.

Hydraulic, solar, and wind energies are the predominant renewable energy sources across the world, and they are distributed unevenly. For instance, Fig. 2.7 shows the world map of Global Horizontal Irradiance. On both sides of the

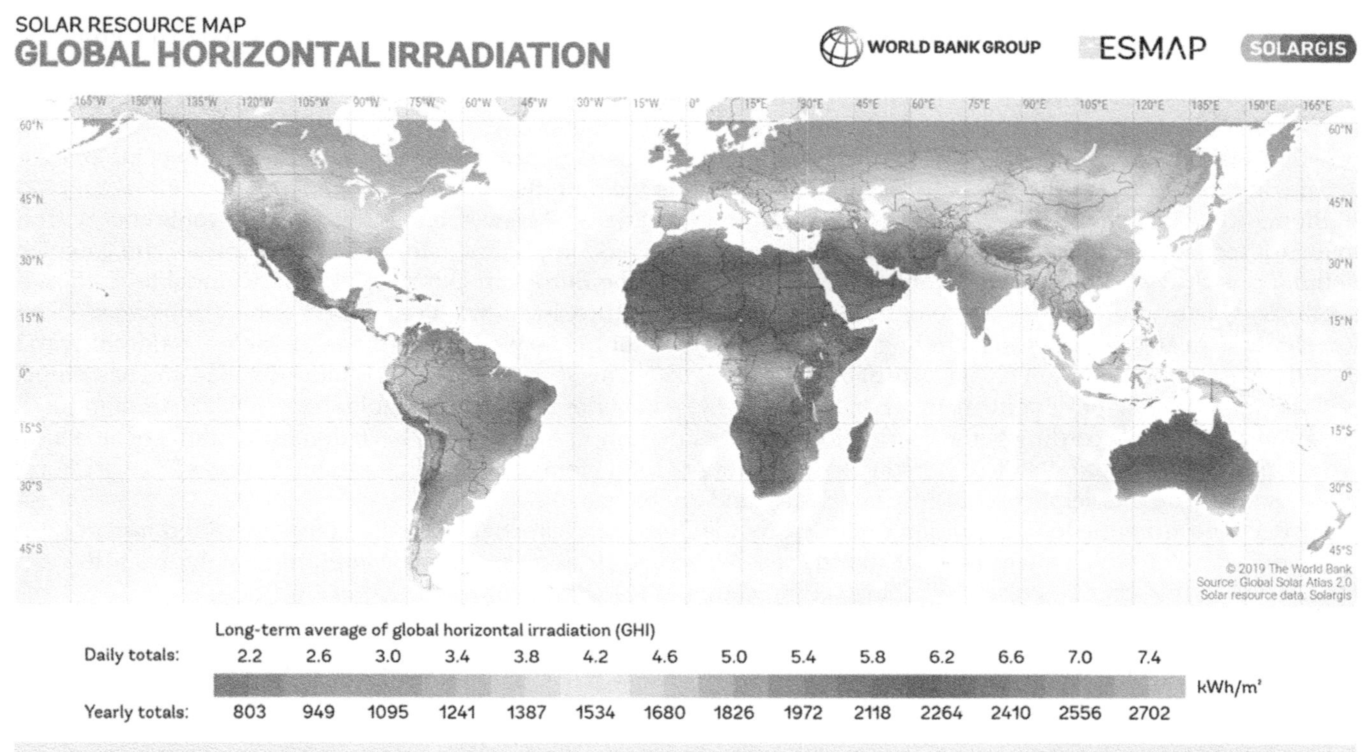

FIG. 2.7 Global horizontal irradiation world map [6].

WIND RESOURCE MAP

WIND POWER DENSITY POTENTIAL

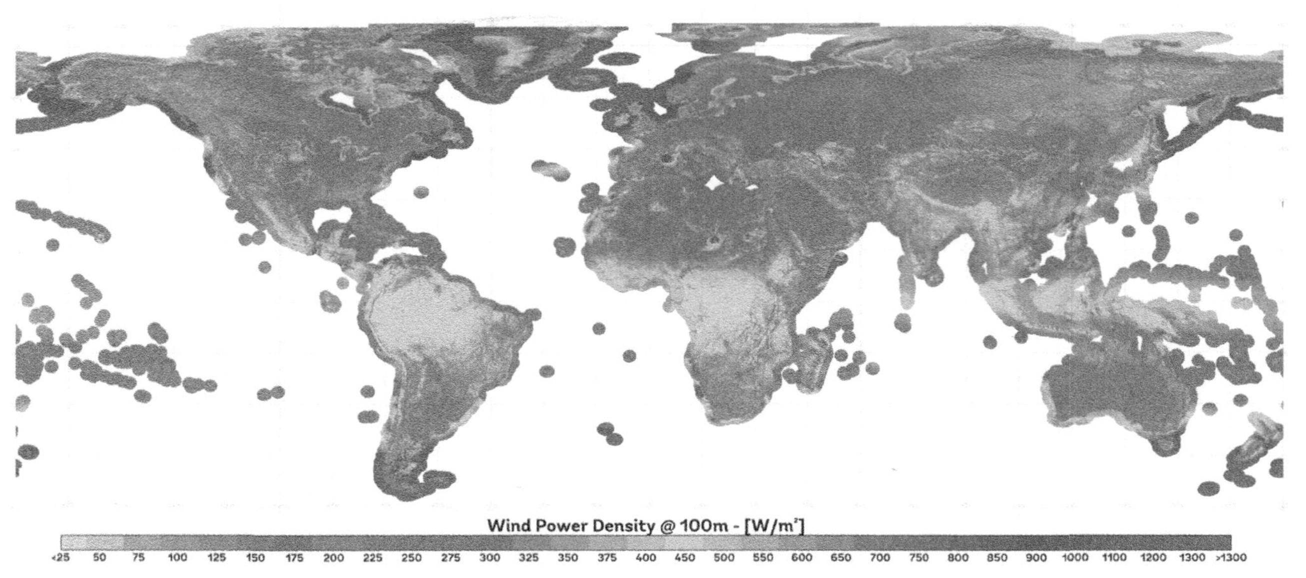

FIG. 2.8 Wind energy world map. *Global Wind Atlas produced by Denmark Technical University (DUT). www.globalwindatlas.info.*

Equator, the availability of solar energy is very high, and it is only the regions in the very north of America, Europe, and Asia that receive significantly lower irradiation. The plot confirms that this technology can be applied across the world.

The situation for wind is similar, with uneven distribution worldwide and, interestingly, a large fraction of the resources found offshore or in coastal areas, in particular in the northern hemisphere (red regions, dark gray in print version, in Fig. 2.8). This explains the very strong interest seen in offshore wind turbines, either fixed to the ground in shallow regions of the sea or floating. Whichever the layout, wind energy poses the need to transport the electric power produced. This is very specific to a particular site; furthermore, for offshore installations, cables to gather electric power from all the wind turbines in a farm and to, then, transport this to the consumers inland with as low a power loss as possible represent a significant portion of the investment. For this, significant R&D efforts are currently being put into optimizing the configuration of electric cables and collectors yielding the lowest cost of energy [7].

Biomass can be defined in several ways. The International Energy Agency defines this as "renewable energy from living (or recently living) plants and animals, e.g., wood chippings, crops, and manure." A more formal and complete definition is provided by the Renewable Energy Directive of the European Union (EU): "biodegradable fraction of products, waste, and residues of biological origin from agriculture, including vegetal and animal substances, from forestry and related industries, including fisheries and aquaculture, as well as the biodegradable fraction of waste, including industrial and municipal waste of biological origin."[d] This second definition is more detailed and has a much wider scope than the former one by the IEA. Interestingly, it includes the biodegradable share of waste from industry and municipalities. According to the World Bioenergy Association, out of the ~155 PWh domestic supply of bioenergy worldwide in 2018, around 85% corresponded to traditional solid biomass (excluding waste), whereas 7% were liquid biofuels, 5% were municipal and industrial wastes, and 3% biogas [8].

As deduced from the long definition provided by the European Directive, biomass comes in very different types of products, all of which share the fact that the energy stored in these materials comes, ultimately, from the sun. In addition to this, biomass resources share certain common features:

- Low energy density: Raw biofuels have a low heating value (for instance, the high heating value of forestry and agricultural waste is 15–20 MJ/kg, whereas this is ~50 MJ/kg for natural gas or ~27 MJ/kg for coal).

[d] Directive (EU) 2018/2001 of the European Parliament and of the Council of December 11, 2018 on the promotion of the use of energy from renewable sources (recast).

- Regional and seasonal dependence: The composition and characteristics of biofuels are very sensitive to climate and, therefore, change across regions and seasonally. Therefore, only rarely is it possible to source biomass with constant characteristics throughout the year.

The low energy density of biomass poses a large challenge for transportation, as this adds significant costs to this raw energy (ranging from 10% to 20% of the total sales price of the end product for large-scale supplies). In order to offset this problem, at least partially, processing biomass with the aim of increasing volumetric energy density is a common intermediate step of the supply chain (except for very short-haul transportation, where the impact of transportation costs is weaker) [9].

As a general rule of thumb and of course dependent on market conditions, it is commonly assumed that transportation of unprocessed biomass by road is uneconomical for distances longer than ~100–200 km, and it also has an impact on the associated net (life cycle) emissions of GHGs [10]. For longer distances, rail and maritime transportation are more cost-effective, although they do not ensure a positive business case for the biomass supply chain for all users.

Based on the counteracting effects introduced in the previous paragraphs, biomass is typically sourced locally. Nevertheless, a changing regulatory framework now incentivizing the consumption of renewable fuels is already changing the economics of biomass transportation for certain end users. For instance, Drax power station in the United Kingdom is rated at 2.6 GWe and runs on biomass sourced in the United States and transported by vessel to a British port and then rail to the power station. This is likely to be more frequent in the future.

Geothermal energy originates inside the Earth and is extracted from the crust, typically from wells drilled for this purpose. There are three main sources of thermal energy (heat) inside the Earth:

- Residual thermal energy is generated from the formation of the Earth (primordial heat) and conserved over time, as stated by the first law of thermodynamics. This thermal energy flows radially outward from the mantle and through the crust; due to the heterogeneous characteristics (thickness and composition) of the latter layer, this heat flow is heterogeneous and yields unevenly distributed geothermal energy resources in the subsurface.
- Radioactive decay of certain elements like uranium, thorium, rubidium, etc. The concentration of these elements is higher in the crust since they are typically displaced from the mantle due to their large atomic radii, which is less compatible with the very high pressure in this layer.
- Gravitational pressure. The very high pressure exerted on the gases and solids in the deeper regions of the Earth brings about a temperature increase similar to that of a gas undergoing adiabatic compression. This thermal energy is trapped due to the good insulation properties of bedrock.

The two first sources are dominant and contribute roughly the same share to the total internal heat flow of the Earth. It is estimated that this amounts to ~47 TW [11], which represents a very large potential to contribute to the primary energy demand of the world (the estimated total electric power capacity worldwide is a little short of 6 TWe). Unfortunately, as it was to be expected from the foregoing considerations about the factors influencing the flow of the Earth's internal heat across the crust, the availability of exploitable geothermal energy in the world is markedly uneven. This is illustrated in Fig. 2.9, showing an *optimal geothermal suitability distribution* produced by Coro and Trumpy, based on geological and environmental data and tested against the sites where geothermal power plants are currently found [12] (darker colors and higher indices show more suitable locations).

Geothermal energy is usually classified according to the temperature level: high enthalpy (>150°C), medium enthalpy (90°C–150°C), and low enthalpy (<90°C); this classification is nevertheless not standardized, and, therefore, the foregoing ranges are likely to change according to the bibliographical context. High-enthalpy geothermal resources are typically used for the production of electricity with organic Rankine cycle (ORC) power systems (power systems conceptually similar to steam turbines but running on different working fluids, specifically suitable for lower temperature applications). Medium-enthalpy geothermal energy can also be used to produce electricity, but the lower efficiency of the power system compromises the economics of this type of application; rather, this energy source is frequently used for district heating or other thermal applications directly. The interest in low-enthalpy geothermal sources is now increasing, in particular for their exploitation through ground-source heat pumps. The economics of these applications are strongly driven by the depth at which the geothermal source is found (capital cost) and by the behavior of energy markets, thus making it possible that medium- or even low-enthalpy geothermal resources can be used for the production of electricity.

Fig. 2.9 shows that the areas with the largest availability of geothermal energy are the west coast of America, in particular, North and Central America (El Salvador and Costa Rica), although this energy source is also available in Chile. In Europe, Turkey, Italy, and Iceland have the largest installed electric capacity, and these assets run with

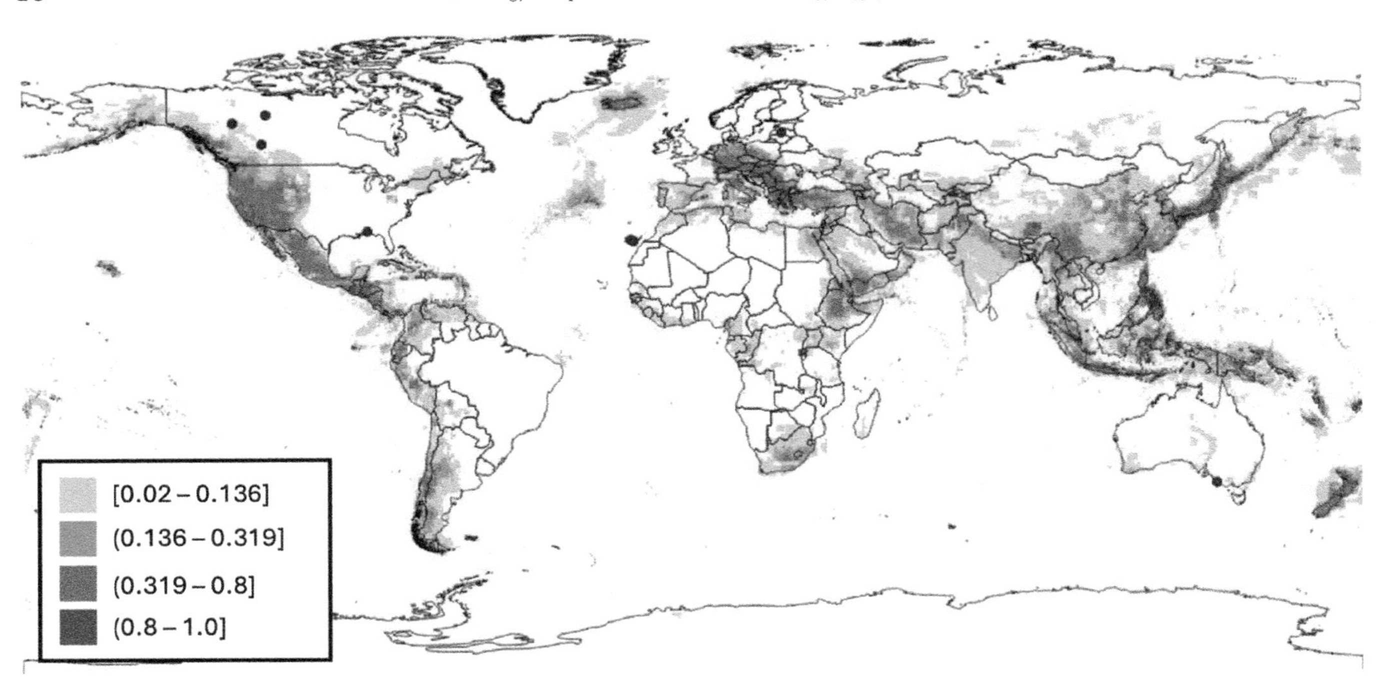

FIG. 2.9 Optimal geothermal suitability distribution. Warmer colors indicate higher suitability scores [10]. *From R.E.H. Sims, R.N. Schock, A. Adegbululgbe, J. Fenhann, I. Konstantinaviciute, W. Moomaw, H.B. Nimir, B. Schlamadinger, J. Torres-Martínez, C. Turner, Y. Uchiyama, S.J.V. Vuori, N. Wamukonya, X. Zhang, Energy supply, in: B. Metz, O.R. Davidson, P.R. Bosch, R. Dave, L.A. Meyer (Eds.), Climate Change 2007: Mitigation, Contribution of Working Group III to the Fourth Assessment Report of the Intergovernmental Panel on Climate Change, Cambridge University Press, Cambridge, United Kingdom and New York, NY, USA (Figure 4.17).*

capacity factors well above 75% [13]. For thermal applications, Germany, France, and the Netherlands add to the foregoing countries for their noteworthy utilization of geothermal resources. Indonesia, the Philippines, Japan in Asia, and New Zealand in Australia also have very large amounts of exploitable geothermal energy available.

The exploitation of geothermal energy implies transportation, as this energy source is extracted from several kilometers deep underground (>3 km). Akin to solar or wind, geothermal energy cannot be transported for long distances in its raw form; rather, it must be converted, typically into electricity, prior to transportation. Distribution (i.e., short-distance transportation) is, on the other hand, possible, and there is a well-established technology for district heating applications and other thermal uses [11], typically relying on pumping stations to circulate a hot heat transfer fluid (most commonly, water) through a closed network of insulated pipes [14].

Lastly, hydropower has not been discussed so far in spite of being an abundant source of renewable energy that is currently exploited efficiently with mature, well-established technology.

Conclusions

This section has illustrated the distribution of energy sources on Earth, showcasing that energy sources are scattered across the globe, in certain cases benefiting from higher concentrations in specific regions (for instance, oil in the Gulf region or solar energy in the sunbelt), but, for the most part, posing the need to harvest this energy from distant places in the world. This applies to all renewable and nonrenewable energy sources.

On the end of the energy supply chain, energy is consumed virtually everywhere, which implies that these dots, energy source and consumer, must be connected by means of a complex energy transportation system. This is typically called the energy supply chain, which can be implemented in multiple ways. From a fundamental standpoint, there are two alternatives to connecting supply and demand: Either energy is transported in raw form (for instance, oil) or it is transformed into an intermediate form of energy. This is commonly called an energy carrier or an energy vector. Of course, any solution in between these two is also possible; for instance, transportation of primary energy in raw form, conversion into an energy carrier at some intermediate location (e.g., refinery), and then distribution of the energy carrier. These different options are discussed in the following sections.

Electrical power transmission and distribution

Current limitations

This section discusses the challenges of electrical transmission and distribution, including a summary of electric grid challenges across major continents, a deeper discussion of technology solutions, and the economics of grid modernization.

Electricity is a fundamental enabler of economic growth, productivity, and modern quality of life. The grid provides the critical links between power generation and end-use loads. However, the transmission and distribution systems connecting suppliers with consumers face pressing challenges worldwide, calling for an in-depth examination of the technical, economic, and policy hurdles impacting grid infrastructure globally. The text below summarizes distinct regional grid issues and highlights new advanced technologies that can modernize aging systems. The substantial costs and financing realities of upgrading grids are analyzed. Recommendations for collaborative action between utilities, regulators, technology firms, and governments are presented to promote progress toward more resilient, efficient, and environmentally sustainable electricity delivery.

The interconnected network of high-voltage transmission lines, lower-voltage distribution wires, substations, transformers, poles, switches, and advanced control systems comprises the electric grid. This complex engineering system allows electricity to be generated at centralized and distributed energy plants and then transmitted and distributed across vast distances to homes, businesses, and factories. Other key grid components include real-time monitoring capabilities to maintain voltage, frequency, and power quality standards and manage flows. The grid provides the essential links between suppliers and consumers that enable reliable access to affordable, quality electricity services that power the economy.

The electrical grid is undergoing transformative change, driven by several key trends, including:

- aging infrastructure, built decades ago and needing modernization and upgrades,
- rising electricity demand, requiring capacity expansion,
- integration of more renewable generation, creating instability,
- shift toward distributed energy resources and two-way power flows,
- growing digitalization, electrification, and power quality needs,
- increasing extreme weather damage and cybersecurity risks,
- market shifts and declining demand in maturing economies,
- consumer desire for more control, choice, and reliability.

These interrelated dynamics are necessitating substantial investments and innovation in grid technologies, markets, business models, and regulations worldwide. The individual features in the aforementioned list are discussed in detail as follows:

Aging infrastructure and equipment. Much of the grid in many nations dates back to the 1950s through the 1980s, with components long exceeding standard lifespan assumptions. For example, 70% of US transformers are over 25 years old, and 65% of circuit breakers are 30+ years old. The average age of UK power infrastructure is 46 years. Outdated equipment leads to voltage fluctuations, more losses, and an increased risk of failure. But replacing aging infrastructure is quite expensive, at US$1 to 5 million per mile of transmission line and US$5 to 10 million for a high-voltage transformer. Frequent repairs also drain utility budgets. Japan spends US$12 billion per year on grid maintenance alone. With infrastructure investment lagging depreciation rates, innovative funding solutions and new asset management approaches are essential.

Rising electricity demand. Global electricity demand grew 2.5% annually over the last decade and is projected to rise nearly 50% by 2050. Developing economies often see very rapid demand growth at 5%–10% annually as shown in Fig. 2.10. Expanding populations, urbanization, industrialization, appliance adoption, and rising living standards drive growth. Even regions with flat demand, like Europe, face increasing peak capacity needs due to cooling demands. Meeting demand requires adequate investment in generation and delivery infrastructure. Steps like demand response, energy efficiency, and distributed energy solutions help minimize grid capacity needs. But substantial new infrastructure is still vital in most regions.

Renewable energy integration. Renewable electricity generation is growing rapidly as costs plummet but presents grid integration challenges. Wind and solar output fluctuate based on weather patterns rather than being directly dispatchable. This complicates balancing supply and demand. Renewables also congest certain corridors, as prime resources are often located far from load centers. More transmission capacity, energy storage, flexible generation,

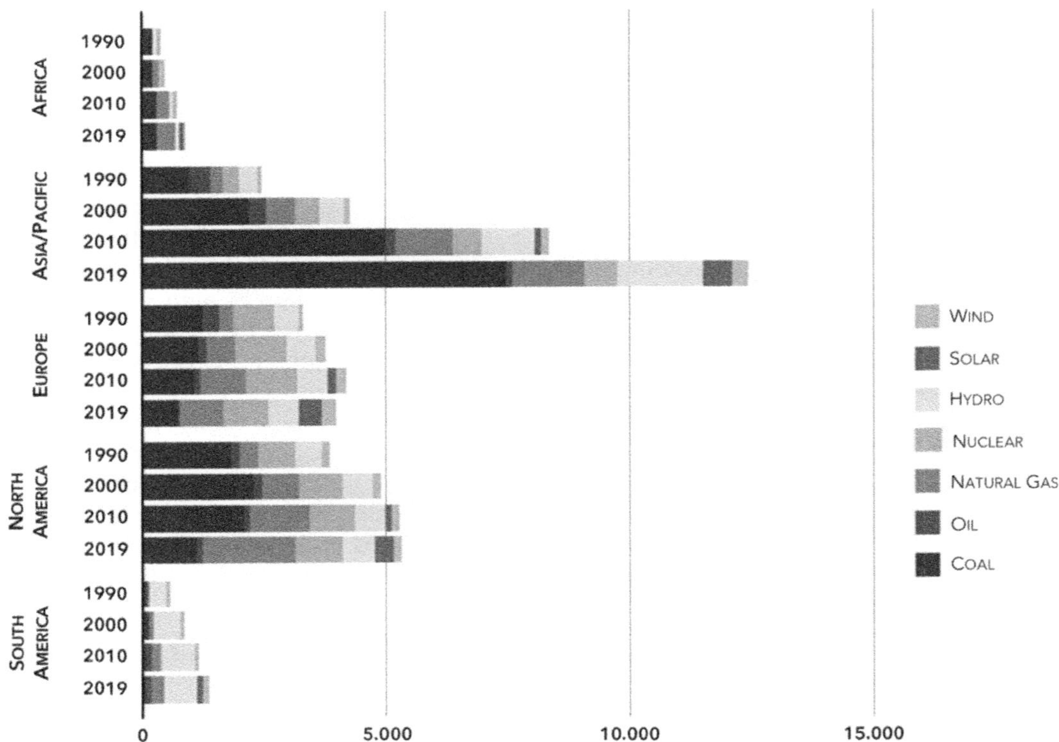

FIG. 2.10 Electricity generation mix over time, by region. The bars reflect terawatt-hours of electricity generation by region and fuel source for 1990, 2000, 2010, and 2019 [2].

advanced forecasting, and grid management capabilities are required to accommodate renewable expansion. However, enabling infrastructure development often lags renewable investments. Streamlined planning and supportive policies are needed.

Distributed energy resources. Rooftop solar, batteries, electric vehicles, microgrids, and other distributed resources (of either primary or transformed energy) are transforming grid architecture. Localized generation reduces stress on the core grid, but high adoption creates complications. Bidirectional and less predictable power flows emerge on local grids. The rise of prosumers alters utility business models. While providing opportunities, distributed resources must be properly integrated with central grid coordination, monitoring, and management. Communication and control capabilities take on heightened importance. New technical standards, markets, and regulatory frameworks are also essential.

Power quality needs. Voltage and frequency fluctuations disrupt sensitive digital loads and manufacturing processes. Modern equipment and advancing technologies necessitate very high-quality and reliable electricity. But variable renewables, extreme events, and aging hardware degrade power quality, Fig. 2.11. Utilities are deploying more advanced sensors, automation, energy storage, and power electronics to enhance quality. However, achieving costly resilience enhancements remains challenging, particularly in developing nations. New reliability metrics and standards matched to emerging technologies are also required.

Physical and cyber threats. Grid infrastructure is prone to storm, flood, and wildfire damage as extreme weather intensifies. Rising sea levels also threaten coastal facilities. The merging of operational and information technologies exposes systems to cyber intrusions. A 2022 climate analysis found 25% of US transmission lines face elevated weather risk. Ukraine's grid has been repeatedly disrupted by cyberattacks. Hardening, redundancy, quicker recovery capabilities, and heightened vigilance are imperative but costly to implement.

The economics of grid infrastructure. Electricity transmission and distribution assets are extremely capital-intensive, coupled with long lifecycles that produce limited direct profitability for owners. High-voltage direct current transmission can cost over US$1 million per mile. New 765 kV lines are over US$3 million per mile, and extra high-voltage lines planned in India top US$5 million per mile. Distribution upgrades cost US$150,000 to US$500,000 per mile. Substation costs range from US$2 to US$6 million. Given typical returns on grid investments in the low single digits, supplementary revenue streams, and creative financing are essential to attract private capital at the scale required.

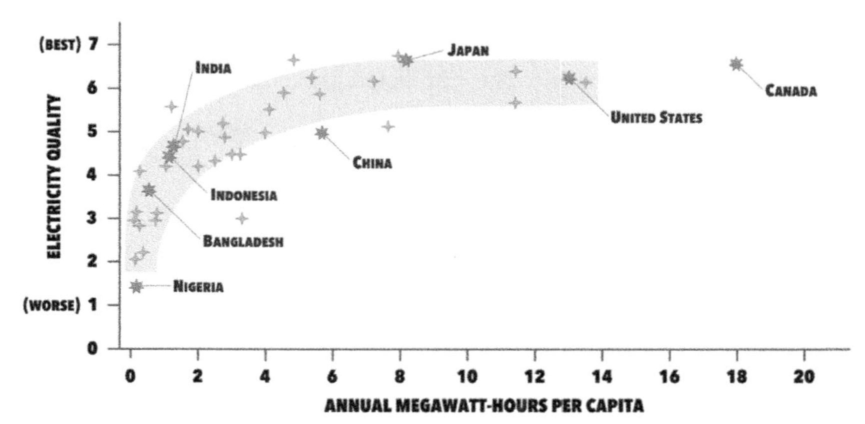

FIG. 2.11 Per capita generation and electricity reliability .

Key regional grid challenges

North America. The North American power grid faces limitations in capacity, efficiency, reliability, and security. Rising electricity demand is stressing aging transmission infrastructure that was not designed for large intermittent renewable energy sources like solar and wind. Congestion leads to bottlenecks that increase costs and inefficiencies. Much of the grid uses aboveground lines that are vulnerable to weather events and physical or cyberattacks. Investments in smart grid technologies have been limited. Regulatory barriers between regions also constrain optimal power flows. The largest blackout in North American history occurred in 2003, cutting power to 50 million people and demonstrating the grid's vulnerability. Climate change is causing more extreme weather events that further threaten reliability.

Europe. Europe faces constraints on adequate interconnection between national grids and renewable energy integration. Cross-border interconnections are limited, hampering efficiency and the ability to balance supply and demand across the continent. The rapid growth of renewable energy has also strained grid flexibility due to intermittency issues. Distribution systems were not designed for decentralized, two-way power flows from sources like rooftop solar. Variable renewable energy sources are driving the need for major transmission investments and grid modernization. Europe also lacks a coordinated continental-scale grid management system. The regulatory environment and differences between national energy policies pose challenges to improving interconnection infrastructure.

Asia. Many parts of Asia lack sufficient transmission infrastructure. Rural electrification rates remain low in some regions. Developing Asian nations are rapidly expanding power generation capacity, mostly from coal, which is straining existing grid infrastructure. Nations like India have seen major blackouts, demonstrating inadequate transmission capabilities. Renewable energy growth throughout Asia also requires substantial new transmission investments to connect sources to load centers. Meanwhile, aging equipment routinely causes disruptions in countries like China. Insufficient financing, inconsistent policies, regulatory issues, and unreliability discourage private investment. Conflicts between national and regional planning also hamper cross-border interconnection projects. Lack of coordination further complicates optimal power trading and dispatch across the region.

Africa. Africa has the lowest electrification rate of any continent, along with chronic power shortages. Grid connectivity is extremely limited, with only a few major transmission links between countries. Most generation is located far from load centers. Running electricity to rural areas is costly over long distances. The continent's inadequate transmission infrastructure also leads to high energy costs. Losses are high from technical issues and electricity theft. Natural gas resources often lack pipelines to power plants. Renewable energy penetration also requires substantial new transmission investment, especially from hydro and geothermal sources to urban areas. Political instability further hinders infrastructure development and maintenance in some nations. Overall, Africa critically lacks high-capacity transmission lines to deliver power reliably over long distances.

South America. South America faces constraints on integrating abundant hydroelectric and renewable resources. Grid interconnection between countries is limited, preventing optimal utilization of generation assets. For example, Brazil's hydroelectric potential remains underdeveloped. Cross-border interties are insufficient to take advantage of complementary solar, wind, and hydro resources across nations. Congestion is increasing on internal transmission networks with growing electricity demand. Much of the continent relies on vulnerable aboveground transmission lines susceptible to natural disasters. Infrastructure is strained by rising electrification. Investments in transmission lag behind power generation. Utilities struggle to recover the costs of major new projects. Variable clean energy integration further exacerbates grid modernization needs and the lack of transmission capacity.

Advanced grid technologies

New grid technologies provide many solutions to the aforelisted grid challenges, but they require substantial investment. Key innovations include the following:

- composite, sensors, drones, and robotics to monitor grids,
- solid-state transformers for voltage regulation,
- storage systems to balance renewable variability,
- advanced power electronics and grid-forming inverters for stability,
- flexible AC transmission devices to control power flows,
- high-temperature superconductors for efficient transmission,
- distributed energy management systems,
- enhanced cybersecurity hardware, software, and services,
- renewable forecasting and predictive analytics,
- resilient smart grid architectures and automation.

Despite proven benefits, adoption of advanced technologies often lags due to financial constraints. Prioritized deployment, guided by cost-benefit analysis, maximizes value. Interoperability and integration with legacy systems during transitions are also crucial.

Grid modernization economics

The overall investment needed to modernize and expand global grid infrastructure is estimated at US$7–10 trillion by 2040. In the United States alone, over US$2 trillion is projected to be required by 2030, a similar investment to that needed in Europe by 2050. Upgrade and resilience investments often conflict with desires to limit consumer cost impacts. But deferring action raises risks and long-run costs. Transmission projects are hampered by conflicting incentives across states and stakeholders. Passing costs to parties proportionally benefiting from enhancements helps secure support.

Innovative utility financing, such as securitization and rate recovery, can fund projects by issuing low-cost bonds to cover costs. Private investors and technology partners are increasingly participating in grid upgrades through public-private partnerships. Contractual returns help attract private capital. Location-based marginal pricing of transmission service charges users based on congestion and losses to incentivize efficient siting decisions.

Upgrades should focus on high-value, high-risk areas based on data-driven assessments of system conditions, vulnerabilities, and interdependency impacts. No regret moves with ancillary benefits beyond grid reliability are prudent starting points. Flexible adaptation pathways adjust plans based on changing conditions vs. overly rigid long-term forecasts.

Policy recommendations

Modernizing aging infrastructure, enhancing sustainability, and strengthening grid resilience against growing threats require targeted policy efforts alongside technology and investment innovations:

- grid enhancement incentives, grants, and loans,
- streamlined regulatory approval and siting processes,
- support for demonstrations and pilot projects,
- favorable tax treatment and accelerated depreciation,
- interconnection standards and data exchange protocols,
- long-term policy clarity and priority grid expansion plans,
- wholesale market design reforms and incentives,
- cost recovery frameworks facilitating investment,
- carbon pricing and clean energy incentives to motivate sustainability upgrades.

Through appropriate financial incentives, clear signals, transparent processes, and proactive planning, substantial progress can be made in addressing the pressing challenges facing electricity transmission and distribution worldwide.

Conclusion

Reliable, efficient, and environmentally sustainable electricity grid infrastructure is foundational to future prosperity and human development. However, aging assets, rising demand, distributed generation, extreme weather, cyber threats, and institutional constraints jeopardize grid resilience and capability. Major investments in technologies, infrastructure, markets, and policies are imperative to address these challenges through a combination of innovation, collaboration, planning, and leadership. By mobilizing the needed resources and expertise, the transmission and distribution networks that power the world can be strengthened to meet the growing needs of the 21st century.

Transportability of energy sources

Need to transport energy: Conversion of primary energy sources into energy carriers. General considerations

The global abundance and widely distributed nature of renewable and nonrenewable primary energy sources are clear and have been discussed in section "Mapping of Energy Sources" of this chapter. It has also been established that meeting the growing and evolving demand for final energy, in particular electricity, has required and will continue to require large efforts to maintain, upgrade, and substantially expand the existing infrastructure, which poses monumental challenges technically and economically. An alternative and/or complementary pathway to meeting the growing demand for final energy is to convert primary energies into energy carriers. The decarbonization of industry further underscores this particular node in the energy value chain.

There are several key aspects of the demand for energy that should be evaluated to understand the viability and appropriateness of a particular energy carrier. This section considers the most relevant aspects of the conversion of primary energy into these varied energy carriers, represented in the second node of the simplified value chain in Fig. 2.12.

Primary energy sources are those derived directly from nature. They are often reactive and therefore not ideal for transportation. Actually, the property of a primary energy source that makes it so attractive for energy production also makes it problematic for transportation since a reaction of a primary energy source during transportation could lead to an adverse event, which could cause damage to equipment and loss of life. For these reasons, it is often desirable to convert these primary energy sources into energy carriers. According to ISO standard 13600, an energy carrier is defined as "a substance or sometimes a phenomenon that contains energy that can be later converted to other forms, such as mechanical work or heat." [16]. Energy carriers "occupy intermediate steps in the energy-supply chain between primary sources and end-use applications. An energy carrier is therefore a transmitter of energy" [17]. In more detail than Fig. 2.12, Fig. 2.13 illustrates how energy carriers are pivotal in the transportation of energy between the primary energy source and the end user.

Energy carriers convert primary energy sources into stable forms of energy that are less susceptible to unintentional reactions during transport. Some primary energy sources are also the energy carrier itself, as is the case with natural gas and woody biomass. Energy carriers have shown a shift from solid to liquid and from liquid to gas in recent years. Interestingly, energy carriers also act as a form of energy storage, and this is very convenient since, with the rise of renewables, the need for storage has grown and outpaced today's conventional storage technology. Therefore, an increase in the utilization of energy carriers and the associated technologies to aid in the energy transition to carbon neutrality will be seen in the near future.

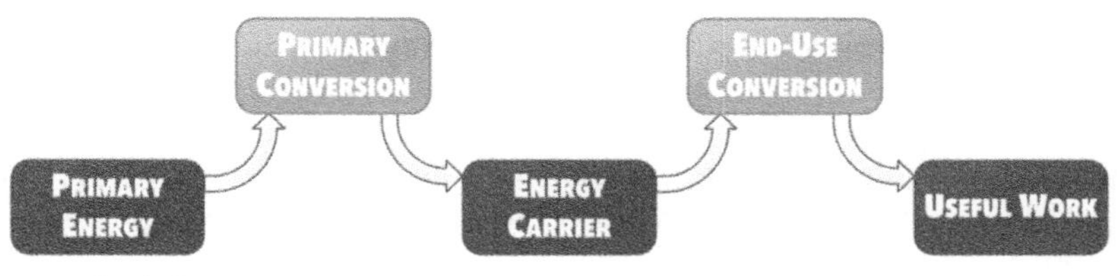

FIG. 2.12 Simplified energy supply chain.

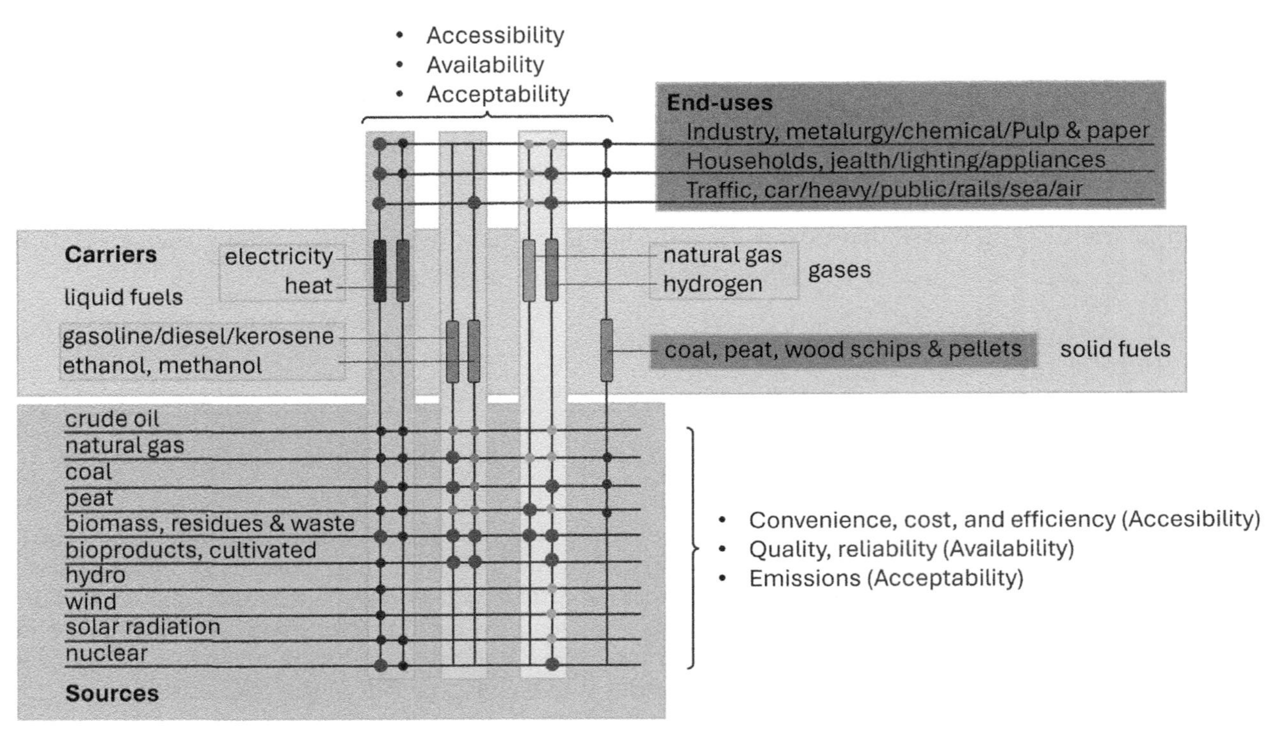

FIG. 2.13 Diagram of the interplay between primary energy sources, energy carriers, and end users [17].

Considerations to assess the viability of conversion into energy carriers

From a demand-side perspective, the form of useful "duty" sought to be done can broadly be defined as of either kinetic (e.g., rotating equipment/machinery, and transportation) or thermal (chemical processes, drying, HVAC, etc.) nature. For thousands of years, all the way through the first industrial revolution, solid fuels (generally in the form of biomass and coal) were combusted on-site for heat. Prior to thermal power generation, flowing water frequently provided the mechanical power to do work such as milling grains [18]. These historical examples are important to note, as the second industrial revolution at the turn of the previous century was in large part enabled by the electrification of mechanical processes. At the heart of this revolution was the ability to use electricity as an energy carrier, which enabled industry growth at various scales and locations [19]. Today, many industrialists refer to the further evolution of energy systems marked by decarbonization, distributed energy resources, and the use of data to improve cost and efficiency as a third industrial revolution [20].

Regardless of the original source of energy, either renewable or nonrenewable energy carriers transport chemical energy in gaseous, liquid, or solid forms. There is a broad range of processes to convert primary energy into these types of energy carriers, and new processes and combinations therein have been and continue to be developed. Key features or dimensions along which the conversion of primary energy to energy carriers can be considered for viability, along with reflections on example conversion steps, with a more detailed examination are provided elsewhere in this volume. The dimensions considered here are scale and scalability, capital cost, operating cost, turn-down or part-load efficiency, and finally setting or location.

Scale and scalability. Depending on the scale of primary energies available, a viable conversion technology must economically match the scale of that primary resource. Often, multiple conversion steps are required due to the scale (or lack of scalability) of the conversion process.

On a commercial scale, electricity producers must convert and condition the power generated into the appropriate voltage and phase in order to connect to the grid and distribute power. Even with residential-scale distributed power generation with solar, voltage conversion of DC to AC can take place at the scale of the home or even as small as the individual solar panel [21]. Renewable electricity (wind, solar, etc.) requires significant processing in the form of voltage regulation and grid tie-in, where storage might occur at the primary energy site, elsewhere on the grid, or at the point of consumption. Viability is again often driven by scale.

Crude oil is another primary energy that requires a certain amount of processing at or near the wellhead to treat the crude with dewatering, solids removal, and even basic fractionation (stabilization) before the crude can be stored in an atmospheric tank [22]. Only then can the crude be aggregated into the sufficient quantity to operate a sophisticated large-scale oil refinery. The oil refinery then produces more fungible energy carriers in the form of gasoline, diesel, and jet fuel. The scales of each of these stages can significantly impact the ultimate cost of energy for the end user.

These are but two examples of how the appropriate scale of a conversion process is mission-critical to the viability of producing an energy carrier.

Capital cost. Often related to scale, the first-time build cost of physical systems to convert primary energy into a useful energy carrier can make or break the viability of a carrier. There are many examples of the high up-front capital cost of a conversion step preventing the conversion of even extremely low-cost primary energy into a viable energy carrier.

One well-established example is the liquefaction of already-processed natural gas into LNG. LNG is a well-regarded energy carrier that is considered completely mature [23]. Yet, the capital intensity of LNG plants can be high enough to prevent the monetization of even the cheapest sources of natural gas. In recent years, there has been renewed interest in floating LNG plants, designed to be implemented at multiple locations during the equipment's useful life, thereby spreading high capital costs over multiple projects [24].

Operating cost. The operating costs of conversion systems can vary as widely as capital costs. Like capital cost, the economic viability of a conversion process can also be made or broken based on operating costs. In the oil and gas industry, many upstream and midstream conversion processes have become increasingly automated (or centrally controlled) in an effort to minimize operating costs. Replacement parts and maintenance costs have been reduced through higher reliability, unit standardization, and the use of predictive maintenance [25]. Often, inputs for a conversion system are among the most significant costs, and this includes energy inputs. When conversion steps occur near stranded resources, like remote wind farms or natural gas monetization, the variability in energy costs is not a significant operational risk over the life of the project. On the contrary, when energy-intensive conversion steps occur closer to commodity energy markets, energy price fluctuations can add significant risk to the operation of a conversion step.

Turn-down ratios, part-load efficiency, and load-following. The conversion of primary energies to a viable energy carrier can be located anywhere between the primary energy source and the end user of the energy, depending on scale, economics, technical requirements, safety, and a wide range of other considerations. In many cases, the availability of the primary energy or the demand for the energy carrier will vary over time. As a result, in many applications, the ability of a conversion technology to perform at high efficiencies while following the variability of either supply (of primary energy) or demand (for the carrier produced) is critical. Historically, upstream conversion of oil and gas products has been relatively insensitive to major variabilities, with a steady supply of primary energy and the ability to send the energy carrier product to storage. In the case of renewable energies like wind and solar, the dynamics are dramatically different. Interestingly, DC power distribution has recently gained tracking in large buildings due to the proliferation of on-site electricity generation and on-site battery energy storage [26].

Setting or location. Most conversion systems will ultimately be located near or closer to the primary energy source, given that energy carriers are often used to address the challenges of variability, transportation, or storage of energy products. The location or setting of a conversion facility becomes much more critical if the conversion process includes either coproducts that require disposal or monetization or critical inputs.

For example, natural gas processing will frequently produce natural gas liquids (NGLs) (C5+) as a coproduct that requires specific infrastructure connected to an NGL fractionation plant. Natural gas processing plants can also produce a pure CO_2 stream ideal for sequestration, but only if the infrastructure to dry, compress, and transport CO_2 to a sequestration facility is viable.

An emerging example case to be considered is the use of electrolyzers to produce hydrogen from renewable power that would otherwise be curtailed due to grid limitations. Of course, the production of hydrogen even from negative-cost electricity still requires the availability of high-purity water, which complicates the infrastructure and economics of this solution.

Direct transportation of primary energy sources and carriers

Primary energy sources can be classified as gaseous, liquid, or solid. This chapter discusses major gaseous, liquid, and solid energy sources. It should be noted that this chapter intends to give the reader a brief introduction to several of these energy carriers, whereas detailed discussions can be found in subsequent chapters.

Gaseous energy carriers

Hydrogen. One of the cleanest forms of gaseous energy sources is hydrogen, since, when used directly in combustion, the by-product is water, which, in its pure form, can be used for day-to-day applications. Hydrogen storage plays a critical role in advancing the utilization of hydrogen as a clean and versatile energy carrier. Nevertheless, given hydrogen's low density and high reactivity, finding efficient and safe methods for its storage is a key challenge.

Various storage technologies have been developed, each with its own advantages and limitations. These include compressed hydrogen gas storage, where hydrogen is pressurized and stored in high-strength tanks; liquid hydrogen storage, where hydrogen is cooled to extremely low temperatures to become a liquid; and solid-state hydrogen storage, which involves adsorbing hydrogen onto porous materials or storing it within chemical compounds. Researchers continuously explore innovative materials and approaches to enhance hydrogen storage capacity, improve energy density, and ensure practicality for applications ranging from fuel cell vehicles to renewable energy storage. Successful hydrogen storage solutions are pivotal to realizing a sustainable energy future and reducing the world's reliance on fossil fuels. Efficient hydrogen storage mechanisms are the linchpin of the hydrogen economy, holding the key to unleashing the potential of hydrogen as a transformative energy source. Given its diverse applications, from fuel cells to industrial processes, devising effective storage methods is crucial. Researchers are investigating a spectrum of approaches, including physical storage through compression and liquefaction and chemical storage via reversible reactions with materials like metal hydrides. Solid-state storage options, like adsorption onto porous materials, are also gaining traction. Pursuing high storage capacity, safety, and economic viability drive innovation in this field. Successfully overcoming the challenges of hydrogen storage will accelerate the shift toward sustainable energy solutions and redefine the landscape of global energy distribution and consumption. Detailed discussions on the advantages and opportunities for hydrogen as an energy carrier are discussed in a following dedicated section and, in much more detail, later in this book.

Natural gas. Natural gas is comprised mainly of methane (CH_4) (>70%) and a few higher alkanes like ethane, propane, and isomers of butane. One of the advantages of natural gas is that it generates fewer pollutants than coal [27] although natural gas purification (due to the presence of H_2S and CO_2), storage, and transportation are energy intensive. Natural gas storage and transport play integral roles in ensuring a reliable and efficient supply of this essential energy resource. Most of the time, the storage of natural gas is in underground reservoirs. These storage facilities serve as a crucial buffer, mitigating the effects of demand fluctuations and supply disruptions. These underground reservoirs can be depleted oil and gas reservoirs, aquifers, or salt caverns. The gas is injected during periods of low demand and withdrawn when needed, maintaining a steady supply for residential, commercial, and industrial consumers. This strategic storage capability enhances energy security and stability. Transporting natural gas involves a diverse range of methods to meet varying needs. Pipelines are the backbone of domestic gas distribution, efficiently transporting large volumes of gas over considerable distances. They offer a continuous and cost-effective means of delivering gas to end users. For international trade, LNG has gained prominence. Natural gas is cooled to around −162°C, converting it into a liquid form that takes up significantly less volume, thus allowing for efficient marine transport. Upon reaching its destination, LNG is regasified, injected into the local distribution network, or used for power generation. A detailed discussion on natural gas and its transport are provided later in this book.

Ammonia. Ammonia has emerged as a promising energy carrier with the potential to address energy storage and transportation challenges sustainably. Ammonia, composed of nitrogen and hydrogen, can be produced using renewable energy sources through a process called "green ammonia" production [28]. It is also considered a hydrogen carrier, allowing for the circumvention of storage and transport issues for hydrogen, and can serve as a versatile fuel for various applications [29]. In this regard, one of the major challenges in utilizing ammonia is the production of nitric oxides during combustion.

Ammonia's properties make it particularly suitable for long-distance transport and storage, overcoming some of the limitations of other renewable energy carriers. Ammonia is typically transported in its liquid form, which requires careful consideration of safety protocols due to its toxicity and potential reactivity. Specialized vessels designed for ammonia transport ensure secure containment and regulated conditions to prevent leaks or accidents. Establishing effective storage methods is essential because of their crucial role in sectors such as agriculture, chemicals, and energy.

Depending on the specific application, ammonia can be stored in various forms, including liquid and pressurized gas. For energy purposes, liquid ammonia is particularly promising due to its high energy density and relative ease of handling. However, its toxic nature necessitates stringent safety measures during storage and handling. As research continues to focus on production efficiency, storage, safety, and integration of ammonia into existing energy systems, it holds the potential to play a significant role in the transition to a cleaner and more sustainable energy future. A detailed discussion on ammonia combustion can be found in subsequent chapters.

Liquid petroleum gas (LPG). LPG is a hydrocarbon fuel mixture primarily composed of propane and butane, both of which are alkanes. Propane is a three-carbon alkane with the chemical formula C_3H_8, while butane is a four-carbon alkane with the formula C_4H_{10}. Both of these compounds are gases at room temperature and atmospheric pressure, but they can be easily liquefied through compression and cooling. This liquefied state enables LPG to be efficiently stored and transported in containers, making it a practical energy source.

LPG has established itself as a versatile and valuable energy carrier, serving a range of domestic, commercial, and industrial applications. As said, its distinctive properties enable it to be stored and transported as a liquid at moderate pressures and low temperatures, facilitating efficient distribution and accessibility. LPG is widely used for cooking, heating, and as a fuel source for vehicles, particularly in areas where natural gas infrastructure is limited. Its relatively low emissions and high energy content make it an attractive transitional energy source as societies move toward cleaner alternatives. The adaptability, portability, and convenience of LPG position it as a significant contributor to global energy systems, supporting diverse needs while playing a role in reducing environmental impacts.

Indeed, LPG transport plays a pivotal role in distributing this versatile energy source to various consumers across the globe. Specialized LPG transport vessels, such as ships and trucks, are designed with safety measures to prevent leaks, fires, and explosions. LPG is stored in sturdy containers to prevent any accidental releases during transit. These vessels have sophisticated pressure and temperature monitoring systems to ensure safe conditions throughout the journey; upon arrival, LPG is easily regasified for use in various applications. LPG storage can take various forms for small-scale use, ranging from small household cylinders to larger tanks at industrial facilities and distribution centers. The efficient and secure transport of LPG enables access to a clean and versatile energy source in regions where natural gas pipelines might not be feasible, contributing to a diversified energy mix and addressing energy needs in a reliable and environmentally responsible manner.

Liquid energy carriers

Crude oil and petroleum products. Crude oil and its refined derivatives, known as petroleum products, are indispensable energy carriers that power modern societies and fuel economic development. Crude oil, a complex mixture of hydrocarbons, is extracted from the Earth's crust and serves as the primary feedstock for a range of products. Refining processes transform crude oil into various petroleum products, such as gasoline, diesel, jet fuel, and petrochemical feedstocks. These products are vital for transportation, industrial processes, and the production of many goods. Their energy density, ease of transport, and established infrastructure make them versatile energy carriers, enabling the movement of people and goods across vast distances and powering industries that underpin modern life. Developments in the transport and storage of crude oil and petroleum products have continuously evolved to meet the demands of a dynamic global energy landscape. Innovations in transportation have led to the construction of larger and more efficient oil tankers and pipelines, enabling the movement of vast quantities of crude oil across continents and oceans. Advanced safety measures, including double-hulled vessels and sophisticated leak detection technologies, have been implemented to mitigate environmental risks. The construction of modern storage tanks with advanced corrosion resistance and containment systems ensures the safe and secure storage of crude oil and petroleum products. Many of the port cities, like Rotterdam and Fujairah (UAE), have their developments revolving around crude oil storage and transportation facilities.

Additionally, strategic storage reserves have been established in some regions to stabilize supply during emergencies. Alongside these developments, the industry has explored digital technologies such as sensors, data analytics, and automation to enhance efficiency, optimize inventory management, and ensure timely delivery. As sustainability becomes a paramount concern, efforts have been directed toward reducing the environmental impact of transport and storage. More details on this are discussed in subsequent chapters.

Biofuels. Biofuels have emerged as a promising and environmentally conscious energy carrier, offering a renewable alternative to fossil fuels. Derived from organic materials such as crops, agricultural residues, and waste, biofuels encompass a range of products like biodiesel, ethanol, and biogas. Their production and use contribute to reduced GHG emissions compared to traditional fossil fuels, as the carbon dioxide (CO_2) released during combustion is offset by the carbon absorbed during the growth of the feedstock. Additionally, some of the oxygenated biofuels help reduce soot emissions by increasing combustion efficiency [30]. As technology advances, biofuels can be seamlessly integrated into existing infrastructure, powering vehicles, and industrial processes with minor modifications. For example, most light-duty vehicles in the United States and EU can operate using a 10% blend of ethanol in gasoline. Transportation of biofuels is done through existing fuel distribution networks, pipelines, and transportation modes such as trucks, trains, and ships. Biodiesel, ethanol, and other biofuels have properties like their fossil counterparts, allowing them to be seamlessly blended with traditional fuels or used as standalone options in vehicles and equipment. However, given the diverse feedstocks used to produce biofuels, careful attention must be paid to the supply chain to prevent any adverse impacts on land use, food production, and ecosystems. Advancements in logistics, sustainable feedstock sourcing, and efficient transportation are pivotal in maximizing the positive environmental benefits of biofuels while minimizing their potential drawbacks. More discussion on biofuels is provided later in this book.

Gas to liquid (GTL). GTL technology has emerged as a transformative energy carrier, converting natural gas into liquid fuels through advanced synthesis processes. This innovation addresses the challenges of transporting and utilizing natural gas, which is often stranded due to remote locations or a lack of infrastructure. GTL technology involves the conversion of natural gas into liquid hydrocarbons like diesel, naphtha, and waxes, effectively creating a transportable and versatile energy source. These GTL products possess lower sulfur and aromatic content than their crude oil counterparts, resulting in cleaner-burning fuels with reduced emissions. GTL fuels can be seamlessly integrated into existing diesel and gasoline supply chains, offering a bridge between traditional fossil fuels and emerging renewable energy sources. More discussion on GTL is provided later in this book.

Vegetable oils. Vegetable oils have gained prominence as a renewable and environmentally friendly energy carrier, contributing to the diversification of energy sources and reducing carbon emissions. Extracted from crops such as soybeans, canola, and palm, these oils can be converted into biodiesel through transesterification processes. Biodiesel, a substitute for conventional diesel fuel, exhibits similar combustion properties but generates fewer harmful pollutants and GHG emissions. This characteristic makes vegetable oils an attractive option for powering diesel engines in various sectors, including transportation, agriculture, and industrial machinery. Moreover, the versatility of vegetable oils extends to their potential use in combined heat and power systems and as feedstocks for producing renewable chemicals. As research advances, optimizing feedstock sourcing, processing methods, and sustainability criteria remains pivotal to ensuring the overall ecological benefits of vegetable oils as an energy carrier.

Solid energy carriers (sources)

Coal. Coal, historically a dominant energy carrier, has played a significant role in powering industrialization and global development. Mined from the Earth's crust, coal is a carbon-rich fossil fuel that has been used for centuries to generate electricity, heat homes, and fuel various industries. Its energy density and widespread availability made coal a key driver of economic growth, particularly during the Industrial Revolution. However, the combustion of coal releases substantial amounts of CO_2 and other pollutants into the atmosphere, contributing to climate change and air pollution. To combat this, clean coal technologies such as carbon capture and storage were developed with the aim of mitigating emissions.

Several countries have historically relied heavily on coal to meet their energy needs, often due to abundant domestic coal reserves or economic considerations. Among these, China and India stand out as the world's largest coal consumers. China, driven by its rapid industrialization and urbanization, has been a major coal consumer for decades, though it has been making efforts to shift toward cleaner energy sources in recent years. With its growing population and energy demands, India also heavily relies on coal to generate electricity and support its expanding industries. Other significant coal consumers include the United States, Russia, and Japan. While these countries have made strides in diversifying their energy mix to include renewable sources, the challenge lies in balancing their energy security needs with environmental concerns, transitioning away from coal, and embracing sustainable alternatives to mitigate the impacts of carbon emissions. This has already been discussed in the foregoing section of this chapter.

The transport of coal is a critical aspect of its supply chain, facilitating the movement of this abundant fossil fuel from mining sites to power plants and industrial facilities. Coal transportation methods vary depending on distance, location, and infrastructure. Railroads are a common mode of transporting coal over longer distances, offering efficient and cost-effective delivery. Conveyor belts and trucks are also employed for shorter hauls, often within mining complexes or nearby power plants. Specialized bulk carrier ships, known as coal carriers or colliers, are used for international coal trade, transporting large volumes of coal across oceans to meet energy demands in different regions. The logistics of coal transport require careful planning to ensure timely and reliable delivery while adhering to safety and environmental regulations.

Coal is typically stored in large piles or in storage facilities near mining sites, power plants, or industrial facilities. The storage methods can vary depending on factors such as coal type, moisture content, and local climate conditions. Covered storage areas protect coal from exposure to the elements, preventing moisture absorption and degradation of its energy content. Additionally, coal stockpiles are managed to prevent spontaneous combustion, a potential risk due to the coal's high carbon content. Careful monitoring, proper ventilation, and temperature control are employed to minimize this risk.

Biomass. Biomass has emerged as a renewable and versatile energy carrier, encompassing diverse organic materials such as wood, agricultural residues, and dedicated energy crops. Biomass can be harnessed through various processes, including combustion, gasification, and fermentation, to produce heat, electricity, and biofuels. Similar to biofuels, its appeal lies in its ability to utilize CO_2 absorbed during plant growth, creating a closed carbon cycle that mitigates GHG emissions. It can also help in utilizing organic wastes from industries and households, which otherwise would contribute to CH_4 emissions and global climate change. Biomass can be readily integrated into existing energy systems, often co-firing with coal in power plants or producing biofuels for vehicles. However, its sustainability hinges on responsible sourcing practices to prevent deforestation, ensure food security, and minimize environmental impacts.

As global efforts intensify to combat climate change, biomass is a promising energy carrier, contributing to a cleaner energy mix while supporting rural economies and land management practices. More discussion on biomass can be found in subsequent chapters.

Hydrogen as an energy carrier

One example of an energy carrier that embodies many of the challenges mentioned earlier is hydrogen. Green hydrogen, in particular, is being generated in places with a high concentration of renewables; however, the infrastructure is not yet in place for the transmission of hydrogen directly in pipelines to end users. One alternative to gas transmission is to cool the hydrogen to liquify and transport it under pressure; this allows for it to be transported by truck and ship. However, this is very costly and energy-intensive while still carrying the risks of the volatility of the primary energy source. This is where ammonia comes in as an ideal energy carrier for hydrogen. Green ammonia can be produced from green hydrogen by pairing one nitrogen atom with three hydrogen atoms at high pressure and temperature through the Haber process. This can be done in a carbon-neutral manner through the utilization of renewables; unfortunately, most ammonia today is formed from natural gas through steam methane reforming, which generates large amounts of GHGs.

Hydrogen, as an energy carrier, exhibits distinct physical properties that set it apart from other common energy sources and carriers. Firstly, it is the lightest element in the periodic table, particularly compared to heavier elements found in carbon-based energy sources. This characteristic gives it a high energy-to-weight ratio, making it suitable for applications where weight is a critical factor. For example, hydrogen has an energy content of 120 MJ/kg compared to the energy content of natural gas and coal, which have an energy content of approximately 46.5 MJ/kg and 21 MJ/kg, respectively.[e] A table of the representative energy content of common energy sources is provided in Table 2.2.

Hydrogen is gaseous at room temperature and requires specialized storage and transportation infrastructure due to its molecular size and high energy content. This can be costly and challenging to implement and requires substantial considerations for safety. High leakage through plumbing fittings and rotating equipment is caused purely by the small size of the hydrogen molecule. The aforecited liquefaction of hydrogen for storage and transport can provide benefits similar to those of liquefied natural gas; this includes a substantial improvement in density, increasing its gaseous density of 0.09 kg/m^3 at standard temperature and pressure to 70.85 kg/m^3 when in liquid state. In other words, the storage capacity of liquefied hydrogen increases by over 78,000% compared to gaseous hydrogen at standard temperature and pressure. However, gaseous hydrogen is liquefied by cooling it to below −253°C (−423°F), which is nearly twice as cold as liquefying natural gas, which remains liquid at −162°C (−260°F). For comparison, using today's

TABLE 2.2 Physical properties of common energy sources.

Fuel	Hydrogen	Compressed natural gas (CNG)	Liquefied natural gas (LNG)	Methane	Propane	Bituminous coal	Crude oil
Chemical structure	H_2	CH_4 (majority), C_2H_6, and inert gases	CH_4 (majority), C_2H_6, and inert gases	CH_4	C_3H_8 (majority) and C_4H_{10} (minority)	C (majority), H_2, and inert gases	C_4 to C_{16} hydrocarbons
Physical state	Compressed gas	Compressed gas	Cryogenic liquid	Compressed gas	Pressurized liquid	Solid	Liquid
Flash point (°C)	N/A	−185	−188	−188	−73 to −101	160–180	−8
Energy content (LHV) (MJ/kg)	120	47	49.5	50	46	29	42
Energy content (HHV) (MJ/kg)	142	52	55	55.5	50	30	47
Density@15°C (kg/m^3)	0.09	0.7–0.9	N/A	0.67	1.7	690–800	900–1000

[e] DOE/GO-102021-5498, https://afdc.energy.gov/fuels/properties, January 2021.

technologies, the energy required to liquefy hydrogen is more than 30% of the energy content of the gas itself.[f] In addition, boil-off from storing liquefied hydrogen can lead to significant product loss if not properly controlled, as opposed to fossil fuels like coal, oil, and natural gas that are denser and easier to transport at room temperature. Another beneficial property of hydrogen is that it is a clean-burning fuel, as its combustion produces only water vapor as a by-product if kept free from contaminants.

Nonelectricity transport as an enabler of energy storage

Introduction

Previous sections of this chapter have illustrated the heterogeneous distribution of energy sources, some of which are scattered across the globe while others concentrate in very specific regions, and the spatial mismatch between these resources and the nodes where final energy is consumed. As a consequence of this, dedicated sections have touched upon the two most common alternatives to transporting energy: using the grid, once converted into electricity, or using other means of transportation in the form of either primary energy sources or energy carriers.

But the mismatch between energy sources and consumption nodes does not take place in the spatial domain only. There is also a mismatch in time since energy sources are harvested at times when the demand for energy is not high enough to consume them completely.

Energy storage is therefore a critical challenge for increasing the utilization of intermittent renewable energy sources and enhancing grid resilience. Batteries are one common approach but remain limited in scale and geographic flexibility, as shown in Fig. 2.14: Nonelectricity transport methods, however, can serve as innovative large-scale storage solutions by using excess energy to move a material to a higher gravitational potential. The stored potential energy can be recaptured through controlled descent. This section explores the technology fundamentals behind using nonelectricity transport of energy for grid-scale energy storage.

Variable and nonprogrammable renewable energy is making an increasing contribution to power generation. In parallel, "electrification of everything" is a fundamental mantra of decarbonization. These drivers combine to mean that long-term, high-capacity energy storage will become essential to balance supply and demand on the power transmission grid [31].

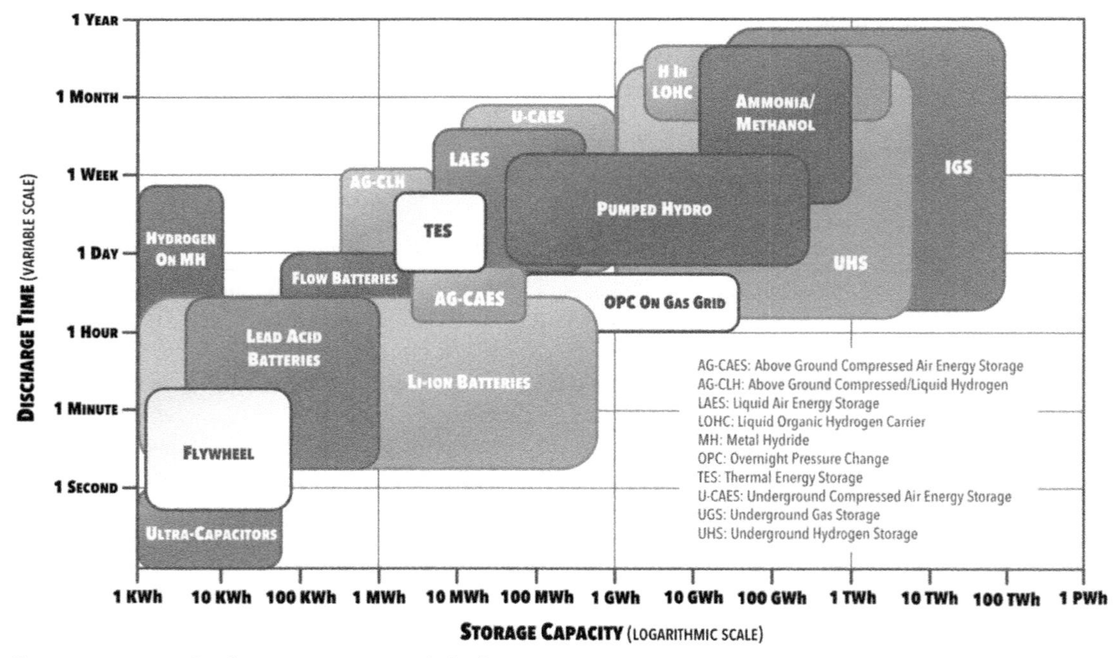

FIG. 2.14 Energy storage technologies: capacities and discharge times compared. *From S.B. Harrison, Compressed and Liquid Air for Long Duration & High Capacity, Modern Power Stations, 2023, pp. 24–25 (June issue).*

[f] DOE/Office of Energy Efficiency & Renewable Energy, https://www.energy.gov/eere/fuelcells/liquid-hydrogen-delivery.

Electric energy storage for grid stability

The electric power grid is undergoing a major transition worldwide, shifting away from centralized fossil fuel generation and toward more distributed renewable energy sources like wind and solar power. In 2020, renewables accounted for 29% of global electricity production, with solar and wind comprising over 80% of new additions. This energy transition aims to reduce GHG emissions and mitigate climate change. However, the variability and intermittency of renewables pose significant challenges for maintaining a stable and reliable electric grid. Energy storage can provide flexibility and help integrate more renewable resources onto the grid. By charging when renewable generation is abundant and discharging when it falls short, storage smooths out the variability and shifts supply to meet demand. Thus grid-scale energy storage is becoming an increasingly critical component of the 21st-century electric grid.

Many different energy storage technologies exist, each with its own strengths and limitations. As the scales of storage deployments grow, the technologies must be evaluated based on metrics such as cycle life, capital costs, round-trip efficiency, discharge duration, geographic constraints, and more. No single solution can meet all grid storage needs, so a mix of technologies is required. This section provides a technology overview and comparison of the most promising utility-scale energy storage options for supporting grid stability as renewable penetration increases. The focus is on mature and emerging storage technologies with capacity scales of tens of megawatts up to gigawatts, and discharge durations spanning minutes to days. After covering the key challenges and metrics for grid-scale energy storage, the section analyzes the current leading storage technology classes. Fig. 2.15 presents a summary of these technologies, alongside their current technology readiness level and commercial maturity.

Grid energy storage challenges and evaluation metrics

Several major challenges exist for storage technologies aimed at grid stability applications. First is the need for low costs, as storage must compete with conventional generators in providing grid services. Capital costs are critical, with desired targets below US$100/kWh for 4h duration systems. Lifetime and cycling also impact overall costs. Technologies should have long lifespans (15+ years) and cycle lives (>5000 cycles). Safety and environmental impacts are also important considerations. In addition, long discharge times from 10 to 12h are needed to cover daily fluctuations or multiday weather events. High round-trip efficiencies[g] (>85%) minimize losses during energy conversions. Location-flexible options that do not require specific geologic formations or terrains can be sited at more points on the grid. Finally, storage must be scalable to capacities of hundreds of MWs or higher to support grid-level applications.

Based on these application challenges, some key criteria for evaluating and comparing storage technologies include the following:

- Maturity level: commercial readiness.
- Capital costs: installed US$/kW and US$/kWh.
- Operating costs: US$/kW-year, including fixed O&M and variable O&M.
- Cycle life and lifetime: number of cycles and calendar years.
- Discharge duration: hours.
- Round-trip efficiency: % energy in versus energy out.
- Energy density: Wh/L or Wh/kg.
- Discharge power rating: ability to charge and discharge at high power levels relative to energy capacity.
- Geographic constraints: location limitations.
- Scalability: feasible size range.
- Safety: potential hazards.
- Environmental impacts: resource usage and emissions.

These metrics cover the technical performance, economic viability, and site flexibility required for a grid-scale energy storage solution to be successful. The following sections evaluate both existing and emerging energy storage technologies against these criteria.

[g] Round-trip efficiency is the ratio from the energy taken from the grid/source during the charging phase to the energy delivered to the end user during discharge.

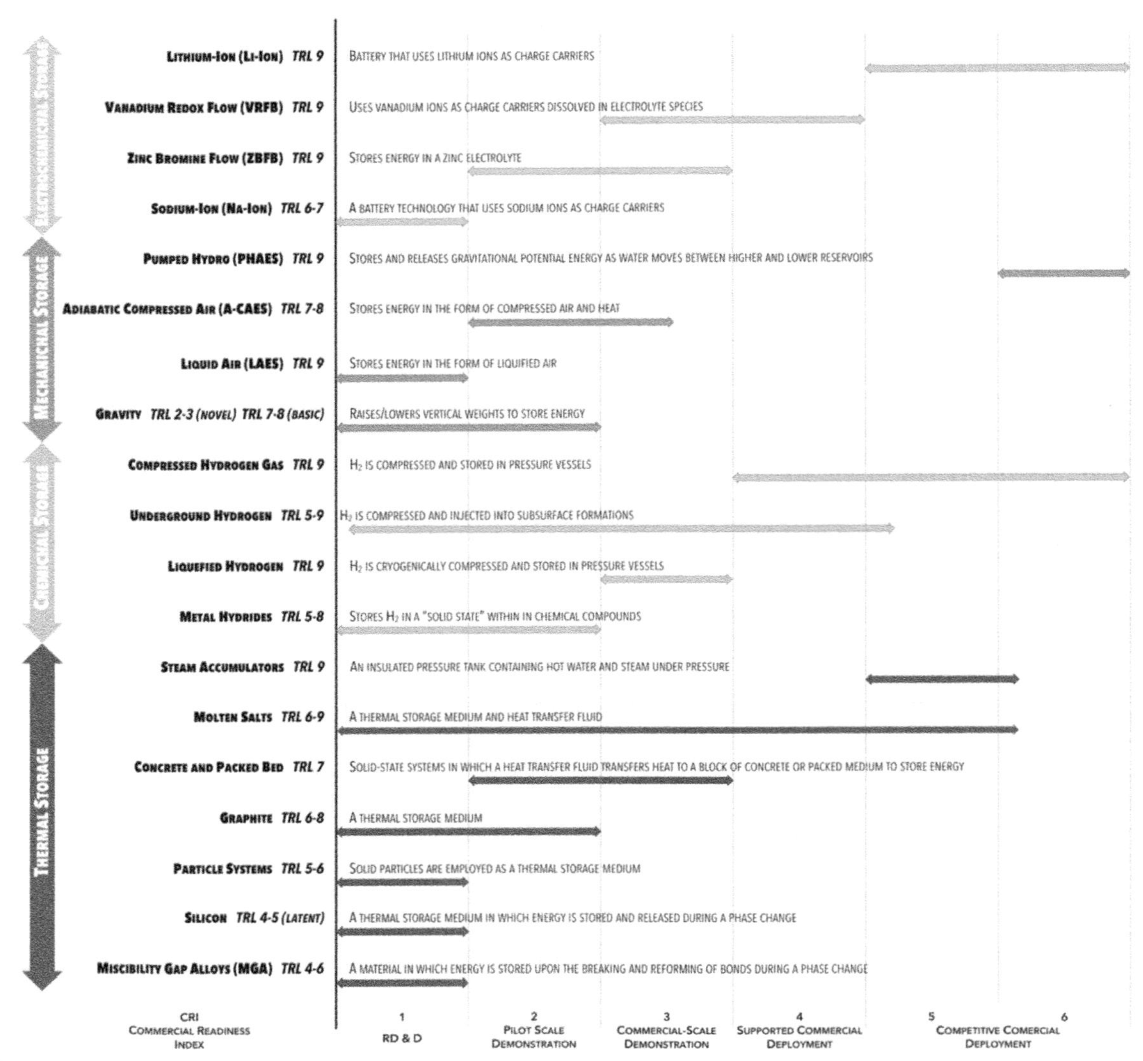

FIG. 2.15 Renewable energy storage roadmap—technology options [32].

Survey of grid energy storage technologies

Pumped hydro storage. Pumped hydro storage (PHS, or simply pumped hydro) is currently the predominant form of grid-scale energy storage, representing over 90% of storage capacity worldwide. In PHS, water is pumped uphill to an upper reservoir during the generation of electricity higher than electricity demand times and then discharged through hydraulic turbines, generating power when electricity demand is higher than power generation. The scale makes it attractive for rated outputs of 500 MW to over 3000 MW, with an energy storage capacity ranging from 8 h to more than 24 h. PHS can also provide valuable ancillary services to the grid beyond energy shifting, such as frequency regulation, spinning reserves, and black start capability.

At present, pumped hydro is the power storage technology with highest capacity. Much has been said about the use of lithium-ion batteries for power storage, but the timescale over which they can be depleted is relatively short, typically between 4 and 8 h. Nevertheless, this is not sufficient to support many requirements, such as continuous power delivery from a solar farm. Internationally, PHES is a commercially competitive technology deployed extensively across the world. Medium- and long-duration large-scale grid-connected systems are commercially competitive, whereas small-scale systems are undergoing supported commercial deployments.

However, PHS faces siting constraints and high capital costs. Suitable reservoirs and geography are required, limiting widespread deployment. PHS costs range from US$1700 to 4500/kW and US$100 to $300/kWh. Round-trip efficiencies are around 70%–80%. Cycling capacity over 25+ years is high at thousands of cycles. Given suitable sites and high storage capacity needs, PHS is a relatively mature and cost-effective option. But geographic restrictions limit growth potential, especially in regions like the United States with fewer mountains and rivers.

Compressed air energy storage. Compressed air energy storage (CAES) is another geologically dependent technology, which uses underground caverns to store compressed air. Like PHS, CAES is a fairly mature technology able to meet long discharge times of 10+ hours. Capital costs range from around US$1200 to 2200/kW and US$60 to 100/kWh. Efficiencies are 70%–89%. Location limitations exist but suitable geology is more common than with PHS, with over 130 candidate sites in the United States. Operating costs are lower than PHS since no water pumping is needed. Scalability is good up to hundreds of MWs. CAES can provide valuable operating reserves along with energy time-shift capabilities. Overall, CAES can serve as a cost-effective and flexible storage option when suitable geology is available nearby.

CAES systems are currently at the commercial-scale demonstration stage. The first commercial plants were installed in Europe (Huntorf, Germany 1978) and the United States (McIntosh, U.S., 1991). Huntorf is rated at 320 MW and can run for 2 h at full load during discharge, whereas McIntosh is rated 110 MW and can run for 26 h at full load.[h] Ore recently, a commercially contracted Hydrostor facility, is operational at 1.75 MW (discharge) and 10 MWh in Ontario, Canada, and two larger-scale projects are planned to commence construction in California, the United States (500 MW), and in North South Wales (200 MW).

Liquid air. Liquid air energy storage operates on the principle that air can be liquefied at low temperatures and ambient pressure, greatly reducing its volume by approximately 700-fold. Off-peak energy is used to liquefy air at −196°C in unpressurized vessels. When needed, the air is heated, causing it to expand and drive a turbine, which generates electricity.

These systems are in the pilot-scale demonstration phase, given that Highview Power has completed a 5-MW/15-MWh pilot plant and commenced construction of a 50-MW/250-MWh LAES system in the UK. Highview Power also plans to construct a 50-MW/500-MWh LAES plant in Chile.

Lithium-ion batteries. Lithium-ion (Li-ion) batteries have become the dominant battery technology thanks to their high cycle life, high round-trip efficiency (90%–95%), high energy density, and ability to scale down to small system sizes. Costs have declined significantly in recent years, reaching around US $175/kWh in 2020, with forecasts of US$100/kWh by 2024. However, degradation and relatively short discharge times limit lifespan. Most Li-ion batteries last 3–15 years with daily cycling. Discharge durations so far have been in the range of 2–6 h. Safety can be a concern, and environmental impacts depend on the materials used.

Overall, Li-ion batteries are the most mature and widely deployed battery technology. But their potential for longer duration storage may be limited. They work well for short-term power smoothing applications and ancillary services, but have not yet been proven at multihour storage scales needed for major renewable integration or seasonal storage. Energy density increase and cost reduction are still needed.

Flow batteries. Flow batteries are an emerging battery technology well-suited for long-duration energy storage. They use liquid electrolytes stored in external tanks and pumped through electrochemical cells to charge/discharge. This allows customizable sizing of power (cell stack size) and energy capacity (tank volume). Common flow battery chemistries include vanadium redox (VRFB), zinc-bromine (ZnBr), and iron-chromium (ICB). Flow batteries can achieve durations beyond 10 h cost-effectively and cycle lives of 10,000–20,000 cycles with 80%–90% efficiency. Capital costs are currently still high, around US$200–350/kWh, but dropping with scale.

Flow batteries are highly promising for sustained, bulk energy storage given their flexible sizing, long life, low degradation, and safe liquid electrolytes. Major limitations are higher upfront capital cost and low energy density. If costs continue declining, flow batteries could become preferred for durations beyond 8 h where Li-ion batteries are less ideal. Their future looks bright.

Hydrogen energy storage. Hydrogen is gaining significant interest for long-term energy storage. It offers a high energy density fuel that can be produced via water electrolysis when power supply exceeds demand. The hydrogen can then be stored in tanks or underground caverns and, later, converted back to power in fuel cells or other power generation devices when needed (reversible electrolyzer-fuel cell systems are very interesting for multiday storage). Storage costs are still high, with capital costs of US$2000–10,000/kW and US$20–$30/kWh, but they are expected to drop as technology matures. Round-trip efficiency ranges from 25% to 45%.

Hydrogen storage scales to large capacities exceeding 100 MW. Location flexibility is high since underground storage does not depend on topology. Environmental impacts vary depending on primary energy source (fossil fuels release CO_2; renewable electrolysis is clean). Safety procedures are critical. Overall hydrogen storage promises

[h] The plant was upgraded to 226 MWe in 1991, therefore reducing the running time during discharge for the same air storage capacity.

emission-free, multiday energy storage scalable to massive capacities. It remains an emerging technology but could fill a critical need for seasonal storage and high-capacity reserves as renewable energy expands.

Hydrogen can be stored in different forms:

- *Compressed hydrogen tanks*. There are three storage applications considered for the use of compressed hydrogen gas tanks, and these are represented at maturity:
  - Grid-connected storage, which involves the use of hydrogen gas tank storage for direct conversion to electricity for power applications, has completed pilot-scale demonstrations.
  - Storage of hydrogen at industrial sites along the export supply chain (e.g., use as a feedstock for ammonia production), which has been commercial for many decades.
  - Integration of hydrogen gas tanks at hydrogen refueling stations (HRSs), which has reached supported commercial deployment. Over 900 HRS and commercial-scale demonstrations have been undertaken worldwide. These deployments are focused on improving the costs of high-pressure compression and refueling speeds for heavy vehicles.
- *Liquefied hydrogen* involves compressing, cooling, and storing hydrogen at −253°C in cryogenic tanks. Commercial and demonstration projects include the Hydrogen Energy Supply Chain Project from Australia to Japan, which was the world's first commercial-scale demonstration of liquefied hydrogen export. Liquefied hydrogen at volumes of up to 100,000 m^3 is being developed internationally. In the transport sector, cryogenic hydrogen for refueling heavy-duty hydrogen vehicles is under investigation due to its potential to speed up refueling times.
- *Underground hydrogen* technology involves compressing and injecting hydrogen gas into subsurface formations via wells. There are four main geological options for underground hydrogen: salt caverns, depleted oil and gas reservoirs, aquifers, and hard rock caverns.

Overall, the maturity of underground hydrogen storage is in early stages. There are different maturity levels for various geologies, and more recent international demonstrations indicate that this maturity differs by end-use application also. Today, underground storage is only commercially used in the context of industrial chemical and refinery industries (e.g., in salt caverns across the United States and UK). Salt cavern systems for grid-connected storage applications are undergoing pilot-scale demonstrations. There are currently no large-scale underground hydrogen projects operating for the purposes of providing storage for electricity grids; however, a few are scheduled to be demonstrated internationally, such as the US Advanced Clean Energy Project. Depleted hydrocarbon reservoirs, aquifers, and engineered caverns are undergoing pilot-scale demonstrations.

- *Hydrogen pipelines*: Line packing is a widely used storage method in gas networks whereby natural gas is stored within the pipeline network. This method can also be used to store renewable gases such as hydrogen. The volume of hydrogen (or other gases) stored in a pipeline can be increased by increasing the pressure of the gas contained within the pipeline, and additional storage capacity can be added by increasing the diameter of the pipeline.

Hydrogen can be transported via a dedicated hydrogen pipeline (100% hydrogen) or injected into existing natural gas networks at lower concentrations. The extent to which hydrogen can be blended depends on the pipeline's material and its resistance to hydrogen embrittlement.

Pure hydrogen pipelines have been successfully demonstrated commercially. Air Products operates a pure hydrogen pipeline over 2500 km in the United States to supply the chemical and refinery sectors, and the European hydrogen pipeline is over 1600 km, partially operated by Air Liquide. In Australia, APA Group plans to convert the Parmelia Gas Pipeline in West Australia to a 100% hydrogen pipeline capable of transporting pure or blended hydrogen.

Blended hydrogen in gas networks has been demonstrated internationally. Blends of up to 20% hydrogen have been demonstrated in the UK, Europe, and the United States, in controlled environments, with the aim being to commence injecting hydrogen into gas networks within the next year.

Other notable technologies. In addition to the aforelisted technologies, the following list provides other technologies that are at a lower level of maturity:

- *Gravity*: In its basic form, gravity energy storage involves mechanically raising weights during the charging phase, which are then released during the discharging phase to drive a generator to produce electricity. Although this describes gravity storage with vertical weight systems, it is worth noting that rail and mountain gravity systems also exist, even though they are not discussed in this section.

Permutations of vertical gravity systems exist and include dry aboveground (using cranes) or underground systems (using deep caverns, such as abandoned mine shafts), as well as wet underground systems that operate under a principle similar to PHES. Some pilot and commercial-scale demonstrations are being deployed in grid applications, industrial parks, and mining applications. More novel designs are less developed.

- *Thermal energy storage*: Stores heat or cold for heating/cooling but it has lower potential to provide electricity grid services.

Thermal storage refers to systems that convert and/or store an energy input (i.e., heat, electricity, or concentrated solar thermal) in the form of thermal energy, which can be used at a later period as direct heat or to generate electricity. These can be steam accumulators, tanks with hot/cold molten salts as in concentrated solar power plants, graphite blocks, miscibility gap alloys (MGAs), silicon, particle systems, concrete storage, and packed bed systems. Unlike some of the other storage processes, thermal storage technologies can generally be "charged" using renewable generated heat, electricity, and waste process heat and can provide a thermal energy output or an electrical energy output if paired with equipment such as a boiler, heat exchanger (heat storage medium to gas/steam, typically), or turbine.

To manage scope, this section has not explored water tank thermal energy storage or the use of cold (low temperature and $<0°C$) thermal energy storage. Both water tanks and cold thermal energy storage have a broad range of existing applications in buildings and refrigeration, such as in vehicles and static chillers.

- *Flywheels*: This technology provides high power output for short-duration storage (seconds to minutes). Costs are in the range of ~US$2300–4000/kW (CapEx) and US$5000–10,000/kWh (OpEx).
- *Supercapacitors*: Supercapacitors are somewhat similar to batteries, conceptually, but with a different working principle and internal architecture. They are more resistant to degradation over time and working cycles and have a much faster response (power management) but lower specific energy storage capacity.
- *Superconducting magnetic energy storage (SMES)*: Advanced high-power output technology based on the creation of magnetic fields to store energy. It has a very fast response capability (~5 ms) and low degradation over time. Scale can be from 100 kW to 10 MW with discharge times ranging from a few minutes to under an hour, depending on rating. Installation costs are very high: ~US$10,000/kW.
- *Power to gas*: Converts electricity to hydrogen (covered above) or CH_4 for gas grid storage.

Conclusions

In reviewing the wide range of storage technologies suitable for grid stability applications, no single solution stands out as superior in all cases. The technology choice depends highly on the desired storage discharge duration, capacity needs, location constraints, capital budget, and grid operator preferences. Location-flexible lithium-ion batteries currently lead the market in terms of commercial installations and can be appropriate for many short-duration needs. But their future potential for longer duration storage services has limits. Flow batteries and hydrogen storage are newer technologies that show significant promise for sustained, bulk energy storage and deep renewable integration. For areas with favorable geology, pumped hydro and compressed air storage offer efficient and mature long-duration capabilities.

Overall, a suite of complementary storage solutions deployed in parallel will likely be needed. Even just for daily load shifting, a mix of lithium-ion batteries, flow batteries, and hydrogen storage could provide an optimal balance of power, short-term smoothing, and bulk energy shifting at night. Looking out to seasonal storage needs that come with high renewable penetration, hydrogen and flow batteries may become critical assets. On the utility side, storage siting, cycling patterns, and smart control strategies can also help maximize the value of any added storage capacity. There is no perfect storage solution but, by combining emerging technologies each optimized for different grid services, the challenges of stabilizing the grid as the transition to renewable energy takes place can be solved.

The global energy market: Current dynamics and the role of natural gas

Introduction

This chapter has illustrated the need to transport energy across the world and the options to implement energy transport, either as an energy source directly (*raw* energy) or with intermediate energy conversion steps (mostly energy carriers or electricity directly). The need to make use of energy storage systems as instrumental elements of this global energy infrastructure in order to ensure the reliability, flexibility, and resilience of the energy supply chain has also been discussed, along with the technologies that are either available today or under development. Nevertheless, all these contents have been technical so far, with the aim of touching base on the key elements of energy transport that will be discussed in detail in this book. This means that only two vertices of the energy trilemma have been discussed: affordability and security. There is another critical element, which is economics and, associated to it, the environment. This is discussed in this section.

Energy underpins the world's economy. Fossil fuels still dominate the global energy system, but the map is shifting as new resources, technologies, and policies reshape markets. This section analyzes the complex, interconnected worldwide energy market with an in-depth examination of natural gas developments across every major region. Current market conditions, demand and supply forces, geopolitics, environmental impacts, and likely trajectories are explored. Challenges, opportunities, and implications for consumers, businesses, and policymakers are discussed.

Natural gas is one of the mainstays of global energy: Worldwide consumption is rising rapidly, and in 2018, gas accounted for almost half of the growth in total global energy demand. Gas plays many different roles in the energy sector, and where it replaces more polluting fuels, it also reduces air pollution and limits emissions of CO_2. The World Energy Outlook team [33] and reports published by the International Energy Agency provide evidence that switching to natural gas from dirtier hydrocarbon fuel sources (coal and fuel oil) has already helped to limit the rise in global emissions since 2010, alongside the deployment of renewables and nuclear energy and improvements in energy efficiency. Indeed, natural gas stands out as a highly strategic fuel for multiple reasons. It has the lowest carbon footprint among fossil fuels, thereby holding a large potential to support and accelerate the energy transition toward full decarbonization by 2050. It is currently sustaining a large share of the primary energy supply to many national power generation systems throughout the world (20% in 2022, according to the International Energy Agency), and it also plays a fundamental role as feedstock for the industry and residential applications. It relies on well-established technologies along the entire supply chain, which brings about lower installation, operation, and maintenance costs in addition to higher availability and reliability. Despite the soaring prices of natural gas triggered by geopolitical strain worldwide, international markets seem to have cooled down in recent times, as evidenced by the 60%–70% price in the first quarter of 2023 [34].

Studies show that the contribution of gas to energy transitions varies widely across regions, between sectors, and over time. They also highlight the limits of this contribution: Gas cannot, of course, do it all. It can bring environmental benefits, but it remains a source of emissions in its own right, and new gas infrastructure can lock in these emissions for the future. The near-term priority is to minimize emissions all along the chain, from gas production to consumption, with a particular focus on CH_4 emissions. For the longer term, the industry needs to seriously explore the possibilities to further reduce the emission intensity of gas supply via biomethane or low-emission hydrogen. Another crucial variable for the future, both for coal and gas, is the extent to which emissions are mitigated by large-scale deployment of carbon capture, utilization, and storage technologies.

Fuel switching, primarily from coal to natural gas, is a means to reduce emissions of CO_2 and air pollutants, and even though switching between unabated consumption of fossil fuels, on its own, does not provide a long-term answer to climate change, there can nonetheless be significant CO_2 and air quality benefits, in specific countries, sectors, and timeframes, from using less emission-intensive fuels. The deployment of carbon capture, utilization, and storage technologies for both coal and gas is another crucial variable for the future.

The clearest example is the "quick win" for emissions from running existing gas-fired plants instead of coal-fired plants to generate electricity. In 2019, the International Energy Agency estimated that up to 1.2 gigatons of CO_2 could be abated in the short term by switching from coal to existing gas-fired plants, if relative prices and regulation are supportive. The vast majority of this potential lies in the United States and in Europe. Doing so would bring down global power sector emissions by 10% and total energy-related CO_2 emissions by 4%.

The environmental case for new gas infrastructure is more complex, but we find that in more carbon-intensive energy systems like China and potentially India, it can play a significant role alongside the rise of renewable energy. Unlike in the United States and Europe, the power sector is not likely to be the main arena for switching. In China, the focus has been on the residential and industrial sectors as part of a strong policy push to improve air quality. In India, much will depend on the pace at which new city gas distribution infrastructure is built out; switching in the Indian context may affect demand for liquids more than coal.

It is critical to take into account both CO_2 and CH_4 emissions. On average, coal-to-gas switching reduces emissions by 50% when producing electricity and by 33% when providing heat. Best practices all along the gas supply chain, especially to reduce CH_4 leaks, are essential to maximizing the climate benefits of switching to gas.

Global energy landscape

World primary energy demand has nearly doubled since 1980 and continues to rise by 1.3% annually, although the pandemic temporarily depressed use. Fossil fuels supply 84% of primary energy, led by oil (31%), coal (27%), and natural gas (24%), as shown in Fig. 2.16: Renewable energy is growing fastest globally at a 30% five-year average pace, but still only meets 5% of demand. Energy intensity has declined, but absolute use rises with population and economic growth.

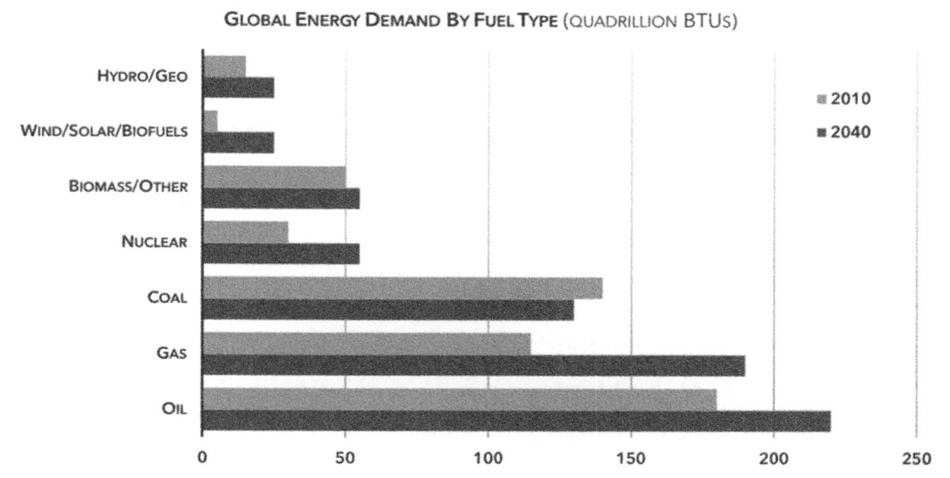

FIG. 2.16 Global energy demand by fuel type [35].

Developing nations account for the most demand growth, led by Asia Pacific. Over 70% of global growth comes from China and India. But per capita energy use in developing regions remains far below that in developed countries. The United States consumes about seven times more energy than India on a per capita basis. Energy security concerns and import dependence motivate increased domestic supply efforts. But meeting rising electricity needs from a growing middle class is paramount across the developing world.

Regional energy profiles

The Americas lead the oil supply, providing 20% of world output, mostly from the United States and Canada. Hydropower supplies half of South America's electricity. North American shale gas has transformed global markets. The United States produces a quarter of global natural gas and leads in coal output. Renewables are expanding, providing 11% of the US energy.

Europe has pioneered policies supporting efficiency and renewable energy adoption. The EU aims for a 55% reduction of greenhouse by 2030, with respect to 1990 levels. Renewables generate 22% of EU energy, led by Germany and Scandinavia. But Russian oil and natural gas imports have been rising. The Ukraine conflict sparked an energy supply crisis and demand reduction efforts.

Asia Pacific dominates energy demand, accounting for half of global use. China consumes a quarter of the world energy. But per capita energy use remains modest outside wealthy nations like South Korea and Japan. Coal provided 60% of China's energy in 2020. India still has 240 million people lacking access to electricity. Both nations are adding renewables but seeking more fossil fuels to meet demand.

Africa's energy poverty constrains growth. Nearly 600 million sub-Saharan Africans lack electricity access. Renewables are expanding, but fossil fuels dominate African energy supply, with natural gas reserves increasingly leveraged. Major new oil and gas finds underscore the continent's potential.

Natural gas: Supply and demand trends

Natural gas represented 24% of the primary energy supply in 2020. Global gas consumption is projected to expand 1.6% per year through 2050, with demand centered in Asia, as shown in Fig. 2.17. Natural gas emits 50%–60% less CO_2 than coal when combusted. This makes gas crucial for managing emissions during the energy transition, although CH_4 leakage risks warrant attention. Half of the growth in global gas use since 2000 is attributable to the power sector. Gas-fired power offers grid flexibility to balance variable renewables.

Production has grown by 50% since 2005. The United States produces 23% of global natural gas, doubling output over the last decade due to shale development. The shale revolution transformed the United States from a major importer to a leading net exporter. Canada, Russia, Iran, and Qatar are other major producers. But new African and Central Asian discoveries signal future supply growth. Global natural gas reserves keep increasing with more shale, subsea, and LNG resources.

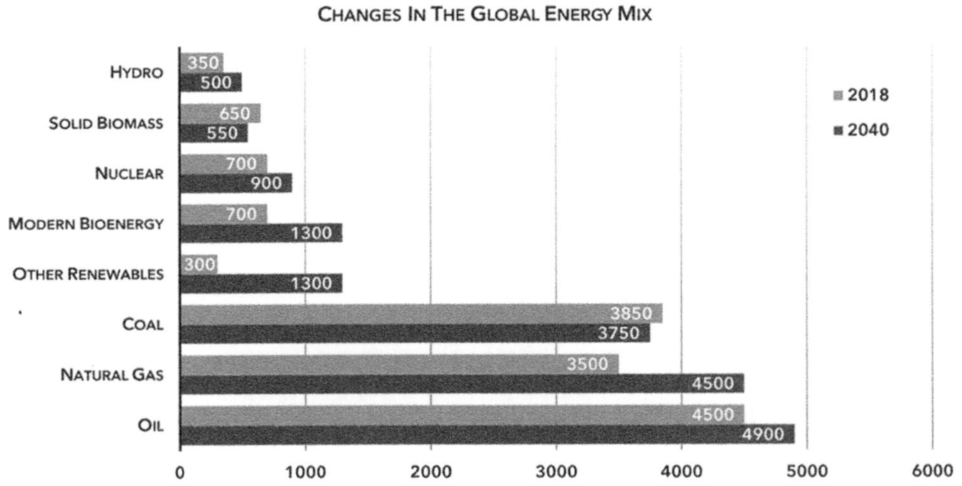

FIG. 2.17 Changes in global energy mix 2018–2040. *Adapted from L. Cozzi, T. Gould,* World Energy Outlook, *International Energy Agency report, 2018.*

The LNG trade represents 35% of internationally traded gas volumes. LNG provides key links between major consumer centers and gas deposits too remote for pipelines. Contract flexibility, falling costs, and ease of shipping underpin LNG's growth. Exports reached 380 million tons in 2021, with Qatar, the United States, and Australia as the top exporters. Japan, China, and South Korea lead imports. Europe's imports are surging amid Ukraine-related Russian supply cuts.

Natural gas outlook by region

North America enjoys abundant, low-cost gas from shale fields like the Marcellus and Permian basins. Shale gas provides over 60% of the US supply, helping retire coal plants. Exports to Asia and Europe are rising. But pipelines face opposition, and CH_4 emission concerns persist. Canada is the world's fifth-largest gas producer and aims to expand LNG exports. Mexico has reserves but lacks infrastructure and has become reliant on US gas imports.

South America holds major reserves in fields like Brazil's presalt deposits. Argentina, Trinidad and Tobago, and Peru also have substantial resources. But development has lagged, and LNG plans have been deferred. Political and fiscal instability hamper gas investments. Yet low-carbon policies position gas for growth by displacing oil and coal.

Europe relies on imports for 90% of its natural gas supply, with Russia providing 40% preinvasion. But the Ukraine crisis accelerated efforts to diversify the European gas supply. Massive new LNG and pipeline projects aim to boost imports from Africa, the Middle East, the United States, and the Azerbaijan/Caspian region. Europe's gas use is forecast to drop long-term absent Russian flows. Renewables, hydrogen, and electrification will displace gas in time per net-zero emissions plans.

Asia leads global gas demand growth, accounting for 75% of the global rise in 2021 alone, driven by China and India. China's gas use has tripled since 2010. Pollution reduction and energy security efforts spur coal-to-gas switching. But pipeline imports and domestic production struggle to keep pace. Asian LNG imports hit record highs in 2022 as Europe's demand recedes. Many nations also push for greener energy.

Africa and the Middle East dominate reserves and new discoveries like Mozambique's Rovuma Basin and Israel's massive Leviathan field. Yet development lags, and gas flaring remains high. Utilizing gas reserves domestically supports economic growth and oil exports. But instability hampers investments despite need and potential.

Implications and future outlook

Natural gas is poised for an extended transitional role in the global energy economy as more nations target net-zero emissions by mid-century. Gas offers cost and emissions advantages relative to coal and oil. Rising electricity needs also drive gas generation demand. But eventually, renewables, hydrogen, and electrification will likely replace gas use in the long-term if carbon capture is not deployed widespread.

LNG's flexibility helps meet demand but can also propagate price spikes. Securing diverse, resilient supply chains is imperative for import-dependent nations. Volatility from geopolitics, extreme weather, and limited price hedging adds risk. Stranded asset concerns for gas infrastructure exist if the transition accelerates.

LNG enables transporting natural gas globally via tanker and has seen rapid growth at 7% annually. It allows non-pipeline countries to import gas. Volumes reached ~400 bcm in 2021, with Australia, Qatar, and the United States as the top exporters. New liquefaction trains keep coming online; however, LNG infrastructure remains capital intensive. Some price de-linkage from oil is evident in new flexible spot-priced contracts. The European gas crisis has spiked Asian LNG prices due to the competition for cargoes.

Natural gas has seen pronounced growth as a cleaner-burning fossil fuel. LNG trade drives increasing global gas interconnectivity. Market liberalization has spurred investment in gas infrastructure and enabled rising demand, especially for power generation. Gas is displacing higher-carbon coal in many countries. Wholesale spot markets are emerging, but regional prices still diverge based on local supply-demand conditions and rigidities. Pipeline gas and LNG remain segmented. Key uncertainties include competition from renewables, CH_4 emission concerns, and the potential for greater electrification.

Global natural gas pricing dynamics

The global scenario depicted in the previous sections yields largely dissimilar gas prices in different regions of the world:

- North America enjoys abundant low-cost shale gas, with spot prices around US$2 to US$6/MMBtu.
- Europe balances LNG imports and Russian piped gas with prices over US$30/MMBtu.
- Asian LNG import reliance keeps prices elevated over US$15/MMBtu.
- Middle East gas is cheap, below US$3/MMBtu, as production is associated with oil.

Pricing is converging somewhat with the LNG trade, but transportation costs prevent full arbitrage. Oil-indexed long-term contracts remain common.

Managing CH_4 leakage across the gas value chain is critical to achieving the promised emission benefits relative to other fossils. Supportive policies, private-sector action, advanced detection technologies, and best practices adoption can strengthen CH_4 management. Sustainability will determine the fuel's long-term prospects.

Overall, natural gas is primed to thrive in the near term amid the global energy transition. Other energy carriers are discussed in short summaries as follows:

Oil market fundamentals. Oil remains the largest global energy source. Markets are shaped by OPEC supply adjustments and strategic petroleum reserves in consumer nations. Prices fluctuate cyclically and spiked over US$100/barrel in 2022 with the Ukraine invasion. Despite environmental concerns, oil demand continues to grow, constraining supply. Transportation reliance and petrochemical demand are key drivers. The US shale boom has been a disruptive factor in increasing non-OPEC production. Trade in crude and products is highly interconnected between regions.

Coal market trends. International coal trade has declined slightly in recent years. China dominates both coal production and consumption, skewing market dynamics. Coal maintains a cost advantage but faces escalating environmental pressures, leading many nations to reduce coal generation. Trade reflects geographical supply-demand imbalances, with Indonesia, Australia, South Africa, Russia, and the United States among the major exporting nations. Steam coal is widely traded, while metallurgical coal relies on bilateral contracts. LNG is displacing some coal demand.

Some comments on the oil market

The global oil demand has increased steadily in the last 20 years and is now close to 100 million barrels a day, representing one-third of the world's annual energy supply. The shrinkage in demand due to the pandemic experienced in 2020 is today more than offset by an arguably rebound effect, which is expected to attenuate in the short term. Nevertheless, oil demand is forecasted to continue an ascending pattern until 2030, even if with an increasingly moderate annual growth rate [36]. The largest contribution to the demand for oil is transportation, with somewhat similar shares for road (land) and aero/maritime transport. Demand for the petrochemical industry follows far behind.

As for any commodity traded in the market, prices are governed by the balance between demand and supply. This effect is particularly strong in the oil market, given that some oil producers do not act independently but rather collectively in the market. In particular, the Organization of the Petroleum Exporting Countries (OPEC) gathers

13 countries whose state-owned oil resources represent one-third of the global oil market (60% if the extended OPEC+ countries are considered). This implies a strong influence on market behavior.

The United States, not an OPEC member, is the largest producer of oil in the world, with a production capacity of some 12 million barrels a day. Saudi Arabia (OPEC) and Russia follow with a production capacity of over 10 million barrels a day, and then, there is a long list of countries whose production ranges from 2.5 to 5 mbbl/d; among the latter, OPEC (Iraq, UAE, and Iran) and non-OPEC (Canada, China, and Brazil) countries are found. Globally, around 45% of the global production of oil is controlled by private companies, while more than half of it is controlled directly by state-owned companies. Regionally, one-third of the oil produced annually comes from the Middle East, another third from North America, and one-sixth from Russia and the Commonwealth of Independent States (CIS). This suggests a high sensitivity to international geopolitics.

Two major events have struck the international oil market in the last 5 years. The very large reduction of oil demand during the pandemic of COVID-19 (−10% in 2020 with respect to 2019) saw prices plummet from US$ 65–70/barrel (barrel of Brent oil) to some US$ 20/barrel in just 3 months (December 2019 to March 2020). The subsequent reduction in global oil production agreed by OPEC+ members reduced supply and brought prices to intermediate, though still historically low, prices (~US$ 50/barrel) for the remainder of 2020. As a second major event, the increasing political stress between Russia and Ukraine, ending with Russia's attack on this latter country in February 2022, brought about a tumultuous reorganization of the international oil market. In particular, Russia's decision to reduce the supply of natural gas to Europe and the EU's reaction to resorting to other suppliers and energy sources in an attempt to revert the situation through political actions (condemnation, sanctions) brought about unforeseen market behaviors that translated into skyrocketing oil (energy) prices exceeding US$ 120/barrel in June 2022.

The energy trilemma is a compound concept representing the need to balance opposing objectives of energy supply: security of supply (energy security), affordability of supply (energy equity), and sustainability of supply (environmental sustainability). All three pillars were strongly affected in the aftermath of the invasion of Ukraine. In addition to soaring prices, the complete reliance of the EU on oil from Russia hit the security of supply hard, given that over 90% of the total oil imports at the time were sourced in Russia. A package consisting of securing oil imports from alternative countries and also switching to other energy sources (for instance, coal) enabled reducing the total monthly imports from Russia into the EU from almost 13 million tons in January 2022 to almost 1 million tons in March 2023, thereby restoring energy security. Regarding sustainability, the rapidly increasing share of renewable energy capacity in Europe in recent years has compensated for the higher emissions of some of the said alternative energy sources, like coal, hence avoiding a deleterious impact on this third pillar of the energy trilemma if not for a slower carbon emission annual reduction rate. Furthermore, the main influence of this third element of the energy trilemma on the energy market has actually more to do with the energy transition toward a decarbonized energy sector. This is discussed in the next section.

Environmental considerations of the energy market. Monetization of emissions

Globally, the GHGs that are most closely tracked, regulated, and reduction incentivized are CO_2, CH_4, and nitrous oxide (N_2O). These are released from the combustion of fuel and fugitive emissions. It is estimated that the energy market contributes to approximately 75% of the total GHG emissions globally.[i]

A North American perspective

The United States and Canada take a number of different approaches to GHG reduction, including regulation, incentives, and funding for technology development.

Carbon dioxide. The reduction of CO_2 emissions has received substantial and increasing attention over the past decade. In recent years, the U.S. Inflation Reduction Act (IRA)[j] and the Bipartisan Infrastructure Law (BIL)[k] have been some of the largest government investments in GHG reductions and include tax credits, technology development funding, and community implementation funding. From a carbon management perspective, these laws work in parallel. The BIL allocates over US$ 12 billion for carbon management research, development, and demonstration over the

[i] "Greenhouse Gas Emissions from Energy Highlights" IEA dataset, August 2023, https://www.iea.org/data-and-statistics/data-product/greenhouse-gas-emissions-from-energy-highlights#.

[j] Building A Clean Energy Economy: Guidebook, https://www.whitehouse.gov/wp-content/uploads/2022/12/Inflation-Reduction-Act-Guidebook.pdf.

[k] An Interactive Diagram for Carbon Management Provisions, Office of Fossil Energy and Carbon Management, https://www.energy.gov/fecm/interactive-diagram-carbon-management-provisions.

next several years, while the IRA research, development, and demonstration funds for CO_2 emission reduction of US$ 5.8 billion are focused on the industrial sector. The IRA expanded the 45Q tax credit for carbon capture, utilization, and sequestration. The monetary potential was substantially increased with base credit amounts of US$17/metric ton CO_2 capture and sequestration and US$12/metric ton CO_2 injected for enhanced oil recovery. For direct air capture facilities, these tax credits are US$36/metric ton and US$26/metric ton, respectively. For facilities that comply with certain wage and apprenticeship requirements, a bonus credit of five times the base credit amount is available. Beyond these tax credit values, the IRA also expanded the entities that qualify for these credits by lowering the threshold requirements to benefit from tax credits.

Methane. Both the United States and Canada have regulatory conditions for CH_4 emissions at two levels: federal and state or province.[l,m,n] At the federal level, fees are imposed for exceeding regulatory CH_4 emission limits. Various states or provinces have additional regulations, which come in the form of taxes, stricter limitations on GHGs, and maximums for trade. These regulations are being adapted regularly. For example, in the past year, new legislation has been drafted in the United States that would impose stricter limits at the federal level, and the Environmental Protection Agency (EPA) has issued new rules.

Beyond regulation in the United States, the IRA allocates US$1.55 billion to the EPA to support CH_4 pollution reduction oil and gas industry operations through research and development and community implementation and imparts a waste emission charge for facilities in excess of specified limits. Additionally, the BIL provides US$4.7 billion to plug and remediate orphaned oil and gas wells on tribal, federal, state, and private lands.

Mexico lags the United States and Canada in terms of CH_4 emission regulation but has made more progress in the past 5 years with the issuance of the "Guidelines for the Prevention and Integral Control of Methane Emissions from the Hydrocarbon Sector".[o] These guidelines provide a regulation frame for companies to follow for planning, prevention, monitoring, and maintenance associated with CH_4 emissions.

Nitrous oxide. Nitrous oxide (NOx) emissions are emitted with the combustion of fuels. The EPA has rules for emission limits and cap and trade regulations for NOx emissions.[p] These are further regulations at the state level in some areas.

A European Union perspective

The Green Deal is Europe's main instrument to fight the global climate crisis on different fronts, with the ultimate goal of becoming the first net-zero carbon emissions region in the world by 2050. This is indeed a very ambitious target, especially for the energy sector, inasmuch as it is responsible for three-quarters of the total emissions of CO_2 in the EU.

The EU Climate Law, adopted by the European Parliament in June 2021, is the legally binding instrument to implement the objectives set forth in the Green Deal. It sets two different timeframes, with the intermediate objective of reducing GHG emissions by 55% (with respect to 1990 levels) by 2030 prior to achieving carbon neutrality in 2050.

For the energy sector, the roadmap to achieve carbon neutrality by 2050 is set out in the Clean Energy for All Europeans legislative package, comprised of eight laws (acts) regulating different areas of energy-related activities:

- Energy Performance in Buildings Directive, providing provisions for more energy-efficient buildings.
- Renewable Energy Directive, setting a binding target to achieve a 32% share of renewable energy in the European energy mix by 2030. A recent proposal aims to increase this target to 42.5%.
- The Energy Efficiency Directive, under the principle of "energy efficiency first" as a cornerstone of EU energy policy, aims to reduce energy consumption in 2030 by 11.7% with respect to the 2020 projections.
- Governance of the EU Regulation, urging each Member State of the EU to develop an integrated 10-year National Energy and Climate Plan for 2021–2030, also incorporating a longer-term view toward 2050.
- Electricity Regulation, setting principles for the internal market of electricity in the EU. This regulation provides a framework for the further integration of renewable energy into the electricity market, sets out new rules on bidding zones and cross-zonal capacity allocation, and reinforces the role of the market in providing price signals for investment.

[l] https://www.whitehouse.gov/wp-content/uploads/2021/11/US-Methane-Emissions-Reduction-Action-Plan-1.pdf.

[m] https://www.epa.gov/controlling-air-pollution-oil-and-natural-gas-industry/epa-issues-supplemental-proposal-reduce.

[n] Greenhouse Gas Pollution Pricing Act (S.C. 2018, c. 12, s. 186), https://laws-lois.justice.gc.ca/eng/acts/g-11.55/.

[o] "Guidelines for the prevention and comprehensive control of methane emissions from the hydrocarbons sector (Mexico)" February 2022, https://www.iea.org/policies/8685-guidelines-for-the-prevention-and-comprehensive-control-of-methane-emissions-from-the-hydrocarbons-sector-mexico.

[p] https://www3.epa.gov/region1/airquality/nox.html.

- The Electricity Directive establishes common rules for the generation, transmission, distribution, energy storage, and supply of electricity, together with consumer protection provisions, with a view to creating truly integrated, competitive, consumer-centered, flexible, fair, and transparent electricity markets in the EU.
- The Risk Preparedness Regulation introduces important rules for cooperation between member states with the aim of preventing, preparing for, and managing electricity crises. Each member state's competent authority must establish a risk-preparedness plan, based on the regional and national electricity crisis scenarios.
- Regulation of the Agency for the Cooperation of Energy Regulators, updating the role and functioning of the European Union Agency for the Cooperation of Energy Regulators (ACER), in particular increasing the competence of the ACER in cross-border cooperation.

As a complement to the aforedescribed legislative package, the European Emission Trading System (ETS) is a key tool of the EU for reducing GHG emissions cost-effectively. Basically, the ETS is the world's first and largest major carbon market, established in 2005 and then operated in four phases, the last one starting in 2021 and running until 2030.

The ETS covers the following sectors and gases, focusing on emissions that can be measured, reported, and verified with a high level of accuracy[q]:

- Carbon dioxide from:
  - electricity and heat generation
  - energy-intensive industry sectors, including oil refineries, steel works, and the production of iron, aluminum, metals, cement, lime, glass, ceramics, pulp, paper, cardboard, acids, and bulk organic chemicals
  - aviation within the European Economic Area and departing flights to Switzerland and the United Kingdom
  - maritime transport
- Nitrous oxide (N_2O) from the production of nitric, adipic, and glyoxylic acids and glyoxal
- Perfluorocarbons (PFCs) from the production of aluminum.

Companies in the sectors listed earlier must participate in the ETS, which is mandatory for companies in said sectors, except for installations below a certain size in some sectors or under a government's decision if other instruments enabling similar emission reductions are put in place.

The Emission Trading System is based on a *cap-and-trade* model where one allowance permits the holder to emit 1 ton of CO_2 (tCO_2). The scheme sets an upper limit (*cap*) for the total amount of GHGs that the participating installations are allowed to emit. For each EU ETS Phase, this quantity to be allocated by each member state is defined in the National Allocation Plan, with the aim of contributing to the emission reduction targets set forth internationally. These total emission allowances are then auctioned off or allocated (for free) and can later be traded. The installations participating in the ETS must monitor and report their CO_2 emissions, ensuring they submit enough allowances to the authorities to cover their emissions. If the allocated emission allowances are not enough, said installation must purchase allowances in the market. If, on the other hand, the emissions of an installation are below their allowance, the leftover credits can be traded in the market (ETS).

In spite of the historically low prices of carbon allowances in the ETS, ~€ 10 per ton in Fig. 2.18, these carbon permits have experienced an almost tenfold increase in price, exceeding € 100 in early 2023 and then remaining at ~€ 90 per ton of CO_2. This, along with the concerted actions implemented through the regulatory framework described earlier, has managed to continue reducing the emissions of CO_2 in the EU, as shown in Table 2.3.

Finally, a few words on EU taxonomy, an instrument focused on sustainable finance that provides a classification system that helps direct investments to the economic activities most needed for the transition, in line with the European Green Deal objectives. This classification and the associated procedure aim to help scale up investments in projects and activities that are necessary to reach the objectives of the European Green Deal, based on the four following criteria to have an economic activity qualified as *environmentally sustainable*:

1. Making a substantial contribution to at least one environmental objective.
2. Doing no significant harm to any of the other five environmental objectives.
3. Complying with minimum safeguards.
4. Complying with the technical screening criteria set out in the taxonomy delegated acts.

[q] Website of the EU Emissions Trading System (EU ETS): https://climate.ec.europa.eu/eu-action/eu-emissions-trading-system-eu-ets_en (accessed September 15, 2023).

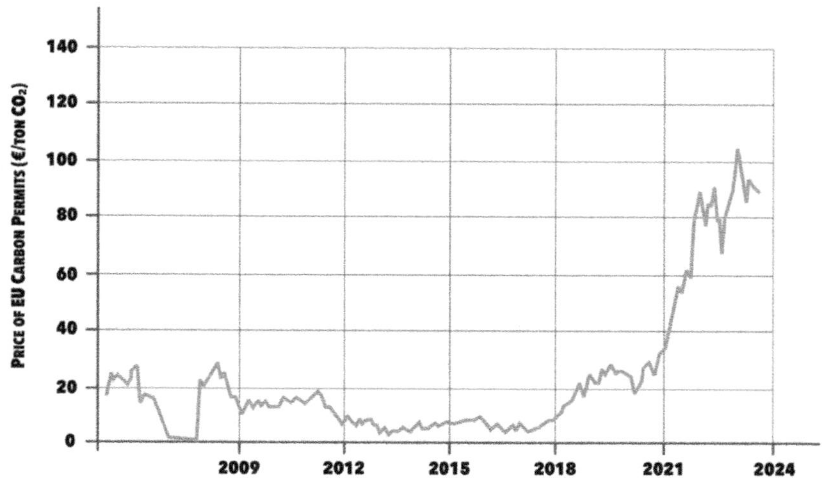

FIG. 2.18 Price of carbon permits (€/tonCO_2) in the emission trading scheme.

TABLE 2.3 Verified emissions from stationary installations [37].

Year	2013	2014	2015	2016	2017	2018	2019	2020	2021
Verified emissions from stationary installations	1908	1814	1803	1751	1755	1683	1530	1356	1335
Annual change (%)	–	−4.9	−0.6	−2.9	0.2	−4.1	−9.1	−11.4	6.6
Verified emissions from electricity and heat generation	1191	1100	1091	1046	1036	964	822	696	707
Annual change (%)	–	−7.7	−0.8	−4.1	−1.0	−7.0	−14.7	−15.3	8.4
Verified emissions from industrial production	717	714	712	705	719	719	708	659	631
Annual change (%)	–	−0.4	−0.3	−1.0	2.0	0.1	−1.6	−6.9	−4.6

Closure

This introductory chapter has aimed to provide a global picture of the energy supply chain from fundamental, technical, economic, and environmental standpoints. The first section presented the scattered distribution of renewable and nonrenewable energy sources across the world as a means to highlight the need to connect these sources with the nodes where energy is consumed, i.e., the need to transport energy in either its raw form or, if possible and sustainable, transform it into energy carriers of different types (including electricity). This spatial mismatch between energy sources and consumers triggers the need for a focus on energy transport, which you, as a reader, now have in your hands.

The most popular energy carrier is electricity. In advanced economies, it is ubiquitous, and anyone has access to it, whether for residential or industrial use, on a small or large scale. This is thanks to a very dense transportation and distribution grid, a monumental and very expensive infrastructure built over the course of one and a half centuries. The features of the electrical power transmission grid are presented in section "Electrical power transmission and distribution."

Access to electricity is nonetheless not as straightforward in underserved economies or economies that are currently in the growth phase, given the lack of infrastructures, so transportation and distribution of electricity are not reliable and universal means to supply energy to the end users. Also, for certain applications or just when the distance traveled is very long, transportation of electricity reveals uneconomical and alternative energy carriers are therefore needed. The different options available or under development today and their main features are introduced in section "Transportability of energy sources", with a special focus on hydrogen technologies for the very advantageous characteristics of this energy vector.

The mismatch between energy supply and consumption is not only regional; on the contrary, there is a secondary or complementary temporal mismatch. In other words, the complex logistics that come about because of the need to transport energy for long distances inevitably require the ability to store energy to avoid the need to devise a sort of *just-in-time* global energy supply chain; such a system would be not only uneconomical but, simply, not possible technically. This is why energy storage is needed to guarantee the viability, availability, and reliability of the energy supply chain, as introduced in section "Nonelectricity transport as an enabler of energy storage."

Sections to are mostly technical, even if with some references to economic and environmental aspects of the technologies presented. Nevertheless, in order to address the three pillars of the energy trilemma, it is mandatory to guarantee that the future energy supply chain ensures environmental sustainability and equity (affordability). These two elements of the discussion are introduced in section "The global energy market: Current dynamics and the role of natural gas," where the global energy market is introduced, with an emphasis on the natural gas and oil markets, alongside the current regulatory framework setup in North America and Europe to handle emissions incurred along the energy supply chain.

The authors hope that this chapter has accomplished the twofold objective of presenting energy transport as a cornerstone of modern civilization and providing a global picture that would help the reader embrace all the different needs and challenges posed by energy transport, as well as the multiple technical solutions that are currently at hand or under development.

References

[1] BP, A global view of gas - in maps and charts, 2017. https://www.bp.com/en/global/corporate/news-and-insights/reimagining-energy/global-view-of-gas-infographic.html. (Accessed 12 September 2023).
[2] International Energy Agency, Key World Energy Statistics, 2021. IEA report.
[3] European Commission, EU Strategy for Liquefied Natural Gas and Gas Storage, COM(2016) 49 Final, 2016.
[4] International Energy Agency, Coal 2020 Analysis and Forecast to 2025, 2020. IEA report.
[5] World Nuclear Association, Transport of Low-Enriched Uranium, 2020. Report No. 2020/007.
[6] The World Bank, Solar resource data: Solargis, 2017, Downloaded from https://solargis.com/es/maps-and-gis-data/download/world. (Accessed 14 June 2023).
[7] V. Timmers, et al., All-DC offshore wind farms: when are they more cost-effective than AC designs? IET Renewable Power Gener. (2022), https://doi.org/10.1049/rpg2.12550.
[8] World Bioenergy Association, Global Bioenergy Statistics 2020, 2020. WBA report.
[9] J. Allen, M. Browne, J. Boyd, A. Hunter, H. Palmer, Transport and Supply Logistics of Biomass Fuels: Volume 1 – Supply Chain Options for Biomass Fuels, ETSU Report ETSU/B/W2/00399/REP/1, Harwell ETSU, 1996.
[10] S.M. Zahraeea, S.R. Golroudbaryb, N. Shiwakotia, P. Stasinopoulosa, A. Kraslawski, Economic and environmental assessment of biomass supply chain for design of transportation modes: strategic and tactical decisions point of view, in: 31st CIRP Design Conference, 2021.
[11] D.R. Boden, Geology and heat architecture of the Earth's interior, in: Geologic Fundamentals and Geothermal Energy, CRC Press, Boca Raton, 2016, https://doi.org/10.1201/9781315371436-4.
[12] G. Coro, E. Trumpy, Predicting geographical suitability of geothermal power plants, J. Clean. Prod. 267 (2020) 121874.
[13] European Geothermal Energy Council, EGEC Geothermal Market Report 2022, 2023. Brussels.
[14] J.C. Robinsx, et al., 2021 U.S. Geothermal Power Production and District Heating Market Report, Report NREL/TP-5700-78291, National Renewable Energy Laboratory, 2021.
[15] K. Schwab (Ed.), Global Competitiveness Report, World Economic Forum, 2015.
[16] International Organization for Standardization, Technical Energy Systems – Basic Concepts, 1997. ISO Standard 13600.
[17] International Panel for Climate Change, IPCC Fourth Assessment Report: Climate Change 2007, 2007. IPCC Report.
[18] T.S. Reynolds, Stronger Than a Hundred Men: A History of the Vertical Water Wheel, Johns Hopkins Univ Press, Boulder, 1983.
[19] A. Atkeson, P.J. Kehow, The Transition to a New Economy After the Second Industrial Revolution, National Bureau of Economic Research, 2001.
[20] J. Rifkin, The Third Industrial Revolution: How Lateral Power Is Transforming Energy, the Economy, and the World, St. Martin's Press, 2011.
[21] A. Benchaib, Advanced Control of AC/DC Power Networks, Wiley, 2015.
[22] J.G. Speight, Dewatering, Desalting, and Distillation in Petroleum Refining, CRC Press, 2023.
[23] S. Mokhatab, et al., Handbook of Liquified Natural Gas, Elsevier, 2016.
[24] C. Stachtiaris, An Evaluation of the Commercial Viability of Offshore Stranded Gas Fields Using LNG Floating Production, Storage and Offloading Platforms (LNG FPSO), Imperial College London, 2010.
[25] A.J. Kidnay, et al., Fundamentals of Natural Gas Processing, third ed., CRC Press, 2020.
[26] D.L. Gerber, et al., A simulation based efficiency comparison of AC and DC power distribution networks in commercial buildings, in: IEEE Second International Conference on DC Microgrids (ICDCM), Nuremberg, Germany, 2017.
[27] E.J. Moniz, et al., The Future of Natural Gas, Massachusetts Institute of Technology, Cambridge, MA, 2011.
[28] G. Chehade, I. Dincer, Progress in green ammonia production as potential carbon-free fuel, Fuel 299 (2021) 120845.
[29] J.B. Baker et al., Experimental ignition delay time measurements and chemical kinetics modeling of hydrogen/ammonia/natural gas fuels, J. Eng. Gas Turbines Power 145 (2023) 041002.
[30] S. Barak, et al., Measuring the effectiveness of high-performance co-optima biofuels on suppressing soot formation at high temperature, Proc. Natl. Acad. Sci. USA 117 (2020) 3451–3460.

[31] S.B. Harrison, Compressed and Liquid Air for Long Duration & High Capacity, Modern Power Stations, 2023, pp. 24–25 (June issue).
[32] V. Srinivasan, et al., Renewable Energy Storage Roadmap, Commonwealth Scientific and Industrial Research Organisation (CSIRO), 2023.
[33] L. Cozzi, T. Gould, World Energy Outlook, 2023. International Energy Agency report.
[34] International Energy Agency, Gas Market Report, Q2 2023, 2023.
[35] Various, Outlook for Energy: A View to 2040, 2018. Exxon Mobile report.
[36] Y. Akizuki, et al., Oil 2023, Analysis and Forecast to 2028, 2023. International Energy Agency report.
[37] European Commission, Report from the Commission to the European Parliament and the Council on the Functioning of the European Carbon Market in 2021 Pursuant to Articles 10(5) and 21(2) of Directive 2003/87/EC (as Amended by Directive 2009/29/EC and Directive (EU) 2018/410), 2022. Brussels, December 2022, COM(2022) 516 final.

CHAPTER

3

Fundamentals

Rainer Kurz[a], *Christoph Meyenberg*[b], *Enver Karakas*[c], *Michael Müller*[d], *Shane Harvey*[c], *Todd Omatick*[c], *Mahmoud Abdellatif*[e], *and Kaartik Palaniappan*[f]

[a]Solar Turbines, San Diego, CA, United States [b]Voith, Crailsheim, Germany [c]Ebara Elliott Energy, Jeannette, PA, United States [d]EagleBurgmann, Wolfratshausen, Bayern, Germany [e]GASCO, Abu Dhabi, United Arab Emirates [f]fIndependent Consultant, Kuala Lumpur, Malaysia

Pipeline transport

Pipeline transport allows the transmission of gases and liquids over large distances. The transport is typically from a source to a market or usage area. In 2014, there were 3,500,000 km of pipeline in 120 countries of the world [1]. About 65% of these are in the United States (Fig. 3.1). Pipelines are among the safest way of transporting materials. For the purpose of this book, we will predominantly discuss the pipeline transport of gases like natural gas, hydrogen, and CO_2, as well as liquids like oil and ammonia. Natural gas pipelines are usually made from carbon steel, often with internal coating to reduce roughness.

When fluids flow through pipelines, friction losses occur. Friction losses generally increase with flow velocity and pipe roughness. In Fig. 3.2, a control volume is shown with gas entering from the left with known flow rate, pressure, and velocity. Friction forces that are proportional to the square of velocity act on the surface of the control volume, and gas exits the control volume with changes to its pressure and velocity.

This yields the universal flow equation, with friction factor f and specific gravity SG:

$$p_1^2 - p_2^2 = Q_{std}^2 \frac{f \cdot Z_{ave} T \cdot SG \cdot L}{C'^2 D^5} \tag{3.1}$$

To calculate the pipeline hydraulic behavior of a piping system, including time-varying (transient) flow pressures, temperatures, and conservation laws have to be applied, and the resulting equations have to be solved.

The time-dependent mass conservation:

$$\frac{\partial(\rho v)}{\partial x} + \frac{\partial \rho}{\partial t} = 0 \tag{3.2}$$

where the through-flow area (A) is constant. The density is calculated from:

$$\frac{p}{\rho} = ZRT \tag{3.3}$$

The momentum balance is satisfied by:

$$\rho\left(\frac{\partial v}{\partial t}\right) + \rho v\left(\frac{\partial v}{\partial x}\right) = -\frac{\partial p}{\partial x} - \frac{\rho \cdot f_{DW} v|v|}{2D} \tag{3.4}$$

for a pipeline with no differences in elevation. The resulting system of equations can be solved numerically, for example, by the method of characteristics, finite differences, or finite volumes. Time steps in the computational domain have

Energy Transport Infrastructure for a Decarbonized Economy
https://doi.org/10.1016/B978-0-443-21893-4.00013-1

Copyright © 2025 Elsevier Inc. All rights are reserved, including those for text and data mining, AI training, and similar technologies.

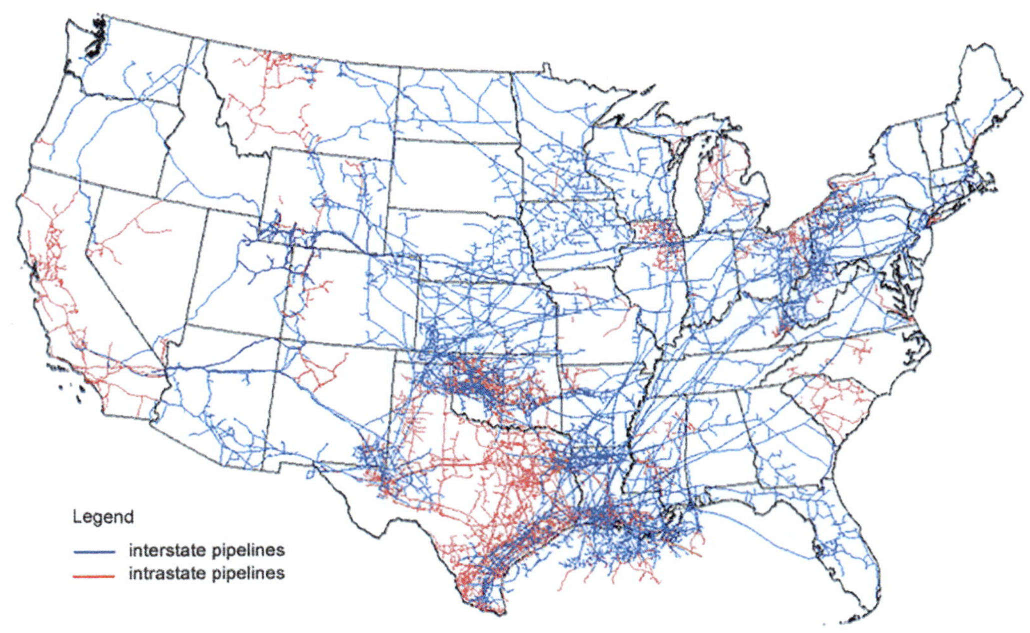

FIG. 3.1 US pipeline network [2].

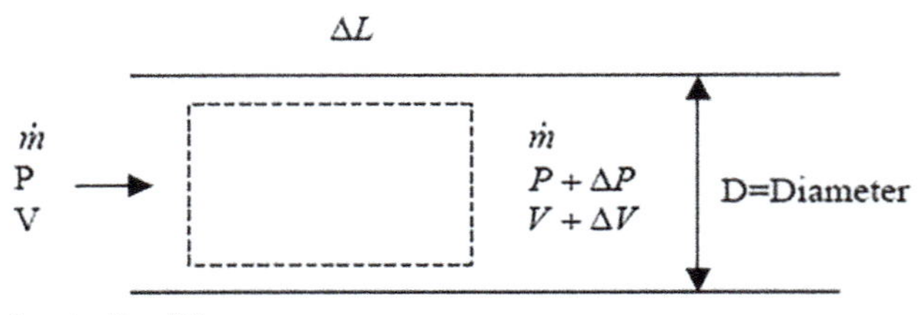

FIG. 3.2 Balance of forces in a straight pipeline [3].

a direct relationship with the wave speed and the length of the pipeline computational segment. The fundamental flow equation,

$$Q = 0.0011493\,\frac{T_b}{P_b}\cdot e\cdot D_i^{2.5}\left[\frac{p_1^2 - p_2^2}{SG\cdot T_a\cdot L\cdot Z_a\cdot f_{DW}}\right]^{0.5} \tag{3.5}$$

also called the general flow equation, for flow (Q) in sm^3/d, and L is the pipe length, D the pipe diameter (in mm), p_b and T_b the base pressure and temperature, T_a the average gas temperature, p_1 and p_2 inlet and outlet pressure, respectively, and all in SI units [4], is widely used in the pipeline industry. The factor e is the pipe efficiency. It allows for the closure of the loss terms in the momentum equation. The equation is computationally efficient, requires fewer resources compared to other equations, and uses the Darcy Weisbach friction factor (f_{DW}) calculated as a function of the Reynolds number. It is very accurate for calculating frictional losses in fluid flow.

Since a pipeline is not an adiabatic system and is subject to heat transfer with the environment, the energy equation must also be solved to capture this effect.

The above applies to any type of pipeline. It omits the impact of pipeline elevation changes, as they are usually not as critical for gas pipelines, but become an important concern for liquid pipelines or dense phase pipelines. In gas pipelines, the friction losses cause a drop in pressure, which further increases the flow velocity. Therefore, the pressure loss in a pipeline is not linear but rather increases along the pipeline. A capacity limit is reached when the flow is choked. To allow for long-distance pipeline transport, compressor stations (Fig. 3.3) are needed that essentially compensate for the pressure losses in the pipeline. Since, for a given flow capacity, a large pipeline diameter yields low flow velocities and low-pressure loss, and a small pipe diameter yields higher flow velocities and higher pressure losses, the preference

FIG. 3.3 Pipeline compressor station with a gas turbine driven centrifugal compressor.

seems to be for larger pipeline diameters. However, larger pipeline diameters cause higher material and construction costs. On the other hand, a smaller pipeline would require more installed power for the compressor stations. Another design parameter is the pressure rating of the pipeline, as for a given mass flow or energy flow requirement a higher operating pressure would allow a smaller diameter pipeline while maintaining flow velocity. However, creating a higher pressure capability will also increase the cost of the pipeline.

This poses an optimization problem. Notwithstanding political considerations, there are considerable variables in trying to optimize gas transportation pipeline size and station spacing. There is a very wide range in the type of variables, such as a right of way costs, permit costs, freight rates, onshore or offshore construction, construction interferences, availability of material, quality of skilled labor, contract costs, and if the pipeline route and station locations are remote, harsh, mountainous, swampy, or have difficult access with lake and river crossings. In some cases, housing has to be provided for the construction crew and pipeline and station operators. Some companies may have gathered considerable data in the area of their operation such as enclosed or partially enclosed station building costs, power supply costs, training, operation and maintenance expenses, taxes, insurance, performance of equipment, contract and service obligations, risk factor in design and operation, and many other factors associated with the business that can be used effectively in a study.

Kurz et al. [3] evaluated the trade-off for a 3220-km (2000 mile) natural gas pipeline for a flow of 500 MMSCFD. Small-diameter pipelines will require more stations at closer distances. In this case, the lower cost of pipe is offset by an increase in station and operation/maintenance expenses. On the other hand, large-diameter pipelines will require fewer stations at farther distances. Thus, the higher cost of pipe is offset by the decrease in station and operating/maintenance expenses. Figs. 3.4–3.6 are created in order to illustrate this trend and use them as visual aids in determining the most theoretically economic pipeline design.

The pressure drop along the pipelines was calculated using the general flow equation (Eq. 3.1). Various line sizes and station spacing were considered at compression ratios of 1.2, 1.3, 1.4, 1.6, 1.8, 2.0, and 2.2. After calculating pipe wall thickness, the weight of the pipeline was determined, which was used in the calculation of pipeline construction cost.

The required shaft power at each station was calculated assuming 83% compressor isentropic efficiency, taking into account the station piping losses mentioned above. This power was used to determine station cost, gas turbine driver heat rate, and consequently, annual operation and maintenance costs for the life of the project (reduced to a present worth sum at 10% interest rate).

Each curve is based on a constant compression ratio with various station spacing. The minima of each curve represent the economic pipe diameter and station spacing for the respective compression ratio.

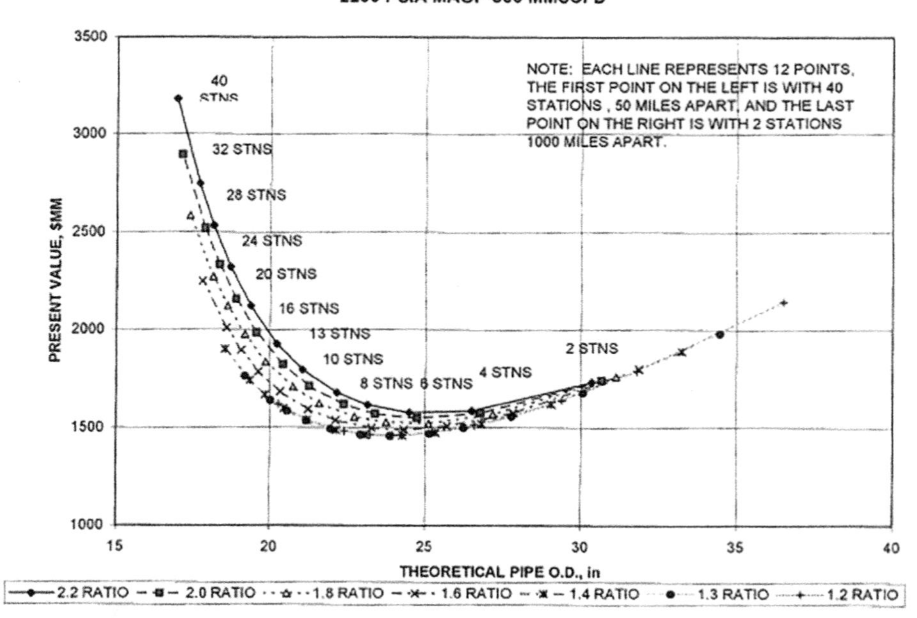

FIG. 3.4 Present value for 152 bar (2200 psi) pipeline for different number of stations and station pressure ratios [3].

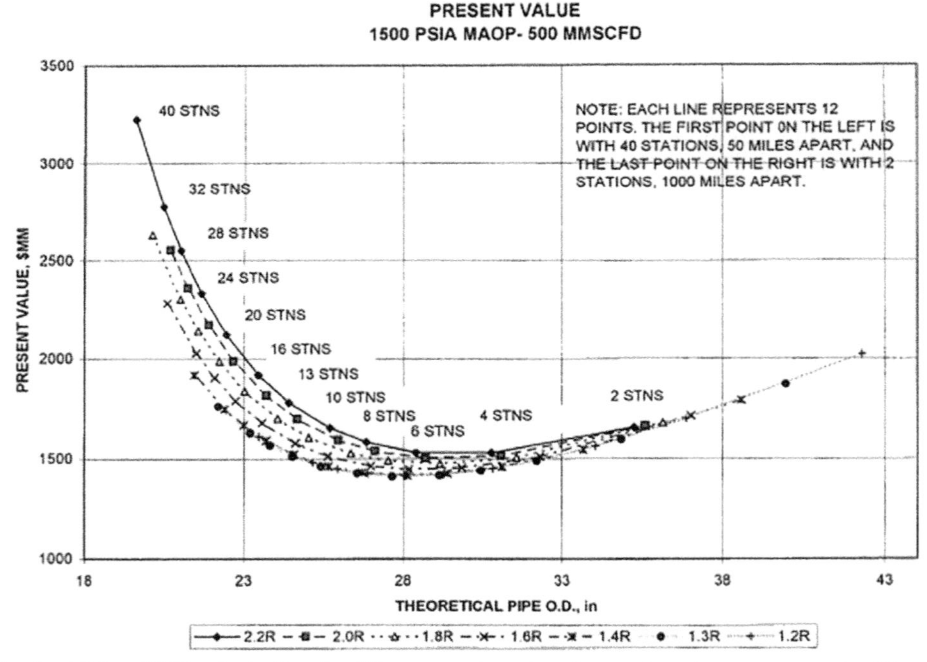

FIG. 3.5 Present value for 103 bar (1500 psi) pipeline for different number of stations and station pressure ratios [3].

As can be seen, there is a close contest between 1.2, 1.3, and 1.4 compression ratios for each pressure level; however, the 1.3 ratio happens to be the minimum in each case. As a further refinement, the impact of fuel consumption along the pipeline was included, using commercially available, pipeline simulation software.

Assuming that pipes will be available in 2″ diameter increments from pipe mills, the nearest even increments of the above-mentioned theoretical diameters were selected (24, 28, and 34″ for 152 bar, 103 bar, and 69 bar (2200, 1500, 1000 psia) pressures, respectively) and analyzed by varying the number of stations along the pipeline. The result of this refinement is shown in Fig. 3.7, where the present value is plotted against a number of stations for each pressure level. The minima for each is shown in the chart with present value, total horsepower, and number of stations. For the

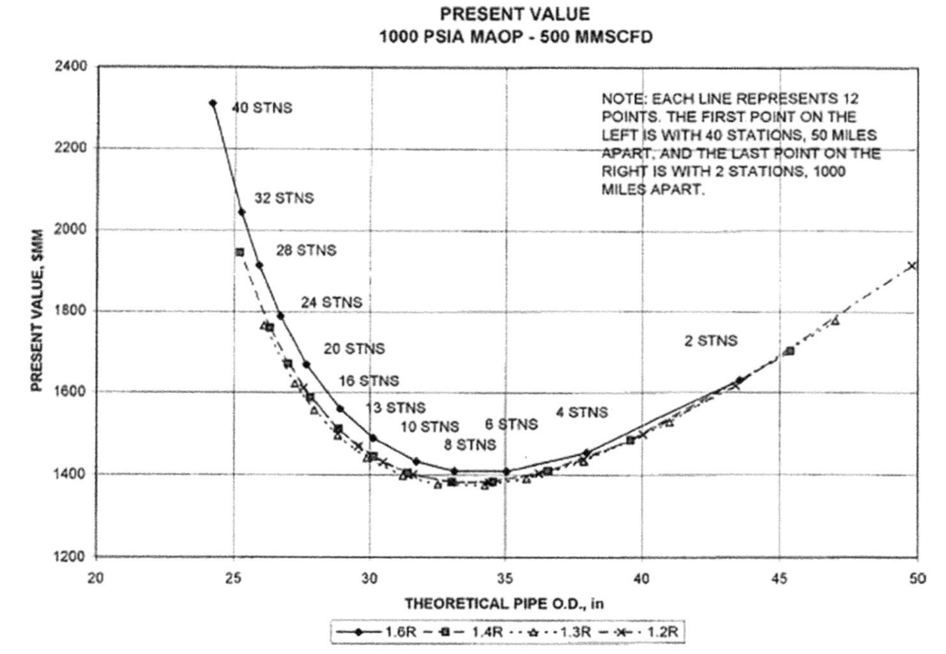

FIG. 3.6 Present value for 69 bar (1000 psi) pipeline for different number of stations and station pressure ratios [3].

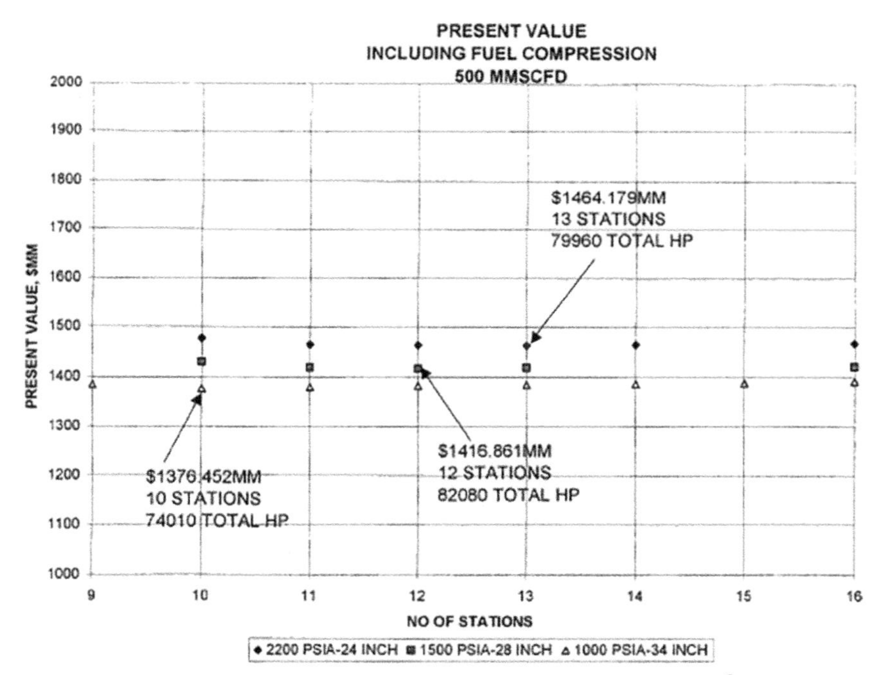

FIG. 3.7 Optimum number of stations and optimum MAOP for the 3220 km (2000 mile), 560,000 m^3_N/h sample pipeline. The lowest cost configurations for each MAOP solution are marked [3].

set of assumptions used, the 69 bar (1000 psia) pipeline has the lowest present value, thus would be the most cost-effective solution.

One of the questions in the discussion on carbon reduction relates to the energy requirements for the transport of the gas involved, i.e., natural gas, CO_2, and hydrogen. To answer the question, pipelines for the transport of these gases are simulated by Kurz et al. [5] (Figs. 3.8 and 3.9).

The simulations include:

1. A dense-phase CO_2 pipeline with a 32″ nominal diameter and three compressor stations equally spaced, transporting about 766 kg/s or 1300 MMSCFD for a distance of 500 miles (800 km). The operating pressure is 2200

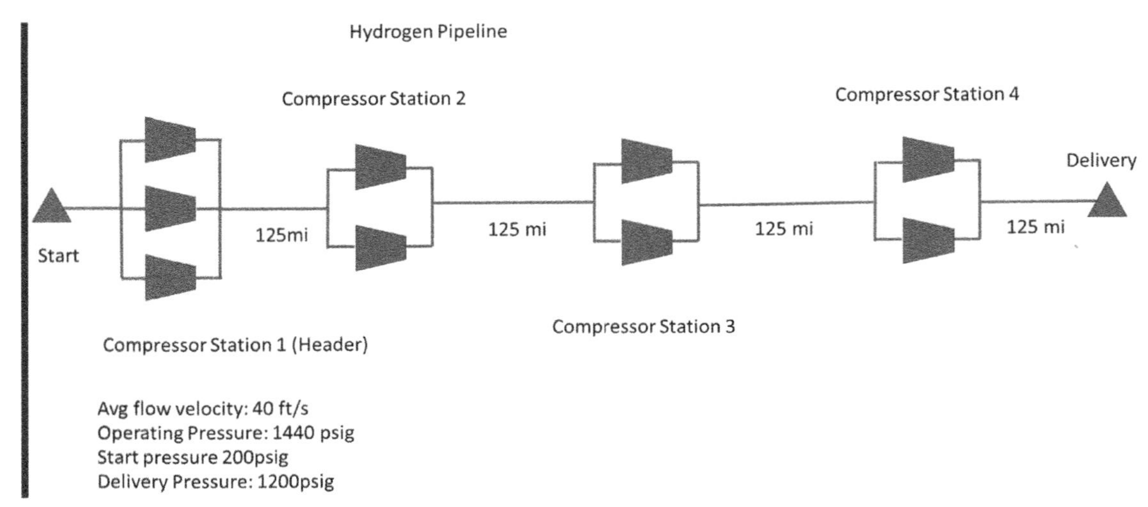

FIG. 3.8 Hydrogen pipeline, 500 miles (800 km), 1000 MMSCFD.

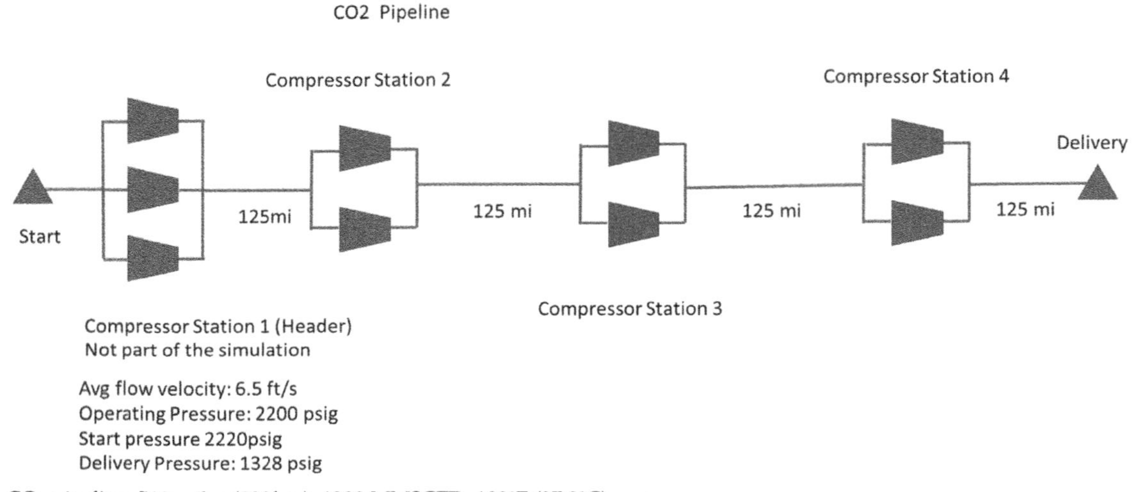

FIG. 3.9 CO_2 pipeline, 500 miles (800 km), 1300 MMSCFD, 100°F (37.8°C).

psig, and the inlet pressure for the compressor station is kept above 1320 psig (91 barg) to avoid two-phase flow at all prevailing ambient temperatures. The calculations shown are for a 37.8°C (100°F) gas temperature at the head station discharge.

2. A hydrogen pipeline for the same distance (500 miles, 800 km) and a 1440 psig (99 barg) operating pressure, transporting 1000 MMSCFD with three compressor stations equally spaced.
3. A 500-mile (800 km) natural gas pipeline, for 700 MMSCFD, 1440 psig (99 barg) operating pressure, natural gas with SG = 0.58.

It should be noted that, although it is not included in these results, the power consumption of the header station is very large relative to the pipeline stations for both the CO_2 and hydrogen cases. This power consumption is very sensitive to the suction pressure, which could be as low as atmospheric pressure for CO_2 or as low as 5 bar for hydrogen (Figs. 3.6 and 3.19). The results (for the pipeline stations only) are summarized in Table 3.1.

The head and power requirements outlined in Table 3.3 would require a 7–8 stage centrifugal compressor, with 585 mm (23 in.) diameter impellers for hydrogen, while the CO_2 application can be covered with a single-stage compressor with a 380 mm (15 in.) impeller. The natural gas compressor would require a single impeller or two impellers with a typical diameter of 530–560 mm (21–22 in.).

To operate a 1 GW power plant, 600 MMSCFD of hydrogen or 180 MMSCFD of natural gas is needed, and 162 MMSCFD CO_2 is produced. To be able to compare the power consumption based on the flows related to the operation

TABLE 3.1 Summarized results for the pipeline simulations.

Gas	Flow (MMSCFD)	Station PR	Head (kJ/kg)	Power (kW)
Hydrogen	1000	1.21	238.9	7946
Natural gas	700	1.45	50.8	9909
CO_2	1300	1.65	6.55	6182

TABLE 3.2 Summarized results for the pipeline simulations corrected to 1 GW plant flow.

Gas	Flow (MMSCFD)	Station PR	Head (kJ/kg)	Power (kW)
Hydrogen	600	1.21	238.9	4768
Natural gas	180	1.45	50.8	2548
CO_2	162	1.65	6.55	770

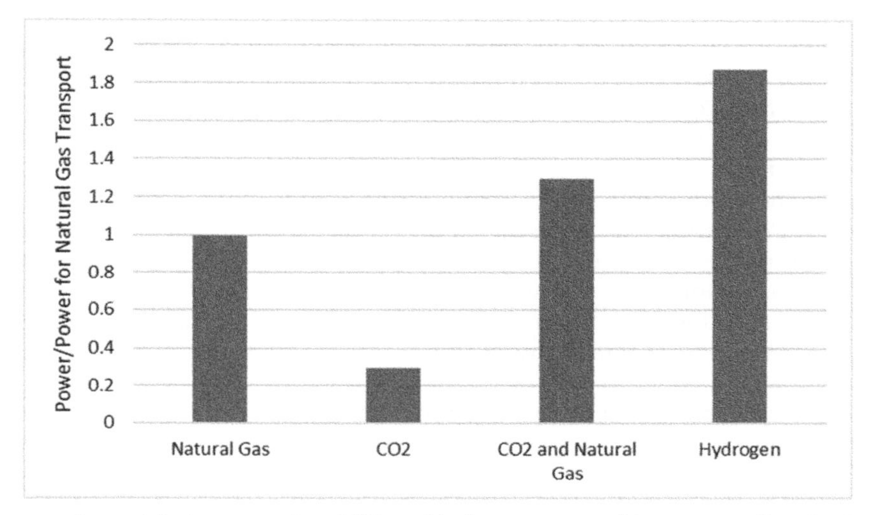

FIG. 3.10 Relative power requirement for transportation of CO_2 and hydrogen compared to transportation of natural gas [5].

of a 1 GW power plant, the flow from Table 3.1 (and with it the power consumption) is scaled to the flow for a 1 GW power plant and shown in Table 3.2.

In Table 3.2, the data are shown by scaling the pipeline to the flow required for a 1 GW power plant in order to compare flow and compression power requirements for a fixed electrical power demand. The pipelines themselves would probably be too small to be economically feasible. This is why the pipeline flows in Table 3.1 were used.

The data in Table 3.2 and Fig. 3.10 show the significant difference in consumed power in these cases. It becomes clear that for a power plant of a given size, transporting hydrogen to the plant requires more energy than the transport of natural gas to the plant and the transport of CO_2 from the plant. Table 3.2 indicates that even if the power for the natural gas and the CO_2 pipeline are combined, they are still significantly lower than the power for the hydrogen transport. Bringing natural gas to a power plant and transporting the generated CO_2 to a sequestration site is more energy efficient than transporting hydrogen, generated elsewhere, over larger distances.

Refrigeration process and liquefaction

The liquefaction process for most gases requires refrigeration of the gas, which is fundamentally cooling the feed gas by heat exchange between the relatively warm feed gas and colder refrigerants. Refrigeration involves the removal of heat from one substance and its transfer to another substance. Heat can be transferred along a temperature differential, in other words, heat will flow from a higher temperature source to a lower temperature sink.

To generate the cold temperatures required for LNG production, work must be put into the refrigeration cycle through compression, and heat must be rejected from the cycle via a heat exchanger. Liquefaction of natural gas requires the temperature of the feed gas to be reduced to a near subcooled (liquid) condition.

For LNG production, any two-phase refrigeration process involves the following fundamental steps:

- A refrigerant (in many cases a hydrocarbon or nitrogen) is compressed to a vapor state at a higher temperature, but still above the saturated vapor pressure at that temperature (1–2).
- The pressurized vapor flows to a condenser that transfers heat out of the vapor typically to the ambient air or to a coolant (water). The condenser cools the hot refrigerant vapor until the vapor condenses the refrigerant vapor into a liquid by removing additional heat (the heat of vaporization) (2–3).
- The next stage in the refrigeration cycle is to expand the high-pressure refrigerant liquid by passing it through a Joule Thomson (J-T) expansion valve or an expander. The expansion produces a mixture of liquid and vapor at a lower temperature and pressure (3–4).
- The cold liquid-vapor refrigerant mixture then flows to an evaporator unit, where the refrigerant fluid reaches a saturated vapor condition by absorbing heat from the working fluid (feed gas) passing from cooling coil tubes within the evaporator (4–1).

From the evaporation, the refrigerant flows to the compressor and the cycle continues. A typical refrigeration cycle is shown in Fig. 3.11.

The thermodynamics cycle of the refrigerant is shown as a temperature versus specific entropy graph in Fig. 3.12 to better understand the refrigerant temperature, work input, phase change, and heat rejection and input for an ideal vapor-compression refrigeration cycle.

It should be noted that Fig. 3.12 illustrates the ideal refrigeration cycle. In reality, points 3 and 1 are at subcooled (liquid only) and superheated (vapor only) parts of the T-s diagram, respectively.

In order to improve the thermodynamic and thermal efficiency of the refrigeration and liquefaction process, an expander can be utilized in the refrigeration cycle in lieu of a J-T expansion valve. For a given pressure differential, an expander can achieve a much greater temperature reduction compared to an expansion valve by recovering work. Expansion via a J-T valve is isenthalpic (no enthalpy change), while the pressure drop across the expander is near isentropic with enthalpy removal from the fluid. This results in colder temperatures at the discharge of the expander with respect to the J-T valve. Fig. 3.13 shows expansion with a J-T valve and a two-phase flashing expander over a pressure versus enthalpy graph.

The following concepts are the most commonly used refrigeration processes in LNG liquefaction:

1. Reverse Brayton Cycle—Nitrogen is typically used as the refrigerant and maintained in vapor phase throughout the process [6]. Methane can also be used in this cycle as a refrigerant. Both N_2 and methane are readily available at LNG

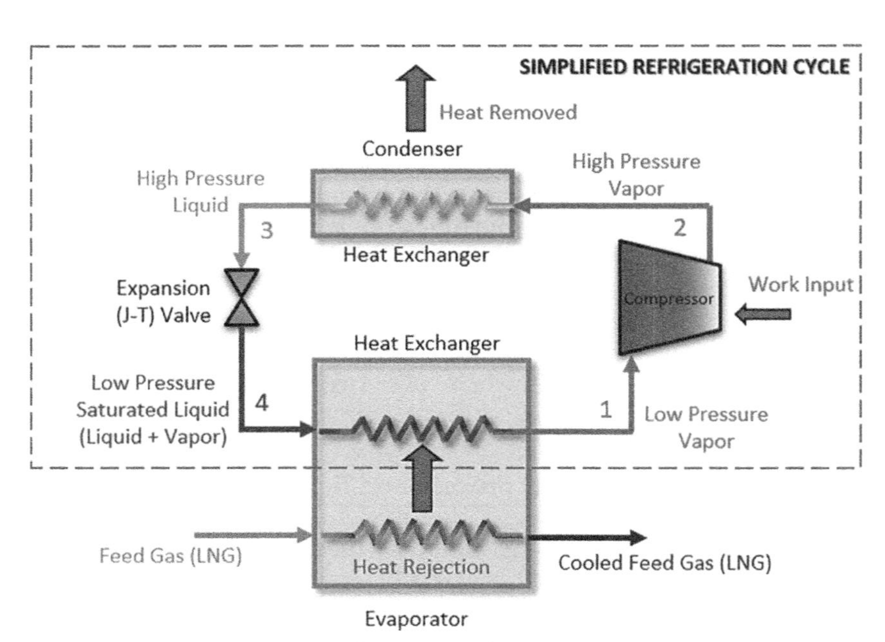

FIG. 3.11 Simplified single-stage compression expansion refrigeration cycle.

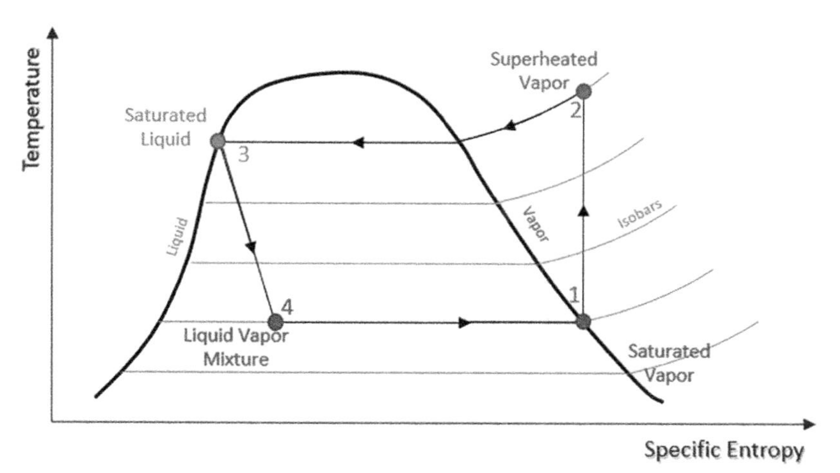

FIG. 3.12 Refrigeration cycle T-s diagram illustrating an ideal case.

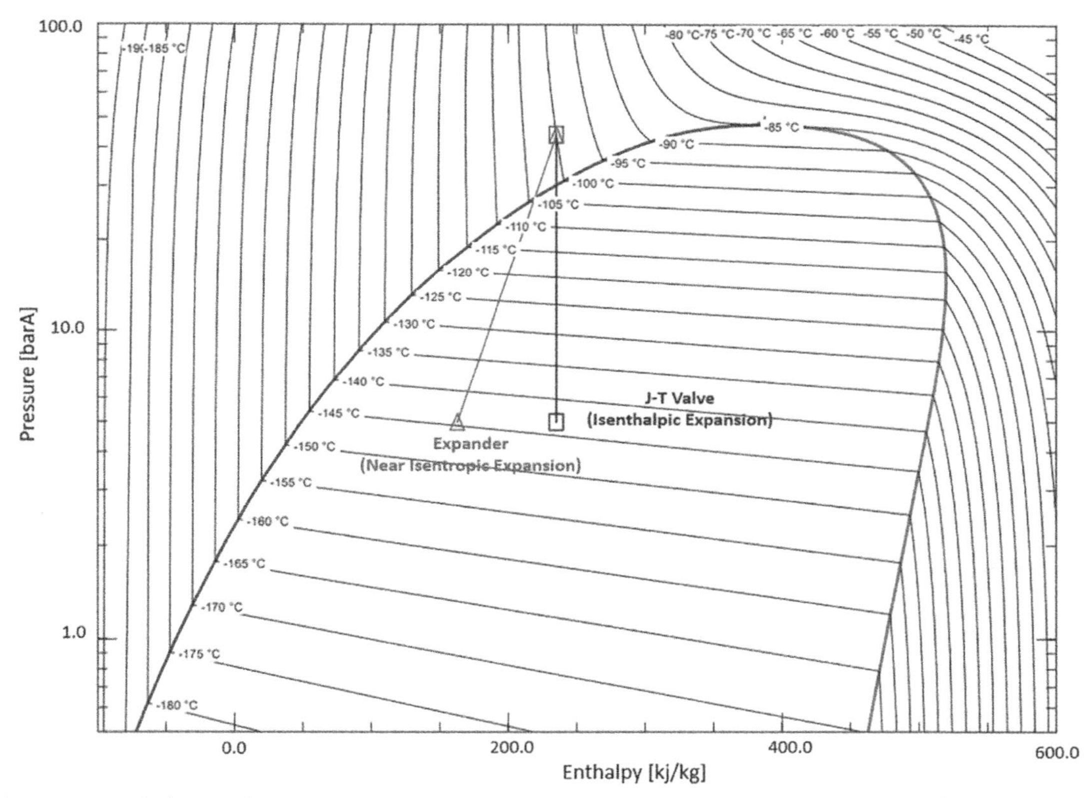

FIG. 3.13 Pressure vs. enthalpy graph showing the expansion via expander and J-T valve for 95% methane and 5% propane fluid mixture.

liquefaction plants for equipment purging and from the feed gas, respectively. Using methane as refrigerant from the feed gas allows a smaller footprint. A reverse Brayton cycle achieves refrigeration by expanding vapor and extracting work. Fig. 3.14 shows a schematic of an ideal reverse Brayton cycle, and the temperature versus entropy plot of an ideal reverse Brayton cycle is given in Fig. 3.15. In this cycle, with the heat from the feed gas (LNG), warm refrigerant gas at low pressure is compressed (1–2). The resulting high-pressure fluid is cooled at constant pressure while rejecting heat to ambient (2–3). The cooled refrigerant is then expanded, and work is extracted to produce a cold refrigerant stream (3–4), which is subsequently warmed while providing refrigeration (4–1).

This cycle is used widely in LNG production due to its simple operation and low capital expense [6,7]. Using N_2 is attractive for small-scale LNG production, since import and storage of hydrocarbon refrigerants are not needed.

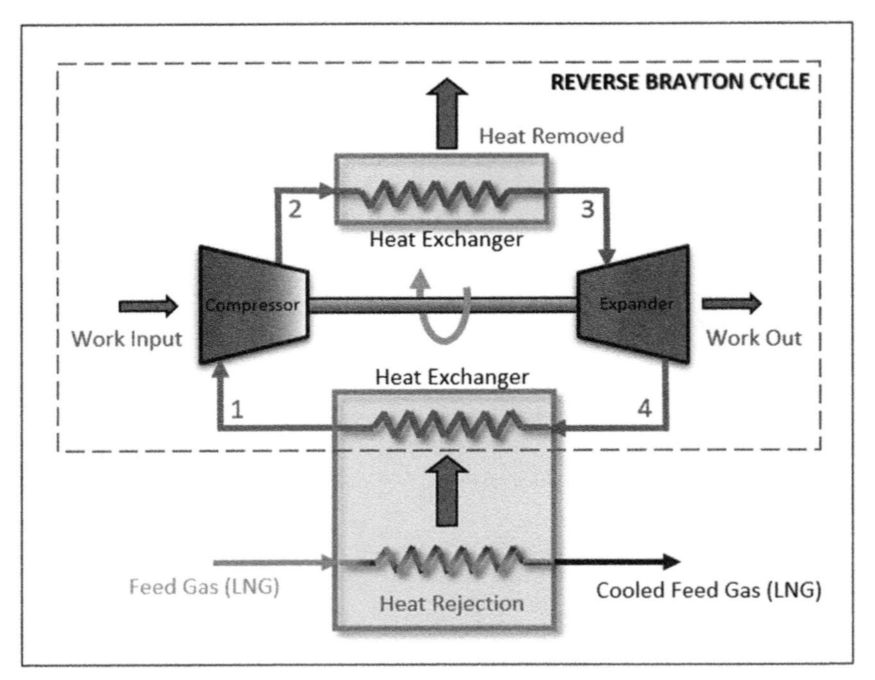

FIG. 3.14 Reverse Brayton Refrigeration Cycle.

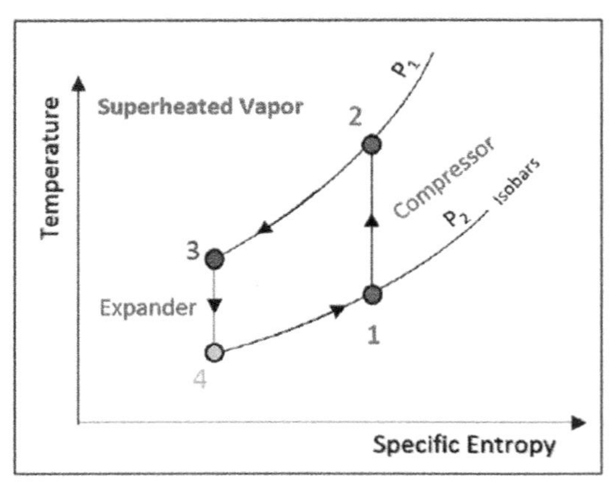

FIG. 3.15 Reverse Brayton Cycle T-s diagram illustrating an ideal case.

As seen in Fig. 3.15, there is no phase change at any of the heat exchangers of the reverse Brayton cycle, and heat transfer is provided by sensibly warming the refrigerant gas (4–1). Therefore, mass flow rate is required to be relatively greater for the reverse Brayton cycle compared to a two-phase cycle to provide sufficient refrigeration. This results in larger equipment and piping. It can be mitigated by the use of high operating pressures, which are subject to design constraints of both the expander and compressor [8]. Therefore, this refrigeration cycle is preferred for small-scale LNG, peak-shaving, and floating LNG (FLNG) applications in the range of 1 MMTPA to 2 MMTPA [8]. It should be noted that in real applications, pressure drop and increase across the expander and compressor is not 100% isentropic, and fluid and heat flow involves losses across piping and heat exchangers.

2. Two-phase Refrigeration Cycle—Involves evaporation and condensation of pure gases or mixed refrigerants (MRs) at heat exchangers. The main difference of two-phase refrigeration with respect to the reverse Brayton cycle is that the cycle provides refrigeration by vaporizing the refrigerant (latent heat) rather than cooling via sensible heat. More refrigeration is provided per mass of refrigerant by latent heat compared to sensible heat.

For LNG liquefaction applications, the choice of pure refrigerant fluids is propane, ethylene, and methane. MRs are often a blend of hydrocarbons with nitrogen. The composition of the MR is determined by matching the cooling curve of the feed gas from ambient to cryogenic temperatures. The basic principle of using MR for cooling and liquefying the gas is to match as closely as possible the cooling/heating curves of the feed gas and the refrigerant, which results in a more efficient liquefaction process requiring a lower power consumption per unit of LNG produced. MR technology can be categorized as single-mixed refrigerant (SMR) and dual-mixed refrigerant (DMR) cycles.

The SMR process provides the benefit of operational simplicity and flexibility, in addition to reduced equipment count; however, it comes at the cost of lower efficiency than the DMR cycle, which better matches the overall MR boiling curve to the feed condensation curve [9]. The SMR cycle uses only one MR circuit for precooling, liquefaction, and subcooling in a single heat exchanger. The precooled MR cycle is the most widely used LNG process for land-based facilities. The feed gas is initially cooled to an intermediate temperature (about −30 °C) by a precooling circuit. Precooling can be provided by utilizing a cold mixed refrigerant (CMR) composed of nitrogen, methane, ethane, and propane. Another and more common option of precooling is providing refrigeration via a propane precooling circuit (C3MR). A schematic view of the basic precooled MR process is shown in Fig. 3.16.

A key component of a precooled MR cycle is the coil-wound heat exchanger (CWHE) where the liquefaction takes place as shown in Fig. 3.16. These heat exchangers incorporate high surface area in a compact form factor and make the process much more efficient than most other liquefaction and refrigeration processes.

Another common process for refrigeration and liquefaction is the cascade process where the feed gas is cooled in multiple stages by using different hydrocarbon refrigerants in stages. Three main refrigeration loops are employed in this process. Typically, a propane cooling circuit is followed by an ethylene loop and finally by a methane loop. The propane loop not only precools the feed gas but also precools ethylene and methane at their respective loops, while the ethylene loop also precools the methane loop [5]. A basic schematic view of the cascade process by ConocoPhillips is given in Fig. 3.17.

The liquefaction process selection is a key activity that starts at an early stage of an LNG project. It should be addressed at the conceptual, feasibility, and pre-FEED stages of development since it has such a large impact on the overall profitability of the project. When a comparison is conducted thoroughly, sufficient process and utility details must be developed to define the capital and operating costs for each licensor. Quotations from the various licensors and main equipment suppliers must be obtained to highlight the differences in the processes and to finally select an optimized design that will best meet the LNG project owner's objectives.

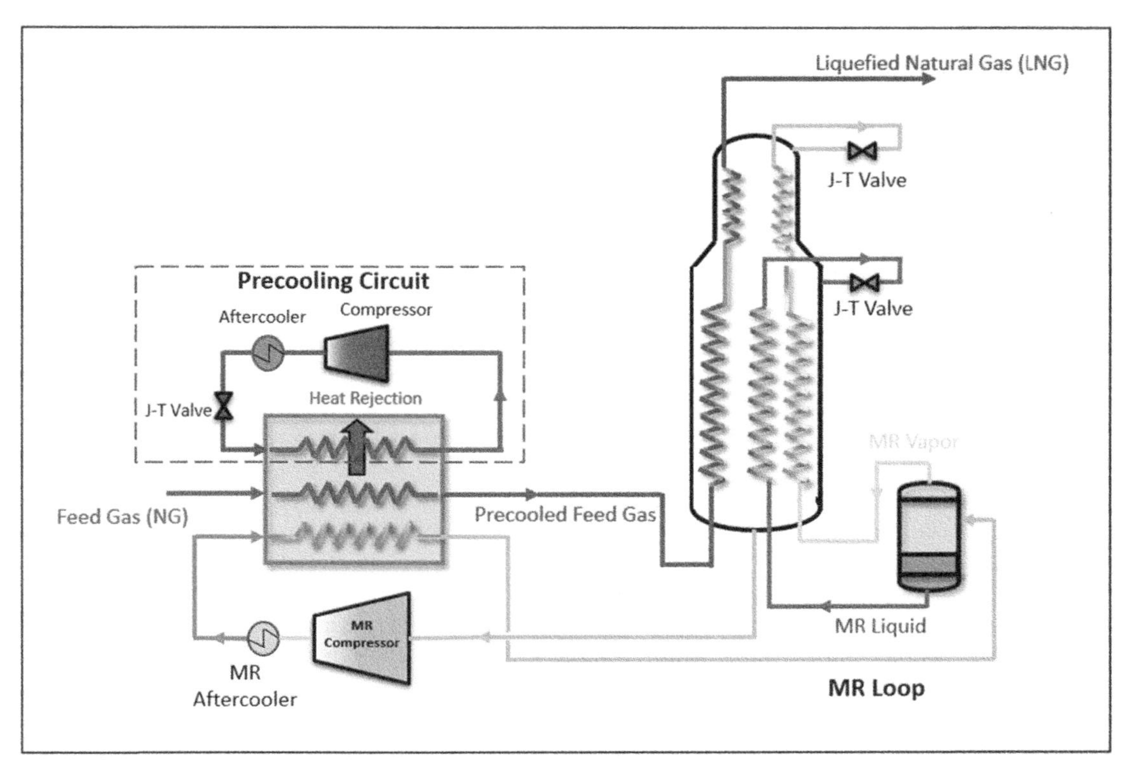

FIG. 3.16 Simplified precooled MR refrigeration process for natural gas liquefaction.

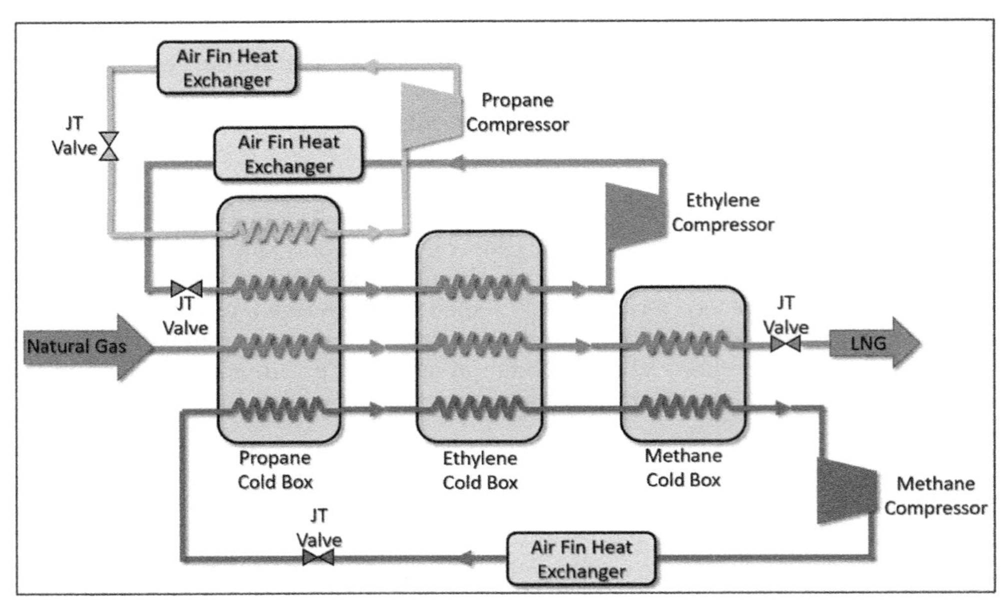

FIG. 3.17 Simplified optimized cascade process by ConocoPhillips [10].

Machinery drivers

Electric motors

Electric motors are utilized to create mechanical work from electrical energy. Electric motors operate based on the interaction of electric current and electromagnetic fields. Lenz's Law states that the direction of an induced current in a conductor will always be such that it opposes the change that produced it. In simpler terms, when a magnetic field through a conductor changes, the induced current in the conductor will flow in a direction that creates a magnetic field opposing the change in the original magnetic field. By applying a changing magnetic field to a conductor, a force resisting the change is created. If the conductor is free to rotate, a motor is created. The same principal in reverse is the basis of generators, where mechanical work is applied to the spinning rotor creating a current in the generator.

Induction motors are simple AC machines. The induction motor is the most common type of large motor due to its simplicity, ruggedness, and low cost. It consists of two main parts: the stator and the rotor. The stator is the stationary part and consists of a laminated iron core with evenly spaced windings or coils. These windings, when connected to an AC power supply, create a rotating magnetic field.

The rotor is the rotating part of the motor. It consists of a laminated iron core with conducting bars placed parallel to the motor's shaft. The bars are shorted at each end by conducting rings, forming a closed loop or cage-like structure. The conducting bars are typically made of aluminum or copper and are typically permanently shorted to provide a closed circuit. When the AC power is applied to the stator windings, it produces a rotating magnetic field. This rotating magnetic field induces currents in the "squirrel cage" rotor bars due to electromagnetic induction. As a result, magnetic fields are generated in the rotor bars, which interact with the magnetic field of the stator. This interaction causes the rotor to rotate, as the magnetic fields in the rotor bars try to align with the rotating magnetic field of the stator.

Squirrel cage induction motors are widely used in various applications including pumps, fans, compressors, conveyors, and many industrial machinery applications. They are known for their robustness, reliability, and ability to provide high torque at low speeds, making them suitable for a wide range of tasks. The squirrel cage rotor design offers several advantages, such as simplicity, durability, and self-starting capability. Since the rotor bars are permanently shorted, there is no need for additional brushes or slip rings to supply current to the rotor, making it maintenance-free. However, the speed of the squirrel cage induction motor is determined by the frequency of the AC power supply and the number of poles in the stator windings.

In theory, an induction motor could be designed for any industrial application and power, but in practice, as the power level of the motor increases beyond around 20 MW, the improved efficiency of the synchronous motor overshadows the increase in complexity.

A synchronous AC motor is a type of alternating current (AC) motor that runs at a constant speed and maintains a fixed relationship, or synchronization, with the frequency of the power supply. Unlike induction motors, which rely on the principle of induction to generate a rotating magnetic field, synchronous motors have a rotor that rotates at the same speed as the rotating magnetic field produced by the stator.

The construction of a synchronous AC motor consists of a stator, which contains the stationary windings connected to the power supply, and a rotor, which is the rotating part of the motor. The stator windings create a rotating magnetic field when energized with AC power. The rotor of a synchronous motor can have different configurations, including salient pole and cylindrical rotor designs.

In a synchronous motor, the rotor is magnetized either by permanent magnets or by DC excitation through a separate DC power supply. The magnetic field of the rotor interacts with the rotating magnetic field of the stator, causing the rotor to rotate at the same speed as the rotating magnetic field. This synchronous speed is directly determined by the frequency of the power supply and the number of poles in the stator windings.

Synchronous motors have fixed stator windings electrically connected to the AC supply with a separate source of excitation connected to a field winding on the rotating shaft. A three-phase stator is similar to that of an induction motor. The rotating field has the same number of poles as the stator and is supplied by an external source of DC. Magnetic flux links the rotor and stator windings causing the motor to operate at synchronous speed. A synchronous motor starts as an induction motor, until the rotor speed is near synchronous speed where it is locked in step with the stator by application of a field excitation. When the synchronous motor is operating at synchronous speed, it is possible to alter the power factor by varying the excitation supplied to the motor field.

An important advantage of a synchronous motor is that the motor power factor can be controlled by adjusting the excitation of the rotating DC field. Unlike AC induction motors, which run at a lagging power factor, a synchronous motor can run at unity or even at a leading power factor. This can help to improve the overall electrical system power factor and voltage drop and also improve the voltage drop at the terminals of the motor (for direct online applications). While synchronous motors can be made at any size, the additional complexity and cost mean they are normally considered only at power levels above 20 MW. The speed of the synchronous speed of AC motors is dependent on the line frequency (Hz) and the number of poles in the stator. Synchronous Speed (in RPM) is equal to (120 × Frequency) divided by the Number of Poles as shown in Table 3.3. Induction motors do nt operate at a synchronous speed. Due to the nature of the induced current in the rotor, a slip exists between the fields in the rotor and stator. Slip is typically 2%–5%; higher slip relates to higher torque generation but also relates to lower efficiencies.

The typical voltage of AC motors is a function of the machine power range and frequency, and general preferred motor voltage ratings are described in Table 3.4.

Meeting the operational speed range of the compressor is important in gas compression systems, because centrifugal compressors and most reciprocating compressors operate most efficiently in terms of capacity control by varying speed [11]. For centrifugal compressors, varying flow rates without speed control involves suction or discharge throttling or recycling gas. Both of these capacity control options are significantly less efficient than changing the rotational speed of the centrifugal compressor. For reciprocating compressors, capacity may be varied by other means besides recycling flow and speed variation, such as opening volume pockets, deactivating the head-end of a cylinder, or delayed valve opening/closing.

However, speed variation provides substantially more range and control of the reciprocating compressor throughput. For these reasons, a large majority of electric motor-driven gas compression systems will require designs for

TABLE 3.3 Typical synchronous motor speeds.

	Synchronous speed	
No. of poles	**50 Hz**	**60 Hz**
2	3000	3600
4	1500	1800
6	1000	1200
8	750	900
12	500	600
16	375	450
18	333	400

TABLE 3.4 Preferred machine power and voltage rating [11].

Horsepower	KW	Voltage rating
a. 60 Hz power supply:		
100–600	75–500	460 or 575
200–5000	150–3500	2300
200–10,000	150–7000	4000
1000–15,000	800–10,000	6600
3500 and up	2500 and up	13,200
b. 50 Hz power supply:		
100–500	75–375	400
600–8000	500–6000	3000–3300
700–15,000	500–12,500	6000–6600
3000 and up	2500 and up	10,000–11,000

adjustable speed, typically accomplished through a variable frequency drive (VFD) controlling the motor or a variable speed hydraulic drive (VSHD) with a fixed speed motor. An alternative that is rarely used is to use a multi-speed motor available in 2-speed, 3-speed, or 4-speed configurations.

In addition to speed matching with a gas compressor, other common electric motor issues must also be considered for the gas compressor application. The cost, complexity, and reliability of the drive train will be impacted as more components are added.

Electric motor configurations include constant speed motors driving the compressor via a variable speed gearbox (VSHD, Fig. 3.18) and variable frequency drive speed (VFD) controlled motors driving the compressor either directly or via a gearbox (Fig. 3.19). The package for a constant-speed motor would look similar to Fig. 3.19.

Of importance for many applications are the performance characteristics of the driver, for example, the power as a function of ambient conditions or the power output at various output speeds. In general, a VFD-controlled motor is a constant torque machine, thus exhibiting a linear drop in power with speed (Fig. 3.20), implemented by maintaining a constant Volts/Hz ratio until the motor becomes power-limited above a certain corner frequency. There are

FIG. 3.18 Electric motor drive package, constant speed motor driving the compressor via a variable speed gearbox. *Courtesy of Solar Turbines.*

FIG. 3.19 Electric motor drive package, VFD controlled motor driving the compressor via a speed increasing gearbox. *Courtesy of Solar Turbines.*

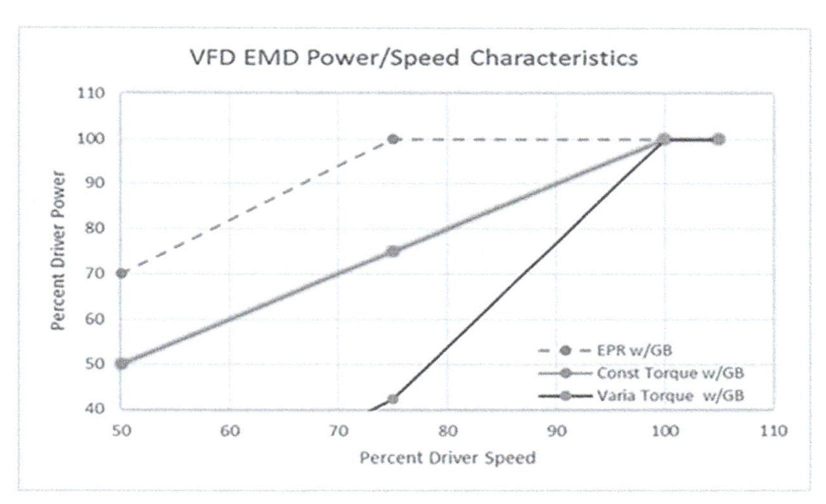

FIG. 3.20 Speed-power relationship for VFD driven electric motors with variable torque, constant torque and expanded range characteristics. *Courtesy of Solar Turbines.*

exceptions to this behavior where the motor is oversized to provide constant power over a wider range (Expanded Power Range, EPR), or often for thermal reasons, the torque is reduced with speed [12]. The speed-power relationship has a significant impact on control concepts for variable speed drives. In particular, the linear reductions of power seen in most VFD-controlled electric motors impose a limit on the flexibility compared to VSHD and two-shaft gas turbine drives.

The starting characteristics, including the amount of torque at low speeds (or, for constant-speed electric motors, the amount of additional current required during starting) must be considered (Fig. 3.21).

A unit's installation location (onshore, offshore, or subsea) determines access to maintenance intervention, as well as the environmental conditions (for example, salt in the air) the equipment has to be designed for. For electric drivers, the question is also whether the electricity can be brought to the site via transmission lines, or whether it has to be generated onsite, usually with gas turbine-driven generators.

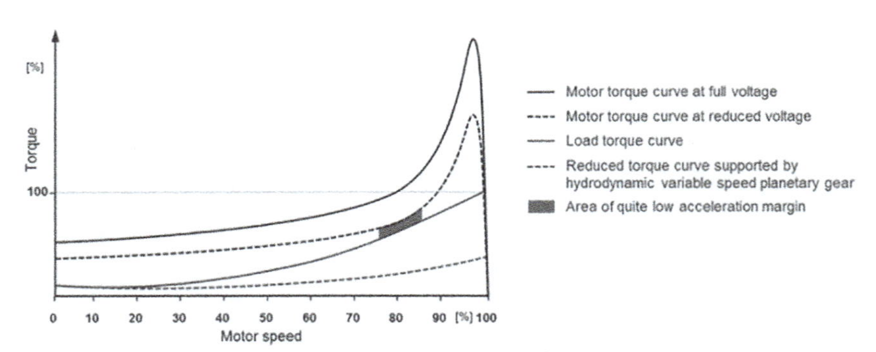

FIG. 3.21 Squirrel cage induction motor capability curve at different voltage levels and load torque curve [13].

Gas turbines

Gas turbines are widely used as drivers for not only compressors and pumps, but also for power generation due to their high power density, low emissions, good efficiency, and attractive operating cost.

Explanations of the working principles of a gas turbine have to start with the thermodynamic principles of the Brayton cycle, which essentially defines the requirements for the gas turbine components. Since the major components of a gas turbine perform based on aerodynamic principles, we will explain these, too [14].

The Brayton or gas turbine cycle (Fig. 3.22) involves compression of air (or another working gas), the subsequent heating of this gas (either by injecting and burning fuel or by indirectly heating the gas) without a change in pressure, followed by the expansion of the hot, pressurized gas. The compression process consumes power, while the expansion process extracts power from the gas. Some of the power from the expansion process can be used to drive the compression process. If the compression and expansion process are performed efficiently enough, the process will produce useable power output. This principle is used for any gas turbine, from early concepts by F. J. Stolze (in 1899), C.G. Curtiss (in 1895), S. Moss (in 1900), Lemale and Armengaud (in 1901) to today's jet engines and industrial gas turbines [15]. The process is thus substantially different from a steam turbine (Rankine) cycle that does not require the compression process but derives the pressure increase from external heating. The process is similar to processes used in Diesel or Otto reciprocating engines that also involve compression, combustion, and expansion. However, in a reciprocating engine, compression, combustion, and expansion occur at the same place (the cylinder), but sequentially, in a gas turbine, they occur in dedicated components all at the same time.

The major components of a gas turbine include the compressor, the combustor, and the turbine. The compressor (usually an axial flow compressor, but some smaller gas turbines also use centrifugal compressors) compresses the air to several times atmospheric pressure (Fig. 3.23). In the combustor, fuel is injected into the pressurized air from the compressor and burned, thus increasing the temperature. In the turbine section, energy is extracted from the hot pressurized gas, thus reducing pressure and temperature. A significant part of the turbine's energy (from 50% to 60%) is used to power the compressor, and the remaining power can be used to drive generators or mechanical

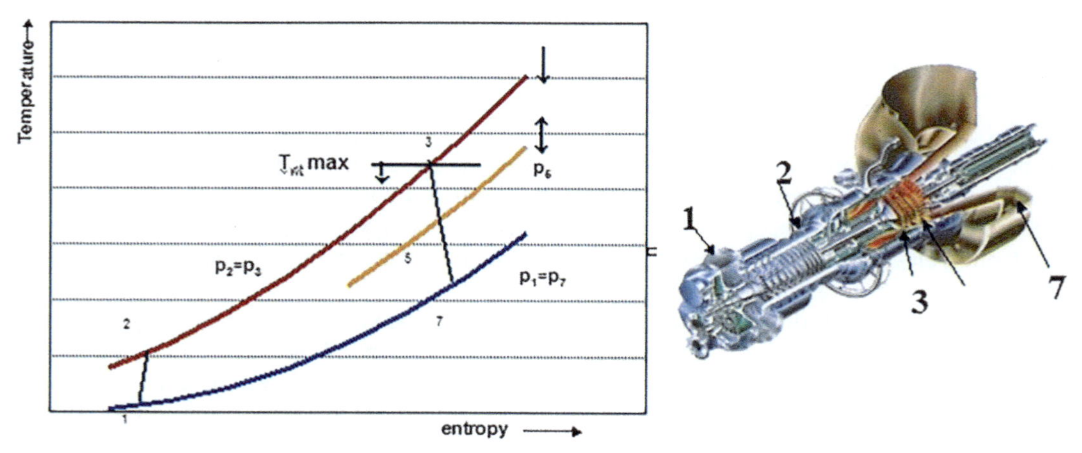

FIG. 3.22 Open Brayton Cycle. *Courtesy of Solar Turbines.*

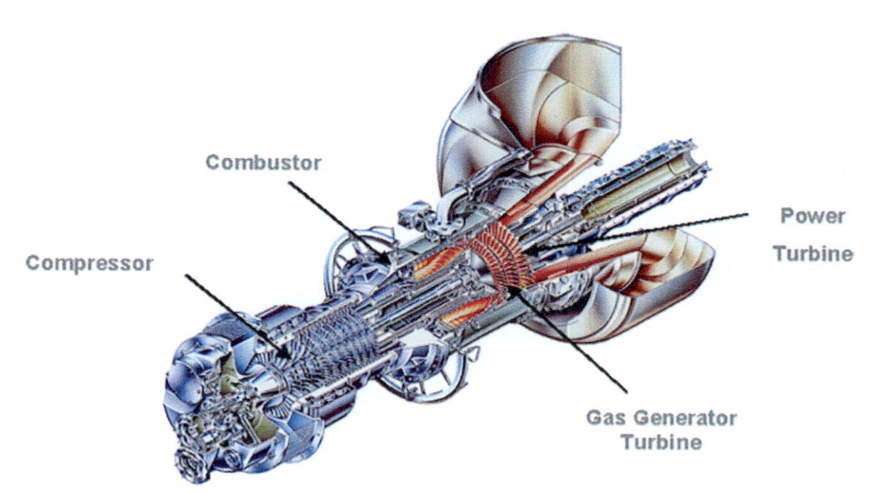

FIG. 3.23 Typical industrial gas turbine. *Courtesy of Solar Turbines.*

equipment (gas compressors and pumps). Industrial gas turbines are built with a number of different arrangements for the major components:

- Single-shaft gas turbines have all compressor and turbine stages running on the same shaft.
- Two-shaft gas turbines consist of two sections: the gas producer (or gas generator) with the gas turbine compressor, the combustor, and the high-pressure portion of the turbine on one shaft and a power turbine on a second shaft (Fig. 3.23). In this configuration, the high-pressure or gas producer turbine only drives the compressor, while the low-pressure or power turbine, working on a separate shaft at speeds independent of the gas producer, can drive mechanical equipment.
- Multiple spool engines: Industrial gas turbines derived from aircraft engines sometimes have two compressor sections (the HP and the LP compressor), each driven by a separate turbine section (the LP compressor is driven by an LP turbine by a shaft that rotates concentrically within the shaft that is used for the HP turbine to drive the HP compressor), and running at different speeds. The energy left in the gas after this process is used to drive a power turbine (on a third, separate shaft), or the LP shaft is used as an output shaft.

The energy conversion from mechanical work into the gas (in the compressor) and from energy in the gas back to mechanical energy (in the turbine) is performed by means of aerodynamics, i.e., by appropriately manipulating gas flows. Leonard Euler (in 1754) equated the torque produced by a turbine wheel to the change of circumferential momentum of a working fluid passing through the wheel. Somewhat earlier (in 1738), Daniel Bernoulli stated the principle that (in inviscid, subsonic flow) an increase in flow velocity is always accompanied by a reduction in static pressure and vice versa, as long as no external energy is introduced.

While Euler's equation applies Newton's principles of action and reaction, Bernoulli's law is an application of the conservation of energy. These two principles explain the energy transfer in a turbomachinery stage (Fig. 3.24).

The compressed air from the compressor enters the gas turbine combustor. Here, the fuel (natural gas, natural gas mixtures, hydrogen mixtures, diesel, kerosene, and many others) is injected into the pressurized air and burns in a continuous flame. The flame temperature resulting from combustion is usually so high that any direct contact between the combustor material and the flame has to be avoided, and the combustor has to be cooled using air from the engine compressor. Additional air from the engine compressor is mixed into the combustion products for further cooling.

Another important topic is the combustion process and emissions control. Unlike reciprocating engines, gas turbine combustion is continuous. This has the advantage that the combustion process can be made very efficient, with very low levels of products of incomplete combustion like carbon monoxide (CO) or unburned hydrocarbons (UHC). The other major emissions component, oxides of nitrogen (NOx), is not related to combustion efficiency, but strictly to the temperature levels in the flame (and the amount of nitrogen in the fuel). The solution to NOx emissions, therefore, lies in lowering the flame temperature. Initially, this was accomplished by injecting massive amounts of steam or water into the flame zone, thus "cooling" the flame. This approach has significant drawbacks, not the least the requirement to provide large amounts (fuel-to-water ratios are approximately around 1) of extremely clean water.

Since the 1990s, combustion technology has focused on systems often referred to as dry low NOx combustion or lean-premix combustion (Fig. 3.25). The idea behind these systems is to make sure that the mixture in the flame zone

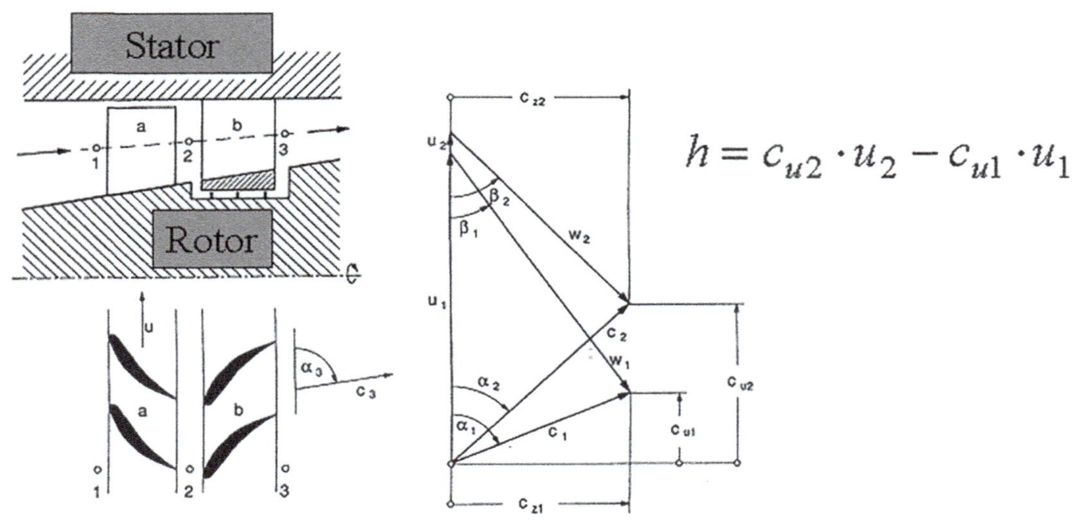

FIG. 3.24 Velocities in a typical compressor stage. Mechanical work transferred to the air is determined by the change in circumferential momentum of the air [14].

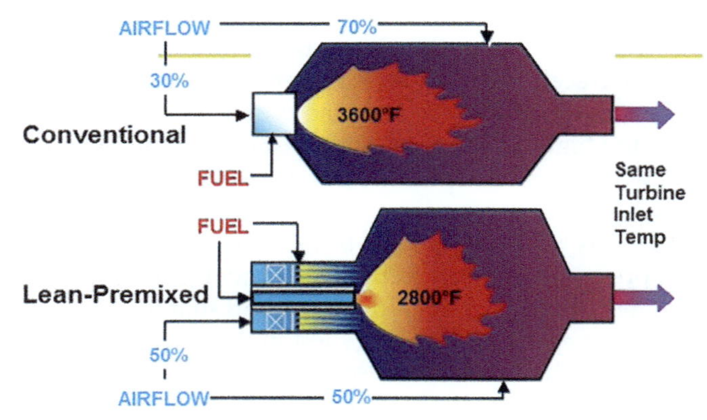

FIG. 3.25 Conventional and lean-premix combustion systems. *Courtesy of Solar Turbines.*

has a surplus of air, rather than allowing the flame to burn under stoichiometric conditions. This lean mixture, assuming the mixing has been done thoroughly, will burn at a lower flame temperature and thus produce less NOx. One of the key requirements is the thorough mixing of fuel and air before the mixture enters the flame zone. Incomplete mixing will create zones where the mixture is stoichiometric (or at least less lean than intended), thus locally creating more NOx. The flame temperature has to be carefully managed in a temperature window that minimizes both NOx and CO. Lean-premix combustion systems allow the NOx, as well as CO and UHC emissions, within prescribed limits for a wide range of loads, usually between full load and about 40% or 50% load. In order to accomplish this, the airflow into the combustion zone has to be manipulated over the load range.

The gas turbine power output is a function of the speed and firing temperature, as well as the position of certain secondary control elements such as adjustable compressor vanes, bleed valves, and in rare cases, adjustable power turbine vanes. The output is primarily controlled by the amount of fuel injected into the combustor. Most single-shaft gas turbines run at a constant speed when they drive generators. In this case, the control system modifies fuel flow (and secondary controls) to keep the speed constant, independent of the generator load. A higher load will, in general, lead to higher firing temperatures. Two-shaft machines are preferably used to drive mechanical equipment, because being able to vary the power turbine speed allows for a very elegant way to adjust the driven equipment to process conditions. Again, the power output is controlled by fuel flow (and secondary controls), and a higher load will lead to higher gas producer speeds and higher firing temperatures.

Fig. 3.26 shows the influence of ambient pressure and ambient temperature on gas turbine power and heat rate. The influence of ambient temperature on gas turbine performance is very distinct. Any industrial gas turbine in production will produce more power when the inlet temperature is lower and less power when the ambient temperature is higher.

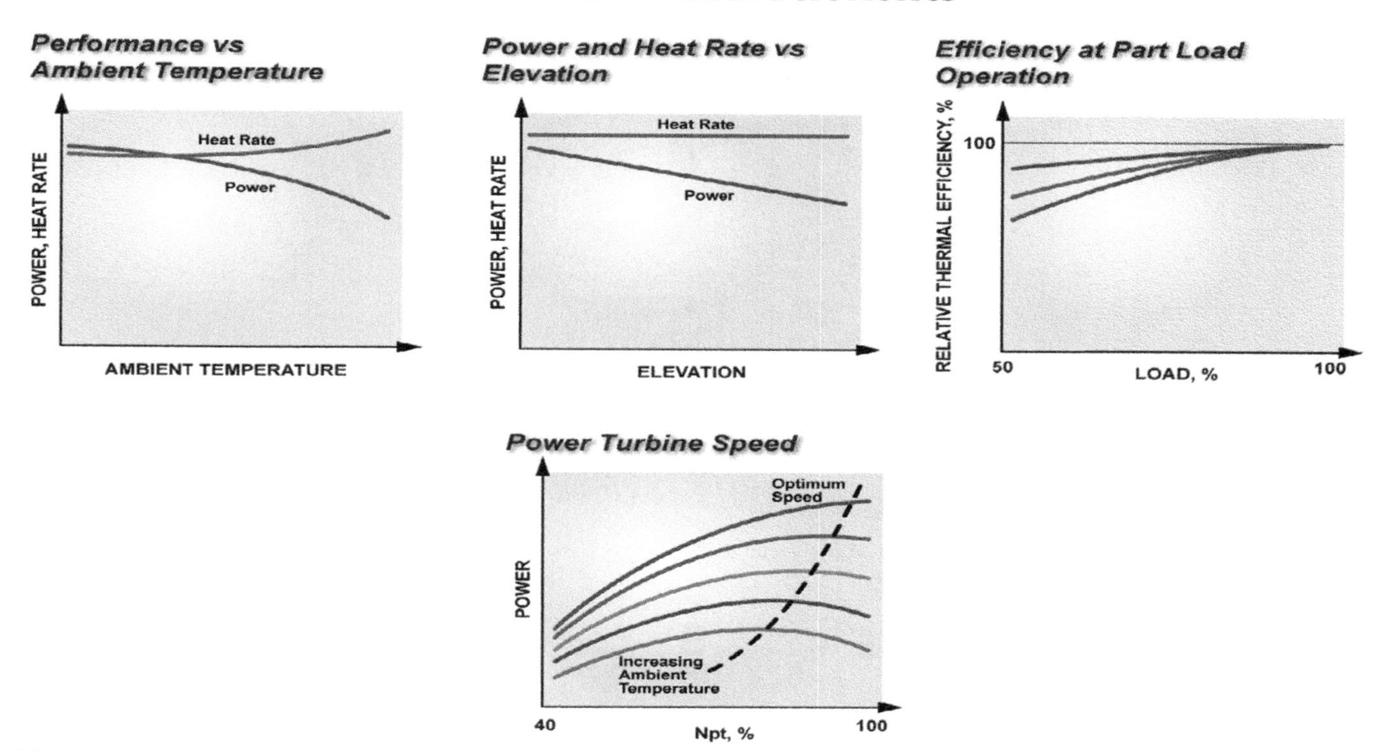

FIG. 3.26 Gas turbine performance characteristics. *Courtesy of Solar Turbines.*

The rate of change cannot be generalized and is different for different gas turbine models. Full-load gas turbine power output is typically limited by the constraints of maximum firing temperature and maximum gas producer speed (or, in twin-spool engines, by one of the gas producer speeds). Gas turbine efficiency is less impacted by the ambient temperature than the power.

The air humidity does impact power output, but only to a small degree (generally not more than 1%–3%, even on hot days). The impact of humidity tends to increase at higher ambient temperatures.

Lower ambient pressure (for example, due to higher site elevation) will lead to lower power output, but has practically no impact on efficiency. It must be noted that the pressure drop, due to the inlet and exhaust systems, impacts power and efficiency negatively with the inlet pressure drop having a more severe impact.

Gas turbines operated in part load will generally lose some efficiency. Again, the reduction in efficiency with part load is very model-specific. Most gas turbines show a very small drop in efficiency for at least the first 10% of drop in load. In two-shaft engines, the power turbine speed impacts available power and efficiency. For any load and ambient temperature, there is an optimum power turbine speed. Usually, lowering the load (or increasing the ambient temperature) will lower the optimum power turbine speed. Small deviations from the optimum (by say ±10%) have very little impact on power and efficiency.

Gas engines

Reciprocating gas engines have been used as drivers for compressors and pumps for more than a century (Fig. 3.27). The first practical gas-fueled internal combustion engine was built by Étienne Lenoir, a Belgian engineer, in 1860. It was further improved by German engineer Nikolaus Otto and later by Rudolf Diesel.

The machines (Fig. 3.28) generally have a reciprocating piston in a cylinder, with the piston connected to a crankshaft. Valves (*V*) allow gas (air or air and fuel) to enter the cylinder and exhaust gas (which is the result of the combustion) to leave the cylinder. The general concept involves the intake of the working gas by the piston (*P*) in a cylinder, the compression of this gas by the piston, the addition of fuel, the ignition of the fuel-air mixture (either by an external source or the heat of compression), and the expansion of the combustion products in the cylinder, thus acting with a

FIG. 3.27 Modern stationary gas engine. *Courtesy of Solar Turbines.*

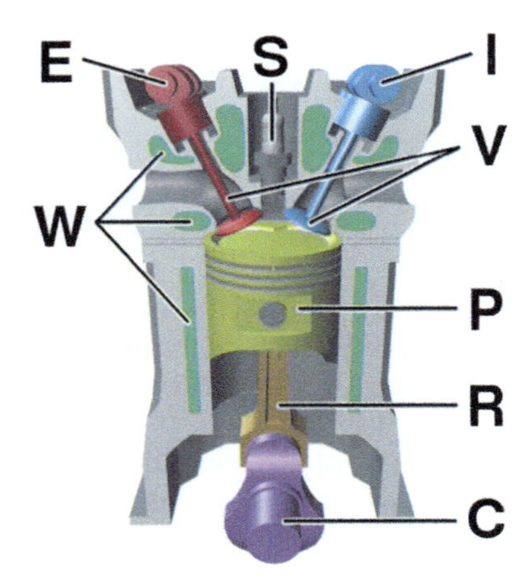

FIG. 3.28 Components [16].

force on the piston, and the exhaust of the combustion products. We thus have four actions: intake-compression-power-exhaust (Fig. 3.29). Forces are transmitted from the piston to the crankshaft (C) via a piston rod (*R*).

The engines can be categorized either as two-stroke or four-stroke engines. In a four-stroke engine, each of the four actions described above is accomplished by a separate stroke of the cylinder. The gas flow into and out of the cylinder is controlled by valves that are usually operated via one or multiple camshafts (E, I). A complete cycle requires two revolutions of the crankshaft. In a two-stroke engine, ports in the cylinder wall are often used to allow entry and exit of the working fluid. The position of the piston will determine the opening and closing of the ports. The 2-stroke engine accomplishes the four actions in just two strokes, thus only requiring one turn of the crankshaft. In particular, it has one power stroke per crankshaft revolution.

While a two-stroke engine is far simpler, a four-stroke design gives better control over the fuel-air mixture through control of the valve timing. Generally, the drivers for high-speed, separable compressors are four-stroke designs.

Any internal combustion reciprocating engine has to ignite the fuel-air mixture for every power stroke. Therefore, the entire combustion process has to be completed very fast, and great care has to be taken to completely consume all fuel. If the fuel is natural gas, the ignition has to be initiated by a spark plug (S), while Diesel cycles for liquid fuels can initiate the ignition simply by the temperature of the compressed air.

Further distinctions are the number of cylinders acting on a crankshaft. The number of cylinders can be as high as 16–24. Another distinction is in the selection of the engine cycle, for example, a Diesel or Otto Cycle.

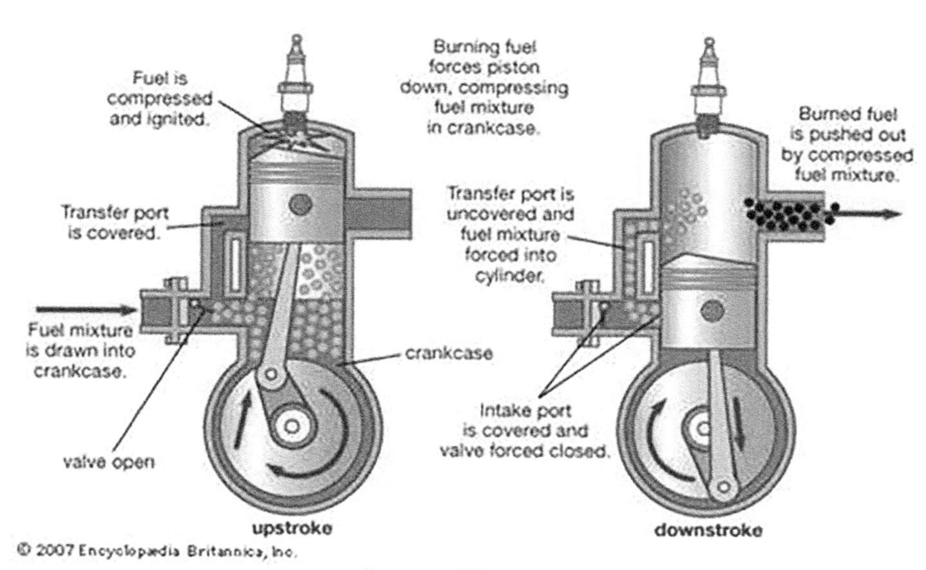

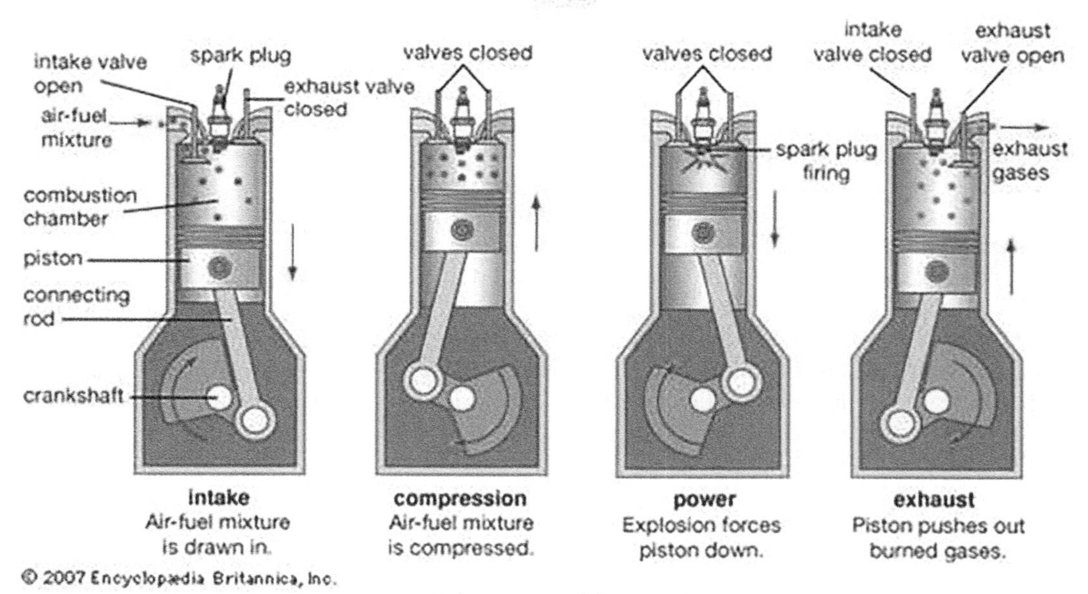

FIG. 3.29 Actions in a 2-stroke versus 4-stroke engine: intake-compression-power-exhaust [17].

In many instances, engines are augmented by turbochargers. A turbocharger extracts energy from the exhaust gas via a turbine, which in turn drives a turbocompressor. This allows an increase in the density of the air flowing into the cylinder. Therefore, the mass flow of air can be increased without changing the volumetric flow.

The cylinders have to be cooled either by air (for smaller engines) or with a liquid via a cooling jacket. The removal of heat is necessary to avoid the overheating of components, but it removes energy from the process cycle.

The force exerted by the expanding gases on the piston causes a torque on the crankshaft. A key parameter of the engine is the cylinder displacement. It is described by the length of the stroke and piston area. Available power and torque depend on the engine speed, and the speed power and speed torque curves depend on the individual engine

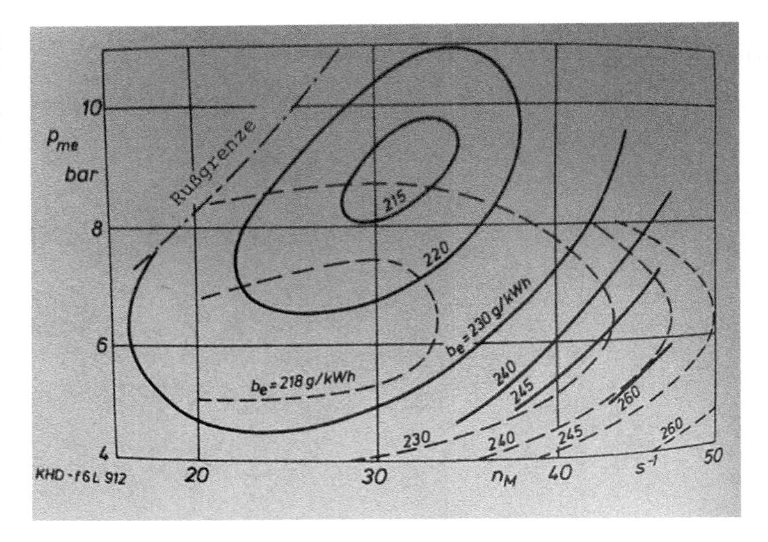

FIG. 3.30 Performance map of a reciprocating engine showing fuel consumption over speed and torque. Comparison is between a turbocharged (*solid lines*) engine and an engine without supercharging (*broken lines*). p_{me}, the average effective gas pressure on the piston, is an indicator for the produced engine torque. b_e is the specific fuel consumption [18].

design. Typical heavy-duty gas engines used in gas compression advertise constant torque across a range of speeds over which the engine can operate continuously. This, in turn, means that the engine power of such an engine increases linearly with speed.

Even with multiple cylinders, the torque at the engine output is subject to fluctuation, thus introducing torsional vibrations into the drivetrain. The design of the drivetrain must avoid natural frequencies that can be excited by these torsional vibrations. Otherwise, high cycle fatigue failures may be experienced. Engines usually employ flywheels as high inertia filters and torsional dampers to dissipate these oscillations.

The starting of the engine is accomplished by a starter motor. This can be an electric motor or a pneumatic motor. The motor has to be sized to bring the gas engine and the driven compressor to a speed where the engine can sustain itself, in other words, where it produces enough torque to overcome friction losses, the power demand of the intake, compression, and exhaust actions, and the power demand of the driven equipment. Unloading the driven equipment by a properly designed recycle loop, for example, can reduce the starting requirements. Auxiliary heaters may be used.

The power and efficiency of gas engines are load- and speed-dependent (Fig. 3.30). In many instances, the engine is supercharged with a turbocharger, which both increases torque and power and improves fuel consumption. The turbocharger extracts energy from the engine exhaust flow via a turbine to power a centrifugal compressor. The compressed air is cooled before it is fed to the engine. This increases the mass flow of air into the engine significantly, and in addition to increasing power and efficiency, it also provides excess air for lean combustion.

Steam turbines

Steam turbines are machines that convert the stored energy in steam into mechanical work. First invented in the 1880s, steam turbines quickly replaced steam engines as the primary driver for industrial and power generation applications. Although they find competition from electric motors, steam turbines still find many applications in industry as prime movers for industrial equipment and are the basis for the majority of global electrical generation.

There are two basic types of steam turbines: impulse and reaction (Fig. 3.31). In impulse turbines, the steam is accelerated through nozzles and directed onto the rotating buckets impacting a force onto the rotor. The steam's pressure is converted into velocity in the nozzle, and no pressure loss occurs in the rotor. In reaction turbines, a reactionary force is created by the steam expanding through the rotor blades. The pressure is lowered within the rotating nozzle. Both sets of blades are shaped more like airfoils, very similar to modern gas turbine expander blading.

Impulse turbines use fewer stages but tend to be less efficient than reaction turbines. In mechanical drive applications, such as pumps and compressor drives, both impulse and reaction turbines can be found, whereas in large power plant turbines, the higher efficiency of the reaction design is key.

Steam turbines are available in various power and speeds, from small industrial drivers below 50 kW to up to 1500 MW in nuclear power stations. Turbines driving high-speed compressors in a refinery can reach speeds over 10,000 rpm, while large power generation turbines run at 1500 rpm. It is very common for turbines to run at speeds

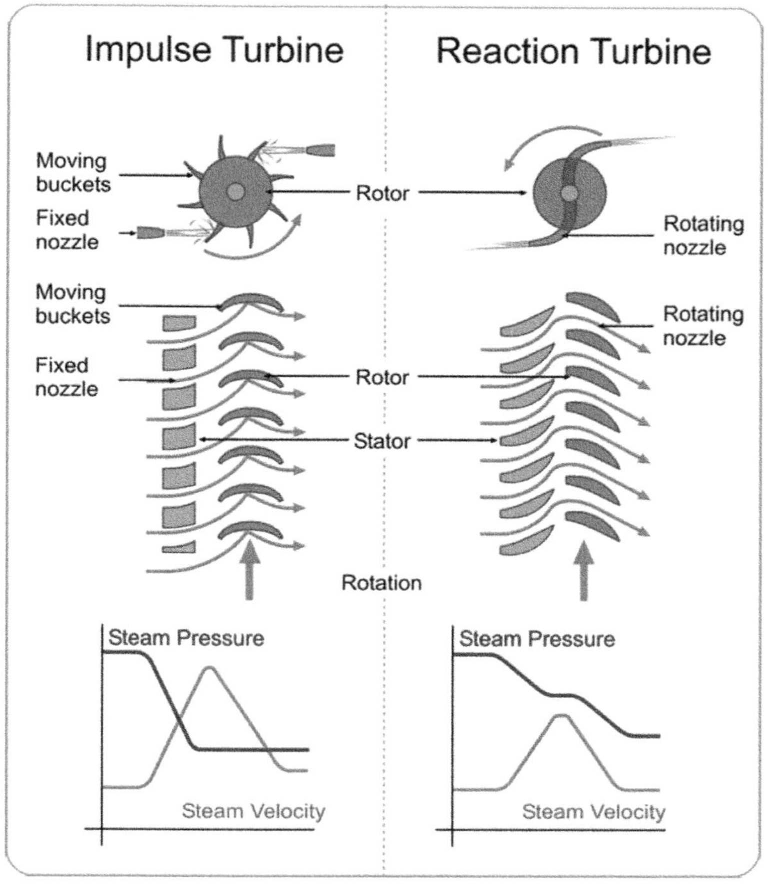

FIG. 3.31 Impulse and reaction turbines [17].

the same as electrical motors (1500, 1800, 3000, or 3600 rpm). Speed-increasing or decreasing gearboxes are also utilized to match the needs of the driven equipment.

Steam turbines can be configured in a number of different ways depending on the plant's available steam and process needs:

- *Backpressure* refers to turbines where the exhaust pressure of the turbine avoids condensation. Steam is exhausted as a vapor into a lower steam pressure header (Fig. 3.32).
- *Condensing* refers to a turbine that the exhaust is subatmospheric and is condensed back into water (Fig. 3.33).

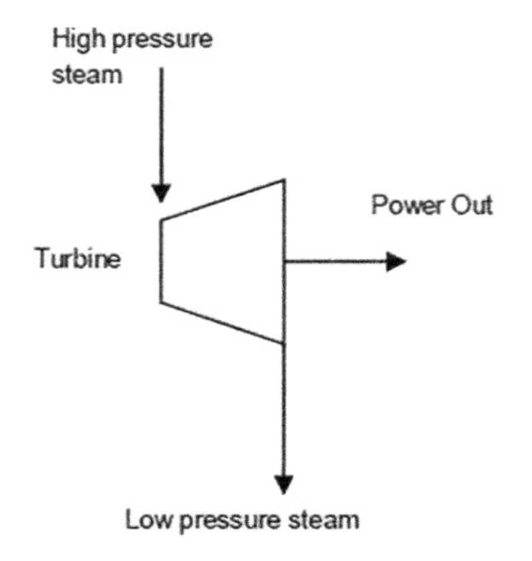

FIG. 3.32 Backpressure turbine.

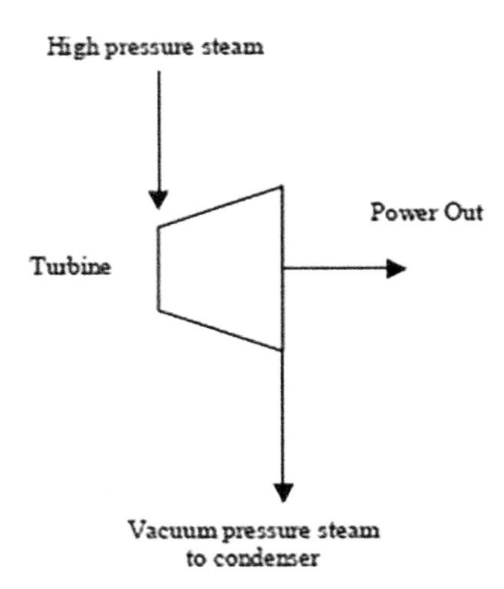

FIG. 3.33 Condensing turbine.

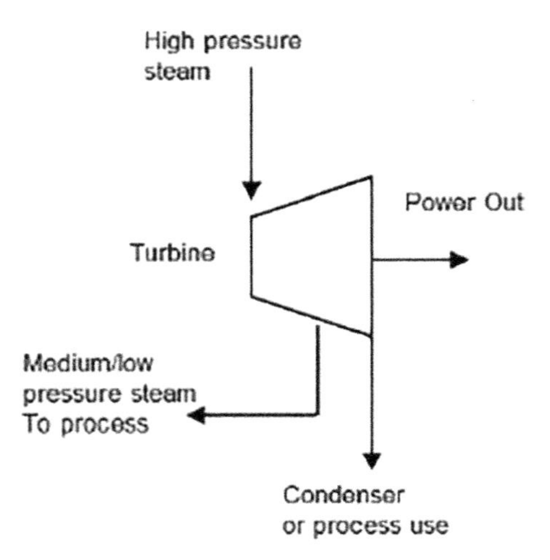

FIG. 3.34 Extraction turbine.

- *Extraction* means that some of the steam is extracted at an intermediate pressure level for other process needs, either heating or other turbines (Fig. 3.34).
- *Induction* turbines introduce steam into the turbine at an intermediate level. It is the same basic configuration as an extraction turbine, except steam is taken into the turbine and condensed. This is common when surplus steam at low-pressure levels is available to maximize overall plant thermal efficiency.
- A *double-flow* exhaust has steam enter the center of the exhaust and has mirrored stages to a common exhaust pressure.

Turbines can be one or many of these at the same time. It is very common to have extraction-condensing turbines for large mechanical drive applications.

The thermodynamic basis of steam turbines is the Rankine cycle where heat is added to high-pressure water to create high-pressure steam. The steam is then expanded through the turbine to produce work. The steam is then condensed (heat rejection), a pump is used to restore the water pressure, and the cycle continues (Fig. 3.35).

Theoretical steam rate (TSR) is the quantity of steam per unit of hourly power for an ideal Rankine cycle heat engine. TSR is equal to the difference in enthalpy of the steam at the inlet and exhaust of the turbine.

$$\text{TSR}\ (\text{kg/kWh}) = 3600/(h1 - h2), \tag{3.6}$$

where $h1$ and $h2$ are enthalpy with units kJ/kg.

Steam turbine efficiency can vary significantly. In very large power generation turbines, efficiency can be over 90%, while for small industrial drivers, it can be as low as 40%.

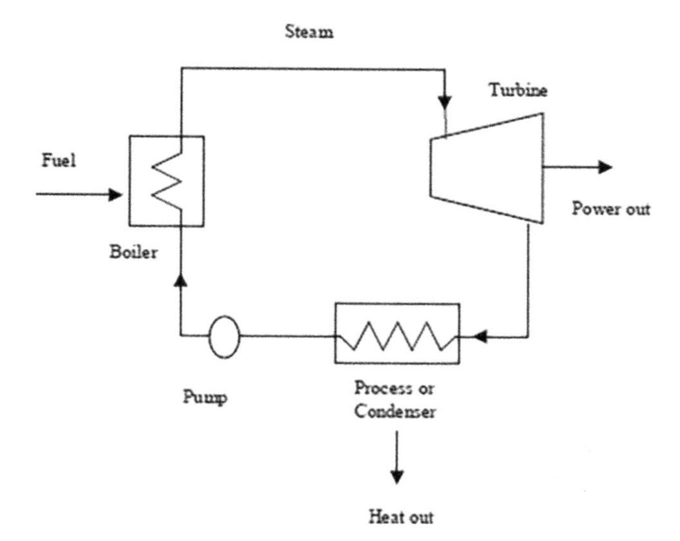

FIG. 3.35 Rankine Cycle.

Speed control

Steam turbines are inherently variable speed drivers, as such their speed must be controlled via governor valves. Governor valves modulate the steam flow into the turbine to maintain a set point (speed, power, header pressure). On multisection turbines, such as an extraction type, a second governor on the extraction valve(s) is utilized to maintain a second set point.

Trip

A sudden loss of load on a turbine can lead to the turbine speed increasing to the point of catastrophic failure. As such, over-speed protection and trip capability of the turbine are needed. Upon a trip signal, valves are closed to stop the introduction of further steam into the turbine. Specialized, fast-acting trip valves are utilized to cut steam flow.

Expanders

An expander (or turboexpander) is fundamentally an expansion turbine that is primarily used in the expansion process to produce work from the pressure of gas, steam, or liquid. Expanders are used across a wide range of operating temperatures and pressures but are most likely to be utilized in large gas or liquid flow applications. In many industrial processes, a considerable amount of valuable energy and gases can go to waste. By installing an expander to a pressure let down or waste gas process, the lost valuable energy and residual gases can be recovered. Recovered energy can be utilized to drive other machinery by mechanical means or by an electrical generator.

Expanders produce work by expanding the pressurized fluid and lowering it to the desired discharge pressure by the process. During the pressure drop, the temperature of the fluid is considerably reduced due to the Joule-Thomson effect. Based on the process requirement, some expanders are used for cooling purposes primarily, and energy recovery or mechanical work becomes a byproduct. If reducing the pressure is the main goal, heat recovery becomes the byproduct of the expander. Expanders produce work by expanding the pressurized fluid and lowering it to the desired discharge pressure by the process.

In refrigeration applications, the use of an expander for pressure reduction has the added benefit that for a given pressure drop, the end temperature is much lower than what is accomplished with a pressure reduction valve (Fig. 3.36).

Centrifugal turbines are usually designed as radial inflow turbines (Fig. 3.37). The gas enters the turbine at station 0, flows through radial guide vanes (1–2), enters a centrifugal impeller (3–4), and leaves the turbine through a diffuser (4–5). Thus, high-pressure fluid at the inlet of the expander first travels through variable-position stationary inlet guide vanes, where approximately 50% of pressure drop occurs. The other half of the pressure drop and enthalpy reduction occurs through the expander wheel.

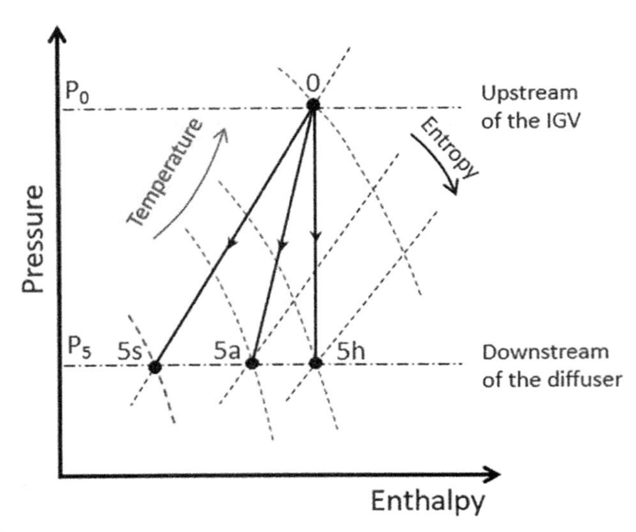

FIG. 3.36 Expansion processes [17].

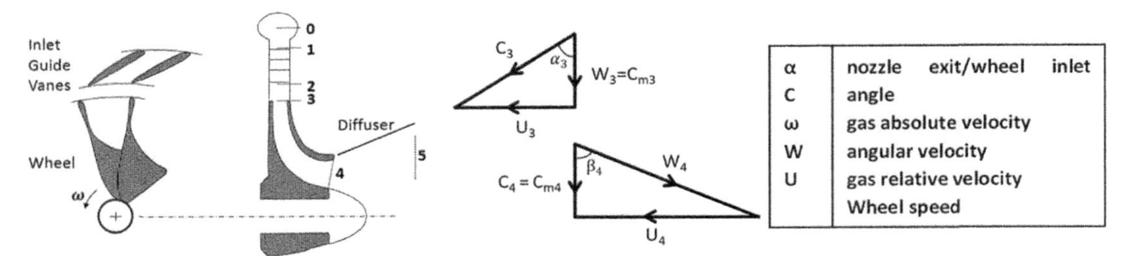

FIG. 3.37 Radial inflow turbine [17].

Expanders are frequently utilized in refrigeration processes. A refrigeration cycle requires the gas to be greatly expanded to reduce its temperature. This is usually achieved via a Joule Thomson (J-T) expansion valve at a simple refrigeration process. The expansion process across a J-T valve is isenthalpic (constant enthalpy) and highly inefficient with no energy recovery. Expanders reduce the pressure of the process fluid similar to J-T valves and additionally extract work and reduce the enthalpy of the gas with a near isentropic (almost constant entropy) process. Reduction in enthalpy, due to extraction of work, results in greater temperature reduction with respect to expansion via a J-T valve (Fig. 3.36).

One of the commonly used refrigeration cycles used in the liquefaction and cooling of process gases is the reversed Brayton cycle, which is shown in Fig. 3.38. Nitrogen and methane gas are commonly used as the refrigerant within the cooling loop and are maintained in gas form throughout the cycle [19,20].

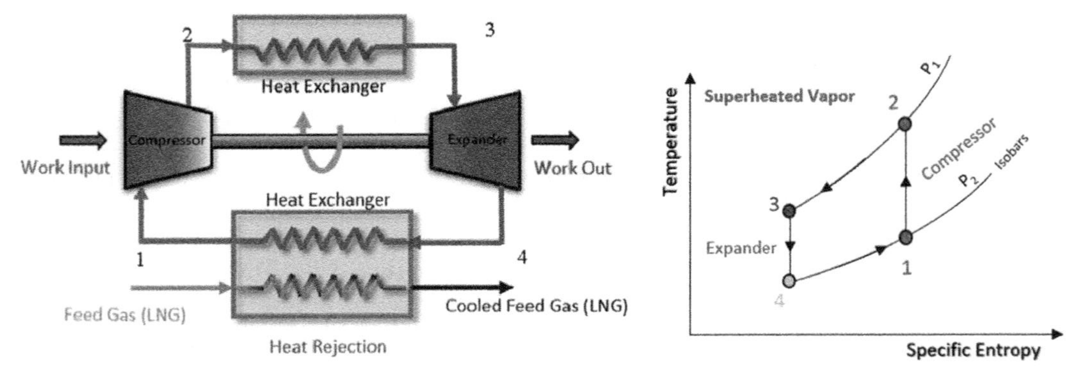

FIG. 3.38 Reversed Brayton Cycle.

This cycle implements a gas expander in lieu of an expansion valve that is mechanically coupled to a compressor. An expander and compressor are coupled together via a common shaft, which are both driven by pressurized fluid. An expander compressor is used as a pressure booster to meet a need in the process that otherwise would have required a separate motor-driven compressor. The expansion stage of these units typically has isentropic efficiencies between 85% and 92%. The compressor stage consists of a compressor wheel with a vaneless diffuser with a typical efficiency range of 75%–80%.

Expanders can also be used in two-phase refrigeration processes that utilize latent heat of vaporization for better and more efficient cooling. In this approach, refrigerant is allowed to evaporate, providing refrigeration duty prior to being condensed once again within the cycle. This cycle is known as the vapor compression cycle, and more details of this cycle can be found in the liquefaction section. This cycle typically consists of two separate heat exchangers and a J-T valve as shown in Fig. 3.39. As mentioned previously, the enthalpy across the J-T valve is the same at the inlet and outlet. The pressure drop at the J-T valve is highly irreversible, providing no ability to extract work. Utilizing an expander in place of a J-T valve enables the return of a refrigerant to its original pressure with a more isentropic path as shown in Figs. 3.39 and 3.40 [1]. There is an enthalpy differential between the inlet and outlet of the expander due to energy recovery. This also results in a cooler fluid temperature exiting the expander. With that, fluid exiting the expander will also have a lower percentage of vapor. These facts are important when considering refrigeration and liquefaction of process gases [19].

Expanders can be coupled to turbomachinery or a generator with the purpose of converting the mechanical energy of the fluid to electricity. Expanders that are coupled to an electrical generator are often called "generator-loaded expanders." These expanders are mostly installed to pressure let down applications to be used in lieu of Joule-Thomson expansion valves in natural gas processes. They provide refrigeration and power recovery in

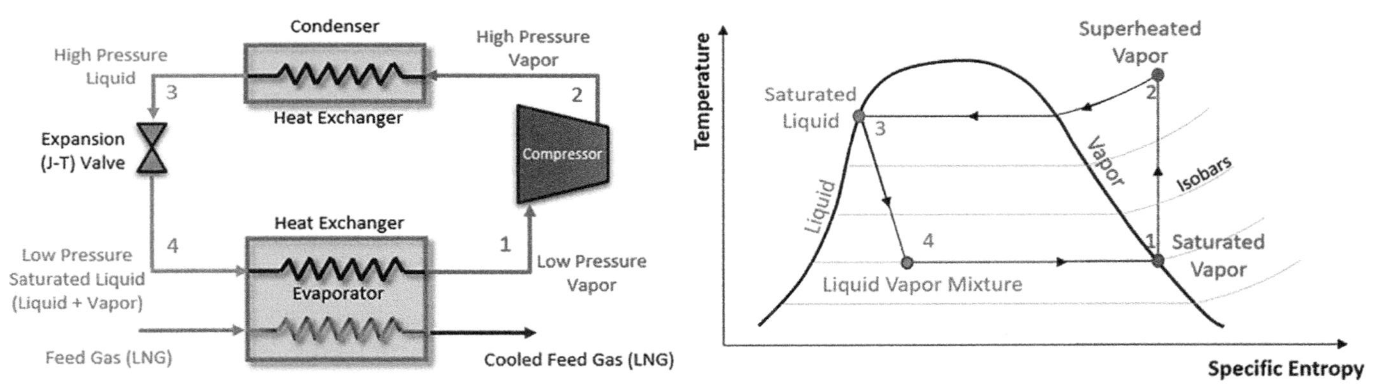

FIG. 3.39 Simple refrigeration cycle with JT valve and no enthalpy change which results in a highly irreversible process (nonisentropic flow) [19].

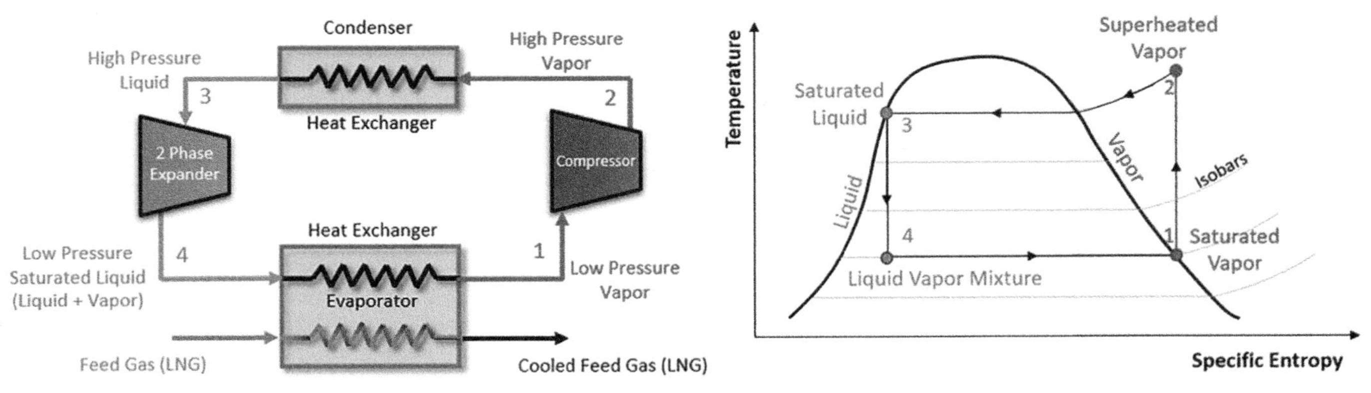

FIG. 3.40 Simple refrigeration cycle with flashing expander, enthalpy change with energy recovery, and near isentropic flow. Enhanced cooling with the enthalpy change results in a higher percentage of liquid at the expander outlet [19].

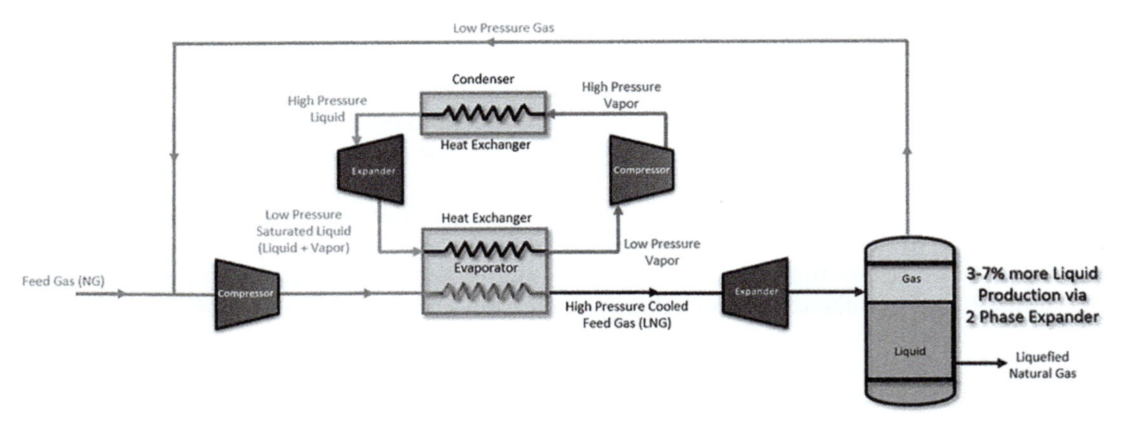

FIG. 3.41 Schematic view of liquefaction and let down loop for LNG with expanders.

liquefaction processes of many cryogenic fluids such as natural gas, carbon dioxide, ammonia, and hydrogen, and in refineries, this includes gas process plants and air separation applications. A typical installation scheme is shown in Fig. 3.41.

A liquid-phase-only LNG expander with a submerged generator is shown in Fig. 3.42. These liquid-only expanders become an important part of liquefaction processes.

Some expanders are used primarily for dissipating excess energy for a high-pressure flow. Instead of utilizing mechanical work for productive purposes, these expanders are coupled with a mechanical brake or load that absorbs the energy and dissipates heat. These expanders are classified as "energy-dissipative expanders" and are used for applications where it is not economical to recover the potential energy.

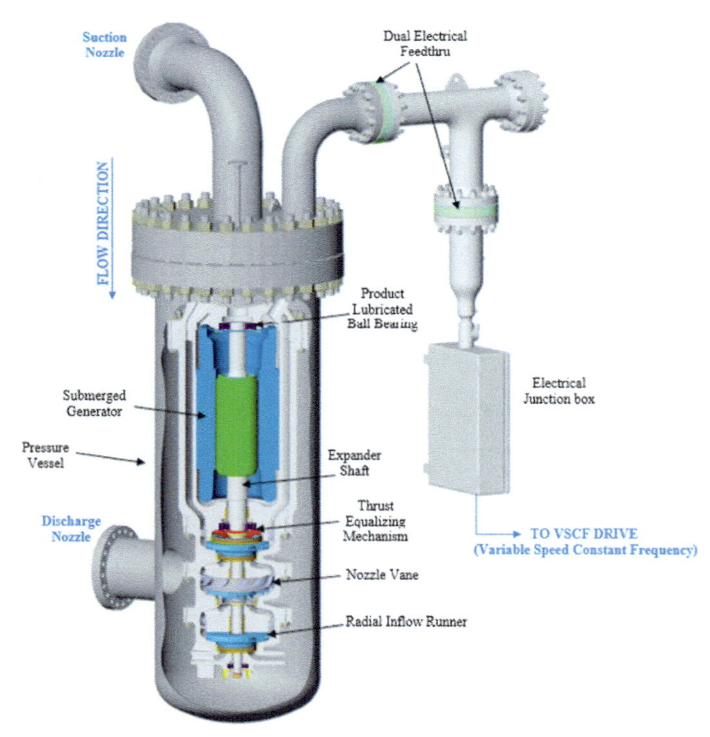

FIG. 3.42 Generator coupled turboexpander with submerged generator design.

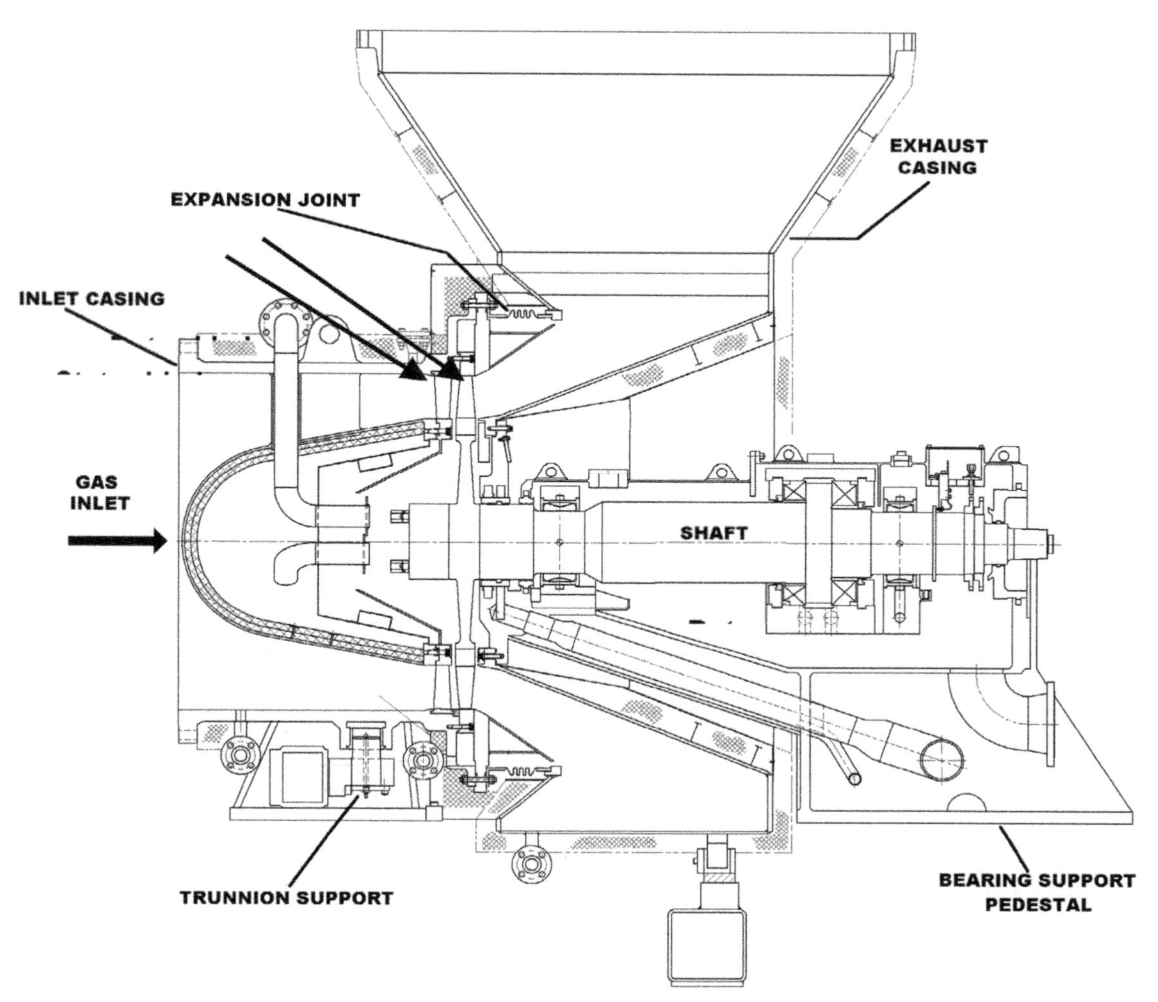

FIG. 3.43 Axial turbine hot gas expander [17].

Expanders, especially hot gas expanders, can also use an axial turbine. An example of an axial flow expander is shown in Fig. 3.43. In low-temperature applications, expanders often use magnetic bearings instead of oil bearings.

Pumping

A pump is a mechanical device used to transport fluids from one location to another by increasing the mechanical energy of the fluid. Pumps are a crucial part of energy transportation and are commonly utilized not only in transportation but also take a part in the process and enhancement of energy resources. Pumps are primarily designed for handling liquids, but there are certain types of pumps that can handle gaseous fluid although they are less common than compressors for this purpose.

Pumps transport fluid by creating a pressure differential that drives the fluid from the pump suction side to discharge. A low-pressure region at the suction is obtained by either increasing the volume or reducing the pressure within the pump internals. This low-pressure region allows fluid to be sucked into the pump inlet. Fluid flows into the pump due to the pressure differential between the pump inlet and the suction pipe or surrounding environment. Once the fluid enters the pump, the volume of the pump chamber is reduced or an external force is applied to the fluid. This will result in an increase in fluid pressure due to compression. Based on how the pressure is increased within the pump's internal passages, pumps are often classified into two main categories, namely centrifugal and positive displacement pumps. These pump types will be covered in the next sections.

Pumping applications generally deal with liquids that are assumed to be incompressible. According to incompressible fluid behavior, the density of the pumped fluid from suction to discharge is assumed to be constant. This assumption simplifies many of the design approaches and pump components. However, depending on the heat input (temperature rise across the pump) and total differential pressure increase or the thermodynamic and transport properties of the process fluid, compressibility may need to be taken into account. For applications with fluid in the supercritical phase, including supercritical hydrocarbons, supercritical carbon dioxide, or even for some liquid hydrogen applications, the assumption of constant density may lead to poor pump performance or even a catastrophic failure. Therefore, fluid properties across pump internals may need to be reviewed and confirmed for these applications [21].

Why pumping?

Pumping liquids requires less energy, because liquid volume does not significantly change during the pumping process. In contrast, compressing gases, which are handled by compressors, requires more energy since gases are highly compressible, and there is a considerable volume reduction during compression. In addition, pumping liquids requires less energy since energy transfer is more direct with respect to the compressors due to the incompressible nature of liquids. Compressing gases often involves more energy losses due to factors such as heat transfer.

Once the gas is condensed and liquefied, its density increases considerably for most fluids. Methane (natural gas), hydrogen, and ammonia (NH_3) density under varying pressure for a given saturation temperature are reviewed to better observe the density difference between liquid and gas phases. This is particularly important to identify and determine if transporting fluid in liquid form is more efficient with regards to density differential between each phase. Fluid properties may be calculated by using published thermal and transport properties by the National Institute of Standards and Technology (NIST) and RefProp software [22]. According to Fig. 3.44, natural gas density in the liquid phase is around 630 times denser with respect to the gas phase under atmospheric pressure. There is a negligible amount of change in density in the liquid phase (~422 kg/m^3) regardless of the pressure increase. This further confirms the assumption of incompressibility under the liquid phase. The same behavior is observed for hydrogen and ammonia based on Figs. 3.45 and 3.46. The liquid-to-gas density difference is 845 and 950 times for hydrogen and ammonia fluids, respectively. While liquid ammonia density is constant under increasing pressure, liquid hydrogen has a tendency to show an increase in density. For applications where the differential pressure is relatively high, the compressibility of the hydrogen may need to be taken into account in the design of hydrogen pumping applications.

Due to the considerable difference in density of phases, it can be concluded that storing fluids in the liquid phase will allow for a significant reduction in footprint and a storage tank's overall size. However, storage and transportation of fluids in the liquid phase often require refrigeration and liquefaction of the process fluid. Hydrocarbons, ammonia, and

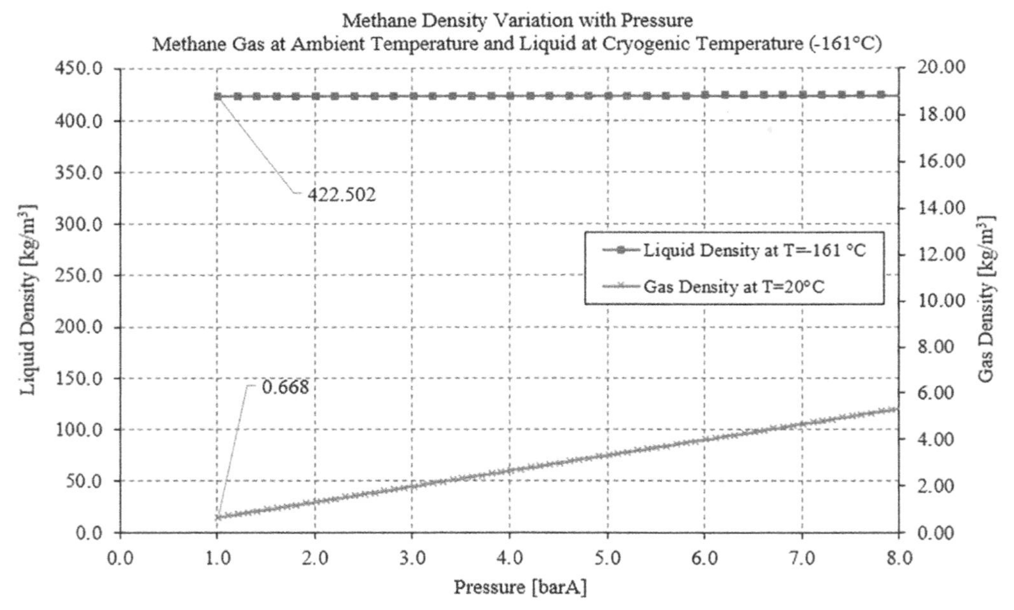

FIG. 3.44 Methane (natural gas) density variation with pressure at cryogenic temperature of −161°C.

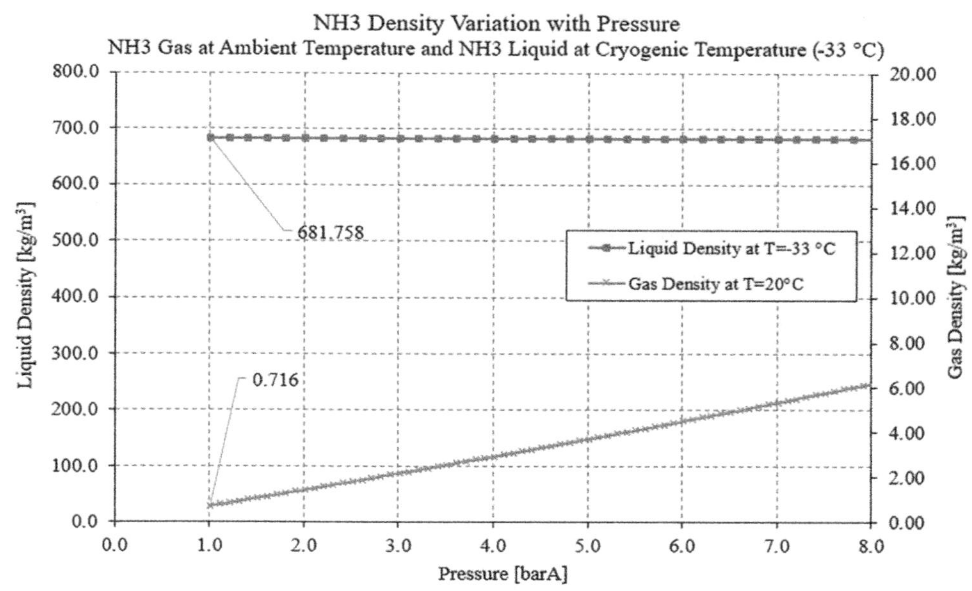

FIG. 3.45 Ammonia (NH_3) density variation with pressure at cryogenic temperature of −33°C.

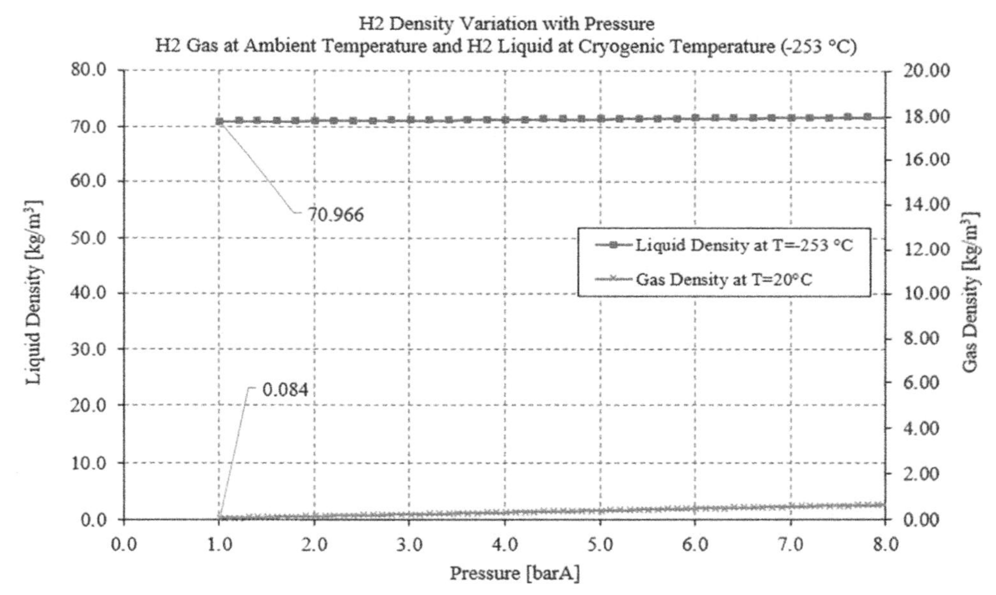

FIG. 3.46 Hydrogen density variation with pressure at cryogenic temperature of −253°C.

hydrogen gases must be cryogenically cooled to be liquefied. Liquefaction of cryogenic fluids involves cyclic compression and cooling via heat exchangers, which can be complex and very costly. Therefore, a feasibility study must be conducted in order to justify the liquefaction and transportation of fluids in liquid form. For applications where pipeline transportation is not possible due to extreme distances or geographic obstacles, transportation by a carrier may be preferred. For transportation of fluids overseas, special ships with cryogenically rated storage tanks are used since transport via pipeline is not possible. Refer to Chapter 4 for more details on these large carriers and liquefaction processes.

Centrifugal pump

In this section, the working principles and design features of centrifugal pumps, areas of application, and pump selection methodology are described.

FIG. 3.47 Single-stage volute centrifugal pump with induction motor drive. *Courtesy of Ebara Corporation.*

Centrifugal pumps are of eminent technical and economic importance in many areas of life and industry. It is reported that the world market volume for centrifugal pumps is in the order of 20 billion US dollars per year [1]. Centrifugal pumps have a range of applications from small cooling or circulation pumps with less than a couple of watts to 60 MW storage pumps and hydraulic recovery pump turbines with more than 250 MW when operating in the pumping mode. Centrifugal pumps are designed for flow rates from 0.001 to 60 m^3/s, heads of 1 to 5000 m, and speeds from a few hundred to 30,000 revolution per minute [21].

Centrifugal pumps utilize rotating components (impellers) that increase the kinetic energy of the fluid by forcing the process fluid radially in the outward direction by centrifugal forces. As the fluid enters a centrifugal pump impeller, it is drawn into eye of the impeller due to the low-pressure region created by the rotating blades. The impeller then imparts a centrifugal force, pushing it radially (or semi-axially) outwards toward the periphery of the impeller. While fluid is traveling radially in the outward direction, its velocity and kinetic energy increases due to the rotational motion. The fluid has the highest flow velocity, hence there is dynamic pressure at the discharge of the impeller. A diffuser or a volute is installed at this location to convert the dynamic pressure to a static head. This is accomplished by deaccelerating the fluid and increasing the static pressure by gradually increasing the effective area of the flow path. An exploded view of a typical centrifugal pump with hydraulic components, such as an impeller and volute, casings, and motor driver is shown in Fig. 3.47.

Basic principles and components

A centrifugal pump, as shown in Fig. 3.47, is fundamentally composed of a casing, a housing for shaft support elements such as bearings or bushings, the pump shaft, an impeller, and a driver. The process fluid first travels through the suction nozzle and into the impeller eye. The overhung impeller is mounted on the shaft via a key and/or a locknut to transmit the torque and accelerate the fluid in a circumferential direction. The pump shaft is driven via a coupling by an electric motor. Across the impeller, dynamic pressure increases with the acceleration in fluid via rotation, and static pressure increases with the curved shape of the meridional section. The fluid exiting the impeller is decelerated at the discharge section with an area increase in the flow path in order to attain the maximum static pressure.

At an external motor application, a dynamic shaft seal is used to prevent any leakage from the pump section to the motor section and/or to the environment. For applications that deal with hazardous and flammable liquids, shaft seals and the motor section are often purged with inert gas to monitor and prevent any leakage to electrical sources and the environment.

The most important component of a centrifugal pump is the impeller, which is also used for subcategorizing the type of centrifugal pump. Regardless of the type or shape of the impeller, impellers are always described by the hub (rear or back shroud), shroud, and blades. Fig. 3.48 shows the three-dimensional, meridional section and blade progression of a typical radial impeller.

Based on the impeller's meridional geometry and flow direction at impeller discharge, centrifugal pumps are classified as radial, semi-axial (mixed flow), and axial flow pumps. Typical geometries of impeller meridional views for these pumps are given in Fig. 3.49. According to the discharge flow direction, semi-axial pumps sometimes have open

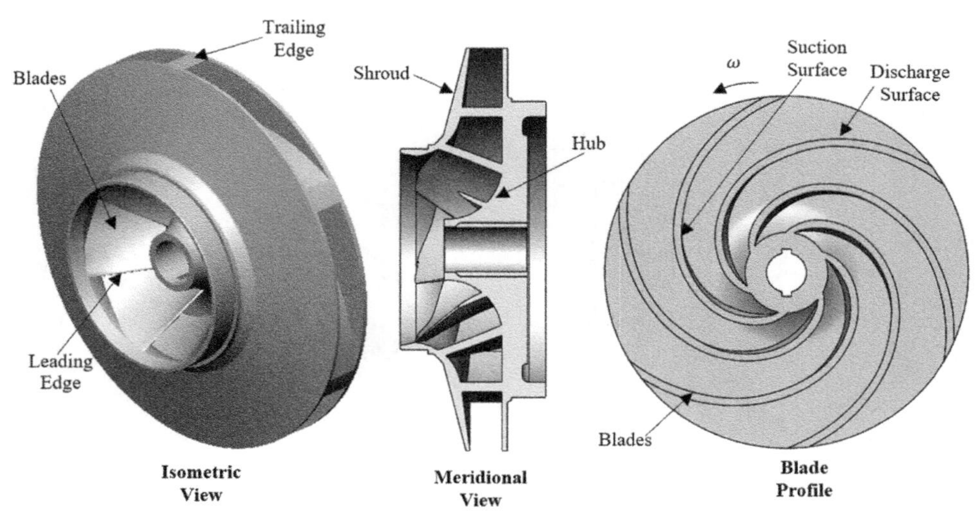

FIG. 3.48 Radial impeller, isometric, meridional, and radial progression of blades (ω: angular velocity).

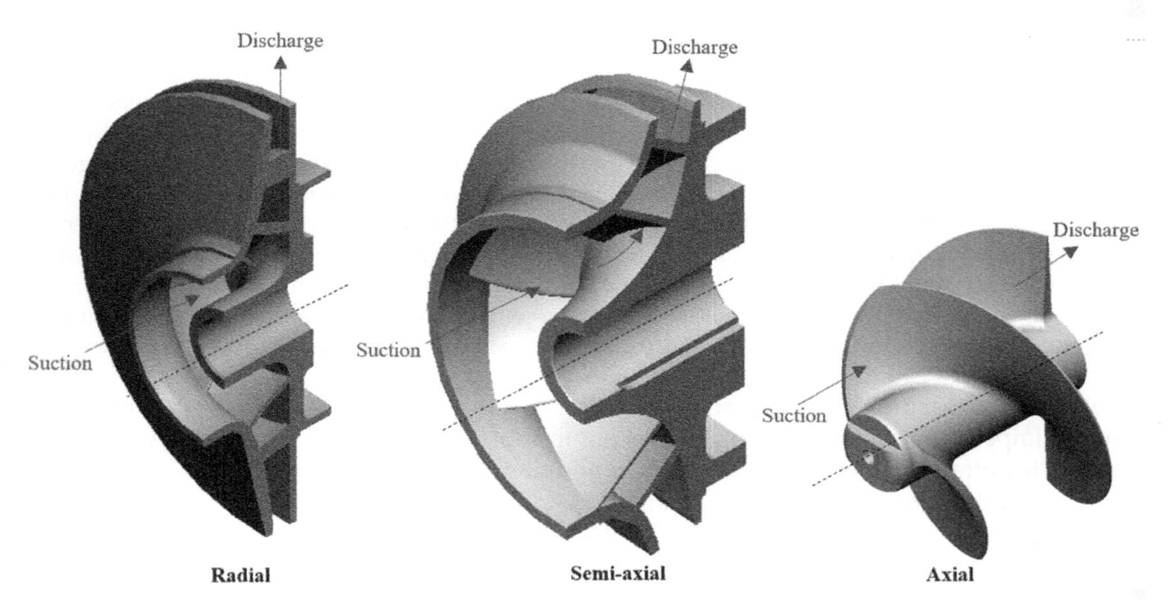

FIG. 3.49 Pump impellers based on flow direction. Radial, semi-axial (mixed flow), and axial flow impellers.

impellers in which the shroud at the front section of the impeller is removed. Axial pumps only have a hub section, and flow is only in the axial direction with no radial component. The geometry and type of each impeller are driven based on the flow, pressure, and efficiency requirements of each application. While axial flow impellers are capable of high flow rates, they are often limited in head production. Radial-type impellers are often used for high-pressure applications, but the flow capacity is limited for these units. It should be noted that if the pressure generated by one impeller is insufficient to meet the process requirement, multiple impellers can be installed in series to increase the pressure for a given flow rate. Unlike compressor applications, each impeller has the same geometry and shape, generating the same amount of differential pressure.

Selection of the volute or diffuser is based on the impeller type and number of stages required to meet the head requirement. Radial-type impellers often utilize a radial-type diffuser vane to keep the stage span short for rotordynamic reasons. Since radial-type impellers are commonly used for applications requiring high differential head, they are installed in a series of stages. This results in a relatively long pump shaft at which rotordynamics can be problematic. Using a radial diffuser vane reduces the stage axial length (span). Radial diffuser vanes have guide vanes to route

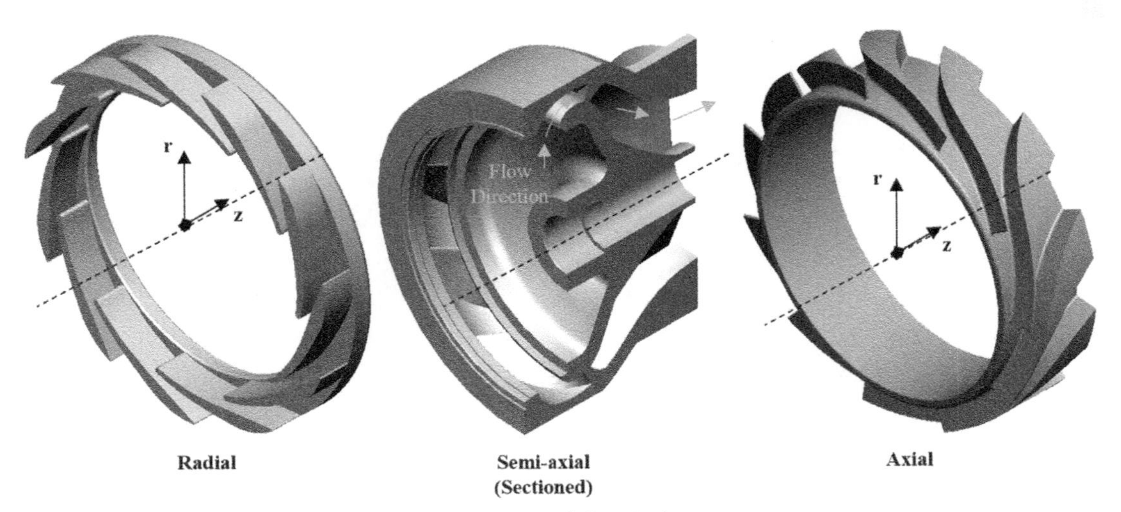

FIG. 3.50 Centrifugal pump diffuser vanes (*r*: radial direction, *z*: axial direction).

the discharge flow from one stage to the next with minimal losses. For single-stage applications, axial diffuser vanes or volutes are preferred. Most of the single-stage machines use a volute due to its simplicity relative to the diffuser vane.

Semi-axial flow impellers are often followed by semi-axial diffuser vanes to attain good efficiency by minimizing transitional losses between stages. These types of diffusers have a similar axial length with respect to axial diffuser vanes. Semi-axial diffuser vanes not only diffuse the fluid but also guide it via vanes to the next stage or impeller.

Axial impellers are often coupled with axial or semi-axial diffuser vanes. Commonly used diffuser vanes for centrifugal pumps are shown in Fig. 3.50. It is important to note that the flow transition from impeller discharge to diffuser vane is very crucial in achieving the best efficiency in centrifugal pumps. Therefore, the blade angles of each component and incidence must be evaluated for design flow rate. With that, streamlined analysis with velocity triangles should also include the flow between the rotating impeller and stationary diffuser vane. To improve the accuracy of streamline analysis, velocities and pressure increase at various radial positions shall be calculated and averaged.

It is worth mentioning that besides having pump impellers in series as in multistage pumps, centrifugal pumps can have double-entry impellers. This configuration allows pump suppliers to increase flow capacity for a given impeller pump diameter as the effective flow area is doubled. In addition, this configuration eliminates the axial thrust load associated with differential pressure at the impeller shroud and hub sections.

Impellers, diffuser vanes, and volutes have complicated blade and flow passages that progress in multiple axes, which are sometimes not possible to manufacture by a machining process. Impellers and semi-axial diffusers with very small flow passages are usually fabricated as castings. The quantity of stages (set of impellers and diffuser vanes) should be considered in the decision of how to manufacture hydraulic components. More importantly, surface finish requirements can determine the manufacturing method as it can have an impact on hydraulic performance. Additive manufacturing is another option to fabricate complicated shapes of impellers, volutes, and diffuser vanes. Postprocessing may be required to attain a desirable surface finish at flow passages.

Performance parameters

In this section, important performance parameters for centrifugal pumps are described. Depending on the specifics of the pumping application, there may be additional performance and operational requirements that are not listed below. The below items are typical of centrifugal pumping applications:

- Differential Head (ΔH), the head difference between discharge and suction locations
- Flow Rate (Q), which is the output volumetric flow at the discharge pipe
- Input Power (P), which is the shaft input power excluding the losses of electrical or mechanical drive
- Net Positive Suction Head required (NPSHr), net positive suction head required at pump suction datum to prevent cavitation (vapor formation)

- Efficiency (η), ratio of output hydraulic power to shaft input power. Efficiency is a calculated value and derived from differential head, flow rate, and shaft input power.

For incompressible flow, pump total differential head is directly proportional to enthalpy rise according to conservation of energy, mass, and momentum. The following equation can be written for the total differential head:

$$\Delta H\, g = \Delta h_{\text{total}} = \frac{p_{d,\text{total}} - p_{s,\text{total}}}{\rho} \tag{3.7}$$

where p_{total} is the total pressure, which is the sum of dynamic and static pressures. Subscripts d and s are used for the discharge and suction locations of the pump, respectively. ρ is the density of the liquid, g is the acceleration of gravity and Δh_{total} is change in total enthalpy from suction to discharge of the pump. The head is independent of fluid density regardless of the fluid or medium. Therefore, pumps generate the same amount of head when transporting oil, liquefied natural gas, or mercury. However, suction and discharge pressures and power consumption are proportional to the density.

Pump differential head can be expressed in terms of total and static pressures as follows:

$$\Delta H = \frac{p_d - p_s}{\rho g} + z_d - z_s + \frac{c_d^2 - c_s^2}{2g} \tag{3.8}$$

The first term in Eq. (3.8) is the static head difference measured at pump discharge and suction. The second term is the elevation difference between the discharge and suction nozzle of the pump where z_d and z_s are the height of the suction and discharge nozzle with respect to pump datum. Pump datum is often a reference to the impeller eye for a vertical unit or shaft centerline for a horizontally suspended pump. The third term is the dynamic head difference, where c_d and c_s are the absolute flow velocity at the discharge and suction nozzle of the pump, respectively. Eq. (3.2) is applicable to pumps only which do not take the plant suction tank or container and final discharge location (tank) into account. Fig. 3.51 shows a schematic view of a horizontal pump located between two tanks. The differential head between each tank should be equal to differential head of the pump

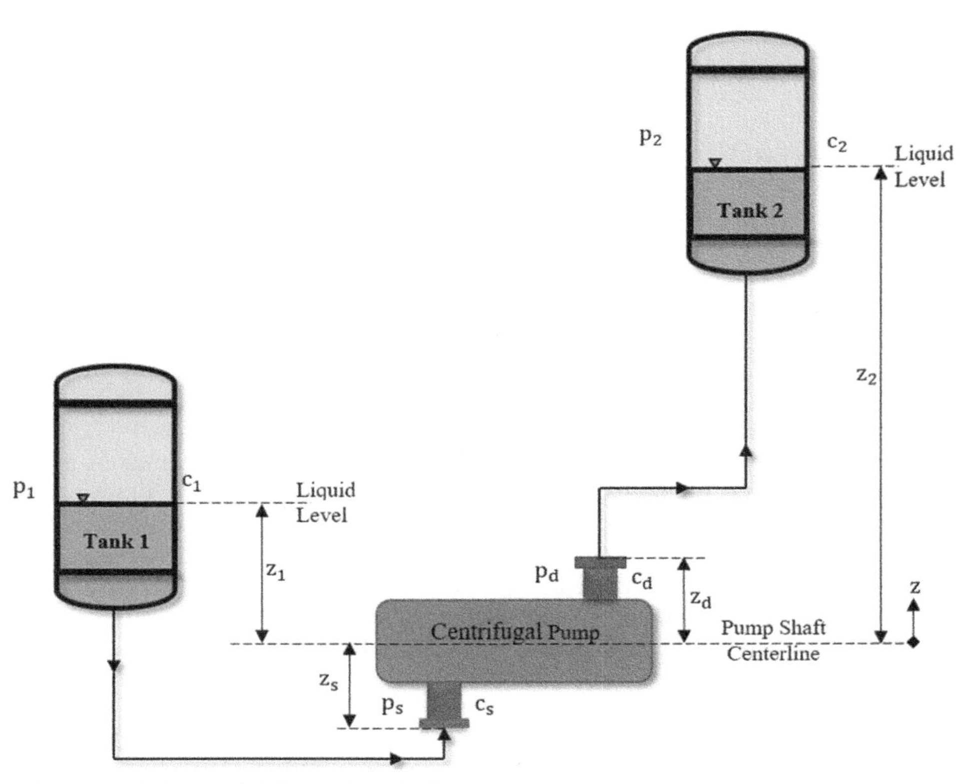

FIG. 3.51 Centrifugal pump and plant total differential head calculation arrangement.

minus any head losses across piping. With that, differential head between each tank at a plant can be written as follows:

$$\Delta H_{\mathrm{plant}} = \Delta H - H_{\mathrm{loss}} = \frac{p_2 - p_1}{\rho g} + z_2 - z_1 + \frac{c_2{}^2 - c_1{}^2}{2g} \tag{3.9}$$

To achieve the head required by the process, losses associated with piping must be calculated and minimized. Pump differential head is usually increased to compensate for these losses. A precise calculation of piping losses is crucial to ensure that the pump is not overpressurizing the discharge piping and/or tank, and more importantly, that it is not oversized and consumes more power.

Power that is generated by pump hydraulics is obtained by multiplying the mass flow rate (ρ Q) with the total enthalpy change ($\Delta h_{\mathrm{total}}$). According to Eq. (3.1), change in enthalpy is proportional to pump differential pressure or head. Following the relationship for hydraulic power, (P_{hyd}) can be written by using Eq. (3.1):

$$P_{\mathrm{hyd}} = \rho\, Q \Delta h_{\mathrm{total}} = Q\, \rho\, g \Delta H \tag{3.10}$$

The efficiency of the pump can be calculated based on the input power and hydraulic power. Input power to the pump shaft (P) is greater than the hydraulic power (P_{hyd}) due to all the losses occurring within the pump hydraulic section. The ratio of hydraulic power to shaft input power is the efficiency of the pump hydraulics, η:

$$\eta = \frac{P_{\mathrm{hyd}}}{P} = \frac{Q\, \rho\, g \Delta H}{P} \tag{3.11}$$

A pump's total efficiency, including the motor or drive losses, can be calculated as follows:

$$\eta_{\mathrm{Total}} = \eta_{\mathrm{motor}}\, \eta \tag{3.12}$$

Net Positive Suction Head (NPSH) is a critical performance parameter for safe pump operation in the event of low suction pressure conditions, which can result in vaporization. A liquid at a given temperature can form vapor bubbles as its pressure is dropped below its saturation point. The vaporization due to low-pressure regions within the flow field of a liquid is called cavitation. The formation of the vapor bubbles starts when the static pressure in the liquid reaches the vapor pressure (p_v) of the liquid for a given temperature [23]. The cavitation inception and the tendency to cavitation is defined in a nondimensionalized form as follows [24]:

$$\sigma = \frac{(p_s - p_v)}{\frac{1}{2}\rho U^2} \tag{3.13}$$

where σ is the cavitation number and U is the reference velocity, which is the impeller tip speed and can be defined as $\omega D_t/2$ where ω is the rotational speed and D_t is the impeller tip diameter. It should be noted that for any flow rate with or without the presence of vapor, there is a corresponding cavitation number. In the context of centrifugal pumps, there are three distinct cavitation numbers that can be related to the pump's cavitation characteristics [24]. The first is the cavitation inception number, σ_i, which corresponds to the initial formation of vapor bubbles within the pump. The second is the critical cavitation number, σ_c, at which the pump's differential pressure drops by 3% with a decrease in suction pressure. Further reduction in suction pressure results in major differential pressure loss at the pump, and this corresponds to the breakdown of the cavitation number, σ_b. In industrial pump applications, the Net Positive Suction Head requirement (NPSHr) of the pump is defined as the suction head that corresponds to the critical cavitation number. It is often referred to as NPSH3, which implies that pump head loss is 3% [25]. For most centrifugal pumping applications, the focus is on the NPSH3 (critical cavitation number) performance, as this is the cavitation performance specification accepted by many industrial norms and guidelines (ANSI, ISO, and API) [25–27]. A typical pump cavitation performance curve is given in Fig. 3.52, which outlines important cavitation numbers.

In order to achieve acceptable performance and reduce the required Net Positive Suction Head (NPSHr), pumps often utilize an inducer to delay the cavitation or enhance the suction performance [23]. Inducer technology was first initiated by NASA due to a lack of suction pressure in rocket turbopump applications. [28,29]. Rocket turbopumps have very little inlet pressure to draw fluid into the pump internals, which are thus prone to cavitate. The main purpose of the inducer is to delay the cavitation inception by producing enough pressure at the impeller eye (suction) section so that the pump can operate at very low liquid levels, hence with minimum suction pressure. Inducers are often installed in pump applications where the available Net Positive Suction Head (NPSHa) is less than ideal and too close to NPSHr by the pump. Inducers are fundamentally axial flow impellers with helical blade geometry that generate relatively low heads with respect to impellers. They are installed at the upstream of the impeller to gradually increase suction pressure. A two-stage, submerged motor LNG in-tank pump with a helical inducer is shown in Fig. 3.53.

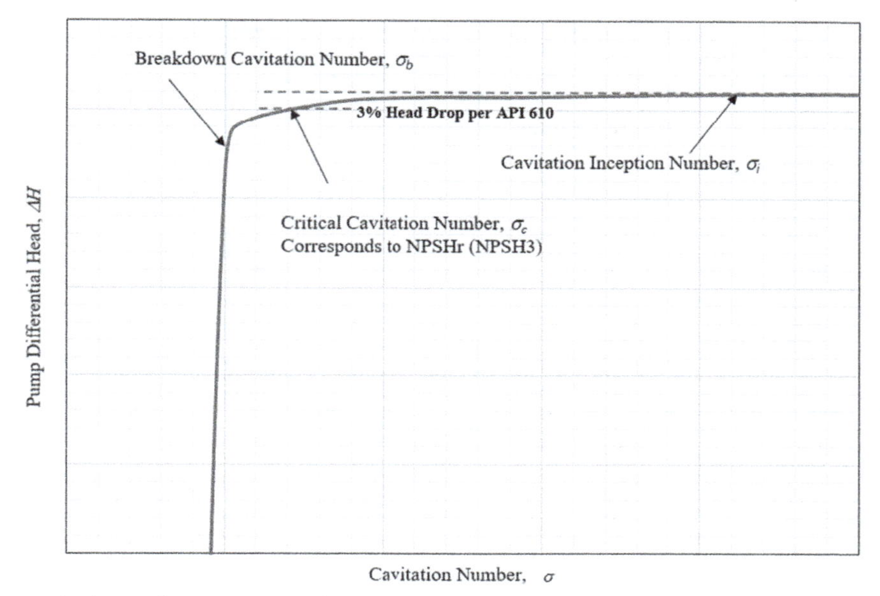

FIG. 3.52 Typical pump cavitation performance curve (drop curve) with inception, critical and breakdown cavitation numbers.

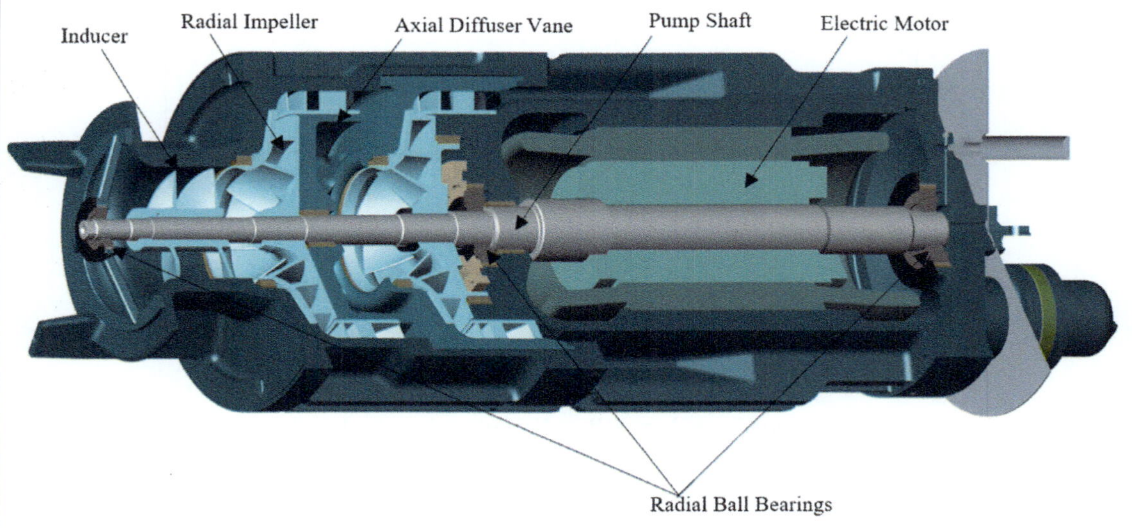

FIG. 3.53 A two-stage centrifugal pump with a helical inducer. *Courtesy of Ebara Corporation.*

Pump performance characteristics and performance curves

Pump performance curves are plotted to better observe and understand how a pump performs and operates as a function of the volumetric flow rate. Pump performance curves show the pump differential head (ΔH), shaft power (P), hydraulic efficiency (η), and NPSH3 as a function of volumetric flow rate for a given operational speed (N). At a certain flow rate, a pump achieves its highest efficiency point, which is described as BEP (Best Efficiency Point). The BEP flow should be closely matched to the required rated flow point of the process to ensure that the pump is operated at its highest efficiency. This will also enable the pump to achieve better reliability and consequently increase the mean time between overhauls. A typical pump performance curve is given in Fig. 3.54.

As seen from the differential head versus flow curve, the head with maximum and minimum impellers are also shown. Per API 610 guidelines, each centrifugal pump shall be capable of a 5% head increase at a rated condition by replacement of the impeller(s) with one(s) of larger diameter or variable speed capability or use of a blank stage [27]. This requirement is intended to prevent a change in selection caused by the refinement of hydraulic requirements after the pump has been purchased. It is not intended to accommodate future expandability. If there is a future operating requirement, it should be specified separately and considered in selection [27].

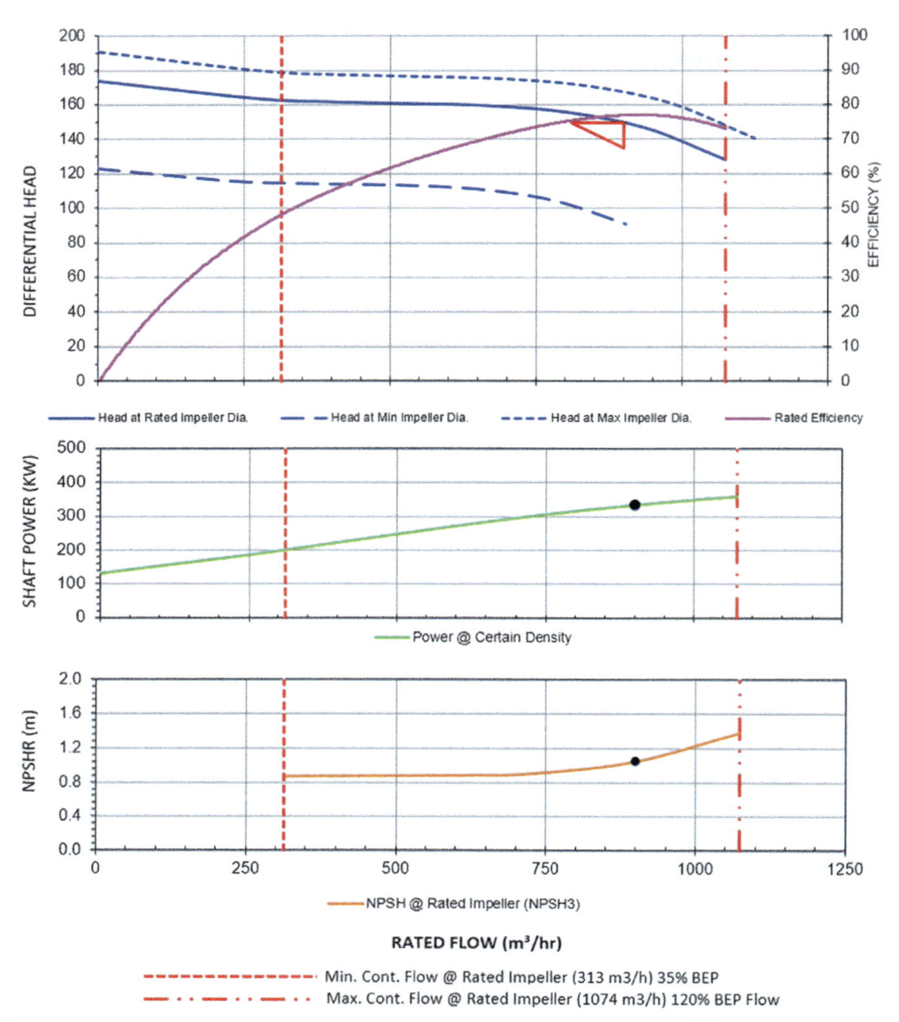

FIG. 3.54 Typical pump performance curves.

Pump performance with different speeds and impeller diameters is calculated by using established affinity laws. This law states that for similar conditions of flow, capacity will vary directly with the ratio of speed and/or impeller diameter and head with the square of this ratio at the BEP. Other flow points to the left or right of the BEP will correspond similarly [30]. Table 3.5 can be used in refiguring pump performance with impeller diameter and speed change.

The affinity laws described above are only applicable to a certain range of impeller diameters. Excessive impeller trim can result in a considerable drop in efficiency and may even create unstable pump performance. Therefore,

TABLE 3.5 Pump affinity laws for refiguring pump performance with impeller or speed change.

Diameter change	Speed change	Diameter and speed change
$Q_2 = Q_1\left(\frac{D_2}{D_1}\right)$	$Q_2 = Q_1\left(\frac{N_2}{N_1}\right)$	$Q_2 = Q_1\left(\frac{D_2}{D_1}\right)\left(\frac{N_2}{N_1}\right)$
$H_2 = H_1\left(\frac{D_2}{D_1}\right)^2$	$H_2 = H_1\left(\frac{N_2}{N_1}\right)^2$	$H_2 = H_1\left(\frac{D_2}{D_1}\right)^2\left(\frac{N_2}{N_1}\right)^2$
$P_2 = P_1\left(\frac{D_2}{D_1}\right)^3$	$P_2 = P_1\left(\frac{N_2}{N_1}\right)^3$	$P_2 = P_1\left(\frac{D_2}{D_1}\right)^3\left(\frac{N_2}{N_1}\right)^3$

Subscripts 1 and 2 indicate original and new flow and head and shaft power, respectively.

impeller trims greater than 80% should be avoided for radial-type impellers. For semi-axial flow (mixed flow) impellers, trimming should be limited to 90% of the original impeller diameter. This is primarily due to the impeller's outside diameter to eye diameter ratio. Variations in blade exit angles for original and new impeller diameters shall be reviewed to avoid any efficiency drop.

Pump types and areas of application

At the beginning of the pumping section, we classified pumps based on their operational principles (centrifugal and positive displacement pumps) and flow direction at the impeller and diffusers/volutes. In this section, two important dimensionless parameters will be introduced, which are widely used in the selection of hydraulic components and in determination of the centrifugal pump type.

Pump specific speed

As described in previous sections, any pump application is characterized by the flow rate (Q), head (H), and rotor speed (N). These important performance parameters assist pump users and designers to choose what type of centrifugal pump to use. These performance parameters are used to define a dimensionless number called pump-specific speed. Pump-specific speed for centrifugal pumps is defined as:

$$N_s = \frac{N\sqrt{Q_{\text{BEP}}}}{H^{0.75}} \tag{3.14}$$

Specific speed is not truly a nondimensional quantity since the acceleration of gravity is intentionally neglected in calculations. Therefore, it is important to calculate the specific speed by using either a set of US customary units or SI units. For US units, rpm, gpm, and ft. shall be used in determining the specific speed. Specific speed should always be calculated at the BEP flow with maximum impeller diameter and single stage only. A comparison of pump hydraulic profiles and the associated specific speed values are published by Hydraulic Institute and given in Fig. 3.55 [31].

Figs. 3.56 and 3.57 show the maximum attainable efficiency of single-stage centrifugal pumps for different flow rates as a function of pump-specific speed [32]. These graphs are helpful in the selection of hydraulics for particular applications based on the required head and flow rate. Pump designers usually adjust the pump number of stages, operational speed, and even number of units to attain maximum efficiency. In most cases, the total process flow is divided into multiple pumps operating in a parallel configuration to improve pump and process efficiency. The same is applicable to high-head applications. If the head requirement is too high for a single, multistage pump to handle, pumps in a series configuration can be utilized.

While specific speed is useful and prominently assists pump designers and end-users to determine what type of impeller geometry and shape will produce better efficiency for a given flow rate, there are still important design considerations and application-specific requirements that need to be considered and met. Pump designers and end-users should consider initial cost, cost to operate each pump, reliability, dependability, availability, mean time between

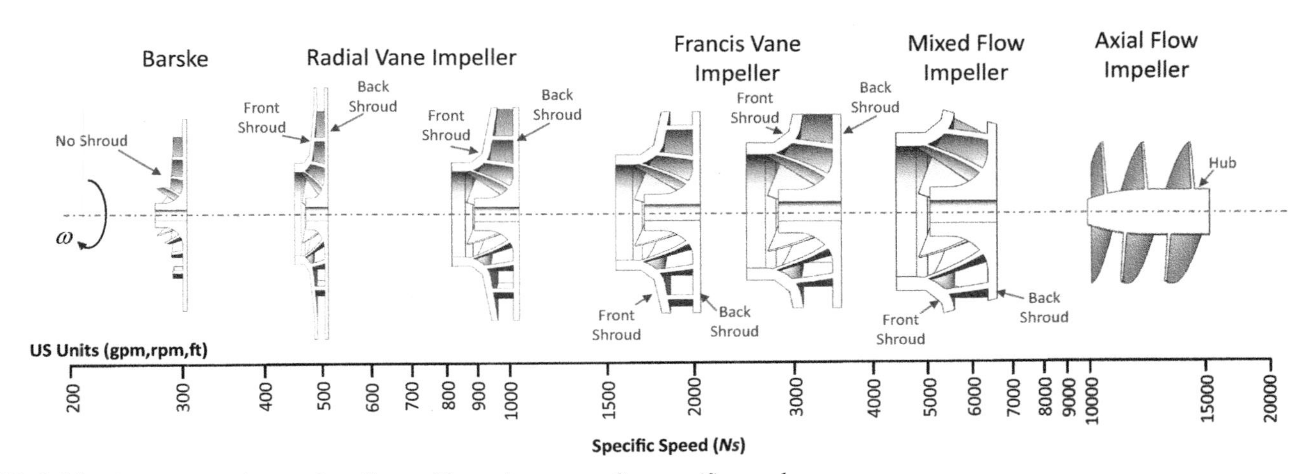

FIG. 3.55 Comparison of pump impeller profiles and corresponding specific speeds.

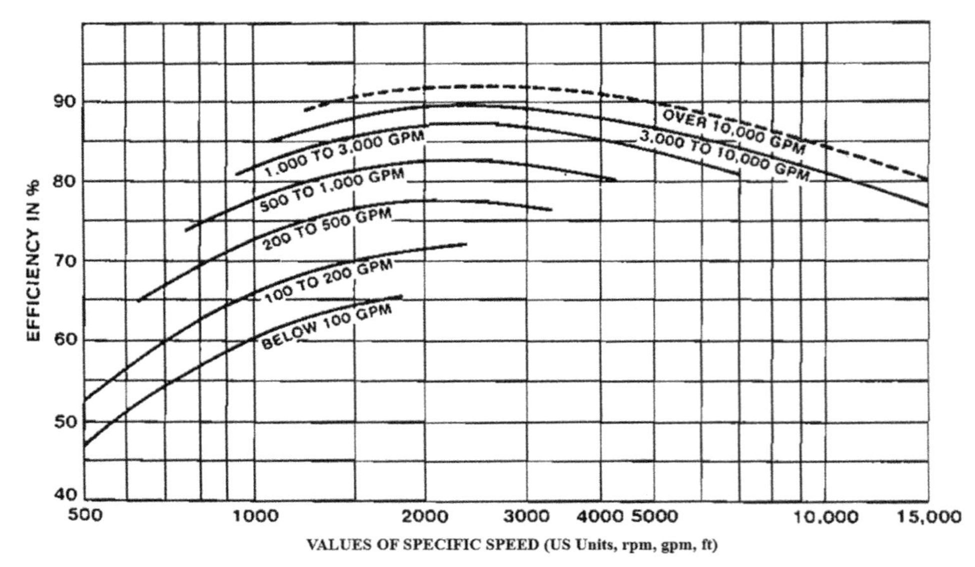

FIG. 3.56 Efficiency of single-stage centrifugal pumps with different flow rates as a function of specific speed [32].

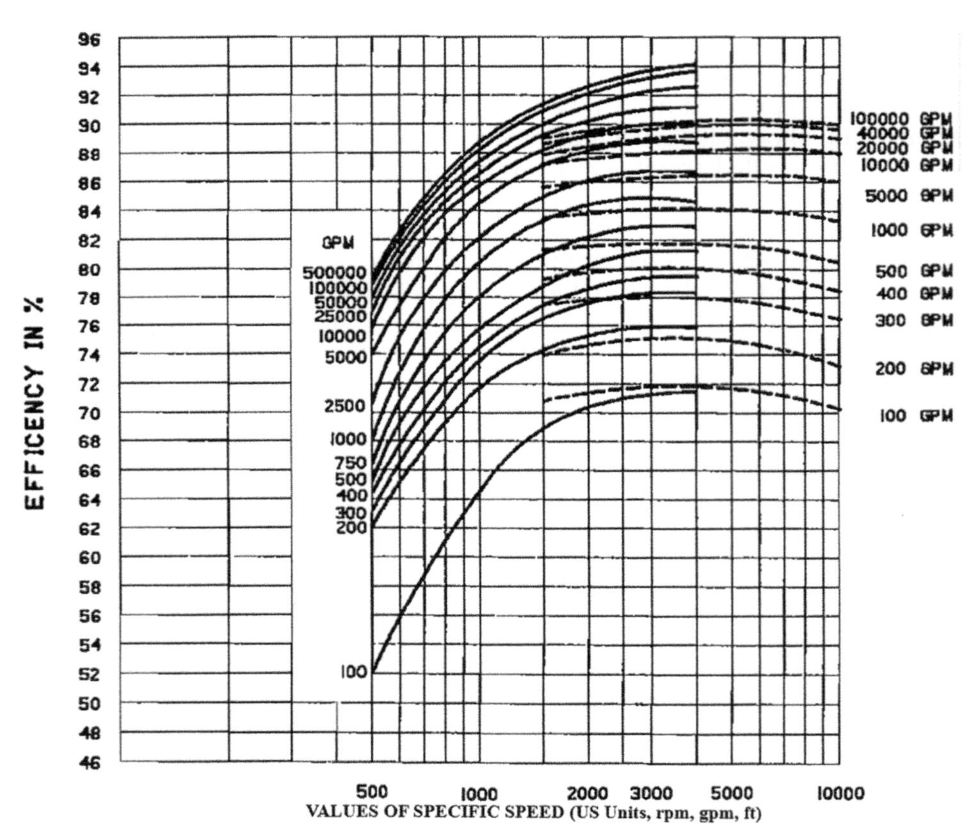

FIG. 3.57 Efficiency of single-stage end suction and double suction centrifugal pumps (–) compared to bowl efficiencies of wet pit pumps (--) at different flow rates as a function of specific speed [32].

overhauls, overall pump footprint, off-design running conditions (if any), ease of installation, and maintenance in the selection of hydraulics and pump type.

A pump type table is provided by Gülich [21], which lists different impeller shapes and corresponding pump type, maximum differential head, attainable efficiency, and specific speed. Table 3.6 is reproduced from Gülich's pump-type table [31].

TABLE 3.6 Pump types.

Ns (rpm, gpm, ft)	Type	Impeller shape	Maximum head/stage	Hydraulic efficiency (%)	Remarks
<25	Piston pumps	Positive displacement pumps	Limited by mechanical constraints	80–95	
<100	Gear pumps			75–90	
100–500	Screw pumps			65–85	Also for gas/liquid mixtures
25–200	Peripheral pumps		400 m	30–35	Single and multi-stage
100–550	Side channel pumps		250 m	34–47	
350–1500	Radial pumps		800 m	40–88	Below $Ns < 500$ usually small pumps only
2500			400 m	70–92	In most cases: Head/stage <250 m
5000			60 m	60–88	$Ns = 5000$ is essentially the upper limit for radial impellers
1750	Semi-axial (mixed flow) pumps		100 m	75–90	For $Ns < 2500$ often multistage For $Ns > 3750$ seldom single stage
8000			20 m	75–90	For $Ns > 5000$, only single stage
8000–20,000	Axial pumps		2–15 m	70–88	Flow rates up to $60\,m^3/s$, only single stage

Pump suction specific speed and cavitation performance

Pump suction performance is critical in the safe operation of a pump under low suction operating conditions. Cavitation at pump inlet can occur due to a lack of suction pressure, which is highly detrimental to the pumps directly and indirectly. Cavitation erosion can destroy inducer and impeller blades due to vapor formation, which can be considered direct cavitation damage. Indirect failure modes are often due to high vibration and noise. Vapor formation cannot be tolerated in the pump internals, especially at locations where a hydrodynamic fluid film is required. The locations that are vulnerable to cavitation are mainly pump shaft support locations such as shaft bushings, hydrodynamic or hydrostatic bearings, and rolling element ball bearings [23,33].

Cavitation performance of a pump is characterized by a pump's suction-specific speed, which is defined as:

$$N_{ss} = \frac{N\sqrt{Q_{\text{BEP}}}}{(\text{NPSH})^{0.75}} \tag{3.15}$$

Similar to pump-specific speed, suction-specific speed is not a true dimensionless parameter and shall be calculated and compared by using either the US or metric unit system. For calculation in US units, rpm, gpm and ft. are used, while rpm, m^3/s, and m are used for determining *Nss* in the metric unit system. N_{ss} shall always be calculated at the

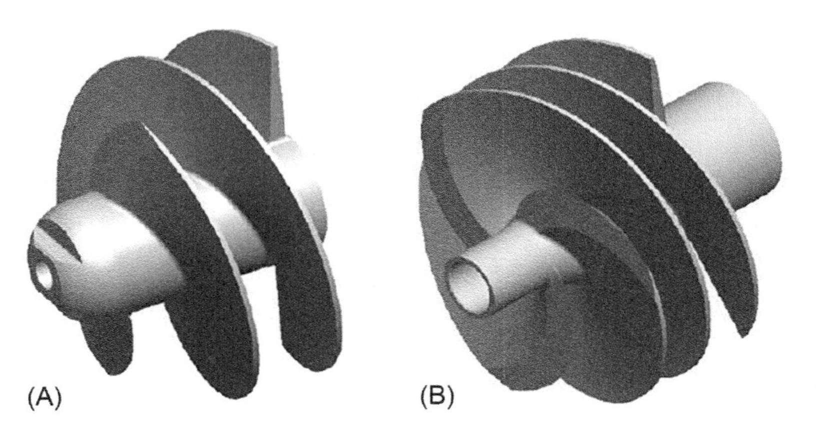

FIG. 3.58 Helical inducers used in centrifugal pumps to improve suction performance. (A) Constant pitch inducer, (B) variable pitch and hub inducer [33].

BEP flow rate (Q_{BEP}) with corresponding Net Positive Suction Head (NPSH). NPSH in Eq. (3.9) is the NPSHr level at 3% pump differential head drop according to applicable API and ISO guidelines [26,27].

Pumps with lower NSPHr have better suction performance. Lower NPSHr values indicate that a pump can be operated with a lower suction head without the presence of any cavitation. According to Eq. (3.15), low NPSHr corresponds to high-suction-specific speed.

Impellers are prone to cavitate with decreased suction head (pressure) and have a N_{ss} less than 10,000 in US units. To improve suction performance pumps, the impeller eye section for the first stage impeller can be enlarged to slow down the meridional flow velocity to prevent cavitation inception. Pumps with suction impellers can achieve a maximum N_{ss} of 14,000 in US units. It should be noted that pump inlet configuration plays an important role in suction performance. Sudden changes in flow velocity and direction, sharp turns, and discontinuation at flow passages can lower suction specific speed of the pump.

Helical inducers with constant pitch geometry are utilized in applications where NPSHa is less than ideal. Helical constant pitch inducers are installed at the suction side of the pump to increase the suction pressure of the first-stage impeller (Fig. 3.54). This type of inducer can attain N_{ss} of 35,000 in US units.

Variable pitch and hub inducers are developed for applications that demand high suction performance as in rocket pump and industrial in-tank pump applications. These inducers are sometimes followed by guide vanes or diffusers at the upstream of the first-stage impeller. N_{ss} can be as high as 80,000 in US units for this type of inducer. The key design consideration for high-specific speed inducers and pumps is to increase the suction pressure gradually to prevent vapor formation within blades and passages. Typical geometric shapes of helical inducers are shown in Fig. 3.58.

Positive displacement pumps

In a positive displacement pump, a movable element is utilized in the form of a piston or diaphragm, which physically pushes and displaces the fluid from inlet to outlet. The movement of the piston or diaphragm results in a decrease in the physical volume of the pump chamber, hence an increase in pressure, causing the fluid to be discharged. Fluid is usually trapped within the pump chamber using closed valves or seals. Constant flow is maintained during operation and at a fixed speed, regardless of changes in pressure. The flow rate can be adjusted via speed control, because flow rate is directly proportional to its speed unlike in a centrifugal pump [34].

As indicated in the previous section, positive displacement pumps have low pump-specific speeds due to their low flow capacity (Table 3.6). Positive displacement pumps are generally preferred for low-flow rate, high pressure, and high viscosity applications. For a high rate of fluid transfer, centrifugal pumps are preferred. High rate of pumping applications for fluids with low viscosity, namely hydrocarbons, diesel, petrol, ethanol, alcohols, and centrifugal pumps have been the choice. The selection of pump types between positive displacement and centrifugal style depends on many other factors including economics, reliability, continuous operation, etc. Each pumping application must be reviewed by process and pump design engineers to determine the best suitable option considering initial and operating costs.

Positive displacement pumps have the capability to operate with vapor at suction and are not required to be primed as in a centrifugal pump application. A positive displacement pump will usually self-prime due to the very small

TABLE 3.7 Positive displacement pumps [34].

Classification	Pump type	Geometric shape
Reciprocating	Piston pump	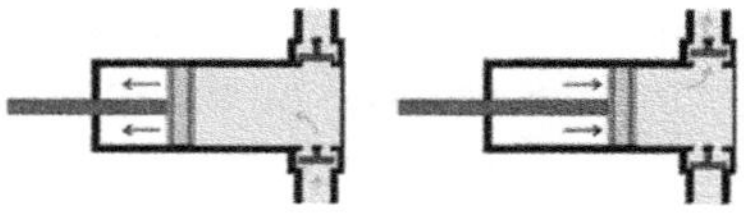
	Plunger pump	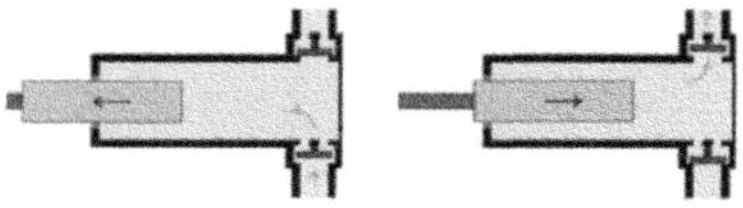
	Diaphragm pump	
Rotary	External gear pump	
	Internal gear pump	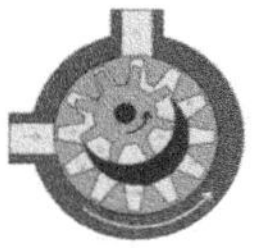
	Lope pump	
	Vane pump	

clearances that exist within the pump. This will help it pull a vacuum and discharge vapor through the pump until the liquid reaches the pump. Positive displacement pumps can also pump fluids that contain a certain amount of gases. These pumps are also ideal for applications with partial solids.

There are two main types of positive displacement pumps, namely reciprocating and rotary pumps. Reciprocating positive displacement pumps operate by making repetitive strokes creating a vacuum inside the pump and pushing the fluid through the discharge side. A rotary-type pump uses a rotating gear system to pump fluid rather than the backward and forward movement of reciprocating pumps. The most common type of positive displacement pumps are shown in Table 3.7.

Compression

Gas compression principles

The working principles of centrifugal gas compressors can be understood by applying some basic laws of physics. Using the first and second laws of thermodynamics together with basic laws of fluid dynamics, such as Bernoulli's law and Euler's law, we can explain the fundamental working principles, and by extension, can increase the understanding of the operational behavior of centrifugal gas compressors. We will first discuss the thermodynamics of gas compression, and then move on to the fluid dynamics.

This general description of the thermodynamics of gas compression applies to any type of compressor, independent of its detailed working principles. Compression fundamentally involves the use of mechanical energy to increase the energy of a gas that flows through a compressor. The increased energy of the gas shows as increased temperature, and if the compression process is done properly, increased pressure and density [35].

Gases have properties that can be observed or measured, such as the pressure, temperature, mass, and volume that contains the gas. A key feature of a gas is that pressure, temperature, and density (density is the mass of gas contained in a given volume) are related (Boyles Law); an increase in pressure (at constant temperature) leads to an increase in density, while an increase in temperature (at constant pressure) leads to a reduction in density.

For any consideration regarding temperature changes, we have to discuss whether heat is lost across system boundaries. Most compressor analyses consider the compressor to be adiabatic, which means that it is assumed that no heat is lost through the cylinder walls. The simplest compression process is a positive displacement process used in reciprocating compressors, easily visualized via a simple bicycle pump example. The idea is to trap an amount of gas in a cylinder with a piston that is able to move inside the cylinder. If we want to push the piston into the gas, we have to expend work. Because the mass of gas is trapped in our contraption, it stays constant while the volume is reduced. Therefore, the density (mass divided by volume) of the gas increases.

To describe the compression process, we can make use of the first and second laws of thermodynamics. The first law tells us that when we transfer one form of energy into another form of energy, no energy is lost. The second law tells us that most of these energy transfers are not reversible. For example, we can convert electrical energy in a motor to create mechanical energy. We can then use the mechanical energy and convert it back to electricity in a generator. However, the electricity generated that way would be less than the electricity originally fed into the motor. This is because there were losses, and these losses caused the creation of heat. This is also referred to as an increase in entropy.

We also need to introduce the concept of enthalpy. Enthalpy describes the total heat content of a gas. Consider a flow process with a flow entering the system at point 1 and leaving at point 2 (Fig. 3.59). We also feed mechanical energy (work), W, into the system, and the system exchanges heat q with the environment.

The first law of thermodynamics, defining the conservation of energy, can be written for a steady-state flow process:

$$\left(h_2 + \frac{w_2^2}{2} + gz_2\right) - \left(h_1 + \frac{w_1^2}{2} + gz_1\right) = q_{12} + W_{t,12} \tag{3.16}$$

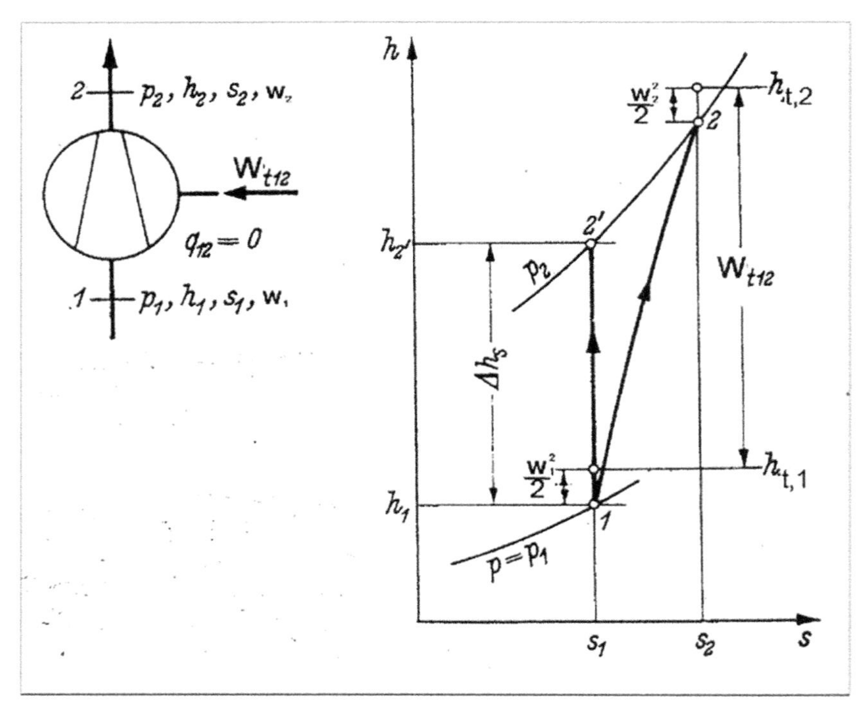

FIG. 3.59 Compression process in an enthalpy-entropy diagram.

with $q=0$ for adiabatic processes, and $gz=0$ because changes in elevation are not significant for gas compressors. We can combine enthalpy and velocity into total enthalpy by:

$$h_t = h + \frac{w^2}{2} \tag{3.17}$$

$W_{t,12}$ is the amount of work[a] we have to apply to effect the change in enthalpy in the gas. The work $W_{t,12}$ is related to the required power, P, by multiplying it with the mass flow.

$$P = \dot{m} W_{t,12} \tag{3.18}$$

Power and enthalpy differences are thus related by:

$$P = \dot{m}(h_{t,2} - h_{t,1}) \tag{3.19}$$

If we can find a relationship that combines enthalpy with the pressure and temperature of a gas, we have found the necessary tools to describe the gas compression process.

Now, let's consider a specific type of compression: the isentropic compression in an adiabatic system. Before we do that, we have to consider the concept of entropy. The second thermodynamics law tells us:

$$\dot{m}(s_2 - s_1) = \int_1^2 \frac{dq}{T} + s_{irr} \tag{3.20}$$

that a change in entropy is either due to irreversible losses, or because heat crosses the system boundaries at a certain temperature (the $\frac{dq}{T}$ term in the equation above). For adiabatic flows, where no heat q enters or leaves, the change in entropy simply describes the losses generated in the compression process. These losses come from the friction of gas with solid surfaces and the mixing of gas of different energy levels.

Therefore, an adiabatic, reversible compression process ($dq=0, s_{irr}=0$) does not change the entropy of the system. It is isentropic. This is an ideal process. This isentropic compression process assumes that no entropy is generated in the compression process. Because the system is adiabatic (that is, no heat can enter or leave the system except with the flowing gas), the only change in entropy can come from losses, which are irreversible.

The performance quality of a compressor can be assessed by comparing the actual head (which directly relates to the amount of power we need to spend for the compression) with the head that the ideal, isentropic compression would require. This defines the isentropic efficiency.

$$\eta_s = \frac{\Delta h_s}{\Delta h} \tag{3.21}$$

For a compressor receiving gas at a certain suction pressure and temperature and delivering it at a certain output pressure, the isentropic head represents the energy input required by a reversible, adiabatic (thus isentropic) compression. The actual compressor will require a higher amount of energy input than needed for the ideal (isentropic) compression (Fig. 3.59).

Fig. 3.60 shows a Mollier diagram for methane. We can see that for low pressures and temperatures, the gas indeed behaves as an ideal gas. That is, the change in enthalpy is only a function of temperature.

For the elevated pressures we see in natural gas compression, this equation becomes inaccurate, and an additional variable, the compressibility factor Z, has to be added:

$$\frac{p}{\rho} = ZRT \tag{3.22}$$

Unfortunately, the compressibility factor Z itself is a function of pressure, temperature, and gas composition.

To calculate enthalpy in a real gas, we get additional terms for the deviation between real gas behavior and ideal gas behavior:

$$\Delta h = \left(h^0 - h(p_1)\right)_{T_1} + \int_{T_1}^{T_2} c_p dT - \left(h^0 - h(p_2)\right)_{T_2} \tag{3.23}$$

[a] Physically, there is no difference between work, head, and enthalpy difference. In systems with consistent units (such as the SI system), work, head, and enthalpy difference have the same unit (e.g., kJ/kg in SI units). Only in inconsistent systems (such as US customary units), we need to consider that the enthalpy difference (e.g., in BTU/lb$_m$) is related to head and work (e.g., in ft. lb$_f$/lb$_m$) by the mechanical equivalent of heat (e.g., in ft. lb$_f$/BTU).

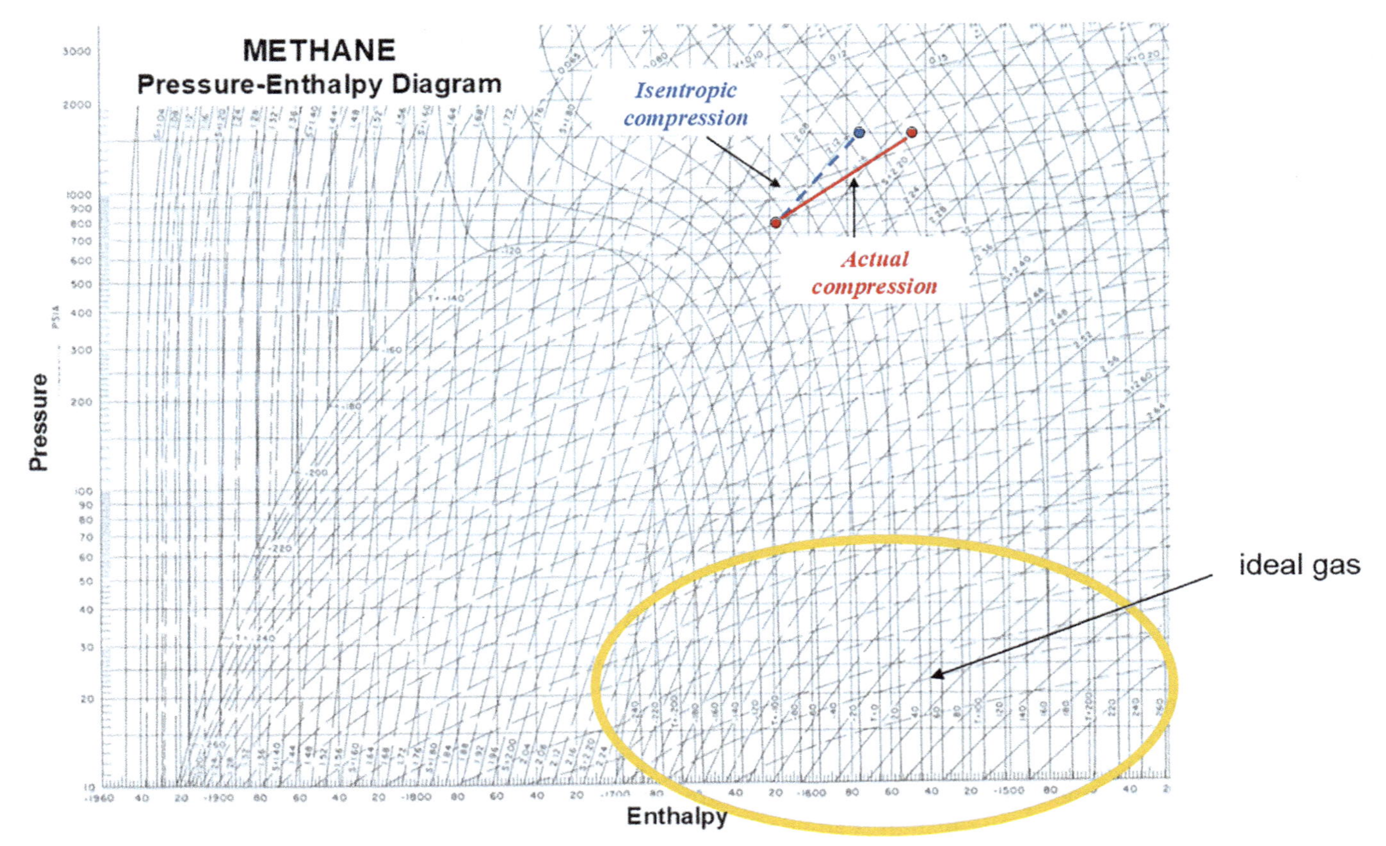

FIG. 3.60 Mollier diagram for a gas (methane): Enthalpy versus pressure with lines of constant temperature, density, and entropy [36].

The terms $(h^0 - h(p_1))_{T1}$ and $(h^0 - h(p_2))_{T2}$ are called departure functions, because they describe the deviation of the real gas behavior from the ideal gas behavior. They relate the enthalpy at some pressure and temperature to a reference state at low pressure but at the same temperature. The departure functions can be calculated solely from an equation of state, while the term $\int c_p dT$ is evaluated in the ideal gas state.

While a Mollier diagram (Fig. 3.60) is perfectly suited for pure gases, we generally have to work with gas mixtures. For gas mixtures, so-called Equations of State (EoS) are used. Equations of State are semi-empirical relationships that allow us to calculate the compressibility factor as well as the departure functions for a given set of pressures and temperatures or pressures and entropies. Equations of State also take into account how the components of gas mixtures influence each other, providing mixing rules for that purpose.

For gas compression applications, the most frequently used Equations of State are Redlich-Kwong, Soave-Redlich-Kwong, Benedict-Webb-Rubin, Benedict-Webb-Rubin-Starling, and Lee-Kessler-Ploecker. Lately, AGA 8 and GERG (REFPROP [22]) are considered frequently [35].

For real gases (where k and c_p in the equations above become functions of temperature and pressure), the enthalpy of a gas $h(p,T)$ is thus calculated in a more complicated way using Equations of State (Poling et al. [37]). These represent relationships that allow us to calculate the enthalpy of gas of a known composition if any two of its pressures, temperatures, or entropy are known.

We therefore can calculate the actual head for the compression by:

$$\Delta h = h(p_2, T_2) - h(p_1, T_1) \tag{3.24}$$

and the isentropic head by

$$\Delta h = h(p_2, s_1) - h(p_1, T_1) \tag{3.25}$$

$$s_1 = s(p_1, T_1) \tag{3.26}$$

Our equation for the actual head implicitly includes the entropy rise Δs, because:

$$\Delta h = h(p_2, T_2) - h(p_1, T_1) = h(p_2, s_1 + \Delta s) - h(p_1, s_1) \tag{3.27}$$

Fig. 3.60 shows this compression process in a Mollier diagram.

Because the enthalpy definition above is on a per mass flow basis, the absorbed gas power P_g (that is, the power that the compressor transferred into the gas), can be calculated as:

$$P_g = \dot{m} \cdot \Delta h \tag{3.28}$$

The mechanical power P necessary to drive the compressor is the gas-absorbed power increased by all mechanical losses (friction in the seals and bearings), expressed by a mechanical efficiency η_m (typically in the order of 1% or 2% of the total absorbed power):

$$P = \frac{1}{\eta_m} \dot{m} \cdot \Delta h = \frac{\dot{m} \cdot \Delta h_s}{\eta_m \eta_s} \tag{3.29}$$

We also encounter energy conservation on a different level in turbomachines. The aerodynamic function of a turbomachine relies on the capability to trade two forms of energy: kinetic energy (velocity energy) and potential energy (pressure energy). This will be discussed in a subsequent section.

If cooling is applied during the compression process (for example, with intercoolers between two compressors in series), then the increase in entropy is smaller than for an uncooled process. Therefore, the compression power requirement will be reduced.

The task of gas compression is to bring gas from a certain suction pressure to a higher discharge pressure by means of mechanical work. The actual compression process is often compared to one of two ideal processes.

The compression process is isentropic if the process is frictionless and no heat is added to or removed from the gas during compression. With these assumptions, the entropy of the gas does not change during the compression process, and the process is reversible. Because there is no heat transfer across the system boundaries, the process is often referred to as reversible adiabatic.

The polytropic compression process is like the isentropic cycle reversible, but it is not adiabatic. It can be described as an infinite number of isentropic steps, each interrupted by isobaric heat transfer, such that the efficiency in each step is the same. The heat addition allows the process to yield the same discharge temperature as the real process.

While the path from the inlet to outlet of the compressor in the isentropic process is defined by following a constant entropy, the isentropic path is defined by infinitesimal steps of constant polytropic efficiency η_p. This means that the actual compression process consists of an infinite (for practical purposes, a large number such as 20 is sufficient) number of steps. Each step consists of an isentropic compression step, followed by an isobaric heat addition.

$$\Delta h = \frac{1}{\eta_p} \int_{p_1}^{p_2} v dp = \frac{\Delta h_p}{\eta_p} \tag{3.30}$$

and

$$\Delta h_p = \int_{p_1}^{p_2} v dp \tag{3.31}$$

The polytropic efficiency η_p is defined as:

$$\eta_p = \frac{\Delta h_p}{\Delta h} \tag{3.32}$$

Using the polytropic process for comparison reasons works fundamentally the same way as using the isentropic process for comparison reasons. The difference lies in the fact that the polytropic process uses the same discharge temperature as the actual process, while the isentropic process has a different (lower) discharge temperature than the actual process for the same compression task. In particular, both the isentropic and the polytropic processes are reversible (and adiabatic) processes. In order to fully define the isentropic compression process for a given gas, suction pressure, suction temperature, and discharge pressure have to be known. To define the polytropic process, either the polytropic compression efficiency or the discharge temperature also has to be known.

For designers of compressors, the polytropic efficiency has an important advantage. If a compressor has five stages, and each stage has the same isentropic efficiency η_s, then the overall compressor efficiency will be lower than η_s. If, for the same example, we assume that each stage has the same polytropic efficiency η_p, then the polytropic efficiency of the entire machine is also η_p. For users of gas compressors, however, the isentropic efficiency has the advantage that the change in isentropic efficiency for given operating conditions is directly related to the change in power consumption.

Centrifugal compressors

Turbocompressors are so-called dynamic compressors (as opposed to positive displacement compressors). The working principle evolves around the deflection of a gas flow in a rotating system.

The basic principle for any aerodynamic machine lies in the fact that if we deflect the flow of a fluid, we have to exert a force. This is the principle the wing of an aircraft uses to keep the plane in the air (Fig. 3.61, Brun and Kurz [17]).

A force applied does not yet relate to a portion of the work applied. Work is done if a force is applied over a distance. Thus, the force has to be applied to a moving object, such that the force (or a least part of the force) is in the direction of the movement. So, if we apply a force (by changing the direction of a fluid or gas) in a rotating (i.e., moving) system, we can transfer work. This is the basic working principle for any type of turbomachinery.

Bernoulli's law (which is strictly true only for incompressible flows, but can be modified for the subsonic compressible flows found in gas compressors) describes the interchangeability of two forms of energy: static pressure and velocity. We consider here the flow of a gas in a duct (Fig. 3.62).

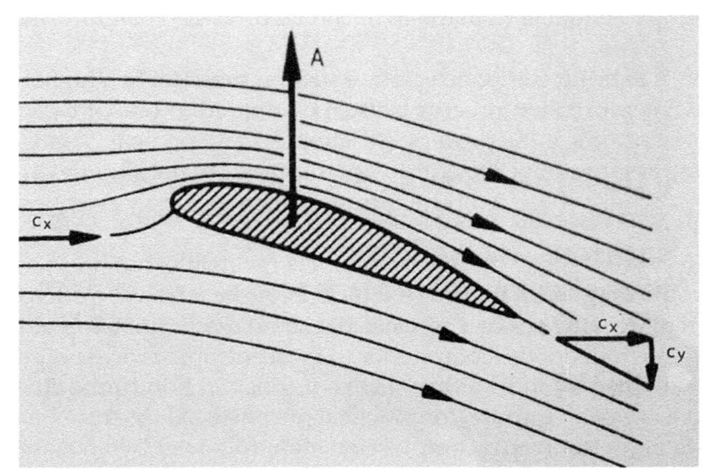

FIG. 3.61 Airfoil creates lift A by deflecting air downwards [35].

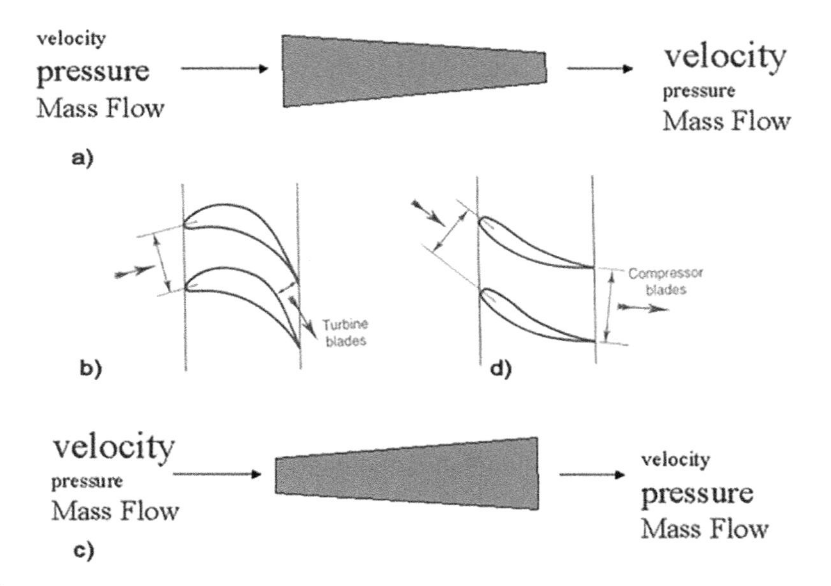

FIG. 3.62 Bernoulli's law.

The incompressible formulation of Bernoulli's law for a frictionless, stationary, adiabatic flow without any work input is:

$$p_t = p + \frac{\rho}{2}c^2 = \text{const} \tag{3.33}$$

For compressible flows, the equation becomes (using the concept of enthalpy discussed earlier):

$$h_t = h + \frac{c^2}{2} = \text{const} \tag{3.34}$$

Another requirement is that mass cannot appear or disappear. Thus, for any flow from a point 1 to a point 2:

$$\begin{gathered} \dot{m}_1 = \rho_1 Q_1 = \dot{m}_2 = \rho_2 Q_2 \\ \rho \cdot Q = \rho \cdot c \cdot A \end{gathered} \tag{3.35}$$

This requirement is valid for compressible and incompressible flows with the caveat that for compressible flows, the density is a function of pressure and temperatures, and thus ultimately a function of the velocity.

These two concepts explain the working principles of the vanes and diffusers used in turbomachines (Fig. 3.62). Due to requirements for mass conservation, any flow channel that has a wider flow area at its inlet and a smaller flow area at its exit will require a velocity increase from inlet to exit. If no energy is introduced to the system, Bernoulli's law requires a drop in static pressure (Fig. 3.62A). Examples of flow channels like this are turbine blades and nozzles and inlet vanes in compressors (Fig. 3.62B). Conversely, any flow channel that has a smaller flow area A at its inlet and a larger flow area at its exit will require a velocity decrease from inlet to exit. If no energy is introduced to the system, Bernoulli's law requires an increase in static pressure (Fig. 3.62C). Examples of flow channels like this are vaned or vaneless diffusers, flow channels in impellers, and rotor and stator blades of axial compressors volutes (Fig. 3.62D).

If these flow channels are in a rotating system (for example, in an impeller), mechanical energy is added to or removed from the system. Nevertheless, if the velocities are considered in a rotating system of coordinates, the above principles are applicable as well.

Another important concept is the conservation of momentum (Fig. 3.63). The change in momentum M of gas flowing from point 1 to point 2 is its mass times its change in velocity (m c) and is also the sum of all forces F acting. The change in momentum is:

$$\frac{d\vec{M}}{dt} = \dot{m}\left(\vec{c}_2 - \vec{c}_1\right) = \vec{F} \tag{3.36}$$

To change the momentum of this gas, either by changing the velocity or the direction of the gas (or both), a force is necessary. Fig. 3.63 outlines this concept for the case of a bent, conical pipe. The gas flows in through the area A_1 with w_1, p_1, and out through the flow area A_2 with w_2, p_2. The differences in the force due to the pressure (p_1A_1 and p_2A_2,

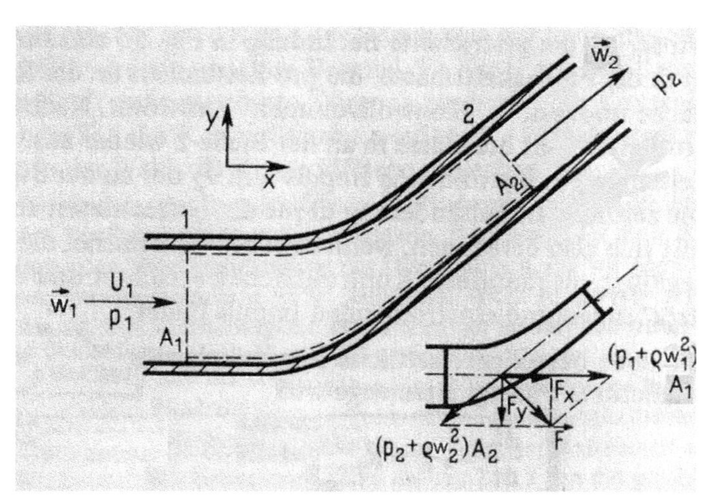

FIG. 3.63 Conservation of momentum.

respectively), and the fact that a certain mass flow of gas is forced to change its direction, generates a reaction force F_R. Split into x and y coordinates, and considering that

$$\dot{m} = \rho_1 A_1 w_1 = \rho_2 A_2 w_2 \tag{3.37}$$

we get (due to the choice of coordinates, $w_{1y} = 0$)

$$\begin{aligned} x : \rho A_1 w_1 (w_{2x} - w_1) &= p_1 A_1 - (p_2 A_2)_x + F_{Rx} \\ y : \rho A_1 w_1 (w_{2y}) &= -(p_2 A_2)_y + F_{Ry} \end{aligned} \tag{3.38}$$

It should also be noted that this formulation is also valid for viscous flows, because the friction forces become internal forces. So, we have identified the force that's created by deflecting the fluid flow. Now, if we would let the channel spin around an axis in an x-direction, at a certain radius r, we get a force that moves at a certain speed. In other words, we either extract power (if our flow channel drives some load connected to the shaft), or absorb power (if we drive the shaft with a motor or hand crank). This works if we have rotating blades or flow channels that are capable of changing the direction of the flow. For a rotating row of vanes (whether they are considered individual airfoils or as in the previous discussion, flow channels), in order to change the velocity of the gas, the vanes have to exert a force upon the gas. This is fundamentally the same force that F_{Ry} acts in the previous example for the pipe. This force has to act in the direction of the circumferential rotation of the vanes in order to do work on the gas. According to the conservation of momentum, the force that the blades exert is balanced by the change in circumferential velocity times the associated mass of the gas.

Fig. 3.64 shows a compressor stage that uses the principles outlined above. Since the blades on the rotor move while the blades in the stator are stationary, we have to find a way to describe velocities in a rotating system and in a stationary system. This works by simply adding velocity vectors. The vanes of the rotating blades "see" the gas in a coordinate system that rotates with the rotor. The transformation of velocity coordinates from an absolute frame of reference (c) to the frame of reference rotating with a velocity (u) is by:

$$\vec{w} = \vec{c} - \vec{u} \tag{3.39}$$

The velocities here are vectors describing both the magnitude of the velocity and its direction.

Fig. 3.64 shows this vector addition in the form of a velocity triangle. If we were to rotate with the rotor, we would experience velocity c_1 (in a stationary system) entering the rotor as velocity w_1, which we can find by subtracting the vector u_1 from the vector c_1. This is called a velocity triangle. Similarly, at the exit from the rotor, we see (while still rotating with the rotor) the air leaving with velocity w_2 and calculate the velocity c_2 entering the stator by adding u_2. In axial machines, where the air flows more or less parallel to the axis of rotation, we can essentially treat this as a two-dimensional problem for every constant diameter. This means that u_1 and u_2 are about the same, and we do not need to worry about the third spatial coordinate.

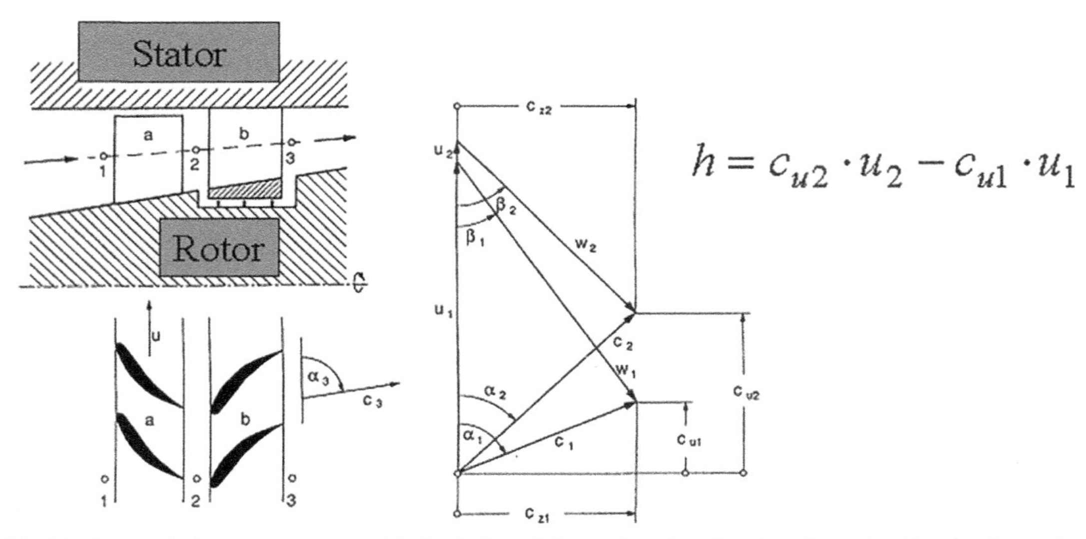

FIG. 3.64 Velocities in a typical compressor stage. Mechanical work h transferred to the air is determined by the change in circumferential momentum of the air.

What we can see in Fig. 3.64 is that the rotor blade deflects the air from w_1 to w_2, and since it rotates, it adds work to the air. It also has reduced the velocity, thus increasing the pressure according to Bernoulli's law (see Fig. 3.62D). The velocity c_2 entering the stator is higher than c_1, but the stator also has a shape to increase the flow passage (Fig. 3.62D). Therefore, the pressure is further increased. The exit velocity c_3 becomes the inlet velocity c_1 for the next stage. The important step is that the change in the circumferential velocity ($cu_2 - cu_1$) multiplied by the speed at which the blade rotates (u_2 and u_1, respectively) gives us the entire amount of work that was transferred to the air, which is also the power per unit mass flow of air absorbed by this compressor:

$$P = \dot{m} \cdot \Delta h = \dot{m} \cdot (u_2 c_{u2} - u_1 c_{u1}) \tag{3.40}$$

This relationship is usually referred to as Euler's Law, after Leonard Euler who formulated it in the 18th century. It describes the conservation of angular momentum at the inlet and outlet of a stage. As mentioned earlier, the only place where work is added is in the rotating part of the machine. All the stator does is convert some of the kinetic energy (velocity) to an additional pressure. The impeller adds energy (expressed as total pressure), but both in the form of more pressure and more velocity. The diffuser converts velocity into pressure but does not add energy to the flow. The important contribution of Euler's law is that it connects thermodynamic properties (like enthalpy, or therefore pressures and temperatures) with aerodynamic properties (velocities).

Fig. 3.64 shows the relationships for an axial compressor. The axial velocity $c_{z1} = c_{z2}$ in this example stays about constant. This is a good assumption for axial machines.

The work input H (also known as the actual head) in the equations above does not tell us anything about the pressure increase of the stage. Only when we know the efficiency of the stage can we gauge the amount of pressure rise.

While in an axial compressor the flow is generally more or less parallel to the axis of the turbomachine, in a centrifugal compressor, we have for each stage a more or less axial flow into the stage, while the gas leaves the stage with a significant radial component [17]. Therefore, in a centrifugal machine, we have to consider all three dimensions. However, just as with the axial machines, the important feature is the force in the direction of the rotational speed, and thus, the changes in velocities in a circumferential direction. While in an axial machine, the blade speed u at the inlet and exit are almost the same, but they are quite different in a centrifugal machine. Fig. 3.65 shows the impeller (i.e., the rotating blades) and the diffusor of a centrifugal compressor.

Fig. 3.66 shows the velocity triangles at the impeller where u is the circumferential blade velocity at the inlet (1) and exit (2) of the impeller. The $c_1 - w_1$ plane has to be imagined perpendicular to the $c_2 - w_2$ plane, because the inlet flow is more or less axial, while the exit flow is more or less radial. However, the circumferential components of c_1 and c_2, respectively, are in the same plane.

Similar to the axial compressor, the relative velocities are found by subtracting the vector of blade velocity u from the gas velocity in the stationary reference frame. The velocity c_1 is the absolute velocity at the inlet to the impeller, and c_2 is the velocity of the gas entering the diffusor. The relative velocity w_1 is often the highest velocity in the machine if it is taken at the tip of the impeller. The relative velocity w_2 is largely determined by the direction of the blades at the exit of the impeller. If w_2 points against the direction of rotation (as in Fig. 3.66), the blades are called backward bent, as

FIG. 3.65 Centrifugal compressor stage showing the spinning impeller feeding the gas into a diffusor.

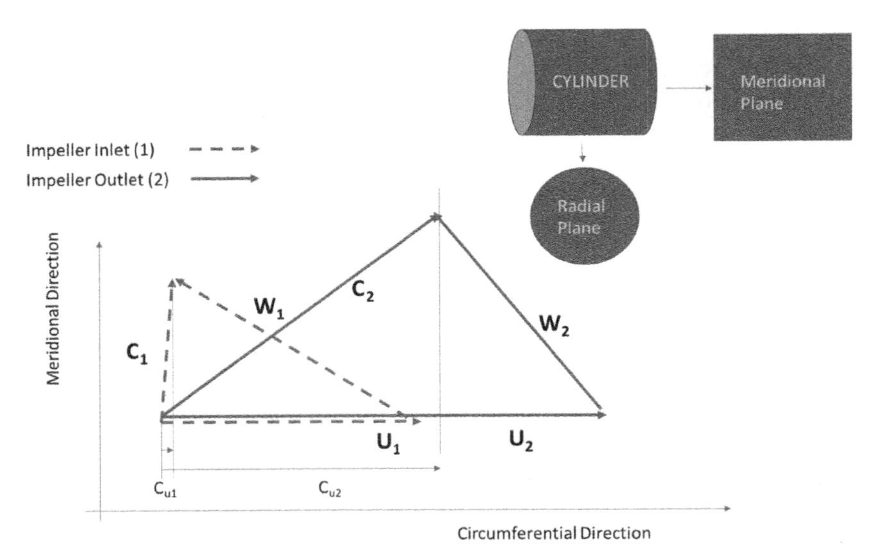

FIG. 3.66 Velocity triangles for a centrifugal compressor.

opposed to forward bent blades and radial blades. Most compressor impellers in oil and gas applications are of the backward bent design, because, as we will later see, this favors a wider operating range. For impellers, rather than using Cartesian coordinates, we have used cylindrical coordinates:

- Meridional coordinate (m) that increases from the inlet to the outlet of the impeller
- Span-wise coordinate (s) that increases from the impeller hub to the shroud
- Circumferential coordinate (u) that increases from the blade pressure side to the suction side of the next blade

c_u is the circumferential component of the gas velocity taken in an absolute reference frame at the inlet (1) and exit (2) (Fig. 3.66). At this point, one of the advantages of centrifugal compressors over axial compressors becomes apparent. In the axial compressor, the entire energy transfer has to come from the turning of the flow imposed by the blade ($c_{u2} - c_{u1}$), while the centrifugal compressor has added support from the centrifugal forces on the gas while flowing from the diameter at the impeller inlet ($u_1 = \pi\ D_i N$) to the higher diameter at the impeller exit ($u_2 = \pi\ D_{\text{tip}}\ N$).

However, like axial machines, Euler's law also applies to centrifugal machines:

$$P = \dot{m} \cdot \Delta h = \dot{m} \cdot (u_2 c_{u2} - u_1 c_{u1}) \tag{3.41}$$

Fig. 3.67 illustrates the velocity build-up in a meridional section of the compressor.

Fig. 3.68 helps to explain the aerodynamics of a vaneless diffusor. The absolute velocity at the impeller exit c_2 has a meridional and a circumferential component. Because the diffuser has no vanes, the circumferential component has to maintain its momentum, which enforces (c_u times r) = const. In other words, c_u is reduced when the flow moves from the diffuser inlet to the diffusor exit.

The meridional velocity component is reduced proportionally to the increase in the diffusor area when the radius is increased. The two effects together mean that the flow path becomes a logarithmic spiral. The more c_2 leans toward the tangential direction, the longer the flow path through the diffusor becomes. The shortest flow path would be for a situation where c_2 is entirely in the radial direction. In the absence of flow separation in the diffusor, the velocity energy difference between the inlet and exit from the diffusor is converted into pressure. Diffusors can also be built with vanes or a flow channel. The general idea is the same, i.e., the flow is decelerated and pressure is built up. Vanes and flow channels allow a more efficient conversion but are more sensitive to off-design conditions.

We have introduced the general principles for the operation of a centrifugal compressor. Now, we need to discuss how a compressor operates when the operating conditions change. We will first discuss the behavior of a compressor stage running at constant speed.

The impeller exit geometry (backsweep) determines the direction of the relative velocity w_2 at the impeller exit. The basic, "ideal" slope of head vs. flow is dictated by the kinematic flow relationship of the compressor, in particular the amount of impeller backsweep (Fig. 3.69).

Any increase in flow at constant speed causes a reduction of the circumferential component of the absolute exit velocity (c_{u2}), as seen in Fig. 3.69. The meridional velocity, c_m, is also impacted. As flow goes up, the meridional velocity

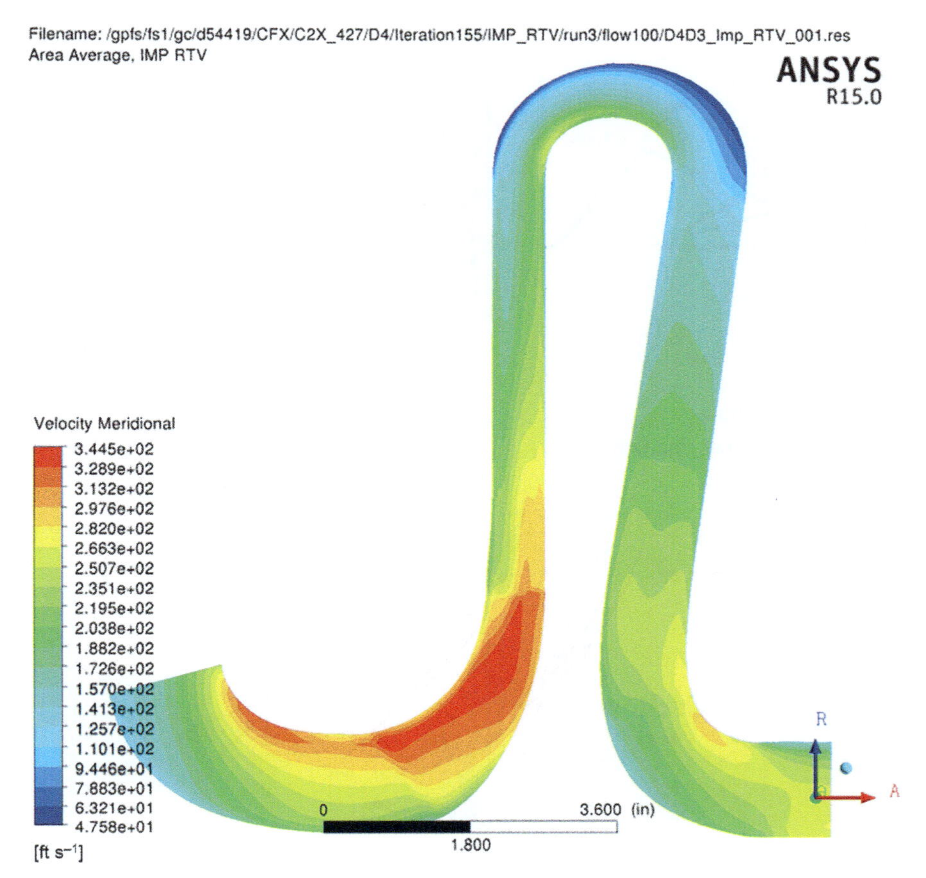

FIG. 3.67 Velocity in a meridional section of a centrifugal compressor stage.

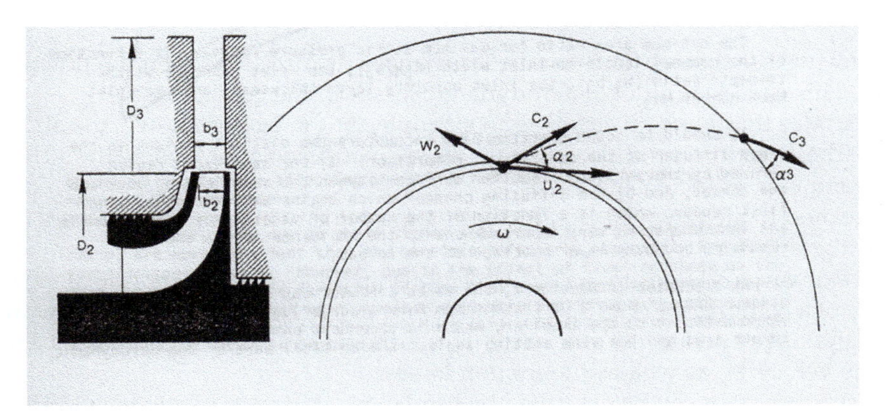

FIG. 3.68 The diffusor.

goes up. The compressor speed is fixed, thus u_1 and u_2 do o't change. We assume the inlet flow is strictly axial, so only the magnitude (not the direction of c_1) changes. This example is valid even for inlet flow that is not axial, as long as the flow direction stays. Further, the direction of the impeller exit flow in the relative frame, which is dictated by the blade geometry, will not change. We can see that the change in circumferential velocity ($cu_2 - cu_1$) is impacted by the change in inlet flow; it is reduced when the inlet flow is increased. It follows Euler's equation above that this causes a reduction in the head. This is shown in Fig. 3.69 on the right. The impeller geometry causes a reduction in the head with an increase in flow (and vice versa).

So far, we have not discussed any losses in the compressor. The losses are lowest at the design point of the compressor, and there are two types of losses to consider. First, there is the incidence loss as seen in Fig. 3.70. The impeller inlet, as well as other components (like the diffuser), are sized under the assumption that the gas comes from a certain

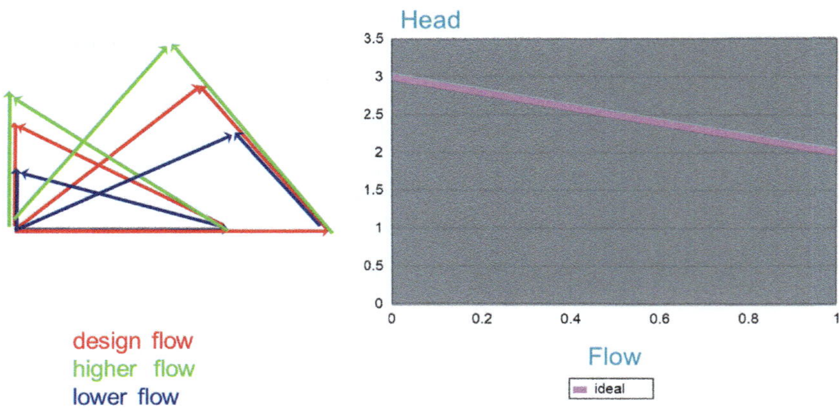

FIG. 3.69 Constant speed operation at changing flow.

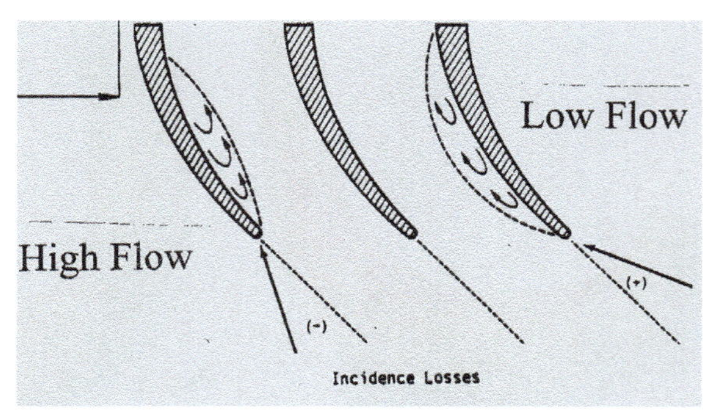

FIG. 3.70 Incidence loss at the impeller inlet.

direction. From Fig. 3.69, we see that w_1 for the impeller changes its direction when the flow is increased or reduced from the design point. This causes additional aerodynamic losses in the compressor.

Fig. 3.71 further illustrates this using an airfoil as an example. At the "design flow," the air follows the contours of the airfoil. If we change the direction of the incoming air, we see increasing zones where the airflow ceases to follow the contours of the airfoil and creates increasing losses.

The second type is related to the flow velocity. Increasing the velocity magnitude increases the friction losses in the impeller channels and the diffuser.

Adding the influence of various losses to the basic relationship developed earlier (Fig. 3.69) shapes the head-flow-efficiency characteristic of a compressor (Fig. 3.72). Whenever the flow deviates from the flow the stage was designed for, the components of the stage operate less efficiently. This is the reason for incidence losses. Furthermore, the higher the flow, the higher the velocities, as well as the friction losses.

Centrifugal compressor behavior can be described by head-flow-efficiency relationships. The basic relationship for a compressor at constant speed is shown in Fig. 3.72. The compressor shows a distinct relationship between head and flow. In the case of machines with backward bent impellers (the type generally used in upstream and midstream compression applications), the head of the compressor increases with reduced flow. Due to the increase in losses when the compressor is operated away from its design point, the curve eventually becomes horizontal and subsequently starts to drop again (Fig. 3.73). The curve section with a positive slope is usually not available for stable operation. When the flow is increased beyond the design flow, the losses also increase, as does the slope of the curve (sometimes to a vertical line).

A compressor, operated at a constant speed, is operated at its best efficiency point. If we reduce the flow through the compressor (for example, because the discharge pressure that the compressor has to overcome is increased), then the compressor efficiency will be gradually reduced. At a certain flow, a stall (probably in the form of a rotating stall) in one or more of the compressor components will occur. At further flow reduction, the compressor will eventually reach its stability limit and go into surge (Fig. 3.73).

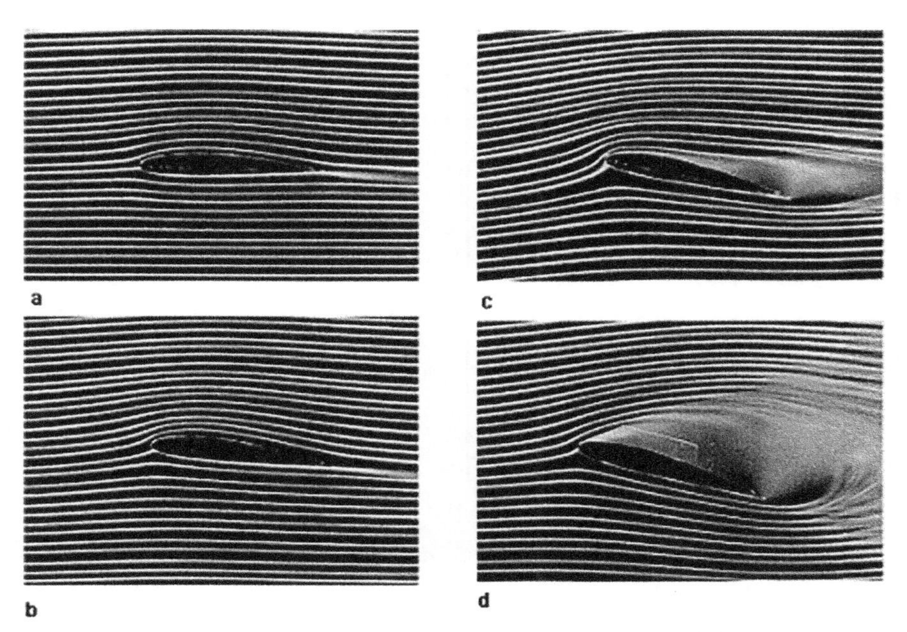

FIG. 3.71 Unseparated (A, B), partially separated (C), and fully separated (D) flow over an airfoil at increasing angle of attack [38].

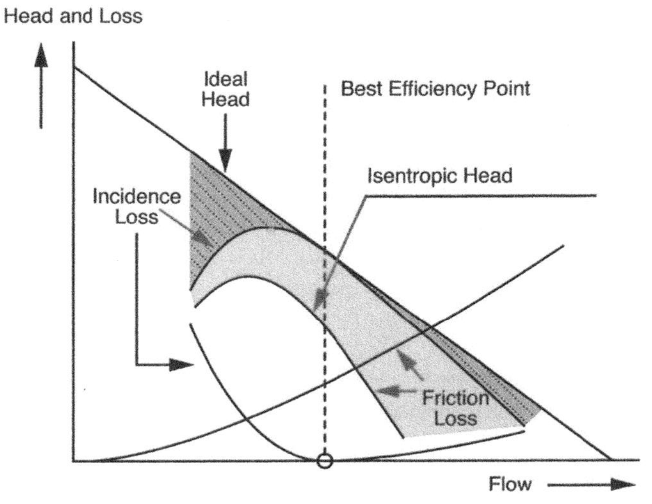

FIG. 3.72 Head versus flow relationship at constant speed: Subtracting the incidence loss and the friction loss from the ideal head yields the isentropic head vs. flow curve [35].

If, again starting from the best efficiency point, the flow is increased, then we also see a reduction in efficiency, accompanied by a reduction in head. Eventually, the head and efficiency will drop steeply until the compressor will not produce any head at all. This operating scenario is called a choke. (For practical applications, the compressor is usually considered to be in choke when the head falls below a certain percentage of the head at the best efficiency point.)

We also see that the resulting curve has a negative slope for the higher flow, but at some point reaches a maximum, followed by a positive slope. The horizontal slope marks the stability limit of the compressor, and operating it at lower flows than this point usually will lead to a surge.

If the flow through a compressor at constant speed is reduced, the losses in all aerodynamic components will increase, because their operating conditions will move away from the design point. Eventually, the flow in one of the aerodynamic components, usually in the diffuser or the impeller inlet, will separate (the last picture in Fig. 3.71 shows such a flow separation for an airfoil). It should be noted that stall usually appears in one stage of a compressor first. The separation can be stationary, or of a propagating (and therefore rotating) nature. Stall and surge are not

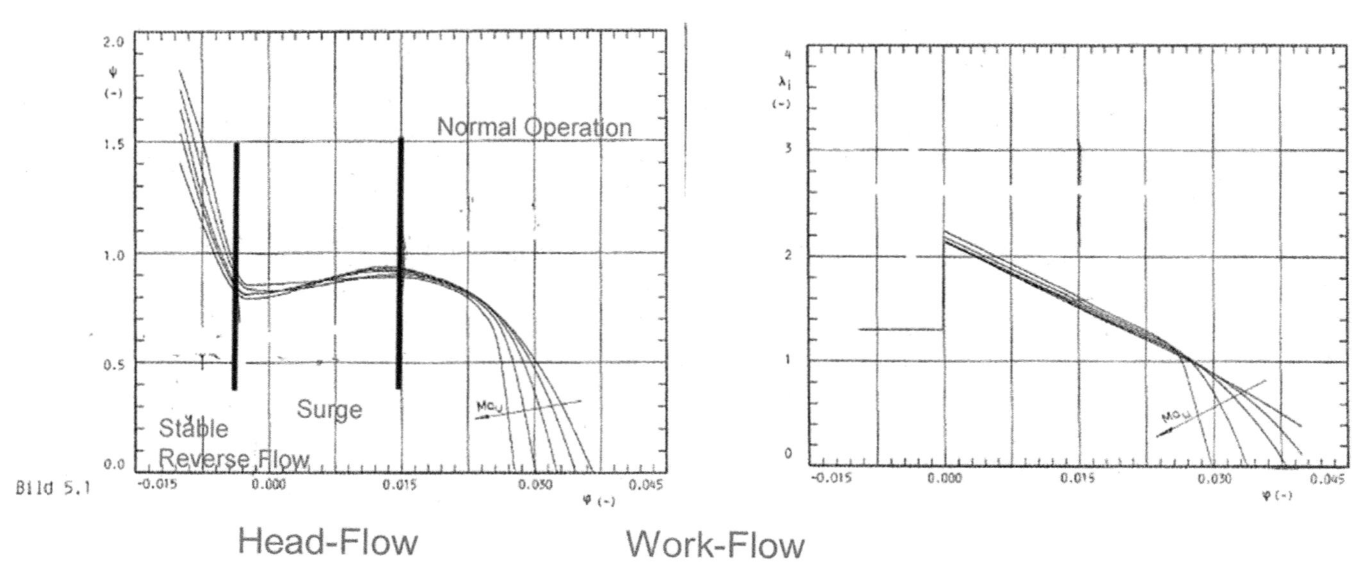

FIG. 3.73 Compressor map showing head vs. flow and work vs. flow including operation in surge [39].

directly related. If we reduce the flow at a constant speed, stall can appear before the compressor actually reaches its maximum head or before it actually surges.

Flow separation and stall in a vaneless diffuser means that all or parts of the flow will not exit the diffuser on its discharge end, but will form areas where the flow stagnates or reverses its direction back to the diffuser inlet (i.e., the impeller exit). This is due to either boundary layer separation or insufficient kinetic energy to overcome the diffuser pressure gradient.

Stall in the impeller inlet or a vaned diffuser is due to the incoming flow (relative to the rotating impeller) changing with the flow rate through the compressor. Therefore, a reduction in flow will lead to an increased mismatch between the direction of the incoming flow the impeller was designed for and the actual direction of the incoming flow. At some point, this mismatch becomes so significant that the flow through the impeller breaks down. Similarly, vanes in the diffuser will reduce the operating range of a stage compared to a vaneless diffuser.

Flow separation can take on the characteristics of a rotating stall. When the flow through the compressor stage is reduced, parts of the diffuser may experience flow separations. Rotating stall occurs if the regions of flow separation are not stationary but move in the direction of the rotating impeller (typically at 15%–30% of the impeller speed). Rotating stalls can often be detected by the increasing vibration signatures in the subsynchronous region but with distinct frequencies. This is different from the ubiquitous increase in flow noise when the compressor operates close to stall, in stationary stall, or in choke, as this noise generally has no distinct frequencies.

Another type of noise is generated by the interaction of the impellers with a vaned diffuser. This noise is not the result of aerodynamic off-design conditions, but rather due to the interaction between the impeller and the diffusor. The onset of the stall does not necessarily constitute an operating limit of the compressor. In fact, in many cases, the flow can be reduced further before the actual stability limit is reached, and a surge may occur.

At flows lower than the flow at the stability limit, practical operation of the compressor is not possible. The stability limit is approximately the flow at which the isentropic head-flow characteristic has a horizontal tangent (Fig. 3.73). At flows to the left of the stability limit, the compressor cannot produce the same head as at the stability limit. It is therefore no longer able to overcome the pressure differential between the suction and discharge side. Because the gas volume upstream (at discharge pressure) is now at a higher pressure than the compressor can achieve, the gas will follow its natural tendency to flow from the higher to the lower pressure. The flow through the compressor is reversed. Due to the flow reversal, the system pressure at the discharge side will be reduced over time, and eventually, the compressor will be able to overcome the pressure on the discharge side again. If no corrective action is taken, the compressor will again operate to the left of the stability limit, and the above-described cycle is repeated. The compressor is in surge. The observer will detect strong oscillations of pressure and flow in the compression system. It must be noted that the violence and onset of the surge are a function of the interaction between the compressor and the piping system.

At high flow, the head and efficiency will drop steeply until the compressor will not produce any head at all. This operating scenario is called a choke. However, for practical purposes, the compressor is usually considered to be in a choke when the head falls below a certain percentage of the head at its best efficiency point. Some compressor manufacturers do not allow the operation of their machines in a deep choke. In these cases, the compressor map has a distinct high-flow limit for each speed line.

The efficiency starts to drop off at higher flows, because a higher flow causes higher internal velocities, and thus higher friction losses. The head reduction is a result of both the increased losses and the basic kinematic relationships in a centrifugal compressor. Even without any losses, a compressor with backward bent blades (as they are used in virtually every industrial centrifugal compressor) will experience a reduction in the head with the increased flow (Fig. 3.69). "Choke" and "stonewall" are different terms for the same phenomenon. There are two distinctly different behaviors in choke: (1) Compressors at low Mach numbers, and in particular single and two-stage machines show a gradual decline in the head. This is mainly due to the increase in losses in the machine. (2) Other machines, especially multistage machines and machines at higher Mach numbers, show an almost vertical drop in the head at a certain flow. This is due to a true choke event, where at some component (often the inlet of an impeller), the flow in the narrowest flow path reaches the speed of sound and thus prevents any increase in flow, regardless of how low the discharge pressure is set. Fig. 3.74 shows both of these behaviors. At low Mach numbers, increased losses show a gradual reduction of head and efficiency, while at a higher machine Mach number, true choke causes an almost vertical drop in head and efficiency.

Once the head-flow efficiency characteristics of a compressor are known for different speeds, a map can be created (Figs. 3.75 and 3.76) that allows one to determine the compressor efficiency, speed, and operating range based on various defined operating conditions.

Centrifugal compressor behavior can be described by their head-flow-efficiency relationships as explained previously. The basic relationship, for a compressor at a constant speed, is shown in Fig. 3.72. The compressor shows a distinct relationship between head and flow. In the case of machines with backward bent impellers (generally used in upstream and midstream compression applications), the head of the compressor increases with reduced flow. Due to the increase in losses when the compressor is operated away from its design point, the curve eventually becomes horizontal and subsequently starts to drop again. The curve section with a positive slope is usually not available for stable operation. When the flow is increased beyond the design flow, the losses also increase, as does the slope of the curve (sometimes to a vertical line).

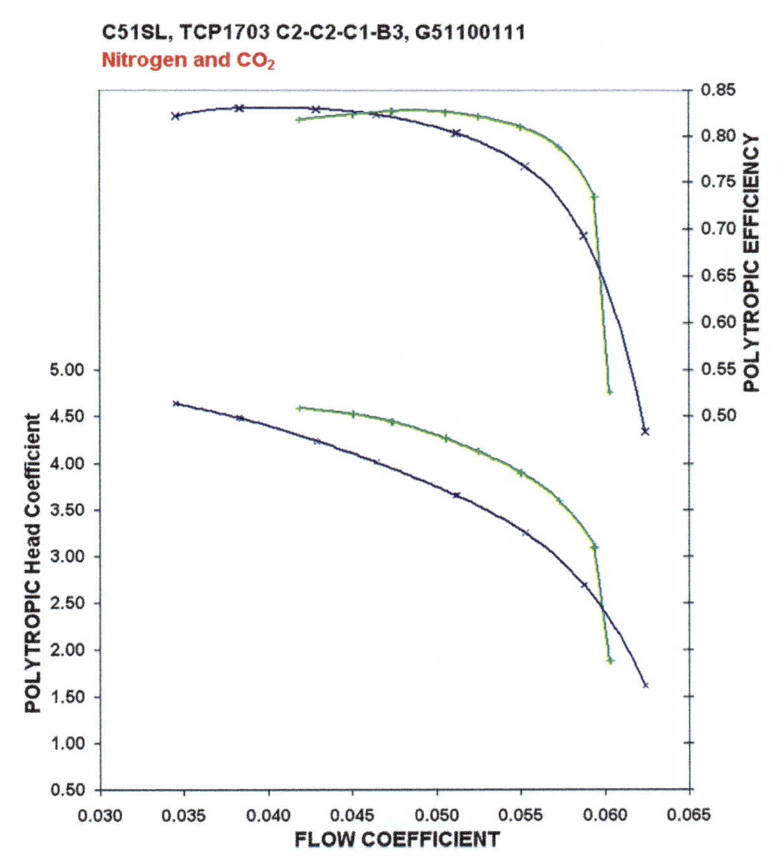

FIG. 3.74 Stage map for $M_n = 0.56$ (*blue*) and $M_n = 0.76$ (*green*).

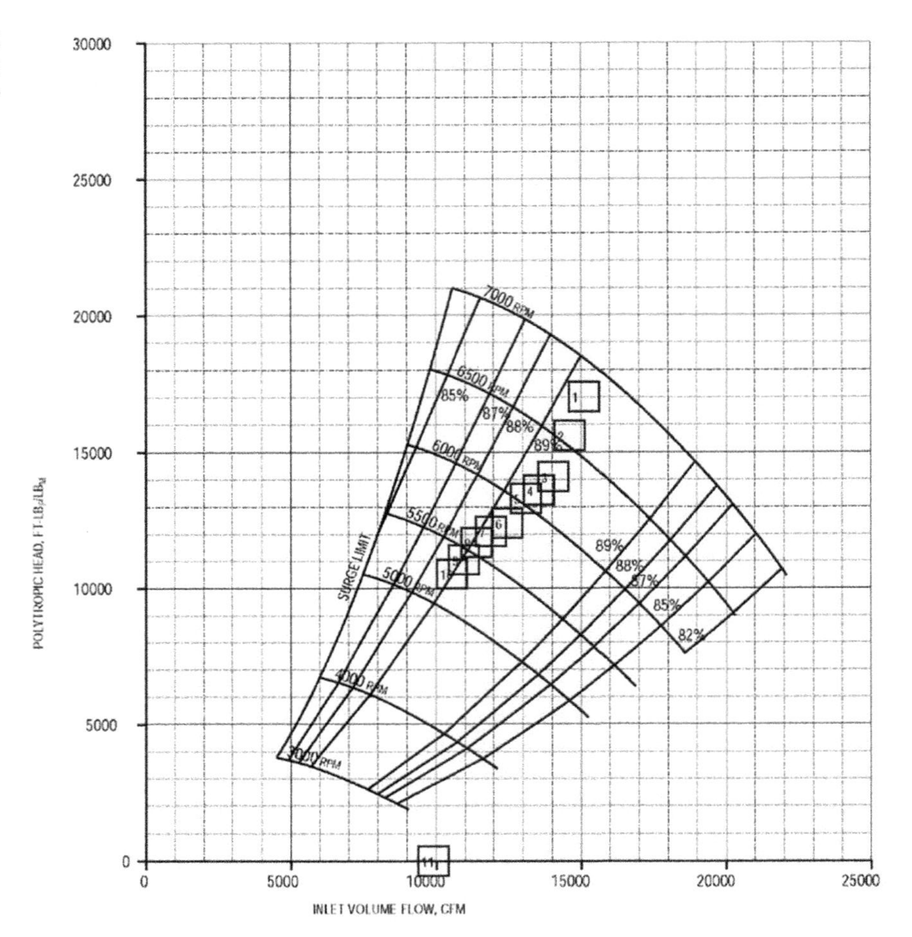

FIG. 3.75 Performance map for a variable speed centrifugal compressor with a number of specified operating conditions typical for a pipeline application [35].

Applying different control mechanisms such as speed variation, adjustable inlet vanes, and adjustable diffuser vanes allow the compressor to operate on a family of curves as seen in Fig. 3.76. Additionally, compressors can be controlled by suction or discharge throttling (Fig. 3.76) and recycling (Fig. 3.77).

Fig. 3.78 indicates the effectiveness and efficiency of different control methods. Of particular importance in upstream and midstream applications is a compressor that can be operated at varying speeds, since this is the most effective and efficient control method. Using a throttle, recycling, or adjustable vanes are very effective in reducing the volumetric flow, but they are not very efficient, because the power consumption is not reduced at the same rate as a speed-controlled machine.

If speed control is not available, the compressor can be equipped with a suction throttle or variable guide vanes. The latter, if available, is rather effective in front of each impeller, but the mechanical complexity proves usually to be prohibitive in pipeline applications. The former is a mechanically simple means of control, but it has a detrimental effect on the overall efficiency.

Within the control system, subsystems protect the compressor as well as its driver. In general, process control will be enabled as long as the compressor and its driver stay within acceptable, predefined boundaries. For the compressor, these boundaries may include maximum and minimum operating speed, a stability limit (aka surge line), choke or overload (on some machines), and pressure, temperature, and torque limits.

The operating point of the compressor is determined by the system characteristic within which it operates (Fig. 3.79). The system imposes a suction and discharge pressure, and the compressor, based on either the defined power input (for gas turbines) or speed input (for electric motors), reacts to that with a certain flow. Control parameters are process focused, i.e., one controls suction pressure, discharge, or flow by adjusting the power input or speed setting.

The essential components of a centrifugal compressor (Figs. 3.80 and 3.81) that accomplish the task of compressing gas are discussed next. The gas enters the compressor at the suction flange and is then directed axially into the impeller with the help of inlet guide vanes. After each impeller, the flow enters a diffuser, followed by a crossover bend. The subsequent turn vane conditions the flow to enter the next impeller in an approximately axial direction.

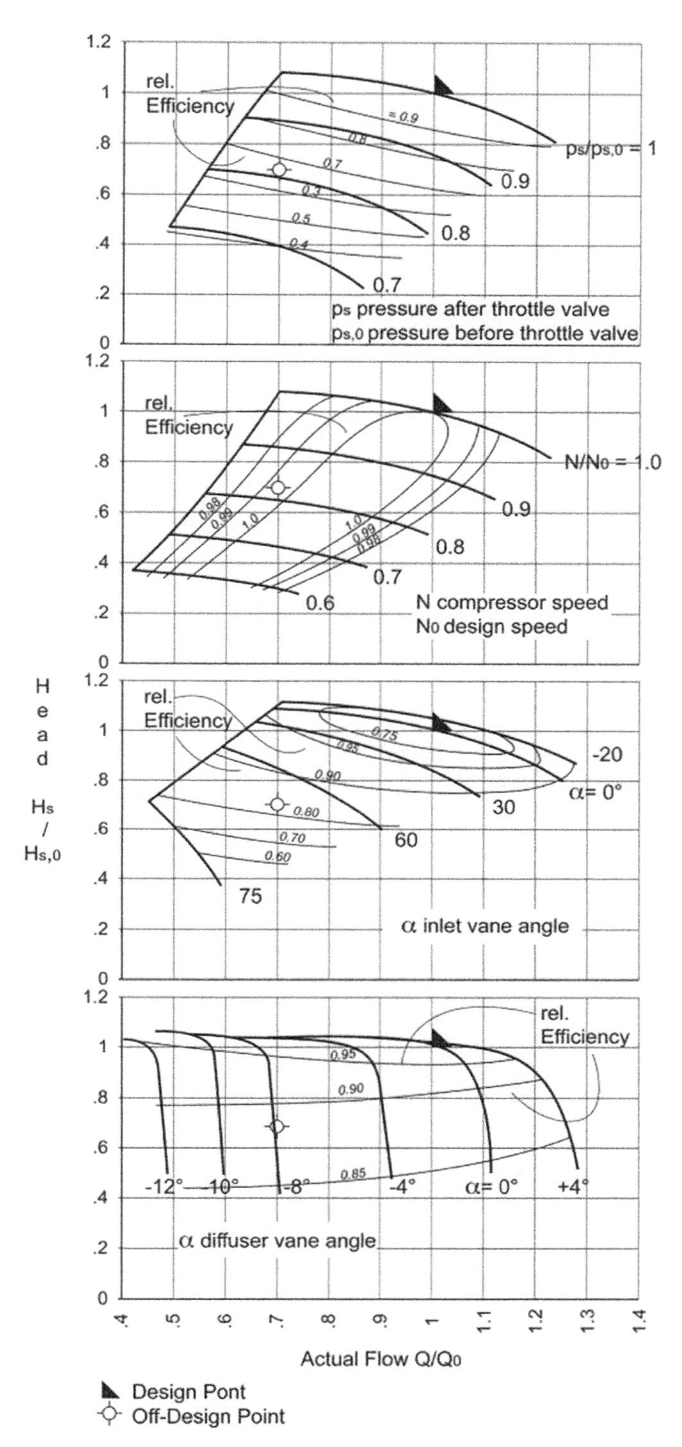

FIG. 3.76 Control methods for centrifugal compressors: Throttling, variable speed, and adjustable guide vanes [40].

After the last diffuser, the gas is gathered using a volute and leaves the compressor now at a higher pressure at the discharge flange.

The only rotating parts of the compressor are the shaft with the impellers, the balance piston, the thrust bearing collar, and the rotating portion of the dry gas seal.

Compressors are typically composed of a module (bundle), casing, end caps, and bearing and seal assemblies. The bundle is composed of the rotor, stators, inlet housing, and discharge volute. The bundle holds all the essential aerodynamic components needed to perform the required duty. The casing is a pressure vessel comprised of a vertically or

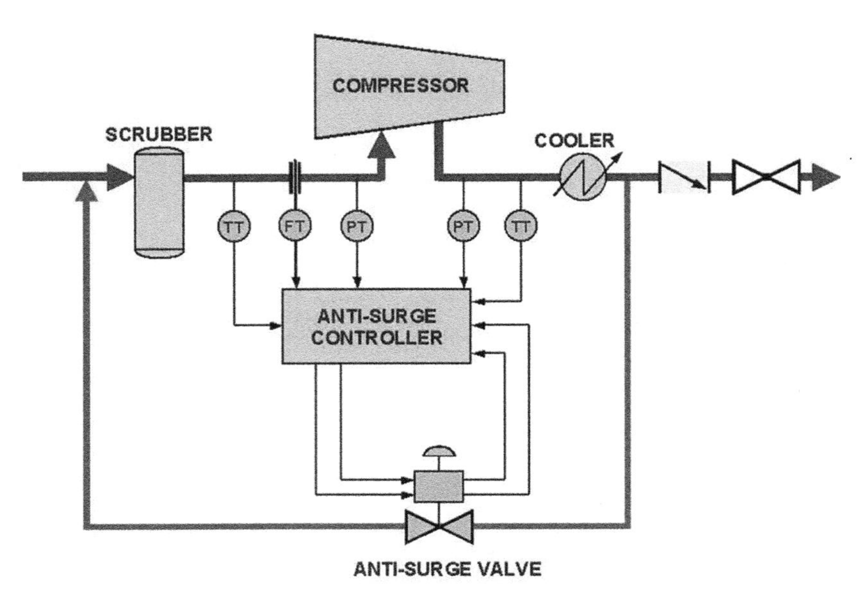

FIG. 3.77 Recycle system.

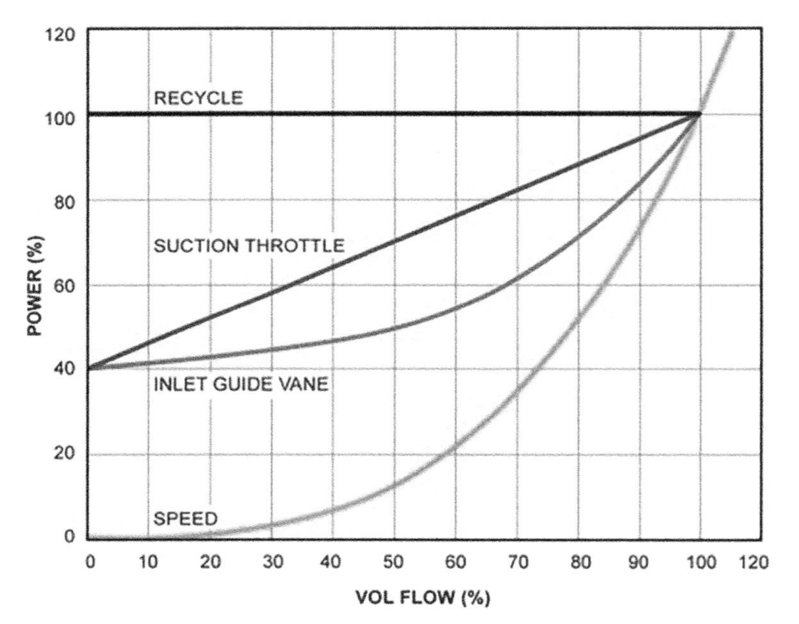

FIG. 3.78 Power consumption for different control methods.

horizontally split barrel that contains the bundle. Suction and discharge endcaps contain the bearings and seal assembles along with service ports for oil and gas. Fig. 3.80 shows a typical centrifugal compressor cross section. Fig. 3.81 shows the major components of the compressor.

Depending on the arrangement of the journal bearings, we can distinguish beam-style compressors and overhung compressors (Fig. 3.82). In a beam-style compressor, all impellers sit between the bearings, while in an overhung design, the impeller sits outside the bearing span. Designs with overhung impellers are typically limited to a single impeller (two impellers if an impeller sits on either end of the shaft), but allow the gas to flow axially into the compressor.

Integrally geared centrifugal compressors

Integrally geared compressors are centrifugal compressors that have their speed-increasing gear integrated into the compressor body. An integrally geared compressor (IGC) is comprised of one or more compressor stages attached to

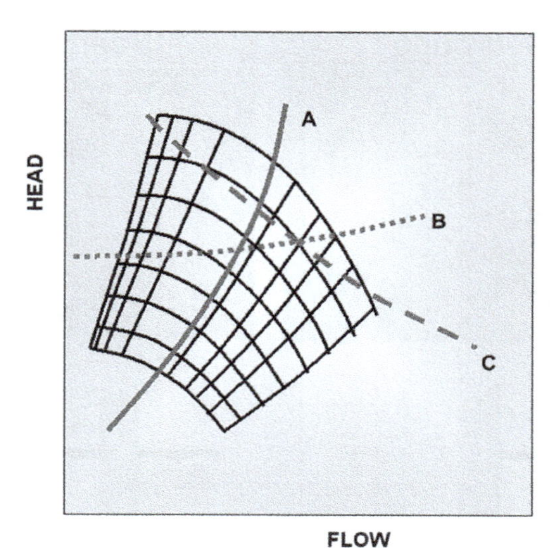

FIG. 3.79 System characteristics and compressor map [35]. Curve *A* is typical for pipeline operations, curve *B* can be seen in plant air compression, curve *C* would be the result of pumping gas into a fixed volume storage cavity.

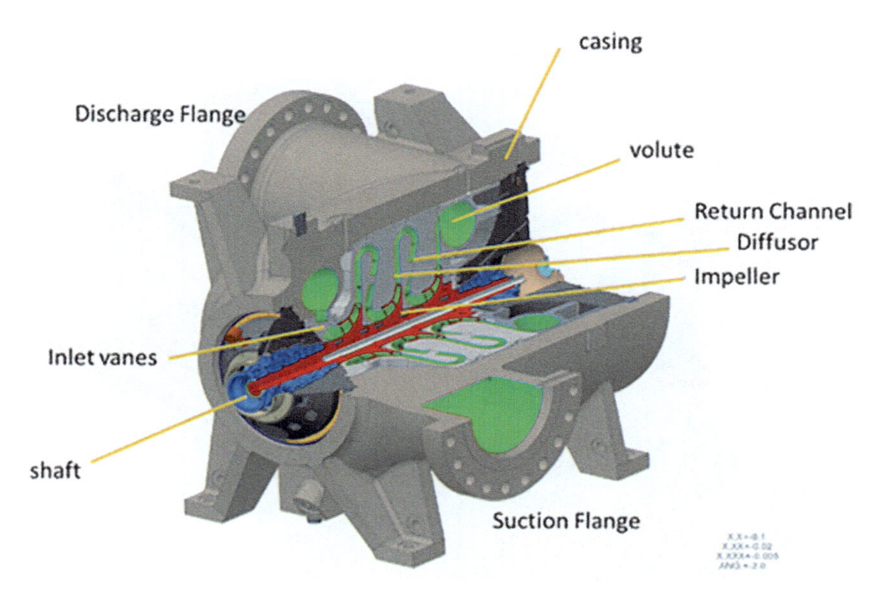

FIG. 3.80 Compressor cross section showing the aerodynamic components [35]. *Courtesy of Solar Turbines.*

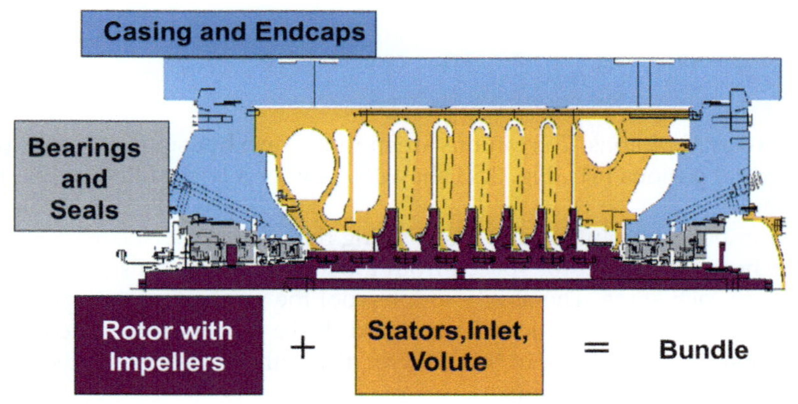

FIG. 3.81 Major mechanical components for a barrel-type compressor. *Courtesy of Solar Turbines.*

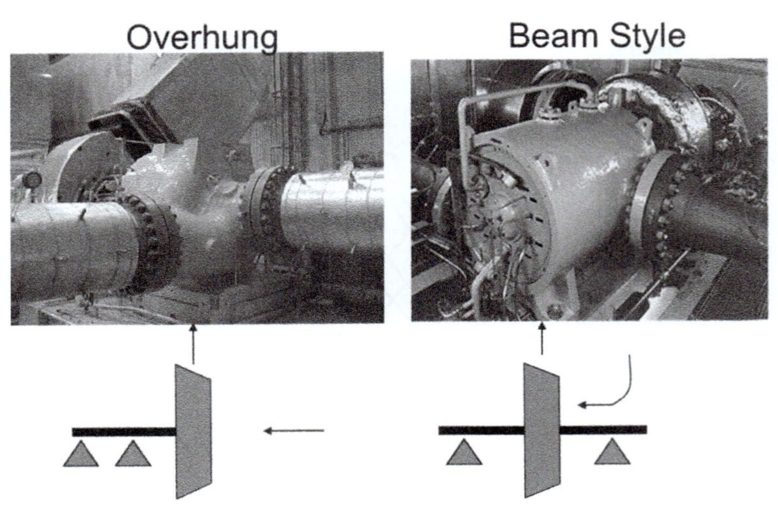

FIG. 3.82 Overhung and beam style design. *Courtesy of Solar Turbines.*

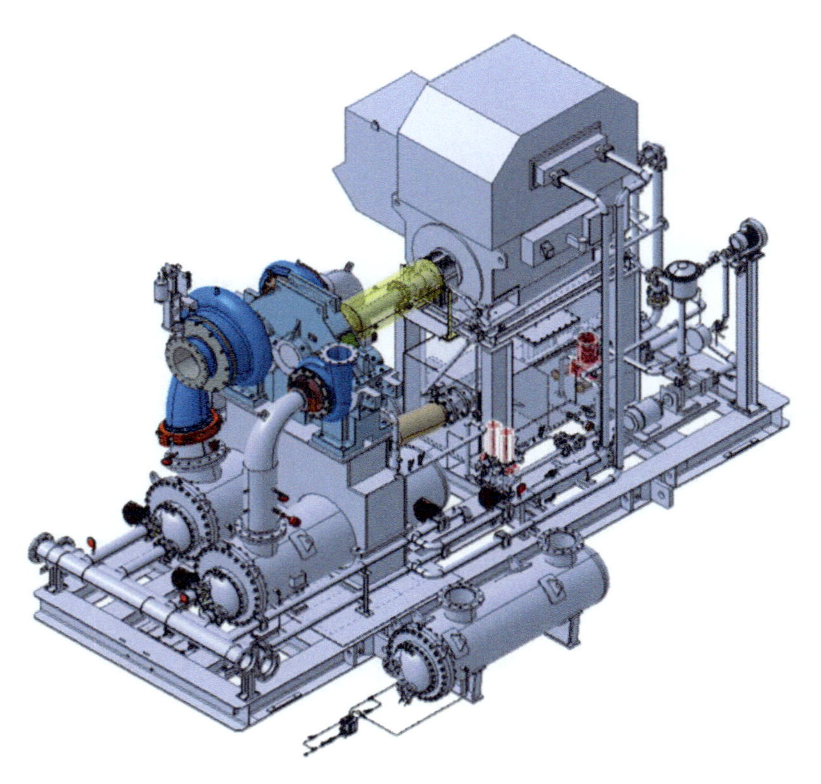

FIG. 3.83 Integrally geared compressor: Arrangement with electric motor drive [41].

the end of one or more high-speed pinions. The pinions are mounted in a housing that contains a large, low-speed bull gear that drives the individual pinions [41]. Fig. 3.83 shows a view of a typical IGC, while Figs. 3.84 and 3.85 show assembly and components.

Srinivasan et al. [42] describe the evolution of IGCs entering the process gas industry with several relevant examples. The author describes how multiple rotor speeds from the individual pinions are applied to achieve an aerodynamically efficient design for each stage. The ability to intercool the gas between every stage of compression using intercoolers benefits the overall thermodynamic efficiencies. Multiple processes can be combined into one unit reducing overall costs and real estate requirements. In this chapter, the authors present the technical challenges associated with designing, building, and testing these units.

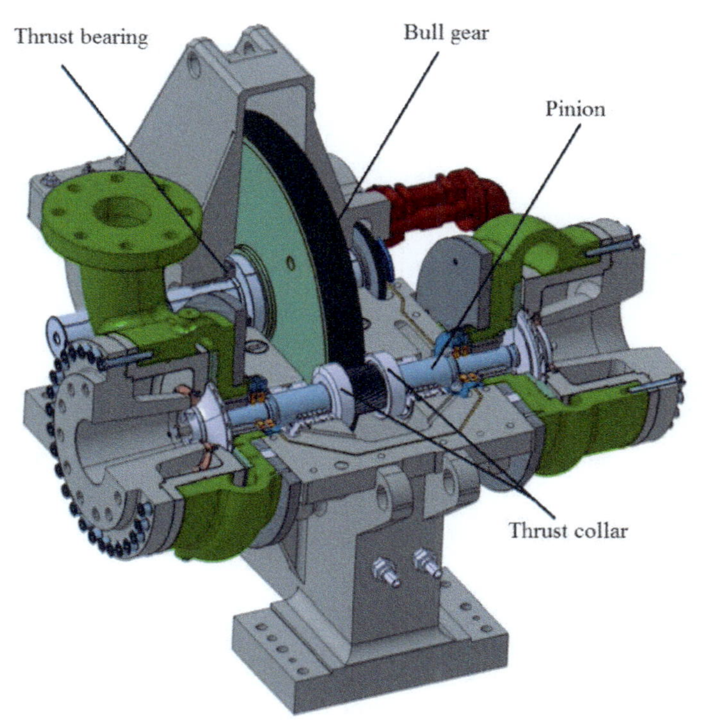

FIG. 3.84 Assembly showing pinion, bull gear, pinion, thrust collar, and thrust bearing on bull gear [41].

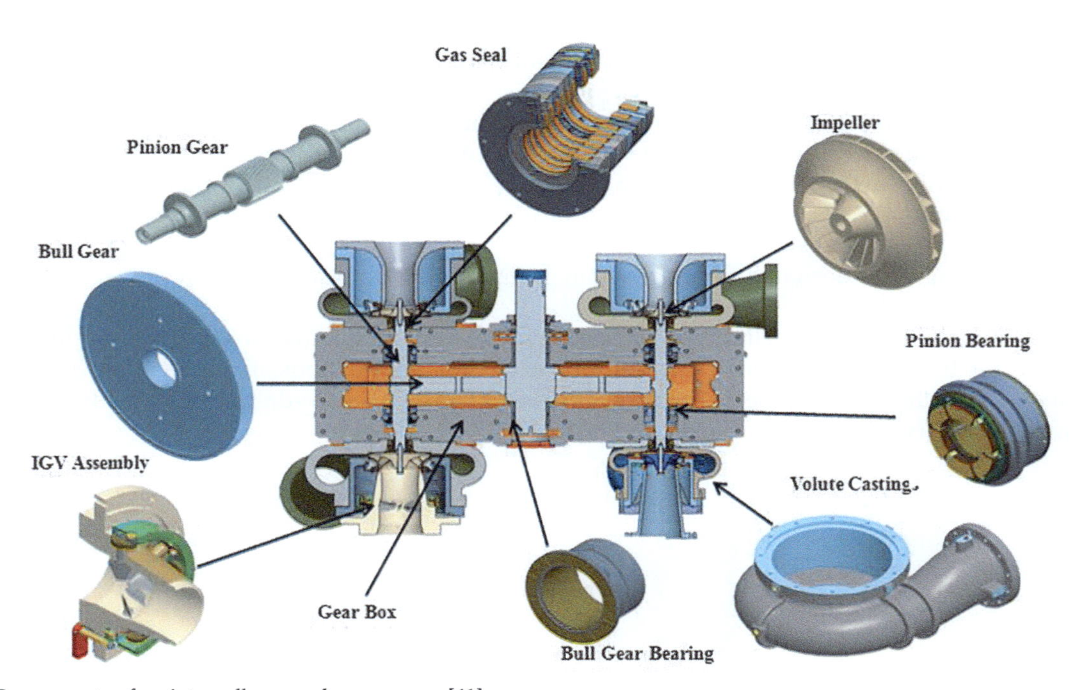

FIG. 3.85 Components of an integrally geared compressor [41].

Since integrally geared type compressors can employ adjustable inlet guide vanes and adjustable diffuser vanes (if necessary for every stage), they can achieve a good operating range even with constant speed drives (Fig. 3.86), such as constant speed electric motors. Nevertheless, there are examples of integral gear-type compressors driven by variable speed, two-shaft gas turbines, expanders, or steam turbines.

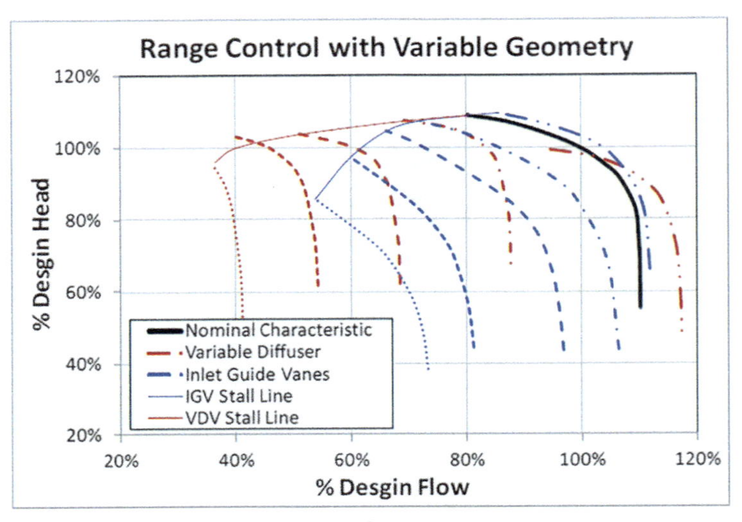

FIG. 3.86 Adjustable inlet vanes versus adjustable diffuser vanes [41].

Reciprocating compressors

Reciprocating compressors are positive displacement compressors. In other words, compression is a result of trapping a gas in a space with a certain volume and subsequently reducing the volume. In a reciprocating compressor, the volume reduction is created by a piston making a reciprocating motion inside a cylinder. Gas change is affected by valves (Fig. 3.87). The piston is connected to the crankshaft via a piston rod (Fig. 3.88). The compressor takes gas in via the suction valve during the downward motion of the piston. Valves are essentially check valves and are operated by the pressure difference on either side of the valve. The suction valve opens when the pressure in the cylinder drops below the suction line pressure. Once the piston moves upward again, the pressure in the cylinder exceeds the suction line pressure, and the suction valve closes. The piston continues to travel upwards, thus reducing the volume and increasing the pressure and temperature of the trapped gas. Once the gas pressure exceeds the discharge line pressure, the discharge valve opens and gas flows to the discharge line. After passing the top dead center, the piston will move downward again, and the discharge valve will close once the cylinder pressure becomes lower than the discharge line pressure. After further travel, when the cylinder pressure becomes lower than the suction pressure, the suction valve will open again, and gas will flow in from the suction line. Then, a new compression cycle begins (Fig. 3.89).

Many compressor stages are designed as double-acting stages. Thus, both sides of the piston contribute to the compression effort (Fig. 3.88).

Since the flow into and out of the cylinder is not continuous, pressure fluctuations are induced in the suction and discharge lines. These have to be dampened using a number of different methods, including orifices, damping bottles with choke tubes, or side branch absorbers. The pressure loss of these pulsation-damping methods has to be included in the performance calculations of the compressor.

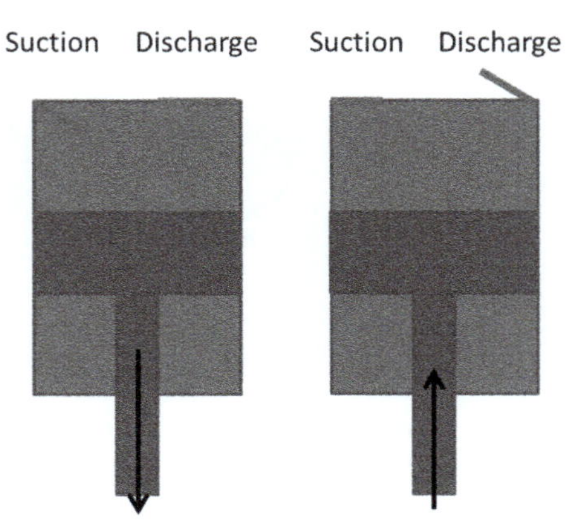

FIG. 3.87 Reciprocating compressor schematic.

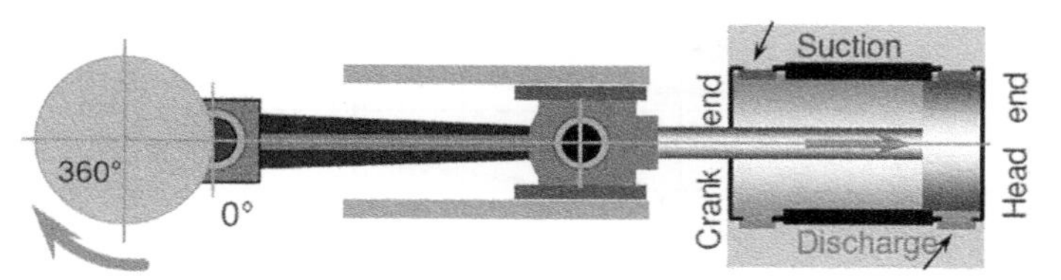

FIG. 3.88 Single-stage, double-acting compressor [43].

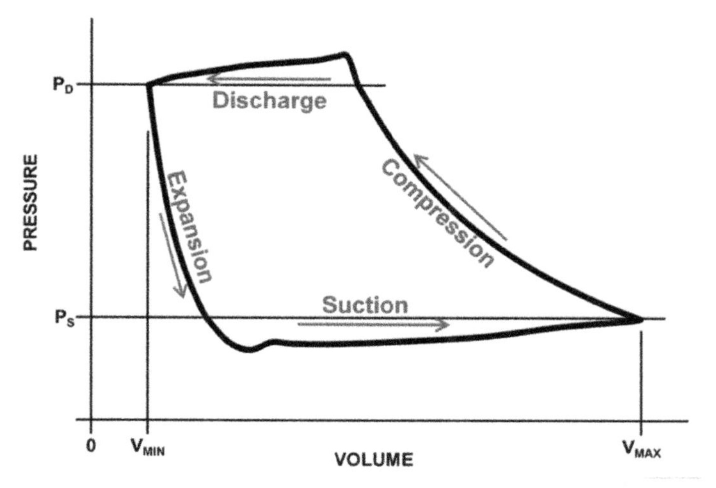

FIG. 3.89 Compression cycle for a reciprocating compressor in a p-V (pressure-volume) diagram [44].

For higher pressure ratios, compressors are built with multiple stages in series, usually with intercoolers in between. This allows for high pressure ratios while limiting the discharge temperature. High-flow, low-ratio compressors can also be designed with multiple stages in parallel to counter inherent flow limitations.

The major source of losses comes from the pressure loss in suction and discharge valves (Fig. 3.90). These losses depend on the pressure, gas density, and flow velocities, as well as the valve opening characteristics. Accordingly, the relative impact of valve losses tends to get reduced for higher pressure ratios and lighter gases (Fig. 3.91).

Commonly used control methods for reciprocating compressors are variable speed, variable clearance, timed valve closing, cylinder end de-activation, throttling, and recycling.

- Increasing the speed simply increases the compression cycles per minute, thus impacting the flow. However, changing the speed will change pulsation frequencies, thus making the damping of these frequencies more difficult. Often, certain speeds have to be avoided. Also, higher speeds tend to reduce compressor efficiency due to dynamic effects. For gas engine drivers, changing the driver speed may impact efficiency and emissions characteristics.

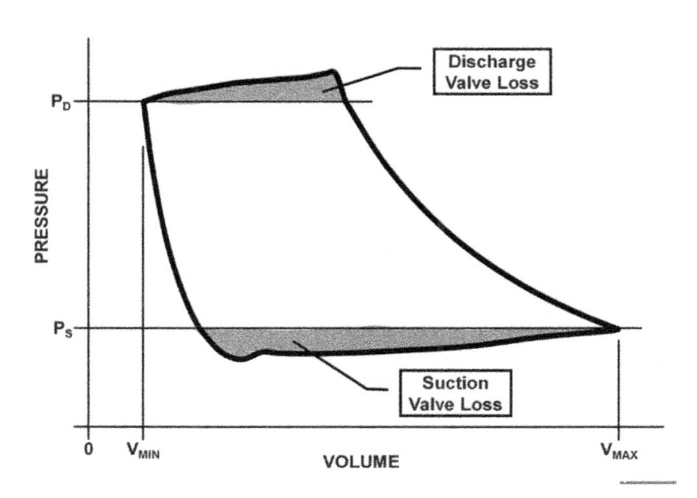

FIG. 3.90 Valve losses compressor [44].

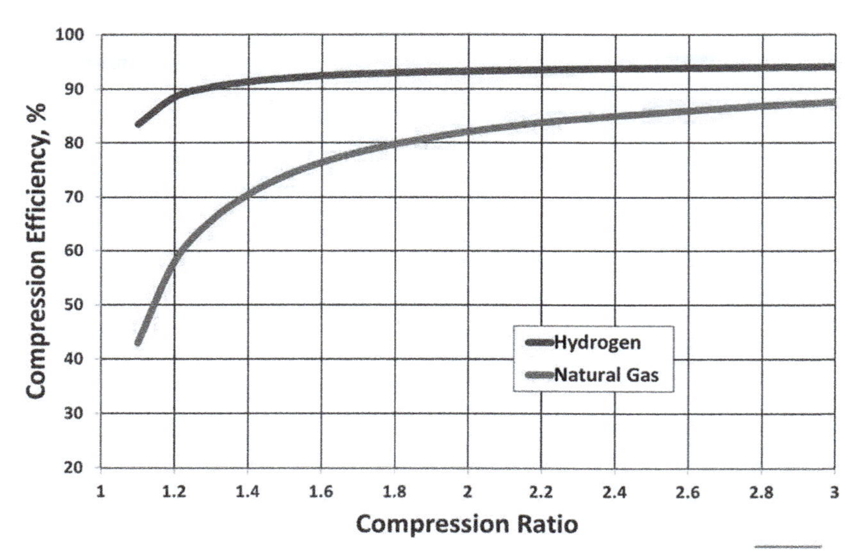

FIG. 3.91 Compression efficiency [44].

- Variable clearance control (either with fixed pockets or variable pockets) allows for the modification of the cylinder volume at the top dead center. The higher this clearance, the lower the volumetric efficiency of the cylinder. Volumetric efficiency denotes the ratio between piston displacement and actual capacity. Since different volumetric efficiencies barely affect the compression efficiency, modifying the clearance of a compressor allows for an elegant way to change the compressor flow capacity. The impact of a certain change in clearance on the compressor flow depends on the stage pressure ratio. Thus, clearance control is most effective for high-pressure ratio stages.
- Timed valve closing allows for a delay in the opening of valves, which can impact the pulsation frequencies of the system, but allow for control of the flow capacity.
- Cylinder end deactivation involves keeping the valves for one or multiple cylinder ends open, thus essentially eliminating the flow contribution of that end. This will impact the pulsation characteristics of the compressors, but allow for fast, significant changes in capacity.
- Suction throttling impacts capacity, but can actually increase the load of the compressor. However, it does not impact pulsations and allows for smooth changes in capacity. This method will reduce the system efficiency.
- Recycling, i.e., bringing all or part of the discharge flow back to suction via a controlled recycle valve, allows for capacity control, but it does not reduce load. Continuous recycling may require cooling of the recycled gas.

Control concepts for reciprocating compressors have to take into account that the torque or power the compressor absorbs has to be lower than the torque or power from the driver to avoid causing the driver to stall. In other words, control devices have to be used both for flow (capacity) control and for load control. Also to be considered is the impact of the line pulsations that are induced by the compressor on the compressor performance itself (Fig. 3.92).

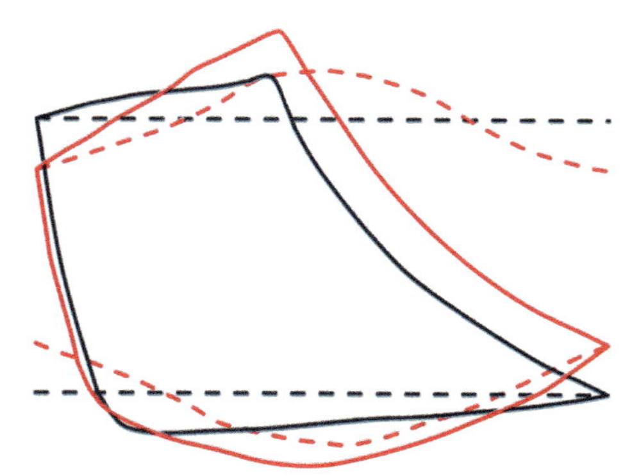

FIG. 3.92 Pressure volume diagram. *Black broken lines* show the situation for constant line pressures, and the *red broken lines* for fluctuating line pressures. The *solid lines* show the respective impact on the *p-V* diagram [44].

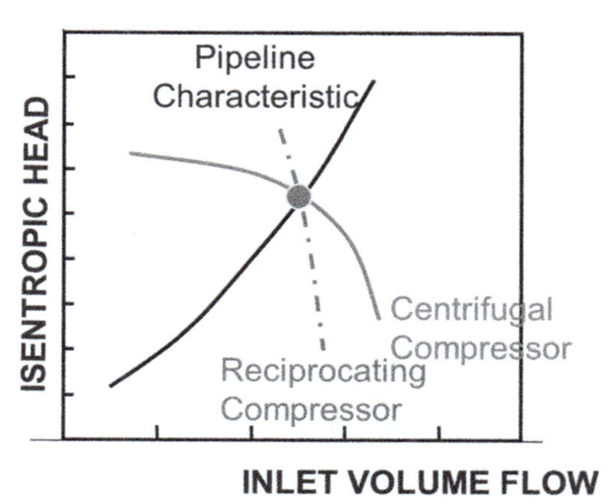

FIG. 3.93 Characteristics of a reciprocating compressor at constant speed and a centrifugal compressor at constant speed compared to a typical load characteristic (for example a pipeline).

FIG. 3.94 Multistage reciprocating compressor.

Without additional control measures, a reciprocation compressor stage shows very little change in actual suction flow, as opposed to a centrifugal compressor without any additional controls (as described in an earlier part of this chapter), where changes in pressure ratio lead to a distinct change in flow (Fig. 3.93).

The general arrangement of a multi-stage reciprocating compressor with pulsation bottles is shown in Fig. 3.94.

Screw compressors

Screw compressors are another type of positive displacement compressor. Rather than a reciprocating piston, the compression task is solved by two rotors in a casing, where the rotors have intermeshing helical lobes and rotate against each other with tight clearances between the rotors and between the rotors and the casing [17]. While rotating, the lobes and casing form compression chambers that steadily decrease their volume as the rotors turn. There are two general types of screw compressors: oil-flooded and dry screws. Dry screw compressors [a synchronizing gear (timing gear)] that synchronize the two rotors, thus avoiding contact (Fig. 3.95). Oil-flooded screw compressors, also called oil-injected or wet screw compressors, have oil continuously injected into the rotor chamber (Fig. 3.96). This oil is discharged with the compressed gas and has to be separated from the gas stream downstream of the compressor.

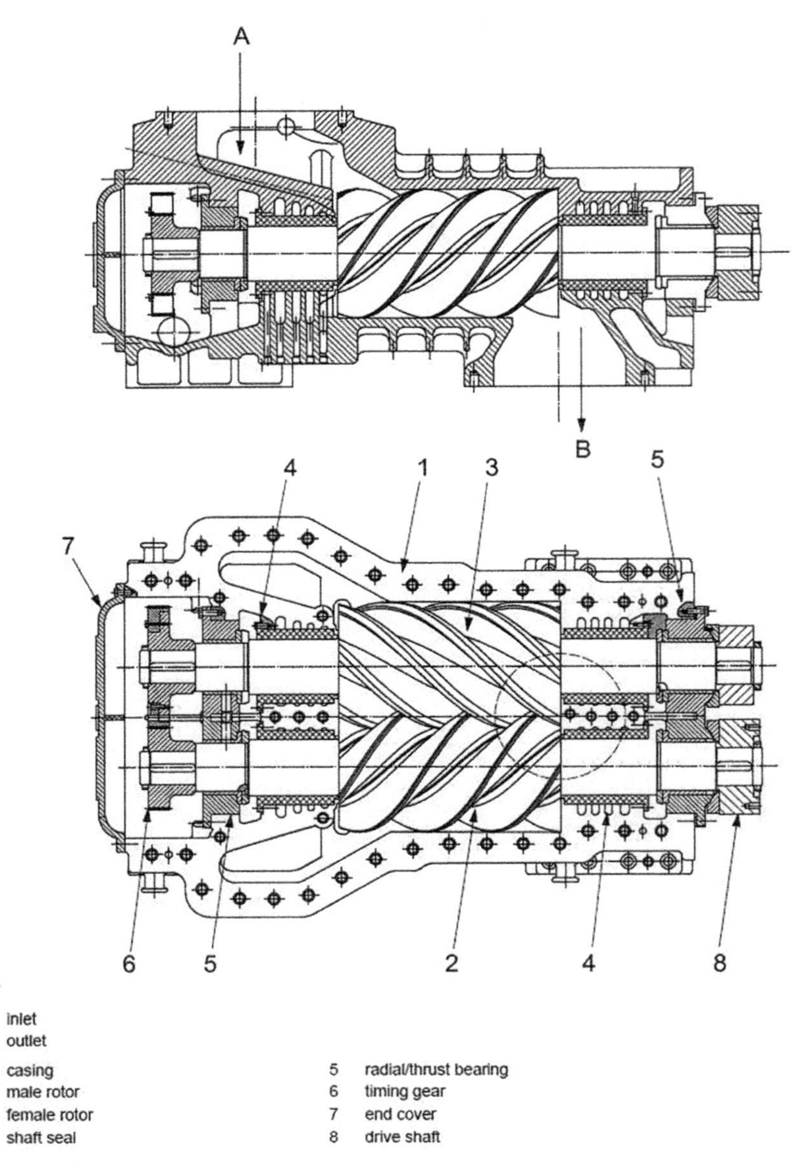

FIG. 3.95 Dry screw compressor [17].

The oil lubricates the male and female rotor, removes some of the compression heat, fills internal clearances to improve the volumetric efficiency, and flushes away contaminants entering the machine. Since the rotors are lubricated, no synchronizing gear is necessary, and only one of the rotors is connected to the driver.

Capacity control is usually via speed control; elsewhere, recycling can be used. Oil-flooded screw compressors can additionally use a sliding valve that can reduce the effective rotor length, thus reducing the volume reduction.

Diaphragm compressors

Diaphragm compressors are hermetically sealed, positive displacement compressors. They use two systems: a gas compression system and a hydraulic system (Fig. 3.97).

Rather than compressing the gas directly with a piston, a metallic diaphragm is used to compress the gas. This diaphragm is brought into oscillating motion by hydraulic fluid, which is in turn moved by a piston that is driven via a piston rod and crankshaft. A housing with specially machined contours forming the gas compression chamber clamps

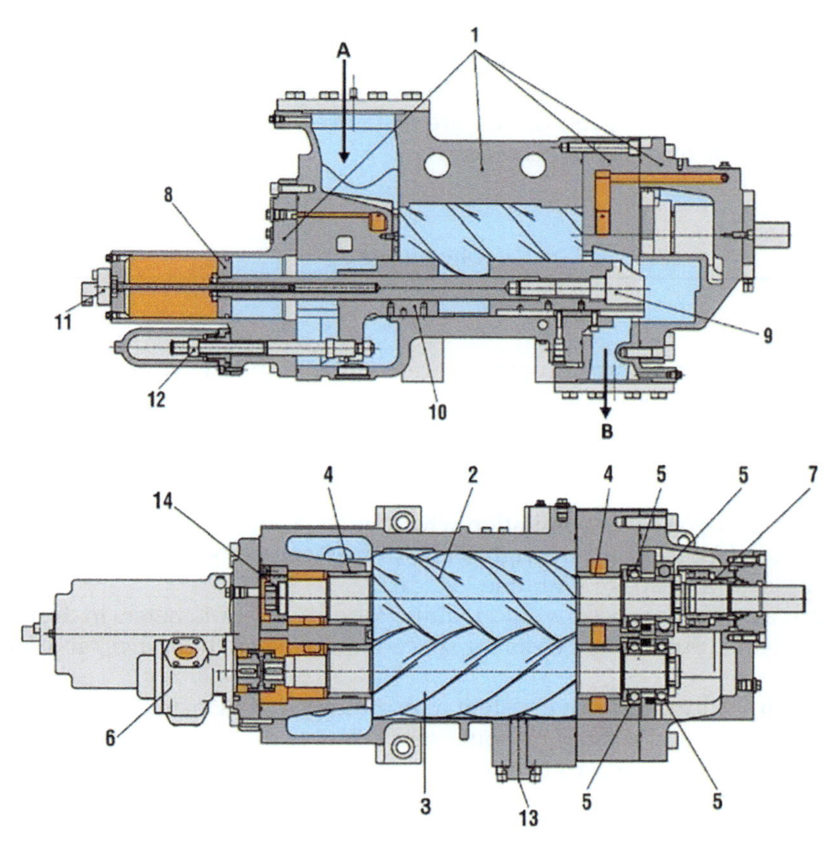

Key

A Inlet
B Discharge

1 Casing
2 Male (Main) Rotor
3 Female (Secondary) Rotor
4 Radial Bearings
5 Axial (Thrust) Bearings
6 Oil Pump (Shaft-Driven, Optional)
7 Shaft Seal
8 Capacity Slide Valve Piston
9 Capacity Slide Valve
10 v_i Slide (For Variable v_i, Optional)
11 Capacity Slide Valve Position Sensor
12 v_i Adjustment Screw (For Variable v_i, Optional)
13 Oil Injection Port
14 Thrust Balancing Piston

FIG. 3.96 Oil-flooded screw compressor [17].

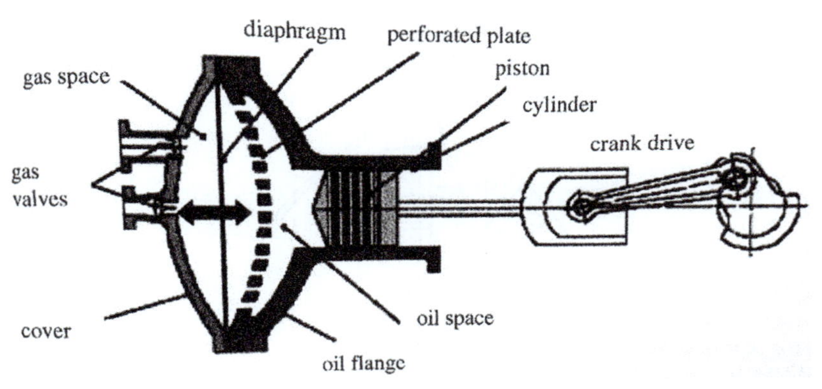

FIG. 3.97 Diaphragm compressor [45].

the diaphragm set. No sliding components on the gas side require lubrication, thus contamination with lube oil is avoided. There is no gas leakage along the piston and rod packing found in conventional reciprocating compressors. Diaphragm compressors are used in low capacity, high-pressure applications or where leak-free and lubrication-free designs are required. For hydrogen applications, they are often used at hydrogen production facilities for cylinder or tube trailer filling applications or for refueling stations.

Gearboxes

Introduction

Gearboxes in centrifugal pumps or compressor trains are essential components of the machinery used to transport a fluid through pipelines. Their purpose is to adapt the driver's speed to the optimum speed of the driven equipment. The optimum speed and use of a gearbox are determined by considerations of train efficiency, cost of energy, rotordynamics, reliability and availability, size and weight of the working machine, and the auxiliary systems. All of these factors affect the gearbox cost.

A lateral vibration analysis often sets the limitation when tuning a rotor to the best efficiency. The application of a gearbox for speed adaptation helps to find the right balance between aerodynamics and rotordynamics and might enable the application engineer to manage with one less casing.

Furthermore, direct drives with low speed would require large impeller diameters in the centrifugal machine without the use of a speed increaser. Subsequent costs for space requirements, buildings, and foundations must be considered in an evaluation.

On the downside, the complexity of the drive system with multiple rotors increases. For the torsional train vibration analysis, an additional two rotors and a coupling could mean double the number of elements., therefore resulting in more torsional natural frequencies of concern to avoid.

In the same way that the number of sensors and monitoring units increases, so does the size of the lubricating oil system.

The above illustrates that the use of a gearbox always requires consideration of the special use case. Gas turbines, as mechanical drives, provide a rather high shaft speed, where compressors can easily be matched to, therefore usually no gearboxes are applied. Pumps, on the other hand, have a lower speed, which is the reason they are often driven by two-pole electric motors. In many cases, however, the introduction of a gearbox pays for itself when comparing the price tag, productive uptime, energy, maintenance, and real estate costs.

Basics of power transmission with gearboxes, design elements, gearing, efficiency, and rotordynamics

The most prominent part of a gearbox is the toothed gearwheels. The gearboxes in use for the machinery considered here are rolling contact gear types with parallel shafts. The most common tooth flank shape is the involute gearing because of its simple manufacturing with straight tools and operational simplicity with a certain insensitiveness to deviations of the center distance (Fig. 3.98). It also offers the possibility to achieve small variations of the shaft offset by shifting the involute profile during design and manufacturing without changing the tooth number.

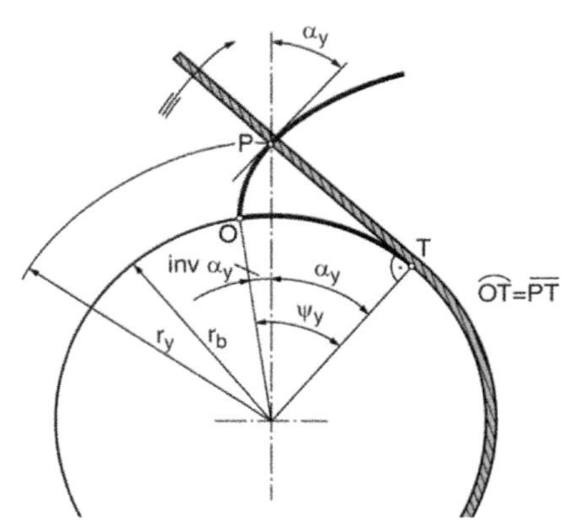

FIG. 3.98 Geometrical generation principle of an involute. *Courtesy of Voith Turbo.*

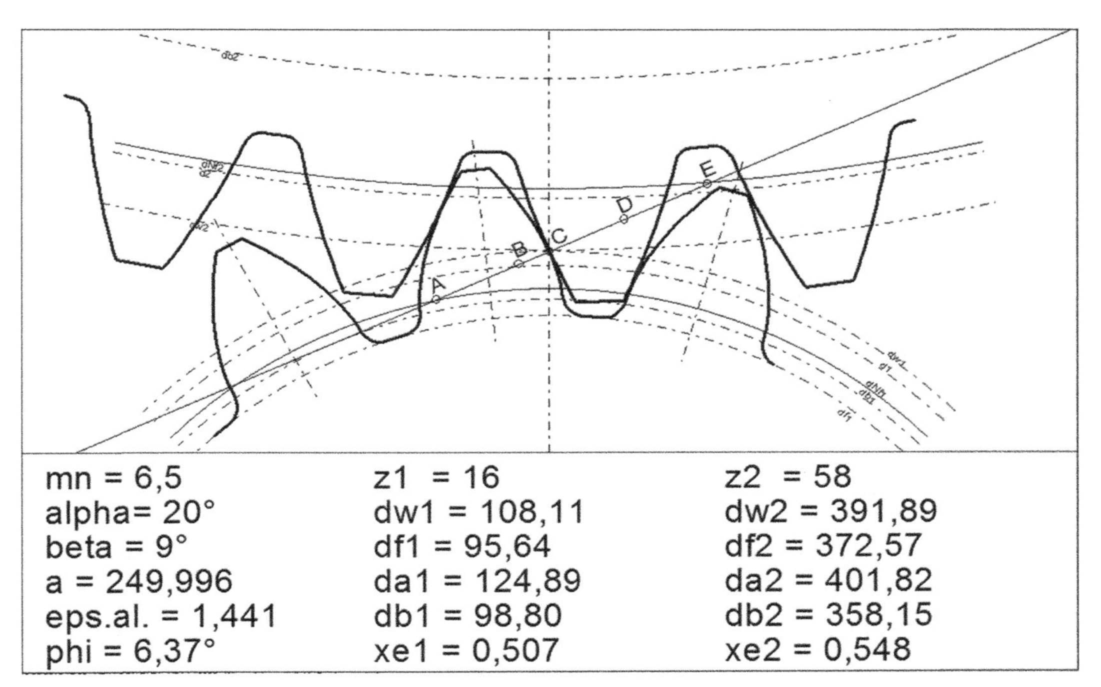

FIG. 3.99 Tooth contact. *Courtesy of Voith Turbo.*

A practical description of the involute shape is given when imagining a knot at point "*P*" in a yarn strand which is unwound from a reel starting at point "*O*." The reel itself represents the base circle of a gear wheel.

By using the involute (Fig. 3.98), the line of contact on which the meshing point of the tooth contact moves continuously is a straight line tangentially touching the base circles of the gears [46]. In the motion, the tooth flanks slide on each other, thus creating a lubricating film with the oil. When the meshing point reaches the rolling circle, which results from the pitch circle plus the profile shift, there is no more sliding, but a rolling motion that displaces the lubricating oil. In the design of the toothing, it must be taken into consideration that the tooth forces do not break the lubricating film on this point (Fig. 3.99).

The simplest gearwheel is a spur gear with straight teeth. The surface of the reel is equivalent to the base circle of the gear wheel. The circle dividing teeth and spaces in equal length is called the pitch circle. A characteristic of an involute tooth geometry is that the tooth flanks slide on each other with the highest relative speed when the tooth root meets the tooth tip and slows down until two teeth meet in the pitch circle with no relative movement. At this point, the teeth roll directly onto each other. For lubrication, the range of slow, relative speed represents the highest requirement due to mixed friction. If damage occurs due to incorrect design, it typically manifests as material chipping on the tooth surface in the mixed friction zone (pitting) or direct tooth contact of the tooth tip and root in the sliding zone (scuffing), caused by inadequate lubrication. When sizing the tooth geometry, the root strength must also be considered to avoid fractures.

Because straight tooth geometry causes rough running and generates a lot of noise, helical tooth geometry was introduced (Fig. 3.100). Due to the helix angle, the teeth of the drive and driven wheel overlap gradually, and therefore, a smoother operation is enabled. On the downside, the helical tooth geometry generates an axial load, which needs to be compensated. This could be done by a well-sized axial bearing or thrust collars, which are common in integral gears. Further development includes double helical tooth geometry, which equalizes the axial thrust.

The transmission ratio of a gearbox is defined by the number of teeth of two gears, which either mesh directly with each other or are connected by an intermediate gear. The smaller gear is called the pinion. If the pinion drives the big gear, the speed is reduced, and the torque increases to the same extent. If the large gear drives the pinion, the speed is increased, and the torque decreases. Except for the friction losses, the transmitted power remains the same. The vector sum of the torques in a gearbox is zero, so the difference between input torque and output torque must be compensated by the gearbox housing.

During operation, the shaft moves within the limits of the bearing clearances and is deformed by the forces acting on it. Especially at the pinion, the tooth flanks are deflecting from the original shape due to bending of the shaft, torsional twist, and thermal distortion resulting from friction losses and the heating up of the oil film flowing from the inner to the outer sides. As the highest forces occur in the rated condition, they stipulate the deformation to be considered.

FIG. 3.100 Double helical toothing at a gear stage. *Courtesy of Voith Turbo.*

A tooth flank correction is necessary to make better use of the teeth and avoid edge or marginal supports. This would mean an increase in stress at the tooth flank and root. By producing a width crowning, the greatest load transfer is placed in the center of the toothing, since the stress peak in the center of the tooth is more tolerable due to the supporting effect from the outside. For smooth meshing of the teeth, a profile modification at the tooth crest compensating for the bending of the tooth is calculated to reduce mechanical shocks and associated noise. This can be calculated over the length and height of the tooth flank by finite element methods. Derived from the calculations, the helix (or lead) and profile modifications will be applied in the finishing process of the gear wheels. As the modifications are based on rated load, the tooth contact pattern can deviate from its optimum in part load operation or test conditions.

Depending on the application, there are many gear manufacturing processes that are adapted to the different service life and accuracy requirements of industries ranging from automotive to critical applications in the oil and gas segment. For gearboxes in the midstream sector, the manufacturing process chain goes from the (1) blank to (2) gear cutting of the basic body by hobbing and (3) heat treatment to (4) hardening of the surfaces and finally (5) hard finishing by profile grinding.

Heat treatment, which in case hardening consists of a process sequence of heating, carburizing, quenching, and tempering, is of particular importance for the load-bearing capacity of the gears. The goal is to obtain a hard surface and a ductile core to withstand the two types of loads, compression and bending, as the toothing is designed to last indefinitely.

Hardening is followed by fine machining, which contributes to a further increase in load capacity and also noise minimization by reducing radial runout and pitch deviations, improving surface structure and quality and, finally, gear modifications to improve the contact pattern.

For the production of double helical teeth (and depending on the helix angle), a separating distance of the two gear rims is required so that the grinding wheels have the necessary space to run out (Fig. 3.101).

FIG. 3.101 Grinding of an involute toothing. *Courtesy of Voith Turbo.*

FIG. 3.102 Various bearing types: Radial tilting pad, journal sleeve, axial tilting pad. *Courtesy of Voith Turbo.*

By transmitting power and motion from the input side to the output side, shafts play a significant role in a gearbox. Typically, there at least two shafts: an input shaft that is driven by an engine or motor and an output shaft that delivers power to the working component. While the pinion often is manufactured as one part with its shaft, the bull gear is made separately from the shaft and connected by a shrink fit. The most common form of shaft end is probably cylindrical with a key, but this requires a temperature differential during assembly and disassembly to produce the required shrink fit. As heating is often not permissible due to hazardous areas, the shaft ends are tapered to allow the removal of the coupling hub by pressurized oil.

The connection of the gear shafts to the housing is preferably designed as plain bearings unless there are special requirements that make rolling bearings necessary. The bearings hold the shafts in a radial and axial direction. For low-speed drive shafts, circular cylindrical bearings, lemon-offset bearings, or 2-face bearings are normally sufficient in the radial direction. At medium speeds, three- or four-surface bearings with fixed segments are used to increase stability and damping and reduce shaft vibrations. Only at very high speeds, such as in integral gears, are tilting pad bearings with even higher damping properties required. Axial bearings can be designed as combined bearings with a supporting bearing, separately with fixed segments or with tilting pads (Fig. 3.102).

For better mounting, bearings are usually divided horizontally into two parts. The lubricating oil is fed under pressure through one or two circumferential channels on the outside and radial bores to the inside into axial grooves, which distribute the oil over the bearing width. From there, it is moved in the circumferential direction by hydrodynamic forces and forms a wedge-shaped bearing pressure profile. The bearing's inner diameter is larger than the shaft diameter in order to form a lubrication gap. After the gear unit has been assembled, the actual bearing clearances, due to manufacturing tolerances, should be measured and followed by a comparison with the intended nominal values. A bearing clearance log should be prepared as part of the quality documentation.

To monitor plain bearings, the temperature is measured at two points on the loaded side in the bearing metal or in the oil drain by thermocouples or resistance thermometers. The shaft vibrations are measured by two eddy current probes offset by 90 degrees. In axial bearings, the position of the thrust collar is monitored with eddy current probes.

Parallel shaft gearboxes

The simplest gearbox comprises of one pinion and one gear wheel as rotating parts plus the bearings and casing as stationary elements (Fig. 3.103). While a speed-decreasing gear with a drive pinion and a driven wheel is sometimes asked for in pump applications, centrifugal compressors mostly require speed increases with a drive wheel and a driven pinion. Applications are overall in the oil and gas and process industries. This is why this kind of gearbox is regulated by American Petroleum Institute (API) standard 613. The key requirements are safety, durability, reliability, and serviceability.

Parallel shaft gearboxes are available up to 140 MW for gear ratios below 5:1. The two shafts are not inline, but have a centerline distance or are offset.

The main limitations in the design of gearboxes are the pitch-line velocity of the gear teeth, the elastic deflection of the pinion and the tooth bases and tips, input and output speeds, bearing velocities and application, and safety factors.

FIG. 3.103 Parallel shaft gear with rotor turning device. The copper plating at the pinion is applied for the run-in period. *Courtesy of Voith Turbo.*

TABLE 3.8 Maximum values for characteristic parameters acc to API 613 [47].

	Characteristic parameters of a parallel shaft gearbox	
Pitch line velocity for shrunk-on forged gears	**Max 150 m/s**	**Max 30,000 ft./min**
Pinion length-to-diameter ratio	1.6 for single helical gears 2.2 for double helical gears including the gap	
Bearing velocity		
Journal bearing load	3.4 MPa	500 lb./in2
Bearing width/diameter ratio		

These parameters act on the centerline distance, shaft diameters, gear module selection, gear width, and bearing width therefore also acting on the overall size.

Taking all values of Table 3.8 into consideration, turbo compressor trains are restricted to approximately 35 MW by API-compliant gear designs.

Integral gearboxes

A special form of parallel shaft gearboxes is used for integrally geared compressors (see above). Casings of this type (Fig. 3.104) can allow up to 10, single-stage compression volute casings to be attached. Typical applications are air separation, PTA, fertilizer, and CCS.

The gearbox is built up around a large, central gear wheel, also called bull gear, distributing the power to several pinions arranged around it in up to three horizontal split lines. Preferred positions of the pinions are 9, 12, and 3 o'clock, while for additional pinions, 6, 11, and 1 o'clock are also possible. Each compression stage includes one radial impeller in its volute, allowing axial gas entry with the lowest aerodynamic losses and radial discharge. Losses are kept low also using three-dimensional impeller blades either in an open or shrouded design. Up to two impellers can be attached to each pinion (one on each end), allowing them to run at their optimum speed while being controlled individually by adjustable inlet guide vanes. This and the possibility to cool down the gas after each stage keeps the polytropic and isentropic efficiency extremely high. The objective for the gearbox layout is to not only maximize efficiency but also to avoid mechanical losses, handle high speeds, and balance axial thrust built up by the impellers.

FIG. 3.104 Integral gearbox with flanges for mounting the compressor volute. *Courtesy of Voith Turbo.*

Epicyclic gearboxes

When the parallel shaft gear units reach their design limits due to physical size or high gear ratio, a different type of gearbox becomes necessary. The approach of epicyclic or planetary gearboxes is to split the power between several planet gears, which move around a central sun gear while also meshing with a surrounding ring gear.

The main advantage of planetary gears is the balanced load distribution over three to six planetary gears. Consequently, planetary gearboxes have a higher power density than parallel shaft gearboxes in terms of both weight and volume, as well as higher efficiency, but they can also be considered more complex. In addition, the input and output shafts of a planetary gearbox are coaxially arranged, which allows for a compact train configuration.

It must be ensured that the load distribution of the gear unit, which is statically overdetermined by the coaxial arrangement, is uniform. Therefore, the permissible elastic deformations must be greater than the dimensional tolerances of the gear components. Accordingly, the ring gear is designed as a floating thin ring, which is held centered on the outside by a serration with sufficient clearance. Wilhelm Stoeckicht first used this principle successfully. The second aspect of this design is the floating nature of the sun wheel, which is centered radially between the planets and held axially by the double-helical toothing (Figs. 3.105 and 3.106). The driven shaft is attached by a flexible coupling.

A high efficiency of up to 99% results from the lower velocities of the toothing and in the bearings compared to a parallel shaft gearbox of equal gear ratio. Because of the smaller tooth forces, a smaller tooth module can be selected. This means that the gear diameters become smaller. The uniformly distributed load application also allows for smaller bearings with corresponding lower bearing velocities, since the radial forces cancel each other out and only the weight forces have to be considered.

The gear ratio is defined by input speed and output speed. As the epicyclic gearbox comprises three elements, one of each can be fixed in the casing. Therefore, the selection of the gear ratio can be affected by the possibilities given. For example, the input can be through the ring gear, with a fixed sun wheel, and the output through the planet carrier. An output speed of more than 60,000 rpm well below the first critical speed can be achieved due to the possible short output shaft driven by the sun wheel.

FIG. 3.105 Floating sun wheel and elastic ring gears according to the Stöckicht principle. *Courtesy of Voith Turbo.*

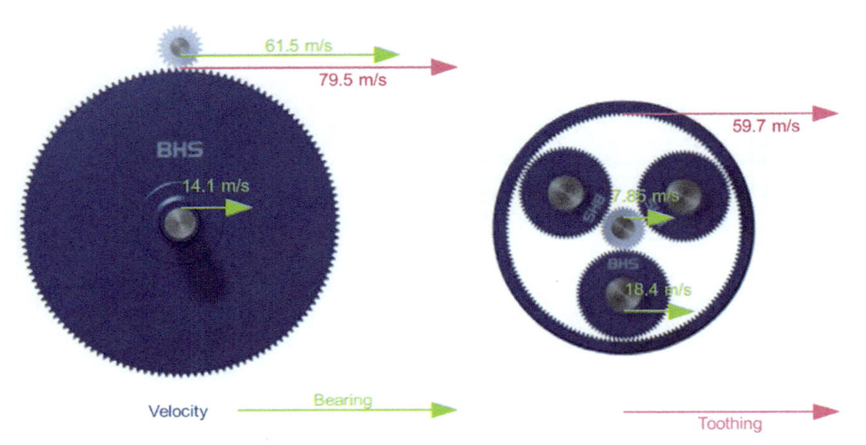

FIG. 3.106 Comparison of velocities in parallel shaft and epicyclic gearboxes showing exemplary values. *Courtesy of Voith Turbo.*

Hydrodynamic variable speed gearboxes

Variable speed is the most efficient form of turbomachinery control as it actsing on all impellers at the same time and therefore provides the best efficiency when operating on the system resistance parabola. Turbines provide this, while electric motors are bound to the grid to which they are connected. One possibility is to vary the frequency of the power supply electronically, but highly efficient, stepless speed converting gearboxes also exist for power transmission of up to 50 MW and beyond.

Epicyclic gearboxes allow superimposition of two drives to generate variable speed. Besides the main driver running at a constant speed, the second power source is a lower-rated controllable drive providing the variable speed. This can be an external motor or an integrated power-splitting component such as a hydrodynamic torque converter. With this, the variable-speed planetary gearbox becomes a purely mechanical unit.

The variable speed gearbox comprises mainly a hydrodynamic torque converter and a revolving planetary gear (Fig. 3.107).

The hydrodynamic torque converter provides the speed variation (Fig. 3.108). Its components include a pump impeller, driven by the main shaft at constant speed and controlled by adjustable guide vanes, and a turbine wheel as output. The power transmitting fluid can be mineral oil, which is also used as a lubricant for bearings and gear toothing. Variations of this configuration are possible for instance, adjustable blades providing variable head in the transmission oil flow at the pump impeller.

The hydrodynamic principle was developed by Hermann Föttinger [48] and patented in 1905 during his search for a wear-free and shiftable 500 HP gearbox for a ferry (Fig. 3.109). He integrated a pump wheel and a turbine wheel in a common housing and thus invented a fluid coupling with very high power density.

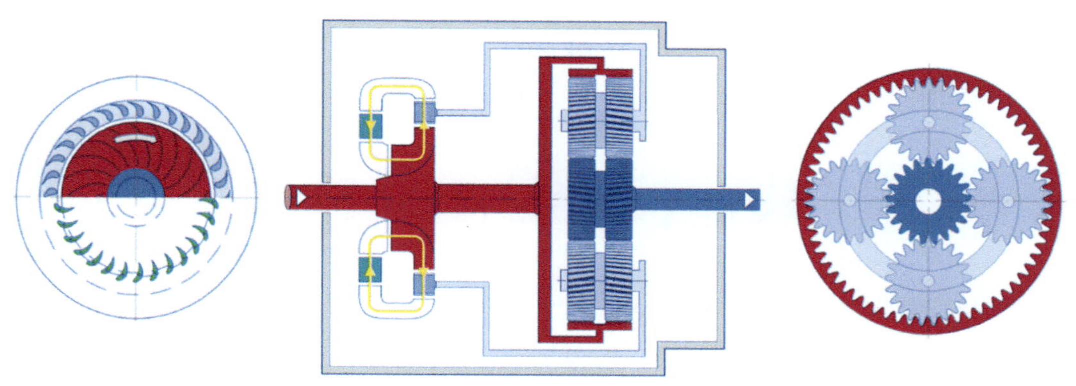

FIG. 3.107 Simplified schematic of a variable speed planetary gearbox with torque converter and planetary gear. *Courtesy of Voith Turbo.*

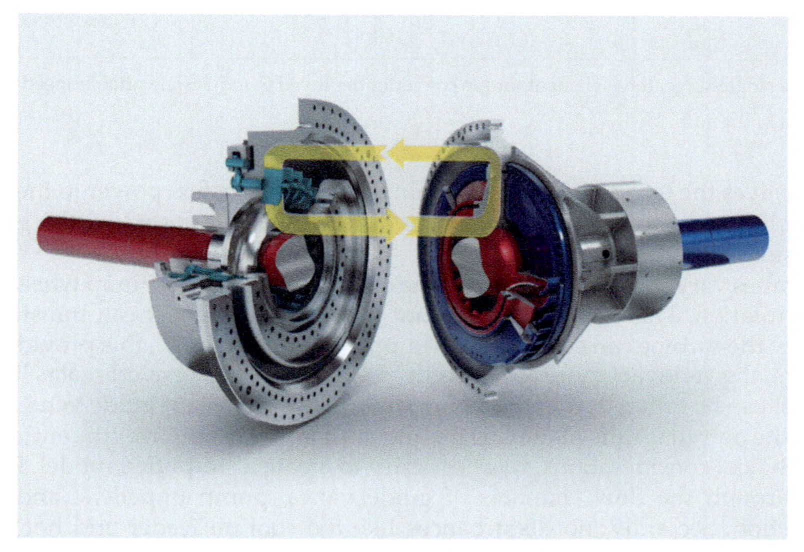

FIG. 3.108 Torque converter components: *Red* = driven shaft and pump impeller, *blue* = turbine wheel and drive shaft, *green* = adjustable guide vanes, *gray* = casing and volute. *Courtesy of Voith Turbo.*

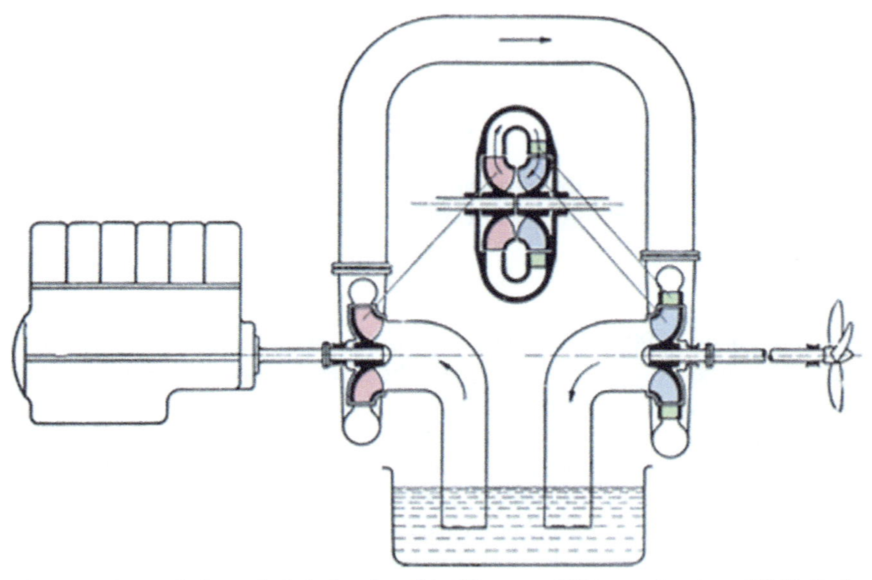

FIG. 3.109 Hydrodynamic power transmission principle developed by Hermann Föttinger. *Courtesy of Voith Turbo.*

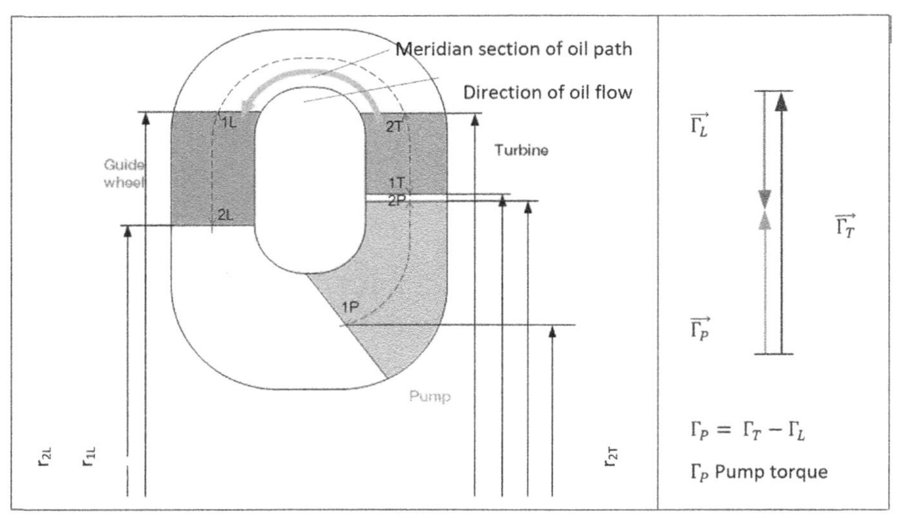

FIG. 3.110 Cross section of a single-stage, three-element torque converter deployed in industrial variable speed planetary gearboxes. *Courtesy of Voith Turbo.*

The pump impeller receives the torque Γ_P from the main drive and transfers power to the oil. The oil is deflected in the turbine wheel and delivers the torque Γ_T. The torque difference is supported by the guide vanes in the casing, causing a torque conversion.

The adjustable guide vanes vary the velocity angle of the oil at the entry of the pump wheel creating a preswirl in the direction of the impeller rotation. This reduces the amount of work the impeller can transfer to the oil and therefore controls the head and flow the turbine can convert. In the upper operating range, this provides a high efficiency, while in the low operating range, the vanes act more as a throttle. Thus, the torque converter acts like a stepless, controllable hydrodynamic gearbox. It can be noted that in this configuration for a constant guide vane angle, the pump torque is not affected too much by the output torque of the turbine and is almost constant over the entire speed range (Fig. 3.110).

For nonnumerical analytical considerations, it is necessary to create a simplified model. First, the oil is assumed to run on the ideal path through the flow channels of guide vanes, pump impellers, and turbine wheels without unwanted angular deflection. Secondly, no disturbances like the cool oil feeder and hot oil drain are considered, and there are no fluid losses through sealing labyrinths.

The torque ratio of the turbine and pump wheel is denoted by the conversion ratio μ.

$$\mu = \frac{\Gamma_T}{\Gamma_P} \tag{3.42}$$

The speed ratio between the turbine and pump wheel is denoted by the speed coefficient ν.

$$\nu = \frac{N_T}{N_P} \tag{3.43}$$

Typical values for ν are 0.3–1.2 depending on the geometry of the components, although higher values can also occur.

The efficiency η is the ratio between turbine power output and pump power input.

$$\eta = \frac{P_T}{P_P} = \mu \cdot \nu \tag{3.44}$$

The maximum efficiency can reach $\eta \approx 90\%$. The speed ratio of the optimum point, which depends on the characteristics of the converter, is in the range $\approx 0.7 < \nu < \approx 0.9$, where the flow enters the pump as well as the turbine with minimum misalignment to the blade edges and exits the turbine free of swirl [49].

The highest turbine torque is generated at maximum deflection in the stall point. In the single-stage, three-element configuration, this is the case when the turbine wheel is at a standstill or even turns in the opposite direction of the pump wheel. The exact position of the stall point depends on the head the pump is able to generate, which is controlled by the guide vanes. As the speed of the turbine wheel increases, the deflection, and thus the torque delivered, finally drops to zero at the run-out point until the oil flows through the turbine wheel in a straight path, and the maximum value for ν is reached (Fig. 3.111).

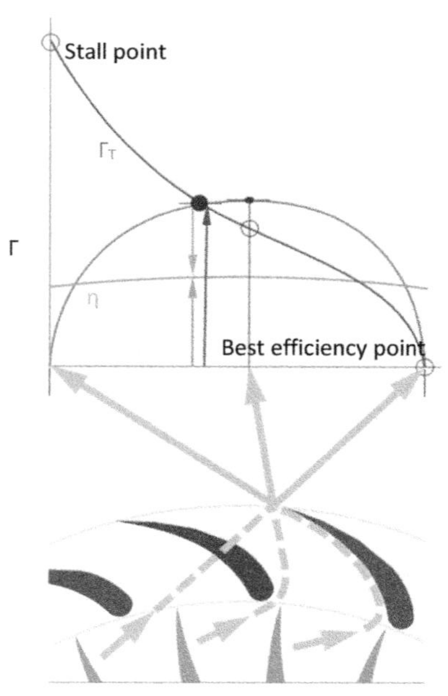

FIG. 3.111 Characteristic curves of the Föttinger torque converter. *Courtesy of Voith Turbo.*

Due to the relation

$$P = \Gamma \cdot N \tag{3.45}$$

the turbine power is zero in two states: at the standstill of the turbine (although the torque is at maximum) and at a maximum speed of the turbine when the torque (and thus the efficiency) is zero.

While μ, ν, and η describe the characteristics of a torque converter, the performance number λ is a parameter used to size the torque converter for a certain speed and power during the selection according to the similarity laws. For geometrical similarity in the sense of mathematical affinity, the Euler number is predominant, which is defined by a pressure difference related to the kinetic energy of a fluid flow.

When scaling the geometry of a turbo machine, dimensions and velocities follow proportionalities to a reference diameter D and the angular velocity $\omega = 2\pi \cdot N$ as

$$\text{Dimension} \quad L \sim D \tag{3.46}$$

$$\text{Velocity } c \sim v \sim \omega \cdot D \tag{3.47}$$

The volume flow rate is derived from the meridian velocity $c_m \sim \omega \cdot D$ in the torque converter channel and its cross-section area $A \sim D^2$.

$$\dot{Q} \sim \omega \cdot D^3 \tag{3.48}$$

Euler's law applies to the pump impeller and the turbine wheel:

$$P = \rho \cdot \dot{Q} \cdot \Delta(c_u \cdot u) \tag{3.49}$$

The power rises with the cube and the torque with the square of the speed:

$$P \sim \rho \cdot \omega^3 \cdot D^5 \tag{3.50}$$

$$\Gamma \sim \rho \cdot \omega^2 \cdot D^3 \tag{3.51}$$

With a proportionality factor λ, a performance number or power absorption factor can be defined, which finally allows to scale the torque converter:

$$\lambda = \frac{P}{\rho \cdot D^5 \cdot 2\pi \cdot N^3} \tag{3.52}$$

$$\lambda = \frac{\Gamma}{\rho \cdot D^5 \cdot 2\pi \cdot N^2} \tag{3.53}$$

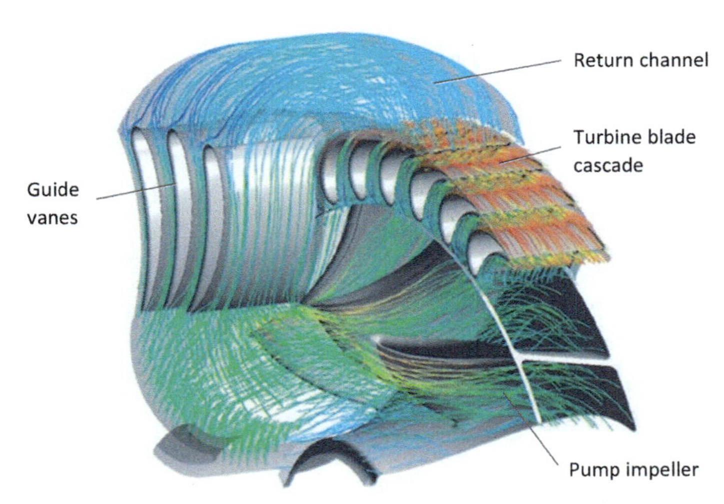

FIG. 3.112 Three-dimensional flow field of a torque converter at the best efficiency point. *Courtesy of Voith Turbo.*

Since the late 1990s, computational fluid dynamics (CFD) made numerical simulations commercially viable and delivered precise and detailed results (Fig. 3.112). Flow deviations like local swirls and misaligned flow against the blade edges can be mitigated. Optimum channel design minimizes the friction of the fluid. Visualization made interpretations increasingly easier. This contributed to today's options for torque converters with selection criteria like high start-up torque, wide operating range, and high overall or high peak efficiency.

In order to adapt the variable speed gearbox to the requirements of pumps and compressors, the entire speed range may be unnecessary in order to achieve desired efficiency increases.

This is achieved by a combination with the rotating planetary gear. As described above, planetary gears consist of three main components: the ring gear, the planet carrier with the planets and the sun wheel. In conventional planetary gears, one component is fixed and serves as a support or intermediate gear. If the system is given a further degree of freedom and the third component is allowed to rotate, such that the result is a superimposed gearbox. The most frequently encountered superimposed transmission in everyday life is the differential gear in the driveline of a vehicle, which has two output shafts. In contrast, in combination with the torque converter, a gearbox is required that combines the main drive with the superimposed branch from the torque converter.

The rating of the torque converter can be selected to split off only a smaller portion of about 20%–30% of the main drive power through the pump impeller. The effect of branching and merging the power flow has two sides. On the positive side, only in the smaller part of the transferred power are there losses of the converter, which are noticeable in the partial load range despite the optimizations by CFD. As 70%–80% of the power is transmitted directly via the revolving gear, the combined efficiency can reach 96%. On the downside, a limitation of the turndown range must be taken into account due to the kinematics described below. However, for compressors, mainly 70%–105% of the rated speed is expected by the industry. For pumps with a larger operating range, a fluid coupling can be integrated to cover the low-speed range, which is also distinguished by low-power transmission.

As for every gearbox, the selection is based on both the maximum torque and the gear ratio needed at maximum continuous speed (Fig. 3.113). For applications in the oil and gas industry, the requirements of API have to be considered as well. That means that the motor power rating is based on the highest specified operating point plus a 10% margin. Since the constant-speed motor is able to provide rated power over the entire compressor speed range, the upper limit of the output torque of the variable-speed gear unit increases as the speed decreases. The torque at the peak determines the design torque of the gearbox. On the other hand, the required effective transmission ratio is determined using actual motor speed and the maximum continuous compressor speed, which is defined by API with a 5% margin added to the rated speed. To obtain the nominal ratio of the revolving epicyclic gear, the superimposed speed must be subtracted.

The transmission oil heats up in the torque converter and must be cooled. The cooler capacity limits the minimum speed in the upper torque range because the efficiency drops from the center of the operating range to the

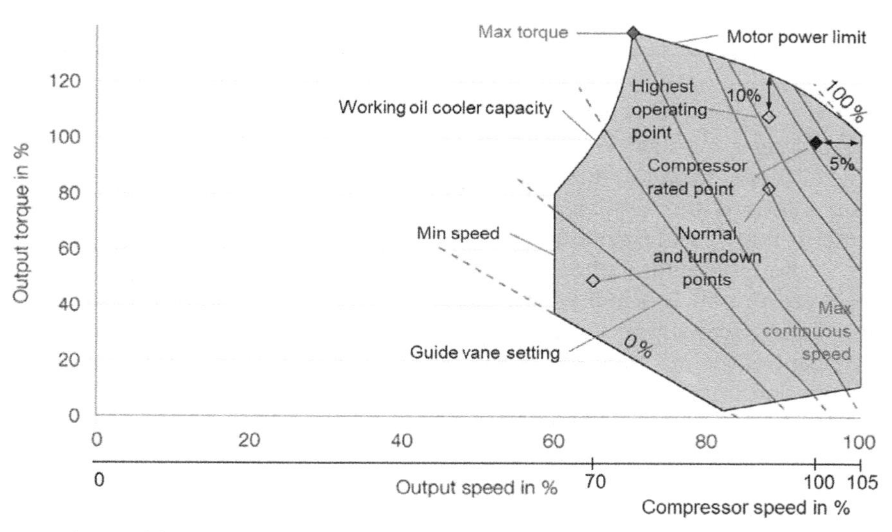

FIG. 3.113 Operating map of a variable speed planetary gearbox as centrifugal compressor drive. *Courtesy of Voith Turbo.*

boundaries causing more heat to be generated. The oil circulation is enabled by a pump, which feeds the oil from a reservoir through the cooler to the torque converter. To maintain a constant supply temperature, a control valve regulates a bypass around the cooler. For safety reasons, the discharge temperature of the torque converter is monitored.

Furthermore, the minimum speed is determined by the limitations of several components inside the variable speed gearbox, such as bearing load under increased temperature, turbine wheel stress and cavitation, and circumferential forces in the planetary gear. The torque converter guide vane settings between 0% (fully closed) and 100% (fully open) also limit the operating range (Figs. 3.113 and 3.114).

Because the Stoeckicht principle requires a floating sun wheel, a minimum torque must be applied depending on the speed. This forms the right lower boundary line of the operating range.

FIG. 3.114 Variable speed planetary gearbox with lube oil system and instrumentation as pipeline compressor drive. *Courtesy of Voith Turbo.*

Shaft sealing

General

Sealing is one of the challenges for all kinds of rotating equipment. The shaft sealing of rotating equipment has an especially wide range of solutions, from low-cost packing solutions in water applications to highly critical machinery applications with backup solutions and highly engineered seal supply systems. This chapter will give a general overview of typical arrangements for the sealing of pumps and turbomachinery in the energy sector using mechanical seals and magnetic couplings. Although liquid- and gas-lubricated mechanical seal solutions can be used in both pumps and compressors, liquid seals are primarily used for pumps and dry gas seals for turbomachinery. The choice of seal arrangements and the referring seal supply systems are strongly dependent on the criticality of the process media and the overall safety concept of the machine and plant. Besides safety, the availability and criticality of the machine for the process make the second important influence on seal specifications and design. The global pressure to reduce emissions within all energy transport processes will be one of the most important drivers.

The given solutions should be understood as general approaches or project-related deviations where experts of the different technical subsections cooperate and find the right solution based not only on accepted standards but also on future-oriented developments. This approach should lead to optimal solutions, especially when based on a thorough exchange of information about all application-related data.

Liquid lubricated mechanical seals

The general design of a mechanical seal is shown in Fig. 3.115 and consists of two rings acting as sliding faces, one rotating with the shaft and the other in standstill. They generate a narrow axial seal gap separating a cavity of higher pressure and one of lower pressure. In order to close the gap, the seal face is spring loaded in the axial direction and pushed against the stationary seat. This setup enables the system to follow axial shaft movements, too.

The closing forces result as a sum of the spring load and hydraulic load when frictional forces are negligible in a first approach

$$F_{\text{Close}} = F_{\text{Spring}} + F_{\text{hydraulic}} \tag{3.54}$$

with the hydraulic force being dependent on the load factor k, describing the ratio between hydraulically loaded area A_H (diameter of the hydraulic sealing point and in this example OD of a sliding face) and the sliding faces area A.

Different arrangements are used to overcome the different challenges of sealing the fluids given. These challenges may include

- Evaporating fluids when passing the seal gap
- Liquid contamination in gas seals
- Low viscosity resulting in wide mixed friction operation
- Solid matter ingredients causing abrasion, clogging

Secondary sealing elements, e.g., O-rings, seal the contact to the shaft and housing. By this description, it can be easily recognized that during normal operation of a mechanical seal, a low leakage will occur through the very narrow seal face gap.

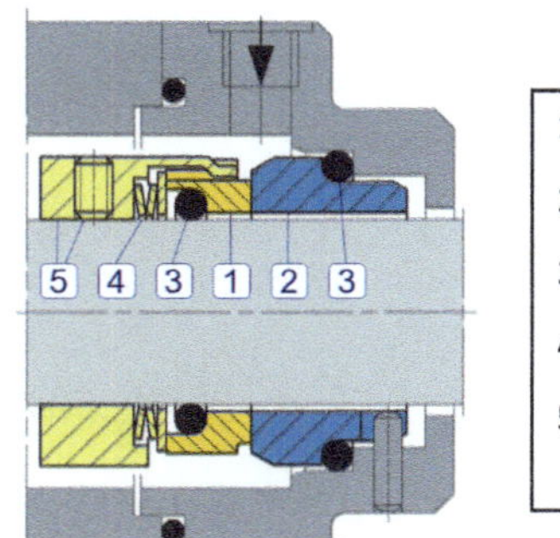

1 (Rotating) Seal Face

2 (Stationary) Seat

3 Stationary Sealing Element (O-Ring)

4 Spring

5 Construction Material and Torque Transmission

FIG. 3.115 Principle schematic of a mechanical seal.

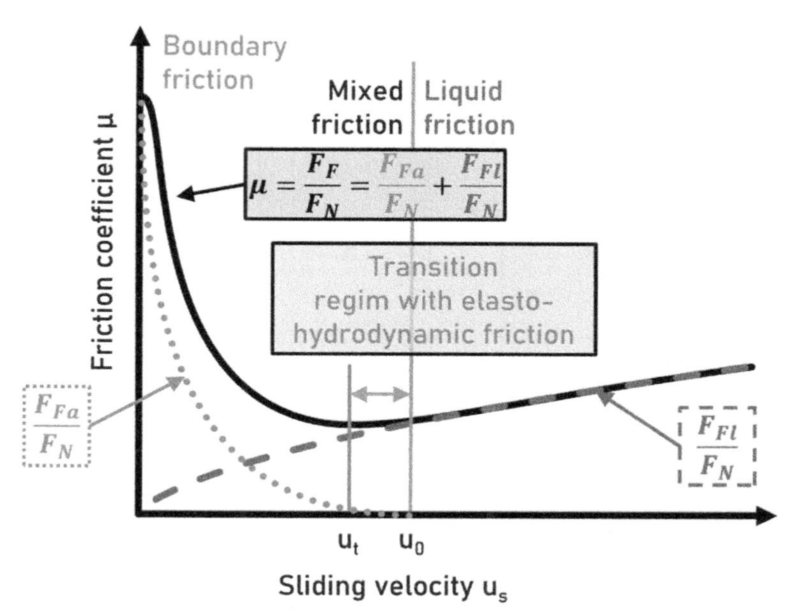

FIG. 3.116 Schematic of Stribeck diagram. F_F, Friction force; F_N, Normal force; F_{Fa}, Friction force due to asperity contact; F_{Fl}, Friction force due to fully trained liquid friction; u_t, Sliding speed starting transition; u_0, Sliding speed starting contactless friction. *Prof. Dr.-Ing Peter Waidner, Gleitringdichtungen, ISBN:978-3-947198-11-5, www.seal-ing.de [seal-ing.de].*

At given operational parameters, it is an optimization issue to balance seal leakage, energy dissipation, and wear of the sliding faces. The frictional behavior in the sealing gap can be explained by analogy to the Stribeck curve (Fig. 3.116), which is well known for bearings.

When at standstill, the sliding faces are in contact with no film or a poor film of liquid between the surfaces. Starting to rotate will induce friction between the surfaces. Increasing rpm brings them into mixed friction with a more or less completely established fluid film and reduces the axial forces to be taken over by the ring surfaces. Furthermore, the fluid inside helps lubrication and partly transfers the dissipated heat, all of this is highly influenced by the fluid properties and those of the chosen sliding faces and their coatings. When continuing to increase rpm, the fluid film compensates increasingly to the axial closing forces in the sliding gap until they lift off. In the region of liquid friction, the faces are in lift-off. Since the leakage is proportional to the cube of the gap height it is easily recognizable that even a small increase of seal gap height h means a significant increase of leakage.

The challenge between low leakages and minimal wear is easily recognizable with the leakage proportional to the 3 of the minimum seal gap height. Geometrically, this lift-off behavior can be positively supported by features like grooves or waviness. As this has to be realized in the micron and submicron area, it becomes evident that unwanted waviness has to be avoided. Besides using the appropriate materials and tolerances for internal parts and assemblies in the seal cartridge, the interface to the machine is highly important.

Desirable properties of sliding face materials are high Young's Modulus, high thermal conductivity, low thermal expansion coefficient, and low coefficient of friction in combination with its mating surface, particularly when operated in mixed friction partial evaporation conditions. Furthermore, highly loaded seals demand high values of yield strength and low density for the high-speed rotating ring (Table 3.9).

O-rings are typically the first option of choice for secondary seals due to their good sealing performance, ease of handling, wide material selection, and low cost. In the energy sector, hydrocarbons in different mixtures often have to be sealed. Different FKM (Fluoroelastomer) or FFKM (Perfluoroeleastomer) qualities are available with good resistance to fluids. Special care needs to be given to the dynamic O-rings sealing the seal face; In static sealing, points swelling can be accepted to a certain level but for a dynamic sealing element it can lead to restriction of the seal movement, and therefore cause unwanted behavior as such as hang-ups of the seal, causing (temporary) increased leakage, or damage to the sliding faces. Operating pressures of far more than 100barg are possible with O-rings, and typical offerings cover temperature ranges from 20°C to 200°C, with special compounds going down to ~−40°C/up to ~230°C.

TABLE 3.9 Sliding materials.

Magnitude / Material	Carbon	TC (Ni)	SiC sintered	Si3N4
Spec. mass (kg/dm³)	2..2.5	14.5	3.1	3.26
Tensile strength (MPa)	80..120	1700	450	850
E-modulus 10^{-4}(N/mm²)	2..4	60-63	30-41	35
Thermal conductivity λ (W/mK)	7..12	80	100...130	30
Thermal coefficient of expansion α (10^{-6}/K).	4..5	4.8	4	2.1

Exceeding these ranges or dealing with more aggressive fluids (or preferring "one-size-fits-all" solutions) leads to the area of nonelastomer sealing elements, typically based on PTFE (Polytetrafluoroethylene) and enriched with further content to improve mechanical properties. Such sealing elements may be used down to less than −200°C and up to 230°C–250°C.

Different stainless steel materials are available and are chosen with regard to the fluid and its corrosion properties, machinability/remaining waviness and it's thermal expansion properties. The latter is of special importance in the event that they are shrink-fitted with sliding faces (Fig. 3.117).

Oil-lubricated seals, similar to the above-described pump seals, have been used in the past and for some applications such as screw compressors and are still beneficial. Sealing pressures of ~100 barg and sliding speeds of ~100 m/s, in general, are possible and in operation. Nevertheless, dry gas seals are the preferred solution (where applicable) with higher rotational speeds and pressures, and dissipate far less energy (Fig. 3.118).

For both oil and gas-lubricated compressor seals, the general arrangement to accommodate high pressures and high RPMs is to design with high pressure at the outer diameter and the seat rotating, with the the stationary seal face supported by springs.

Dry gas seals (DGS) operate in lift-off during normal operation. The sealing gap of 3–30 μm allows for contact-free, and thus wear-free, operation of the sliding faces. Some seals have a minimum operating speed of approximately 3 m/s to achieve liftoff, but this is highly dependent on seal design, operating fluid, and conditions (temperature and pressure). Some seal designs incorporate pressure balancing to lift off statically before dynamic operation.

Depending on the train driver and size, startup isn't the only point of interest. Large trains, especially when driven by electric drivers with high rotating mass, perform long, coast-down curves causing the DGS to be in high-speed contacting operation for a long time. When using steam turbines, heating up and cooling down is necessary for bigger

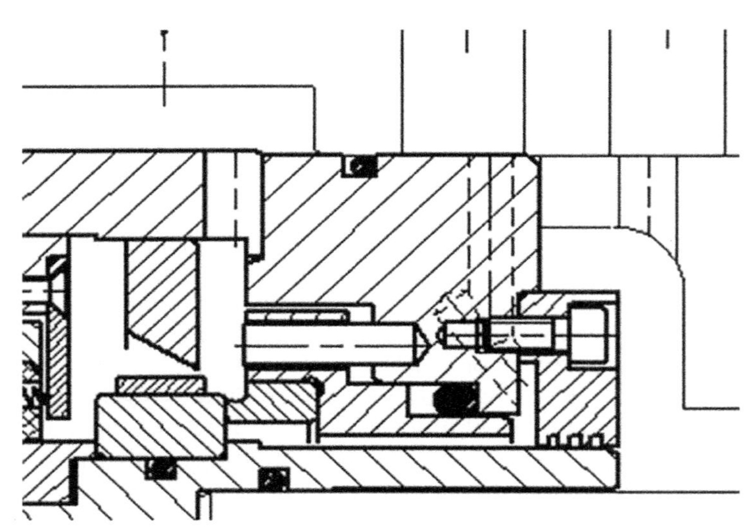

FIG. 3.117 Seal for pump applications.

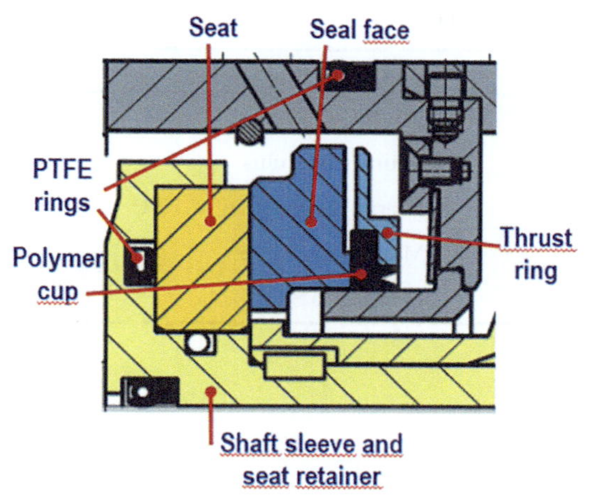

FIG. 3.118 Dry gas seal for compressor applications.

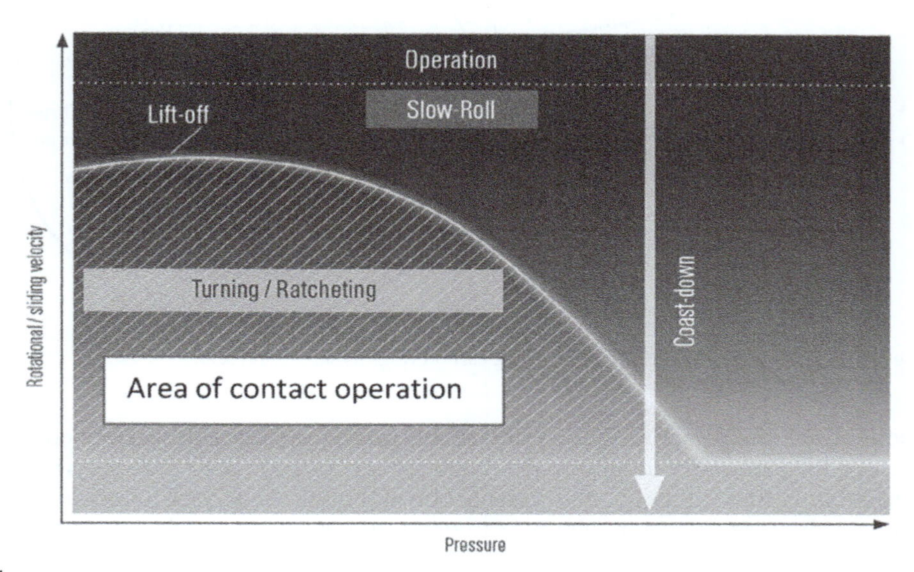

FIG. 3.119 Slow roll.

machine sizes. With regard to DGS, the best operation mode in this case is to slow roll at some 100 RPM above lift-off (Fig. 3.119). Where not possible, continuous/discontinuous turning/ratcheting can be accommodated with satisfactory design life using standard coatings with hard-hard material combinations. The most common coatings for this purpose are amorphous carbon coatings, also known as diamond-like carbon (DLC). These coatings are applied in the range of several microns. Hard-soft material combinations typically have one sliding face made out of carbon material and have life advantages; however, these typically have noticeably higher leakage during normal operation.

When more abrasion resistance or more safety margin is required, for example with a low leakage application (low seal gap), industrial, crystalline diamond can be applied. It helps to increase operation duration tremendously for turning/ratcheting. This is one possibility when turning speeds are significantly above 0.2 rpm but still below contact-free operation. Due to its high thermal conductivity, diamond coating also helps to avoid hot spots in short-term contacts and thus helps to avoid thermally induced cracks [50]. Other typical damage scenarios are well known and are addressed in API 692 [51].

A comparison of DLC and crystalline diamond coating is shown in Table 3.10.

Similar to pump applications, there are different compressor seal arrangements, but in general, the variety is less.

A typical example of a compressor tandem seal with an intermediate labyrinth and a floating carbon ring separation seal is shown in Fig. 3.120. This is the most common one used in the industry and allows for highly efficient shaft sealing without process air coming into the atmosphere during normal operation when seal pressures are above approximately 5 barg.

TABLE 3.10 Characteristics of DLC and diamond coating.

DLC	Property	Cryst. diamond
1000 … 4000HV	Hardness	10,000HV
30 … 220GPa	Young's modulus	1150GPa
0.1 … 10W/mK	Therm, conductivity @ 20°C	2000W/mK
≈250°C	Max. temperature	≈600°C
Adhesive (fairly good bonding)	Bonding to SiC	Covalent (excellent bonding)

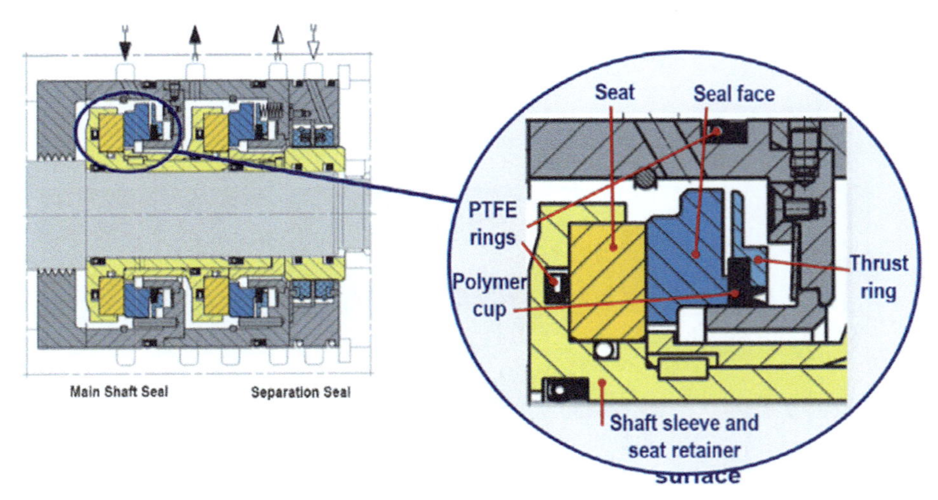

FIG. 3.120 Tandem seal with intermediate labyrinth and floating carbon rings as separation seal.

In all operating modes (including standstill), the primary seal to the products side is supplied with clean and dry gas by a seal supply system, which ensures a minimum flow velocity across the process side labyrinth seal in order to keep dirty process gas from entering the seal cavity. As condensation of the gas needs to be avoided during depressurization in the sealing gap, the supply system must maintain a safety margin of 20K to the dew point line needs to be ensured during the whole depressurization process. Higher Cs, respectively, other higher mole weight molecules should be considered (even if they are low mole fractions) for the calculation of the dew point line for DGS operation, even though they can often be neglected for compressor performance calculations and operation.

In order to avoid hazardous process gas leakage to the atmosphere, a secondary seal gas (mainly N_2) is injected and ducted via an internal labyrinth and buffers the primary seal gas leakage to the primary vent. The secondary seal is operated with pressure above vent pressure and high enough to give good operational stability to the seal, which is of vital importance at higher sliding speeds, especially above approximately 140m/s. Secondary seal leakage and part of the separation gas ducted to the secondary vent are typically ducted to the atmosphere. The second part of the separation gas is ducted through the gap toward the bearing side. Supported by mechanical barriers, this part of the seal prevents oil from migrating into the DGS, which is of vital importance for the safe operation of the DGS.

The upcoming challenges in the energy market with regard to the industrial generation of H_2 and the increasing use of sCO_2 will extend sliding velocities and temperatures.

Aiming for more safe installations, vent studies will lead to the need for higher safety margins for foreseeable damages such as secondary vent pressures. End users, compressor manufacturers, and seal manufacturers are all working on these challenges.

The need for emissions reduction has already led to new products. For example, coaxial barrier seals, following the demand of end users for reduced N_2 consumption, have been available for more than a decade. Based on this technology, the demand for reducing/avoiding the venting and flaring of greenhouse gases such as methane might increase considerably.

Maintenance

Proactive maintenance

A plant can utilize different maintenance strategies or a combination of strategies to maintain plant equipment. Universally, maintenance strategies have progressed from reactive maintenance to preventive maintenance to proactive maintenance.

Reactive maintenance is performing corrective maintenance on equipment as a reaction to a finding or a failure, for example, if a pump runs and experiences a seal failure, then seal replacement work would be performed. Another example is if there is an audible noise from the pump because of bearing rub, then a bearing replacement is completed. This is also called the 'run-to-failure' philosophy.

Preventive maintenance is a scheduled maintenance program at a predetermined frequency, which could be either time or utilization based that is consistently performed to keep a machine running. The only data needed to perform this type of maintenance would be run hours or calendar duration. An example of this would be the replacement of engine oil every month or 700 h of machine operation.

Proactive maintenance on the other hand anticipates and identifies a problem before it occurs. It involves gathering information or data from the equipment and using the data to evaluate, analyze, and determine what maintenance is needed. This a three-step process where data from equipment operation is extracted, gathered data are processed, and the results analyzed to determine what maintenance should be done to proactively prevent a possible failure.

The three main categories for proactive maintenance are:

(i) Condition based maintenance (CBM)—Monitor and track changes in specific equipment metrics against set thresholds using defined screening technologies. Equipment data is collected from sensors or samples.

(ii) Predictive maintenance (PdM)—Monitor and track changes in one or more metrics using sensors with the incorporation of some analytics to detect anomalies in trending that indicate deterioration of equipment health or impending failure. From the trend information, analytics are used to estimate a predicted failure point in time.

(iii) Prescriptive maintenance (RxM)—Monitor and track changes in a wide range of equipment and operational metrics using sensors and the Internet of Things (IoT). Extensive usage of analytics and proprietary algorithm modeling (AI) is used to determine potential causes of trending anomalies. This may include the identification of subcomponents that could cause the issue. The identification uses a machine learning library (ML) to match an anomaly against relevant to-do actions and inspection or rectification tasks where the predicted failure point not only can be identified, but also be deferred out further in time via adjustments in operating mode.

Proactive maintenance has evolved over time; with more data extracted from the equipment, a more knowledge-based, intelligent analysis can be performed to identify the maintenance activity needed to rectify a fault condition that is occurring.

The Potential Failure (*P-F*) Curve is a curve that depicts the progression of equipment health from potential failure (*P*) toward a failure (*F*) over time. It starts off with some form of equipment degradation denoted as "failure starts" where prescriptive maintenance could be used to rectify the situation. If nothing is done, the progression will indicate attributes that can be identified and rectified using predictive maintenance, followed by condition-based maintenance. If still nothing is done, failure will eventually occur where only reactive maintenance can resolve the issue.

Preventive maintenance that is done on a prefixed interval might be able to identify and rectify a possible failure, but the downside is that there is also the possibility that it could be performed too early in the *P-F* cycle generating no meaningful value or is too late/too close to the failure that irreversible, significant damage to the equipment has already occurred.

The cost of maintenance increases the closer we get to the failure (*F*) since there is less time to mitigate or eliminate the failure. However, the further we are away from the failure, the maintenance tools used for detection and maintenance are more sophisticated and expensive (equipment, software, and training).

With more data gathered from equipment, the identification of potential anomalies or fault conditions can be done early in the *P-F* cycle. The earlier these can be identified, the more flexibility there is for timeline and production planning to execute the maintenance activity. Besides identifying and rectifying a fault early, with more data (specifically for predictive and prescriptive maintenance), there are additional benefits in that it allows for the estimation of the failure point and a prescription of what can be done to avoid or delay the estimated failure point [52].

Condition monitoring

The performance aspect of condition monitoring generally involves the comparison of a measured performance parameter with an expected value [14,53]. Trending then becomes the logging of deviations between measured and expected parameters.

This identifies the two key ingredients of condition monitoring:

- A (digital) performance model for the machine (or pipeline), including its components.
- Sensors and algorithms that allow the measurement and processing of relevant performance data.

In addition, one must have ways to exclude invalid data, be it from sensor failures, sensor inaccuracy, or nonsteady state behavior, in order to get a sense of the accuracy of the data. Also, the digital performance models can be improved by the use of test data on individual engines. These improvements can come from tests on an individual machine to customize the digital model, as well as usage of data from a number of the same machines where available.

The intersection of machine learning methods and sensor data has expanded rapidly in the last decade to include numerous applications of regression, clustering, and even neural network algorithms. Learning algorithms has pushed traditional engine health management into the realm of prognostic health management. The focus is generally placed on industrial gas turbines with an industry standard monitoring system. Allen et al. [54], for example, explore beyond gas path analysis with a novel use of machine learning algorithms for engine component classification. Other applications can involve the optimization of maintenance schedules [55] by estimating degradation rates and the economic impact of delaying maintenance. One of the key challenges is the capability to distinguish data that is valid and suitable for analysis versus data that does not lend itself to be used, for example, because it was taken during transient conditions or because it comes from a faulty sensor [56]. Another example may include a situation where the driven compressor is used to measure the power output of a gas turbine, and the gas composition for said driven compression is unknown or has changed.

The latest buzzword in the industry is the digital twin [57]. It's being praised as the holy grail solution for everything from failure prevention and maintenance reduction to life extension, improved operation, and parts forecasting. While the digital twin concept offers definite value for machinery operation, some of these claims may be overstated. It's worthwhile to discuss what a digital twin really is, what it can do and where its limitations lie.

A digital twin is a computer virtual representation and simulation of a real physical object or process. Although the idea has been around for a long time, the basic concept and definition were introduced in 2002 by Michael Grieves, then at the University of Michigan, as a model for Product Lifecycle Management. NASA then coined the term in 2010 in a Roadmap Report. The digital twin concept consists of three parts: the physical object, the digital or virtual product, and the connections between the two. The concept has branched out into different types, such as the digital twin aggregate (DTA), where multiple digital twins operate in parallel to simulate a complex system.

Implementing a simple digital twin or a more complex DTA requires data about the real-time status of the real-life object, as well as the ability to operate, maintain, or repair the real system remotely via a parallel digital twin representation. Think of Apollo 13, where the ground-based engineers were able to determine how to rescue the mission remotely by simulating all operations and failure scenarios to identify a viable solution. Thus, "the ultimate vision for a digital twin is to be able to create, test, and build [our] equipment in a virtual environment" (John Vickers, NASA, 2010).

In its purest form, the digital twin is a product and technology development tool. That means it is also an alternative to big data approaches, where correlation is assumed to be a proxy for causation. The digital twin is essentially a computer simulation program that uses real-life data to create simulations that can predict how a physical asset will perform. Artificial intelligence, self-learning, and software data analytics can also be utilized to enhance and refine the output.

The distinction between a simulation model and a digital twin is not really clear. The basic components of a digital twin are a model and some real-life operational data. That's nothing new. For decades, we have used performance models—in the form of performance curves or compressor maps, for example—to determine the health of equipment by simply comparing the expected behavior (from the curves) to measured performance data. The difference between a model and a digital twin (i.e., "a digital twin without a physical twin is a model") does not help because the performance curves actually reflect the behavior of a real machine.

The added feature of a true digital twin versus a simple model is that a digital twin can adapt to changes in the physical object over time. For example, if a machine shows wear or has performance degradation, a true digital twin can adjust, whereas a simple model simulation cannot. The digital twin thus provides added value in that it provides a more accurate representation of the current state of the machine. This can be of significant value when scheduling maintenance and repairs or predicting remaining life since the digital twin's updated parameters directly related to known physical failure mechanisms.

One of the key insights of digital twin operations is that the simulation model used in the background does not have to be complicated. The use of performance curves noted above can be part of a digital twin, but updating the model to the current state of the machine using measured data is not trivial. Measured data is always subject to condition, measurement, and statistical uncertainties. The digital twin correction using actual data (from operations, field and factory testing) must be treated as a statistical process—the more data, the better the model improvement. The sourcing of test data from multiple (more or less) identical machines significantly helps when correcting the model. This means that a manufacturer with a large number of machines in operation (gas turbines, steam turbines, compressors, etc.) has an advantage when building a digital twin.

The minimum requirements for the model used in a digital twin are simply that it is sufficiently physics-based, accurate, and quick to run. The concept of a digital twin, leaving all the hype aside, is based on existing and mature technologies. Overuse of the term may lead to underutilization when potential adopters become cynical of the buzzwords and shiny presentations. Using the concept requires trust in the model, the data, and the algorithms used to update the model. Supervision by human experts is still required for checking the plausibility of the model outcomes and digital twin recommendations.

The requirements for sensors and algorithms that allow the measurement and processing of relevant performance data are essentially the same as for a performance test. In general, the following parameters, or a subset thereof, should be measured:

- Speed of all shafts
- Pressures at the inlet and exit of all components
- Temperatures at the inlet and exit of all components
- Output power
- Flows, composition, and temperature
- Relative humidity of the inlet air
- Bleed flows
- Emissions can also provide input about the health state of an engine.

In some instances, the process in question is not in a steady state condition. Many pipelines fall into this category. The problem arising is that all conservation equations are time-dependent. Allen et al. [58] have discussed this problem for pipelines. Quantitative predictions of fouling [59] are also available, and the impact of component fouling on overall engine performance is well understood [60] (Fig. 3.121).

A particular difficulty arises from determining the output power of a driver or the absorbed power of a compressor. If the gas turbine drives a generator, the power output can be easily determined by the electrical output at the generator terminal and then corrected by known generator and gearbox efficiencies. If the gas turbine drives one compressor or multiple compressors, the power measurement requires one to fully instrument the driven compressors and calculate

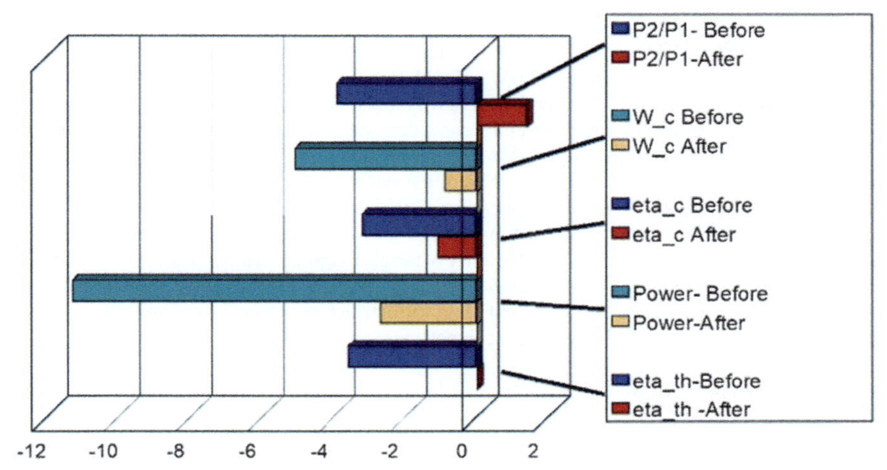

FIG. 3.121 Impact of degradation as a result of compressor fouling on gas turbine parameters (compressor pressure ratio $p2/p1$, compressor air flow W_c, compressor efficiency eta_c, power and thermal efficiency eta_th), before and after removal of compressor foulants by water washing [60].

the absorbed power by thermodynamic and flow measurements. This can be avoided if the torque at the power turbine is measured with a torque-sensing coupling. If the gas turbine drives a pump, thermodynamic measurements are not possible, and either the pump efficiency is assumed, or a torque-measuring coupling is installed. Of importance is the fact that the uncertainty is not just determined by the instrumentation, but also by the operating condition of the equipment [61].

References

[1] The World Factbook — Central Intelligence Agency. n.d. www.cia.gov. Archived from the original on August 21, 2016. Retrieved September 6, 2016.
[2] U.S. Energy Information Administration. n.d. Natural Gas Pipelines. Available at: https://www.eia.gov/energyexplained/natural-gas/natural-gas-pipelines.php (Accessed 19 March 2023).
[3] R. Kurz, S. Ohanian, K. Brun, Compressors in high pressure pipeline applications, in: ASME Paper GT2010-22018, 2010.
[4] S.I. Hyman, M.A. Stoner, M.A. Karnitz, Gas flow formulas - strengths, weaknesses and practical applications, in: Proceedings of the American Gas Association, Operating Section. Los Angeles, USA, 5–7 May 1975, 1975, pp. d125–d132.
[5] R. Kurz, T. Allison, J. Moore, M. McBain, Compression turbomachinery for the decarbonizing world, in: Proc. Turbosymposium 2022, Houston, TX, 2022.
[6] J.C. Bronfenbrenner, Selecting a Suitable Process, LNG Industry, 2009. Summer.
[7] L. Balascak, D. Healey, W. Miller, S. Trautmann, Air Products' Technologies for Small Scale LNG, Small Scale LNG, Oslo, Norway, 2011.
[8] M. Roberts, F. Chen, Ö. Saygı-Arslan, Brayton Refrigeration Cycles for Small-Scale LNG, Gas Processing, 2015.
[9] S. Mokhatab, D. Messersmith, Liquefaction technology selection for baseload LNG plants, Hydrocarbon Processing Magazine (July 2018).
[10] D.L. Andress, R.J. Watkins, Beauty of simplicity: Phillips optimized Cascade LNG liquefaction process, Am. Inst. Phys. Conf. Proc. 710 (1) (2004).
[11] M.G. Nored, J.R. Hollingsworth, K. Brun, Application Guideline for Electric Motor Drive Equipment for Natural Gas Compressors, Gas Machinery Research Council, Dallas, TX, 2009.
[12] R. Kurz, J. Mistry, P. Davis, G.-J. Cole, Application and control of variable speed centrifugal compressors in the oil and gas industry, in: IEEE/PCIC Conference, San Antonio, TX, 2021.
[13] M. Glasbrenner, B. Venkataraman, R. Kurz, G.J. Cole, Electric-motor driven gas compressor packages: starting methods for large electrical motors and torsional integrity, in: Proc. 46th Turbosymposium, Houston, TX, 2017.
[14] R. Kurz, K. Brun, C. Meher-Homji, J. Moore, F. Gonzalez, Gas turbine performance and maintenance, in: Proc 42nd Turbomachinery Symposium, Houston, TX, 2013.
[15] C.B. Meher-Homji, The historical evolution of turbomachinery, Proceedings of the 29th Turbomachinery Symposium, Houston, TX, September 2000, 2000.
[16] Wikimedia Commons. n.d. Four Stroke Engine Diagram. Available at: https://commons.wikimedia.org/wiki/File:Four_stroke_engine_diagram.jpg.
[17] K. Brun, R. Kurz, Compression Machinery for Oil and Gas, Gulf Professional Publishing, Cambridge, MA, 2019.
[18] F. Pischinger, Verbrennungsmotoren, RWTH Aachen, 1975.
[19] E.S. Karakas, H. Watanabe, J. Crutcher, Cryogenic liquid and two phase expanders in liquefaction and cooling processes of natural gas, in: Turbomachinery & Pump Symposia, Houston, TX, September 2023, 2023.
[20] R. Kurz, M. Ji, G. Beck, T. Allison, On small scale LNG concepts, ASME Paper GT2021-58989, 2021.
[21] J.H. Gülich, Centrifugal Pumps, Springer-Verlag Berlin Heidelberg, 2008.
[22] E.W. Lemmon, I.H. Bell, M.L. Huber, M.O. McLinden, NIST Standard Reference Database 23: Reference Fluid Thermodynamic and Transport Properties. REFPROP, Version 10.0, National Institute of Standards and Technology, Standard Reference Data Program, Gaithersburg, 2018.
[23] E.S. Karakas, H. Watanabe, M. Aureli, C.A. Evrensel, Cavitation performance of constant and variable pitch inducers for centrifugal pumps: effect of tip clearance, ASME J. Fluids Eng. 142 (2) (2020).
[24] C.E. Brennen, Hydrodynamics of Pumps, Concepts ETI, Inc, Norwich Vermont, USA, Oxford University Press, Oxford, 1994.
[25] ANSI/IHI 9.6.1, Rotordynamic Pumps Guideline for NPSH Margin, Hydraulic Institute, 2012. ISBN: 19357621102.
[26] ISO 13709:2009, Centrifugal Pumps for Petroleum Petrochemical and Natural Gas Industries, 2009.
[27] API 610 11th Edition. Standard for Centrifugal Pumps for Petroleum, Petrochemical and Natural Gas Industries.
[28] J.K. Jakobsen, Liquid Rocket Engine Turbopump Inducers, NASA Technical Report, SP-8052, May 1971.
[29] D.D. Scheer, M.C. Huppert, F. Viteri, J. Farquhar, Liquid Rocket Engine Axial-Flow Turbopumps, NASA Technical Report, SP-8125, April 1978.
[30] V.S. Lobanoff, R.R. Ross, Centrifugal Pumps, Design & Application, Gulf Publishing Company, Houston, TX, 1985.
[31] ANSI/IHI 14.3 n.d. American National Standard for Rotordynamic Pumps for Design and Application. ISBN: 978-1-935762-81-2.
[32] E.P. Sabini, W.H. Fraser, The Effect of Specific Speed on the Efficiency of Single Stage Centrifugal Pumps, Turbomachinery Laboratories, Department of Mechanical Engineering, Texas A&M University, 1986.
[33] E.S. Karakas, R. Mollath, Cavitation Performance Improvement of an Industrial Cryogenic Centrifugal Pump by Implementing Variable Pitch Inducer, Turbomachinery Laboratory, Texas A&M Engineering Experiment Station; Texas A & M University. Libraries; Texas A & M University; Libraries, 2021.
[34] Michael Smith Engineers. n.d. Positive Displacement Pumps. Available at: https://www.michael-smith-engineers.co.uk/resources/useful-info/positive-displacement-pumps.
[35] R. Kurz, Introduction to Centrifugal Compressors, Solar Turbines Incorporated, San Diego, CA, 2022.
[36] GPSA, Engineering Handbook, ninth ed., 1974.
[37] B.E. Poling, J.M. Prausnitz, J.P. O'Connell, The Properties of Gases and Liquids, McGraw-Hill, 2001.
[38] Y. Nakayama, Visualized Flow, Pergamon, 1988.

[39] N. Aust, Ein Verfahren zur digitalen Simulation instationaerer Vorgaenge in Verdichteranlagen, Diss HS Bw, Hamburg, 1988.
[40] P.C. Rasmussen, R. Kurz, Centrifugal compressor applications-upstream and midstream, Proc. 38th Turbosymposium, Houston, TX, 2009.
[41] K. Wygant, J. Bygrave, W. Bosen, R. Pelton, Tutorial on the application and design of integrally geared compressors, in: Asia Turbomachinery and Pump Symposium, Singapore, 2016.
[42] A. Srinivasan, C. Impasto, Application of integrally geared compressors in the process gas industry, ASME Paper GT2013-95870, 2013.
[43] B. Boutin, P. Taylor, Basics of compressor valve design and troubleshooting, in: GMC 2015, Austin, TX, 2015.
[44] P.G. Basic, Thermodynamics of reciprocating compression, in: GMC 2015, Austin, TX, 2015.
[45] K. Brun, T. Allison, Machinery and Energy Systems for the Hydrogen Economy, Elsevier, 2022.
[46] Roloff/Matek, Maschinenelemente, sixteenth ed., Friedrich Vieweg & Sohn Verlag, 2003.
[47] American Petroleum Institute, API Std 613 6th Edition. Special-purpose Gears for Petroleum, Chemical, and Gas Industry Services, 2021.
[48] Voith Power Transmission, 100 Years of the Foettinger Principle, Springer, 2005.
[49] VDI-Richtlinie 2153, Hydrodynamische Leistungsübertragung, VDI, 1994.
[50] D.I. Müller, D.A. Schrüfer, M. Höfer, L. Schäfer, Crystalline Diamond Coated Seal Faces for Dry Gas Seals in Slow Roll Operation. Technical Paper IREC 2012, Copyright: Compressors, Compressed Air and Vacuum Technology Association within VDMA e. V, 2012.
[51] API 692 Part 2-Dry Gas Seals.
[52] Clark K. n.d. The P-F curve: One of the first, yet hardest, things to learn. Available at: https://www.isa.org/intech-home/2019/march-april/features/improving-maintenance-by-adopting-a-p-f-curve-meth. (Accessed 9 April 2023).
[53] R. Kurz, D. Burnes, D. Reitz, Performance of industrial gas turbines, in: Proc. 51st Turbomachinery and Pump Symposia, Houston, TX, 2022.
[54] C.W. Allen, C.M. Holcomb, M. de Oliveira, Gas turbine machinery diagnostics: a brief review and a sample application, in: ASME Paper GT2017-64755, 2017.
[55] C.W. Allen, C.M. Holcomb, M. de Oliveira, Estimating recoverable performance degradation rates and optimizing maintenance scheduling, in: ASME Paper GT2018-75267, 2018.
[56] M. Venturini, D. Therkorn, Application of a statistical methodology for gas turbine degradation prognostics to alstom field data, in: ASME Paper GT2013-94407, 2013.
[57] K. Brun, R. Kurz, Myth: The Digital Twin Knows It All, Turbomachinery International, 2022.
[58] C.W. Allen, C. Holcomb, R. Zamotorin, R. Kurz, Optimal Parameter Estimation for Efficient Pipeline Simulation, PSIG 2118, Pipeline Simulation Interest Group, 2021.
[59] A. Suman, M. Morini, R. Kurz, N. Aldi, K. Brun, M. Pinelli, P.R. Spina, Quantitative CFD analyses of particle deposition on a transonic axial compressor blade, part II—impact kinematics and particle sticking analysis, J. Turbomach. 137 (2) (2014) 021010.
[60] R. Kurz, K. Brun, Degradation in gas turbine systems, Trans. ASME J. Eng. Gas Turbines Power 123 (2001), https://doi.org/10.1115/1.1340629.
[61] R. Kurz, K. Davis, R. Kaiser, M. McBain, K. Brun, Site performance testing of centrifugal compressors and gas turbine drivers, in: Proc. 50th Turbomachinery and Pump Symposium, Houston, TX, 2021.

CHAPTER

4

Applications

Sebastian Freund[a], *Matthew Holland*[b], *Rainer Kurz*[c], *James Underwood*[c], *and Bernhard Winkelmann*[d]

[a]Energyfreund Consulting, Munich, Germany [b]Southwest Research Institute, San Antonio, TX, United States [c]Solar Turbines, San Diego, CA, United States [d]S&B Enterprises, La Jolla, CA, United States

Introduction

Emissions from fossil fuel combustion have reached new record highs in almost every year of recent history as of 2023. This alarming development is driving regulatory efforts and incentive policies in many jurisdictions around the world to reduce emissions and transform all sectors of the economy toward lower and ultimately zero reliance on fossil fuel, oil, gas, and coal, a process known as decarbonization. Inevitably, the transformation will bring increasingly drastic changes in the kinds of energy used and the ways energy is transported and transformed in almost all applications in the near future (Fig. 4.1).

Most greenhouse gas emissions are CO_2 linked to fossil fuel combustion related to energy supply. Second are methane emissions, which are to a large fraction a result of oil and gas production and also connected to our energy supply: primary energy, the basic supply of energy for any application in all sectors of the economy, is still dominated by fossil fuels that emit CO_2 (Fig. 4.2).

The total global supply of primary energy in terms of heating value was about 600 Exajoule (10^{18} J or 166,670 TWh) in the year 2020 [1]. To illustrate this amount of energy in terms of mass, let us imagine all of this would be oil (heating value 42 MJ/kg): the equivalent energy supply would be about 14 billion tons or 100 billion bbl per year. This corresponds to about 5.5 L of oil per day on average for each of the 8 billion inhabitants. Given that most of the energy, except for locally sourced electricity, needs to be transported to the end user as fuel, this is an incredible amount to transport. Also note that despite all efforts of saving energy and increasing efficiency over the past decades, the global supply has risen constantly over the years and there is no discernible trend to the contrary for the near future as of 2023.

Energy is transported to the end user in various forms and used eventually, e.g., as heat, power, light, or transportation fuel. When dividing the total energy used in various forms by four sectors of the economy, one finds that in typically developed countries, about one-third of the energy is used in industry, one-third in residential and commercial buildings, and one-third for transportation (Fig. 4.3).

Over the last two centuries, mankind has built an impressive amount of infrastructure from roads to railroad tracks, airports but also electrical networks as well as pipelines for liquids and gases. Those investments have consumed not just many years to fully cover very large territories but also cost trillions of Dollars to complete. Whatever future energy mix will be pursued, it will be an advantage to efficiently utilize these assets and build upon the existing infrastructure: Those who currently own and operate infrastructure, and those whose products are sold via this infrastructure both have vested interests in keeping their business. These include large corporations, state-run enterprises, and public companies, all of which have strong political influence. It is almost guaranteed therefore that the inevitable transition toward low-carbon energy will be firstly gradual and nondisruptive, and secondly start with products using existing infrastructure such as renewable electricity, liquid renewable fuels, or biogas and hydrogen, perhaps blended with natural gas.

Examples of infrastructure to be reused and expanded include the electricity grid upgraded to cope with an increasing influx of renewables and natural gas pipelines converted to hydrogen, which are discussed in this book in more detail in dedicated chapters. Road, rail, and shipping lines may transport an increasing amount of renewable liquid fuels such as ammonia, methanol, and synthetic fuels while the amount of petroleum products may decrease.

Energy Transport Infrastructure for a Decarbonized Economy
https://doi.org/10.1016/B978-0-443-21893-4.00005-2

Copyright © 2025 Elsevier Inc. All rights are reserved, including those for text and data mining, AI training, and similar technologies.

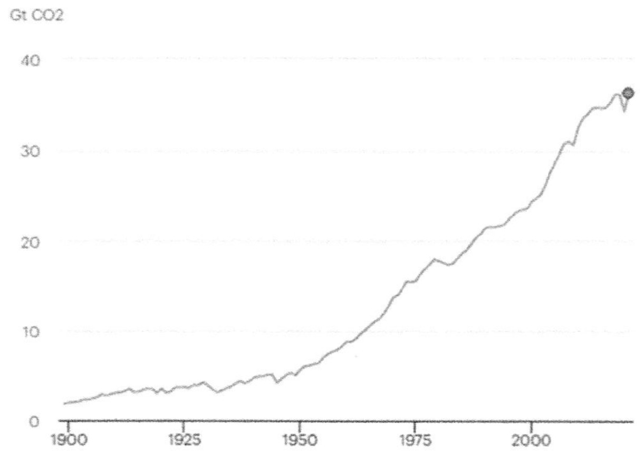

FIG. 4.1 CO_2 emissions reaching a record of 36 billion tons in 2021. *Source: IEA, Global Energy Review, International Energy Agency, Paris, 2021. https://www.iea.org/reports/global-energy-review-co2-emissions-in-2021-2.*

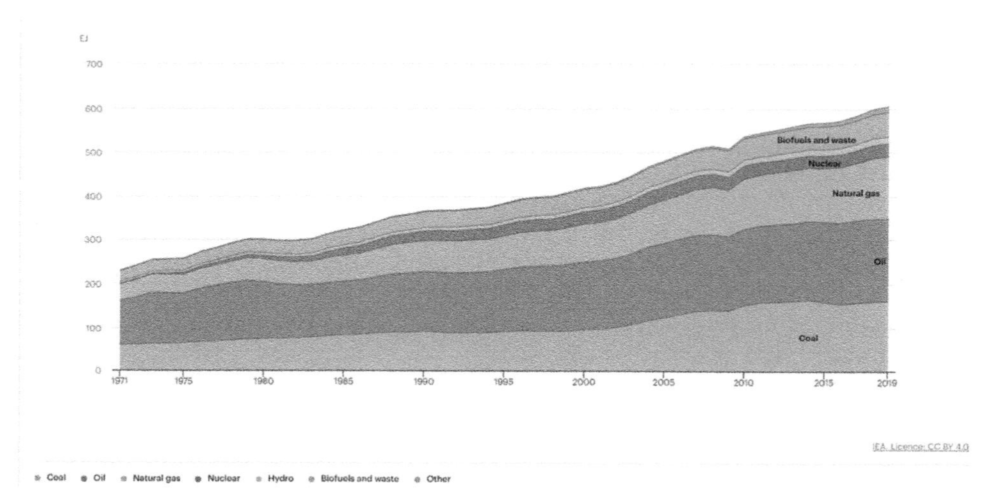

FIG. 4.2 Global primary energy supply by type over the past 50 years [2].

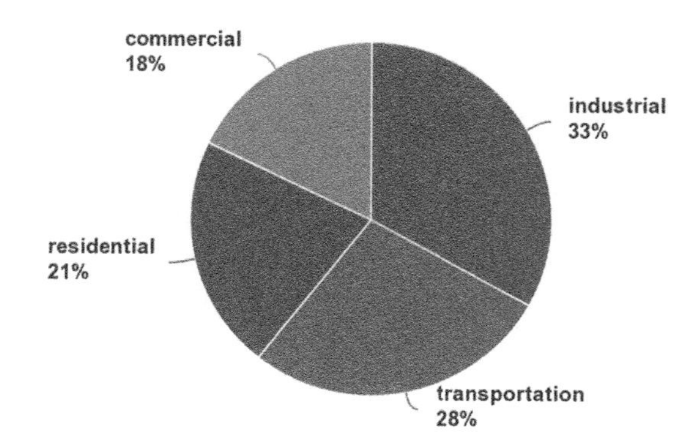

FIG. 4.3 Breakdown of energy consumption in the United States as a typical example for developed industrialized countries [3].

Hydrogen and renewable electricity for the decarbonization of all sectors

To decarbonize all sectors, the forerunners for transitioning away from fossil fuel are currently renewable electricity, biofuels, and low-carbon hydrogen with its derived products such as ammonia, methanol, and synthetic liquid fuels. Renewable electricity is expected not only to replace more and more electricity generated from fossil fuel but also to replace some fossil fuels used for heat through electric heating and for transportation through batteries. There is a clear efficiency and cost advantage of using renewable electricity directly as a fuel substitute rather than indirectly by producing an electrolysis-derived fuel from it. However, the availability of renewable electricity is usually fluctuating and electricity storage is costly and has a limited efficiency too, which is where electrolysis fits in and mitigates the efficiency penalty as less renewable electricity may be curtailed when generated in excess of direct electricity demand. Low carbon hydrogen and its derived products, generated either from renewable electricity or from combustible fuel with carbon capture, are expected to replace fossil fuels in so-called "hard to abate" sectors including long-range transportation, chemical and metallurgical processes. Direct electrification is difficult in long-range transportation because of the low energy density of batteries and may be impossible in other hard to abate sectors, for instance, where fuels are used as a chemical reduction agent.

For development toward a global energy system based on renewable electricity and low carbon hydrogen, the required upscaling of solar and wind power is a multiple of what it was throughout the 2020s. Renewable electricity generation (excl. hydro) reached 3150 TWh/a in 2020 after growth rates of more than 10%/a [4]. Hydroelectricity adds about 4400 TWh/a of renewable energy, but its growth rate is much lower than the other renewables (solar, wind, biomass, geothermal). The additional amount of hydrogen to substitute fossil fuel in hard-to-abate sectors is a multiple of the 2020 production of about 75Mt/a [5], but without carbon dioxide emissions, i.e., requiring even more renewable electricity as well as carbon capture. Biofuels will continue to play a role but their scaling potential seems to be more limited than that of renewable electricity, evident by the low growth rate for biofuels of 4%/a from 2010 to 2020 [4]. An estimate for the amount of renewable electricity required for direct electricity use in 2050 is 90,000 TWh/a (320 EJ/a), or more than three times the current total electricity generation [6]. This includes the replacement of much of today's fossil fuels, which account for up to 70% of total energy consumption. Of the remainder, 20% or about 25,000 TWh/a (90 EJ) may be low-carbon hydrogen used in hard-to-abate sectors or about 10× today's production. Mostly made through electrolysis, hydrogen production will require an additional 35,000 TWh/a of renewable electricity [7]. Together, 125,000 TWh/a of renewable electricity may be needed for decarbonization in 2050, which is almost 40× the amount of renewable electricity generated in 2020. In terms of a growth rate for renewable electricity generation to meet this target within 30 years from 2020 to 2050, this requires an annual increase of 13% or a doubling of renewable generation every 6 years. The historic growth rates of 10% or less have been too low to achieve these numbers and maintaining a high growth rate over a prolonged period of time is very challenging unless technology leaps occur (Fig. 4.4).

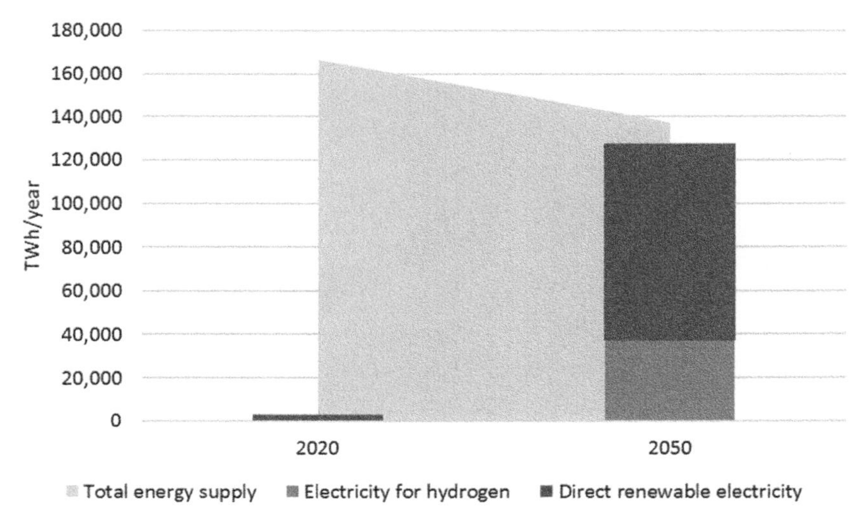

FIG. 4.4 Annual energy supply 2020 and 2050 in a decarbonization scenario using mostly renewable electricity [6] and hydrogen [7].

In summary, a decarbonization of the world's primary energy consumption with a 20% share of hydrogen from renewable electricity in the scenario outlined by the Energy Transition Commission (2021) requires quadrupling current electricity generation, all renewable meaning 40× the renewables installed in 2020, and 10× the current hydrogen generation, mostly from water electrolysis. Total energy consumption decreases until 2050 because of the higher efficiency of direct electricity compared to fossil fuel use and the remainder after direct electricity and hydrogen is covered largely by biofuels.

Hydrogen and water consumption

When forecasting such hydrogen demand, one might wonder if the water consumption for hydrogen production may become a constraint. Fortunately, the oil and gas industry, whose products hydrogen from water electrolysis might one day replace, has answered this question indirectly already as not an issue. In the oil and gas industry, water is consumed for production, e.g., during drilling, for well stimulation, and for enhancing recovery, and further water is consumed in refineries for processing. The global water consumption for oil and gas production was estimated at 21,000 million m^3 per annum in 2008 [8]. When breaking this down per unit of energy, with about 300 billion GJ of oil and gas produced in 2008 [4], the specific water consumption of oil and gas production becomes $0.07\,m^3/GJ$. In hydrogen production by electrolysis, 9 L of water is split for 1 kg of hydrogen, or $0.075\,m^3/GJ$ of lower heating value. Coincidentally, this is a very similar value. In particular, hydrogen and its derivatives can replace petroleum fuel in many applications, and a comparison with oil seems fair. Also, a hydrogen electrolysis process consumes a bit more water than what is finally contained in the gas since water purification and cooling take a share. When accounting for additional water consumption in the electrolysis process by adding 25% retentate from purification and 20% for evaporation losses, the amount of water required per kg of hydrogen becomes 14.4 kg and the specific consumption per unit of lower heating value becomes 120 L/GJ. The production of crude oil may on average consume 80 L of water per GJ of oil and refining may require an additional 40 L/GJ according to the analysis of Spang et al. [8], which totals 120 L/GJ of fuel. Hence the amount of water consumed for making hydrogen is about the same as for petroleum fuel: if all oil would be replaced by hydrogen, the water consumption would be equivalent.

Natural gas applications and LNG growth

While hydrogen is being discussed as the fuel of the future, expected to substitute many other fuels and complement renewable electricity in a zero-emissions scenario, today natural gas is considered a cleaner fuel than coal or oil, and its consumption has increased to about 140 EJ (40,000 TWh) or about 4000 billion cubic meters in 2021, representing about 24% of primary energy supply. The historically rising share of natural gas in the primary energy mix is shown in Fig. 4.2 above. Applications of natural gas are chiefly heating and power generation followed by industrial use in the petrochemical sector (other energy) and a bit of transport fuel and miscellaneous usage. In all scenarios for natural gas demand, the historic growth trend is declining and its future demand depends on how and when global ambitions for decarbonization are implemented (Fig. 4.5).

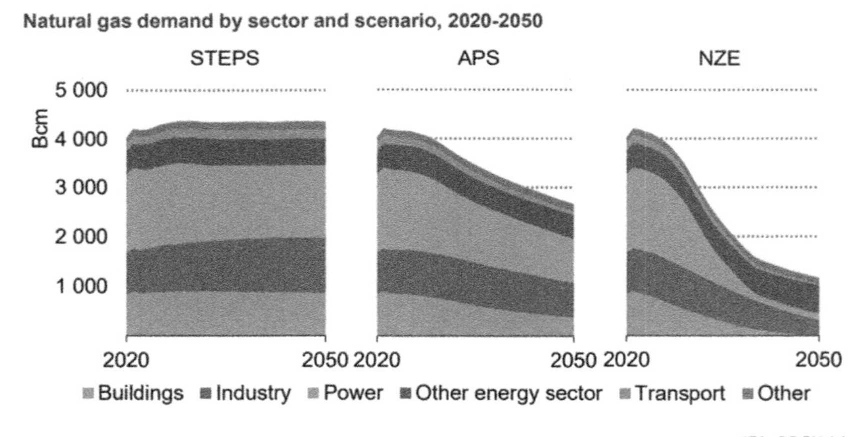

IEA. CC BY 4.0.

FIG. 4.5 Natural gas demand by sector in IEA scenarios Stated Policies, Announced Pledges, Net-Zero Emission. *Source: IEA (2023).*

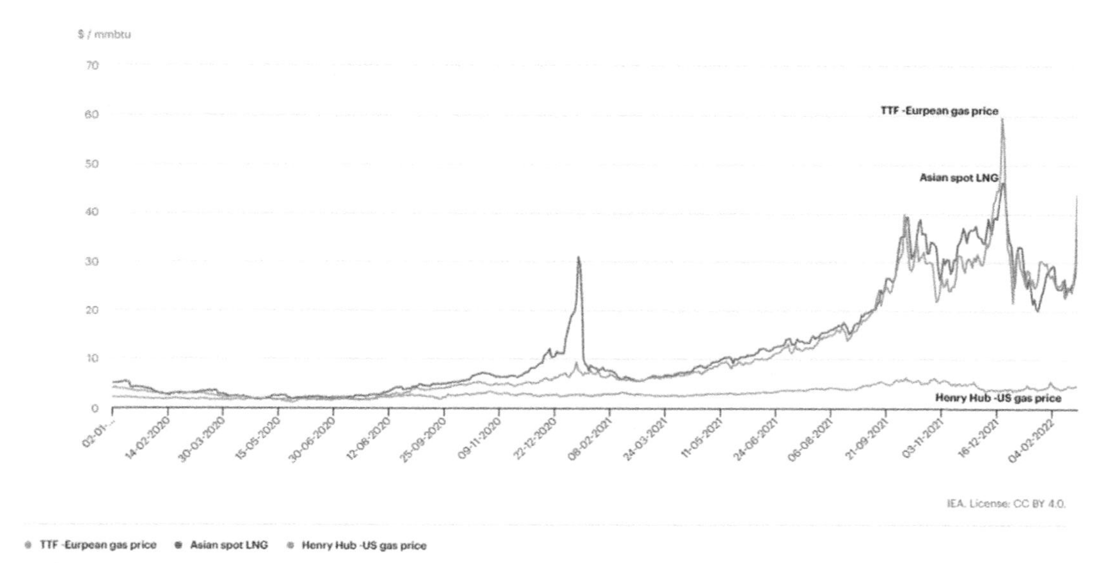

FIG. 4.6 Natural gas spot market prices in the United States, Europe, and Asian LNG import from 2020 to 2022 [9].

Besides decarbonization through a reduction of natural gas use, all climate protection scenarios call for an immediate reduction of flaring and gas leaks. The IEA [10] estimates more than 260 bcm of natural gas is wasted this way (6%), and the latest figures from UNFCCC indicate that fossil methane over a 20-year time period is 87× as potent as carbon dioxide in trapping radiation in the atmosphere and total methane emissions account for more than 25% of global warming.

Unlike oil, which is easily shipped and traded around the world at a global benchmark price, natural gas prices differ greatly by location. Natural gas tends to be inexpensive in regions supplied by pipelines from several gas fields, where availability and competitive markets set prices near the production cost, e.g., at the US Henry Hub. Where pipeline capacity is missing, or when production declines below demand, prices skyrocket. During the European energy price crisis in 2022, natural gas prices at the central European spot markets spiked up to 30× over the previous year (Fig. 4.6).

LNG, traded internationally, on one the hand takes advantage of high spot market prices, but on the other hand can alleviate peak prices by adding additional supply to tight markets when import terminal capacities exist. Some Asian countries, e.g., China, Japan, and Korea have long been relying on LNG imports, and adjusted to high gas prices, while central Europe was well supplied by pipelines from gas fields in the North Sea region and in Russia. Given the substantial geopolitical changes in Europe since Russia's full invasion of Ukraine in 2022, an alternative for Russian natural gas and oil had to be found that was previously delivered by pipelines. As a consequence, Europe had to adjust the natural gas strategy and governments got actively involved in building LNG terminals to become available in the early to mid-2020s. New import terminals have an impact on the pipeline network. As an example, Spain has been receiving pipeline natural gas originating in Northern Europe or Russia. With large LNG volumes anticipated to arrive in the south of Spain, it is likely that Spain will have to reverse pipeline flows to send natural gas to France and possibly further central European countries and pipeline systems may need upgrades for changing volumes and flow directions (Fig. 4.7).

LNG trade and transport capacity has been growing even steeper than natural gas demand and doubled within the decade from 2015 to 2020. This growth may help to dampen price spikes in import markets where competing with pipeline gas. However, LNG with its higher cost compared to pipeline gas will also keep spot market prices higher for pipeline gas, as seen in the convergence of Asian LNG spot price and the European gas price compared to the Henry Hub.

After the energy price crises in Europe in 2022, energy demand was predominantly driven by supply security and availability with a decreased emphasis on emissions: European natural gas demand decreased while coal was temporarily revived. This trend will reverse toward low carbon fuels but an increase in natural gas consumption much above 2019 levels in Europe is unlikely and any trajectory that is not swiftly decreasing within the late 2020s and 2030s is incompatible with the EU's 2050 net zero target. If ambitious policies for climate protection are implemented globally and natural gas demand declines along Net-Zero-Emission scenarios, there will be huge overcapacity in production, processing, and distribution infrastructure including pipelines and LNG terminals and companies now developing new projects may be left with stranded assets. However, some of the infrastructure built for natural gas and LNG might be reused with green or blue hydrogen and ammonia.

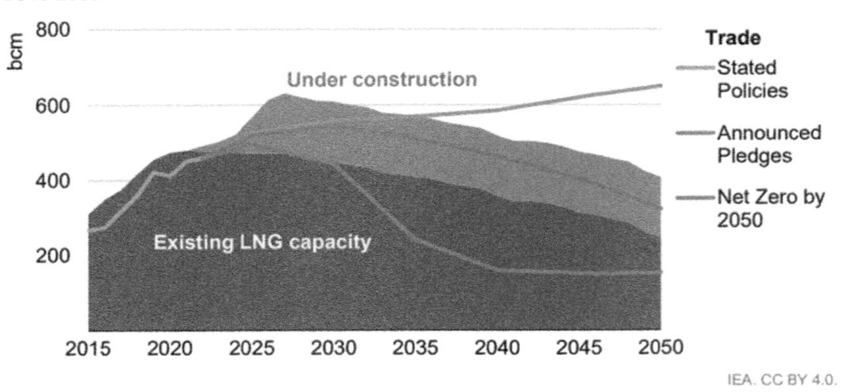

FIG. 4.7 Projection of LNG trade volume in billion normal cubic meters in different IEA scenarios. *Source: IEA, Outlooks for Gas Markets and Investment, International Energy Agency, Paris, 2023. https://www.iea.org/reports/outlooks-for-gas-markets-and-investment.*

Regional cost of energy, regulation, and carbon leakage

Due to transport bottlenecks, taxes, and subsidies, fossil fuel prices and consequently also electricity prices differ substantially around the globe. As a proxy for energy prices, retail gasoline prices are shown in Fig. 4.8 to vary considerably between different parts of the world: Western Europe has the highest prices, followed by China, India, Japan, Korea, and Australia, while the Middle East and many developing countries in North Africa and Central Asia who may not have taxes and even subsidize fuel have the lowest prices. Prices in the United States are middle but appear rather inexpensive relative to income.

High fossil fuel prices combined with increasing costs of CO_2 production through regulations including cap and trade schemes or a carbon tax are a burden for industries with high carbon intensity, such as petrochemistry, steel, cement, glass, of which many produce globally traded goods and thus are facing global competition. Regions with low prices and no carbon regulation have a competitive advantage. In Europe for instance, prices for gas and electricity are higher than in many other parts of the world. The cap and trade scheme adds an additional cost for CO_2 emission

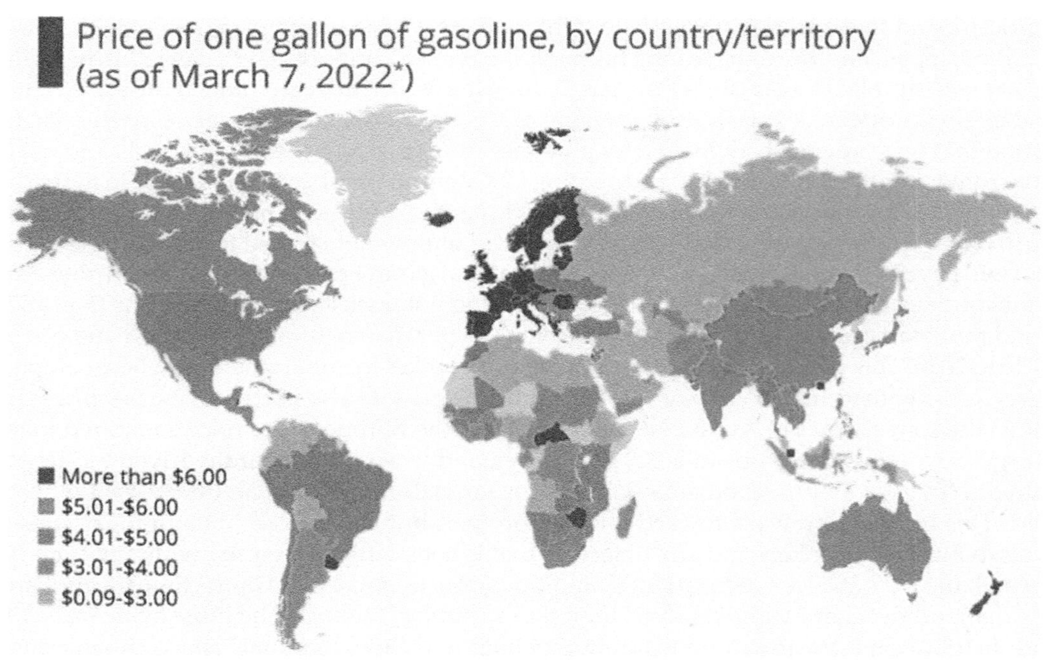

FIG. 4.8 Gasoline prices in US$/gal (3.78 L) around the world in March 2022. *Source: Statista, "How Gas Prices Compare Around the World", by K. Buchholz. 2022. Available online: https://www.statista.com/chart/5316/petrol-prices-around-the-world-visualised/.*

on the order of \$100/ton. Given the new energy situation in Europe, since Russian imports are to be avoided, this disadvantage of fossil fuel prices will probably be persistent. One of the consequences could be the relocation of heavily fossil energy-dependent industries into countries with lower energy costs. In many cases that may mean a shift of production to countries that have lower standards for efficiency and less stringent if any rules for minimizing CO_2 emissions. Ironically, a strong regional regulation of CO_2 emissions could lead to globally higher emissions for the same product volume by shifting production to another legislation without restrictions. This would obviously counter the original goal of emission reduction and outlines one of the difficulties of carbon regulation: Strong opposition of regional industry and unions fearing a shift of production elsewhere, and limited effectiveness in case this shift materializes, a phenomenon called "carbon leakage." To mitigate carbon leakage and protect regional industries, "carbon tariffs" on goods of high CO_2-intensity imported from regions with relatively lax regulations are being considered by jurisdictions like the EU who are advancing carbon regulations. This phenomenon also exemplifies the need to ultimately approach carbon regulation internationally through agreements that implement harmonized regulations around the globe. Efforts to that end include the United Nations Framework Convention on Climate Change and the Paris Agreement of 2015. The potential for carbon leakage can be used by industrial lobbyists as an argument to counter any regional efforts of carbon regulation, halting and delaying the reduction of CO_2 emissions. However, such behavior can be viewed as irresponsible and is actually hampering both, international efforts by not setting leadership examples, and regional economic progress by not becoming less carbon-intensive and developing a competitive advantage for a low-carbon future.

In summary, the adoption of low-carbon energy in industrial sectors heavily reliant on fossil fuels may be lower than expected when production is moved to jurisdictions without decarbonization mandates and lower energy costs: this may delay the energy transition toward net zero and reshape energy and product trade flows.

Stationary applications

Approximately two-thirds of primary energy is used in stationary applications while the other third fuels all our transport. The stationary applications using roughly 400 EJ of primary energy include electricity generation consuming about 250 EJ [4], industrial, residential, and commercial heating, and the use of oil, gas, and coal in the steel and petrochemical industry. Of these applications, power generation and the three industrial sectors with the largest fossil fuel consumption account for 60% of global CO_2 emissions and are considered in this chapter with regard to their decarbonization potential with energy flows and transport issues.

Power generation

Global electric power demand has been rising for the past decades with a rate increasing from 2%/year to more than 3% in 2020–2025, exceeding 25,000 TWh in 2020 [11] (Fig. 4.9).

To satisfy the ever-growing global demand for electricity, power plants are generating electricity from copious amounts of fossil fuels and nuclear fuel in addition to an increasing rate of renewable energies such as hydro, solar, and wind. Most concerning from a global warming perspective for power generation is the use of coal and natural gas with combined emissions of 13.2 Gt in 2022 [12], or about 40% of total CO_2 emissions. Globally in 2020, power plants have consumed about 4500 Mt of coal (about 60% of production) and about 13,000 TWh (1300 billion cubic meters) of natural gas (about 1/3 of production) for generating 15,800 TWh of fossil electricity [4]. Much of the future growth of electricity generation will come from renewables, driven by policies to limit global warming and by cost degression of wind and PV power. This leads first to a relative and over time also to an absolute decline of fossil fuel use for power generation. This shift is illustrated in the diagram below of the share of installed capacity by power plant type and the share of generation in a forecast by the IEA [13] for the Net Zero Emission scenario. Here the world economy is gradually phasing out fossil fuel at a rate consistent with the goal of becoming climate neutral in 2050 by reaching net zero greenhouse gas emissions. This scenario, as of 2023, is more ambitious than scenarios following current trends or announced policies, both of which lead to slower phaseout of fossil fuels and less drastic growth of renewables, electrification, and emission reductions (Fig. 4.10).

The transformation of the electricity generation from fossil fuel toward a mix of fossil, nuclear, and renewables with decreasing CO_2 emissions have started around the turn of the century in most OECD countries. Fig. 4.11 shows the generation by source in the United States for the past 70 years with a trend away from coal since 2005 and a rise of natural gas since the 1980s and renewables since 2000. A very similar trend with a declining share of coal, a rise of

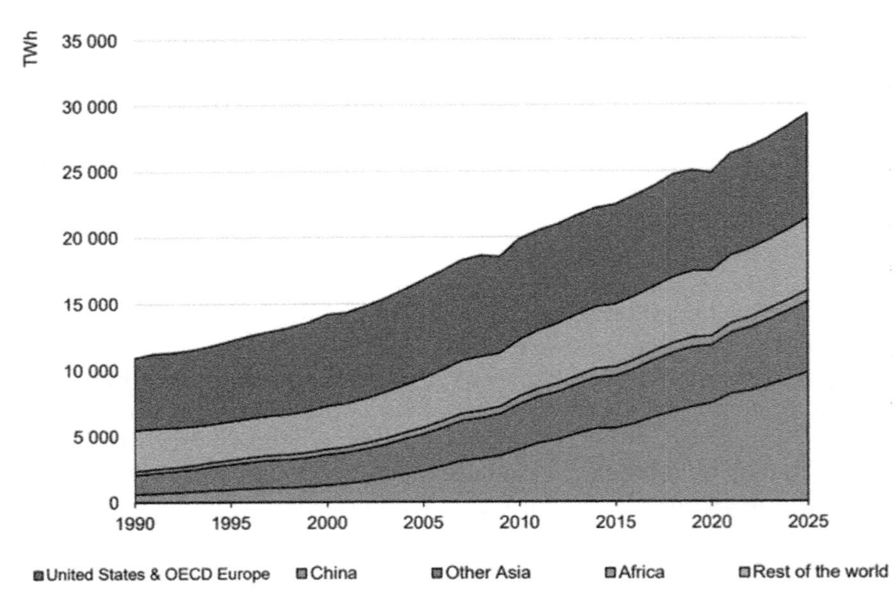

FIG. 4.9 Global electricity demand [11].

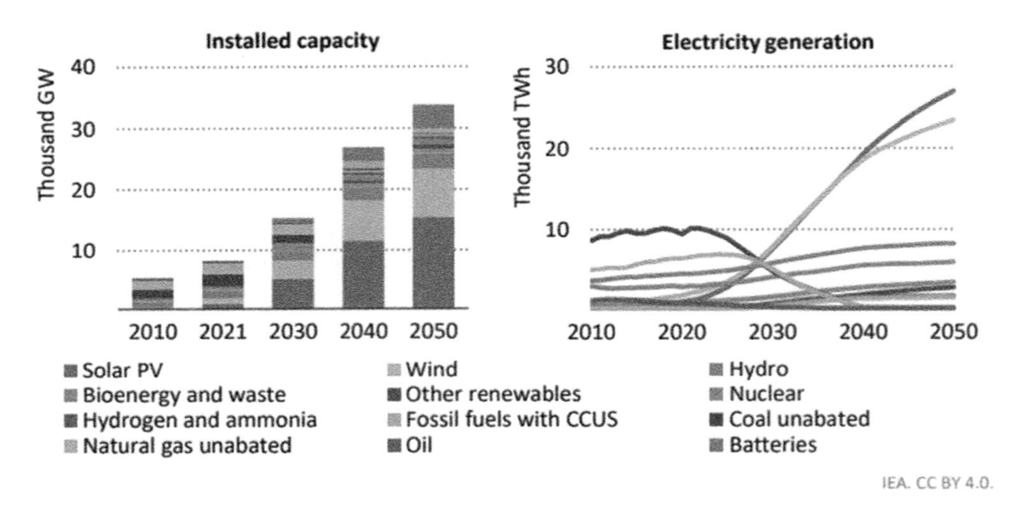

FIG. 4.10 Global installed electric power capacity and generation in a scenario for net zero emissions by 2050 [13].

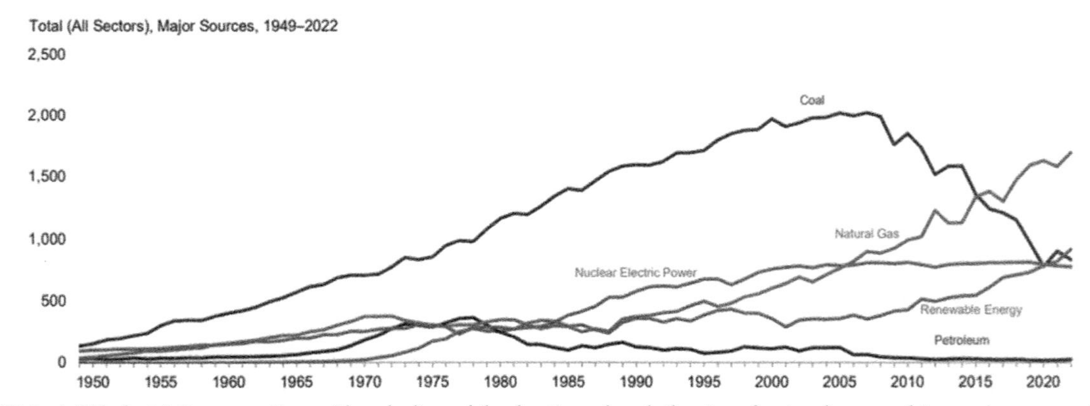

FIG. 4.11 Historic US electricity generation with a decline of the fraction of coal, the rise of natural gas, and increasing renewables [14].

natural gas but lately a drastic increase in renewable generation, even more so than in the United States, is found in Europe and generally in most OECD countries. It can be expected that developing nations will follow this trend eventually although coal is still playing a major role in India, China, and south-east Asia to meet the fast-growing electricity demand.

Gas and oil fired power plants

Gas-fired power plants chiefly include combustion turbine plants, also known as gas turbine generators, in which a gas turbine burns natural gas to drive the generator. These plants are characterized by high power density and fast startup, they are installed by utilities for their flexibility and low investment cost. When a gas turbine plant is combined with a steam turbine cycle using the turbine's waste heat, they become gas turbine combined cycle power plant. These, of all common power plants, have the highest efficiency and with their low specific cost have become the most popular type of utility power plant. A typical gas turbine generator may have an efficiency of electric power output over fuel lower heating value input of around 40%: in a combined cycle power plant, the remaining 60% of waste heat is converted by a steam turbine generator with an efficiency of around 30% into additional electricity, resulting in about 58% overall efficiency. Modern combined cycle plants can reach even higher values. Natural gas-fired steam boilers with steam turbines have traditionally been used in power generation but lost competitiveness to gas turbines because of lower efficiency compared to gas turbine combined cycles and higher costs.

Worldwide as of 2022, about 7418 gas turbines and combined cycle plants as well as some gas and oil boilers and large reciprocating engines are installed in 3293 plants. Their total capacity is 1840 GW of power, producing 6520 TWh (gas-fired) and 720 TWh (oil-fired) of electricity per year [4,15].

Gas turbines are often configured with dual fuel combustors for fuel flexibility to burn liquid fuel, mostly fuel oil, as back-up or as main fuel. Such back-up fuel can be a decarbonized liquid fuel such as a biofuel, methanol and other alcohols (based on green hydrogen or from fermentation), synthetic hydrocarbons (based on green hydrogen and recycled carbon dioxide), or hydrocarbons from pyrolytic processing of waste or other organic matter. Likely future options of gaseous main fuels to replace natural gas are hydrogen and ammonia, delivered via pipeline. Manufacturers have been working on new combustors and retrofit options for firing gas mixtures or pure hydrogen while keeping NOx emissions within regulatory limits, details, and examples can be found in literature, e.g., Schmitt et al. [16] and Omatick et al. [17]. One gas turbine manufacturer building a combined heat and power to be operating by 2023 for the municipal utility in Leipzig, Germany, announces that the two 62 MW industrial F-class units with dry low-emission combustors could run on 100% hydrogen and switch from natural gas to hydrogen as soon as this becomes available [18]. Table 4.1 shows a comparison of the alternative fuel flow rates for a large gas turbine and the substantial increase in fuel volume when switching from natural gas to hydrogen.

Diesel and natural gas fired engines

Internal combustion engines are increasingly used for power generation, often as combined heat and power, and substitute gas turbines typically for distributed generation when power demand is small, kW to single-digit MW-scale. Recently, their relatively low specific cost, good partload efficiency, and fast ramping characteristics have led to power plant projects with multiple large engines reaching a total power on the order of 100 MW and competing with gas

TABLE 4.1 Mass and volume flow in an H-Class single 1–1 configuration gas turbine combined cycle power plant.

Capacity	800	MW electric gross
Efficiency	60	%
Heat rate	6000	kJ/kWh
Natural gas consumption	102,000	kg/h @ LHV = 47 MJ/kg
Natural gas consumption	126,000	m^3/h @ 15°C, 1 atm
Natural gas CO_2 emissions	290	tons/h
Hydrogen consumption	40,000	kg/h @ LHV = 120 MJ/kg
Hydrogen consumption	465,000	m^3/h @ 15°C, 1 atm
Ammonia consumption	242,000	kg/h @ LHV = 19 MJ/kg

Comparison between natural gas, hydrogen, and ammonia.

turbines. Stationary Diesel engines are also used for back-up power as emergency generators and often for temporary installations and off-grid or island power.

Engines for natural gas or biogas are turbocharged Otto cycle engines with various injection systems and mostly spark-ignition. These engines can be redesigned for using decarbonized fuel. Carbon-free or carbon-neutral gases suitable for gas engine generators include

- Biogas or biomethane (similar to natural gas, from anaerobic digestion of plant matter)
- Synthetic natural gas (same as natural gas, Sabatier process of green hydrogen and recycled CO_2)
- Hydrogen (green or blue)
- Ammonia (e.g., Haber-Bosch process of green hydrogen and air)

While biogas, biomethane, and synthetic natural gas are drop-in fuel options for existing natural gas engines, hydrogen, and ammonia require modifications of the combustion and air-intake system. Owing to the large volume of hydrogen, the maximum (rich burn) power density decreases and the engine power at a given displacement is only about 80% of the value reached with methane or gasoline. Because of hydrogen's combustion characteristics, flame speed, and ignitability, very lean burn combustion systems, EGR, and high-pressure turbocharging are required for NOx emission compliance and acceptable power density of hydrogen gas engines [17,19].

Stationary gas engines for power generation may be dual-fuel engines derived from Diesel engines and equipped to burn liquid fuel as backup or as main fuel. Instead of common petroleum-derived fuels such as fuel oil, decarbonized liquid fuels can be used. These fuels include

- Biofuels (plant oils including the diesel alternative fatty-oil methyl ester)
- Methanol and other alcohols (based on green hydrogen or from fermentation)
- Synthetic hydrocarbons (based on green hydrogen and recycled carbon dioxide)
- Hydrocarbons from pyrolytic processing of waste or other organic matter
- Dimethylether (based on green hydrogen)

Some of these liquids may be drop-in replacements for diesel or fuel oil, while others may require modifications of the combustion system or aftertreatment systems because their combustion characteristics or transport properties are too different. Fuel mass flows may be higher, e.g., for alcohols, as their heating value is lower than that of fuel oils, requiring larger storage and transport volumes and subsequently larger pumps and pipes.

Similar to dual-fuel gas engines, diesel engines may be redesigned or retrofitted with additional injection systems for using ammonia and hydrogen in addition to a fraction of diesel or biofuel for ignition through a pilot injection. For fuel flexibility, their liquid injectors may be designed to run also on liquid fuel alone as a back-up. This option is interesting for the decarbonization of stationary diesel engines when green gaseous fuels are available.

Solid fuel-fired power plants

Coal-fired power plants burning hard coal or lignite are the traditionally dominant type of thermal power plant for electricity generation. Worldwide, as of 2022, about 2400 coal-fired plants with 2100 GW of installed power are in operation, producing 9450 TWh of electricity per year and causing 9700 million tons of CO_2 emissions in 2021 [20], or about 27% of total worldwide CO_2 emissions. Actually, more new coal capacity is planned or under construction than what is being retired [21] and the current path of coal capacity is incompatible with any climate protection goal.

In a coal plant, hard coal or lignite-fired steam boilers supply steam turbines sized up to 1 GW of electric power per unit. These plants involve enormous mass flows of coal and combustion products (flue gas including CO_2 and other pollutants like sulfur dioxide, nitrous oxides, etc.) and subsequently large mass flows of byproducts for cleaning the flue gases like gypsum from desulphurization and ammonia for catalytic denitrification (SCR Denox) and fly ash from electrostatic precipitators or filters. Most interesting in terms of energy transport are the mass flows of coal and CO_2. In a decarbonized scenario, if these plants are kept alive, either coal is to be replaced by alternative carbon-free or carbon-neutral fuels, or the CO_2 needs to be captured and sequestered.

With respect to transport, typical mass and volume flows for a modern 2×800 MW supercritical coal plant with 45% gross efficiency operating in mid-merit schedule of 16 full load hours per day are quite substantial:

- Coal (LHV 24 MJ/kg): 8500 tons/day, corresponding to a train with 85 cars and 1380 m in length
- Fly Ash (8% ash content): 680 tons/day
- Gypsum from flue gas desulphurization (1% sulfur content): 460 tons/day
- CO_2: 18,800 tons/day

TABLE 4.2 Comparison of volumes for a large coal plant with bituminous coal and wood pellets.

Item	Unit	Hard coal	Wood pellets
Gross power	MW	2 × 800	2 × 800
Gross efficiency	%	45%	45%
Daily full load hours	h	16	16
Heating value	MJ/kg	24	17
Density	kg/m^3	850	675
Consumption	tons/day	8533	12,047
Supply railcars @ 100 t coal	#/day	85	152
Supply train length @ 16.2 m per car	m	1382	2458

If this plant were equipped with a 90% effective CCS system, a daily volume of 16,900 tons or 24,700 m^3 pressurized liquid (25°C) would have to be removed from the plant and sent to a sequestration site. This could be done best by a pipeline. Alternatively, after chilling the CO_2 to lower the pressure, it can be transported by cryogenic tank rail cars (e.g., VTG type G92.062D with 62 m3), requiring 272 cars or a train with a length of 3900 m each day (note that this train has almost 3× the length of the coal supply train).

Though coal dominates power generation from the combustion of solid fuels, there are other substitute fuels for such boilers. Carbon-free or neutral fuels considered for coal-fired power plants are for example wood chips, wood pellets, and fuel made from combustible refuse ranging from sewage sludge, and dried organic waste to processed domestic or industrial waste including plastic scraps. A coal boiler can also be converted to burning gases, including carbon-free or neutral gases like hydrogen. However, gases may be used with higher efficiency in gas turbine combined cycles and present a niche solution for coal plant retrofit or life extension for use as backup or as a peaking plant rather than baseload.

An example of the scale at which biomass can replace coal is Drax Power Station in Selby, UK. Here 4 of 6 units of 660 MW each have been converted from coal to wood pellets since 2012. Drax has been burning about 20,000 tons of pellets per day according to their website [22] that were sourced from international markets and delivered by train from the nearby port of Immingham (Table 4.2).

Pipelines for natural gas, hydrogen, and captured CO_2

The volume flows required for decarbonized power generation involving hydrogen, natural gas, and captured CO_2, possibly ammonia and other fuels may be larger even than with fossil fuels and one solution for the transport problem are pipelines. Generally, pipelines provide by far the most economical way to transport gas and liquids under a wide range of pressures over short and medium distances. Only for very long distances, or where no continuous demand and rather flexibility in time or location is required, rail or ships may be more economical.

Carbon neutrality in a power plant fueled with gas can be achieved by either bringing decarbonized energy into the plant via a hydrogen or ammonia pipeline, or by traditional natural gas brought by pipeline with added carbon capture. In the latter case, one would either create hydrogen (at the plant site) from natural gas (blue hydrogen) or use the natural gas as a fuel, and subsequently capture the CO_2 from the exhaust [23]. In the scenarios where CO_2 is generated, decarbonation happens by transporting and sequestering the CO_2 captured from the exhaust gas of fossil-fired power plants or from the process gases of blue hydrogen or ammonia generation.

With pipelines and large volume flows, gas compression technology is key for economical operation, even more so when dealing with additional volumes of CO_2 needing compression after capture, e.g., from atmospheric pressure to pipeline pressure for transportation and eventually even higher pressure for geologic sequestration [24]. Long pipelines need intermediate boost compressor stations to keep up the gas flow. Analysis and optimization will be needed to evaluate the best options for building or upgrading such infrastructure. For instance, for a gas turbine plant, supplying blue hydrogen from a hydrogen plant through the pipeline to the power plant, with the compression energy required for transporting hydrogen, versus local production of hydrogen from natural gas reformation at the power plant or even postcombustion CO_2 capture, and using one pipeline with much lower compression power for transporting natural gas and one for transporting captured CO_2 back to a sequestration site.

Industrial

Globally the industrial sector has been estimated to account for 54% of all end-use energy consumption. In developed economies, the share of industrial energy consumption is a bit lower as private consumption levels are higher. In the United States for instance, the industrial sector accounts for 33% of all energy consumption making it the largest consuming sector followed by transportation (28%), residential (21%), and commercial (18%). Fig. 4.12. below shows the breakdown of the energy consumption by primary energy source type.

Most of the industrial sector energy use is in the form of fuel, mostly natural gas and oil, that is used for heating but also as a feedstock for nearly all chemical products. Globally, the industrial sectors with the highest consumption of fuels include petrochemical, cement, and steel. Those three sectors are described in subchapters here in the following, analyzing their current and future energy requirements.

Besides fuel, the industry consumes electricity. The primary loads in the industrial sector include large motors for driving pumps, compressors, and other machinery. Additional loads include a rising amount of electric heating and cooling for processes and building HVAC. Most industrial sites are connected to the electrical grid and purchase their power from local utilities; however, a large portion has onsite generation of various forms. These onsite power generators often use byproducts of the industrial processes as fuels. The fuel flexibility of these onsite generators allows switching between the primary energy sources such as natural gas, oil, or coal and industrial byproducts. One example of this is an industrial gas turbine used in the steel industry fired with coke oven gas, a byproduct, which is rich in hydrogen, methane, and carbon monoxide. The turbine can switch to natural gas or fuel oil operation when the industrial process conditions are unable to meet the fuel demands. In some regions where byproducts are not fully exploited locally, markets are evolving to transport and sell these industrial byproducts as fuels, separating out the methane and hydrogen. These industries are seeing an evolution from flaring, to onsite consumption, to exporting of what was once considered process waste streams (Fig. 4.13).

The largest consumer of energy in the U.S. industrial sector is manufacturing at an astounding 80%, followed by mining (9%), construction (6%), and agriculture (3%) per the EIA's annual energy outlook. The manufacturing sector is dominated by chemical production followed closely by petroleum and coal products, together they account for almost 60% of the energy consumed in the industrial sector which equates to almost 20% of all US energy consumption. As we make the transition toward carbon neutrality, industrial processes will need to be modified to accommodate the shift from fossil fuel to renewable fuels or electricity and alternative sources of carbon such as biomass and captured CO_2 will need to be used in the petrochemical industry to enable a sustainable future.

Decarbonized steel production

Steel is one of the most important globally made commodities with an annual trade volume in 2020 of 1860 million tons. The traditional production of primary steel, i.e., steel made from iron ore rather than secondary steel made through recycling, is an energy-intensive process using coal and fossil energy and emitting on average 1.85 tons of CO_2 per ton of steel, accounting for about 8% of the world's CO_2 emissions [25]. Traditional steel production happens in two steps: First, sintered iron ore is turned into pig iron in a blast furnace using coke as a heating and reduction agent. Second, the iron is turned into steel through the basic oxygen converter process. Both are energy-intensive

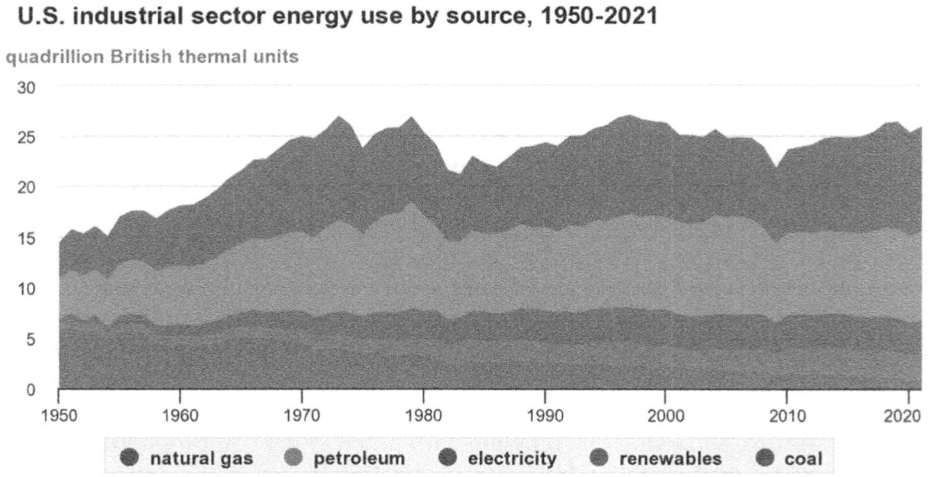

U.S. industrial sector energy use by source, 1950-2021

quadrillion British thermal units

natural gas · petroleum · electricity · renewables · coal

FIG. 4.12 Energy use by type in the US industrial sector. Note: 1 Quadrillion BTU = 293 TWh [26].

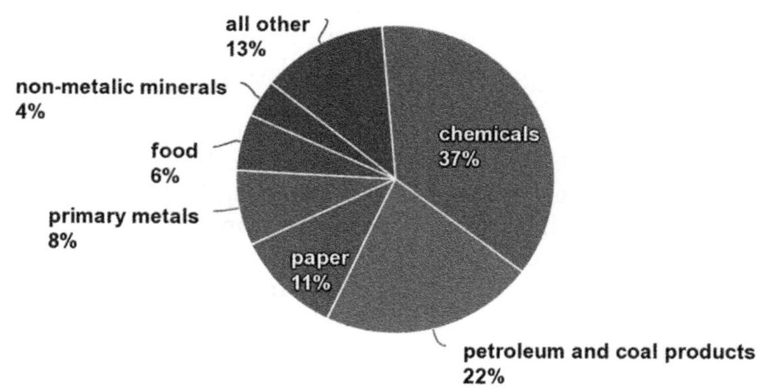

FIG. 4.13 Breakdown of industrial energy use share in the manufacturing sector in the US including fuel and electricity [26].

high-temperature processes that release CO_2 and substantially add to greenhouse gas emissions when using fossil energy. Blast furnaces in particular through the reaction of iron oxide with carbon monoxide from coal into iron and CO_2. The coke production and the iron ore processing are also CO_2 intensive. The basic oxygen converter produces CO_2 by consuming fossil energy and oxidizing some remaining carbon as well as requiring slag building minerals such as CaO that liberate mineral CO_2 from carbonate. Both steps can be decarbonized in various ways by changing to renewable energy and alternative processes. Carbon capture and sequestration is a possibility that has been tried for partial decarbonization but appears to be too costly and complex for general implementation in steel making. Iron ore reduction in a blast furnace produces less CO_2 when coal is partially replaced with natural gas or especially with biogas or hydrogen. Natural gas is easily available for steelmakers and relatively inexpensive; it is already used to a limited degree in steel production. Biogas can be substituted for natural gas but is typically more costly. Hydrogen from coke oven gas as well as from coal or natural gas reformation is also used already in some plants with an annual volume of about 9 Mt [5]. Further decarbonization can be achieved in the "direct reduction" process, which is an alternative to the blast furnace process. Direct reduction uses a different furnace and a reduction gas and no coal, the product is sponge iron, a low-carbon raw iron, rather than traditional pig iron. Natural gas or coal-derived syngas has been used in the direct reduction process but the use of green hydrogen seems to be the current path to decarbonization. Biogas has not been available to steel makers in economic quantities or costs. The amount of hydrogen required for direct reduction is 43 kg H2/t of iron or up to about 50 kg H_2/t of steel. 4 Mt of dedicated hydrogen is used already for direct reduction steel making [5], with estimates for H_2 use increasing to 8 Mt in 2030 and 45 to 60 Mt in 2050. If this hydrogen was made from water electrolysis and consumed 55 kWh/kg, the green hydrogen for steel production would require 3300 TWh of renewable electricity in 2050. Demand for huge quantities of hydrogen will change the way energy is transported to steel plants, unlike coal arriving by train or barge, either hydrogen or copious amounts of electricity need to be provided. Hydrogen production through electrolyzers near off-shore wind sites is one possibility, and pipelines may be used where grid capacity would be insufficient. In fact, hydrogen pipelines are planned for several steel factories in Germany to connect with large electrolyzer projects on the coast until 2030.

Sponge iron from direct reduction furnaces can readily be converted in the electric arc process. Steel conversion is already increasingly done through the electric steel process in electric arc furnaces with graphite electrodes that can run on renewable electricity to reduce CO_2 emissions. Thus, using green hydrogen and renewable electricity makes it possible to completely decarbonize the steel production processes, and several steel manufacturers have announced or started pilot projects.

The cost of decarbonized steel will be higher than traditional production unless substantial carbon taxes are implemented. A cost comparison in very simple terms between the blast furnace and basic oxygen route and the direct reduction and electric arc furnace route for primary steel production is given as follows (neglecting various opex and capex cost differences of the processes): Electric arc furnaces consume perhaps 900 kWh/t steel. The green hydrogen required for direct reduction is about 50 kg/t and may cost $200/t of steel. Together, at $50/MWh for green electricity, these costs may be $245/t. This offsets about 400 kg of coke, costing perhaps $120, resulting in an increase of

\$125/t but avoiding more than one ton of CO_2 emission. For crude steel traded at, e.g., \$500/t, this is a significant price increase; for high-performance steel alloys trading at a multiple of the crude steel price, the green cost premium may be about 10%.

An example of the initial implementation of zero-carbon steel production is the "tkH2Steel" program from ThyssenKrupp Steel in Duisburg, Germany, who are developing a direct reduction pathway for iron and steel production based on hydrogen and electrification with a volume of more than Eur 3B, of which Eur 2B are government funded. Bei 2027, the new production should annually make 2.5 Mt of steel, consume 140,000 t of H2 delivered by pipeline from an electrolysis project, and save 3.5 Mt of CO_2 emissions [27].

Similarly, Salzgitter AG, another steel producer specialized in high-quality automotive sheet metal and pipes in Germany, is undertaking the "SALCOS" project to decarbonize steel production using hydrogen and renewable electricity for direct reduction iron making. They announced obtaining Eur 1B of government funding for the pilot to be commissioned by 2025 to produce 1.9Mt of steel [28]. Salzgitter plans, after full implementation, to save around 7Mt or about 1% of current total German CO2 emissions.

In Austria, Voestalpine has been undertaking several research projects using natural gas, hydrogen, and electricity for steel decarbonization under the "Greentec Steel" brand, including research with a 6 MW electrolyzer in Linz, an innovative prototype of a hydrogen plasma direct steel process in Donawitz, and direct reduction test facilities [29]. Through electric arc furnaces and less raw iron from blast furnaces, they plan to save 5% of Austrias annual CO_2 emissions after 2027.

Another pilot plant is announced by Sweden's SSAB Steel with starting the "Hybrit" project, where together with mining firm LKAB and utility Vattenfall a fully decarbonized steel production value chain is being demonstrated. The goal produces 1.35 Mt of steel annually using a 500 MW fossil-free electrolysis with underground hydrogen storage, thereby reducing 10% of Sweden's current CO_2 production [30].

Decarbonized cement production

Thirty billion tons per annum and growing-concrete is the most copiously used material. Its most important ingredient is cement, which at an annual production of over 4 billion tons [31] accounts for about 9% of global CO_2 emissions and 3% of worldwide energy use [32]. CO_2 is a byproduct of cement production both from the calcination of limestone for Portland cement, which is the most frequently used type of cement, and from the heating of the reaction kiln to high temperatures using fossil fuels. Portland cement requires limestone (calcium carbonate) and other minerals to be heated to 1450°C (2640°F) in a kiln to remove its carbon, producing lime (calcium oxide) in a calcination reaction, and reacting further to form a mixture of calcium silicates and aluminates, called clinker. Normal Portland cement mainly contains ground clinker and a fraction of gypsum plus minor additives. The calcination reaction is responsible for about 60% of Portland cement's CO_2 emissions. The heating of the kiln requires significant amounts of energy, usually from burning fossil fuels, causing an additional 40% of CO_2 emissions. A reduction of CO_2 emissions for Portland cement can be obtained by producing clinker from raw materials mixtures emitting less carbon or requiring lower kiln temperatures, by reducing clinker through blending with cementitious substances such as blast furnace slag from iron production or fly ash from coal-fired power plants, or, as researchers have recently found, through electrolytic production [33]. Besides Portland, there are other types of cement using different raw materials, such as pozzolanic minerals (silica and alumina) as well as entirely different chemistries and processes, even of biological origin from cementitious plant ash or microbial precipitation of limestone similar to corals. However, these substitutes with potentially lower CO_2-impact have not yet achieved a scale to solve the CO_2-problem of Portland cement and currently appear to be of limited supply (Fig. 4.14).

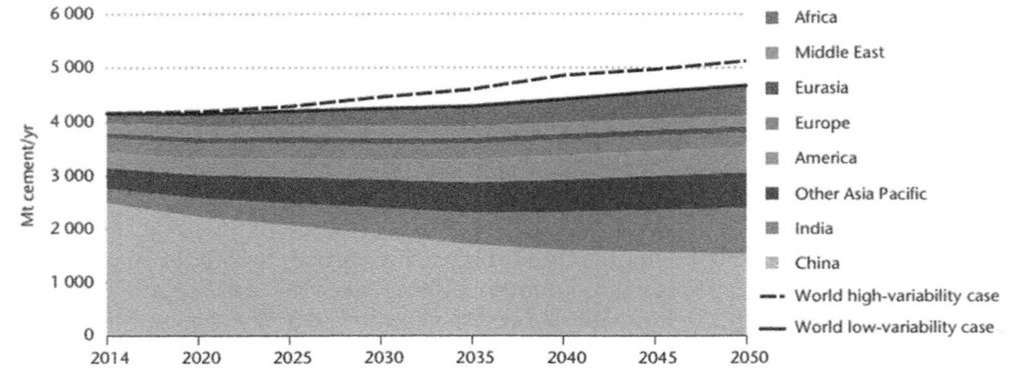

FIG. 4.14 Cement production forecast by region. 10% to 20% growth expected until 2025. *Source: IEA, Technology Roadmap: Low-Carbon Transition in the Cement Industry, IEA, Paris, 2018. Available online: https://www.iea.org/reports/technology-roadmap-low-carbon-transition-in-the-cement-industry.*

A modern cement kiln consumes around 3 to 3.5 GJ of fuel for producing one ton of clinker, depending on the efficiency of the process, the fuel type, and the raw materials [34]. Predominantly fossil fuels such as coal and petroleum coke are burned, to a lesser extent natural gas and fuel oil. On a global average, of the 3.5 GJ thermal energy per ton that clinker kilns today use for heating, still only 7% is provided by renewable and nonrenewable waste fuels [35]. The calcination process and the thermal energy use together account for a CO_2 intensity of about 0.6 tons CO_2 per ton of clinker on a global average. Besides improving the energy efficiency of the process, the use of alternative fuels (e.g., biomass, refuse-derived fuel, hydrogen, and green electricity) would reduce the carbon footprint. The amount of hydrogen when substituting 90% of the fuel for current cement production would be about 44 Mt in China, 7.7 Mt in India, 2.5 Mt in Vietnam, and 2 Mt in the US to list the four largest producers by country. However, the clinker process lends itself quite well to CO_2 capture and sequestration: the flue gases contain a high concentration of CO_2, and the waste heat from the kiln can be used for the regeneration of a CO_2-absorbent such as an amine solution. Potentially about 90% of 2600 Mt CO_2 per annum could be captured from the annual production of 4300 Mt of Portland cement [35]. This would necessitate large-scale CO_2 infrastructure including pipeline transport and geological sequestration. The potential amount of CO_2, captured, compressed, and transported to be sequestered would be about 960 Mt in China, 170 Mt in India, 55 Mt in Vietnam, and 44 Mt in the United States.

Decarbonized petrochemical production

The petrochemical industry supplies industrial chemicals that form the basis for many products of everyday life in developed economies. Major products of the petrochemical industry with their approximate annual volume in 2018 are given in Table 4.3 [36]. The total product output reached almost 700 Mt now and has seen annual growth rates of about 2...3%. Most feedstocks for the primary chemicals, crude oil fractions, originate at refineries. Most of the primary chemicals themselves are feedstocks for the chemical industry. Thus, the petrochemical industry bridges and merges, upstream, the oil and gas industry, with the chemical industry, downstream.

With these seven primary chemicals, their direct derivatives, and many other products produced in Megaton volumes, the petrochemical industry supplies the material for almost all chemical-based industrial and consumer goods.

For making these basic chemicals, the industry accounted for consumption of about 660 Mt crude oil fractions per annum in 2018 (14% of oil consumption). This means most of the crude oil or its fractions that are not used as transportation and heating fuel is used in the petrochemical industry. Besides oil, the petrochemical industry is also a large consumer of natural gas with about 300 billion cubic meters per annum or 8% of total production [36]. The direct CO_2 emission from the production of the primary chemicals can be estimated at almost 1Gt/year or about 3% of the world's total. In particular, the production of the plastic precursors ethylene, propylene, and BTX aromatics, together known as "high value chemicals", emits about 1.2 tons of CO_2 per ton of product, whereby ethylene typically requires the least and BTX the highest energy input but all can be derived from crude oil fractions in coupled processes (Fig. 4.15).

About 40% of the oil and gas consumption of the petrochemical industry is combusted for heat and power generation during production, the remainder is turned into various chemical products. Since most products will release their carbon after the end of their life, the petrochemical industry is one of the major CO_2 emitters. Options for mitigation of CO_2 emissions include three strategies:

1. Replacement of crude oil and natural gas-based production with green hydrogen and climate-neutral, recycled CO_2 or substitution with biomass.
2. Process heat decarbonization through renewable electricity and carbon capture
3. Recycling of products after the end of life rather than incineration or landfill

TABLE 4.3 Primary petrochemicals production in 2018 [36].

Product	Amount per year, Mt
Ethylene and propylene	255
Ammonia	185
BTX aromatics (benzene, toluene, xylenes)	110
Methanol	100

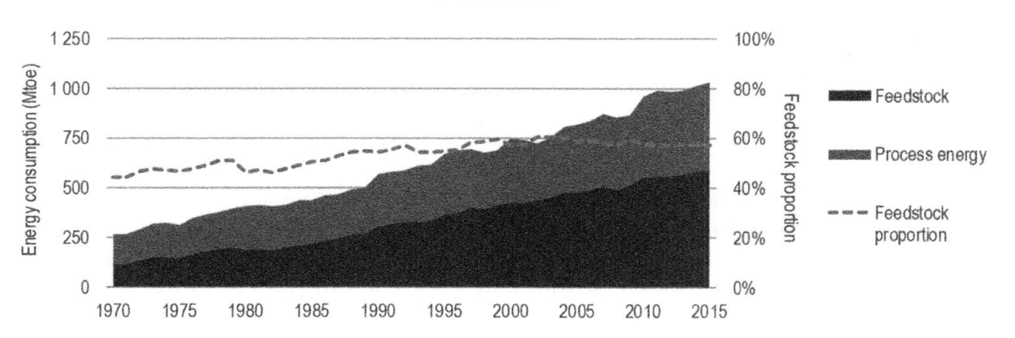

FIG. 4.15 Growth of Feedstock and energy demand of the petrochemical industry [36].

Step 1 is interesting in the context of this book as it involves building hydrogen infrastructure and supply chains as well as CO_2 capture with at least regional transport to the petrochemical plants. Pipelines may be used to connect CO_2 and hydrogen production with the plants turning these into primary chemicals such as methanol, ethylene, propylene, etc. Further transport of the products, currently about 700 Mt/annum just for the seven primary chemicals, mostly as feedstock for plastic and fertilizer, is provided largely by rail in tank cars and by ship in chemical carriers. Final products are containerized and shipped by truck, rail, or container ships around the globe. As the product volume keeps increasing with the economical development of the world and population growth, the transport demand keeps growing as well. At the current trajectory, global demand for the primary seven petrochemical products may increase by more than 50% to reach 1000 Mt per year by 2050 (not accounting for the potential increasing use of methanol and ammonia as alternative "green hydrogen" fuels).

Step 2 of process heat electrification may reduce some oil and gas transport and replace this with a growing electrical transmission system but relies on a huge buildup of renewable electricity sources. IEA [37] estimates 12,000 to 17,000 TWh of electricity is required to replace fossil fuel feedstock and energy for an electrical pathway to primary chemicals in 2050. To be put in perspective, this amount of electricity corresponds to about half of the current total production, or almost 5× the amount generated by PV and wind in 2021.

For step 3, a growing fraction of recycling may add to the overall transport volume when collected plastic scrap is shipped back to central plants by truck, rail, and ships rather than dumped and incinerated near local garbage collectors. However, recycling will also offset some oil and gas transport by replacing these feedstocks. Generally, plastic recycling is key to significantly reduce fuel consumption, emissions, and pollution in the petrochemical industry. On a global average of plastic waste, only 9% is recycled [38] but more than twice as much is "mismanaged" and pollutes the environment. Interestingly, with the exception of the EU, Korea, and Japan, most plastic waste is not even incinerated and used as fuel for power and heat but simply landfilled. North America is particularly concerned because of the high per capita consumption of plastic (Fig. 4.16). Mechanical recycling of postconsumer plastic waste such as PET, HDPE, and PP used, e.g., in packaging into new plastic products can save 80% of the energy and 70% of the emissions compared to the production of new resin, according to a life-cycle plastic recycling study by Franklin Associates [39]. These numbers include collection, cleaning, sorting, and processing. For polyethylene resin as a proxy for such consumer plastics, average literature values for the energy of the feedstock (e.g., crude oil fractions) is about 55 MJ/kg and the energy for processing (oil, gas, and electricity) is about 15 MJ/kg primary energy [40], together about 70 MJ/kg or equivalently 1.6 kg crude oil/kg plastic. When incinerated, plastic releases a lower heating value of 43 MJ/kg, comparable to fuel oil, although typically the net efficiency of an incinerator is much less than that of a heating or power plant, such that, e.g., a credit of only half the heating value should be taken for the substitution of fossil fuel (unless plastic waste is processed into fuel pellets (RDF) and co-fired). Thus, if landfilling and littering of annually 350 Mt of plastic waste [38] were avoided, considerable savings in oil and natural gas would be achieved, besides emissions and pollution reduction. If, for instance, 50% instead of just 9% would be recycled into new plastic, and the remaining 50% instead of just 19% were incinerated for heat and power production, the savings would amount to: (1) the equivalent of 190 Mt of crude oil for new resin production, which is 3.8Mbbl/d or 4% of oil production; (2) 650 TWh of natural gas for heat and power generation, which is about 1.6% of global production.

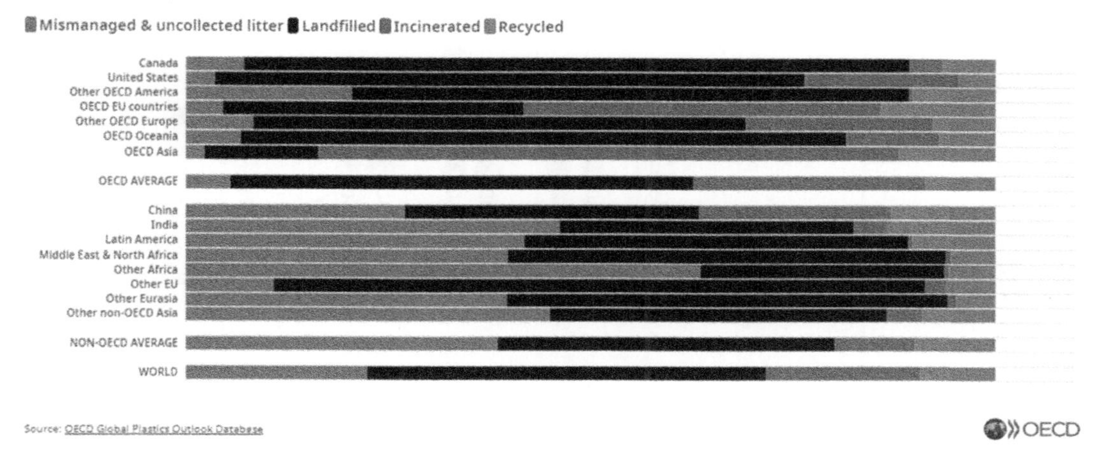

FIG. 4.16 Summary of plastic waste treatment around the world. *Source: OECD, Global Plastics Outlook: Economic Drivers, Environmental Impacts and Policy Options, OECD Publishing, Paris. 2022. 10.1787/de747aef-en. Available online: https://www.oecd.org/environment/plastic-pollution-is-growing-relentlessly-as-waste-management-and-recycling-fall-short.htm.*

Mobility applications

Mobility applications chiefly include the transport of goods and passengers. Currently, 90% of transport energy of more than 2500 million tons of oil equivalent per year is fueled by petroleum derivatives such as Diesel, gasoline, fuel oil, and jet fuel [41].

The enormous amount of oil consumed by transport means requires an international trade network of petroleum and products between a few producing regions and globally distributed consumers. This trade volume across national borders in the year 2020 was about 2100 million tons of crude oil and 1100 million tons of derived products. Most of it was transported by large oil tankers also known as crude carriers.

Within the frame of this book, more interesting than the oil currently fueling the majority of transport, are those 10% of alternative energies that will need to grow substantially in the immediate future and become "green" to decarbonize this sector. According to the IEA statistics, these include in order of volume natural gas, biofuels, and electricity. Lately, biofuels and electricity (incl. green hydrogen) for transportation have been growing at an increasing annual rate. Given the worldwide trend to decarbonize, these fuels must be expected to accelerate their absolute annual growth to achieve a meaningful CO_2 reduction soon.

Natural gas, as a fossil fuel with associated CO_2 and fugitive methane emissions, is not a climate-friendly alternative to oil and is unlikely to grow much in the future. It is used in transportation as compressed natural gas in pressure cylinders or as liquified natural gas (LNG) in vacuum-insulated tanks for Otto engines in cars, trucks, and ships. Reasons for using natural gas instead of gasoline or Diesel include regionally or temporally lower prices and its relatively low combustion emissions.

The share of renewable fuels in the transportation sector has been historically very low and practically zero in regions without regulations mandating certain shares of biofuels or incentivizing electrification. In the EU, "the average share of energy from renewable sources in transport increased from 1.6% in 2004 to 8.9% in 2019" according to Eurostat [42]. Although the growth rate on average was only 0.5% per year, individual countries have reached a multiple and show what is possible with targeted policies and incentives. The share of renewable energy in transport fuel reached 30% in Sweden, 28% in Norway, and 21% in Finland. Note that renewable electricity is accounted here.

The global outlook shows that a dramatic acceleration of substituting fossil fuel in this sector is to be expected. The IEA estimates in all scenarios a growing fraction of both biofuels and hydrogen derivatives made from green electricity (4.17–4.19).

The transport challenge of decarbonized fuels and electricity replacing petroleum products will be that transport from the source to the customer is on the same order of magnitude in terms of energy as today, but requires different

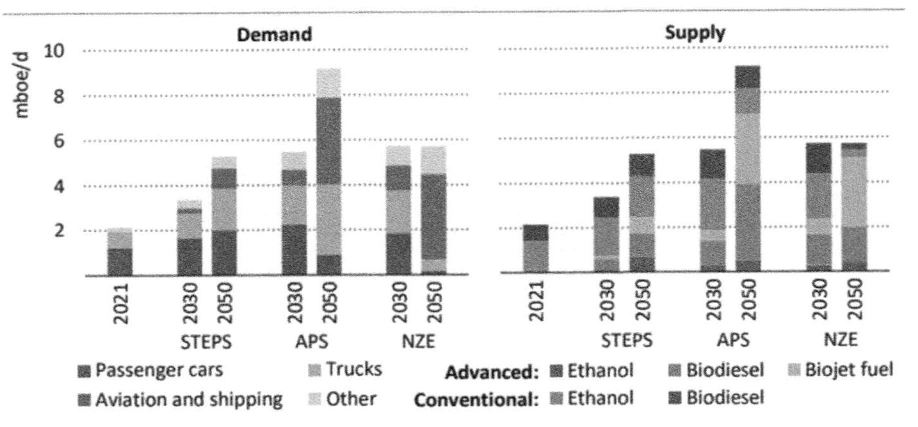

FIG. 4.17 Demand and supply sources of biofuels in IEA scenarios of different greenhouse gas reduction ambition with NZE corresponding to a consistent reduction to net zero emissions by 2050 [43].

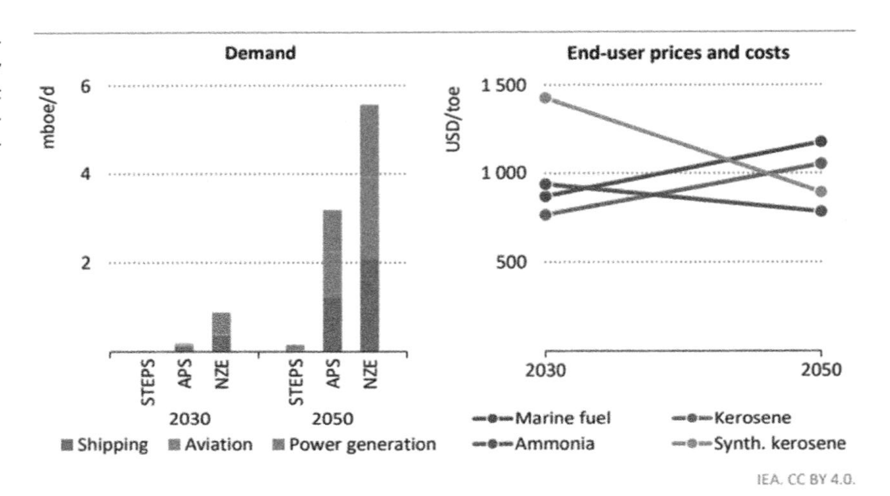

FIG. 4.18 Demand of hydrogen-based low-emission fuels replacing petroleum products and their respective prices forecasted in scenarios of different greenhouse gas reduction ambition with NZE corresponding to a consistent reduction to net zero emissions by 2050 [43].

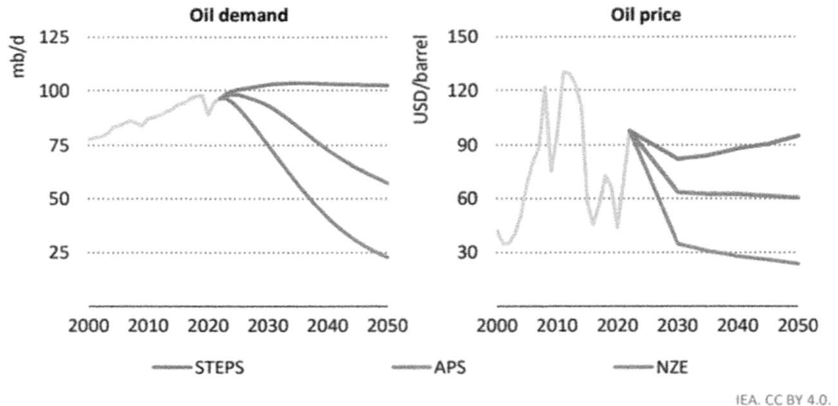

FIG. 4.19 IEA forecasts of global oil demand and crude oil price for scenarios of different greenhouse gas reduction ambition with NZE corresponding to a consistent reduction to net zero emissions by 2050 [43].

means. The current capacity of oil tankers and oil pipelines may no longer be needed. Instead, other transport networks will grow: Gas (including hydrogen) using converted pipelines and gas tankers, electricity requiring additional capacity in transmission and distribution lines, and biofuels or decarbonized fuels with decentralized production requiring transport in smaller scale and smaller batches than crude oil. Since the transport of petroleum and its liquid products in pipelines and large tankers is fundamentally the most efficient and after decades of optimization very cost-effective, any alternative fuel and transport network will incur higher costs but also new business opportunities and engineering demand for efficiency improvements. As an efficiency baseline for transporting energy, consider the example of an oil tanker. A very large crude carrier uses about 100 t per day laden and 80 t per day on ballast for a 42-day roundtrip voyage from the Gulf to Asia carrying 300,000 t of crude oil [44]. The total fuel consumption of 3780 t equals about 1.3% of the energy content of the crude oil carried; hence, the transport efficiency of such an oil tanker is very high, about 98.7%. Any other means of transport, except short pipelines, will likely have a lower transport efficiency. All IEA forecasts, however, show Peak Oil happening in the 2020s and for any climate protection scenario, a steep decline in petroleum production in the coming decades [43]. Low-carbon alternatives such as derivatives or renewable hydrogen are possible replacements. These, unless produced locally, will need considerable transport capacities for global trade as all alternatives to petroleum have a larger volume per unit of energy.

Automotive

Within the transportation sector, the group of passenger cars, trucks, buses, and off-highway vehicles, is the largest fuel consumer. Besides biofuels, renewable electricity and green hydrogen and its derivatives are future contenders for decarbonizing this sector. With a major shift in transportation energy from fossil fuels to green electricity, the existing energy transportation and distribution will have to change: Fewer gas stations, more charging points, and hydrogen refueling dispensers will be required, resulting in dramatically more renewable electricity demand. In the United States in 2020, 90% of transportation fuel was petroleum based, 6% were biofuels. Gasoline and Diesel for road transport accounted for approximately 200 billion gallons in the United States. To put this in perspective, this energy of about 7000 TWh (lower heating value) equates to 1.7 times the total US electricity generation (about 4100 TWh in 2021). This huge consumption of fossil fuels is clearly not compatible with any scenario of climate change mitigation and has to change soon, foremost through electrification with additional renewable generation.

In 2022, the global electric vehicle fleet including cars, trucks, and buses consumed about 110 TWh of electricity, which accounted for less than 0.5% of total final electricity consumption worldwide. The use of EVs displaced on average 700,000 barrels of oil per day in 2022 [45], which is about 0.8% of 90 million barrels of daily production. For passenger cars, 10 million of those sold were electric, or about 14% of global sales; for vans, buses, and trucks, that figure was 310,000 (3.6%), 66,000 (4.5%), and 60,000 (1.2%), respectively [45]. The growth of electric vehicles (EVs), according to the increase in annual sales, averaging about 30% in the time frame from 2016 to 2023, leads to an exponential increase in the number of EVs on the road. In recent years, the total fleet even grew by about 50% in 2021, 60% in 2022, and about 50% in 2023 (Figs. 4.20 and 4.21).

Along with the increase in EV numbers, their electricity consumption will grow. Throughout the 2020s, an annual growth rate of 20% to 25% can be estimated based on EV number estimates from IEA scenarios, resulting in a growth of annual electricity consumption for charging vehicles from 80 TWh in 2020 to between 500 and 900 TWh in 2030, or from 0.4% of global electricity demand to approximately 2% in 2030 (Fig. 4.22).

For individual cars and vans, no electricity grid constraints for charging are envisioned in the near future. However, for fleets, especially truck fleets, grid constraints for chargers present a major hurdle for electrification.

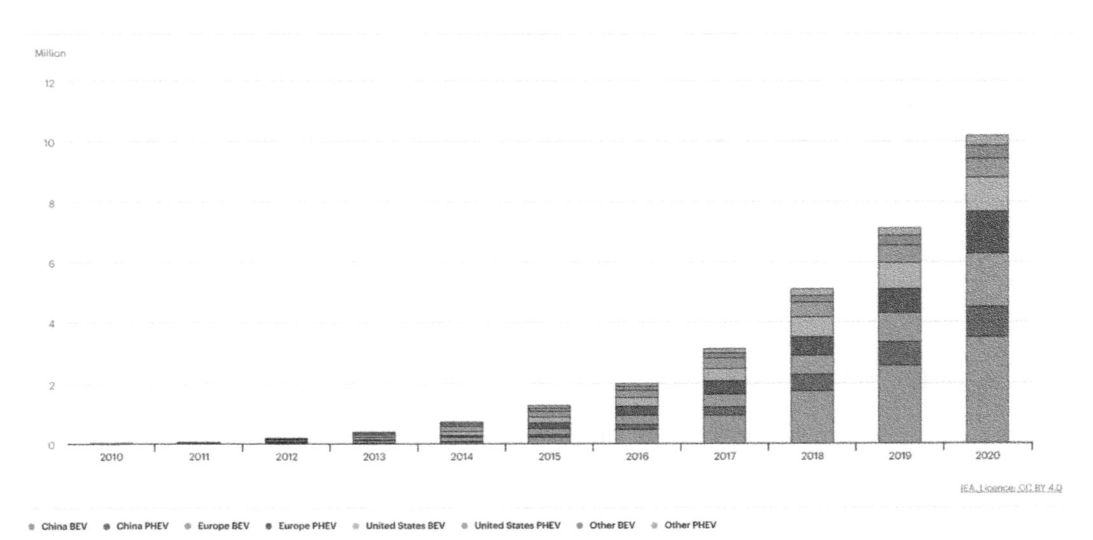

FIG. 4.20 Global historic electric passenger car stock and breakdown by region [46].

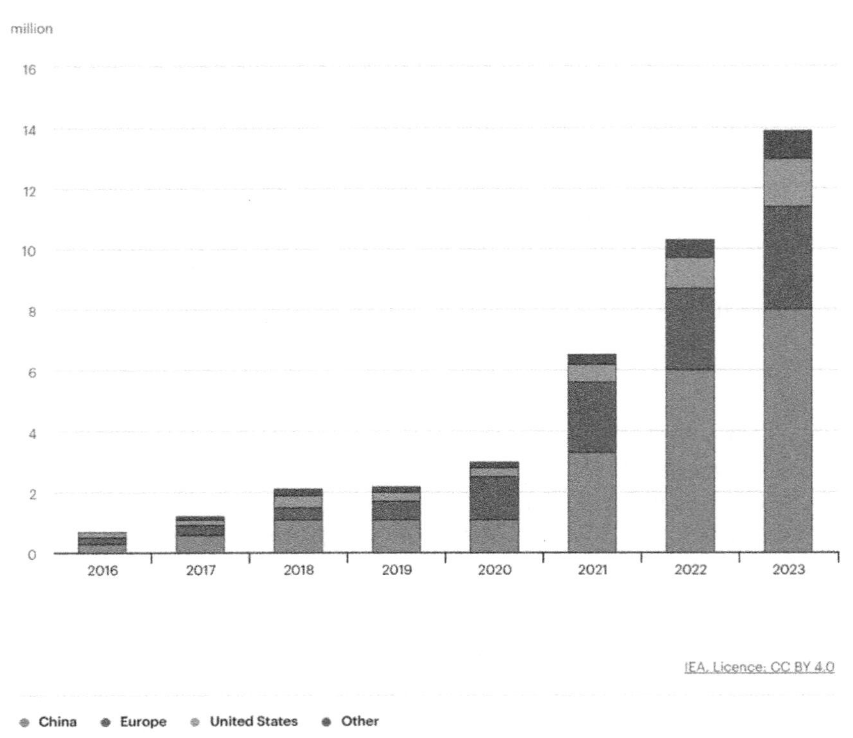

FIG. 4.21 Electric car sales in million units showing substantial growth on track to make electricity the dominant fuel in the 2030+ decades. *Source: IEA, www.iea.org/data-and-statistics/charts/electric-car-sales-2016-2023.*

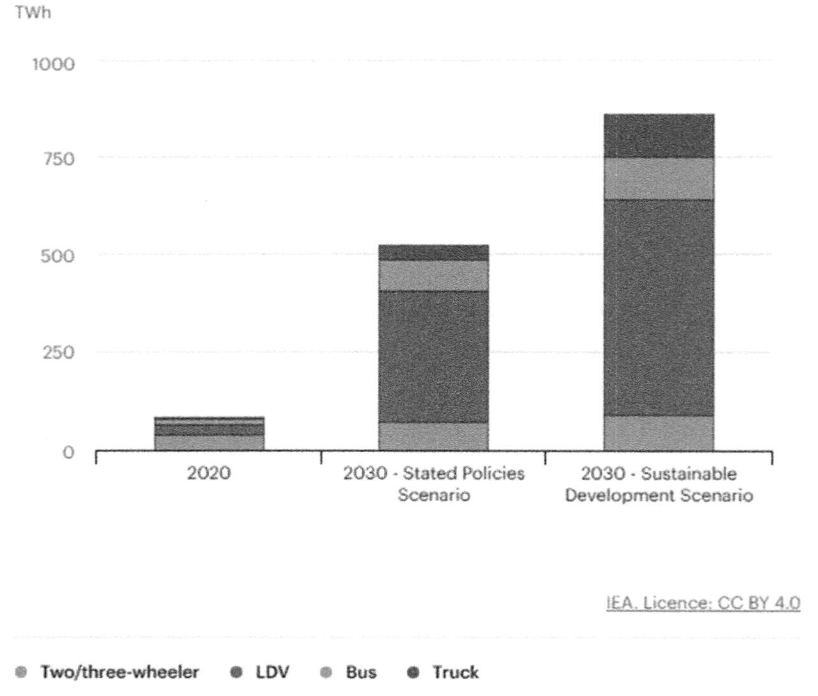

FIG. 4.22 Electricity consumption estimate for charging electric vehicles not including e-fuels [47].

Together with faster refueling times compared to electric battery charging, hydrogen is the obvious albeit more costly solution.

Hydrogen has long been an alternative to battery electricity for cars, vans, and trucks, either for EVs with fuel cells, or for internal combustion engines. For the year 2020, about 35,000 vehicles using hydrogen fuel cells were registered globally, while hydrogen engines remain rare. The growth of fuel-cell vehicles has been orders of magnitude lower than that of battery-electric vehicles. For passenger cars, this is certainly due to the higher cost of the vehicle, higher cost of fuel and fewer opportunities to refuel hydrogen than to charge the battery. For commercial fleets, especially for heavy-duty trucks, substantial growth in the numbers of hydrogen-fueled vehicles can be expected for two reasons: Legislation requiring ever more stringent CO_2-reduction for new registrations, and the issue with fleet charging taking a long time whereas hydrogen refueling is quick. Hydrogen trucks and buses may also offer a higher range and more payload than battery-electric versions. Already thousands of trucks and buses with fuel cells operate in regions in China where the necessary refueling infrastructure has been extended and government support has been in place to develop the market. Europe and the United States can be expected to follow with several thousand annual new registrations of hydrogen-fueled heavy commercial vehicles toward 2030. In the EU, legislation has been passed that stipulates 15% reduction in CO_2 emission in the fleet of every OEM's new heavy vehicles by 2025 and 30% reduction in 2030 [48]. This means that about 15% of the annually registered 400,000 heavy trucks will need to be emission-free after 2025. Some of those are expected to run on hydrogen, others on batteries. The cumulative number of trucks on the road using electricity and hydrogen will demand a power plant capacity of several GW for charging and several thousand tons per day of hydrogen by 2030, depending on the market development and the share of battery electric vs. hydrogen vehicles numbers. In the United States, several states are developing similar legislation to force CO_2 reduction, now targeting not just passenger cars as in the past but also trucks with the aim to phase out Diesel in the long term and replace this with electrification, hydrogen, or other carbon-neutral fuels with an annually rising percentage target for new medium and heavy trucks to be emission-free from 2024 (at least 5%) onwards to 2045 (all trucks "feasible" to be zero emission) [49]. This measure is estimated to put 300,000 zero-emission trucks on the road by 2035, leading to a hydrogen demand of about 7500 t per day and about 6 GW of power plant capacity for charging if the share of hydrogen vs. battery electric trucks was 50:50. Given California's track record of piloting regulations that soon become adopted in many further jurisdictions, these numbers may scale proportionally with the economy of all OECD countries and possibly China and others (Fig. 4.23).

These developments of an increasingly large-scale decarbonization in light and heavy transport are expected to start reducing Diesel and gasoline consumption first in developed economies and soon also in developing economies despite an overall growth in the number of vehicles and total miles driven and goods transported. The IEA has analyzed several future scenarios with different ambitions to reduce road transport CO_2 emissions and concluded a reduction in oil usage. This means, however, an increasing demand for production and transport or alternative fuels that have, except for electricity, higher cost and lower energy density (Fig. 4.24).

Just as trucks have been pivotal to supply fuel to gas stations and heating oil customers for many decades, trucks and road transport will be pivotal for delivering decarbonized energy such as hydrogen or renewable fuels to consumers where no pipeline or gas grid is available. Since the volumes of decarbonized energy carriers per unit of energy are greater than fuel oil or LNG, the transport cost may increase. Hydrogen is an extreme example of inefficient

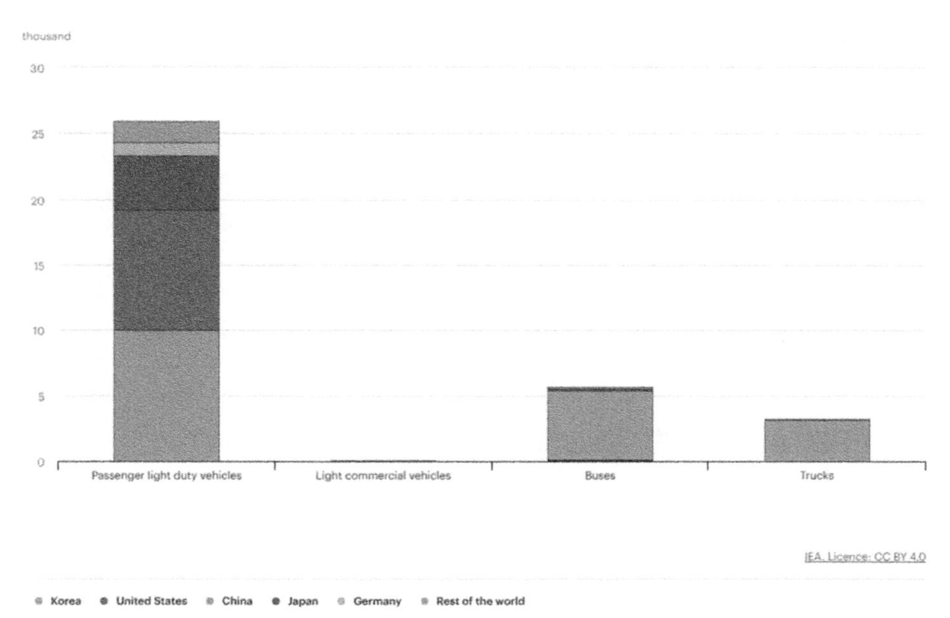

FIG. 4.23 Number of hydrogen fuel cell vehicles by type and country in 2020, measured in the thousands in contrast to battery electric vehicles measured in the millions (IEA statistics: https://www.iea.org/data-and-statistics/charts/fuel-cell-electric-vehicles-stock-by-region-and-by-mode-2020 accessed 2023).

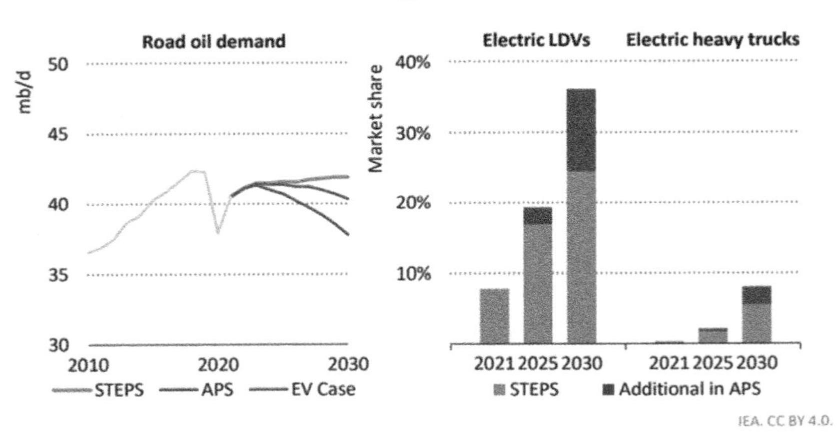

FIG. 4.24 Global road transport oil demand and share of alternative drive train vehicles for scenarios of different CO_2 reduction and future vehicle fleets [43].

transport [50]. There are commonly two modes of hydrogen transportation on the road, either in gaseous form or in liquid form. Both alternatives have advantages and disadvantages, such as an increased density of about 70 kg/m^3 for liquid hydrogen compared to about 10–30 kg/m^2 for gaseous, or the lack of boil-off losses and savings of energy for liquefaction when hydrogen is transported as a gas.

For liquid hydrogen, trailers with vacuum-insulated tanks can carry up to 4 tons at a temperature around −250°C and a pressure of just a few bars with relief valves for boil-off gas to maintain low pressure and cryogenic temperature. Unloading can commence with cryogenic pumps or gravity/pressure-driven drain into the customer's storage vessel. Liquid hydrogen trucking is preferred for higher capacity and where liquid hydrogen is available from liquefaction plants.

For gas transportation up to about 200 bar, steel tube trailers that have been used for many years and many different gases, including hydrogen. These trailers can carry about 300 kg of H2 in up to 10 tubes mounted horizontally on a standard-length trailer. For a higher capacity, new trailers have been developed with smaller and lighter cylindrical composite tanks of type 4. Such cylinders consist of a carbon fiber-wrapped hydrogen-tight plastic liner and steel fittings and are available for 380, 500, or even 640 bar for road transport of hydrogen (EN17339). Mounted on trailers or in containers with lengths of about 20, 30, 40, and 45 ft. (EU) to 53 ft. (US), type 4 trailers can hold up to about 1200 kg of H_2 or four times more than steel tube trailers. However, filling them requires production facilities to have new, high-pressure compressors compared to the 200 bar traditionally used for steel tube trailers (Fig. 4.25).

FIG. 4.25 Trailer with vertically oriented composite cylinders. *Courtesy: Chart https://www.chartindustries.com.*

Rail

Besides road transport, rail has a significant share in the transport sector both for passengers and freight. In the United States, rail transports in ton-miles account for about 25% of all freight, second after road transport (55%). In the EU, rail freight accounts for more than 5% of all ton-kilometers after marine (67%) and road (25%). Railroads are a very efficient mode of transport compared to road transport: on a Diesel consumption per ton-mile base in the United States, trains consume only about 8% as much as trucks according to the Bureau of Transportation Statistics [51], meaning that the railroad is about 12 times as energy efficient on average per unit of freight and distance. Thus, railroads present an efficient alternative for cargo in a world aiming for decarbonization where energy becomes increasingly expensive. However, the IEA estimated that railroads globally consume about 300 TWh of electricity and 30 Mt of Diesel. These led to total CO_2 emissions from Diesel and electric trains of about 100 Mt CO_2 in 2019. This is far less than marine and aviation, let alone trucks, but still efforts are anticipated to decarbonize this sector as well. Electrification is relatively far developed and delivery of green electricity is a matter mostly of contracting and expanding renewable generation. Diesel accounts still for more than 50% of railroad energy consumption in the global total although to much less than half everywhere except North America. To reduce CO_2 emissions, the alternatives are, besides electrification with renewables, batteries for short distances of passenger trains with frequent recharging, climate-neutral liquid fuels, green ammonia, and hydrogen. Quite similar to road transport or marine decarbonization, these fuels, used with internal combustion engines or fuel cells, will cost more, require extra space for tanks, and necessitate new or upgraded refueling infrastructure (Fig. 4.26).

Hydrogen has been explored commercially as a climate-friendly alternative fuel for diesel shunting locomotives and multiunit commuter trains [52]. Hydrogen can provide a larger range than battery-electric trains, on par with Diesel. Similar to electric trains, hydrogen fuel cell electric trains are emission-free and quieter than diesel trains. Compared to diesel multiunit commuter trains, which cost about Eur 3…5M with 2 or 3 units for 120 to 150 seats, the fuel cell and battery-powered versions cost about 30% more with Eur 4 to 6 M. The market potential for hydrogen-powered trains to substitute diesel is substantial; in Germany, for instance, with a well-developed railroad network, still more than one-third of the network was not electrified in 2020, and 2800 diesel commuter trains and more than 120 mainline diesel locomotives in addition to hundreds of shunters are in operation [53].

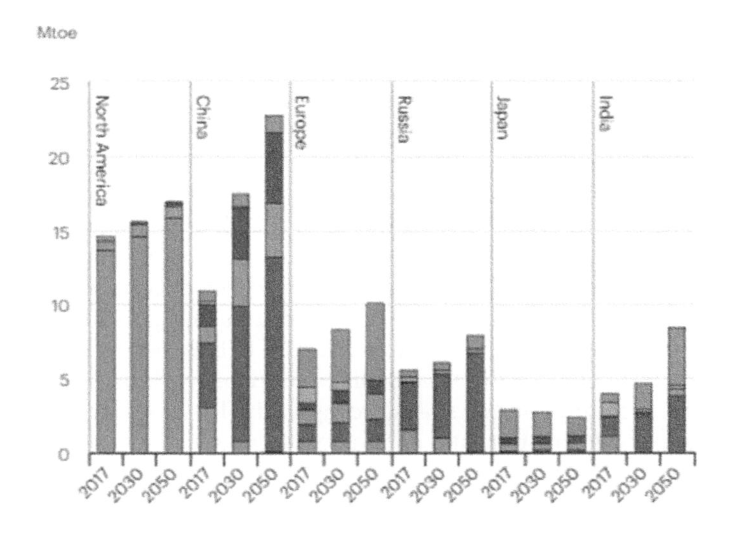

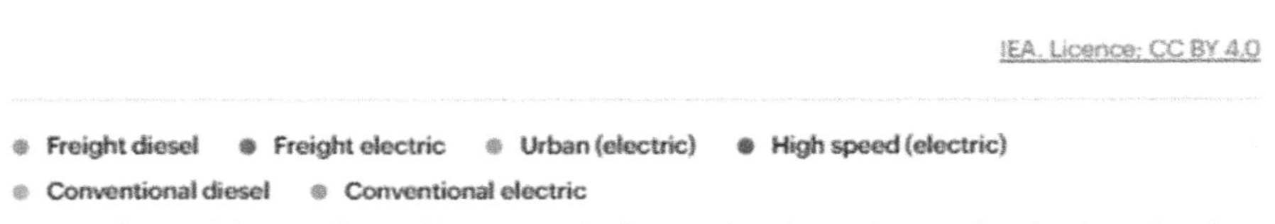

FIG. 4.26 Forecast of energy demand from rail in million tons of oil equivalent by regions and technology in a base scenario with minimum growth [54].

FIG. 4.27 Hydrogen fuel cell commuter train. *Credit: Alstom: https://www.alstom.com/press-releases-news/2020/3/alstoms-hydrogen-train-coradia-ilint-completes-successful-tests.*

As one of the first, train manufacturer Alstom has developed a hydrogen fuel-cell electric commuter train "Coradia iLint", that entered service in 2018 in Lower-Saxony, Germany. More of these trains are on order as of 2021 for lines in Germany, Austria, and the Netherlands. The 150 seater (+ 150 standing places) 2-unit train is equipped with 350 bar tanks on the roof of each unit containing at least 90 kg of hydrogen to supply 200 kW fuel cells driving 360 kW traction motors, supported by a battery for acceleration and breaking recuperation, for achieving a range up to 1000 km and a top speed of 140 km/h (Fig. 4.27). Siemens, another large manufacturer of trains, is developing a similar commuter train with hydrogen fuel cells ("Mireo") and so does Stadler in Switzerland with its "Flirt". Stadler's hydrogen trains were selected in 2018 for the Zillertalbahn narrow-gage line in Austria [55] and in 2019 for Southern California's San Bernardino County Transportation Authority to enter service in 2024 [56]. Besides commuter trains that are light, travel relatively slow and return to the same station each night, also shunting locomotives can be converted from Diesel to hydrogen without major logistical challenges as expected for mainline locomotives with much higher power and tank volume requirements. Among several prototypes is a project of Ballard and BNSF, in which a Diesel-electric shunter was converted to hydrogen for operation in switching yards in Los Angeles, CA [57]. More hydrogen as well as ammonia-fueled trains can be expected in the future with converted Diesel engines.

More so than consuming fuel, freight railroads have been the backbone for land transportation of fuels including crude oil, petroleum products, and coal. In the United States, freight trains moved about 300 Mt of coal (more than half of the consumption) and more than 100 Mt of oil and chemicals in 2021 [58]. With an increasing volume of decarbonized energy carriers that replace coal and oil, and a renewed focus on transport efficiency with increasing energy prices, the role of freight trains in the future energy system should be expected to grow. On the passenger side, a growing population, increasing wealth, and demand for travel with extensions of high-speed rail networks, all will lead to growth in railroad service: from 4 trillion passenger kilometers in 2016 to between 9 and 15 trillion by 2050 [54]. Overall, the outlook is bright for freight and passenger railroads alike: Train networks are both to benefit from and to contribute to the energy transition in the coming decades. Accordingly, the IEA sees large growth potential for efficient freight networks as an alternative to road transport and convenient high-speed rail as an alternative to flying, in both cases decarbonizing the transport sector.

Marine

The marine economy is doubly affected by changes in the energy sector from decarbonization efforts: first, the fuel for ship engines, known as bunker, needs to change in the future from fossil fuel to alternatives with lower CO_2 emissions. Second, a large fraction of cargo shipped globally may change, since most of the oil, a large fraction of the coal, and an increasing share of natural gas production is moved across the oceans in ships. Bunker fuels traditionally include varieties of heavy fuel oil and marine gas oil, the first is less expensive and most widely used in cargo ships, and the latter is a distillate with lower sulfur content and lower combustion emissions but a higher price that is increasingly used in areas with emission regulations. The global shipping industry in 2018 accounted for more than 1 Gt of CO_2 emissions or 3% of the total. International efforts are underway to reduce these emissions, e.g., by the International Maritime Organization (IMO) with the 2023 IMO GHG Strategy that sets guidelines for CO_2 reduction in line with the Paris Agreement from 2015 with the goals of 30% and 70% total reduction by 2030 and 2040, respectively, to reach net zero by 2050 [59].

Renewable marine fuels

There are several options becoming available for low-carbon, climate neutral, or renewable alternative fuels and it seems clear that no single fuel will replace all current bunker fuel oil, rather a variety of fuels is expected with application and market-specific choices. An overview of potential low-carbon fuels with a comparison of their relevant specifics is given in a report on hydrogen as a future fuel for shipping by the American Bureau of Shipping [60]. Table 4.4 shows that the nine alternatives to diesel (MGO) and fuel oil (HFO) all have a significantly larger volume per unit of energy, requiring larger fuel tanks or reducing the range before refueling in the future. Most of them also require either refrigeration or pressurized tanks. What is not shown in the table is how all of them are currently more expensive in the absence of high carbon taxes with the possible exception of LNG.

LNG has been used besides on LNG carriers and also on ferries and cruise ships as well as for smaller coastal or inland boats because it enables low-emission combustion without as much aftertreatment as required for Diesel engines and is generally perceived as a clean fuel. However, LNG is by no means a climate-neutral fuel especially given the fugitive methane emissions of the natural gas supply chain and the methane slip of engines that add considerable to the total global warming potential of natural gas compared to petroleum products.

Two of the fuels in the table can be truly zero carbon when produced with renewable energy: Hydrogen and ammonia. In comparison, hydrogen may be less expensive than renewable ammonia, which is a derivative of it but is much harder to store and transport due to its high volume or very low temperatures in liquid form. Hydrogen is a fuel suitable particularly for smaller vessels and those where fuel costs are less critical. H_2 and has been used in a number of lighthouse projects for "zero emission" ferries, harbor tug boats, or cargo vessels and pushboats on inland waterways, where range requirements are low. These boats have either fuel cells and electric drivetrains or internal combustion engines.

Hydrogen can be burned in modified diesel engines alone in a spark-ignited Otto-cycle or, specifically with Diesel dual-fuel engines, mixed with diesel and compression ignition of pilot injections. Most current hydrogen engines are derived from Diesel engines. Using hydrogen reduces the displacement-specific power of engines compared to diesel or natural gas to between 50% and 80%, thus requiring larger engines with higher displacement or

TABLE 4.4 Characteristics of low-carbon alternatives compared to current bunker fuels [60].

	UNIT	HYDROGEN	MGO	HEAVY FUEL OIL (HFO)	METHANE (LNG)	ETHANE	PROPANE	BUTANE	DIMETHYL-ETHER (DME)	METHANOL	ETHANOL	AMMONIA
Boiling Point	°C	-253	180-360	180-360	-161	-89	-43	-1	-25	65	78	-33
Density	kg/m³	70.8	900	991	430	570	500	600	670	790	790	696
Lower Heating Value	MJ/kg	120.2	42.7	40.2	48	47.8	46.3	45.7	28.7	19.9	26.8	22.5
Auto Ignition Temp	°C	585	250	250	537	515	470	365	350	450	420	630
Flashpoint	°C	-	> 60	> 60	-188	-135	-104	-60	-41	11	16	132
Energy Density Liquid (H_2 Gas at 700 bar)	MJ/L	8.51 (4.8)	38.4	39.8	20.6	27.2	23.2	27.4	19.2	15.7	21.2	15.7
Compared Volume to MGO (H_2 Gas at 700 bar)		4.51 (7.98)	1.00	0.96	1.86	1.41	1.66	1.40	2.00	2.45	1.81	2.45

higher engine speed to reach the same power as with traditional fuel. Challenges besides reduced power chiefly include avoidance of preignition, high intake pressure and turbocharging, spark plug life, and injector availability. With ongoing development of hydrogen internal combustion engines, it is expected that ultimately the same efficiency can be reached than with natural gas engines. Also, strategies for NOx reduction have been developed for hydrogen engines to comply with all emission regulations without or with minimal aftertreatment such as EGR or catalyst [19].

Ammonia can be made "green" and climate neutral from green hydrogen and nitrogen from an air separation unit using renewable electricity in the Haber-Bosch-Process. When cooled to −33°C or compressed to about 16 bar at ambient temperature, ammonia becomes a liquid that is much easier transported than hydrogen and is considered a future hydrogen carrier. The demand is potentially reaching several 100 Mt and requires multiplying annual production volume when becoming the dominant replacement of heavy fuel oil for cargo ships. Ammonia can be thermal-catalytically cracked into hydrogen (and nitrogen) and purified for use in PEM fuel cells. It can also be used straight or partially cracked in internal combustion engines, similar to hydrogen as described above, such as modified Diesel or dual-fuel engines. This allows retrofit of current Diesel engines for ammonia in the future. Both leading designers of two-stroke ship engines, MAN and WinGD have announced plans for having ammonia dual-fuel engines ready for cargo ships in the mid-2020s. Storage and handling, liquid or compressed, or in a combination of both with lower pressure and higher temperature is clearly a disadvantage compared to liquid fuels, but significantly easier than LNG or hydrogen.

Methanol is another potentially green fuel option when made from green hydrogen and climate neutral CO_2-source like biogas or biomass. Methanol is an easily handled liquid and a commodity produced and traded worldwide on a 100 Mt scale already. It has been used as fuel and can be almost a drop-in replacement with some changes to the engine ignition system although its energy density is less than half of fuel oil, reducing range or necessitating larger tanks as with most other alternative bunkers. MAN Energy Solutions [61] as a leading engine manufacturer offers engine retrofit upgrades for methanol/Diesel dual fuel operation, claiming significant emission reductions. Methanol is considered a future energy carrier for green hydrogen, like ammonia, because the transport is easiest and requires no significant changes to handling and storing infrastructure. Unlike ammonia, it requires a carbon source for synthesis and produces CO_2 when burned which adds to global warming unless the carbon source is e.g. biomass or direct air capture. The estimated future cost range for green methanol is similar to that of green ammonia on the order of $500/ton or more depending on the carbon source. Although "blue" production from fossil fuel with carbon dioxide sequestration seems far less costly, this is not clearly providing any environmental benefit over the straight use of a lesser amount of natural gas (such as LNG) than what would be required for methanol production, besides being much less expensive. However, its current use as a fuel and ease of handling is likely giving green methanol a head start for long-distance transport decarbonization.

Renewable fuel transport in cargo ships

New decarbonized fuels that substitute oil, coal, and even LNG that will be traded globally include hydrogen, ammonia, methanol, synthetic fuels, biofuels, and biomass. Biomass, e.g., wood pellets or chips, can be shipped similar to coal in bulk carriers. Methanol, liquid synthetic fuels, and biofuels, will be shipped in tankers, chemical tankers, or even crude oil carriers when volumes increase, with no technological challenges. The same or slightly retrofitted ships can be used today for these cargoes without an economical disruption. Ammonia, however, in its decarbonized variety as a green energy carrier, may become a commodity traded in several 100 million tons per year, a multiple of the trading volume in 2020 of 25Mt/year. According to a market scenario reported by IRENA and AEA [62], future global demand for renewable ammonia may increase by 45–500 Mt between 2030 and 2050 through the use as a fuel and as carrier for renewable hydrogen in addition to growing demand for fertilizer and chemicals. Today, most of the 180 Mt of current annual ammonia production is captive and immediately used to make easily transported fertilizers (urea, ammonium nitrate, -sulfate or -phosphate) or a range of other chemicals before shipping. About 20Mt/year of ammonia are currently shipped in ca. 170 vessels [62] and consequently hundreds more of these may be needed in the future. These vessels are gas tankers with up to about 100,000 m^3 (type VLGC) of refrigerated, heavily insulated tanks that are only slightly pressurized and hold ammonia as a liquefied gas (down to −33°C for ambient pressure). Boil-off gas is reliquefied to maintain pressure or used as fuel in dual-fuel engines, together with fuel oil. Most of these carriers are also used to transport liquefied petroleum gas, vinyl chloride, or other liquefied gases. When using ammonia boil-off as the main fuel for the engines, the transport is decarbonized as well (Fig. 4.28).

Shipping liquid hydrogen (LH2) is fundamentally problematic because of unavoidable heat gain leading to pressure increase and eventually to discharge of boil-off gas. While shipping LNG in ocean-going vessels in principle suffers

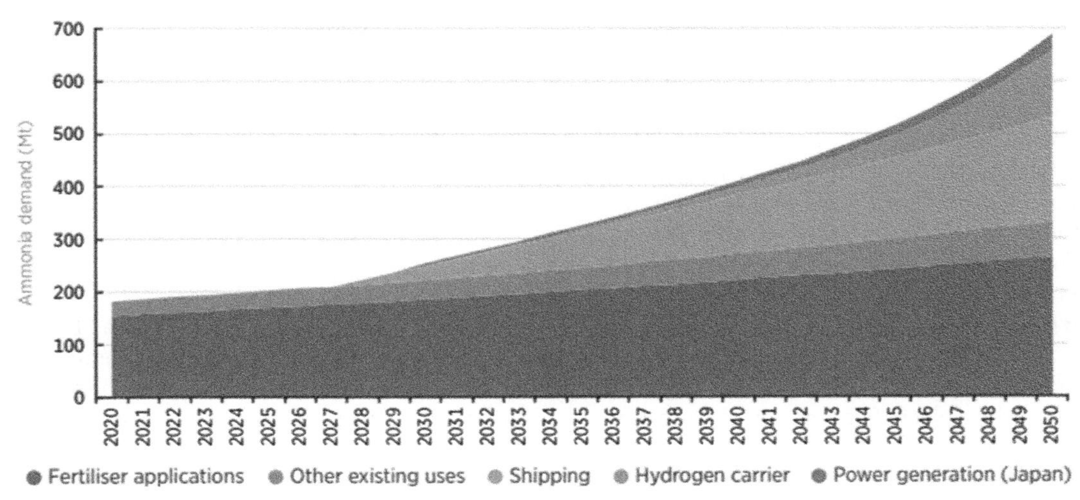

FIG. 4.28 Forecasted ammonia demand up to 2050 by IRENA and the Ammonia Energy Association for a scenario to comply with the Paris Agreement. *From IRENA and AEA, Innovation Outlook: Renewable Ammonia, International Renewable Energy Agency, Abu Dhabi, Ammonia Energy Association, Brooklyn, 2022. https://www.irena.org/Publications. © IRENA- AEA 2022.*

from the same issue, this has been economically resolved through sufficient insulation and effective use of boil-off gas in the ship's engine or alternatively compression and re-liquefaction on board. For LH2 the heat gain through the same tank and volume would be twice as high because of the lower temperature. Moreover, the heat gain per unit energy transported would be approximately 5× higher for the same tank since LNG has a lower heating value of about 22.5MJ/L and LH2 of only 8.4MJ/L. Despite these disadvantages, the first small LH2 carrier, the Suiso Frontier, was built by Kawasaki Heavy Industries in 2019. She is scheduled to transport 1250m3 (about 90t) of LH2 in a vacuum-insulated 4bar tank from a liquefaction plant at the Port of Hastings, Australia to consumers at the port of Kobe, Japan. The ship has Diesel engines installed and can flare boil-off gas (Fig. 4.29).

Overall, the diversification of future energy carriers with dramatic growth in the 2030s and 40s can be expected to increase demand for shipping volumes despite a partial substitution of current fossil energies because of the lower energy density of decarbonized fuel. The exact volumes depend on the future demand and degree of electrification as well as the decline of fossil fuel transports.

FIG. 4.29 The Suiso Frontier is the first liquid hydrogen carrier. A vacuum insulated 1250m3 tank is being installed during constructing in 2019 by Kawasaki Heavy Industries. *Courtesy of Kawasaki, 2022.*

Aviation

The aviation sector is dominated by commercial airlines operating about 31,000 planes for transporting 4.5 billion passengers and 58 Mt of cargo in 2019 [63], while consuming about 225 Mt of jet fuel and emitting about 720 Mt of CO_2. After the recovery from the COVID-19 travel restrictions in the early 2020s, the sector is expected to continue on its historic growth trajectory of about 4% per year along with an ever-growing amount of jet fuel consumption, reaching about 300 Mt by 2025 and emitting about 1000 Mt of CO_2. This leads to a share of global warming of 2% to 5% depending on accounting for secondary climate effects from various emissions at great altitude. An analysis from the ICAO projects the future jet fuel demand based upon current trends of capacity growth and efficiency improvements to almost triple by 2050 [64].

In recognizing the responsibility of the aviation sector to reduce its climate impact, in 2021, the IATA member airlines passed a resolution committing them to achieving net-zero carbon emissions from their operations by 2050 [65]. The strategy involves, by order of contribution, sustainable aviation fuel (SAF) (65%), carbon offsetting (19%), new technologies including electrification and clean hydrogen (13%), and operational efficiency improvements (3%). After the industry's resolution in 2021, also the member states of the International Civil Aviation Organization (ICAO) agreed in 2022 to a net-zero CO_2 target by 2050, and states now must implement policies including incentives and mandates to reduce emissions. In the EU, legislation was passed in 2023 that requires aviation fuel used in member states to contain a fraction of SAF starting from 2% in 2025 and rising to 70% in 2050 [66].

Hydrogen and alternative aviation fuels

The potential for hydrogen to reduce the climate impact of aviation has been estimated to reduce the total sector emissions by half in a study for the EU Hydrogen and Fuel Cell Joint Undertaking [67]. It was cautioned that the technological hurdles and cost increases of converting from jet fuel to hydrogen would be challenging and other jet fuel alternatives like battery electric propulsion and liquid SAF might be more economical than hydrogen in many cases. Hydrogen use with combustion in turbine engines could reduce the climate impact of a plane by up to 75% and with fuel cells by up to 90% compared to jet fuel. Remaining climate impact stems from water vapor and NOx causing cloud formation and is generally not well understood, hard to quantify and rather short-lived compared to CO_2.

The main challenges for introduction of hydrogen to mitigate emissions are the following:

(1) The aviation sector is characterized by heavy regulation, long life cycles, and incremental technological progress.
(2) Hydrogen has a large volume and liquid hydrogen tanks need to be about 4× larger than current jet fuel tanks.
(3) Fuel costs matter in the competitive international commercial environment.

The technical and economic challenges of using hydrogen for aircraft propulsion have prevented commercial success as of 2022. Historically hydrogen has been used in prototype and research aircraft, such as the Tupolev T-155 prototype that was successfully flown in the Soviet Union in 1988. The plane's three Kuznetsov turbofan engines burned hydrogen from a liquid storage tank placed in the unpressurized and air-cooled rear end of the fuselage [68]. Political pressure for reducing emissions has spawned some new developments toward hydrogen-powered aircraft. One example is Airbus' R&D program on hydrogen-fueled planes, of which the company published images of several conceptual regional and short-range aircraft with hybrid propulsion and a liquid hydrogen tank in the rear of the fuselage. For presumably longer range, Airbus showed a blended wing body aircraft concept with distributed tanks and small electric turbo fans [69].

Several aircraft propulsion concepts for using hydrogen have been developed, including:

(1) Electric concepts including "distributed" small electric motors with propellers or more conventional, large electrically-driven propellers or turbo-fans. A fuel cell together with batteries would provide the electric power, fueled by, e.g., liquid hydrogen.
(2) Combustion concepts are based on conventional engines with modified combustors for burning hydrogen with low NO_x emissions.
(3) Hybrid propulsion concepts, where an electric motor supplied by a fuel cell drives the turbofan in conjunction with a combustion turbine, which together boost power during take-off but either one can be turned off during the cruise.

Of these concepts, (1) and (2) have been at least partially employed and tested in the past. Hybrid concepts are promising for larger planes to reduce the maximum electric power and associated weight but require novel engines with electric motor integration. The technology for gas turbine engines burning hydrogen with acceptable emissions in modified combustors is under development for power generation and is described in detail in other chapters of this book. For the tanks, liquid hydrogen is preferred over compressed hydrogen because of its higher storage density and potentially lower weight without a thick-walled pressure vessel. Having a central fuel tank outside the center of gravity of the plane, e.g., in the tail end of the fuselage, leads to weight distribution issues when the tank level changes, resulting in increasing induced drag and higher fuel consumption of the plane. The volume of the tank also reduces the room available for passengers or makes the plane longer, heavier, and costlier to operate. Blended wing body concepts provide more room for distributing hydrogen tanks to avoid these disadvantages but have not been built.

Encouraging examples of progress on hydrogen flight include the development of two US startup firms. Zero Avia, founded in 2017 has since developed fuel cell electric propulsion systems for turboprop commuter planes and in January 2023 made a successful test flight with a retrofitted Do-228. They aim for commercializing this 19-seater with a range of 500 miles within a few years [70]. Similarly, startup firm Universal Hydrogen has converted a De Havilland Dash 8-300 turboprop commuter plane to use liquid hydrogen with a fuel cell electric drive and made a successful test flight with one hydrogen engine in March 2023 [71].

The alternative to concepts using liquid hydrogen for CO2 mitigation is SAF. Though the climate impact reduction is higher with direct hydrogen use than with those fuels due to their supply chain emissions and combustion properties, they are a drop-in replacement and require no changes of engines or infrastructure. These fuels can be made, e.g., through Fischer-Tropsch synthesis with synthetic methane from clean hydrogen and a renewable carbon source. SAF also includes biofuels, clean hydrogen-based methanol or mixtures, and derivatives thereof. Considering that the heating value efficiency of going from gaseous hydrogen to liquid hydrogen (about 30% loss plus boil-off) is similar to the production efficiency of synthetic hydrocarbons made from hydrogen (about 40% loss), liquid hydrogen including its handling may not be much less expensive than SAF in the end. Consequently, the cost difference between SAF and liquid green hydrogen may not be expected to be large enough to justify the disadvantages, complexity, and investments in new aircraft technology and airport infrastructure for liquid hydrogen anytime soon.

Since the required volume and size of the fuel tank increases proportional to the range of an aircraft, alternative fuels, and propulsion concepts requiring larger mass and volume are suited differently well for different categories:

- General aviation and short-range commuter planes, including taxi drones, can be powered electrically by batteries with fuel cells and liquid hydrogen as range extenders.
- Small regional aircraft up to 600 miles would be similarly electric-powered, with fuel cells, larger hydrogen tanks, and less batteries.
- Short-range aircraft, up to about 2000 miles, could use hybrid-electric propulsion with combustion turbines and fuel cells.
- For medium and long-range aircraft, liquid hydrogen is rather impractical because of the required volume and tank weight. Liquid SAF and conventional combustion turbines appear to be a more likely solution to reduce the climate impact.

Airport infrastructure will need to be extended or upgraded in the coming decades to receive and store SAF and hydrogen or ammonia by rail or pipeline.

Sustainable aviation fuel

Given the issue of volumetric energy density and the long timelines and costs for development and certification of novel engines and storage systems for hydrogen in aviation, it may be faster, safer, and less costly to use renewable liquid fuels that can be handled and burned exactly like current jet fuel. Such liquid fuels have become known as SAF. They include synthetic, hydrogen-CO_2 (Fischer Tropsch) or methanol-based fuels ("e-fuels"), Fischer-Tropsch fuel from solid waste or biomass gasification, fuels made through hydroprocessing of pyrolyzed biomass and biofuels made from alcohols, hydrotreated refined plant fatty acid esters and waste oils. SAFs are typically blended within a certain %-range with petroleum distillates or other SAFs and to meet the critical specifications of jet fuel (i.e. Jet-A); some SAFs can be tailored to meet specs without fossil blending to be directly substitutable and carbon neutral. The use of SAF would enable the continuation of current engine technology and aircraft fueling infrastructure, in fact, SAFs and blends are drop-in fuels for instantaneous decarbonization without any changes required for engines, fueling system, and ground infrastructure.

In 2022, the SAF market was estimated at just 300′000 tons [72]. However, the EU proposed legislation to increase a fraction of SAF in aviation fuel made available at EU airports starting in 2025 with 2% and rising to 6% in 2030, 20% in 2035 and gradually to 70% in 2050. This, according to a study published by EASA [73], will lead to a demand for SAF in the EU of more than 2 Mt in 2030, 15 Mt in 2040, and almost 30 Mt in 2050 (see Fig. 4.30). The study also assesses a likely share of SAF types made in Europe, with synthetic, hydrogen-CO_2 (PtL), and alcohol-based (AtJ) as well as biomass

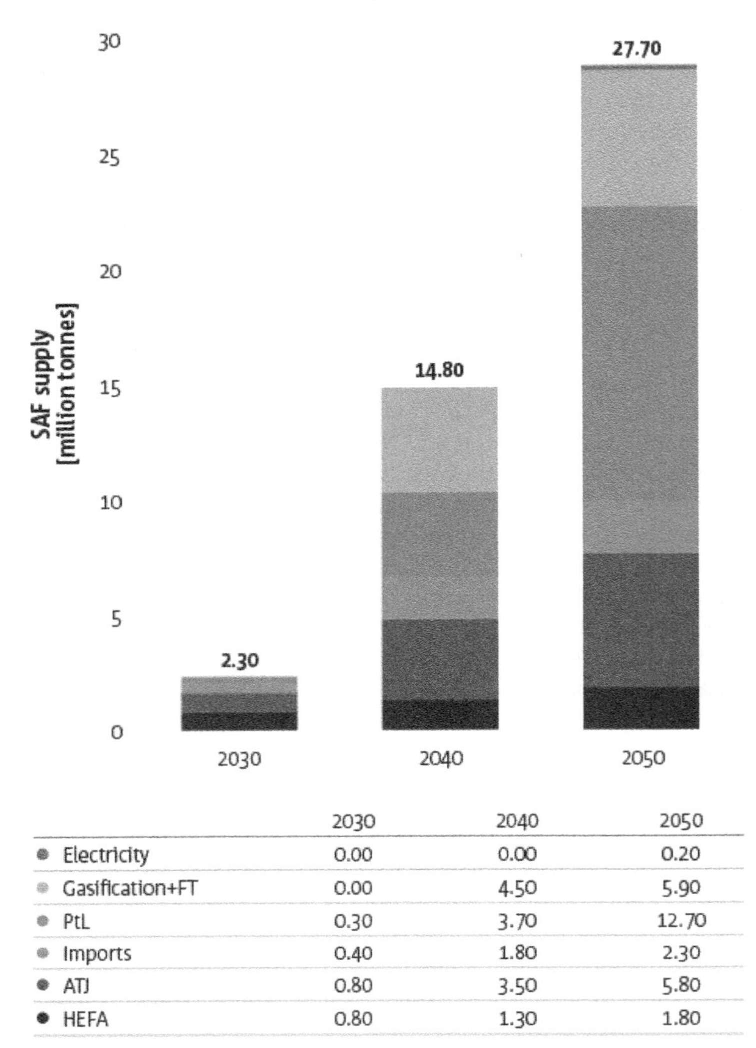

	2030	2040	2050
Electricity	0.00	0.00	0.20
Gasification+FT	0.00	4.50	5.90
PtL	0.30	3.70	12.70
Imports	0.40	1.80	2.30
ATJ	0.80	3.50	5.80
HEFA	0.80	1.30	1.80

FIG. 4.30 Forecast of SAF demand by type and origin in the EU [73]. *Proposal for a Regulation of the European Parliament and of the Council on ensuring a level playing field for sustainable air transport. COM/2021/561 final.*

gasification (FT) SAF having the largest fraction and growth, while fatty acid-based SAF (HEFA) with a limited feedstock base (waste fats) doesn't grow much. Biomass potentially available from agriculture or forestry residues and other underutilized biogenic waste products could provide for about 160 Mt of SAF in the US according to the Bioenergy Technologies Office [74]. In the EU, this number has been estimated with conservative assumptions about biomass utilization to be about 12 Mt and with constraints on technology deployment until 2030 to be 3.4 Mt or 5.5% of jet fuel consumption [75]. Clearly, other feedstocks and green hydrogen are required in the future for scaling production to meet demand. The IATA estimates that 65% of aviation emission reductions by 2050 need to come through the use of SAF with an amount of 360 Mt per annum [72]. The required production growth rates are in excess of 10% per year until 2050 and initially far higher. For the near term, SAF as a drop-in solution for existing technology and infrastructure seems the most practical decarbonization route for commercial airliners and mandates can be expected to be rolled out across further regions after the EU initiated this.

Ammonia

Another option for zero CO_2 emission fuel is ammonia. Based on green hydrogen, ammonia has no carbon emissions and is relatively inexpensive to make, while it has a two to three times higher energy density than liquid or compressed gaseous hydrogen. Unlike SAF, ammonia is not a liquid at ambient conditions and harder to handle, requiring new infrastructure, pressurized and insulated tanks, and safety protocols. Ammonia is combustible in engines and turbines or special high-temperature fuel cells. Compared to SAF, ammonia is more than two times heavier per unit of energy and requires about 3× the volume. These are disadvantages that can be traded against the higher cost of SAF. Ammonia can be used directly in turbofan engines with modifications of the combustion system to account for ammonia's heating value, flame speed, and volume flow rate. An issue with ammonia combustion is the high NOx emission from both the thermal (Zeldovich) mechanism due to combustion temperatures and the prompt NOx resulting from the high nitrogen content of the fuel itself. While NOx abatement through aftertreatment of exhaust gases with catalytic converters is common practice in internal combustion engines and even in stationary gas turbines to meet ever-more stringent emission regulations, for aircraft engines, especially turbofan engines, aftertreatment has not been an option. NOx abatement has been accomplished in modern engines running with jet fuel through various combustor designs with strategies lowering the flame temperature including premixed lean burn, fuel staged lean burn, and rich burn quenched - lean burn. For ammonia combustion, research is ongoing to find a suitable strategy. One possibility with ammonia combustion to lower NOx is fuel processing through catalytic partial reformation into a mixture of ammonia, water, nitrogen, and hydrogen by using high-temperature engine heat. Once the NOx issue is resolved, ammonia may find application in aviation as an alternative to hydrogen for its easier handling and storing if SAF turns out to be too expensive. However, at the time of writing, the use of ammonia as an aviation fuel seems rather far away.

Rockets and spacecraft

Rockets, or launch vehicles, bringing spacecraft into orbit, are a comparatively small sector by fuel consumption within aviation: each typical rocket launch consumes about as much fuel as an intercontinental flight. However, with the promise of commercial space travel on the horizon and increasing demand for spacecraft, even manned missions, this sector may continue to grow fast.

There has been a large growth in the launch rate of rockets across all nations since the beginning of the 2000s, reaching 187 launches in 2022. The two leading nations are the United States and China. Commercial satellites for communication are the most common payload. A Falcon 9 rocket from Space-X, the most active provider of satellite launch services in 2022, carries about 300 t to 500 t of propellant depending on the version, and can deliver a payload up to 13 t into low earth orbit. Of the top 10 American companies [76], seven use kerosene-fueled engines, one uses methane-fueled engines, and two use hydrogen-fueled engines. There are also two LNG vehicles in development. All launch vehicles from these top launch providers use liquid oxygen for their oxidizer. Of the companies who build vehicles that operate in space, most use hydrazine-nitrogen tetraoxide combinations for control thrusters, and about half use ethanol for larger thrusters.

Rocket fuels and propellants

Propellant selection is based on two things. Primarily is the specific energy of a propellant combination. Propellant includes both fuel and oxidizer, the latter may outweigh the mass of fuel. A flight vehicle design requires a minimal mass to take its payload to orbit, a large portion of this mass is the propellant to power the thrusting of the vehicle [77]. Thus, a rocket or spacecraft would require reacting as much energy to turn into thrust in as little mass as possible, leading to needing to store energy in the chemicals that are the rocket's propellants. As of the time of writing, there are no in-space refueling methods in use; however, this is likely to change in the future with lunar missions being carried out with reusable vehicles [78]. Second, the propellants need to be selected for a specific need. For example, a low-cost booster by a small launch provider will likely not utilize liquid hydrogen as a fuel due to its high development and sustainment costs. Kerosenes provide a lower-cost alternative in this situation.

Hydrocarbons

A variety of hydrocarbon fuels have been used historically [79,p. 260], and an even wider variety has been experimented with [80]. The first liquid-fuel rocket built by Robert Goddard was fueled by gasoline [81], and the V-2 rocket built by Wernher von Braun's team was fueled by ethanol [82]. Kerosene is used in multiple specifications, but the most common one for spaceflight is the highly refined RP-1 [80,p. 28]. RP-1 is commonly used due to its relatively low cost.

Hypergolic fuels

Much effort has been put into developing high-performing hypergolic propellants dating back to the 1940s [80, p. 20]. Hypergolic simply refers to a combination of chemicals that ignite on contact with no additional energy required [80,p. 11]. The two main hypergolic oxidizers are nitric acid (HNO_3) and nitrogen tetroxide (N_2O_4 or NTO) [79,p. 258]. The main hypergolic fuels are the three main forms of hydrazine: hydrazine (N_2H_4), unsymmetrical dimethylhydrazine (UDMH) ($(CH_3)_2NNH_2$), monomethylhydrazine (MMH) (CH_3NHNH_2) [79,p. 262]. The challenge with working with these propellants is the material compatibility and safety of personnel, although they do offer advantages in the function of rockets and spacecraft in their ignition properties and that they can be stored for long durations stably, with examples of some probes storing propellant for 40 years [79,p. 262].

Liquid oxygen

Liquid oxygen, commonly referred to as LOX, is the most common oxidizer, owing to its high accessibility and relatively low cost of developing and operating vehicles that use LOX. It is also a good oxidizer for many fuels that could be used, including fuels like RP-1 that are easier to work with and more common [79].

Liquid methane & liquified natural gas

Liquid methane and LNG are relatively new fuels that have only been matured for use in recent years. Sutton [79] in his correspondence with Morehart in 2015 had not seen the use of methane or LNG in flight vehicles. As of the time of writing, there are two engines in development that use methane or LNG [83,84], and one that has been flown on an orbital class rocket [85]. NASA has identified methane and LNG for future use due to its high performance, semi-storability, and relative ease of use [86].

Liquid hydrogen

Much like the other applications in this chapter, hydrogen presents advantages in that it is a very high-performing propellant, better than all stable chemicals [80,p. 159], owing to the extremely low molecular mass of the combustion products and engine exhaust. To achieve this performance, the vehicle and propulsion systems are more complicated and expensive.

Monopropellants

Another way to store chemical energy is in the chemical bonds of a molecule. This molecule can be decomposed to release energy to expand a fluid. Using this method to produce thrust requires specific propellants referred to as monopropellants. There are two main monopropellants that have been used. First, that was used is hydrogen peroxide (H_2O_2), which only requires a catalyst to produce the decomposition into steam and oxygen [80,p. 59]. This leads to hydrogen peroxide also being used as an oxidizer. An early use for hydrogen peroxide was to drive turbines for the turbopumps of early engines [82,87]. Hydrazine has become more prevalent relatively recently in being used for small thrusters [79,p. 264].

There have been alternative monopropellants developed, however not used outside of the experimental stage [79,p. 264] [80,ch 11].

Pressurized gases

There are two uses for high-pressure gas on a launch vehicle or spacecraft. The first and primary use is to maintain pressure in the propellant tanks. This can be either used to drive propellants into an engine or to maintain pressure at the inlet of a turbopump [79,p. 206]. There are multiple methods of doing so, including using heated gases from the engine to pressurize the tanks. When the same propellant is heated and returned to the propellant tanks via a process in the engine, this is referred to as autogenous pressurization and is an effective way to re-use waste heat in the engine [88]. When operating turbopumps, the energy used to pressurize the tank is vital to prevent cavitation, where the energy in the fluid cannot hold the propellant in its liquid form, and more energy is required to move the propellant into the combustion chamber [89].

The second use is as a cold gas thruster. This is not as common since monopropellant and hypergolic thrusters are also available but do provide a simple way to produce thrust [79,p. 266]. The mechanism is essentially storing mechanical energy as pressure and strain energy of the tank to release when needed in the vehicle's use.

Solid fuels

Not all fuels are in the form of a fluid. The first rockets were solid-fueled rocket motors. While there is not as much energy stored in a unit mass of propellant, solid motors are simpler with fewer parts that can serve a wide range of thrust levels and with more energy per unit volume than liquid fuels [90].

Solid fuel motors are used in boosters rather than a main stage for a rocket in modern applications for spaceflight. The only main-stage solid motor rocket was going to be the OmegA by Northrop Grumman, which was canceled in September 2020 [91]. However, there are three American rockets in operation and development that use solid rocket boosters [92–94].

FIG. 4.31 A liquid hydrogen delivery for the NASA Space Shuttle. *Source: US Department of Energy, https://www.energy.gov/sites/default/files/liquid_tanker.jpg.*

Propellant transport and loading

Propellants are typically transported via freight train and truck to their location of use, whether that be a test site or launch site. While the frequency of launches has been increasing, there is likely still not a use case for a direct pipeline for liquid propellants to launch sites from production facilities. Propellants are stored in tanks on the site where they are to be used. In the case of launch sites, they will remain there until loaded onto a vehicle except for some propellant forms. Solid fuel is integral to the vehicle and needs to be manufactured in the rocket [95]. Storable propellants can be loaded onto the vehicle when assembled. Cryogenic propellants, liquid methane, hydrogen, and oxygen, will need to be loaded shortly before use, or launch, and the vehicle prechilled to those temperatures and insulated to minimize boil-off losses. Loading propellants onto a vehicle is done by a series of piping leading to the launch vehicle which then detaches before or at lift-off of the vehicle (Fig. 4.31).

References

[1] IEA, Global Energy Review, International Energy Agency, Paris, 2021. https://www.iea.org/reports/global-energy-review-co2-emissions-in-2021-2.

[2] IEA, World Total Energy Supply by Source, 1971–2019, International Energy Agency, Paris, 2021. https://www.iea.org/data-and-statistics/charts/world-total-energy-supply-by-source-1971-2019.

[3] EIA, Monthly Energy Review, US Energy Information Agency, 2022. Available online: www.eia.gov/mer.

[4] BP, BP Statistical Review of World Energy 2022, British Petroleum, 2022. Available online: https://www.bp.com/content/dam/bp/business-sites/en/global/corporate/pdfs/energy-economics/statistical-review/bp-stats-review-2022-full-report.pdf.

[5] IEA, The Future of Hydrogen, International Energy Agency, Paris, 2019. https://www.iea.org/reports/the-future-of-hydrogen.

[6] Energy Transitions Commission, Making Clean Electrification Possible: 30 Years to Electrify the Global Economy, Energy Transitions Commission, 2021.

[7] Energy Transitions Commission, Making the Hydrogen Economy Possible: Accelerating Clean Hydrogen in an Electrified Economy, Energy Transitions Commission, 2021. https://energy-transitions.org/wp-content/uploads/2021/04/ETC-Global-Hydrogen-Report.pdf.

[8] E.S. Spang, W.R. Moomaw, K.S. Gallagher, P.H. Kirshen, D.H. Marks, The water consumption of energy production: an international comparison, Environ. Res. Lett. 9 (10) (2014).

[9] IEA, Natural Gas Prices in Europe, Asia and the United States, Jan 2020-February 2022, IEA, Paris, 2022. https://www.iea.org/data-and-statistics/charts/natural-gas-prices-in-europe-asia-and-the-united-states-jan-2020-february-2022.

[10] IEA, Outlooks for Gas Markets and Investment, International Energy Agency, Paris, 2023. https://www.iea.org/reports/outlooks-for-gas-markets-and-investment.
[11] IEA, ElectricityMarketReport 2023. International Energy Agency, 2023. https://iea.blob.core.windows.net/assets/255e9cba-da84-4681-8c1f-458ca1a3d9ca/ElectricityMarketReport2023.pdf.
[12] IEA Publications, CO_2 Emissions in 2022, International Energy Agency, 2023. https://www.iea.org/reports/co2-emissions-in-2022.
[13] IEA, World Energy Outlook, IEA, Paris, 2022.
[14] US Energy Information Administration, 2023: Monthly Energy Review, 2023, Available online https://www.eia.gov/totalenergy/data/monthly/pdf/sec7_4.pdf.
[15] Global Energy Monitor, 2023. Available online https://globalenergymonitor.org/.
[16] J. Schmitt, T. Briggs, T. Callahan, S. Freund, R. Kurz, A. Neil, G. Paniagua, D. Sánchez, Chapter 4: Heat engines, in: K. Brun, T. Allison (Eds.), Machinery and Energy Systems for the Hydrogen Economy, Elsevier, 2022, pp. 95–188, https://doi.org/10.1016/B978-0-323-90394-3.00014-X.
[17] T. Omatick, D. Zhang, M.S. Blais, M. Mossolly, P. Renzi, R. Kurz, S. Freund, S. Harvey, Chapter 7: Usage, in: K. Brun, T. Allison (Eds.), Machinery and Energy Systems for the Hydrogen Economy, Elsevier, 2022, pp. 251–304, https://doi.org/10.1016/B978-0-323-90394-3.00009-6.
[18] L. Zeitung, 2020. https://www.l-iz.de/wirtschaft/wirtschaft-leipzig/2020/05/Kohleausstieg-Ende-2022-soll-das-neue-Gasheizkraftwerk-Leipzig-Sued-in-Betrieb-gehen-329779.
[19] J. Wilkes, M. McBain, R. Kurz, J. Goldmeer, T. Callahan, K. Wygant, J. Singh, B. Geswein, S. Freund, Chapter 10: Power generation and mechanical drivers, in: K. Brun, T. Allison (Eds.), Machinery and Energy Systems for the Hydrogen Economy, Elsevier, 2022, pp. 425–473, https://doi.org/10.1016/B978-0-323-90394-3.00006-0.
[20] IEA, Coal-Fired Electricity, IEA, Paris, 2022. https://www.iea.org/reports/coal-fired-electricity.
[21] The Guardian, 2022. Available online: https://www.theguardian.com/environment/2022/apr/26/too-many-new-coal-fired-plants-planned-for-15c-climate-goal-report-concludes.
[22] Drax Group plc, 2023. https://www.drax.com/sustainable-bioenergy/5-incredible-numbers-worlds-largest-biomass-port/.
[23] P. Saxena, D. Burnes, R. Kurz, Study to Adapt Industrial Gas Turbines for Significant and Viable CO_2 Emissions Reduction, 2022. ASME Paper GT2022-83472.
[24] R. Kurz, M. Lubomirsky, G. McLorg, Compression Applications for Carbon Reduction, 2022. ASME Paper GT2022-80080.
[25] World Steel Association, Public policy paper 17.05.2021, 2021, Available online https://worldsteel.org/wp-content/uploads/Climate-change-and-the-production-of-iron-and-steel.pdf.
[26] US Energy Information Administration, Use of Energy Explained: Energy Use in Industry, 2023, Available online: https://www.eia.gov/energyexplained/use-of-energy/industry.php.
[27] ThyssenKrupp Steel Europe AG, Press release 27.04.2023, 2023, Available online: https://www.thyssenkrupp.com/en/newsroom/press-releases/pressdetailpage/eu-commission-approves-german-federal-and-state-government-fund-ing-for-thyssenkrupp-steels-tkh2steel-decarbonization-project-228875.
[28] A.G. Salzgitter, Press release 18.04.2023, 2023, Available online https://www.salzgitter-ag.com/en/newsroom/press-releases/details/salzgitter-ag-receives-official-notice-of-government-funding-for-the-salcosr-low-co2-steel-production-program-20702.html.
[29] A.G. Voestalpine, Press release 23.03.2023, 2023, Available online: https://www.voestalpine.com/group/de/media/presseaussendungen/2023-03-22-voestalpine-aufsichtsrat-genehmigt-1-5-milliarden-euro-fuer-weitere-dekarbonisierung/.
[30] SSAB, Press release 01.04.2022, 2022, Available online https://www.ssab.com/en/news/2022/04/hybrit-receives-support-from-the-eu-innovation-fund.
[31] USGS, US Geological Survey, 2023, Available online https://pubs.usgs.gov/periodicals/mcs2023/mcs2023-cement.pdf.
[32] P. Monteiro, S. Miller, A. Horvath, Towards sustainable concrete, Nat. Mater. 16 (2017) 698–699, https://doi.org/10.1038/nmat4930.
[33] L.D. Ellis, A.F. Badel, M.L. Chiang, R.J.-Y. Park, Y.-M. Chiang, Proc. Natl. Acad. Sci. USA 117 (2020) 12584–12591. https://www.pnas.org/doi/full/10.1073/pnas.1821673116.
[34] Joint Research Centre, Institute for Prospective Technological Studies, I. Kourti, L. Delgado Sancho, F. Schorcht, et al., Best available techniques (BAT) reference document for the production of cement, lime and magnesium oxide – Industrial Emissions Directive 2010/75/EU (integrated pollution prevention and control), Publications Office, 2013. https://data.europa.eu/doi/10.2788/12850.
[35] IEA, Cement Report, IEA, Paris, 2022. Available online: https://www.iea.org/reports/cement.
[36] IEA, Technology Roadmap: Low-Carbon Transition in the Cement Industry, IEA, Paris, 2018. Available online: https://www.iea.org/reports/technology-roadmap-low-carbon-transition-in-the-cement-industry.
[37] IEA, The Future of Petrochemicals, IEA, Paris, 2018. Available online: https://www.iea.org/reports/the-future-of-petrochemicals.
[38] OECD, Global Plastics Outlook: Economic Drivers, Environmental Impacts and Policy Options, OECD Publishing, Paris, 2022, https://doi.org/10.1787/de747aef-en. Available online: https://www.oecd.org/environment/plastic-pollution-is-growing-relentlessly-as-waste-management-and-recycling-fall-short.htm.
[39] Franklin Associates, Life Cycle Impacts For Postconsumer Recycled Resins: PET, HDPE, and PP, Association of Plastic Recyclers, Washington, DC, 2018. Available online: https://plasticsrecycling.org/images/library/2018-APR-LCI-report.pdf.
[40] H. Marczak, Energy inputs on the production of plastic products, J. Ecol. Eng. 2022 23 (9) (2022) 146–156, https://doi.org/10.12911/22998993/151815.
[41] IEA, Sankey Diagrams of Energy Statistics, 2023. https://www.iea.org/sankey/.
[42] Eurostat, Renewable Energy Used in Transport Increasing, 2020, Available online https://ec.europa.eu/eurostat/web/products-eurostat-news/-/ddn-20201223-1.
[43] IEA, World Energy Outlook 2022, IEA, Paris, 2022. Available online: https://www.iea.org/reports/world-energy-outlook-2022.
[44] K. Gkonis, H. Psaraftis, Modeling tankers' optimal speed and emissions, Trans. Soc. Nav. Archit. Mar. Eng. 120 (2013) 90–109.
[45] IEA, Electric Vehicles, 2023, Available online: https://www.iea.org/energy-system/transport/electric-vehicles.
[46] IEA, Global EV Outlook 2021, IEA, Paris, 2021. https://www.iea.org/reports/global-ev-outlook-2021.
[47] IEA, Prospects for Electric Vehicle Deployment, 2021. https://www.iea.org/reports/global-ev-outlook-2021/prospects-for-electric-vehicle-deployment.

[48] EU 2019/1242, European Parliament Regulation, 2019, Available online https://eur-lex.europa.eu/eli/reg/2019/1242/oj.
[49] Reuters, California Passes Landmark Mandate for Zero Emission Trucks, D. Shepardson, N. Groom, 2020, Available online: https://www.reuters.com/article/us-california-trucks-electric-idUSKBN23W31N.
[50] R. Kurz, B. Winkelmann, S. Freund, M. McBain, M. Keith, D. Zhang, S. Cich, P. Renzi, J. Schmitt, Chapter 6: Transport and storage, Section 6.5, in: K. Brun, T. Allison (Eds.), Machinery and Energy Systems for the Hydrogen Economy, Elsevier, 2022, pp. 232–237, https://doi.org/10.1016/B978-0-323-90394-3.00003-5.
[51] Bureau of Transportation Statistics, 2021. Available online https://www.bts.gov/content/fuel-consumption-mode-transportation-1.
[52] S. Freund, et al., Chapter 7: Usage, Section 7.4: Train/Rail, in: K. Brun, T. Allison (Eds.), Machinery and Energy Systems for the Hydrogen Economy, Elsevier, 2022, pp. 283–286, https://doi.org/10.1016/B978-0-323-90394-3.00009-6.
[53] J. Pagenkopf, T. Schirmer, M. Böhm, et al., Marktanalyse alternativer Antriebe im deutschen Schienenpersonennahverkehr, DLR/Now GmbH, 2020. . https://elib.dlr.de/134615/1/DLR_2020_Marktanalyse%20alternative%20Antriebe%20SPNV.pdf.
[54] IEA, The Future of Rail, IEA, Paris, 2019. https://www.iea.org/reports/the-future-of-rail.
[55] Railwaygazette, 2018. Available online: https://www.railwaygazette.com/traction-and-rolling-stock/zillertalbahn-plans-switch-to-hydrogen-power/45986.article.
[56] Railwaygazette, 2019. Available online: https://www.railwaygazette.com/traction-and-rolling-stock/us-hydrogen-train-contract-awarded/55124.article.
[57] Ballard, 2021. https://blog.ballard.com/hydrogen-powered-trains.
[58] AAR (Association of American Railroads), 2023. Available online https://www.aar.org/data-center/railroads-states/.
[59] International Maritime Organization, IMO GHG Strategy, 2023, Available online: https://www.imo.org/en/OurWork/Environment/Pages/2023-IMO-Strategy-on-Reduction-of-GHG-Emissions-from-Ships.aspx.
[60] The American Bureau of Shipping, Sustainability Whitepaper: Hydrogen as Marine Fuel, 2021, Available online: https://absinfo.eagle.org/acton/media/16130/hydrogen-as-marine-fuel-whitepaper.
[61] MAN Energy Solutions, Technical Paper: Methanol in Shipping, 2023, Available online https://www.man-es.com/campaigns/download-Q3-2023.
[62] IRENA and AEA, Innovation Outlook: Renewable Ammonia, International Renewable Energy Agency/Ammonia Energy Association, Abu Dhabi/Brooklyn, 2022. https://www.irena.org/Publications.
[63] ICAO, Annual Report 2019, International Civil Aviation Organization, 2020. Available online: https://www.icao.int/annual-report-2019/Pages/the-world-of-air-transport-in-2019.aspx.
[64] G.G. Fleming, I. de Lépinay, R. Schaufele, Environmental Trends in Aviation to 2050. ICAO 2022. https://www.icao.int/environmental-protection/Documents/EnvironmentalReports/2022/ENVReport2022_Art7.pdf [icao.int].
[65] IATA, Fact Sheet Net Zero Carbon 2050 Resolution, International Air Transport Association, 2023. Available online: https://www.iata.org/en/iata-repository/pressroom/fact-sheets/fact-sheet- - - -iata-net-zero-resolutin/.
[66] Reuters, 2023. Available online: https://www.reuters.com/sustainability/eu-lawmakers-approve-binding-green-fuel-targets-aviation-2023-09-13/.
[67] McKinsey & Company, Hydrogen-powered aviation. A fact-based study of hydrogen technology, economics, and climate impact by 2050, Report for the EU Clean Sky 2 JU and EU Fuel Cells and Hydrogen 2 JU, European Union, 2020, https://doi.org/10.2843/471510. https://www.fch.europa.eu/sites/default/files/FCH%20Docs/20200507_Hydrogen%20Powered%20Aviation%20report_FINAL%20web%20%28ID%208706035%29.pdf.
[68] D. Komissarov, Tupolev Tu-154, The USSR's Medium-Range Jet Airliner, Hinckley, UK, 2007. ISBN 1-85780-241-1.
[69] Airbus, Aviation & Climate Change Fact Sheet, 2021, Available online https://www.airbus.com/innovation/zero-emission/hydrogen/zeroe.html.
[70] ZeroAvia, Company website accessed in 2021, 2021. https://www.zeroavia.com/.
[71] S. Pfeifer, Aviation start-ups test potential of green hydrogen, Financial Times (2023). Available online: https://www.ft.com/content/aa1fb5bb-6393-427a-9450-4ea02f9969d8.
[72] IATA, SAF Deployment Policy, International Air Transport Association, 2023. Available online: https://www.iata.org/contentassets/d13875e9ed784f75bac90f000760e998/saf-policy-2023.pdf.
[73] EASA, European Aviation Environmental Report 2022, European Union Aviation Safety Agency, 2022. https://www.easa.europa.eu/eco/eaer/topics/sustainable-aviation-fuels/current-landscape-future-saf-industry.
[74] Bioenergy Technologies Office, Sustainable Aviation Fuels, US Office of Energy Efficiency & Renewable Energy, 2023. Available online: https://www.energy.gov/eere/bioenergy/sustainable-aviation-fuels.
[75] ICCT, WORKING PAPER 2021–13, International Council on Clean Transportation, 2021. Available online: https://theicct.org/sites/default/files/publications/Sustainable-aviation-fuel-feedstock-eu-mar2021.pdf.
[76] https://aerospace.csis.org/data/space-environment-total-launches-by-country/.
[77] https://arstechnica.com/science/2022/12/top-us-launch-companies-of-2022-the-ars-technica-power-ranking/.
[78] https://ntrs.nasa.gov/api/citations/20090037584/downloads/20090037584.pdf.
[79] G.P. Sutton, O. Biblarz, Rocket Propulsion Elements, ninth ed., 2017.
[80] J.D. Clark, Ignition! An Informal History of Liquid Rocket Propellants, 2017.
[81] https://airandspace.si.edu/stories/editorial/robert-goddard-and-first-liquid-propellant-rocket.
[82] https://www.nationalmuseum.af.mil/Visit/Museum-Exhibits/Fact-Sheets/Display/Article/195894/v-2-rocket/.
[83] https://spaceflight101.com/spx/spacex-raptor/.
[84] https://www.blueorigin.com/engines/be-4.
[85] https://www.nasaspaceflight.com/2023/03/maiden-terran-1/.
[86] https://ntrs.nasa.gov/api/citations/20170005557/downloads/20170005557.pdf.
[87] http://heroicrelics.org/info/redstone/a-7-turbopump.html.
[88] https://trace.tennessee.edu/utk_gradthes/6328/.

[89] https://ocw.mit.edu/courses/16-512-rocket-propulsion-fall-2005/b1f6e25184cb98de9c147c9b4aabbd63_lecture_26.pdf.
[90] https://seitzman.gatech.edu/classes/ae4451/solid_propellant_motors.pdf.
[91] https://spaceflightnow.com/2020/09/14/northrop-grumman-ends-omega-rocket-program-after-losing-military-launch-competition/.
[92] https://www.nasa.gov/exploration/systems/sls/fs/solid-rocket-booster.html.
[93] https://www.ulalaunch.com/rockets/atlas-v.
[94] https://www.ulalaunch.com/rockets/vulcan-centaur.
[95] https://www.nasa.gov/mission_pages/shuttle/vehicle/rsrm_srbs.html.

CHAPTER

5

Liquefied natural gas

Marybeth McBain[a], *Enver Karakas*[a], *Todd Omatick*[a], *Brian Pettinato*[a], *Stephen Ross*[a], *Sterling Scavo-Fulk*[a], *Brian Hantz*[a], *Derrick Bauer*[a], *Matt Taher*[b], *and Rainer Kurz*[c]

[a]Ebara Elliott Energy, Jeannette, PA, United States [b]Bechtel Energy Inc., Houston, TX, United States [c]Solar Turbines, San Diego, CA, United States

The worldwide demand for LNG has dramatically risen over the last 20 years, and due to geopolitical instability, this trend is expected to continue in the foreseeable future. LNG plants often use a wide range of turbomachines, including centrifugal compressors, gas turbines, steam turbines, cryogenic pumps, and expanders. For example, centrifugal compressors driven by gas turbines are commonly used for refrigeration duty in most LNG production plants. Similarly, cryogenic pumps are used in LNG plants to transfer LNG from storage and the movement of LNG at the terminal.

LNG is a liquefied form of natural gas that primarily consists of methane and some smaller fractions of ethane, propane, and butane, all in the liquid phase below its condensing point. LNG is odorless, colorless, nontoxic, and noncorrosive. The liquefaction process involves first the removal of acid gases, inert gases, water, and heavy hydrocarbons. Then, the remaining gas is condensed into a liquid at roughly atmospheric pressure by cooling it to approximately −162°C. At this temperature and near atmospheric pressure, the LNG has 600 times lower volume than in its gas state. It can be pumped and transported using maritime vessels, highway trucks, and short liquid pipelines.

Various types of complex refrigeration cycles are common in modern LNG plants. The majority of the liquefaction processes are done by cascaded vapor-compression refrigeration. In a simple, pure component, cascaded LNG process, the compressors operate sequentially arranged, closed-refrigeration cycles using methane, ethylene, propane, or other hydrocarbons as the refrigerant process fluid where the feed gas is cooled using heat exchangers to lower-temperature steps by each cycle. Different cycles use mixed refrigerant compressors where the feed gas is cooled using multiple Joule-Thomson expansion processes and is then utilized as the refrigerant for precooling of the incoming feed gas. These processes require compression of either the feed/process gas or secondary refrigerant gases in the cooling cycles. In several processes, the Joule-Thomson valves can be replaced using cryogenic expanders to make refrigeration more efficient.

Typical power ranges for LNG turbomachines range from single digits (in small-scale distributed LNG plants) up to well above a hundred thousand horsepower in large LNG export facilities. Many of the compressors are driven by natural gas-fueled gas turbines. Gas turbines are usually the logical compressor mechanical driver choice since natural gas as a fuel is abundantly available in an LNG production plant. Optimized operation of the gas turbine-driven compressors requires matching the compressor and gas turbine driver's speed, power, and utility requirements while maintaining operational flexibility for startup, shutdown, and during system excursions. Pumps are often driven by induction motors, which are submerged into pumped fluid for enhanced cooling.

Introduction

Global natural gas consumption has been rising over the past 20 years. In many cases, increasing regional demand for natural gas can be linked to population growth and motivation for cleaner power sources. Strong forecast market demand for natural gas tends to occur in areas remotely located from the most economical natural gas fields with less pipeline infrastructure. Liquefied natural gas, or LNG, offers the capability to convert natural gas to higher energy concentration and transport its value to higher-demand markets where it can be stored and re-gasified.

Energy Transport Infrastructure for a Decarbonized Economy
https://doi.org/10.1016/B978-0-443-21893-4.00008-8

Copyright © 2025 Elsevier Inc. All rights are reserved, including those for text and data mining, AI training, and similar technologies.

LNG has adapted historically to advances in turbomachinery, from steam turbine drives to gas turbine drivers (both frame units and aero derivatives), and currently, even large synchronous electric motor drivers for the main refrigerant compressors. Additionally, developers have continued to gain efficiencies and cost improvements by considering variations to the refrigerant mixtures and cycle adaptations to most economically and efficiently convert natural gas to its liquefied form. This chapter describes the current processes, various cycles utilized, and specific turbomachinery required to produce and ultimately deliver LNG to consumers.

Background

According to Shell, global LNG demand will continue to rise, exceeding 600 million tons per annum (MTPA) by 2030 [1]. Additionally, LNG offers the opportunity to store natural gas energy value in liquid form and exploit the energy capacity to new markets for various means—power generation, direct heating, industrial processes, and as a supplement to renewable power worldwide. In 2019, LNG represented 9.8% of the global gas supply. This percentage is likely to increase as the primary gas-producing areas develop greater LNG production to capture higher market value for natural gas as a primary and supplemental (to renewables) means of power generation.

Liquefaction is the process of cooling natural gas to approximately −162°C (−259°F) until it forms a liquid. The conversion of natural gas to a liquid allows for the transport of greater energy content since LNG takes up about 1/600th the volume of natural gas (i.e., liquefied natural gas offers an energy density 600 times that of standard natural gas at atmospheric pressure, 28 lbm/scf vs. 0.045 lbm/scf). LNG must then be converted back into gaseous form for commercial use, and this is done at regasification plants at the receiving port terminal where the LNG is delivered.

The LNG value chain

The LNG Value Chain includes the production, processing, and conversion of natural gas to LNG, transportation of the liquefied gas, and regasification as it travels from extraction at the wellhead to final delivery to end-users [2].

Production: Extraction of natural gas

Gas is separated from condensate and/or crude oil and contains impurities, water, and other associated fluids, so before liquefaction can occur, the gas must undergo a treatment process to remove undesired substances. This raw (feed) gas is transported (gas gathering) to processing facilities for the removal of contaminants and natural gas liquids. Natural gas can contain nonhydrocarbons, including nitrogen, hydrogen sulfide, carbon dioxide, mercury, and water. Natural gas liquids, like propane or butane, are also extracted and sold separately.

Gas processing

The second step in the value chain is cleaning and treating the natural gas. Frequently, this takes place at the liquefaction plant. A series of processing steps separates and removes the various other compounds and impurities from the natural gas prior to liquefaction.

One of the main purposes of the liquefaction plant is to provide a consistent composition and reliable combustion characteristics, which are critically important to obtain pipeline-quality gas for end-users, which typically contains 85% to 95% methane. The remaining mixture typically contains heavy hydrocarbons like ethane, butane, or propane and other substances, which are not removed during the processing.

The gas goes through a series of steps involving separation, treatment, dehydration, and compression, where impurities are removed and heavier fluids are separated. First, water and condensate are removed, followed by the removal of carbon dioxide (CO_2), hydrogen sulfide (H_2S), and mercury (Hg). Water and CO_2 may cause ice forming during liquefaction, and H_2S, mercury, and other contaminants could cause corrosion in pipelines and LNG heat exchangers. The remaining mix is precooled, and other heavier natural gas liquids are separated from the mix before liquefaction occurs. Heavy hydrocarbons like propane or butane are sometimes also removed and sold as raw materials to the petrochemical industry or used as fuel. The remaining gas mostly consists of methane and some ethane, which is brought to liquefaction.

Liquefaction: LNG plant

After the gas treatment, the natural gas feedstock is about 85% to 90% methane (CH_4), with the rest being a mix of ethane (C_2H_6), propane (C_3H_8), and butane (C_4H_{10}). The clean, dry natural gas is then ready for liquefaction. Like most gases, natural gas will condense into a liquid when its temperature drops below its boiling point, which is −162°C for methane. To produce LNG, an industrial-scale cryogenic process is required.

The liquefaction of natural gas is achieved through several different refrigeration processes, which can be classified into three groups: pure component cascade liquefaction processes, mixed refrigerant processes, and gas expansion-based processes. These processes will be reviewed in greater detail in later sections. The method to power the plant and to provide motive power for the refrigeration compressors can also vary, and this will be further discussed later. LNG plants are most commonly land based; however, production, treatment, liquefaction, storage, and transportation can also take place using floating liquefied natural gas (FLNG) facilities. The liquefaction and LNG cooling process are based on thermodynamic refrigeration cycles and take place in cryogenic heat exchangers, which absorb heat from the natural gas.

Production capacity and plant size vary, but key infrastructure, in addition to the specific gas treatment and liquefaction equipment, includes equipment for handling boil-off gas (BOG) like venting, flaring or re-liquefaction, send-out systems, and cryogenic storage tanks. Flash gas and BOG can be used as fuel gas for plant consumption, for turbines used for onsite power generation, or recycled to the plant inlet feed gas. The individual systems are often closely integrated.

Liquefaction plants are typically constructed as many parallel liquefaction units, which are commonly referred to as trains. Each train is a complete stand-alone processing unit. LNG process advancements have generally followed an evolutionary path rather than one of radical design changes. Over the last 30 years, the size of LNG plants has grown from approximately 2 MTPA to as much as 7.8 MTPA focused on economies of scale; however, the design of the specialized turbomachinery, heat exchangers, and large gas reserves required for the larger size plant may exceed feasibilities of the economy of scale. Fig. 5.1 shows a current trend of projects that are being evaluated that also include small- and mid-scale LNG process technologies with train sizes ranging from 0.25 MTPA to 2.0 MTPA.

The train size often steers the design toward certain LNG refrigeration cycles since some of these plants scale better than others in terms of the capital cost of equipment. It is interesting to note the emergence of the smaller scale (less than 2.0 MTPA) train size—suggesting some of the LNG operators favor the smaller sized equipment and cycles that favor these train capacities due in part to capital cost, availability of equipment, and more condensed project schedules. Many of these smaller-scale LNG trains plan to add identical trains to eventually reach a similar LNG export plant size to the larger trains, typically over 10 MTPA for the total plant capacity. The smaller-scale LNG designs have considered other cycles, such as the Chart IPSMR with nitrogen expansion and the Black and Veatch PRICO/PRICO plus process. Finally, many of the smaller-scale LNG trains are being considered primarily in North America due to the ability to quickly develop projects with a smaller footprint and smaller power range of equipment.

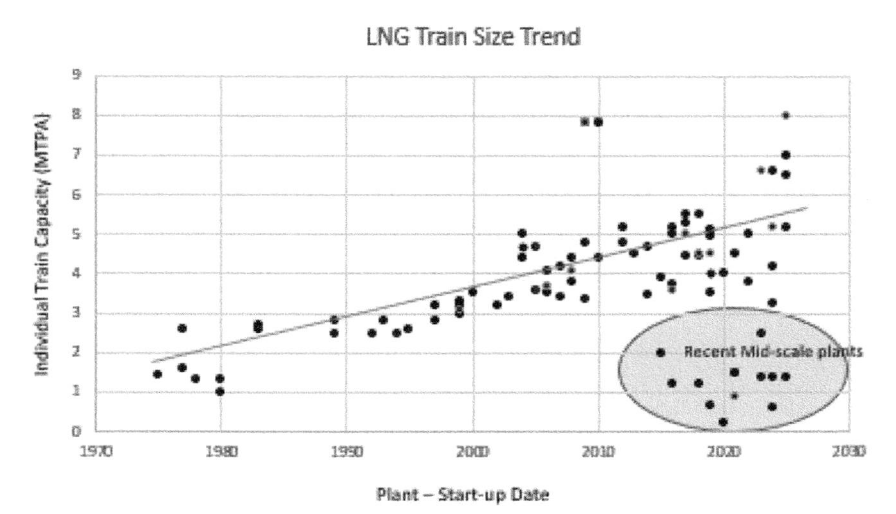

FIG. 5.1 LNG train size by year including plants under construction [3].

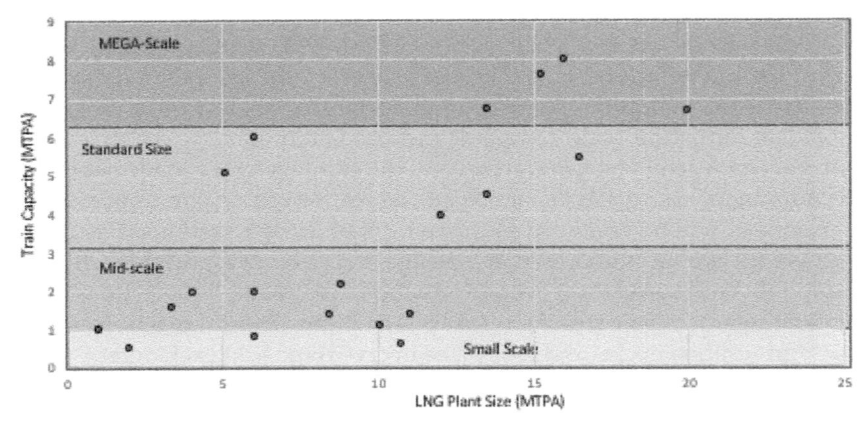

FIG. 5.2 Selected proposed LNG train size MPTA vs. project size MTPA [3].

These small- to mid-scale projects have plant capacities that range from 2.5 MTPA to 20 MTPA as shown in Fig. 5.2. These projects seek to achieve lower unit technical costs, faster permitting, or reduced project risk through strategies involving liquefaction process technologies, modularization, and diversified equipment suppliers.

Additionally, LNG storage from the US natural gas grid is a common practice to supplement mainline natural gas power generation, especially in seasonal demand periods when electricity pricing is high. LNG peak shavers utilize price arbitrage by storing low-cost natural gas in the form of LNG and then selling it back into the market to benefit from higher electric prices (hot summers, cold winters). Most facilities use either a single mixed refrigerant process or nitrogen (less efficient but simple to operate).

LNG transportation

The next step in the LNG value chain is transporting the liquefied natural gas to the regasification terminal. The most economical way to transport LNG over great distances is by sea in specialized LNG carriers, which are double-hulled ships designed to keep the LNG at or near atmospheric pressure at a cryogenic temperature of approximately −162°C. LNG carriers contain specialized materials and advanced systems for handling cryogenic cargoes, including layers of insulation that isolate the LNG cargo from the hull. The insulation system restricts the amount of LNG that evaporates or boils off during the voyages. On many LNG vessels, boil-off gas is captured and compressed to use as supplementary fuel during the voyage. In rare cases, when the liquefaction plant is in the vicinity of the regasification facilities, an LNG can be transported by truck in smaller quantities.

LNG receiving and regasification terminal

The next step in the LNG process chain entails unloading, storage, and regasification at the import terminals, which are typically located near a port. LNG carriers deliver the LNG to a marine terminal where the LNG is stored before undergoing regasification, which converts the LNG back into its gaseous form by heating it to above the boiling point. The LNG is transported at just below its boiling point, so it does not take much of a temperature change to force the phase transition. LNG can also be delivered to floating terminals, which are LNG ships or barges constructed to function as Floating Storage Units (FSU), Floating Regasification Units (FRU), or Floating Storage and Regasification Units (FSRU). Floating facilities allow LNG terminals to be sited offshore and deployed rapidly, often at a lower capital cost than conventional terminals.

Gas utilities/pipeline

Once the natural gas returns to its gaseous phase, it is ready to enter the local distribution system. Pipeline operators generally run their distribution systems at 30–80 bar (435–1160 psia), so the output of the regasification plant has to match the pipeline specification. Rather than compressing the output gas from the regasification plant, pressurizing the LNG using pumps is often easier and more efficient.

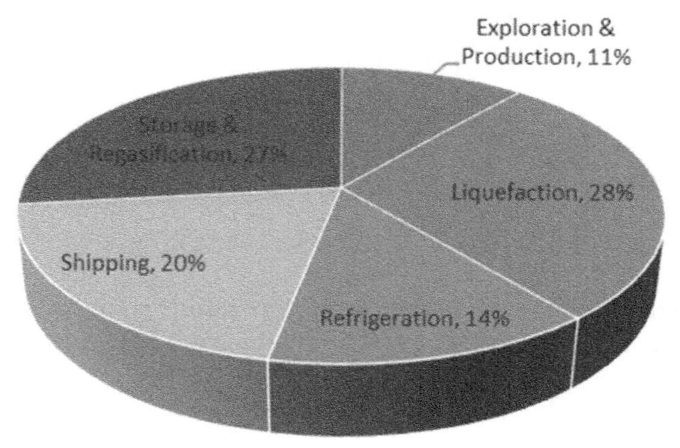

FIG. 5.3 Approximate breakdown of LNG project costs [4].

LNG peak shaving

Another type of facility that may receive LNG is a peak-shaving facility. These plants, which may be operated by utility companies, store LNG in tanks until it is needed at peak demand. An LNG peak-shaving facility is typically connected to the gas supply system and may consist of a small LNG plant to convert the natural gas into LNG. LNG storage tanks, compressors, pumps, and regasification equipment turn the LNG back from liquid to natural gas. In some cases, peak-shaving facilities are filled using cryogenic trucks from an import terminal.

LNG capital investment

LNG projects require considerable capital investment and involve many project partners looking at the entire LNG value chain from production through delivery. Liquefaction is one of the most expensive parts of the value chain, and the turbomachinery that supports this process is critical to the overall value chain. The capital and operating expenditure of the liquefaction facilities is an area that LNG operators have the most control over and can optimize their development costs through process selection and capital equipment. The percentages of overall project costs can vary based on a number of factors; however, a typical breakdown of costs is shown in Fig. 5.3.

The LNG value chain is relatively complex and requires many different types of turbomachinery to complete the thermodynamic processes required. The value chain cycle is fairly efficient, with most of the energy consumed by the liquefaction and transportation steps. Optimizing the liquefaction process and employing energy recovery systems throughout the chain is important to maximizing the return on investment. In most cases, the LNG value chain is the most efficient way to transport natural gas over vast distances.

Overview of refrigeration and liquefaction

The liquefaction process requires refrigeration of natural gas, which is fundamentally cooling the feed gas by heat exchange between the relatively warm feed gas and colder refrigerants. Refrigeration involves the removal of heat from one substance and its transfer to another substance. Heat can be transferred along a temperature differential; in other words, heat will flow from a higher-temperature source to a lower-temperature sink. To generate the cold temperatures required for LNG production, work must be put into the refrigeration cycle through compression, and heat must be rejected from the cycle via a heat exchanger. Liquefaction of natural gas requires the temperature of the feed gas to be reduced to near subcooled (liquid) conditions.

For LNG production, the vapor-compression refrigeration cycle involves the following fundamental steps:

- A refrigerant (in many cases a hydrocarbon) at or slightly above its saturated vapor point is compressed to be a vapor at a higher temperature in its superheated region (1–2).
- The pressurized vapor flows to a condenser that transfers heat out of the vapor typically to the ambient air or to a coolant (water). The condenser cools the hot refrigerant vapor until the vapor condenses the refrigerant vapor into a liquid by removing additional heat (the heat of vaporization) (2–3).

- The next stage in the refrigeration cycle is to expand the high-pressure refrigerant liquid by passing it through a Joule-Thomson (J-T) expansion valve or an expander. The expansion produces a mixture of liquid and vapor at a lower temperature and pressure (3–4).
- The cold liquid-vapor refrigerant mixture then flows to an evaporator unit, where the refrigerant fluid reaches a saturated vapor condition by absorbing heat from the working fluid (feed gas) passing from cooling coil tubes within the evaporator (4–1).

From the evaporation, the refrigerant flows to the compressor, and the cycle continues. A typical refrigeration cycle is shown in Fig. 5.4.

The thermodynamics cycle of the refrigerant is shown as a temperature vs. specific entropy graph in Fig. 5.5 to better understand the refrigerant temperature, work input, phase change, and heat rejection and input for an ideal vapor-compression refrigeration cycle.

It should be noted that Fig. 5.5 is illustrating the ideal refrigeration cycle. In reality, points 3 and 1 may be at the subcooled (liquid only) and superheated (vapor only) part of the T-s diagram, respectively.

To improve the thermodynamic and thermal efficiency of the refrigeration and liquefaction process, an expander can be utilized into the refrigeration cycle in lieu of a J-T expansion valve. For a given pressure differential, an expander

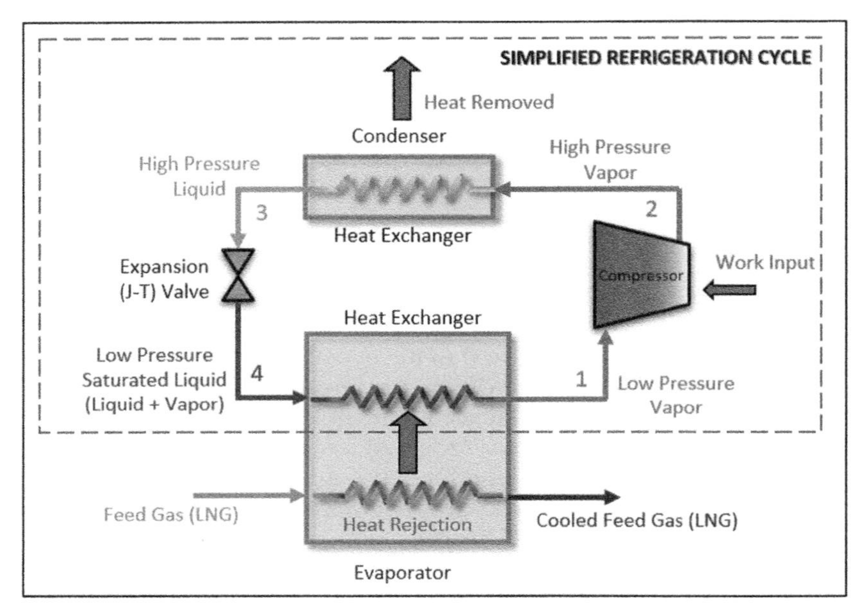

FIG. 5.4 Simplified single-stage compression expansion refrigeration cycle.

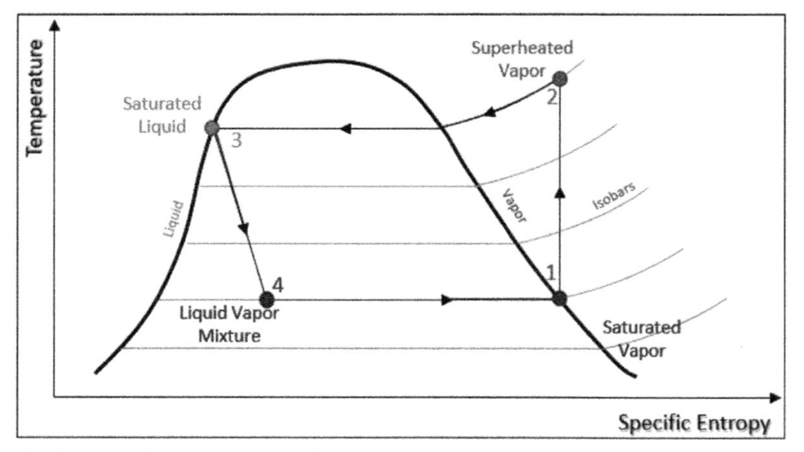

FIG. 5.5 Refrigeration cycle T-s diagram illustrating an ideal case.

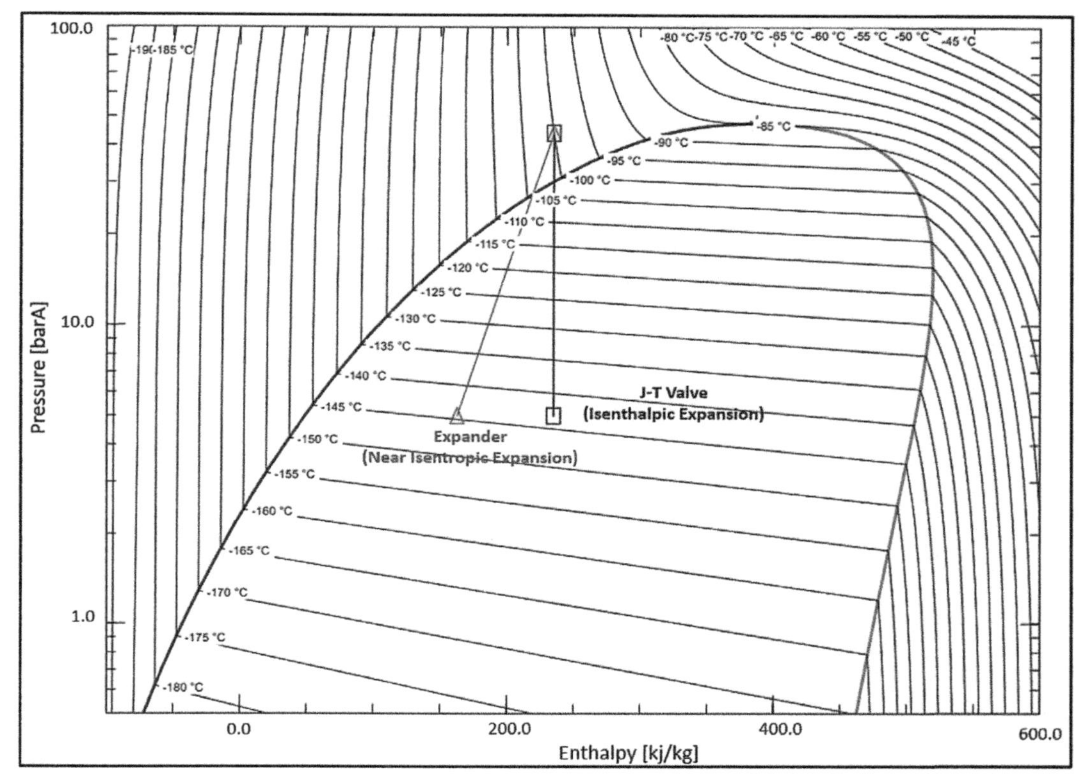

FIG. 5.6 Pressure vs. enthalpy graph showing the expansion via expander and J-T valve for 95% methane and 5% propane fluid mixture.

can achieve a greater temperature reduction in comparison with an expansion valve by recovering work. Expansion via a J-T valve is isenthalpic (no enthalpy change), while the pressure drop across the expander approaches isentropic condition with enthalpy removal from the fluid. This results in colder temperatures at the discharge of the expander with respect to the J-T valve. Fig. 5.6 shows expansion with a J-T valve and a two-phase flashing expander over a pressure vs. enthalpy graph.

The following concepts are the most commonly used refrigeration processes in LNG liquefaction.

1. *Reverse Brayton Cycle*—Nitrogen is typically used as the refrigerant and maintained in a vapor phase throughout the process [5]. Methane can also be used in this cycle as the refrigerant. Both N_2 and methane are readily available at LNG liquefaction plants from equipment purging and from the feed gas, respectively. Using methane as a refrigerant from the feed gas allows a smaller footprint. It should be noted that methane is not able to achieve temperatures as low as nitrogen. This is best illustrated with a vapor pressure curve comparing the different refrigerants. The smaller physical size is related to the greater enthalpy change with methane. Using only methane for an LNG plant could impact overall effectiveness due to decreased temperature differentials. In other words, the enthalpy difference benefit is reduced by reduced heat exchange temperature differential. A reverse Brayton cycle achieves refrigeration by expanding vapor and extracting work. Fig. 5.7 shows a schematic of an ideal reverse Brayton cycle. In this cycle, with the heat from the feed gas (LNG), warm refrigerant gas is compressed at a low pressure (1–2). The resulting high-pressure fluid is cooled at constant pressure while rejecting heat to ambient (2–3). The cooled refrigerant is then expanded, and work is extracted to produce a cold refrigerant stream (3–4), which is subsequently warmed while providing refrigeration (4–1). A temperature vs. entropy plot of an ideal reverse Brayton cycle is given in Fig. 5.8.

   This cycle is used widely in LNG production due to its simple operation and low capital expense [6,7]. Using N_2 is attractive for small-scale LNG production, since import and storage of hydrocarbon refrigerants are not needed.

As seen in Fig. 5.8, there is no phase change at any of the reverse Brayton cycle heat exchangers, and the heat transfer is provided by sensibly warming the refrigerant gas (4–1). Therefore, mass flow rate is required to be relatively greater at reverse Brayton cycle to provide sufficient refrigeration. This results in larger equipment and piping. It can be mitigated by the use of high operating pressures, which are subject to design constraints of both the expander and

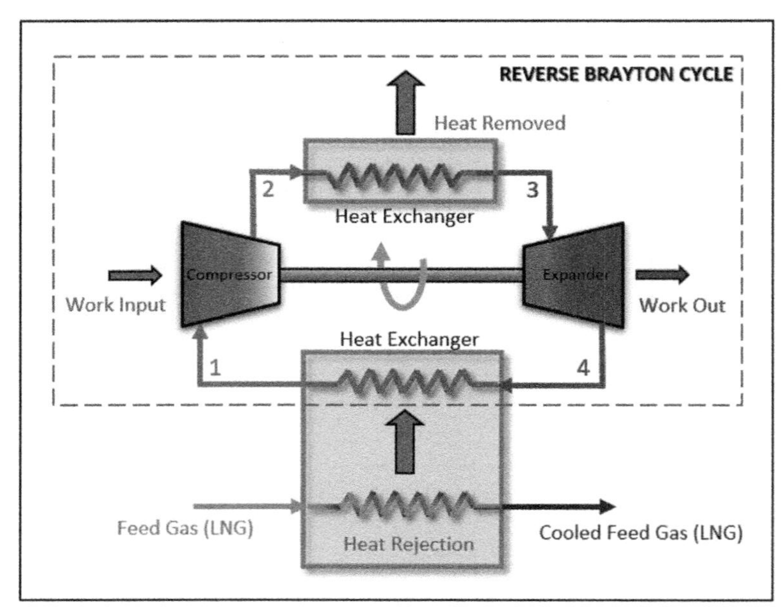

FIG. 5.7 Reverse Brayton refrigeration cycle.

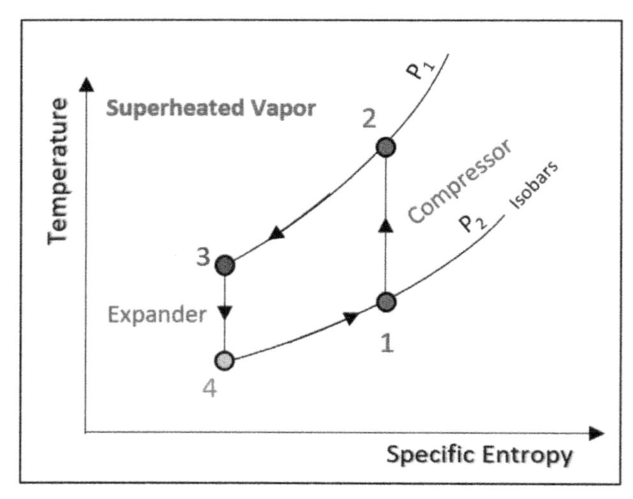

FIG. 5.8 Reverse Brayton cycle T-s diagram illustrating ideal case.

compressor [5]. Therefore, this refrigeration cycle is preferred for small-scale LNG, peak-shaving, and floating LNG (FLNG) applications in the range of 1 MTPA to 2 MTPA [8]. It should be noted that in real applications, pressure drop and increase across the expander and compressor is not 100% isentropic, and fluid and heat flow involves losses across piping and heat exchangers.

2. *Vapor-Compression (Two-Phase) Refrigeration Cycle*—This involves evaporation and condensation of pure gases or mixed refrigerants at heat exchangers. The main difference of the vapor-compression refrigeration with respect to the reverse Brayton cycle is that the cycle provides refrigeration by vaporizing the refrigerant (latent heat) rather than cooling via sensible heat. More refrigeration is provided per mass of refrigerant by latent heat compared to sensible heat.

   For LNG liquefaction, typical choice of pure refrigerant fluids is propane, ethylene, and methane. Mixed refrigerants (MR) are often a blend of hydrocarbons with nitrogen. The composition of the MR is determined with the emphasis of closely matching the cooling curve of the feed gas from feed gas supply to cryogenic temperatures. The basic principle of using MR for cooling and liquefying the gas is to match as closely as possible

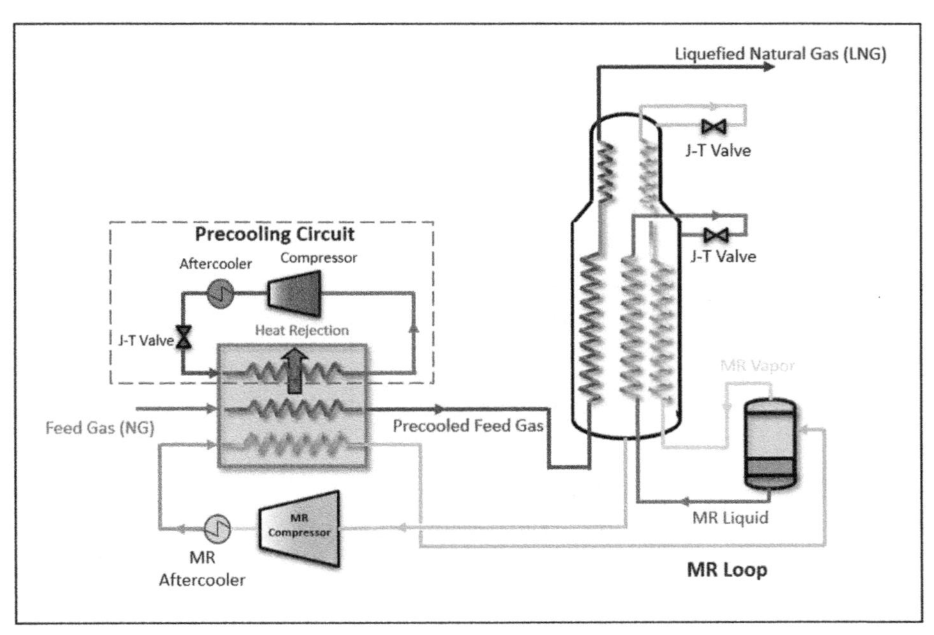

FIG. 5.9 Simplified precooled MR refrigeration process for natural gas liquefaction.

the cooling/heating curves of the feed gas and the refrigerant, which results in a more efficient liquefaction process requiring a lower power consumption per unit of LNG produced. MR technology can be categorized as single mixed refrigerant (SMR) and dual-mixed refrigerant (DMR) cycles.

The SMR process provides the benefit of operational simplicity and flexibility, in addition to reduced equipment count; however, it comes at the cost of lower efficiency than the DMR cycle, which better matches the overall mixed refrigerant boiling curve to the feed condensation curve [8]. The SMR cycle uses only one MR circuit for precooling, liquefaction, and subcooling in a single heat exchanger. The precooled MR cycle is the most widely used LNG process for land-based facilities. The feed gas is initially cooled to an intermediate temperature (about −30°C) by a precooling circuit. Precooling can be provided by utilizing a cold mixed refrigerant (CMR) composed of nitrogen, methane, ethane, and propane. Another and more common option of precooling is providing refrigeration via a propane precooling circuit (C3MR). A schematic view of the basic precooled MR process is shown in Fig. 5.9.

The key component of a precooled MR cycle is the coil-wound heat exchanger (CWHE), where the liquefaction takes place. These heat exchangers make the process much more efficient than most other liquefaction and refrigeration processes.

Another common process for refrigeration and liquefaction is the pure component cascade process where the feed gas is cooled in multiple stages using different hydrocarbon refrigerants in stages. Three main refrigeration loops are employed in this process. Typically, a propane cooling circuit is followed by an ethylene loop and finally by a methane loop. A propane loop not only precools the feed gas, but it also precools ethylene and methane at their respective loops, while the ethylene loop also precools the methane loop [9]. A basic schematic view of the cascade process is given in Fig. 5.10.

The liquefaction process selection is a key activity that starts at an early stage of an LNG project. It should be addressed at the conceptual, feasibility, and prefeed stages of development, since it has such a large impact on the overall profitability of the project. When a comparison is conducted thoroughly, sufficient process and utility details must be developed to define the capital and operating costs for each licensor. Generally, proprietary process simulations are also completed by the respective licensors to predict process performance for a defined feed gas composition or range of compositions and site conditions. Quotations from the various licensors and main equipment suppliers must be obtained to highlight the differences in the processes and finally to select an optimized design that will best meet the LNG project owner's objectives.

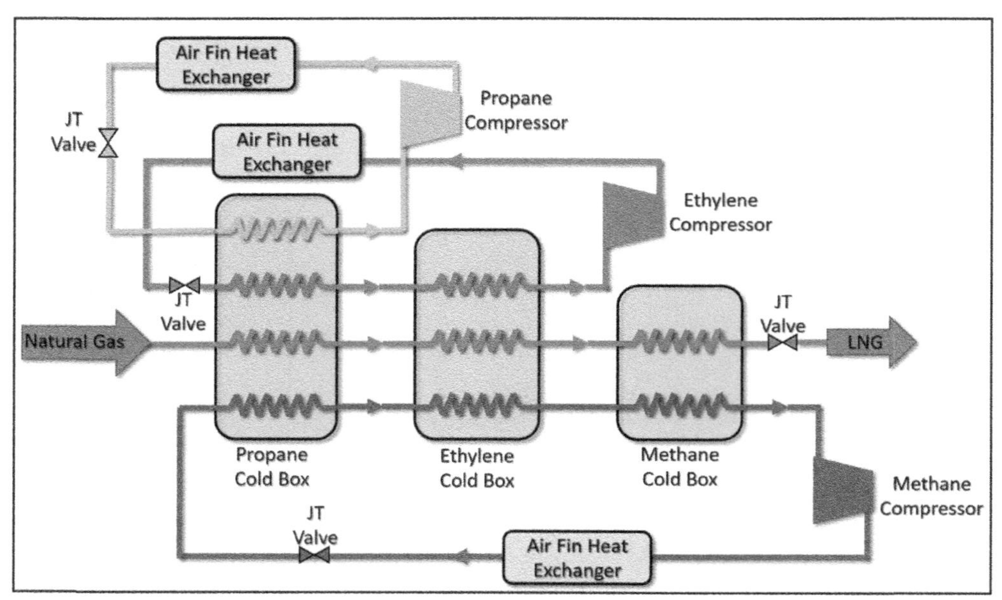

FIG. 5.10 Simplified optimized cascade process.

Compressors for LNG applications

Compressors for LNG applications are used to transfer and pressurize the fluid in gaseous form. Raw natural gas from wells varies in composition and contains some amount of contaminants, such as water, CO_2, and H_2S. Processing is required to remove the contaminants and adjust the composition to meet specifications for sale. If the gas is not transported via a pipeline, it is typically liquefied to reduce its volume for easy transportation and storage. Processing and the liquefaction process, as well as storage of natural gas, requires compressors at various locations. These different services require different design features specific to the compressed fluid and operating conditions. In this section, these design concepts are briefly reviewed, and compressor applications for different processes are detailed.

Natural gas processing compression applications

Whether raw gas is delivered to a pipeline or liquefied for transport or storage, some processing is required. Some common applications for compressors in a gas processing plant are shown in Fig. 5.11.

Feed gas or booster compressor

If the pressure in the gas reservoir or pipeline is too low to pass through the processing plant, a feed gas or booster compressor will be employed to increase the pressure. The raw gas composition for this compressor can vary greatly. Major constituents would be 68%–99% methane, 0.2%–14% ethane, 0%–9% propane and butane, 0.5%–18% nitrogen, 0.1%–2% carbon dioxide, and water vapor. Hydrogen sulfide may also be present. Materials need to be selected to resist corrosion from H_2S and wet CO_2.

Flash gas compressor/off gas compressor

The first processing step is to remove liquid oil and water from the raw gas in a separator. Gas will be "flashed" from a hydrocarbon liquid when the liquid flows from a higher pressure to a lower pressure separator or to a stabilizer column where the light ends (methane, ethane, some propane, and butane) are removed. The gas from the column overhead will be compressed and returned to the feed stream [10]. Flash gas compressors typically handle low flow rates and produce high compression ratios. These may be reciprocating, screw-type, or barrel-type centrifugal compressors. For a centrifugal compressor, the variable mole weight of the feed gas and variable flow rates would require variable speed control and/or a need to recycle flow around the unit.

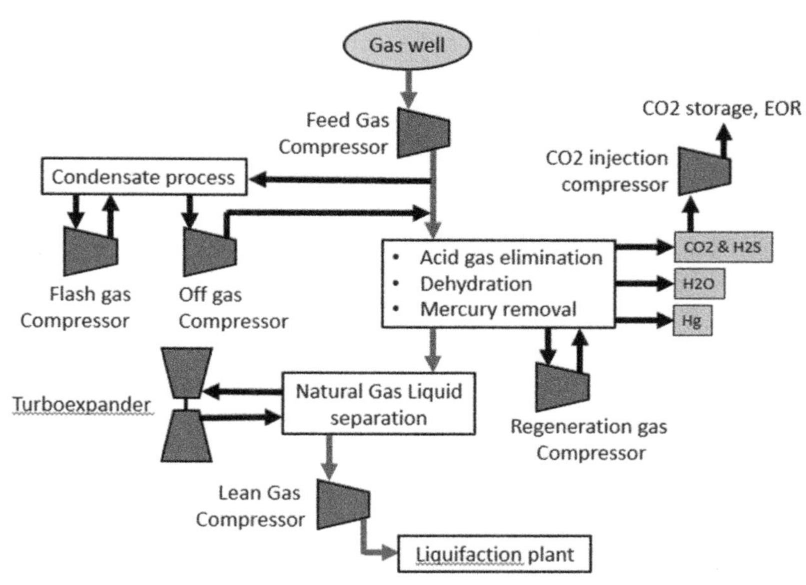

FIG. 5.11 Gas processing plant.

Regeneration gas compressor

The next step is processing "sweetening" or the removal of acid gases, H_2S and CO_2, followed by dehydration. The acid gases are removed by either amine treatment (scrubbing) or membrane separation. The membrane process has larger pressure drops. A solid desiccant dehydrator or molecular sieve typically uses a slipstream of the dried gas to drive off the adsorbed water and regenerate the desiccant. If pressure drops are too great, a regeneration gas compressor may be required. This may be a small, barrel-type centrifugal compressor or a single-stage, integrally geared design.

CO_2 injection compressor

Carbon dioxide removed from natural gas can be reinjected into the well to provide enhanced oil recovery (EOR) or enhanced gas recovery (EGR), as well as carbon capture and storage (CCS). The elevated injection pressure is governed by the reservoir formation pressure and depth of the well. This pressure can be 175bar or more. These pressures can be achieved with integrally geared compressors or barrel-type centrifugals [11] or a compressor in series with a dense-phase pump as was applied at a Middle Eastern EOR installation [12,13]. Discussion of the CO_2 injection pump is beyond the scope of this chapter.

Turboexpander

The first turboexpander was introduced in 1964. Along with a low-temperature distillation column (demethanizer), they are used to remove heavier hydrocarbons from the gas. The overall cryogenic separation plant is shown in Fig. 5.12. More than 90% of the propane and 80% of the ethane can be removed in this manner. In the expander section, gas will enter at 6.2bar and −51°C. As it expands down to 300psi, the temperature will drop isentropically to −90°C. The ethane, propane, and butane will condense at this temperature. These natural gas liquids (NGLs) will be moved to a fractionation plant by light hydrocarbon recycle pumps. The remaining lighter molecular weight gas mixture passes through the cold box, where it is warmed as it cools the inlet gas stream. Its pressure is then increased by the compressor section of the turboexpander, which is powered by the expander side.

Lean gas compressor

The lean gas compressor raises the pipeline-quality natural gas pressure to a level suitable for the liquefaction plant. The various liquefaction processes will be discussed later in the chapter. A general view of a liquefaction plant and LNG tanker with common compressor applications is shown in Fig. 5.13.

Refrigeration compressors

Different processes use different gases for refrigeration. Propane, ethylene, methane, nitrogen, and mixed refrigerants are common. These processes and compressors will be described in more detail later in the chapter.

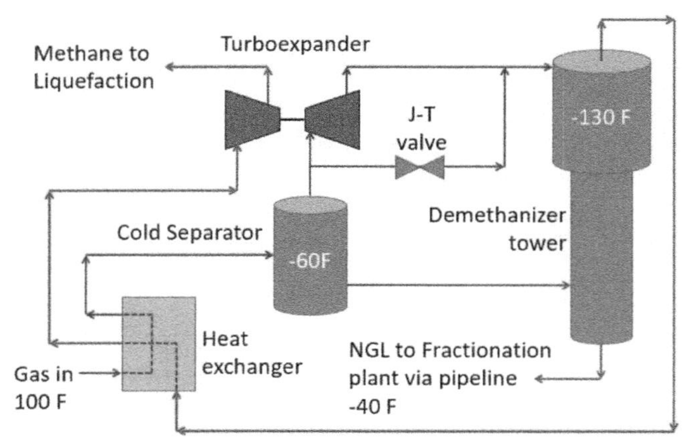

FIG. 5.12 Cryogenic separation plant.

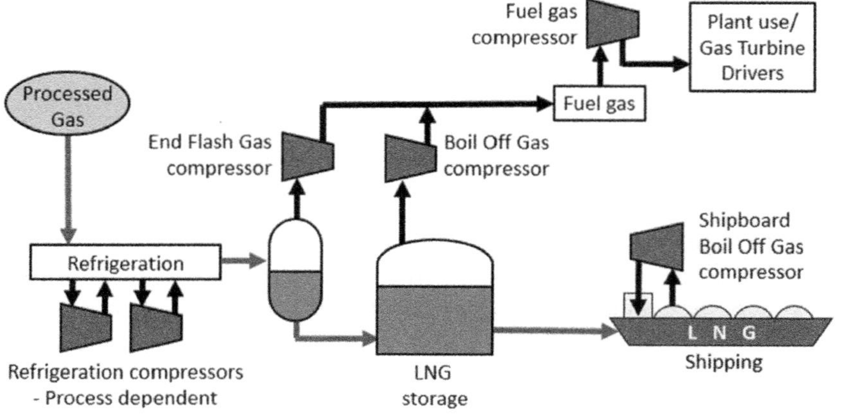

FIG. 5.13 Liquefaction plant compressors.

End flash gas compressor

An end flash gas compressor will compress the low-pressure vapor that comes off the vapor-liquid separator at the end of the liquefaction process. This gas will be used as fuel gas for the plant, which may include gas turbine drivers. The compressor inlet temperature will be near −155°C, so low-temperature materials and design considerations are needed.

Boil-off gas compressors

A boil-off gas (BOG) compressor will compress the low-pressure vapor that comes off the LNG storage tanks. This gas will also be used as fuel gas for plant consumption. Some ships also burn boil-off gas from the cargo tanks as fuel for their engines or recompress and re-liquefy the gas. Like the end flash gas compressors, boil-off compressors will have a similar low inlet temperature and require special materials and designs.

The primary purpose of the BOG compression system is to maintain the LNG storage tank operating pressure. Better control of the LNG tank pressure can lead to better control of the boil-off gas rate.

Broadly speaking, three different types of compressors can be used in a BOG service:

- Reciprocating Compressors (API 618)
- Dry Rotary Screw Compressors (API 619)
- Centrifugal Compressors (API 617)
- Integrally geared with variable IGVs control
- Single shaft (inline) with variable speed or variable IGVs control

While the required pressure ratio of BOG compressors varies in a very narrow range, BOG flow rate swings significantly with the plant operating modes (e.g., holding and ship-loading modes), which is a major factor to select the optimum number of compressor trains. As a result, operating (and upset) conditions must be very carefully considered to define the boundaries of compressor performance, as well as the normal operating point for which the highest efficiency of compressors is desired. The optimal design involves a trade-off between the overall turn-down ratio and the compression efficiency [14].

The machine "reference" Mach number, which is defined as the ratio of impeller tip speed over the sonic speed at the impeller eye (Eq. 5.1), plays a key role in selecting the control method for a BOG centrifugal compressor.

$$\mathrm{Ma}_{ref} = \frac{N\pi D_2}{a} = \frac{N\pi D_2}{\sqrt{RZTn_s}} \tag{5.1}$$

where

a = speed of sound at the first impeller eye
D_2 = impeller tip diameter
Ma_{ref} = machine reference Mach number
N = rotational speed, rps
n_s = isentropic volume exponent
R = gas constant
T = temperature, K
Z = compressibility factor

Due to a low inlet temperature, the speed of sound, a in Eq. (5.1), at the eye of the first impeller of a BOG compressor is approximately in the range of $275 \pm 5\,\mathrm{m/s}$, which results in a high machine reference Mach number that amplifies the effect of aerodynamic stage mismatching as gas flow through multiple stages of the compressor [14]. The control method for a single-shaft BOG centrifugal compressor can be either an adjustable inlet guide vane (AIGV) or variable speed, which are compared by Taher et al. as illustrated in Fig. 5.14.

Special attention must be given to factory acceptance testing of the LP BOG compressor. Dissimilar specific volume reduction between the test and design conditions, which is further amplified by the effect of multiple stages running at the same speed, often leads to major deviation from the limits of ASME PTC-10 Table 3.2 [14].

The polytropic efficiency of centrifugal compressors in a BOG compression application can be precisely evaluated using the Taher-Evans cubic polynomial methods [15]. In the following example, a multistage, single-shaft centrifugal compressor is used to compress a mixture of 96% methane and 4% nitrogen from 1.05 bara at −160°C to 12 bara at −17°C. The polytropic efficiency is calculated using both the cubic polynomial endpoint method and the linear endpoint method. The shape of the polytropic path as approximated by the two methods and the results of the calculation are shown in Fig. 5.15 and Table 5.1, respectively. Thermodynamic properties are calculated by using RefProp version 10.0 by National Institute of Standards and Technology (NIST) [16].

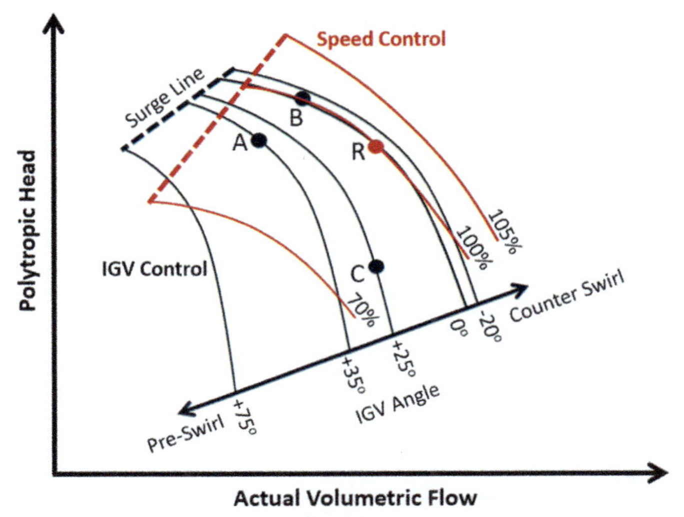

FIG. 5.14 AIGV control method vs. variable speed control method for a single-shaft multistage LP BOG centrifugal compressor operating points are shown as A, B, C. The rated point is denoted with R [14].

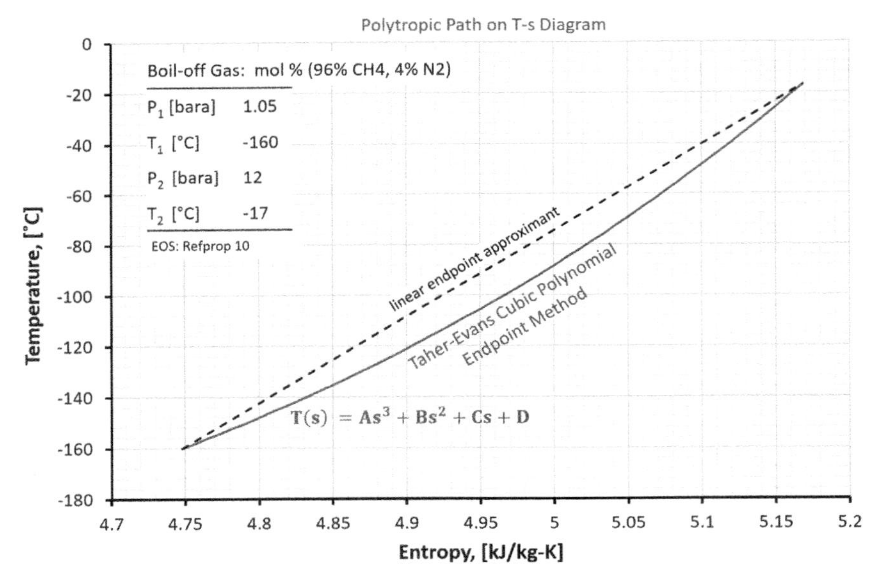

FIG. 5.15 The shape of the polytropic path on T-s diagram as approximated by both the Taher-Evan cubic polynomial endpoint method and the linear endpoint method. The polynomial coefficients A, B, C, and D are given in Table 5.1.

TABLE 5.1 Taher-Evans cubic polynomial endpoint method as applied to a multistage single-shaft centrifugal compressor in a BOG application.

Compositions (mol %): 96% CH_4, 4% N_2			
Phase		**Inlet Superheated gas**	**Discharge Superheated gas**
Pressure	[bara]	1.05	12
Temperature	[°C]	−160	−17
Specific volume	[m^3/kg]	0.52295	0.10379
Z		0.964	0.966
c_p/c_v		1.373	1.372
Specific enthalpy	[kJ/kg]	486.467	767.082
Specific entropy	[kJ/kg-K]	4.748	5.169
Sonic speed	[m/s]	269.4	406.2
$(c_p - c_v)/R$		1.170	1.171
x		0.122	0.120
E	[kg-K^2/kJ]	217.02	488.74
Cubic Polynomial Endpoint Method			
Polytropic efficiency		0.73766	
Polytropic work	[kJ/kg]	206.9998	
A	[kg^3-K^4/kJ^3]	143.56583	
B	[kg^2-K^3/kJ^2]	−1812.357	
C	[kg-K^2/kJ]	7717.617	
D	[K]	−11,040.07	
T-s shape category [15]		I (concave upward along the path on T-s diagram)	
Linear Endpoint Method			
Polytropic efficiency		0.72341	
Polytropic work	[kJ/kg]	202.9989	

The endpoint conditions are shown and calculation results for polytropic efficiency for both the cubic polynomial endpoint method and the linear endpoint method are compared.

As shown in Fig. 5.15, the Taher-Evans cubic polynomial endpoint method accurately predicts the polytropic path's shape. When required, the accuracy of calculation for the polytropic efficiency can be further improved beyond that of the Taher-Evans cubic polynomial endpoint method by using the Taher-Evans cubic polynomial multisegment method [17].

Fuel gas compressors

These compressors raise the pressure of the supply line to the fuel inlet pressure required for plant consumption.

LNG compressor technologies

Each compressor discussed previously can be driven by an electric motor, gas turbine, or steam turbine. Regardless of the driver configuration, each compressor is designed to attain the best performance in terms of efficiency, while maintaining reliability and dependability in a cost-effective manner. Besides the performance requirements, the following aspects must be considered in the selection of compressor type and compressor components:

- **Selection of Aerodynamic Components:** The selection of aerodynamic components is based on the specific performance requirements of the compressor. Operating speed, off-design running conditions, and dynamic behavior, such as rotordynamics of the machinery, should also be considered at the selection of aerodynamics.
- **Shaft Support Configuration:** The main shaft support elements are oil-lubricated, tilting pad journal bearings. Labyrinth-style seals made of aluminum or polymer material are installed between each stage to minimize parasitic leakage. Self-equalizing tilting pad thrust bearings are used to absorb excess axial loading.
- **Shaft End Seals:** Dry gas seals are used in newer compressors. Separation seals keep the lubricating oil away from the gas seal cavity. They can be labyrinth or carbon ring type seals. Buffer gas systems condition the process gas as the main seal gas and inert gas for the separation seals.
- **Thrust Equalizing Mechanism:** Balance pistons are utilized to overcome the axial load. The chamber behind the balance piston is normally equalized back to the inlet of the compressor.
- **Materials of Construction:** Discussed in later sections.
- **Control System, Condition Monitoring, and Safe Operation:** Bearing temperature monitoring, vibration monitoring, and an antisurge system are typically supplied.

Other systems relevant to the compression system beyond the above are discussed in the compressor design section.

Key LNG processes

There are two basic refrigeration cycles used to liquefy purified natural gas (NG) by cooling it down to roughly −160°C: expansion and vapor-compression. The simplest way to cool a gas is to expand it adiabatically. Expansion can be accomplished using a Joule-Thomson (J-T) valve or an expander. The following Eq. (5.2) defines the adiabatic process for an ideal gas. Note that temperature is directly related to pressure and inversely related to volume.

$$\frac{T_1}{T_2} = \frac{p_2}{p_1}^{(k-1)/k} = \frac{v_1}{v_2}^{(k-1)} \tag{5.2}$$

Expansion-based NG liquefaction cycles follow the reverse Brayton cycle and rely on high-pressure refrigerant gas expansion to cool and ultimately liquefy the feed natural gas. The product of the reverse Brayton cycle is the transfer of heat from a cold body to a hot body consuming external work. Fig. 5.7 in the section "Overview of Refrigeration and Liquefaction" shows a simple process diagram of the reverse Brayton cycle. External work, typically an electric motor or gas turbine, drives a compressor, which compresses the refrigerant and drives the cycle (1–2). A heat exchanger cools the high-pressure refrigerant at a constant pressure (2–3). Then, the refrigerant is expanded isentropically, which reduces the temperature following Eq. (5.2) previously (3–4). This expanded refrigerant flows through a heat exchanger extracting heat from the natural gas and ultimately liquefying it (4–1). The working fluid for this cycle in LNG can be methane but is typically nitrogen, so the process is aptly referred to as either nitrogen expansion, reverse Brayton expansion, or reverse Brayton nitrogen expansion.

Vapor-compression, on the other hand, follows the reverse Carnot cycle. This relies on a liquid absorbing heat when changing from a liquid to a gas. Fig. 5.4 in the section "Overview of Refrigeration and Liquefaction" shows a basic vapor-compression refrigeration schematic. Following the reverse Carnot cycle, the cooling fluid is compressed adiabatically, driving the cycle (1–2). Compression adds heat to the gaseous cooling fluid, which is then rejected during isobaric condensing (2–3). The now liquid cooling fluid is then expanded adiabatically resulting in a cold liquid-vapor (3–4). Next, within NG liquefaction, the cold liquid–vapor cooling fluid absorbs heat from the NG feed gas within a heat exchanger, or evaporator, in isobaric expansion and becomes a saturated vapor (4–1). As heat is pulled from the feed gas, the NG cools and ultimately liquefies resulting in LNG at roughly −160°C. The resulting evaporated cooling fluid from isobaric expansion returns to the compressor and the cycle repeats. The expansion process shown in the process diagram may use a Joule-Thomson (J-T) valve or an expander. Refrigerants are used for the cooling fluid due to their heat capacities, vapor pressure behavior, and critical temperatures and pressures among other properties. The refrigerant gases are typically single hydrocarbons, such as propane, or a mixture of different hydrocarbons, such as methane and ethane. Vapor-compression refrigeration is the basic refrigeration process around which most base-load LNG processes are designed.

Air products and chemicals' propane mixed refrigerant (AP-C3MR)

Air Products and Chemicals' (APCI) propane mixed refrigerant (AP-C3MR) cycle is one of the oldest natural gas liquefaction processes. It was first applied in 1972 at Brunei LNG and follows the vapor-compression cycle (Fig. 5.16). Licensed to over 20 sites worldwide, AP-C3MR is the most widely used process producing over 170 MTPA of LNG. APCI incorporates their unique coil-wound design into the main cryogenic heat exchanger (MCHE). These can be fabricated in sizes up to 6m in diameter and 55m tall, using larger tube bundles to accommodate higher capacities [18].

In the C3MR process, propane provides the precooling duty of the feed gas, as well as the mixed refrigerant gas, cooling them down to roughly −30°C. The process diagram previously shows the feed gas on the left going into a precooling heat exchanger before going to the much larger MCHE with the mixed refrigerant. Here, the mixed refrigerant liquefies the natural gas sequentially within the heat exchanger by vapor-liquid separation steps. At each step, the liquid portion is auto-cooled through expansion then reintroduced to the shell side of the MCHE similar to the process described in the AP-SMR section next. This yields LNG at roughly −160°C.

Figs. 5.17 and 5.18 show possible C3MR train configurations with steam turbines and gas turbines, respectively. These examples are from the Lumut, Brunei, and Donggi-Senoro, Indonesia LNG sites, respectively, and show typical older propane compressors. These examples only had three sections, while newer propane compressors will have four sections. As shown in the figures, there are typically three different compressors in a C3MR train: propane, low-pressure mixed refrigerant, and high-pressure mixed refrigerant.

In many cases, you will also find a starter/helper motor with the HP MR train. Note that the equipment diagrams are intended to illustrate string layouts and do not reflect the actual power output of each driver. Although the figures next both show three separate turbomachinery strings, C3MR strings can be variations of this, such as having the high- and low-pressure mixed refrigerant compressors driven by the same turbine. Another variation adds a mid-pressure mixed refrigerant compressor, totaling four compressors in the train. In this case, the propane and high-pressure compressor might be driven by one turbine, and the mid- and low-pressure compressors might be driven by another turbine. Typical capacities for the AP-C3MR process currently range from 1.3 to 6.4 MTPA.

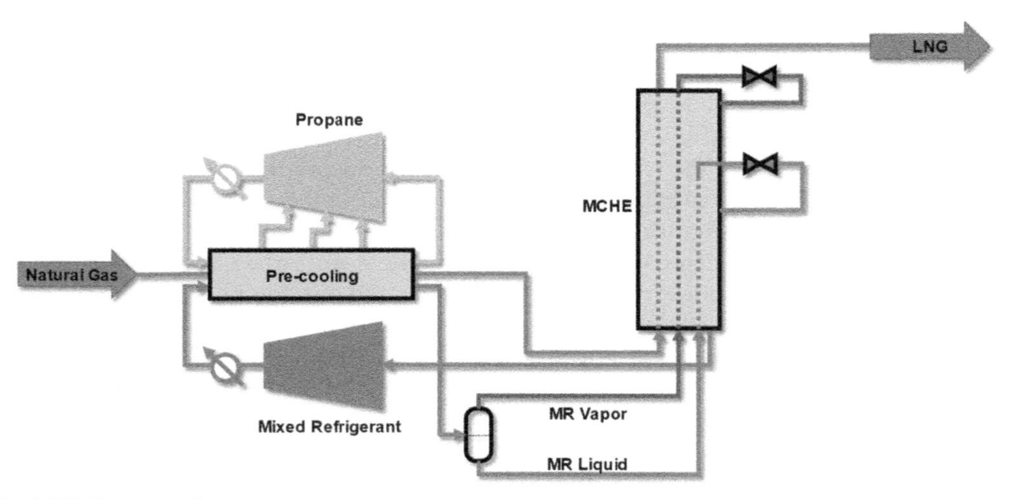

FIG. 5.16 APCI AP-C3MR process diagram.

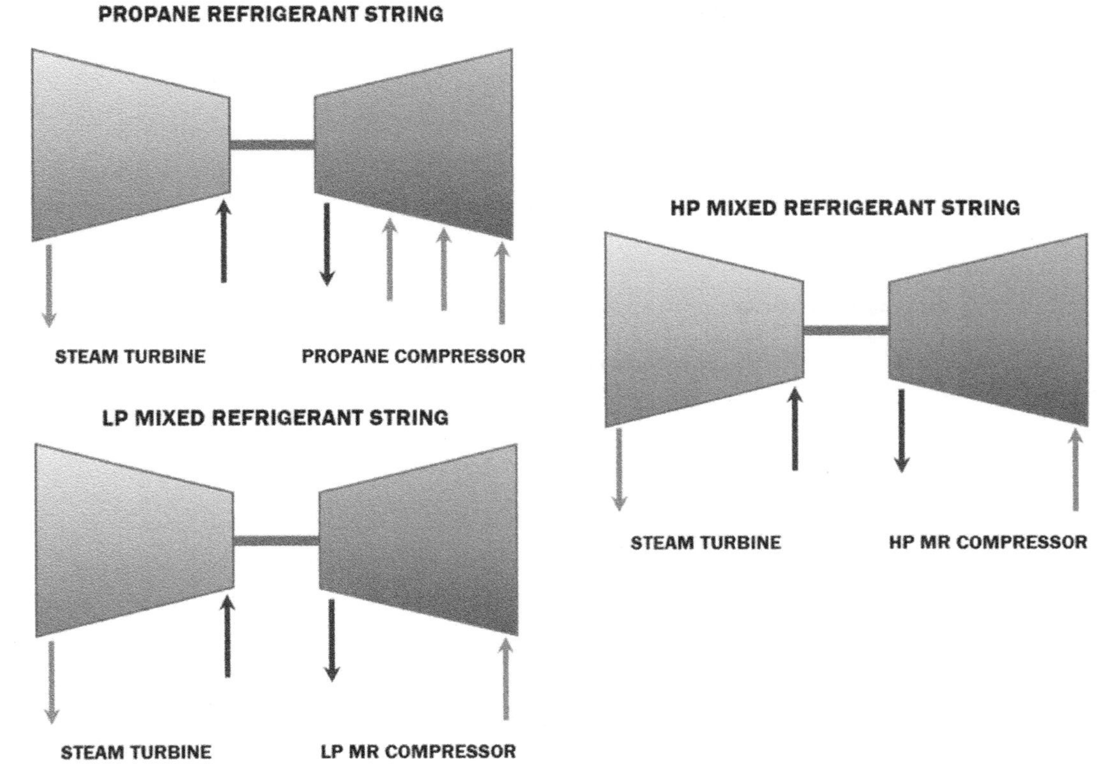

FIG. 5.17 AP-C3MR turbomachinery string with steam turbine drivers.

Air products and chemicals' propane and split mixed refrigerant (AP-C3MR/SplitMR)

The C3MR/SplitMR configuration is a modification to the turbomachinery arrangement of the popular C3MR process. The process itself is unchanged. In this configuration, power is split between the propane and MR strings with two gas turbines of equal power output to improve the power distribution between the fluid cycles. As shown in Fig. 5.19, the propane and HP MR compressors are both driven by a gas turbine in the same string, while a second gas turbine is utilized to drive the LP and MP MR compressors. By placing the propane and HP MR compressors in the same string, the power is equally balanced across both gas turbines. In the RasGas 2 plant, for example, the HP MR compressor utilizes roughly 30% of the load from the gas turbine driver [19]. The configuration is intended to offer more efficient operation for a wide range of ambient temperatures, as the load can be better shifted from precooling and liquefaction loops as needed. Current train capacity for the C3MR/SplitMR process ranges from 3.6 to 5.2 MTPA.

All of the published C3MR/SplitMR trains are currently utilizing heavy-duty gas turbines to drive the turbomachinery strings. Depending on the performance requirements, the compression may be done over one or two compressor frames. Both strings include a helper motor for starting; however, the motors are typically rated for continuous use to aid the gas turbines depending on ambient conditions. The Tangguh LNG trains in Indonesia are unique in their utilization of steam turbine helpers.

ConocoPhillips optimized cascade

Originally developed in the 1960s, the early cascade process was first applied for LNG production at the Kenai, Alaska facility in 1969. After several decades of experience and refinements, the Optimized Cascade process was then first applied at the Trinidad Atlantic LNG facility in 1999. The process is currently utilized in many current and planned LNG facilities around the world as a large-scale LNG liquefaction technology.

The Optimized Cascade process (Fig. 5.10 in the section "Overview of Refrigeration and Liquefaction") consists of three, multistage refrigeration loops using propane, ethylene, and methane as pure refrigerants as discussed by the licensor [20]. Core-in-kettle-type and brazed aluminum heat exchangers with insulated cold boxes are used for cooling the feed gas. In the precooling cycle, the propane refrigerant cools the natural gas, ethylene, and methane to roughly

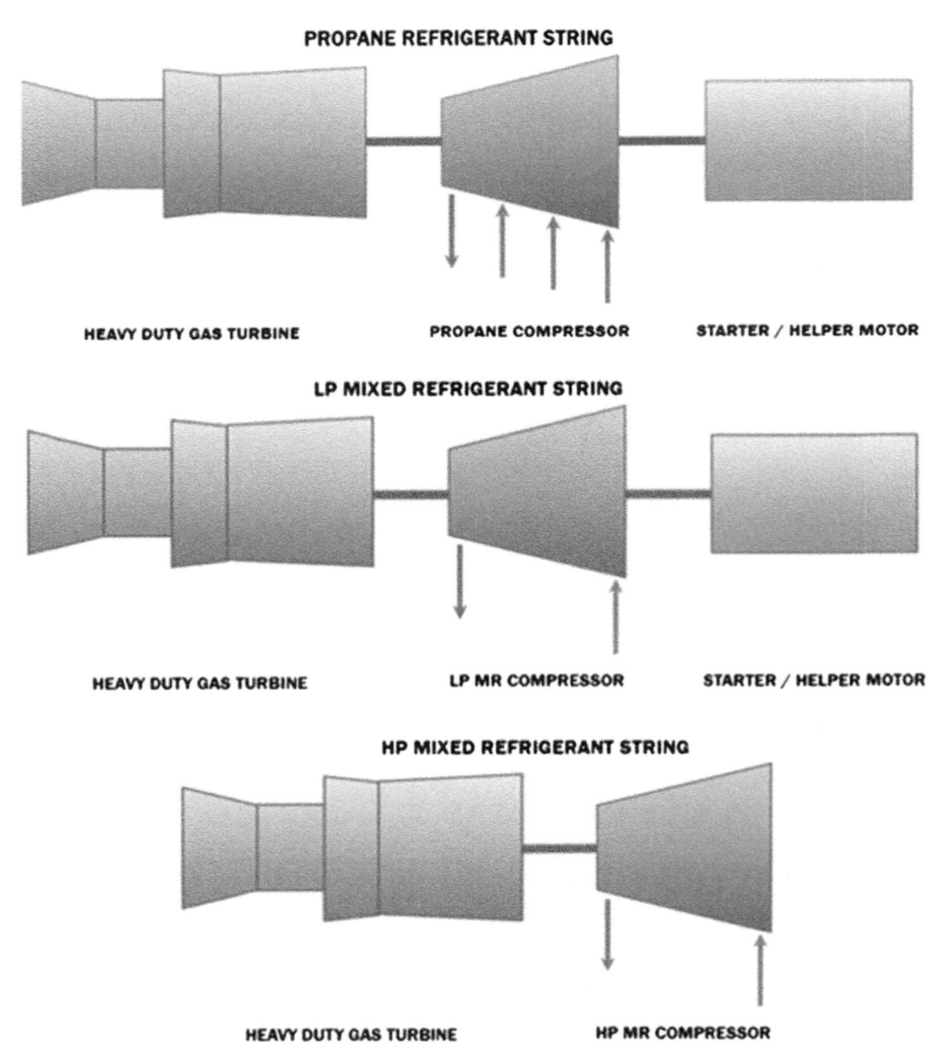

FIG. 5.18 AP-C3MR turbomachinery string with gas turbine drivers.

−30°C, followed by the ethylene loop to condense the natural gas and methane. Finally, the natural gas is subcooled by the methane loop to approximately −160°C for storage and transportation [21]. While the Kenai, Alaska facility produced 1.5 MTPA of LNG using the early cascade process, the Optimized Cascade process is being utilized for train capacities ranging from 3.3 to 5.2 MTPA.

The Optimized Cascade process is typically designed as "two trains in one." Each compressor station consists of two turbomachinery strings working at 50% in parallel to meet the production capacity. This division of labor is intended to improve reliability and availability of the train while providing maintenance flexibility. If any of the refrigerant compressor strings are not in operation, the train can continue production at a reduced rate rather than shutting down completely.

Typical applications of the Optimized Cascade process have involved two-shaft, heavy-duty gas turbines driving centrifugal compressors for each refrigerant loop, as shown in Fig. 5.20. Each train would include two strings in parallel for each fluid at 50% each. While midrange heavy-duty gas turbines are typical drivers, higher power gas turbines were utilized for the Angola LNG train to meet its increased capacity of 5.2 MTPA. The only other train with similar capacity is Trinidad's Atlantic LNG Train 4. Although it uses midrange heavy-duty gas turbines, the compressor stations are further divided into three turbomachinery strings working at 33% in parallel for the ethylene and propane to increase total capacity.

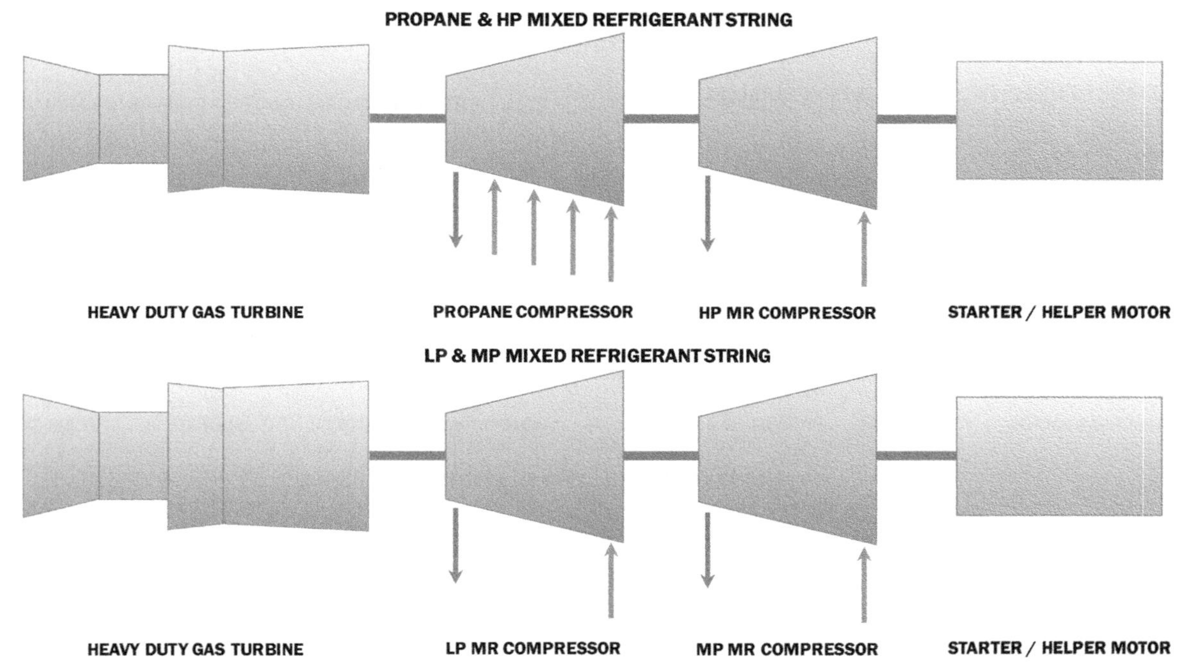

FIG. 5.19 Typical C3MR/SplitMR propane and HP MR string *(top)* and LP/MP MR string *(bottom)*.

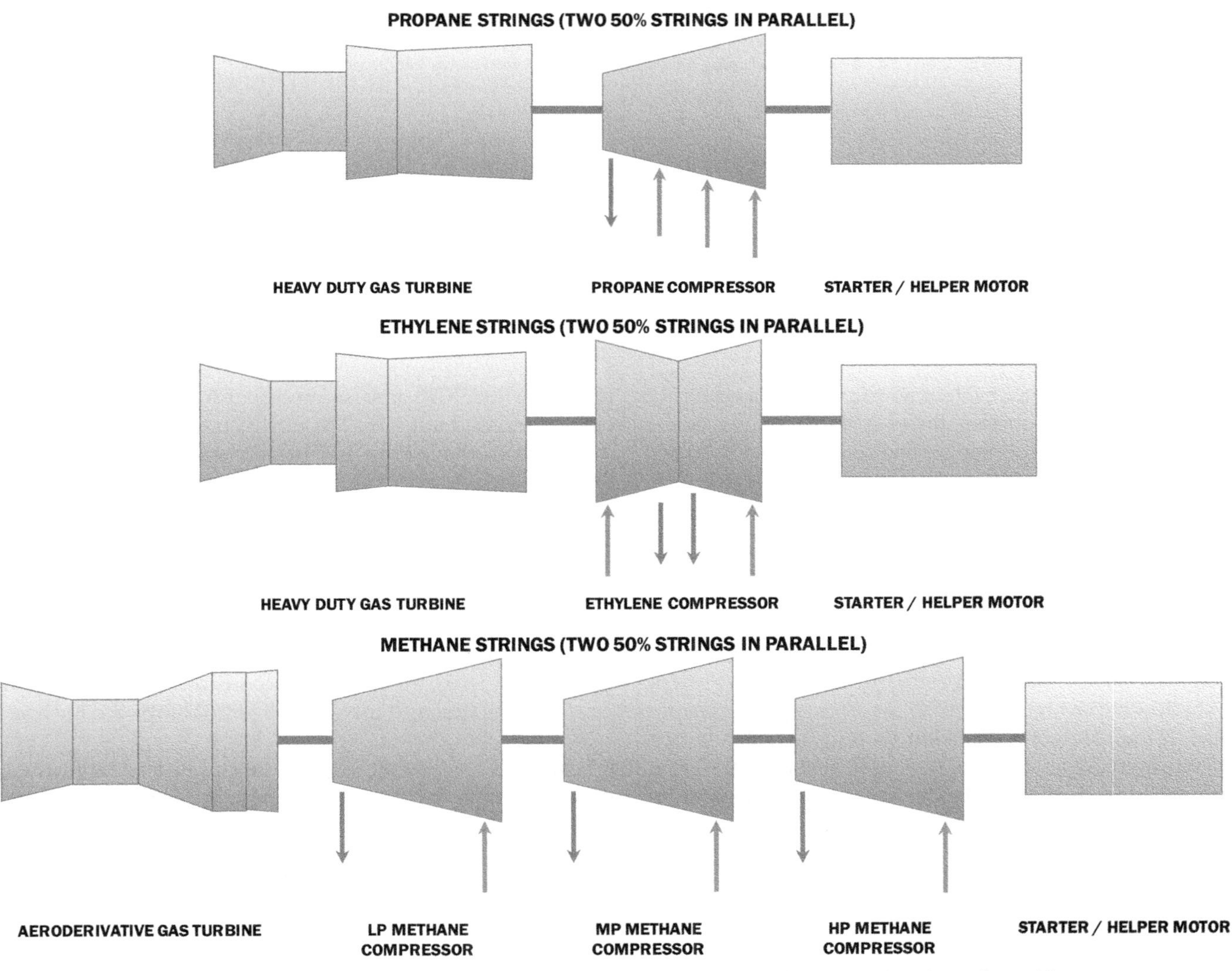

FIG. 5.20 Typical optimized cascade turbomachinery strings for propane *(top)*, ethylene *(middle)*, and methane *(bottom)* loops.

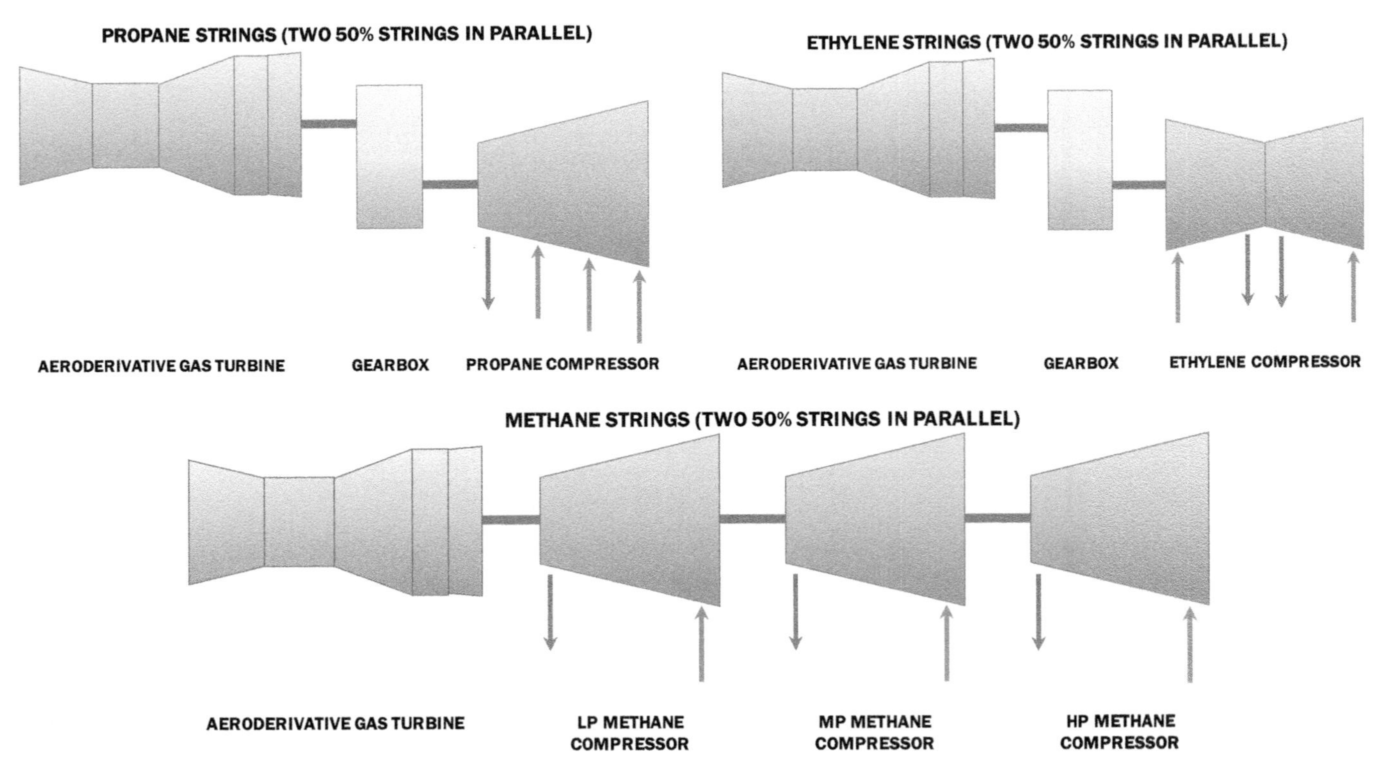

FIG. 5.21 Darwin LNG optimized cascade turbomachinery strings utilizing aeroderivative gas turbines.

More recently, several Australian LNG projects employing the Optimized Cascade process have designed the turbomachinery strings with aeroderivative gas turbines. The Darwin LNG facility was noted as the first such liquefaction facility to use aeroderivative gas turbines for the refrigeration compressors, which also employ speed-reducing gearboxes for the propane and ethylene compressors as shown in Fig. 5.21 [22].

Air products and chemicals "AP-X"

APCI's AP-X LNG cycle was first introduced in 2001. It is an AP-C3MR process modified to include a nitrogen expander cycle, combining the two basic refrigeration processes discussed initially. The nitrogen expansion cycle provides separate subcooling for the feed gas, eliminating that duty from the MR cycle. This allows for higher feed gas flow through the C3MR portion and results in increased train LNG capacities above 5 MTPA despite the physical size limitations of the MCHE. Fig. 5.22 next shows a simplified process diagram. Here, we see a basic combination of a C3MR cycle and a nitrogen expansion cycle both described previously. AP-X employs APCI's coil-wound heat exchangers and can be implemented in various train configurations. The first applications of the AP-X process; Qatargas 2, 3, and 4; and RasGas 3, each had a 7.8 MTPA train capacity. This was over 50% more than the standard AP-C3MR process at the time.

Identical to the AP-C3MR process, the feed gas and mixed refrigerant are precooled with a propane loop down to roughly −30°C. The mixed refrigerant loop then liquefies the feed gas. Then, the nitrogen expansion loop subcools the LNG down to roughly −162°C for storage and transportation. The propane and mixed refrigerant loops are independent of the nitrogen expansion loop in the AP-X process.

Nitrogen expansion loops employ turboexpanders rather than J-T valves to more efficiently produce low nitrogen temperatures and provide the additional refrigerating capacity. Turboexpanders follow a near isentropic process, more efficiently reducing the temperature of the gas as compared to isenthalpic throttling with a J-T valve. With a turboexpander, the cycle can also recover some power to run a generator or assist in driving the nitrogen compressor

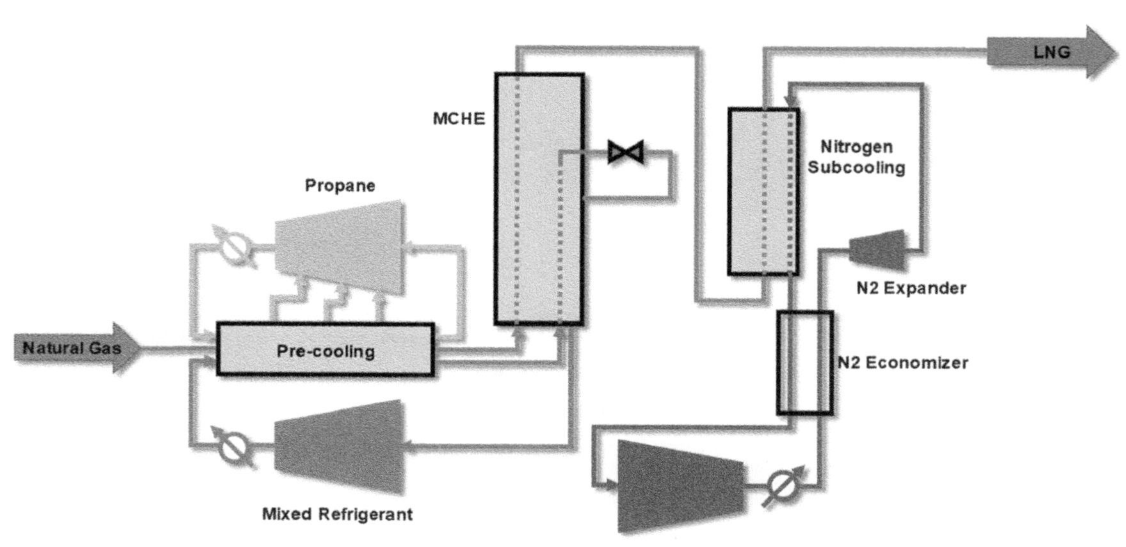

FIG. 5.22 APCI AP-X process diagram.

further increasing its efficiency. While the AP-X process trains in place now have 7.8 MTPA capacities per train, APCI lists that the process has a range of 5–12 MTPA per train.

Similar to the C3MR cycle, the main equipment in the cycles are the heat exchangers and propane, mixed refrigerant, and nitrogen compressors along with their drivers.

A typical refrigeration train within the AP-X process is shown in Fig. 5.23 from RasGas 3, one of the first AP-X applications. This shows that each train has three turbomachinery strings: the propane string, the mixed refrigerant string, and the nitrogen string. All three of these are driven by heavy-duty gas turbines for RasGas 3. The AP-X process has also been applied to three additional Qatargas projects, with all AP-X applications providing a train capacity of 7.8 MTPA. APCI markets their AP-X process as a large- to mega-scale LNG process with a capacity of 5–12 MTPA per train. The RasGas 3 train configurations are not the only AP-X configuration APCI offers. Air products describe a few other train configurations of AP-C3MR in a publication in the LNG journal [23]. One utilizes parallel compression and aeroderivative gas turbines having two 50% strings for the propane, three 33% strings for the mixed refrigerant, and two 50% strings for the nitrogen service. Another utilizes a parallel C3MR configuration with a single string with a back-to-back compressor for nitrogen subcooling, all with heavy-duty gas turbines and helper motors. The division of labor in both of these configurations improves reliability and availability of the train and provides maintenance flexibility.

Dual-mixed refrigerant

The dual-mixed refrigerant (DMR) process, shown in Fig. 5.24, employs two mixed refrigerant loops: one for precooling the gas (replacing the typical propane loop) and the other for liquefaction and subcooling. This process has been successfully proven in Sakhalin, Russia, with two 4.8 MTPA trains. The DMR process uses mixed refrigerant (mainly ethane and propane) for precooling in an additional spiral wound heat exchanger typical of an SMR process [24]. The lower molecular weight of this mixed refrigerant, compared to propane, allows for a smaller condenser and removes the propane compressor bottleneck. Typical capacities for the AP-DMR process currently range from 3.4 to 3.6 MTPA, but the process has a similar range capacity to AP-C3MR.

The DMR process is generally selected on the basis of being able to adjust the composition of the mixed refrigerant, such that pressure limitations associated with propane can be avoided. This is particularly suitable for colder climates as proven by the Sakhalin facility.

A heavy-duty gas turbine is typically used to drive each turbomachinery string in the precooling and liquefaction loops, with starter/helper motors for each. An example string layout is shown in Fig. 5.25.

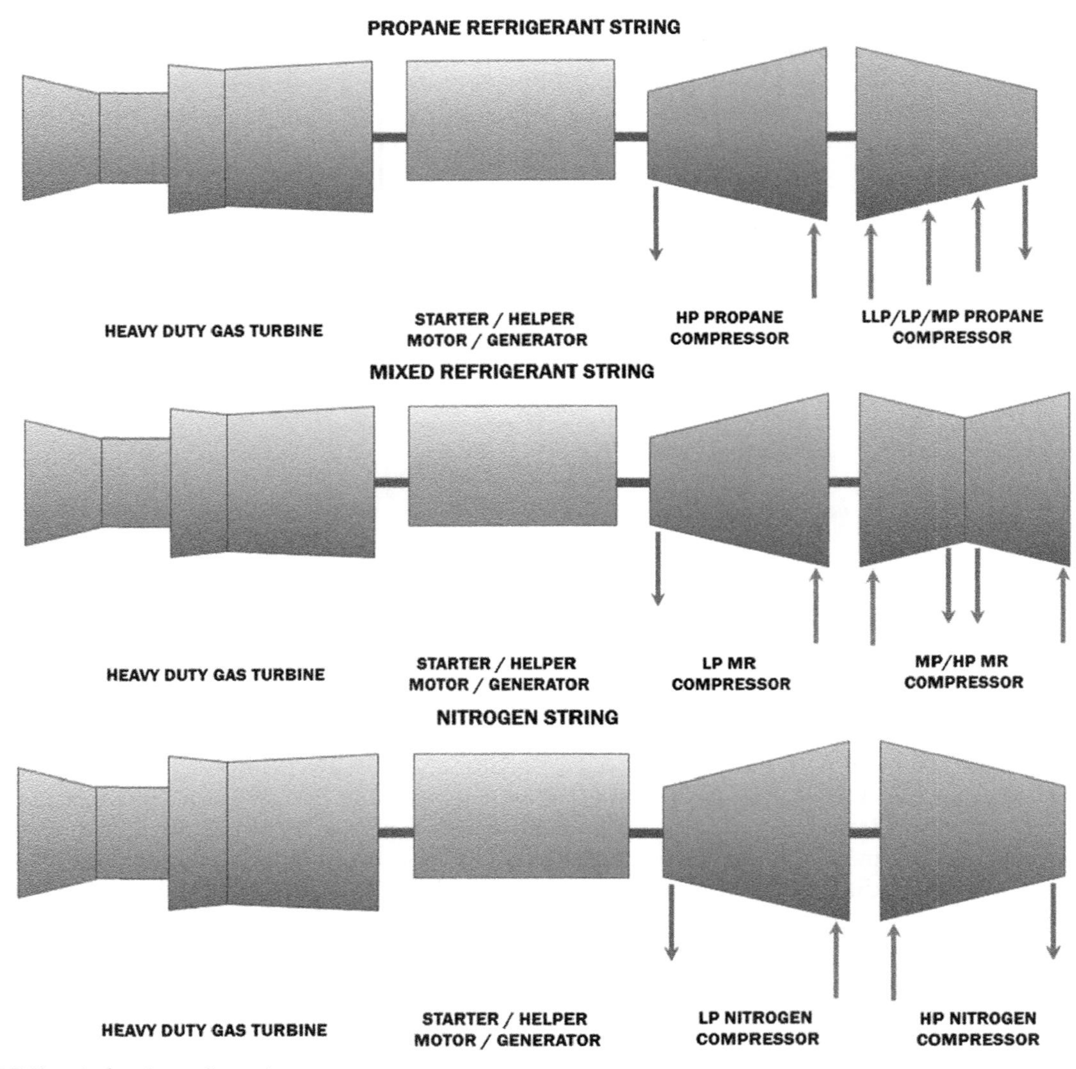

FIG. 5.23 AP-X typical train configuration.

Air products and chemicals' single mixed refrigerant (AP-SMR)

APCI's single mixed refrigerant cycle (AP-SMR) achieves all of the process gas cooling with one mixed refrigerant gas loop. This is the most basic type of mixed refrigerant vapor-compression refrigeration cycles. APCI differentiates their SMR process from others by incorporating their fully modular unique CWHE. It is part of how their SMR process is 88% as efficient as their AP-C3MR and AP-DMR processes [25]. Fig. 5.26 shows the AP-SMR process.

The AP-SMR cycle follows the basic vapor-compression process description at the beginning of this section. The CWHE adds efficiency over other SMR processes. The SMR process involves vapor-liquid separation, auto-cooling through expansion, and reintroduction at each step of cooling within the MCHE with an additional step as compared to C3MR. This additional step is required since there is no separate feed gas precooling. The liquid portions are subcooled alongside the feed gas then are throttled and introduced to the shell of the heat exchanger as the cooling fluid. The vapor portions are cooled and condensed alongside the feed gas. The condensed vapor is then separated. The liquid portion separated from the condensed vapor is further cooled alongside the feed gas in the next section before being throttled and introduced to the shell. The vapor portion is cooled then subcooled alongside the feed gas in two sections. It then exits the top, is throttled, and is introduced to the shell. The evaporated refrigerant is recombined at the bottom of the shell side of the MCHE where it then flows to the compressor inlet to maintain the cycle. This yields LNG at roughly −160°C.

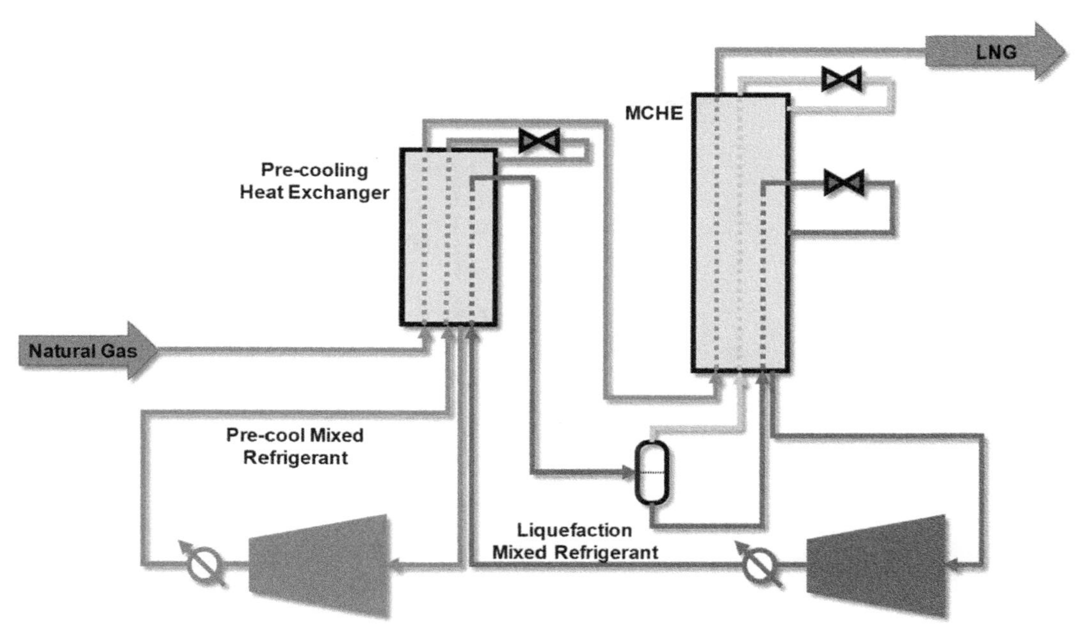

FIG. 5.24 Typical DMR process diagram.

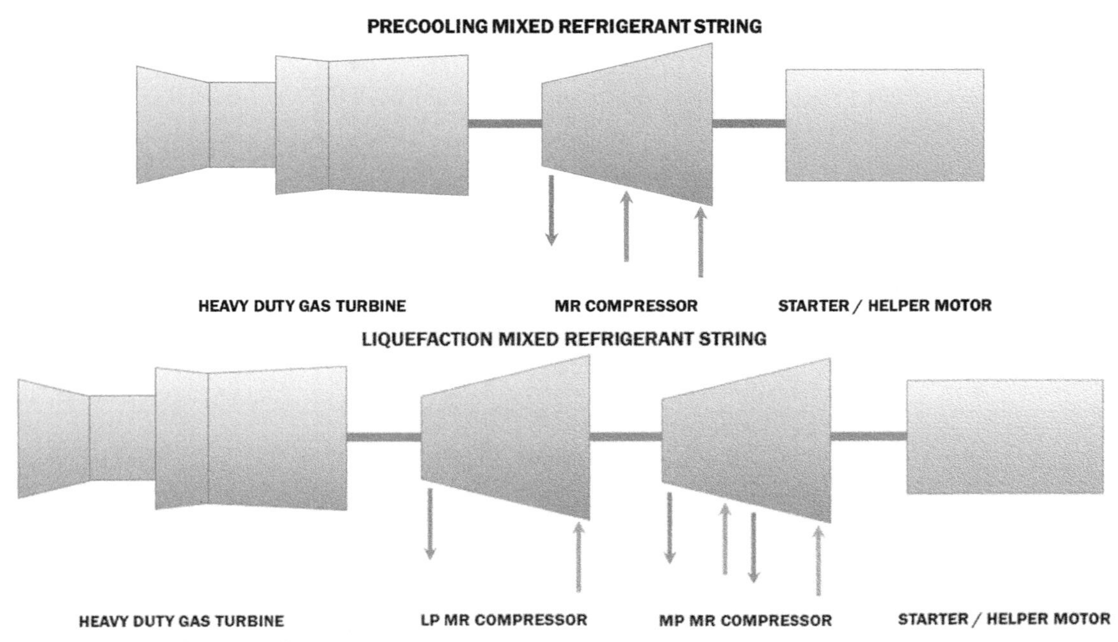

FIG. 5.25 Typical string configuration for DMR process.

The main equipment in the SMR cycle is the refrigeration compressors, turbomachinery drivers, and heat exchangers. A typical refrigeration train within the AP-SMR cycle is shown in Fig. 5.27. This is a simple depiction of the first application of the process, Marsa El Brega in Libya, showing that each train has two turbomachinery strings: a steam turbine combined with a low-pressure mixed refrigerant compressor and a steam turbine combined with a high-pressure, mixed refrigerant compressor. The strings at Marsa El Brega were roughly 32,000 HP and 31,000 HP for the low-pressure and high-pressure compressors, respectively. While this LNG site had an output of 0.8 MTPA per train, more typical capacities for this cycle are 0.3–0.5 MTPA as shown in APCI's experience documentation.

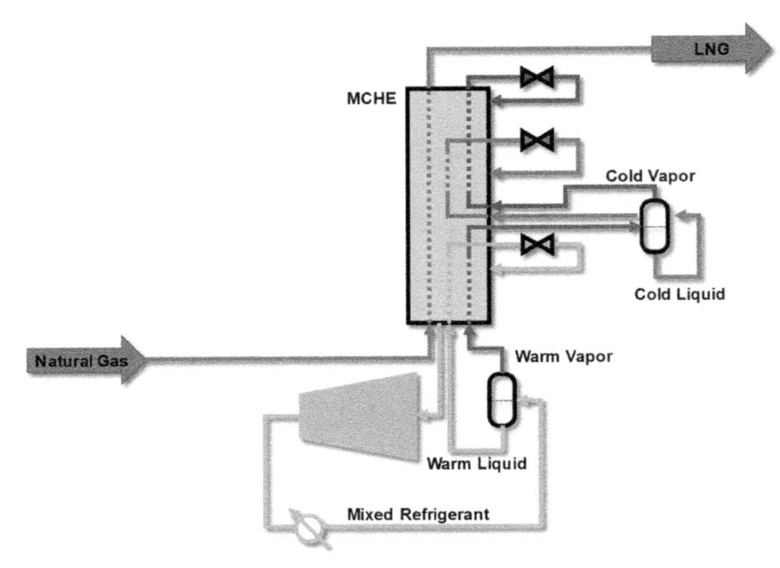

FIG. 5.26 AP-SMR process diagram.

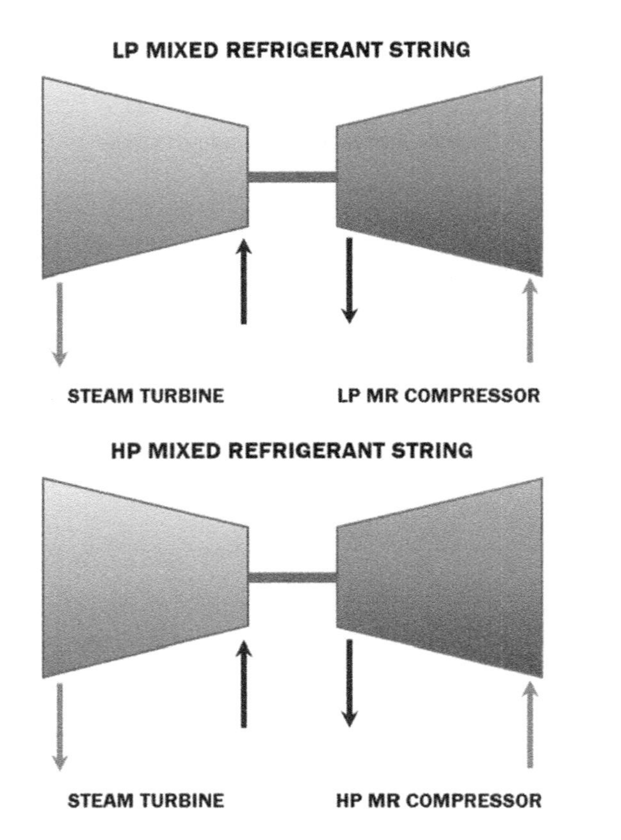

FIG. 5.27 Typical AP-SMR train.

Compressor selection, configuration, and design

Large capacity gas compressors are typically of the axial or centrifugal design having multiple aerodynamic stages of compression. Centrifugal compressors tend to be the most popular for these applications with axial compressors seeing only limited application in mixed refrigerant service. API 617 outlines the requirements and design guidelines. End-users may also have their own specifications in addition to those of API. A cross-section of a basic API 617

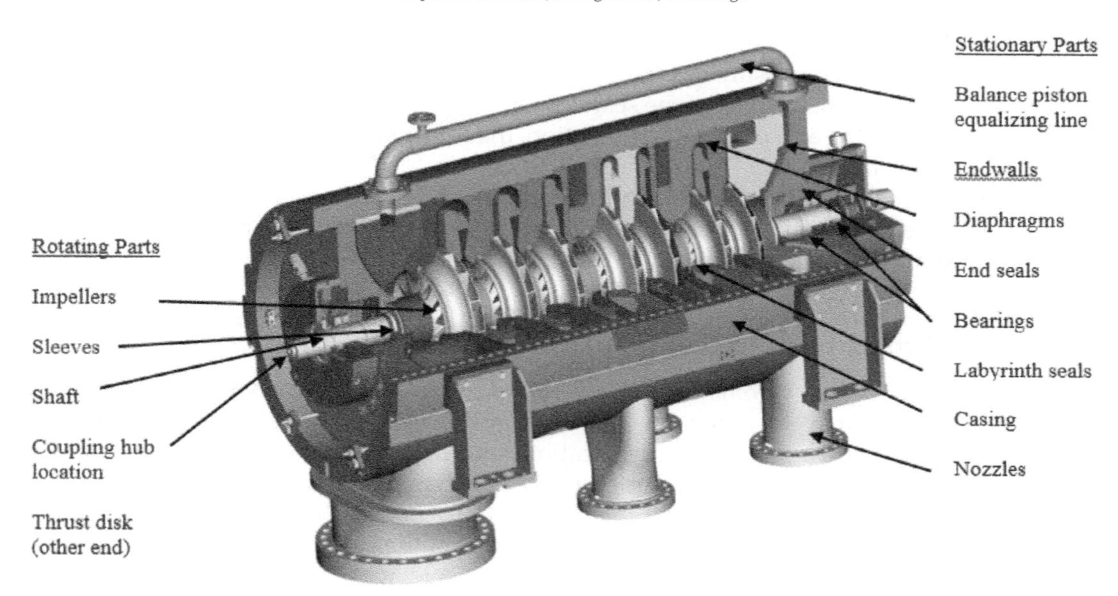

FIG. 5.28 Compressor cross-section. *Courtesy of Elliott Group, Ebara Corporation.*

centrifugal compressor is shown in Fig. 5.28. It is comprised of rotating and stationary parts. Rotating parts are mounted on a shaft, which is free to rotate. Stationary parts act as a pressure vessel, seal in the gas, support the shaft, and define a flow path along with the shaft.

Selection of the compressor string begins with the process requirements, and these usually include gas composition, flow rate, inlet and discharge conditions. The starting point of the compressor selection is typically the amount of flow required. In large-scale LNG processes (>2 MMTPA), the flows are quite large. In general, large compressors are required to handle large flows. However, there are some methods for minimizing the compressor size of straight-through compressors such as operating at a very high speed while using a high-flow coefficient [26] and very high-flow coefficient wheels [27], which continues a trend in LNG centrifugal compressors that started in the 1990s. Iterative compressor selections are made in an effort to maximize the efficiency and range while ensuring reliability and minimizing cost.

Nozzles and nozzle configuration

The flowrate will define the nozzle diameter, which must be sized to maintain acceptable gas velocity. The nozzle size and number of nozzles can dictate the casing size, which must be big enough to accommodate the largest nozzle while maintaining an acceptable flowpath and long enough to prevent nozzle interference. Canting the nozzles can help reduce the casing size, but this adds complexity to the piping installation, which is best avoided whenever possible.

Another important parameter that is considered during the early stages of a project is the compressor nozzle configuration—upward vs. downward nozzles. The upward nozzle configuration will result in a compressor structure at a lower elevation than downward nozzles, but it does require more down time when rotor replacement is required on horizontally split machines. Cumbersome piping disassembly and realignment may take 3–4 weeks vs. about a week otherwise, and this can result in LNG production loss.

Effect of settle-out pressure on selection

After a compressor shutdown, gas becomes trapped between the upstream and downstream check valves, and the pressure eventually equalizes at the settle-out pressure. The higher the settle-out pressure, the higher the starting torque. Judiciously locating the check valve close to the compressor discharge will reduce the high-pressure downstream volume and result in a lower settle-out pressure after machine trip or shutdown.

Regardless of what the settle-out pressure is, the dry gas seal system must prevent leakage. If the settle-out pressure exceeds the static liftoff pressure, then the seal faces will separate, and unfiltered process gas will begin to leak into the primary seal vent. To prevent this leakage at settle-out conditions, the seal is buffered with process gas pulled from the blocked-in compressor, filtered, and boosted through a small reciprocating compressor. A low settle-out pressure that is below the dry gas seal static liftoff pressure has the advantage of not requiring buffer gas recirculation. The dry gas seal static liftoff pressure is mainly a function of the seal design, which is heavily determined by the dry gas seal vendor. When the compressor is started and placed into operation, the buffer gas is provided by the compressor discharge after passing through a set of filters.

Economic and environmental factors favor eliminating compressor blowdown or at the very least minimizing it as much as possible. The drivers, including any starter/helper motor and couplings, need to be sized to handle the startup torque from settle-out pressure, which can be significant.

How the driver effects' compressor selection

In large-scale LNG processes, a gas turbine driver may be the first item selected. When this occurs, the type of gas turbine will dictate the compressor speed range. The first gas turbines used for large-scale LNG were of a single-shaft design derived from industrial power generation applications and designed to operate at either 3000 RPM or 3600 RPM. The allowable speed range of these gas turbines is fairly tight having a total speed variance of about 5%. These types of gas turbines would be direct coupled to the compressor and are still in heavy use today. More recently, multi-shaft aeroderivative gas turbines and industrial gas turbines have been used in LNG production [22,28]. These designs do provide some starting torque and also have a very wide speed range, which make them attractive as a mechanical drive for compressors.

Aerodynamic flowpath and staging

Inline centrifugal compressors can be described as a combination of one or more centrifugal compressor stages operating about a common shaft. Each stage is comprised of rotating impellers that impart kinetic energy into the gas that is then converted to a rise in static pressure as the gas decelerates in the stationary components. Fig. 5.29 illustrates a compressor stage comprised of an impeller, diffuser, return bend, and return channel.

The impeller spins on a shaft and is the means by which energy (work) is imparted on the gas. The impeller is paired with a diffuser, which is a stationary component that primarily converts velocity (kinetic energy) to pressure (static energy). The diffuser is a radial passage that is roughly the same width as the impeller blade, but the radial area expands and provides the desired diffusing effect. As the gas exits the diffuser and enters the return bend, the gas has obtained the majority of the static pressure rise for the centrifugal stage. The return bend is found in multistage centrifugal compressors and is a stationary component, which redirects the gas from a radially outward flow in the diffuser to a radially inward flow into the return channel, which contains guide vanes leading into the next stage. Impeller eye and shaft seals minimize the leakage loss that can occur in the secondary flow passages.

Stages are selected based on aerodynamic performance, then arranged to develop head from compressor inlet to discharge. Fig. 5.30 shows a 1990s vintage compressor prior to being rerated with new aerodynamic stages. The rerate of this compressor was performed in 2004 and involved all new aerodynamic internals. Impellers were upgraded from 3D, straight-line, semi-inducer designs to 3D, straight-line, full-inducer designs, which represented the technology at the time. This rerate provided both a capacity and efficiency improvement.

The industry has continued to progress in the area of aerodynamics [26,27]. New aerodynamic stages are available from compressor OEMs with improved efficiency that are capable of operating at a high Mach number while

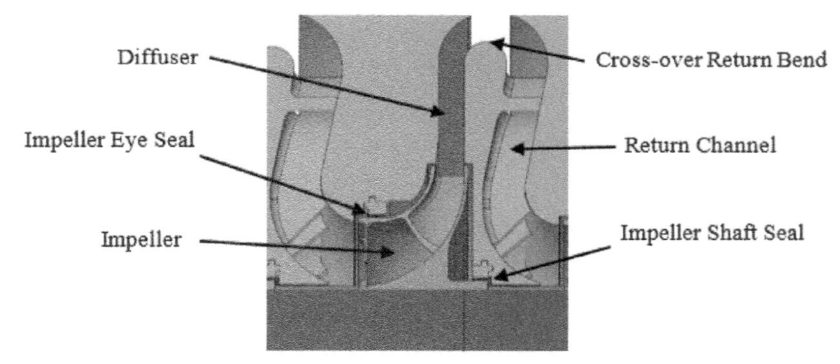

FIG. 5.29 Compressor aerodynamic stage. *Courtesy of Elliott Group, Ebara Corporation.*

FIG. 5.30 1990s Vintage propane compressor prior to rerate. *Courtesy of Elliott Group, Ebara Corporation.*

maintaining stage space that is comparable to previous designs. New aerodynamic stages have also been extended out in flow coefficient with flow coefficients ranging from 0.12 to over 0.24 in covered impellers and utilizing mixed flow geometry. High-flow coefficient impellers have larger throat area and longer blades, which tend to have relatively low natural frequency. Blade strength and natural frequency can be improved by having sculpted blades with a customized thickness profile.

The aerodynamic selection will yield a map for each section. These maps provide the permissible operational envelope of the entire compressor. Increasing the Mach number will shorten the range and also produce a steeper falloff in head as stonewall is approached, but this observation is only true within a given technology. For example, a modern compressor aerodynamic stage operating at a high Mach number may actually have the same or even better range than an older technology stage operating at a lower Mach number. In extreme cases where the curve is essentially vertical at choke, the avoidance of choke is extremely important. The Mach number can be reduced by slowing the compressor down and adding more stages.

The aerodynamic selection and resulting flow path will also factor into the bearing span. High-flow coefficient stages tend to require large axial stage space, which can push the bearing span up leading to lower critical speeds and lower rotordynamic stability. Reducing speed can require additional stages, which also factors into the bearing span. In all cases, a proper rotordynamic analysis must be performed.

Compressor configuration

The compressor type (straight through, side-load, isocooled, etc.) is determined by the process. Introduction of a single component refrigerant in a series of progressively lower pressure and temperature streams will improve the overall efficiency of a refrigeration process. These multiple streams of a single gas can be handled by a single compressor arranged with multiple sections having incoming side-loads fed by kettle-type evaporators. Per design, the side-load compressor is well suited for refrigeration applications where the refrigerant is introduced at progressively lower temperatures to the lower pressure sections. Gas temperature increases in a given section are reduced by the mixing of the side-stream flow into the suction of the following section of the compressor. This maintains gas temperature throughout the machine at reasonable levels [29]. Fig. 5.31 shows the cross-section of a compressor flowpath having two side-loads.

The presence of side-loads makes the aerodynamic design a bit more challenging. Research into side streams (Fig. 5.32A) has led to enhancements with respect to performance, mixing, flow distribution, and harmonic reduction. Computational flow path analyses and proprietary models have confirmed an enhanced design (Fig. 5.32B) of the merging flow paths to minimize losses.

Lateral rotordynamics

LNG plant operation follows service intervals are dictated by the gas turbine. The centrifugal compressor is typically expected to go several gas turbine service intervals without a major overhaul spanning more than 6 years without being overhauled and in some cases going well beyond this. One of the key parameters that effects long-term operation is the lateral vibration level of the equipment. Lateral vibration is of such importance that nearly all critical service turbomachinery applications apply a vibration monitoring system to identify dangerous levels.

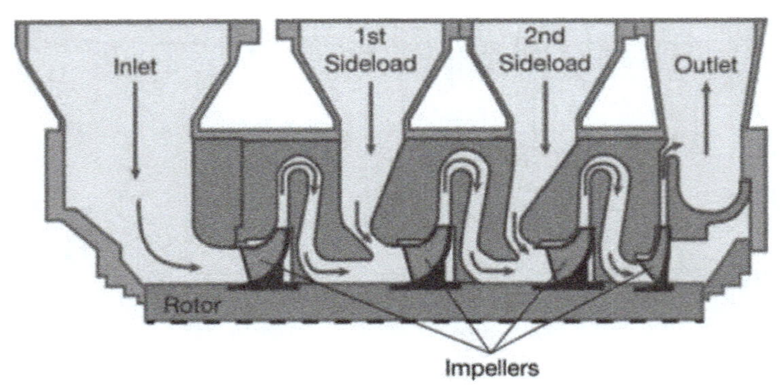

FIG. 5.31 Compressor flowpath with side-loads. *Courtesy of Elliott Group, Ebara Corporation.*

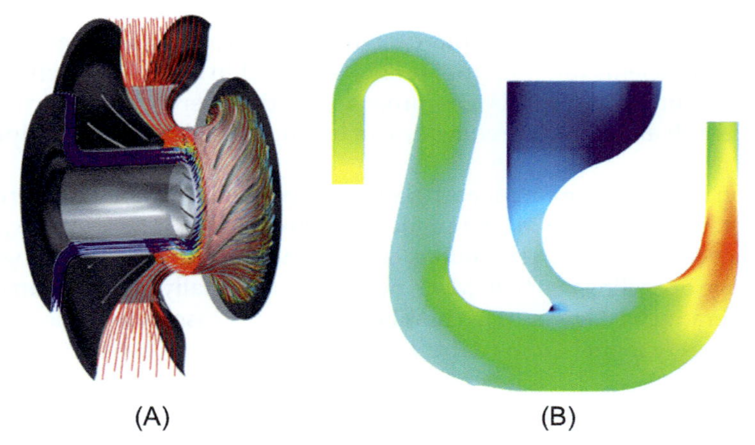

FIG. 5.32 Side-load design (A) previous generation (B) latest generation. *Courtesy of Elliott Group, Ebara Corporation.*

The main causes of lateral vibration are unbalance, misalignment, and instability. The primary methods for ensuring acceptable vibration are to eliminate sources of excitation such as unbalance and misalignment and to desensitize the rotor to sources of excitation through proper design and rotordynamic analysis. Rotordynamic requirements are set by API standards and include achieving separation margin from critical speeds and achieving required logarithmic decrement stability values. Separation margins from lateral natural frequencies are largely determined by rotor sizing, bearing selection, and coupling selection. Logarithmic decrement stability is dependent on the same parameters, primarily the rotor sizing, which determines the critical speed ratio (operating speed/critical speed on stiff supports) and the bearing selection. High gas density, which is an important stability risk factor, is generally not seen in LNG applications. Design options such as hole pattern seals or squeeze film dampers are available to improve rotordynamic stability, but these are seldom required.

Centrifugal compressor rotordynamics are heavily dependent on the aerodynamic path. Most compressor aerodynamicists prefer a small shaft diameter and long stage space for purposes of aerodynamic efficiency, but this combination results in a long bearing span and depressed first critical speed that can quickly create rotordynamic problems. Most OEMs offer a series of aerodynamic stage solutions to solve problems such that the required compression can be performed in a small number of compressor bodies. Rotordynamic problems can be fairly easy to avoid provided there are sufficient stage options for compressor selection. Fig. 5.33 shows an unbalance response plot for a propane compressor. The speed range was 3420 to 3636 RPM, and critical speeds were easily avoided.

Torsional rotordynamics

Torsional rotordynamic concerns are heavily dependent on driver selection for the compressors. In the case of gas turbine drives, there is very little torsional excitation from the gas turbine. A single-shaft gas turbine must be paired with a variable speed starter motor. The starter motor accelerates the string of equipment up to the gas turbine firing

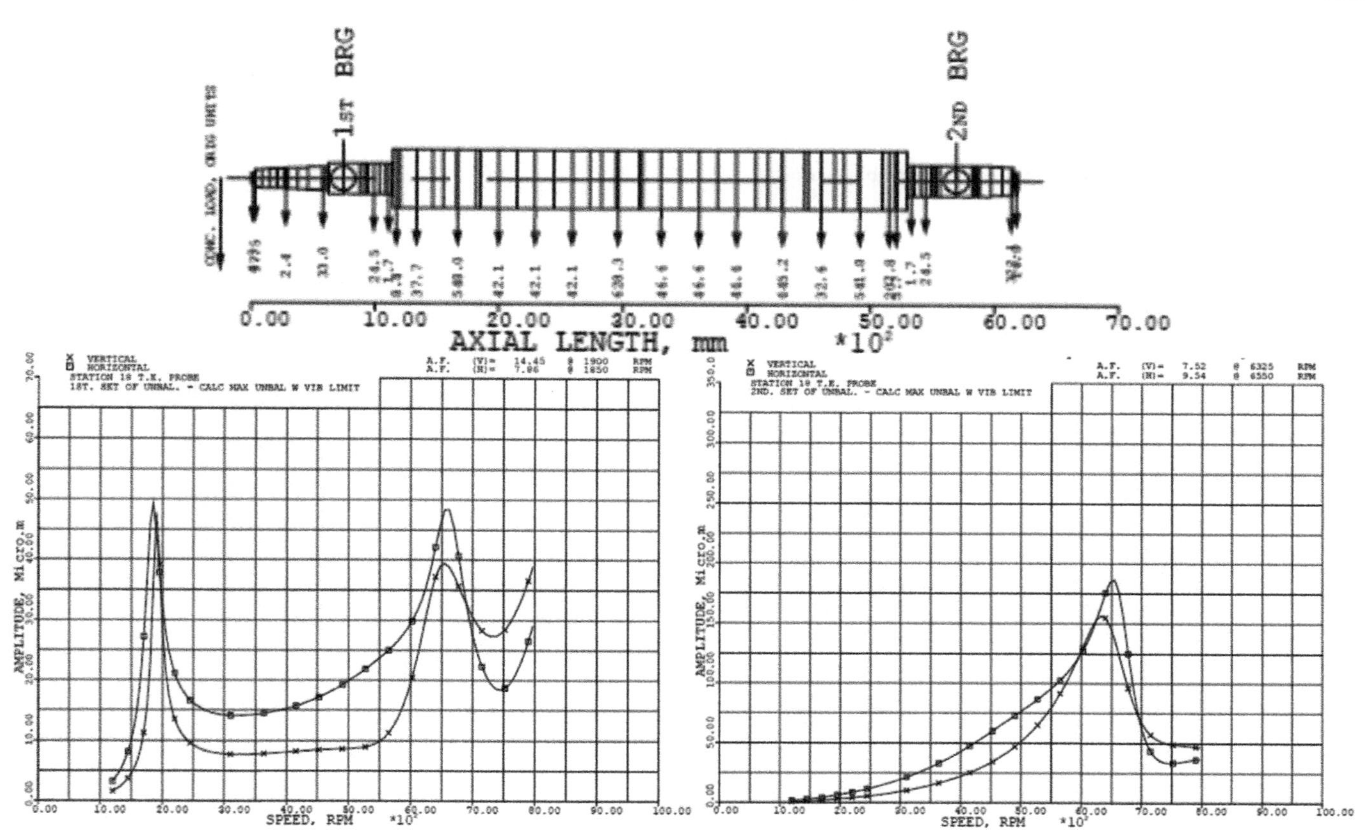

FIG. 5.33 Propane compressor mass-elastic drawing and unbalance response plots.

speed where the gas turbine begins to provide torque. The starter motor continues to assist with string acceleration augmenting the gas turbine torque until the gas turbine fully takes over at the minimum continuous speed. The starter motor is sometimes used as a helper motor to provide additional power. Variable frequency drives (VFDs) smooth out motor startup transients that would be present in a direct-on-line startup. VFDs are used to add variable speed flexibility to motors, but they also add further complication to torsional rotordynamics due to their potential for generating torsional excitation at the motor over a greater number of frequencies, including broadband excitation and a potential for feedback instability [30–32].

Achieving acceptable torsional rotordynamics is essentially an exercise in keeping torsional stresses below threshold values. Nearly all machinery trains lack a torsional measurement system. Therefore, the avoidance of torsional rotordynamic problems is primarily dependent upon a proper torsional analysis and system design. Requirements generally include achieving API separation margins from torsional natural frequencies via coupling selection and tuning and properly sizing components to handle imposed stresses.

An example case [33] is described for a gas turbine-compressor-compressor-starter/helper motor string (Fig. 5.34). The gas turbine is rated at 117,600 HP (87.7 MW) having a rated torque of 171,573 lb-ft (232,621 N-m), and the LCI synchronous starter motor is rated at 16,092 HP (12 MW) having a rated torque of 23,476 lb-ft (31,830 N-m). Due to the gas turbine imparting a narrow speed range, required separation margins from operating speed were easy to achieve; however, difference frequency harmonics from the LCI motor were a concern and needed to be avoided. The difference frequency of concern was the motor operating frequency minus the grid frequency. Since the grid was at 50 Hz (3000 CPM) and the motor operated from 57 Hz (3420 RPM) to 60.6 Hz (3636 RPM), the difference frequency ranged from 7 to 10.6 Hz. The torsional natural frequency range to avoid was a 6× multiple of this difference frequency or 42 Hz to 63.6 Hz. Couplings were selected to ensure proper separation margin in the operating speed range of the primary modes. An LCI synchronous motor also produces startup transient excitation. Fig. 5.35A shows the predicted transient torque at the compressor-to-gas turbine coupling. The oscillatory transient torque levels should become negligible in the continuous speed range, but the measurement during full-load testing (Fig. 5.35B) did not confirm this. Tuning of the drive was required.

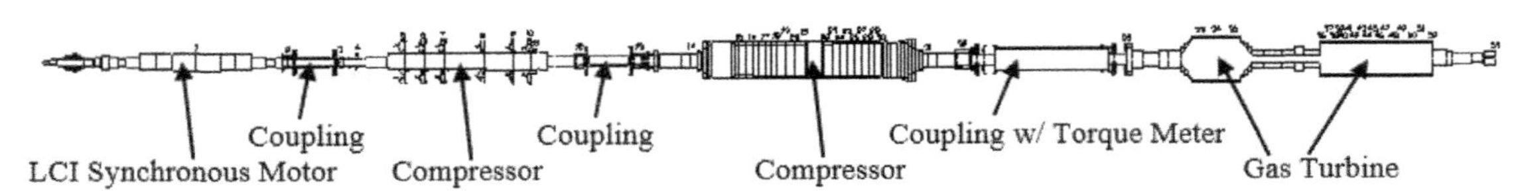

FIG. 5.34 Motor-compressor-compressor-gas turbine string.

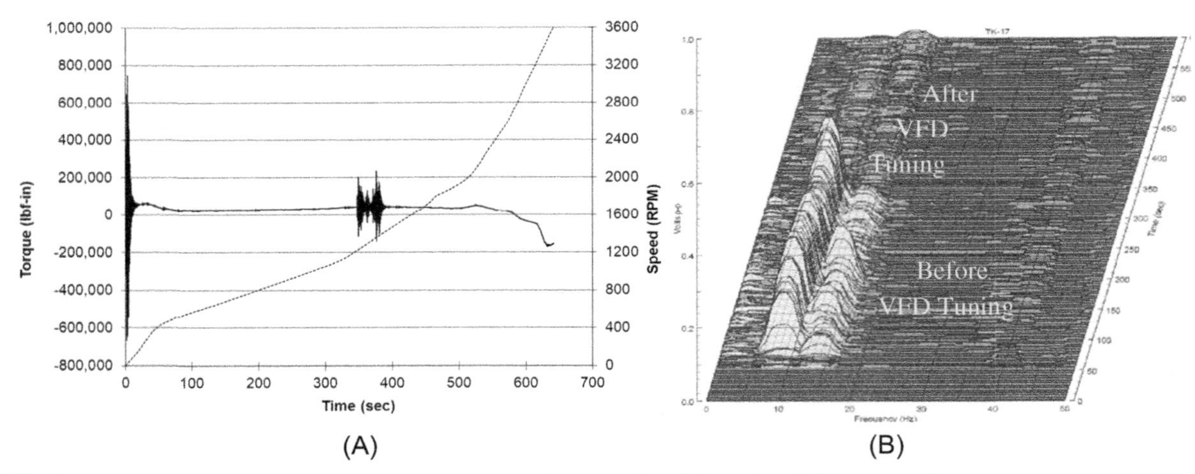

FIG. 5.35 Startup transient torque at the compressor-to-gas turbine coupling (A) predicted (B) measured.

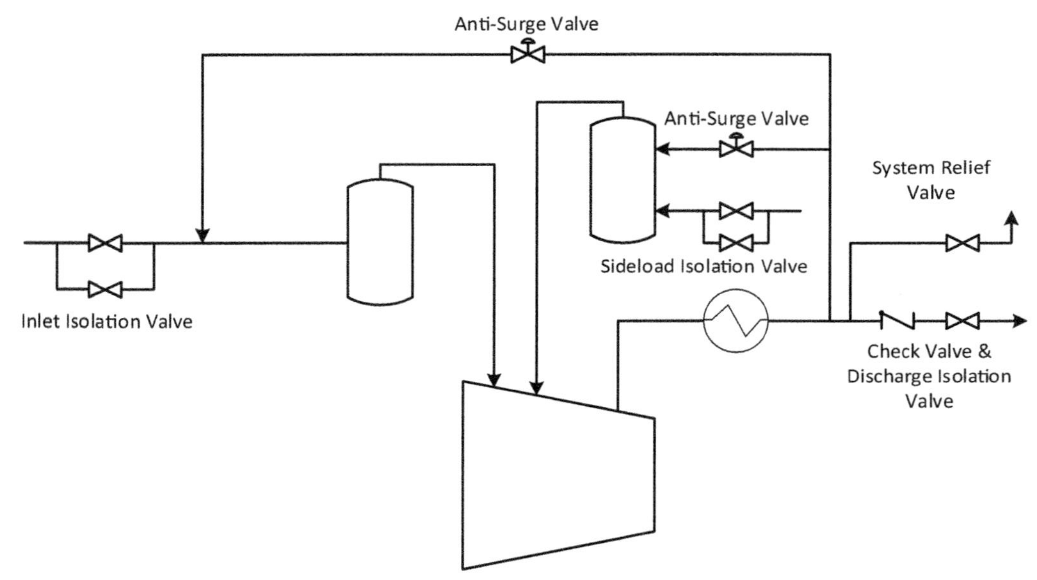

FIG. 5.36 Antisurge recycle loop.

Antisurge system

Each section of the compressor will have an antisurge system to assist with keeping that section out of surge similar to Fig. 5.36. A map for each section of compression is provided for purposes of control. Fig. 5.37 shows a typical compressor first section map. Operational limits are set by surge +10% on the leftmost part of each curve and by stonewall on the rightmost part of each curve. Compressor operation is controlled by inlet throttling, adjustable inlet guide vanes, or speed variation. Applications having inlet pressures that are close to atmospheric will avoid inlet throttle valves to ensure that inlet pressures do not drop below atmospheric, which could lead to air entering the system thereby creating a safety concern.

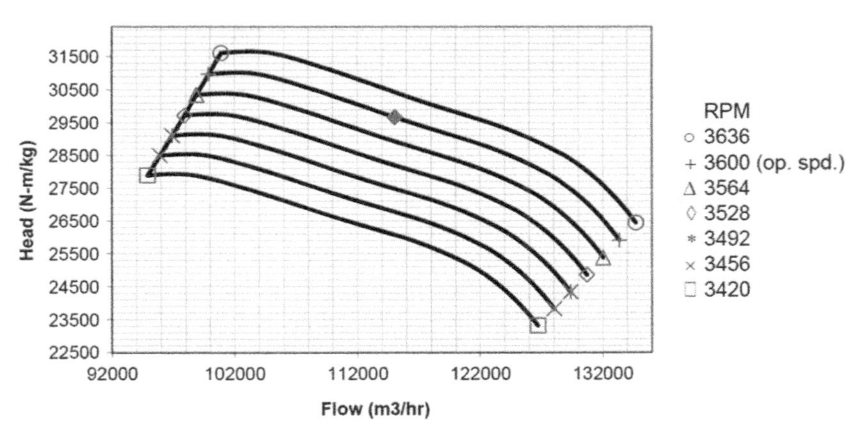

FIG. 5.37 Compressor map, section 1 propane compressor.

FIG. 5.38 Multi-side-load propane compressor. *Courtesy of Elliott Group, Ebara Corporation.*

Compressor selection examples

Propane compressor

LNG propane compression is one of the more difficult compressor applications, typically accommodating wide operational conditions with multiple side-load inlets, a high molecular weight gas (44 MW), and low temperatures. Fig. 5.38 shows a ¾ cutout of such a compressor having three side-loads (one inlet is not pictured due to the cutout). This compressor is used in a 4.7 MTPA plant to precool the natural gas to about −35°C.

The American Petroleum Institute (API) 617 standard defines cold temperature service as −29°C. Low-temperature service requires design and material selection for the prevention of brittle fracture. Steel materials require assessment by impact testing in accordance with ASME BPVC Section I requirements.

The compressor shown in Fig. 5.38 has the following features:

- Gas turbine at the drive end and LCI synchronous helper motor at the nondrive end of the string
- More than 70,000 shaft HP (52.2 MW) at the compressor design condition
- 3420–3636 RPM speed range matching the gas turbine
- Impeller flow coefficients ranging from 0.096 to 0.158
- More than 65,000 ICFM at the first inlet at −37°C (−35°F)
- Propane (C_3H_8) process gas
- Horizontal split casing for simplified maintenance (compressor resides between a gas turbine and other bodies)
- Five nozzles: one main inlet, three side-load, and one discharge

- Internal side-load mixing
- Up nozzle arrangement necessitates piping removal for maintenance. Optional mezzanine mounting with down nozzles eliminates this requirement
- Tandem dry gas seals
- Balance piston recycle line
- Not pictured: piping arrangement with antisurge recycle lines

Control systems can be particularly difficult for this multi-side-load arrangement. The aerodynamics between each nozzle is known as a section, and this compressor has four sections. Each section must be monitored and controlled. Flow monitoring is critical and is performed by at least four flowmeters (one for each section). A fifth flowmeter would be costly, but would enable confirmation of the flows thus determining calibration errors and drift, and this can be quite valuable in maximizing the range. Each stage has an antisurge controller with each controlling one of four recycle valves. A map for each section of compression is provided for purposes of control. Operational limits are set by surge +10% on the leftmost part of each curve and by stonewall on the rightmost part of each curve. Compressor operation is typically controlled by speed variation.

Mixed refrigerant compressor(s)

The LNG mixed refrigerant compressor is not as aerodynamically challenging as the propane compressor, but it does have some unique challenges that exist in the form of large heavy casings operating at higher pressures that often require large-frame barrel designs (Fig. 5.39). Here again, continuous improvements and gradual step outs in experience occurring over long spans of time are allowing new offerings in large-scale, high-pressure casings.

As the size of the compressor casing increases, so too do the deflections that occur under pressure, such that the compressor size and its pressure handling capability are inversely related (Fig. 5.40). Handling either the large frame size or the high pressure can be quite challenging, but the primary original equipment manufacturers (OEMs) for LNG equipment have been doing both routinely, offering large compressors at relatively high pressures.

End flash gas and boil-off gas compressors

After liquefaction, boil-off gas and flashed vapors are produced through pressure drops and through heating. Processing these cryogenic gases is required to avoid flaring or venting. Compressors are used to transport and pressurize these gases, which can either be used as fuel gas, sent back for re-liquefaction, or transported for other distribution.

FIG. 5.39 High-pressure mixed refrigerant (HPMR) barrel compressor. *Courtesy of Elliott Group, Ebara Corporation.*

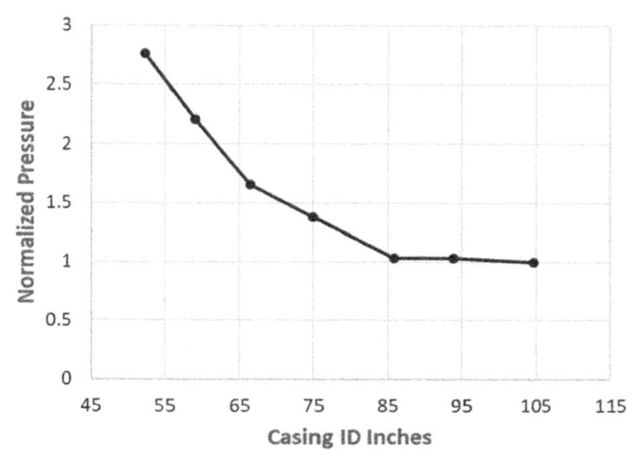

FIG. 5.40 Relative pressure capability vs. casing inner diameter.

FIG. 5.41 End flash gas compressors being prepared for shipment. *Courtesy of Elliott Group.*

End flash gas compressor(s)

The liquefied gas stream is sent to a separator where end flash gas is produced, consisting mostly of methane, but also removing most of the nitrogen and other trace constituents, thereby ensuring the quality of the remaining LNG stream. An end flash gas compressor (Fig. 5.41) is used to pull this end flash gas off the top of the separator, pressurize, and transport it. The compressor inlet gas temperature can be close to the liquefaction temperature. Such cold service applications require stringent material specifications, certification, and testing.

The compressor shown in Fig. 5.41 has the following features:

- Variable frequency drive electric motor and gear box drive
- Two compressors having a total of 15 aerodynamic stages
- Over 4700 shaft HP (3.5 MW) at compressor design condition
- 9007–11,127 RPM speed range
- More than 7700 ICFM at the first inlet
- Methane (CH_4) gas at less than −129°C (−200°F)

- Horizontal split casing for simplified maintenance (middle compressor resides between a gear box and another compressor)
- Two nozzles: one main inlet and one discharge
- Up nozzle arrangement necessitates piping removal for maintenance. Optional mezzanine mounting with down nozzles eliminates this requirement
- Tandem dry gas seals

Boil-off gas compressor(s)

After separation, the remaining gas stream is pumped to storage tanks or loaded onto a ship. Boil-off gas (BOG) compressors are used to pull boil-off gas from the top of storage tanks, thereby keeping pressures within allowances. The amount of boil-off gas can vary considerably, and operational flexibility is extremely important for this application. BOG compressors generally fall into one of the following types: inline centrifugal, integrally geared centrifugal or reciprocating. Since roughly 2010, BOG has been handled with multicasing centrifugal compressors having dry gas seals designed to handle the cryogenic temperatures by employing special static seals within the design. Adjustable inlet guide vanes must also be designed for the range of operating temperatures that will be seen in service. Differences in the coefficient of thermal expansion must be accounted for in the design to prevent lockup of the adjustable inlet guide vanes.

Centrifugal compressors typically employ either a variable speed drive or adjustable inlet guide vanes (AIGVs) for control (Fig. 5.42). Suction throttling is avoided to keep pressures from dropping below atmospheric. Inlet temperatures are similar to the end flash gas compressor. Fig. 5.43 shows a compressor map having adjustable inlet guide vanes. Operational limits are set by surge +10% on the leftmost part of each curve and by stonewall on the rightmost part of each curve.

Testing

The API 617 standard requires a mechanical running test (MRT). The MRT is typically performed under vacuum with the inlet and discharge flanges blocked off. During the MRT, the compressor is operated from zero up through the trip speed and back down. The compressor is then brought up to the maximum continuous speed and held for 4 h. Lube-oil inlet pressure and temperatures are varied through the operating range during this four-hour run test. Axial position and both axial and radial vibration are monitored throughout the test. Dry gas seal primary vent flow and separation gas flow are also measured for each dry gas seal. Radial vibration must meet contractual limits. After the test, bearings are to be removed, inspected, and reassembled. Any evidence of oil migrating past the dry gas seal's separation seal is cause for rejection.

Per API 617, when a factory performance test is specified, the test must be performed in accordance with either ASME PTC 10, or if specified, ISO 5389. A minimum of five points, including surge and overload, must be taken. The purpose of a shop performance test is to determine the aerodynamic performance of a centrifugal compressor (under controlled conditions) so that its aerodynamic performance on a specified gas of known properties under specified conditions may be predicted. Shop performance tests are instrumented per ASME PTC-10 requirements. For LNG

FIG. 5.42 Adjustable inlet guide vane mechanism. *Courtesy of Elliott Group, Ebara Corporation.*

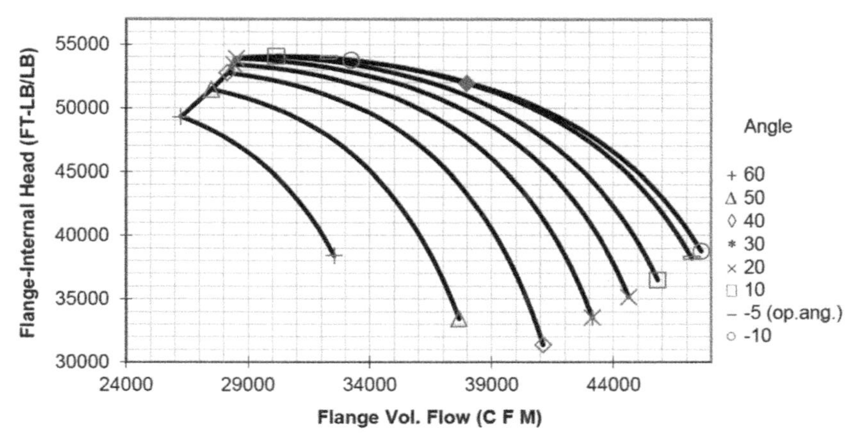

FIG. 5.43 Compressor map with adjustable inlet guide.

applications, a Type 2 test is normally conducted where a substitute inert gas is used, and the compressor performance is derived using the principals of equivalence. For an equivalent test, similarity must be maintained within PTC-10 allowances for the following quantities:

- Volume ratio
- Flow coefficient
- Machine Mach number
- Machine Reynolds number
- Side-stream inlet volume flow rate vs. discharge volume flow rate

A full-load/full-pressure inert gas test may be specified for LNG main refrigeration compressors as a mechanical integrity test of each equipment string to minimize startup problems in the field. This test consists of a fully assembled string of equipment and includes most of the auxiliaries. This test is normally 4 h in duration with the compressor/s at design discharge pressure and design horsepower at Nmc (max continuous speed). As noted by API, "In some cases, matching discharge pressure and horsepower simultaneously is not possible. In these cases, the discharge pressure is matched for 2 h, then the horsepower is matched for the remaining 2 h." Vibration and bearing temperatures can be examined at full load and full speed.

The compressors depicted in Fig. 5.44 were subjected to a full-load mechanical test to demonstrate the overall mechanical design, prove the gas turbine and starting motor integrated control system, and verify the engineering procedures utilized to size the starter motor [34].

FIG. 5.44 Compressor full-load string test-motor *(left)*-compressor-gas turbine (right). *Courtesy of Elliott Group, Ebara Corporation.*

Customized solutions

The value solution for LNG liquefaction is often determined by operational and market realities that effect plant size and LNG train arrangements. OEMs look to provide a customized solution for each quotation balancing range, efficiency, price, and experience in accordance with the customer's needs and expectations. Technology plays an important role in the ability to provide such a solution. Improvement in aerodynamic technology has led to the development of high-flow staging that pushes well beyond previous flow coefficients. Other new staging has also been developed that has achieved some modest efficiency gains, but with rather significant space reduction, thus allowing application of smaller compressors that can achieve better results. In other cases, a single compressor body can be offered up where two were required before, while in other cases, parallel compressor arrangements are required in accordance with the operational needs of the plant to provide greater operational flexibility. All of these changes have been the result of continuous development by both the OEMs and process licensors with numerous step outs adding up to significant gains over time.

Pumps for LNG applications

Pumps for LNG applications are used to transfer and pressurize the fluid in liquid form. As previously mentioned, natural gas is liquefied for easy transportation and storage. The liquefaction process, along with transportation, storage, and refinery processes of natural gas, requires pumps at various locations. Due to the very cold operating temperatures of LNG pumping applications (around −165°C for liquid methane), these pumps are classified as cryogenic pumps. A cold pumping environment requires design features specific to the pumped fluid and operating conditions. In this section, these design concepts are briefly reviewed, and pump applications for different processes are detailed.

LNG pumping applications

The most common applications for LNG pumps are the import and export terminals. Fig. 5.45 is a typical import (receiving) and regasification terminal outlining location and type of LNG pumps within the process.

LNG is often transported by LNG tankers to receiving terminals where the cryogenic liquid is unloaded from the shore to land via the ship's cargo and spray pumps. Cargo pumps are usually operated in parallel to meet the required flow rate and are assisted by smaller-size spray pumps to empty the cargo efficiently. A single cargo pump can handle flow rates around 2000 m^3/h. The maximum discharge pressure is kept below the pipeline rating, which is around 18.9 barg (275 psig) with rated differential pressures ranging between 5 and 8 barg (72.5 and 116 psig).

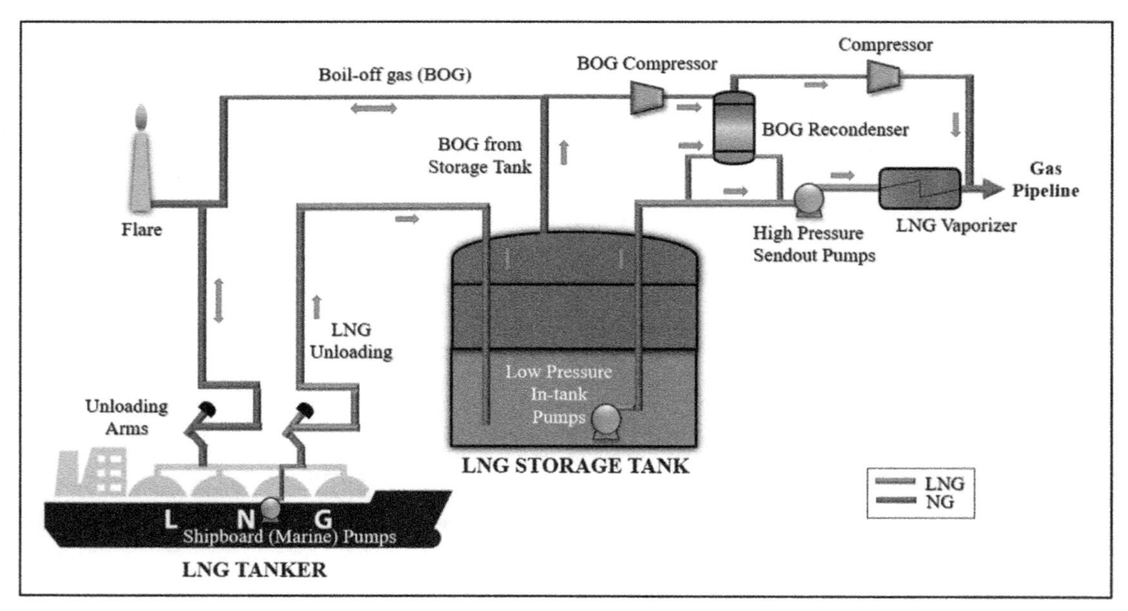

FIG. 5.45 Typical simplified LNG import and regasification terminal with various LNG pumps.

LNG is pumped to LNG storage tanks which are designed to handle very little pressure (less than 300 mbarg) due to their large size and construction. For the purpose of regasification, LNG is pumped out of the storage tank via retractable in-tank pumps that are located at the bottom of the tank. The special retraction system assists operators to safely remove and install the cryogenic in-tank pumps into the discharge column of the storage tank. In-tank pumps are often low-pressure pumps with rated discharge pressures at around 7 barg (101 psig) with flow rates ranging from 400 to 2000 m^3/h. depending on the application. The main purpose of cargo pumps and in-tank pumps is to transport LNG from the ship's tank or the storage tank to the next point in the process without high discharge pressure. Discharge pressures from in-tank pumps are boosted to the required pressures of 100 barg using vessel mounted (canned) send-out pumps. High-pressure send-out pumps have relatively less flow output due to rotordynamics and structural concerns. Therefore, to satisfy the total flow rate required by the process, multiple high-pressure pumps are operated in parallel to each other. It should be noted that cargo and in-tank pumps are also operated in parallel to satisfy the total flow rate required by the plant.

LNG bunkering and peak-shaving plants also require pumps to safely supply LNG to tankers and transport LNG for smaller scale applications. These pumps are much smaller in terms of flow rate and are not required to be operated at all times.

Cryogenic LNG pumping technologies

In this section, LNG cryogenic pump technologies are divided into three main categories based on the type of the motor and driver.

Regardless of the motor and driver configuration, each pump is designed to attain the best performance in terms of efficiency, while maintaining reliability and dependability in a cost-effective manner. In addition to the performance requirements, the following aspects must be considered in the selection of pump type and pump components:

- **Selection of Hydraulic Components:** The selection of hydraulic components is based on the specific performance requirements of the pump. As in any pumping application, pump-specific speed (N_s) is determined to attain the best efficiency. Hydraulic component selection should involve the suction performance requirement in terms of NPSHr. Helical type constants or variable pitch inducers are installed at the downstream of the first stage of each pump to enhance the suction performance and lower NPSHr. Operating speed, off-design running conditions, and dynamic behavior, such as rotordynamics of the machinery, should also be considered at the selection of hydraulics.
- **Shaft Support Configuration:** The main shaft support elements are the radial-type, deep groove, and angular contact ball bearings. Bushings and labyrinth-style wear rings are installed between each stage to support the shaft and minimize back leakage. Submerged, motor-type pumps use ball bearings with little thrust capability, which requires a balance device. Ball bearings are product lubricated and cooled. Due to a relatively low viscosity of LNG, ball bearing races and balls are designed and manufactured to maintain sufficient EHL (elastohydrodynamic lubrication). This requires enhanced surface finish (improved surface roughness) and better controlled curvature at ball bearing components, namely, races and balls. Ceramic ball bearings provide the best surface finish and act as an insulator. Ball bearings with ceramic balls are often preferred for applications with low viscosity and for applications where electrical arcing is a concern. If the fluid properties of the pumped fluid are less than ideal and the cleanliness of the pumped fluid is questionable, hydrodynamic bearings can be utilized. Hydrodynamic bearings have a greater speed capability and can provide greater mean time between pump maintenance, but they require a discharge fluid supply, which can negatively impact overall pump efficiency. For LNG pumps with external motors, lubrication of ball bearings is controlled and achieved with a specific lubrication and cooling system. Thrust bearings can be used at these applications, if required.
- **Thrust Equalizing Mechanism:** Submerged, motor-driven pumps cannot utilize thrust radial ball bearings due to the lack of viscosity of pumped fluid and cryogenic temperatures. LNG has relatively low viscosity to allow elastohydrodynamic film to be built under excess axial load. Although product-lubricated ball bearings are designed to have an angular contact angle to carry both radial and axial loads, the axial load capacity is limited to rotating assembly weight and cannot handle the axial thrust load of a pump. Therefore, the resulting axial load, due to differential pressure of each stage, needs to be equalized by a thrust mechanism. Balance pistons, or thrust mechanisms based on a variable geometry, such as TEM, are utilized to overcome the axial load.
- **Materials of Construction:** Due to very low pumping temperatures ($\sim -165°C$, $-265°F$), materials are selected to ensure adequate strength for each application. While the yield strength is often improved at colder temperatures, impact strength is significantly lower at cryogenic temperatures. With that, materials with sufficient impact and yield strength must be used. The 300-series stainless steel is preferred, while carbon steel is avoided due to lack of impact strength at cold temperatures. Impellers, diffuser vanes, and casings with complicated geometries are cast

aluminum alloy per ASTM A356.0 T6 temper. Preferred shaft material is 17–4 or 15–5 pH magnetized stainless steel since the motor is directly sleeved to the shaft. Aerospace grade aluminum alloys, such as 6061-T6 and 7075-T6, are used for casings, which provide a very good strength-to-weight ratio. Bushings and wear rings are selected to provide an adequate material hardness difference in accordance with API 610 guidelines with enough wear properties. Bronze alloys are commonly used as wear rings and bushing material and are proven to work well at cryogenic conditions. Gaskets and O-rings are usually Teflon or metal based due to low temperature.

- **Footprint, Physical Size, and Weight:** Pumps are vertically suspended, which help with the footprint. Physical size and weight are important design considerations for marine and offshore applications due to limited space and weight restrictions. For in-tank pump applications, overall pump diameter is critical in the determination of the discharge column. Therefore, pumps that have a relatively smaller diameter are preferred to reduce the overall diameter of the discharge column.
- **Motor Cooling and Lubrication:** For submerged, motor-driven pumps, motor cooling is crucial for stable operation of the pump. Lack of cooling can result in vaporization due to boiling of the process fluid at the motor cavity, which can result in a premature pump failure. Sufficient cooling flow is provided via a cooling loop using the process fluid. Since the cooling flow is obtained from diverting the pumped fluid and circulating it to the motor cavity, it should be well controlled and optimized to ensure pump flow output and efficiency are not highly impacted.
- **Condition Monitoring and Safe Operation:** Condition monitoring is conducted via accelerometers installed in close proximity of radial ball bearings. Overall vibration levels are monitored and recorded using accelerometers specifically designed for cryogenic fluids and temperatures. Since the physical size of each pump varies quite significantly, it is advised to establish a vibration and dynamic signature for each pump at initial installation and perform trending analysis to identify whether there is any degradation over time. Pump process parameters should also be monitored to ensure there are no operational issues or upset conditions that may result in premature pump failure. API 610 has guidelines regarding maximum allowable vibration levels in terms of velocity. These levels may be too conservative for small-scale pumps or too restrictive for large-, high-pressure, or high-flow pumps. It is crucial to establish condition monitoring and a safe operation philosophy (including alarm and trip algorithms) with the end-user and pump suppliers for the best practice.
- **Electrical Components:** Due to the hazardous nature of LNG, electrical components and any instrumentation (sensors) located at ex-tanks within certain proximity of process fluid must be certified according to the local and international hazardous area codes. NEC, NFPA, and IECEX regulations mandate that electrical connections (such as electrical terminals and junction boxes) that are not rated as "Intrinsically Safe" are to have a dual-seal feed-through with a chamber between two seals that are continuously purged with nitrogen gas to detect and monitor any hazardous gas leakage to the electrical ignition sources. Once a leak is detected, machinery must be shut down to replace any seals. Nitrogen gas helps not only in detection of leakage but also assists in removing the flammable gas from the hazardous area location via continuous purging.
- **Capital and Operational Cost:** Capital cost and operational cost should be considered in the selection of a pump type. Reliability and availability studies are conducted with the pump suppliers to better understand and plan for operational expenses. It is important to minimize the downtime due to scheduled or unscheduled repairs. Spare pumps may be installed to the process for redundancy, or a spare pump can be made available for immediate install.

Submerged motor-driven LNG pumps

Since LNG is highly flammable and explosive, the design of any electrical component will require consideration of applicable local and international safety regulations and guidelines. Conventional LNG pumps with external motors require complex seal design, complex initial start and cool-down procedures, alignment issues, and higher maintenance. In addition, each electrical component must be designed, manufactured, tested, and certified according to the applicable hazardous area classification and code. As a cryogenic fluid, LNG is an excellent insulator. Submerging the induction motor into a process fluid allows the induction motor to be completely isolated from oxygen in the atmosphere. As a result, there is no requirement for hazardous area certification for the induction motor.

This design feature approach also eliminates the need for rotating dynamic seal(s) between the external motor section and hydraulics section of the pump, which are prone to fail due to a harsh, cryogenic environment.

Conventional, external motors normally have a mechanical coupling located between the motor and the hydraulic components for driving the motor. This requires precise alignment during installation, which is critical for reliability. The submerged design has a common shaft, and alignment or concentricity is accomplished at the factory when the machine is manufactured and assembled. A common single-shaft design eliminates the alignment concerns.

Conventional cryogenic LNG pumps with external motors often require an external lubrication system for the motor bearings. This adds complexity to the system and requires maintenance. The bearings in the submerged motor design are cooled and lubricated by the process fluid in a subcooled condition. It should be noted that at the motor cavity, fluid must be in a subcooled condition, as vapor formation cannot be tolerated by motor and radial ball bearings. Another advantage of a submerged motor design is that the motor itself is cooled by the cryogenic process fluid via cooling return lines and fluid traveling through the gap between the motor rotor and stator [35].

There are three main types of submerged motor pumps utilized in LNG plants and processes, which are, namely, suction vessel (canned) mounted, retractable in-tank pumps and marine pumps.

Suction vessel (canned) mounted pumps

Suction vessel mounted pumps are often preferred for send-out or booster pump applications. There are some applications in which these types of pumps are implemented for loading applications with high flow rates (~2000 m^3/h) and low differential pressure (<10 barg). These pumps are vertically suspended and installed inside a pressure vessel that consists of the necessary process piping, such as suction, discharge, and vent nozzles. A typical, high-pressure suction vessel mounted pump is shown in Fig. 5.46. Pressure vessels are designed and constructed per the ASME Boiler and Pressure Vessel Code (BPVC) and can be certified and registered to additional applicable codes, such as PED and AS1210.

Send-out and booster pumps require multistages to attain the high pressure required by the vaporizers during the regasification process. Discharge pressures can be in excess of 100 barg. As many as 20 stages can be utilized to achieve the required discharge pressure, while rotational speeds do not exceed 3600 RPM. Rotordynamics of the machinery

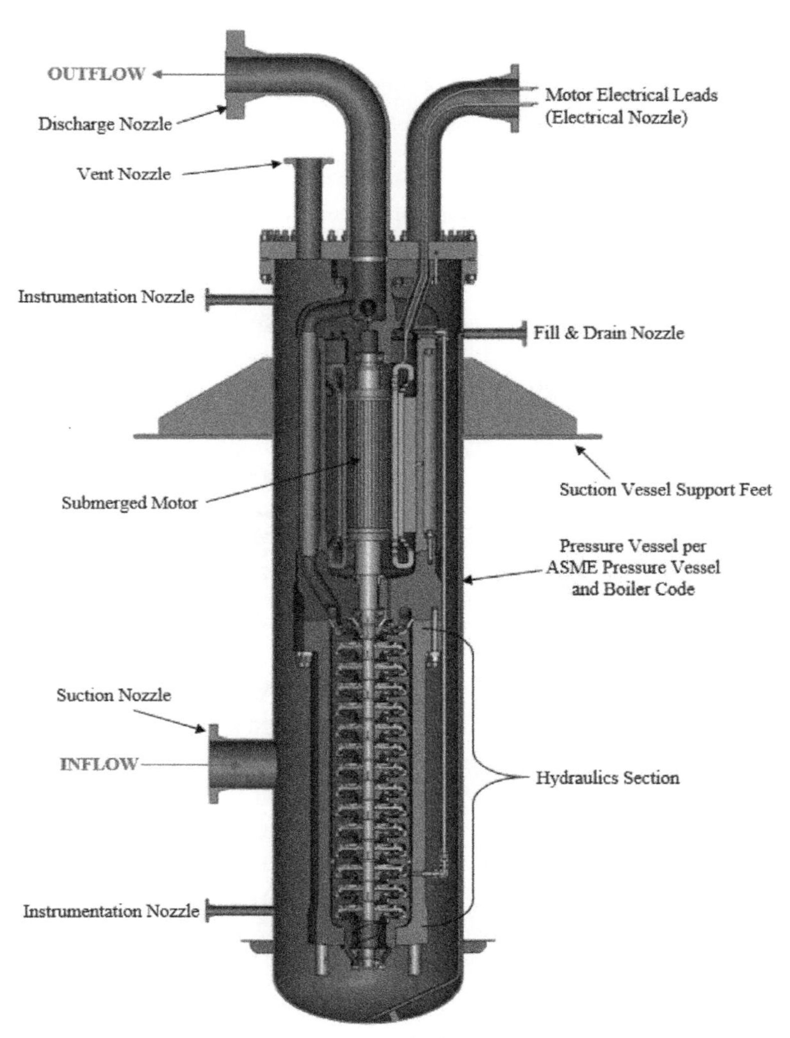

FIG. 5.46 Suction vessel mounted submerged motor pump. *Courtesy of Elliott Group, Ebara Corporation.*

should be considered in determination of the stage count and operational speed. To keep the stage span shorter, radial-type diffuser vanes are used at high-pressure pumps. An example of a radial diffuser stack is shown in the cross-sectional view of a high-pressure cryogenic pump in Fig. 5.47. Since the diffusion of the fluid is achieved in the radial direction of flow, stage overall axial length is much shorter. However, greater radial spacing from the shaft center is necessary, which increases the overall diameter of the outer pump casings. As it can been observed from Fig. 5.45, each pump stage is identical and provides the same amount of head increase. Since the process fluid is incompressible in pumping applications, there is no concern of change in density, and the pressure ratio is constant across each stage for a given flow rate and fluid density.

Loading pumps are single-stage, cryogenic units that provide high-flow output. Similar to high-pressure applications, these pumps are installed vertically into a pressure vessel with identical piping configurations. As in high-pressure pumping applications, an inducer can be installed at the downstream of the first impeller's suction side to lower NPSHr. The main difference is the type of diffuser vanes. Instead of radial diffusers, axial-type diffuser vanes are implemented. Ball bearings at the lower most location can be eliminated according to the rotordynamics. A typical suction vessel mounted loading pump is shown in Fig. 5.48.

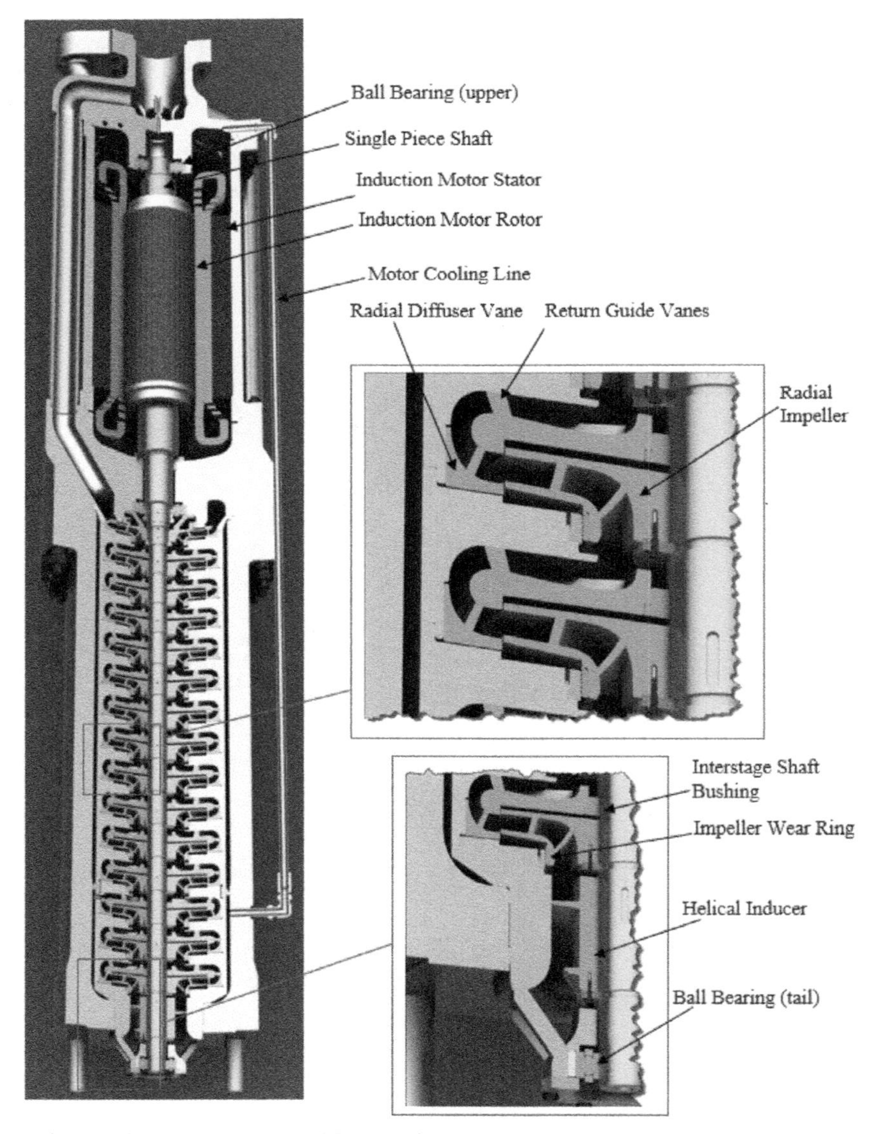

FIG. 5.47 High-pressure submerged motor cryogenic LNG pump for send-out applications. *Courtesy of Elliott Group, Ebara Corporation.*

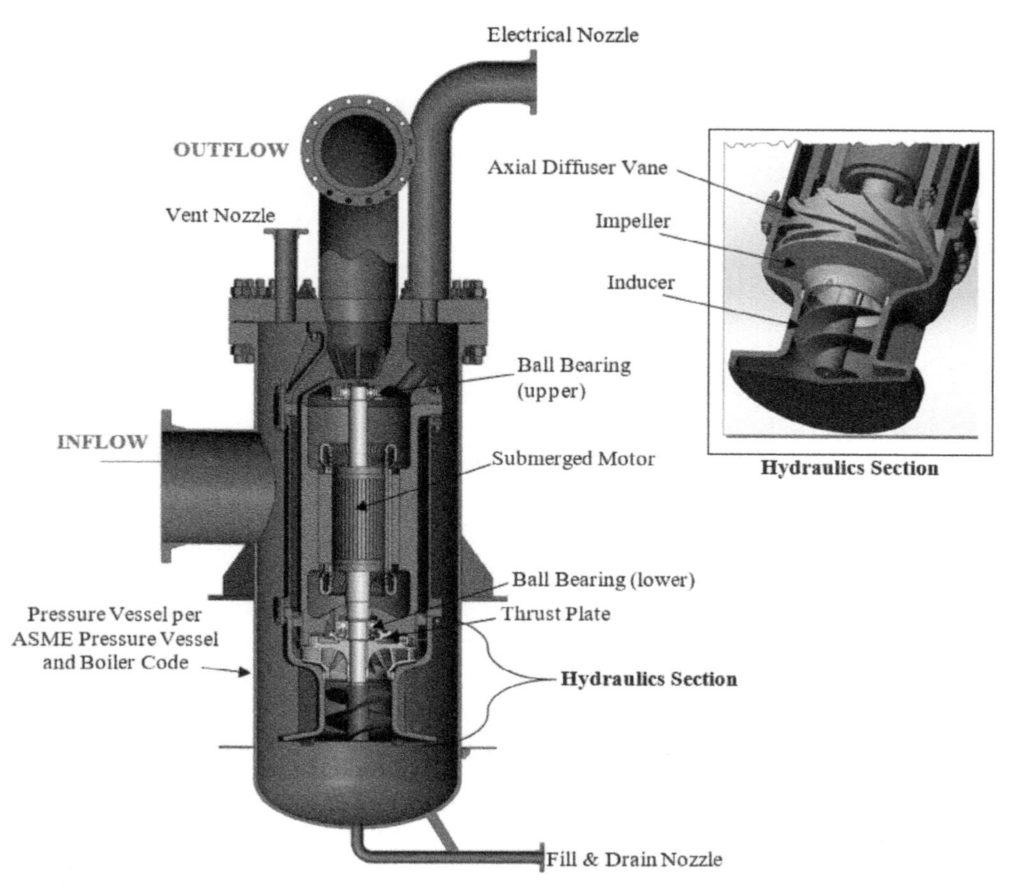

FIG. 5.48 Typical single-stage suction vessel mounted high flow loading pump for LNG applications. *Courtesy of Elliott Group, Ebara Corporation.*

Retractable in-tank pumps

Retractable in-tank pumps are vertically suspended and installed inside the discharge column of LNG storage tanks via a special retraction system. The main purpose of these types of pumps is to transfer LNG from storage tanks to booster pumps for regasification purposes. They are also used to load carriers at LNG export terminals. Pumps are installed inside a discharge column and over to a suction (foot) valve, which closes once the pump is lifted off for maintenance and removed from the column. The suction valve has a spring-loaded mechanism to seal and minimize any leakage from the storage tank to the discharge column once the pump is removed.

A typical in-tank retractable pump is shown in Fig. 5.49 along with a suction valve and LNG storage tank. Every in-tank pump application is furnished with a helical inducer to delay the cavitation inception so that it can safely be operated at very low liquid levels, hence with reduced suction pressure. LNG tank construction for any hydrocarbon application (methane, propane, butane, etc.) can only handle a maximum tank pressure of 300 mbar (4.35 PSI) or less due to cost and manufacturing constraints. With that, the suction pressure of an in-tank pump is often dictated by the static height of the liquid. Inducer design is very crucial for these applications, as it ultimately determines the minimum liquid level in the tank required for safe and stable operation. This minimum liquid level (height) is known as NPSHr, which also determines the nonusable height of a liquid inside a storage tank. Since the construction of the storage tank limits the available suction pressure to the pump, high suction specific inducers are utilized in these applications to increase the suction pressure to the pump to reduce the minimum required liquid level [36]. It should be noted that the overall footprint of a storage tank can be as large as 200 m (656 ft.) in diameter [37]. Therefore, any little improvement in the magnitude of few centimeters can result in substantial improvement in usable amounts of liquid.

Marine pumps

Marine pumps are installed in ships with the purpose of unloading the storage and supplying fuel for the carrier. There are mainly four different types of marine pumps, which are categorized based on their duty: cargo, spray (stripping), emergency, and fuel pumps.

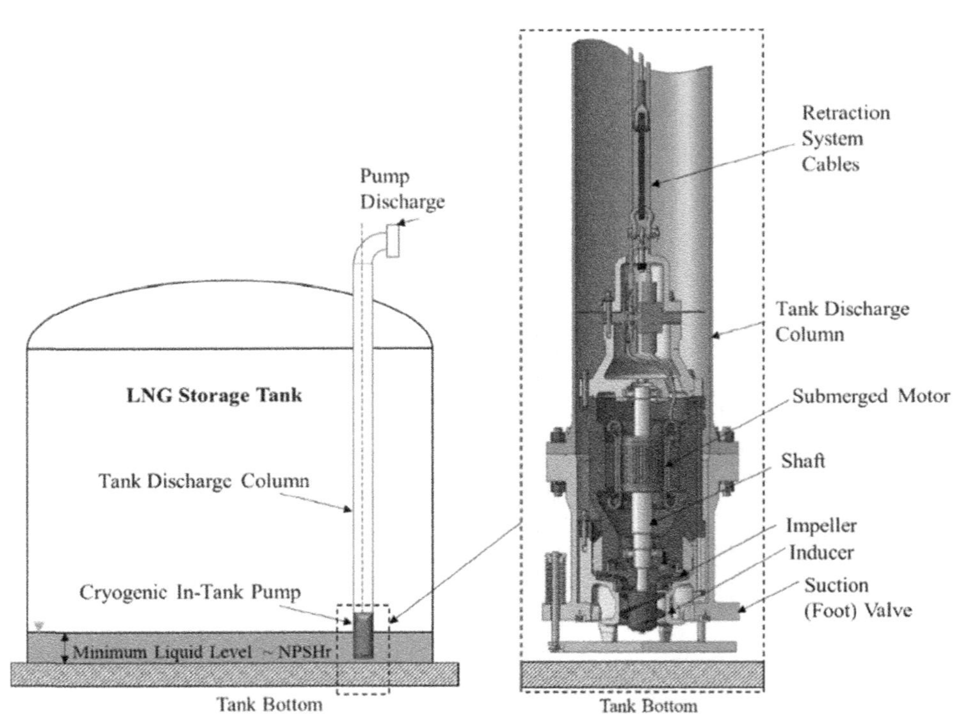

FIG. 5.49 Typical configuration of a retractable in-tank LNG cryogenic pump inside LNG storage tank. *Courtesy of Elliott Group, Ebara Corporation.*

Each tank on a liquefied gas carrier typically has two main cargo pumps installed. These pumps can unload the entire cargo in approximately 12 h. A single-stage design is most often used, although multistage pumps can also be provided depending on the differential pressure requirement. Cargo pumps have very similar design features to single-stage suction vessel mounted loading pumps. The main difference is that cargo pumps are mounted to a support plate inside the carrier's tank, which is closely located near the bottom of the tank for efficient unloading. The discharge flange of the cargo pump is directly connected to the tank's discharge pipe. Because of this design feature, cargo pumps are also categorized as "fixed" pumps. A typical main cargo pump cross-sectional view is shown in Fig. 5.50.

Spray (stripping) pumps spray LNG onto the inside top of the cargo tanks to help keep them cool and reduce boil-off gas vapor. Spray (stripping) pump designs have relatively lower Net Positive Suction Head (NPSHr) required since they are much smaller in size with low-duty flow rates. This allows cargo tanks to be offloaded with minimum to no liquid levels. Similar to retractable in-tank and cargo pumps, these pumps are also furnished with a high suction specific speed inducer. With some applications, these pumps are used to provide fuel to the ship and can be used as a dual-duty pump. A typical configuration of a spray pump is given in Fig. 5.51.

Emergency pumps are identical in design to in-tank retractable pumps. They are only installed and used for emergency cases, where the offloading of the cargo tank cannot be done by main cargo pumps due to a malfunction. Similar to retractable in-tank pumps, they are installed in the cargo tank column via a retraction system during an emergency to quickly empty the cargo tank. These pumps are often stored inside a sealed container and only installed if necessary.

Pumps for supporting functional units in LNG plants

In addition to the core cryogenic and liquefaction units of an LNG plant, there are many functional units within LNG plants. Gas pretreatment is one of the important, functional units in LNG plants. It takes place prior to the chilling, condensation, liquefaction, and refrigeration processes. Gas treatment involves acid gas removal, precooling, dehydration, and mercury removal as shown in Fig. 5.11.

Acid gases are mainly CO_2 and H_2S. These must be removed from the process stream before the process gas is introduced into the cryogenic unit. Both CO_2 and H_2S will freeze and solidify if they are not removed to an acceptable level. For CO_2, a typical acceptable level is between 50 and 100 ppm by volume (ppmv). For H_2S, the amount is determined based on the pipeline gas composition specifications. A typical acceptable level of H_2S is less than 4 ppmv. To effectively remove acid gases, solvent-based processes are used in the LNG industry. The alkanolamines are the most commonly used solvents for removal of H_2S and CO_2.

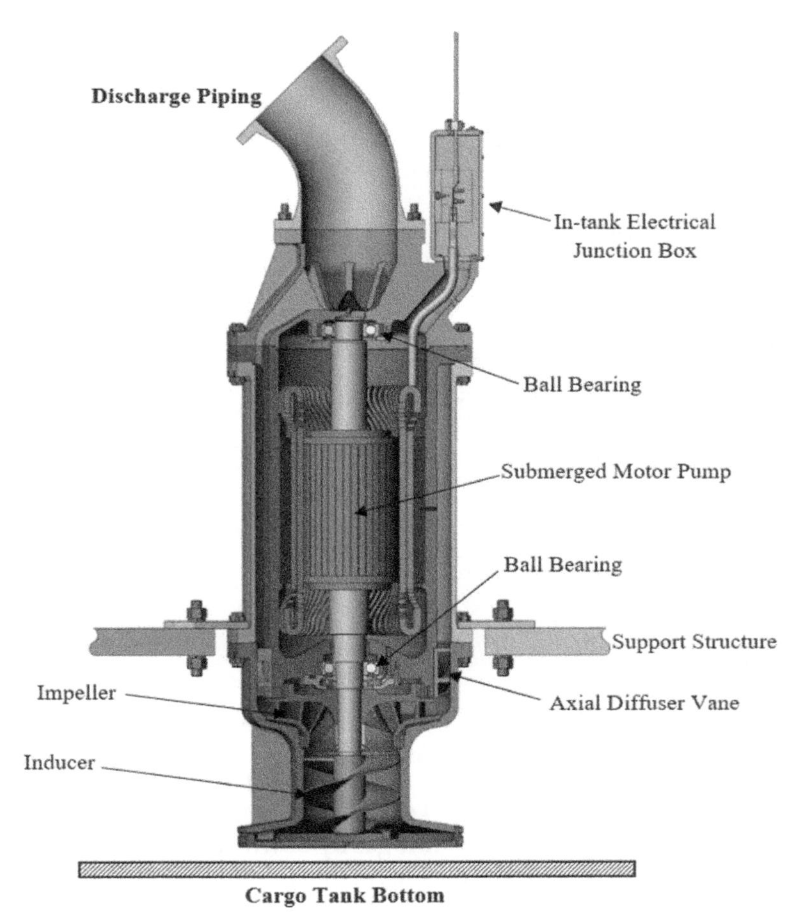

FIG. 5.50 Typical cargo pump used in LNG carriers. *Courtesy of Elliott Group, Ebara Corporation.*

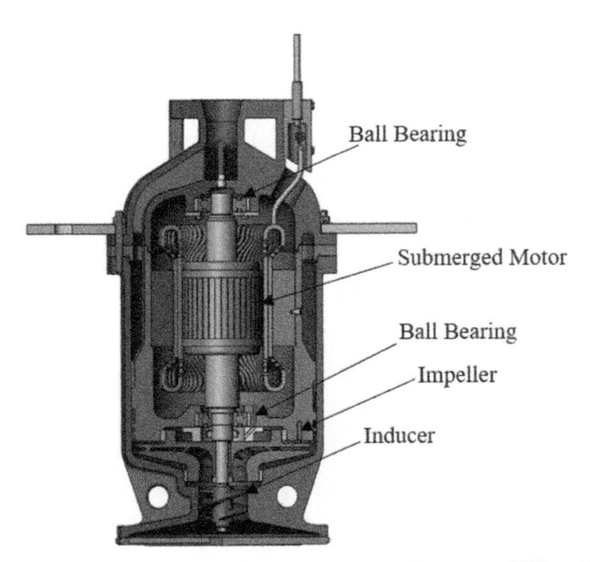

FIG. 5.51 Cross-sectional view of a spray/stripping pump used in LNG carriers. *Courtesy of Elliott Group, Ebara Corporation.*

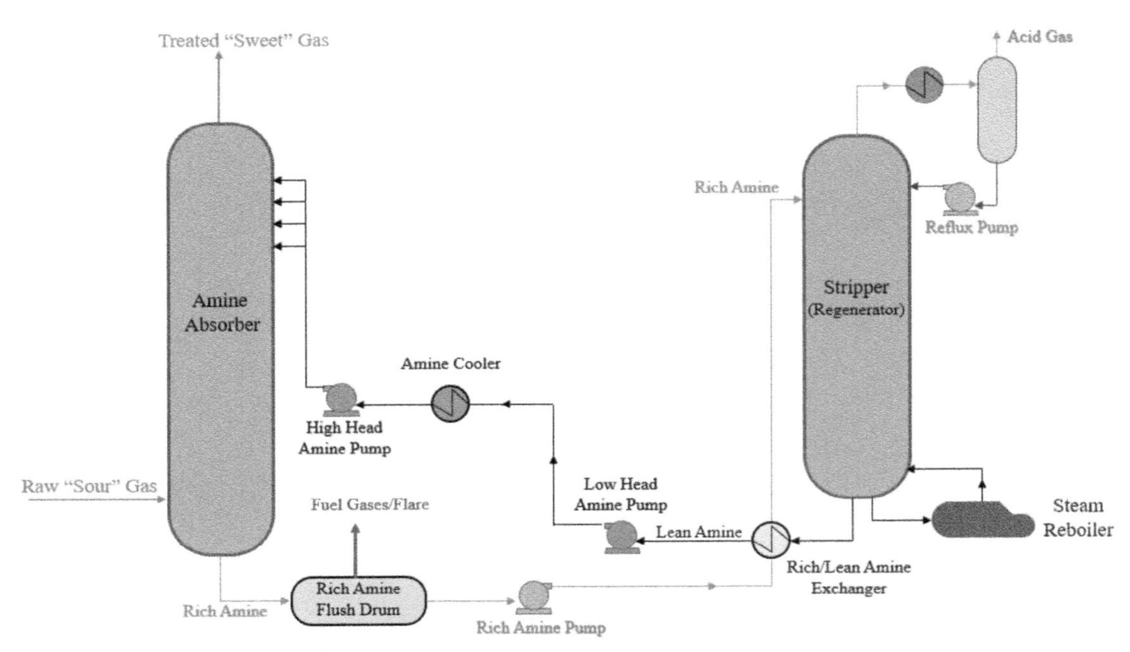

FIG. 5.52 Acid gas removal process flow diagram.

Fig. 5.52 shows a schematic view of a solvent-based, acid gas removal unit. The process consists of two important columns. The first one is the absorber column which is used for removing acid gases from the raw "sour" gas. The second column is the stripper column, which is used for stripping acid gases from "rich" amine solution. There are multiple pumps utilized in this process to pressurize and transfer "lean" amine solution to the high-pressure absorber column where acid gases are absorbed from the "sour" raw gas by downward flowing amine. "Lean" amine solvent is injected into the amine absorber column and makes contact with natural gas and absorbs acid gas components, thus becoming "rich" amine solution.

Amine pumping to the absorber is often handled by two separate pumps with an amine cooler as shown in Fig. 5.53. This is to prevent the possibility of cavitation due to high-temperature pumping. A low differential head amine pump operating at a high temperature is followed by a higher head pump operating at near-ambient temperature [38]. High temperature makes the high differential head pump susceptible to cavitation. Therefore, a cooler is implemented to cool down the amine solution to increase the NPSH available. Another concern is the residual dissolved CO_2 within the amine solution. CO_2 can come out of the solution at a low pump suction pressure and high temperature, which can be highly detrimental to pumps. Therefore, three to four times the available NPSH is reduced to prevent separation of dissolved CO_2. A low, head-high-temperature pump is often a double-suction, single-stage unit with a low NPSH requirement. High-head pumps are typically multistage ring section, horizontal, electric motor-driven units. These units have external motors with mechanical seals. A typical high-head amine pump is shown in Fig. 5.53.

Amine solutions are not highly corrosive fluids. Therefore, standard cast iron or ductile iron is usually acceptable for amine scrubbing applications. However, it is very crucial to select the correct elastomers for the mechanical seal and

FIG. 5.53 High-head ring section horizontal pump. *Courtesy of Elliott Group, Ebara Corporation.*

O-rings that can stand up to the amine. Amine leakage to lube oil for bearings can result in a breakdown of the oil and causes a considerable reduction in viscosity. This can result in premature bearing failure.

Flow rates for amine solution pumping depend on the amount of acid gases. Typically, a flow rate for amine pumping is around 2000 m^3/h. for a 5 MMTPA of LNG containing 15% CO_2 [38]. Total flow is often handled by multiple pumps operating in parallel. The pressure at the absorber column should be high enough to assist the absorption process. With that, high-pressure, lean-amine pumps have a differential head greater than 600 m. Differential head for low head is often between 80 and 120 m.

Expanders for LNG applications

As discussed under the refrigeration and liquefaction sections, there are several methods for liquefying natural gas. Each process shares a common problem of high-pressure LNG leaving the compression to be let down to near atmospheric tank pressure. The common solution is a Joule-Thomson (J-T) valve, which utilizes a highly reversible isenthalpic process of reducing pressure with no energy recovery. In essence, a portion of the gas is lost due to pressure relief and required to be recompressed and sometimes flared-off to the atmosphere. Another option to reduce the pressure is through LNG expanders, which have the ability to reduce enthalpy and recover the lost energy of the fluid during the let-down process. In this section, LNG expanders that are utilized in LNG liquefaction plants will be discussed.

LNG liquid expanders

The first LNG expander was installed at a LNG liquefaction plant in the mid-1990s. All liquefied natural gas (LNG) plants that are commissioned before 1996 are operating with an inefficient expansion valve [39]. In recent years, LNG liquid (hydraulic) expanders have become an important machinery for almost every modern LNG liquefaction plant. All liquid expanders mentioned next are installed in C3/MR and DMR liquefaction processes. Liquid expanders have been mentioned by the cascade process licensor, but the first example is yet to be demonstrated [40]. There are a handful of examples of LNG liquid expander installations to FLNG and FSRU applications.

LNG and MR expanders are shown in Fig. 5.54 in a simplified C3MR/DMR liquefaction process.

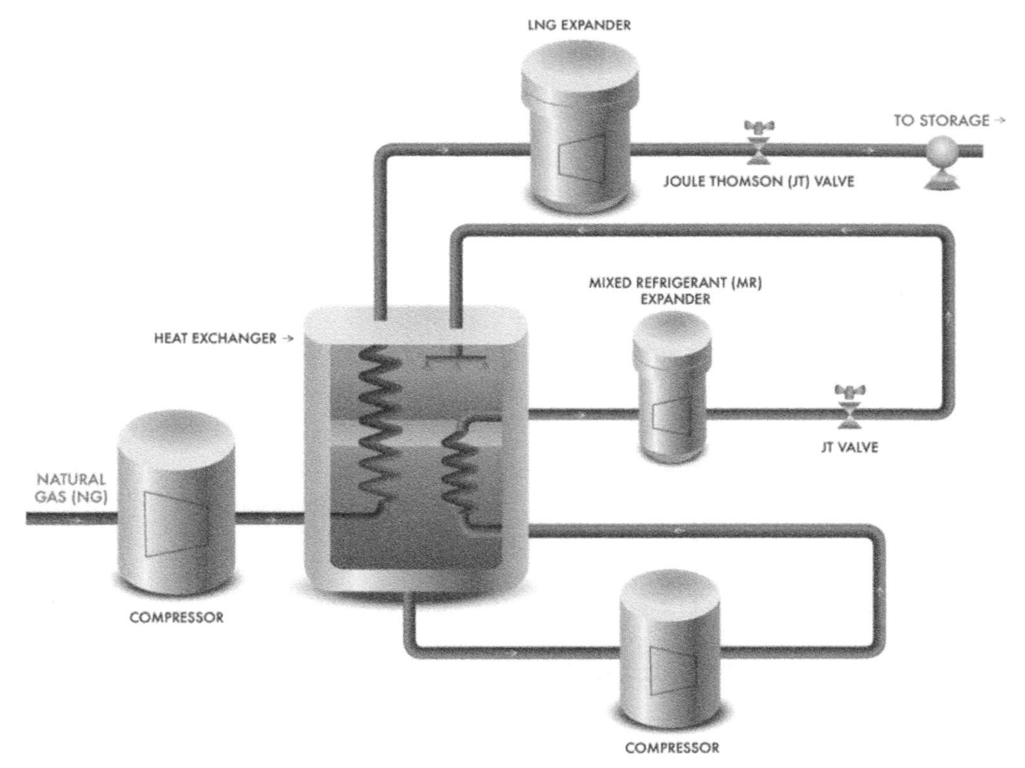

FIG. 5.54 LNG and MR liquid expanders at liquefaction process.

As shown in Fig. 5.54, liquid expanders require a J-T valve at the downstream to achieve the pressure drop required by the process. Since the final pressure at the downstream of the J-T valve results in a vapor-liquid mixture (saturated liquid), liquid expanders cannot be operated at that final condition. Liquid expanders cannot tolerate vaporization within the machine internals especially at the locations of liquid lubrication (elastohydrodynamic lubrication—EHL) and cooling. Shaft supports and thrust equalizing components rely on liquid film. Any vapor formation within these areas can result in unstable operation and premature failure. With that, the liquid expander outlet condition is always maintained to be above the saturation point (bubble point) of the process fluid.

Replacing the existing J-T valve with a cryogenic LNG liquid expander would considerably improve the thermodynamic efficiency of the liquefaction process [41]. Utilizing a liquid expander in the LNG main stream provides:

- Energy recovery and reduction in compression requirements by mechanical integration or electrical feed to the compressor.
- Enhanced cooling by enthalpy change, which improves LNG production by reducing vapor percentage at the flash process, which in turn, directly reduces the vapor recompression requirement.

Change in pressure and enthalpy across a liquid expander and a J-T valve is shown in Fig. 5.55. While the expansion across the J-T valve is isenthalpic (constant enthalpy), the expander has the ability to reduce the enthalpy as indicated previously. Enthalpy reduction also results in additional cooling of the liquid at the process. Therefore, it is not only the energy recovery, but also the cooling enhancement that make expanders very ideal for LNG liquefaction plants. Independent studies have proven that a 1%–3% increase in liquid production gain is attainable by utilizing LNG liquid expanders in the main LNG stream of a liquefaction plant in lieu of J-T valves [42–44]. In addition, the compression power requirement can be lowered by the same amount for both the precooling loop (MR or propone cooling refrigeration cycle) and the main LNG stream, which overall allows a reduction in heat exchanger and compressor sizes [42].

Early designs of expanders have three basic sections, which are, namely, the hydraulic turbine section, the cryogenic shaft seal, and the air-cooled, explosion-proof induction generator. The hydraulic section consists of turbine wheel(s) and various nozzles that assist in converting the kinetic energy of the fluid to mechanical energy at the turbine shaft. The cryogenic shaft seal is the boundary of the liquid that seals and separates the hazardous flame fluid from the surroundings and generator section. The generator section is where the mechanical energy of the shaft is converted to electrical energy.

In 1996, Dr. Hans Kimmel proposed a cryogenic hydraulic turbine with a variable speed to adjust the performance of the expander for different process requirements in terms of pressure drop and flow capacity [45]. At this new design concept, the hydraulic section utilizes axial nozzle vanes with radial inflow reaction type turbine wheels, which are capable of exceeding an isentropic turbine efficiency of 85%. The entire rotating assembly, including the hydraulic turbine section and the generator, is completely submerged similar to LNG pumps with submerged motors. To achieve

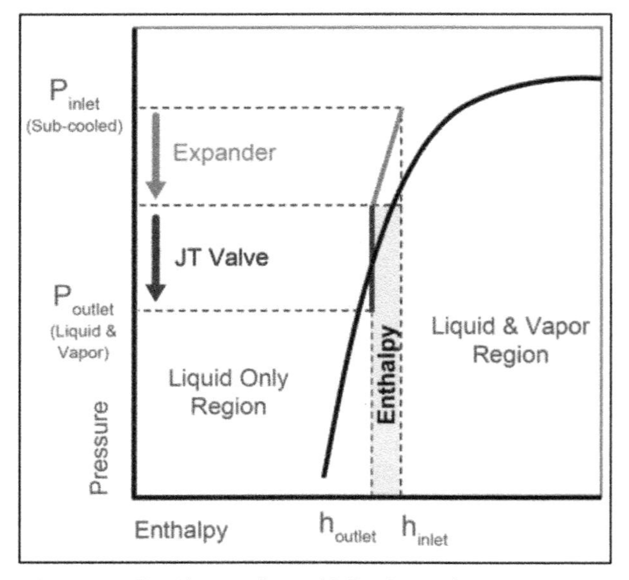

FIG. 5.55 P-h diagram showing expansion across liquid expander and J-T valve at downstream.

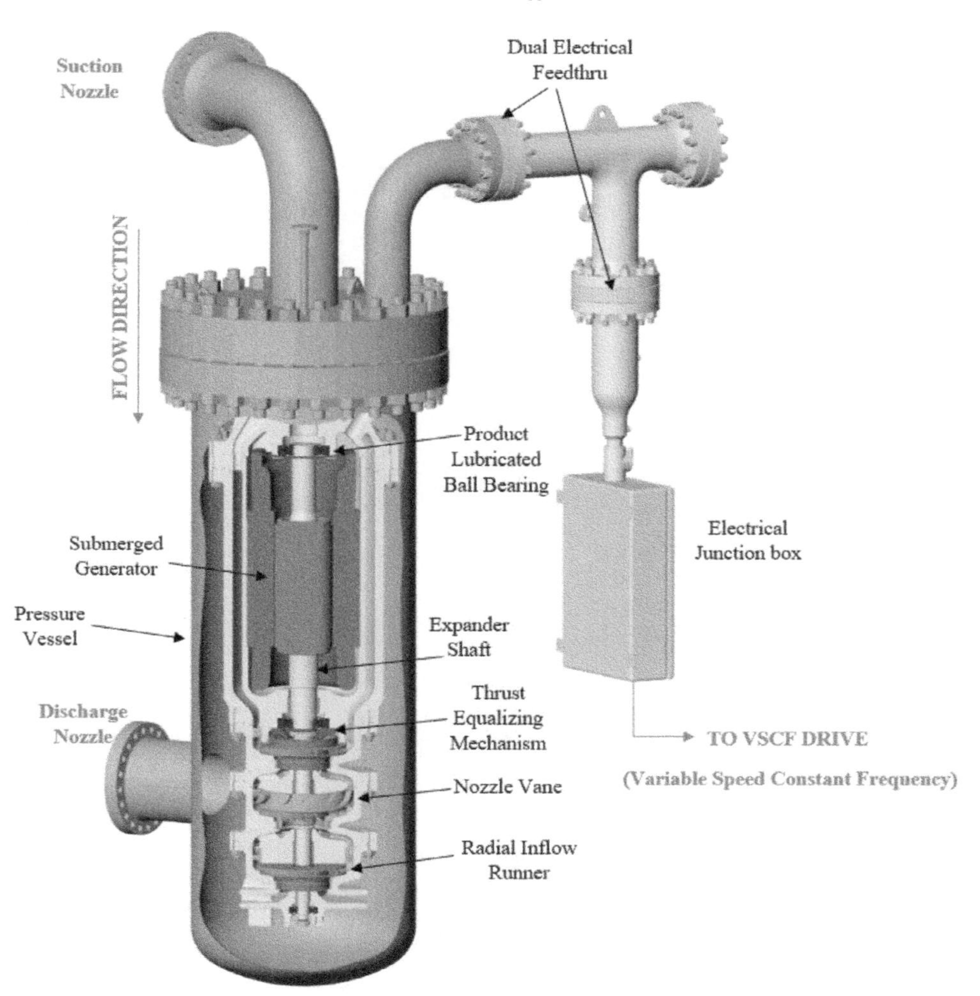

FIG. 5.56 Cross-sectional view of the variable speed submerged-generator hydraulic expander. *Courtesy of Elliott Group, Ebara Corporation.*

a high differential pressure, multistages with multiple hydraulic components are utilized. Fig. 5.56 shows the cross-sectional view of this variable-speed, submerged-generator, LNG liquid expander. With the entire unit submerged in LNG, there is no need for a dynamic rotating shaft seal or any kind of coupling between the hydraulic and generator sections. The expander assembly operates safely within a stainless steel containment vessel designed and tested in accordance with applicable pressure vessel codes.

Since the ball bearings are LNG lubricated and cooled with cryogenic liquid, they cannot handle excess axial thrust load. Therefore, each expander is equipped with a thrust equalizing mechanism to balance the hydraulic thrust to extend the lifetime of the LNG lubricated ball bearings [46].

A small percentage of the process inflow is required to cool the submerged generator, lubricate the ball bearings, and operate the thrust equalizing mechanism. With a submerged-generator design, there is no regular daily or monthly maintenance to undertake (as there are no rotating seals or seal systems), no oil-lubricated bearings or lube oil system, and no actuators or similar ancillaries.

In the advanced design of LNG expanders, flow direction is reversed from the downward direction to the upward direction as seen in Fig. 5.55. In the case of downward flow, thrust force and the rotating assembly weight add together, resulting in a greater axial force. If the flow is in the reverse direction, the axial force can be greatly reduced, which also results in a reduction in the required flow thrust equalizing mechanism. Fig. 5.57 is the expander assembly of an upward flow unit. The most important benefit of an upward flow unit is the physical size reduction of the casings and housing. At upward flow units, casings are subjected to external pressure, whereas in downward style expanders, casings are subjected to internal pressure. Internal pressure requires thicker walls, larger flange sizes, bigger hardware, etc., which increase the overall footprint of the expander. Under external pressure, casing sizes (volume and weight) are reduced considerably.

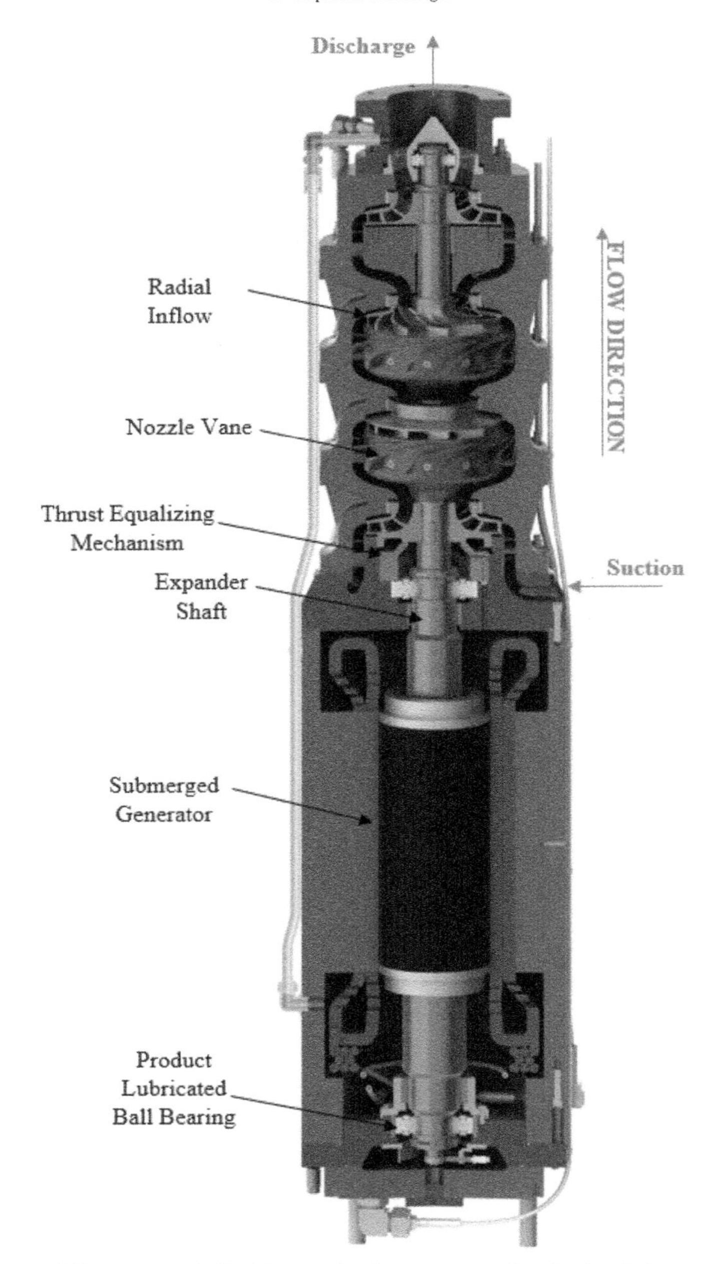

FIG. 5.57 Typical three-stage upward flow cryogenic liquid expander (pressure vessel and related piping not shown). *Courtesy of Elliott Group, Ebara Corporation.*

To extend the maintenance intervals, stainless steel ball bearings can be replaced with hybrid bearings with steel inner and outer races, silicon nitride balls, and molded resin separators. Their greater resistance to debris and reduced wear in low lubricity liquids shows a significant extension of life.

For harsh conditions such as low-viscosity applications with a high content of debris, and to further extend the life expectancy, hydrostatic bearings have been utilized with a touchdown secondary ball bearing for startup and shutdown. Hydrostatic bearings prove to increase the maintenance intervals from 24,000h to 40,000h.

LNG flashing two-phase expanders

Two-phase expander technology is very similar to existing cryogenic liquid turbine and expander technologies. The hydraulic energy of the pressurized fluid is converted by first transforming it into kinetic energy, then into mechanical shaft power, and finally to electrical energy through the use of an electrical power generator. The generator is submerged in the cryogenic liquid and mounted integrally with the expander. Similar to liquid expanders, two-phase expanders use a common shaft that consists of hydraulic and aero components with a submerged generator. The main difference between liquid and two-phase expanders is the additional hydraulic components utilized to handle the vapor formation within the machine internals.

The first two-phase flashing expander was installed in late 1990s, and two more units were installed in early 2000 and 2002 [47]. All of these units have been operational and used in a Nitrogen Rejection Unit (NRU) of a liquefaction plant. Similar to liquid expanders, two-phase expanders are also utilized in lieu of J-T valves to reduce the pressure of LNG. These expanders can take greater energy out of the process bringing greater cooling to the main LNG stream. They also process efficiency even more compared to liquid expanders [48]. A simplified NRU process is shown in Fig. 5.58 with two-phase expanders.

Change in pressure and enthalpy across a two-phase expander and a J-T valve is shown in Fig. 5.59. According to Fig. 5.59, the inlet condition is subcooled (liquid only), and flashing occurs within the machine internals, allowing liquid to vaporize by reducing the pressure below the saturation point. This results in a greater enthalpy change, which consequently results in greater temperature reduction and energy recovery with respect to a liquid expander with a J-T valve at the downstream.

An LNG production gain of 2%, along with a 0.2% thermodynamic cooling gain, is reported for NRU processes with two-phase expanders in operation with J-T valves [49]. There is mention of utilizing two-phase expanders in a cascade process by ConocoPhillips. Studies show that a 3% LNG production gain with 3% savings in overall compression in horsepower can be attained [40]. With an assumed 75% isentropic turbine efficiency for two-phase expanders in high and intermediate stages of the Phillips Optimized Cascade LNG process, a 5%–7% production gain with a 0.5%–0.6% thermal efficiency increase is reported [50]. More common C3MR liquefaction processes can also benefit from this new technology, but there are no real applications, yet.

LNG two-phase expanders are also vertically suspended inside a pressure vessel, and the flow is in the direction of buoyancy so that vapor does not get trapped at the internals of the machinery. A typical two-phase expander for LNG applications is shown in Fig. 5.60. LNG two-phase expanders have two additional hydraulic and aerodynamic components for handling the fluid in saturated conditions. Vapor formation is closely monitored and ensured that it only occurs at the downstream of the runner (turbine wheels) at a component called the jet exducer. The jet exducer is an axial flow rotating component that is capable of reversing the angular momentum of the fluid to achieve high isentropic

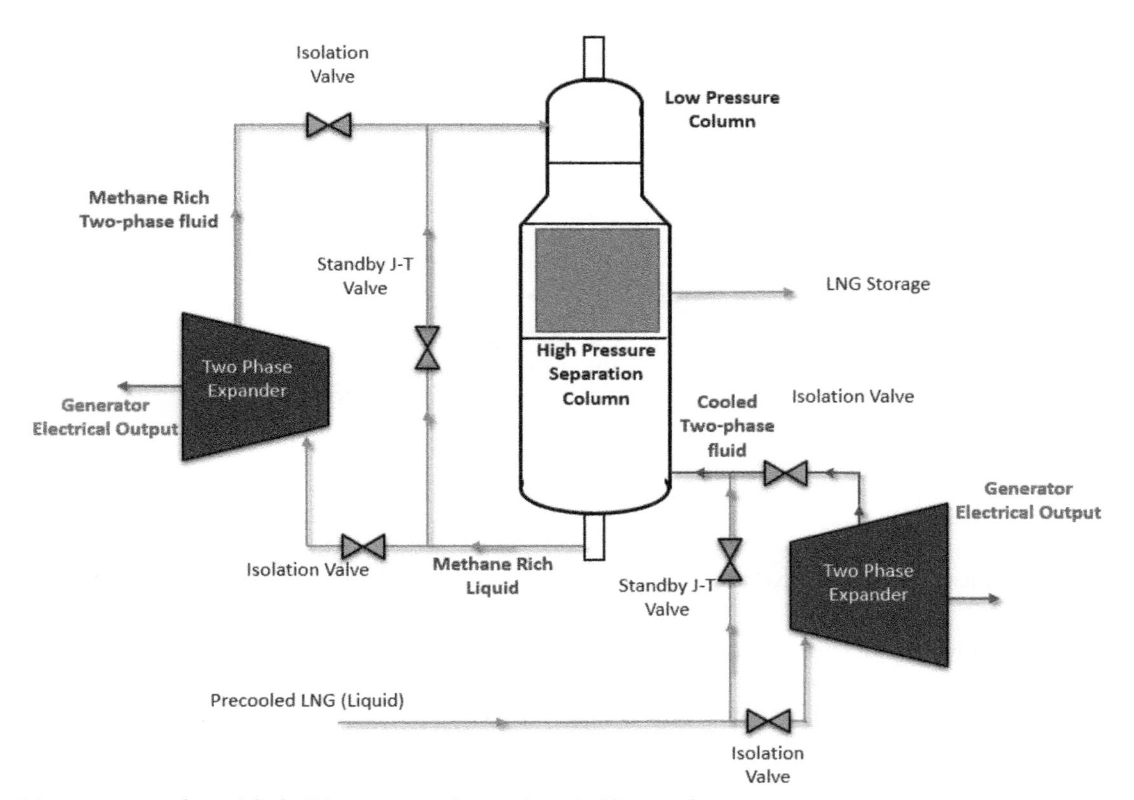

FIG. 5.58 Schematic view of simplified NRU process with two-phase LNG expanders.

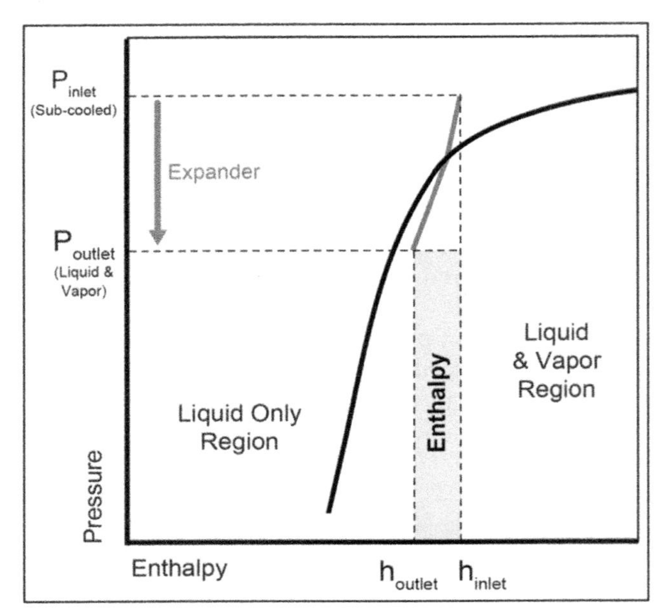

FIG. 5.59 P-h diagram showing expansion across two-phase expander.

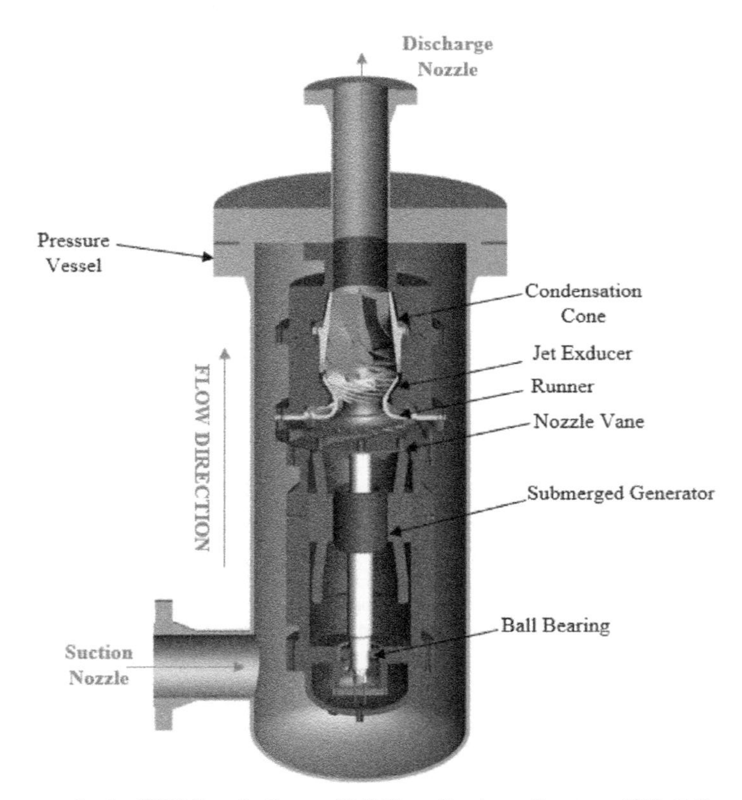

FIG. 5.60 Two-phase flashing expander for LNG liquefaction and NRU applications. *Courtesy of Elliott Group, Ebara Corporation.*

efficiency. Vapor formation results in acceleration of the fluid due to a change in density. The velocity change increases the angular momentum and consequently allows the jet exducer to recover the fluid's angular momentum to mechanical torque to the turbine shaft. A condensation cone is installed at the downstream of the jet exducer, which is utilized to recover the static pressure of the fluid. While the condensation cone is a stationary component, it still improves the overall efficiency of the process by recovering some of the static pressure by deaccelerating the vapor mixture. Fig. 5.61 shows the hydraulic and aerodynamic components utilized in two-phase expanders.

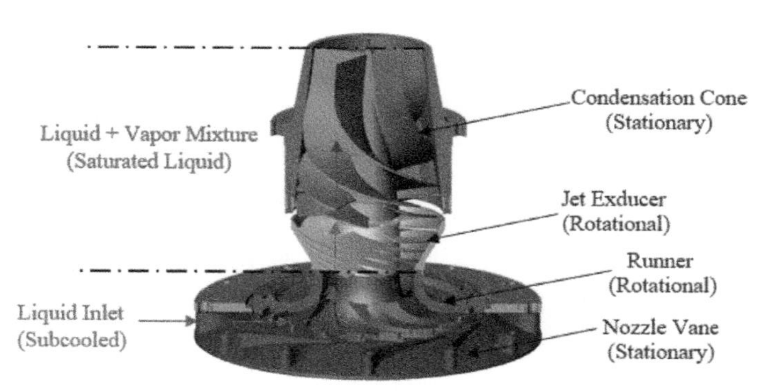

FIG. 5.61 Hydraulic and aerodynamic components of a two-phase flashing expander. *Courtesy of Elliott Group, Ebara Corporation.*

LNG drivers

LNG compressors can be driven by a variety of drivers using different sources of energy: electricity, natural gas, or steam. Some of the drivers allow speed variations over a wide range of speeds, such as variable speed electric drives, two-shaft gas turbine drivers, and steam turbines, while other drivers can only operate at constant speeds, such as constant speed electric drives, or with very limited speed flexibility, such as single-shaft gas turbines.

Today, variable speed drives are favored, not only because variable speeds allow for efficient and effective process control, but also because variable speed drives tend to make the equipment starting process simple. Variable speed drives may also eliminate the need to partially or fully blow down the process system around the compressor to be able to start with limited starter capabilities.

There are applications for small-scale LNG that use constant speed electric motors, especially in nitrogen refrigeration cycles with integrally geared compressors.

Steam turbines were the dominant drivers for LNG refrigeration compressors until the 1960s [51]. Thereafter, the driver market for large LNG plants was dominated by single-shaft industrial gas turbines, until in the mid-2000s, aeroderivative gas turbines (two-shaft design) became used more often. In the same time frame, large VFD-driven electric motors began to appear in the discussion. While not derived from aircraft engines, modern two-shaft industrial gas turbines show the same advantages, such as high efficiency and very short overhaul times (engine exchange, as for aeroderivative designs), relative to single-shaft designs but with a very high fuel flexibility.

Two-shaft gas turbines and aero derivative gas turbines are available mostly in smaller power ranges than single-shaft industrial gas turbines, which thus requires a larger number of (albeit identical) machines to be installed.

Electric motor configurations include constant speed motors driving the compressor via a variable speed gearbox and variable frequency drive speed (VFD) (i.e., controlled motors driving the compressors either directly or via a gearbox).

The electric motor speed (for induction type motors, synchronous motors, and permanent magnet motors alike) is determined by the frequency of the electric current and voltage supplied to the motor. A convenient way to vary the motor speed is to modify the power supply frequency by using a variable frequency drive [52].

Another way to create a variable speed drive is by using a constant speed motor and a variable output speed gearbox [52], often referred to as variable speed hydrodynamic drive (VSHD). VSHD drives are offered up to about 50 MW. The basic components of a VSHD are a hydrodynamic torque converter coupled with a planetary gear. The planetary gear is designed as a superimposing gear, with the torque converter driving one of planetary gear components. The torque converter acts as the control unit for the output speed. The VSHD uses a constant speed electric motor. Both the VHSD and the VFD are controlled by setting their speed until the available power becomes a limit.

Of importance for many applications are the performance characteristics of the driver, for example, the power as a function of ambient conditions or the power output at various output speeds. In general, a VFD-controlled motor is a constant torque machine, thus exhibiting a linear drop in power with speed (Fig. 5.62), implemented by maintaining a constant Volts/Hz ratio until above a certain corner frequency the motor becomes power limited. There are exceptions to this behavior, where the motor is oversized to provide constant power over a wider range, or where often for thermal reasons, the torque is reduced with speed. The speed-power relationship has a significant impact on control concepts for variable speed drives. In particular, the linear reductions of power seen in most VFD-controlled electric motors impose a limit on the flexibility compared to VSHD and two-shaft gas turbine drives.

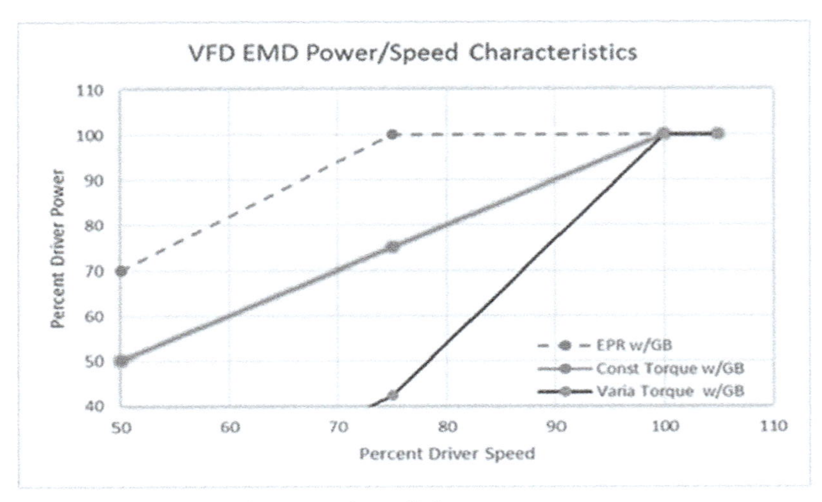

FIG. 5.62 Speed-power characteristics for variable frequency drives [52].

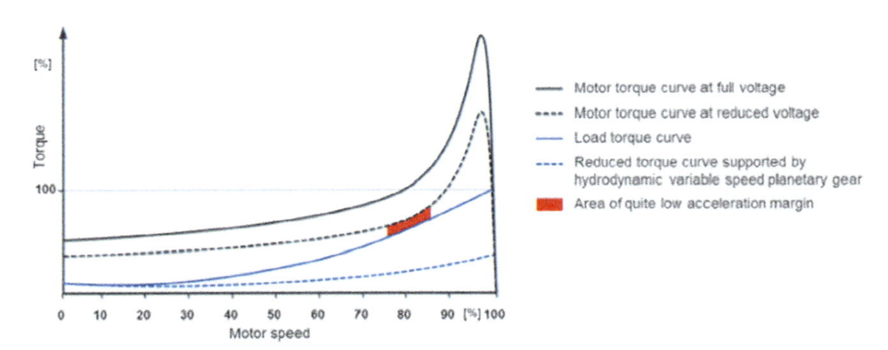

FIG. 5.63 Squirrel cage induction motor capability curve at different voltage levels and load torque curve [52].

Another feature of importance is the fact that the electric motor output is not subject to changes in ambient temperature (within limits). This means that it lacks the convenience of providing more power at lower ambient temperatures, but it has the advantage that the same power is available on hot and cold days. This can be important in applications where the load demand is dependent on ambient conditions.

Furthermore, the starting characteristics must be considered (Fig. 5.63), including the amount of torque at low speeds, or for constant speed electric motors, the amount of additional current.

The location where the units are installed (onshore, offshore, or subsea) determines access to maintenance intervention and the environmental conditions (e.g., salt in the air) that the equipment has to be designed for. For electric drivers, the question is also whether the electricity can be brought to site via transmission lines or if it has to be generated on site, usually with gas turbine-driven generators.

Industrial gas turbines produce mechanical power and serve to drive generators, pumps or gas compressors. They share the working principles with gas turbines that are used for aircraft propulsion [53]. The major components of a gas turbine include the compressor, the combustor, and the turbine (Fig. 5.64). The compressor (usually an axial flow compressor, but some smaller gas turbines also use centrifugal compressors) compresses the air to several times atmospheric pressure. In the combustor, fuel is injected into the pressurized air from the compressor and burned, thus increasing the temperature. In the turbine section, energy is extracted from the hot pressurized gas, thus reducing pressure and temperature. A significant part of the turbine's energy (from 50% to 60%) is used to power the compressor, and the remaining power can be used to drive generators or mechanical equipment (gas compressors and pumps). Industrial gas turbines are built with a number of different arrangements for the major components:

- Single-shaft gas turbines have all compressor and turbine stages running on the same shaft (Fig. 5.65).
- Two-shaft gas turbines consist of the following sections: the gas producer (or gas generator) with the gas turbine compressor, the combustor, and the high-pressure portion of the turbine on one shaft and a power turbine on a

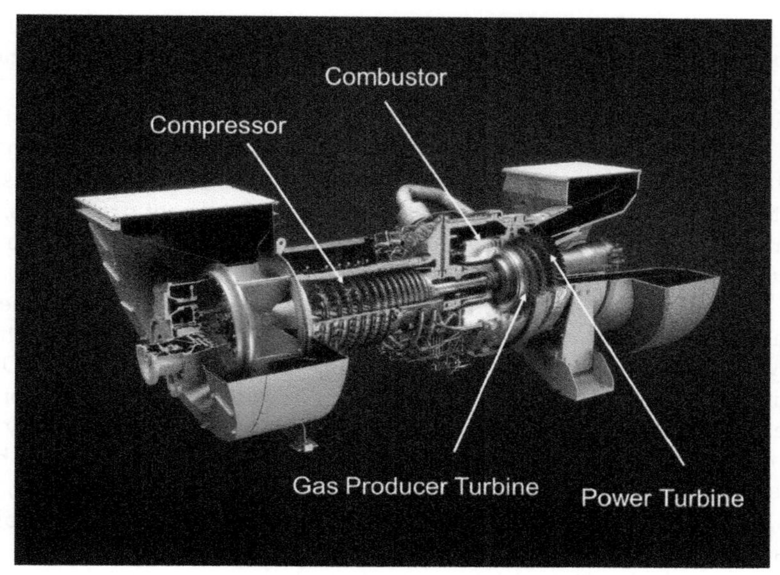

FIG. 5.64 Components of a typical industrial gas turbine [53].

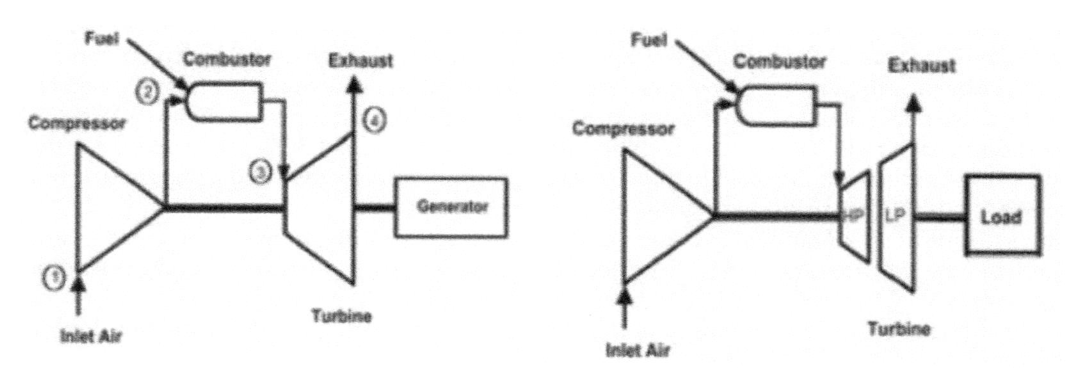

FIG. 5.65 Single- and two-shaft gas turbines. *Courtesy of Solar Turbines.*

second shaft (Fig. 5.65). In this configuration, the high pressure or gas producer turbine only drives the compressor, while the low pressure or power turbine, working on a separate shaft at speeds independent of the gas producer, can drive mechanical equipment.

- Multiple spool engines, which are industrial gas turbines, are derived from aircraft engines and sometimes have two compressor sections (the HP and the LP compressor). Each is driven by a separate turbine section (the LP compressor is driven by an LP turbine connected to a shaft that rotates concentric within another shaft that is used for the HP turbine to drive the HP compressor) and run at different speeds. The energy left in the gas after this process is used to drive a power turbine (on a third, separate shaft), or the LP shaft is used as an output shaft.

The engine configuration has a significant impact on the starting process; the starter motor in a two-shaft gas turbine, or a multispool turbine with a free power turbine, only has to accelerate the gas producer and bring it to self-sustaining idle speed. This requires only small starter motors with less than 1% of the GT power rating. In a multispool gas turbine where the load is driven from the LP spool, the starter brings the HP spool to speed. In a single-shaft machine, the starter does not only have to accelerate the gas producer, but also the driven compressor. This requires very large, VFD-driven starter motors with power requirements in the range of 20% of the driver power.

Another important topic is the combustion process and emissions control. The compressed air from the compressor enters the gas turbine combustor. Here, the fuel (natural gas, natural gas mixtures, hydrogen mixtures, diesel, kerosene, and many others) is injected into the pressurized air and burns in a continuous flame. The flame temperature is usually so high that any direct contact between the combustor material and the flame has to be avoided, and the combustor has to be cooled using air from the engine compressor. Additional air from the engine compressor is mixed into the combustion products for further cooling. The gas turbine combustion is continuous. This has the advantage that the combustion process can be made very efficient, with very low levels of products of incomplete combustion like carbon monoxide (CO) or unburned hydrocarbons (UHC).

The other major emissions' component, oxides of nitrogen (NO_x), is not related to combustion efficiency, but strictly to the temperature levels in the flame (and the amount of nitrogen in the fuel). The solution to NO_x emissions, therefore, lies in lowering the flame temperature. Initially, this was accomplished by injecting massive amounts of steam or water in the flame zone, thus "cooling" the flame. This approach has significant drawbacks, not the least the requirement to provide large amounts (fuel to water ratios are approximately around 1) of extremely clean water.

Since the 1990s, combustion technology has focused on systems often referred to as dry low NO_x combustion or lean-premix combustion. The idea behind these systems is to make sure that the mixture in the flame zone has a surplus of air, rather than allowing the flame to burn under stoichiometric conditions. This lean mixture, assuming the mixing has been done thoroughly, will burn at a lower flame temperature and thus produce less NO_x. One of the key requirements is the thorough mixing of fuel and air before the mixture enters the flame zone. Incomplete mixing will create zones where the mixture is stoichiometric (or at least less lean than intended), thus locally creating more NO_x. The flame temperature has to be carefully managed in a temperature window that minimizes both NO_x and CO. Lean-premix combustion systems allow to keep the NO_x, as well as CO and UHC emissions, within prescribed limits for a wide range of loads, usually between full load and about 40% or 50% load. To accomplish this, the air flow into the combustion zone has to be manipulated over the load range (Fig. 5.66).

The gas turbine power output is a function of the speed and the firing temperature, as well as the position of certain secondary control elements, like adjustable compressor vanes, bleed valves, and in rare cases, adjustable power turbine vanes. The output is primarily controlled by the amount of fuel injected into the combustor. Most single-shaft gas turbines run at constant speed when they drive generators, and their speed range for compressor drives is very limited (from about 95% to 100% speed), because the gas generator is directly coupled to the driven compressors. The gas generator power drops rapidly when the speed is reduced. Two-shaft machines are preferably used to drive mechanical equipment, because being able to vary the power turbine speed allows for a very elegant way to adjust the driven equipment to process conditions. Two-shaft machines allow a speed range of 50%–100% speed for the driven equipment. Again, the power output is controlled by fuel flow (and secondary controls), and a higher load will lead to higher gas producer speeds and higher firing temperatures.

Fig. 5.67 shows the influence of ambient pressure and ambient temperature on gas turbine power and heat rate. The influence of ambient temperature on gas turbine performance is very distinct. Any industrial gas turbine in production will produce more power when the inlet temperature is lower and less power when the ambient temperature gets higher. The rate of change cannot be generalized and is different for different gas turbine models. Full-load gas turbine power output is typically limited by the constraints of maximum firing temperature and maximum gas producer speed (or, in twin spool engines, by one of the gas producer speeds). Gas turbine efficiency is less impacted by the ambient temperature than the power.

The air humidity does impact power output, but only to a small degree (generally not more than 1% to 3%, even on hot days). The impact of humidity tends to increase at higher ambient conditions.

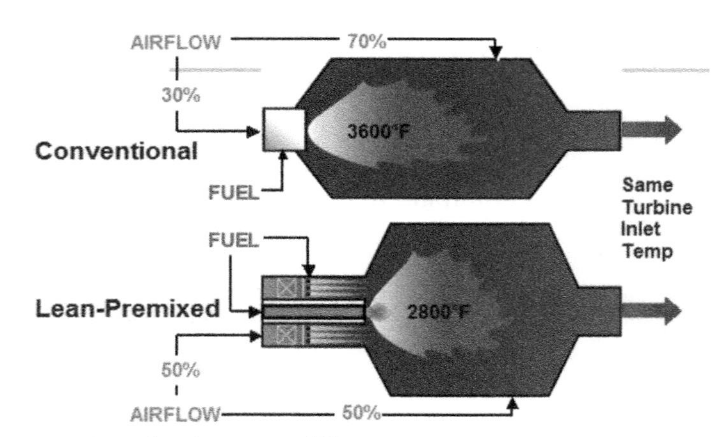

FIG. 5.66 Conventional and lean-premix combustion systems [54].

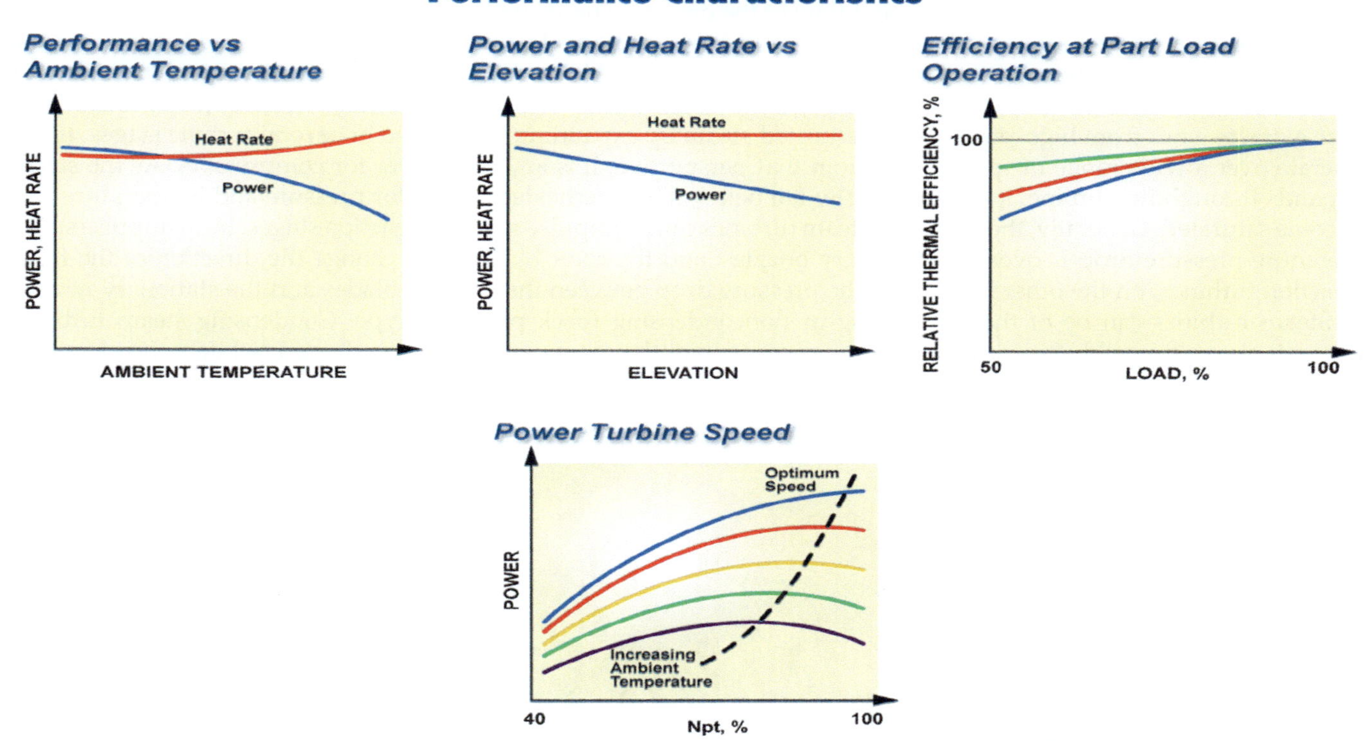

FIG. 5.67 Performance characteristics—Single-shaft and two-shaft designs *(top three figures)* and speed-power correlation for the power turbine in two-shaft designs *(bottom figure)* [53].

Lower ambient pressure (e.g., due to higher site elevation) will lead to lower power output, but has practically no impact on efficiency. It must be noted that the pressure drop (due to the inlet and exhaust systems) impacts power and efficiency negatively with the inlet pressure drop having a more severe impact.

Gas turbines operated in part load will generally lose some efficiency. Again, the reduction in efficiency with part load is very model specific. Most gas turbines show a very small drop in efficiency for at least the first 10% of drop in load. In two-shaft engines, the power turbine speed impacts available power and efficiency. For any load and ambient temperature, there is an optimum power turbine speed. Usually, lowering the load (or increasing the ambient temperature) will lower the optimum power turbine speed. Small deviations from the optimum (by say ± 10%) have very little impact on power and efficiency.

While the speed-power characteristic of a two-shaft gas turbine allows a wide speed range for the driven equipment (Fig. 5.67), a single-shaft gas turbine only allows for a very small speed range (Fig. 5.68), due to the rapid drop in output power when the speed is reduced.

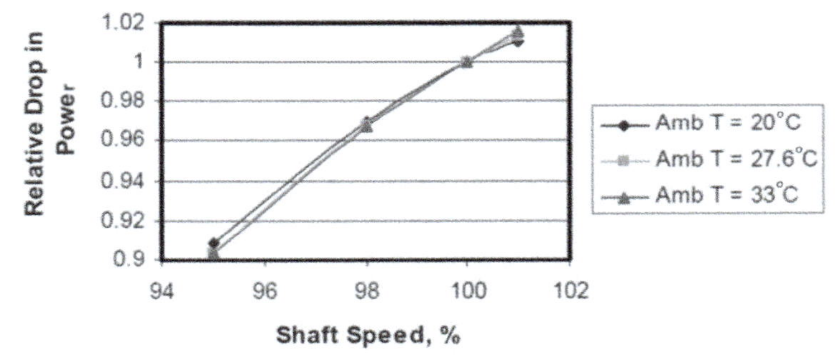

FIG. 5.68 Single-shaft power vs. speed.

Steam turbines (Fig. 5.69) are usually custom designed, so in most configurations, the centrifugal compressor is sized and optimized for the required pressure and flow of the process. The steam turbine is subsequently designed with the defined speed and power requirements of the compressor as operational points. The compressor speed is selected at the most efficient point for the compression duty; however, this speed is rarely the ideal speed for the steam turbine, and the efficiency of the driver is usually sacrificed to ensure optimum compressor operation (Fig. 5.70) [55]. A steam turbine extracts work from high-pressure, superheated steam by expanding it over one or several turbine stages. It can operate over a wide range of speeds, and from that perspective, it is an ideal driver for compressors. As the steam expands through the turbine, it will eventually fall below the superheated region for pressure and temperature and become saturated. Generally, the stages for steam turbines can be impulse stages or reaction stages. In an impulse stage, the entire pressure drop is over the stationary nozzles, and the rotor blades only change the direction of the flow. Reaction turbines, on the other hand, split the pressure drop between the rotating blades and the stationary nozzle.

Steam turbines can be of the condensing or noncondensing (back pressure) type. Condensing steam turbines expands steam to well below atmospheric pressure, where the steam is partially condensed, while noncondensing

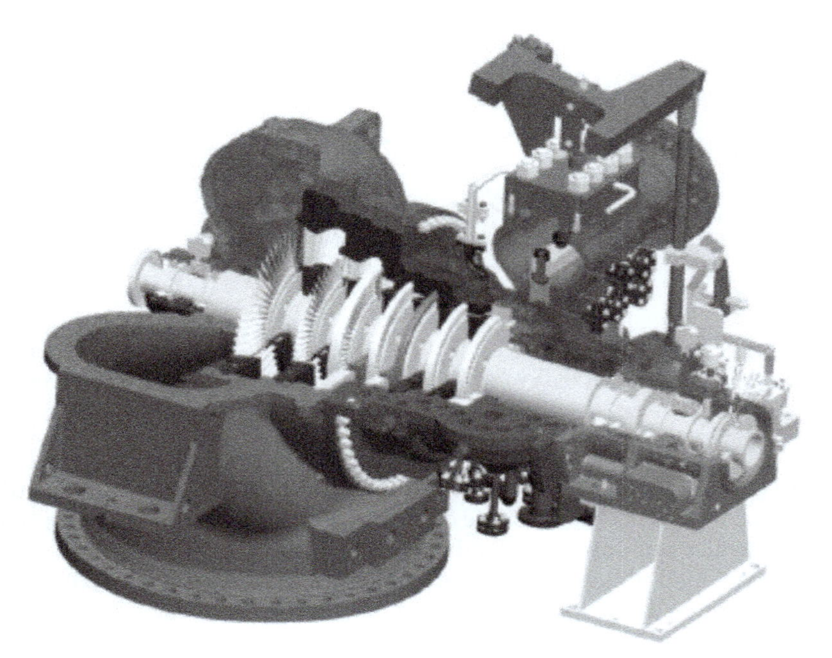

FIG. 5.69 Steam turbine [55].

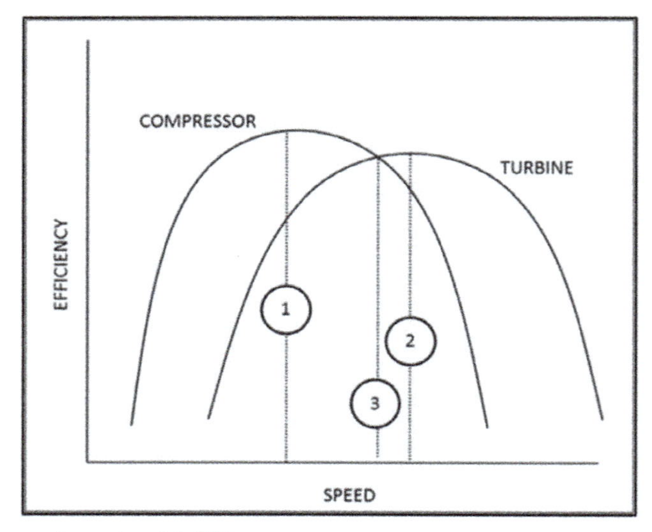

FIG. 5.70 Steam turbine and driven equipment match [55].

(back pressure) steam turbines control the turbine back pressure and use the steam for additional process purposes. Designs also may use cycles where the steam is reheated after some expansion or where steam is extracted for process purposes.

Material considerations for LNG machinery

The allowable casing material selection for centrifugal compressors is heavily influenced by the minimum design metal temperature (MDMT). Centrifugal compressor casings in LNG service may have an MDMT that ranges between −29°C and −196°C. Centrifugal compressor casings are manufactured per the requirements of API 617. This code references ASME BPVC Section VIII for general rules of construction; however, it should be noted that compressor casings are not considered to be a pressure vessel. There are typical materials of construction listed in Annex D of API 617-8th Edition, and each material has a typical low-temperature limit associated with it.

Carbon steels, low alloy steels, and stainless steels are the basic materials utilized for the compressor casings. The suitability of different steels for various temperatures is dictated by the microstructure of each grade. Carbon steels and low alloy steels have a ferritic microstructure, which is the result of atoms that are arranged in a body-centered cubic (BCC) structure. Austenitic stainless steels, nickel-based alloys, and stainless steels have austenitic microstructures where the atoms are arranged in a face-centered cubic structure.

The ductility of a material is based upon the "layers" of atoms sliding past each other, referred to as slip planes. Energy is required to allow the atoms within a slip plane to break over the saddle point of the adjacent layer when an external force, such as a tensile stress or impact, is applied. Body-centered cubic materials, such as carbon steel and low alloy steel, have a number of different slip planes, allowing the material to have excellent ductility regardless of the direction of the applied force. While there are many different slip systems available, they are thermally activated in BCC steels. As the temperature decreases, less slip systems are available, and a material will go through a ductile-to-brittle transition over a small temperature range of approximately 10°C (Fig. 5.71). The temperature which the material switches from failing in a ductile manner to a brittle manner is called the ductile-to-brittle transition temperature.

This transition will occur for every BCC steel, regardless of the steel cleanliness and heat treated condition. The ductile behavior of carbon steel without alloy additions can occur slightly below −46°C, which is why API sets this limit for carbon steels at this temperature. The addition of nickel to the steel will help to push this ductile-to-brittle transition temperature to lower temperatures (Fig. 5.72). The reason this works is that nickel acts as an austenite stabilizer within the steel, resulting in a small percentage of an austenite phase within the steel.

Austenitic alloys, such as austenitic stainless steels (300 series stainless steels), have slip systems in this microstructure arrangement and are always present and not thermally activated, which allows materials with this microstructure to retain toughness at temperatures down to −196°C. The material selection for any centrifugal compressor casing should ensure that the MDMT is below the ductile-to-brittle transition temperature to avoid a sudden and brittle fracture of the casing.

The amount of nickel that is added to the carbon steel influences how low of a temperature is reached for the ductile-to-brittle transition temperature. Carbon steel with no special alloy additions can retain toughness at −46°C. Steel alloys with 2.5% nickel can be rated for temperatures as low as −75°C, alloys with 3.5% nickel are commonly utilized to −101°C, and 9% nickel steels can be used at cryogenic temperatures. The 3.5% nickel steels are the most commonly utilized grades for compressor casings, which have an MDMT between −46°C and −101°C. When the MDMT

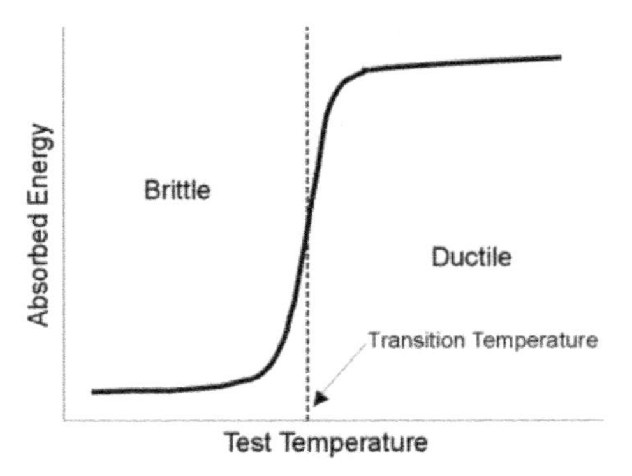

FIG. 5.71 Graph showing ductile-to-brittle transition curve of BCC metals [56].

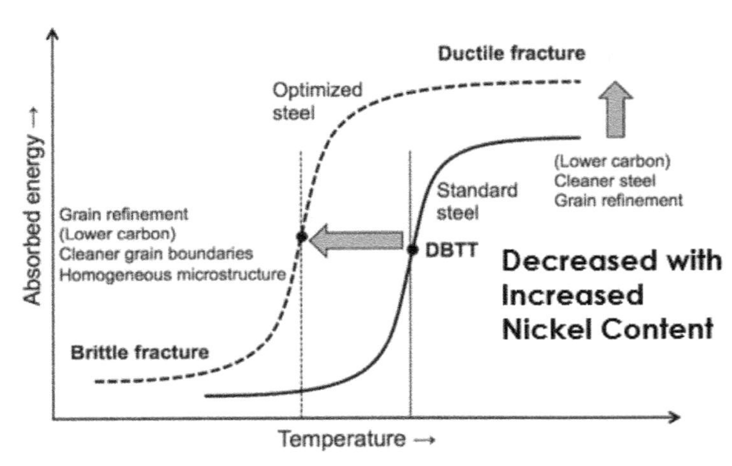

FIG. 5.72 Graph showing the ductile-to-brittle transition temperature is decreased with increased nickel content.

TABLE 5.2 Commonly utilized compressor-casing materials for low-temperature service.

Minimum temperature	Material	Plate specification	Forging specification	Casting specification
−46°C	Carbon Steel	ASTM A516 Grade 60	ASTM A350 Grade LF2	ASTM A352 Grade LCB
−75°C	2.5% Nickel Steel	ASTM A203 Grade B	Not Available	ASTM A352 Grade LC2
−101°C	3.5% Nickel Steel	ASTM A203 Grade E	ASTM A350 Grade LF3	ASTM A352 Grade LC3
−115°C	4.5% Nickel Steel	ASTM A203 Grade E Modified	ASTM A350 Grade LF3 Modified	ASTM A352 Grade LC4
−196°C	304L Stainless Steel	ASTM A204 Type 304L	ASTM A182 Grade F304L	ASTM A351 Grade CF3

temperature is below −101°C, the material utilized is often an austenitic stainless steel, such as 304L (UNS S30403) or 316L (UNS S31603). These stainless steels contain at least 8% nickel, which is sufficient to stabilize a fully austenitic microstructure at temperatures down to −196°C, and this is why the toughness of austenitic stainless steels remains high even at cryogenic temperatures. The austenitic stainless steels are significantly more expensive in comparison to nickel alloy steels; however, they may be necessary due to the MDMT requirements and are readily available for centrifugal compressor-casing fabrication. Table 5.2 shows the materials that are commonly utilized for compressor casings in low-temperature service.

For the rotating compressor components, including the compressor shaft and impellers, higher strength materials are required due to the applied stress during operation. The compressor-casing materials listed in Table 5.1 have yield strengths that are only slightly above 30 ksi (206 MPa), and yield strengths above 80 ksi (551 MPa) are necessary to achieve the required efficiency. While these are commonly more expensive in comparison with alloys utilized for the compressor casings, these higher strength alloys allow the compressor to achieve the required design criteria in less stages by allowing the spin speed to be increased. While the selection of materials is more limited for materials that offer both high yield strength and sufficient ductility at low temperatures, there are options available that meet this requirement. Commonly utilized materials for compressor shafts and impellers are given in Table 5.3.

TABLE 5.3 Commonly utilized materials for rotating compressor equipment for low-temperature service.

Minimum temperature	Material	Forging specification
−101°C	4330 Steel	Not Available
−128°C	13% Cr-4% Ni Steel	ASTM A182 Grade 6FNM
−196°C	9% Nickel Steel	ASTM A522 Type 1
−196°C	Alloy A286	ASTM A638 Grade 660
−196°C	Inconel 718	ASTM B637

Centrifugal compressor casings and rotating components are manufactured to meet the MDMT requirements, ensuring that the material retains adequate toughness to avoid brittle fracture at these temperatures. Compressor OEMs select materials to meet these minimum temperature requirements with consideration to the cost of the material, available material options per API 617, and fabrication requirements. The selection of the MDMT will influence the material that is selected for the casing fabrication and directly influences the cost to the end-user.

Capacity control considerations for LNG

There are two objectives for compressor control: meeting the external process requirements and keeping the compressor within its operational boundaries. Typical control scenarios that have to be considered are process control, starting and stopping of units, and fast or emergency shutdowns [54,57]. The control of centrifugal compressors has to be considered from both the perspective of the compressor and the perspective of the process. Regarding the compressor, it is necessary to discuss the different control devices, such as variable speed, guide vanes, throttles, or recycle valves. It is also important whether a steady-state compressor map is still valid in the case of fast transients. Furthermore, the different operating conditions of the compressor, such as surge, stall, and choke, have to be explained. The control system has to be addressed with regard to instrumentation and device requirements, as well as the control methods of the drivers.

For the process, one must understand the relationship between the flow through the system and the pressures imposed on the compressor. These relationships are different depending on their rate of change. In other words, one must expect different system responses for fast and slow changes, as well as steady-state conditions. Ultimately, the key process variables that need to be controlled to a setpoint have to be understood.

On a very fundamental basis, the key process variable in an LNG process is to bring the feed gas to a temperature where it forms a liquid. This requires to move enough heat from the feed gas to the refrigerant. Therefore, there is a refrigerant temperature that needs to be maintained at the cold point. Fig. 5.73 shows the general concept, and Fig. 5.74 shows these relationships for a nitrogen (reverse Brayton) cycle [58].

For a reverse Brayton cycle, the mass flow through the entire refrigeration loop is the same, so the expander work and the compressor expander work have to balance. To keep the end temperature at −160°C (−255°F), one has to maintain the pressure drop through the expander, regardless of the refrigerant mass flow. This can be done by a combination of speed variation and expander vane settings. Therefore, the expander work extraction is approximately constant. The driven expander compressor will also keep the same pressure ratio, as does the main compressor.

For two-phase refrigeration processes, the refrigerant composition, the available refrigerant condenser temperature, and the required evaporator temperature determine the suction and discharge pressure for the compressor.

More generally, for a given feed gas composition and given refrigerant compositions, the control concept for each refrigeration loop requires to maintain a certain refrigerant temperature. Thus,

- Keep the end temperature at the desired level (e.g., at −160°C (−255°F), i.e., the point on the lower left corner of Fig. 5.72).
- Control the refrigerant mass flow such that the LNG is cooled to the desired temperature (e.g., −156°C (−250°F) in Fig. 5.72), so that it liquefies when going through the final expansion (e.g., through a Joule-Thomson valve) [58].

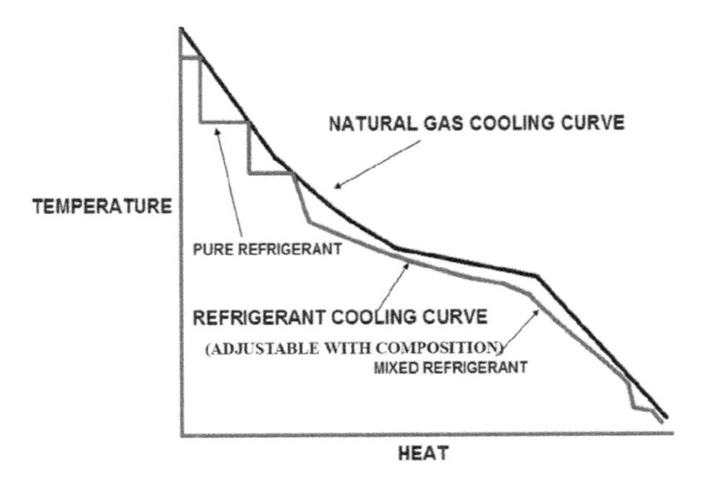

FIG. 5.73 Cooling curve for refrigerant and natural gas (feed gas).

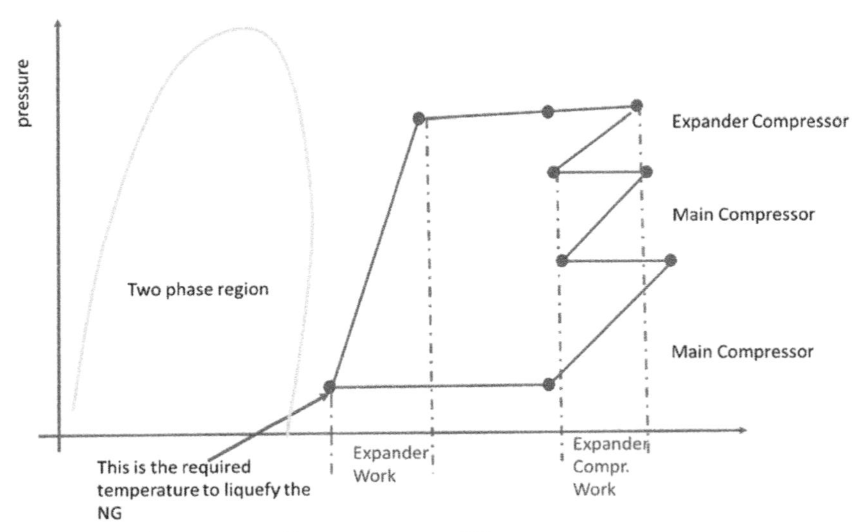

FIG. 5.74 Pressure enthalpy diagram for nitrogen refrigeration process [58].

The final pressure drop of the feed gas is determined by the feed gas pressure available at the inlet of the process and the pressure loss of the feed gas within the process. Thus, the process gas inlet pressure influences, to a small degree, the temperature to which the feed gas has to be cooled prior to the final pressure drop.

Therefore, the pressures and pressure drops within the refrigeration loops have to be maintained to keep the required process temperatures. For a given feed gas mass flow to be set, the (more or less) only other variable in the refrigeration loops is the refrigerant mass flow. This also means that the achievable refrigerant mass flow (limited either by the compressor operating range or the available driver power) determines the maximum amount of LNG that can be produced. Secondary effects to be considered are compressor efficiency for a given operating point and the variability of flow losses with mass flow. A control concern is the availability of certain power levels from the driver. If, as in most cases, the refrigeration compressors are driven by a gas turbine, the available power will be determined by ambient conditions, as discussed previously.

Since the refrigeration compressors work in closed loops, the mass flow in the refrigeration loop can be set by inventory control (i.e., the mass of refrigerant in the loop is modified), by the speed of the compressor or by adjustable compressor vanes. In all cases, the compressor operates at a more or less constant head (because as mentioned previously, the condenser and evaporator temperatures determine the compressor pressure ratio), while the compressor flow changes. Generally, the operating point of the compressor can be controlled either by their speed or by recycling gas. The compressor operating points will change, and thus, compressor sizing has to include consideration of compressor operating range (both surge margin and margin to choke), compressor speed range, and operation with recycle (surge control). In other words, operability has to be evaluated for low ambient temperature cases, as well as for high (and unusually high)-temperature cases. The primary control variable is thus usually the feed gas flow rate, which is varied to maintain a relatively constant suction pressure of the propane (or heavy MR) compressor with some flexibility due to the limited compressor speed.

In constant speed operation, the operation of a propane compressor loop (or the first mixed refrigerant loop in DMR applications) is significantly affected by ambient temperature, with high ambient operation at or near the surge control line and low ambient operation more toward overload. The discharge pressure is set by the condensing pressure of the refrigerant at existing ambient conditions. Given the very limited speed range afforded by the single-shaft gas turbine drivers, this results in operation across the compressor curve and variable head production. The subsequent compression loops (e.g., mixed refrigerant or ethylene and methane) see much less variation of operating pressures and pressure ratios. The largest variation in these loops is the circulation rate due to the variable refrigeration duty demand.

Capacity control for reverse Brayton cycles (Fig. 5.74) is somewhat different since the refrigerant is always a gas, and the expansion process provides some compressor power [58]. The mass flow through the entire refrigeration loop is the same, so the expander work and the compressor expander work have to balance. To keep the end temperature at −160°C (−255°F), one has to maintain the pressure drop through the expander regardless of the refrigerant mass flow.

This can be done by a combination of speed variation and expander vane settings. The expander work extraction is therefore about constant. The driven expander compressor will also keep the same pressure ratio, as does the main compressor. This does not take into account that efficiency of the component will change somewhat. When the refrigerant mass flow is changed, all components will still approximately run at the same head, regardless of mass flow. Therefore, the main compressor speed will drop to maintain head. To reduce the mass flow in the refrigeration loop (to accommodate a lower LNG feed), one simply reduces the speed of the main compressor and uses the expander guide vanes to maintain the expander pressure ratio (or the expander discharge temperature) [58].

Flare in the LNG process

In general, the "flare" is a device or process of safely combusting hydrocarbons into the atmosphere (as opposed to just venting a hydrocarbon gas to the atmosphere). The flare itself is normally a pipe stack or boom that is set at a height normally 40 to 100 ft above the ground to safely combust and disperse a hydrocarbon gas. The flare stack is also located onsite but away from other equipment and buildings. In some cases, ground flares or flare pits may be used as well to combust hydrocarbon gases or liquids. Flares are used and are common in all types of industries, including petroleum refineries, chemical plants, natural gas processes, oil and gas extractions, oil wells, offshore oil and gas rigs, gas wells, and landfills. Fig. 5.75 next shows a flare in operation.

The focus of this section will be the reason for flaring in the LNG process or in an LNG plant. In most typical LNG-producing facilities, a main methane loop exists where the gas is cooled and converted to LNG. To provide cooling, typically depending upon the LNG process, one or more closed-loop hydrocarbon or nitrogen refrigeration loops are used to cool the methane and produce LNG. See the section on LNG production for additional information.

During the operation of an LNG plant, some process fluid or hydrocarbon gas escapes or leaks from the main loops or cooling loops. A good example of this is compressor dry gas seal leakage (the seal is designed to leak gas). These gases are collected and sent to the plant's gas flare system where these hydrocarbon gases are safely combusted. Since the LNG plant process is a continuous process, the plant flare system is always in operation. When passing by an LNG plant, the flare system can typically be seen (especially at night).

Reasons for flaring in LNG plants

Normal Operation—As stated previously, certain equipment in an LNG plant is designed to leak hydrocarbon or other gases in normal operation. This normal leakage is collected and sent to the LNG plant's flare gas system for safe disposal or combustion.

Emergency—Occurs in some type of plant emergency conditions such as a fire or explosion (not very common). During an emergency, plant operations or a safety control system will open flare valves and send large volumes of hydrocarbon gas to the plant flare system. The object is to remove flammable gas from the main LNG loop and

FIG. 5.75 Natural gas being flared at Hess corporation gas plant in Tioga, North Dakota, due to maintenance issues.

hydrocarbon refrigeration loops to safely combust the gas in the flare system. The goal is to remove the hydrocarbons from the plant and safely combust these hydrocarbons in the flare system. As these gas loops hold a considerable amount of hydrocarbon gas, the flare stacks can be quite a sight (especially at night).

Gas Leak—If a gas leak is detected somewhere in the plant, this section or specific gas loop will be sent to flare. For example, if a hydrocarbon refrigeration loop gas leak is detected, this specific gas loop will be sent to flare without necessarily sending the complete plant inventory to flare.

Over Pressure—Within an LNG plant, many safety systems and safety valves exist. If a particular gas loop pressure exceeds the design pressure, typically a safety valve will open and send gas to the flare reducing the pressure in the loop and avoiding an over-pressure situation which could result in a mechanical failure or a rupture releasing hydrocarbon gas.

Normal Shutdown—In a normal plant shutdown, the LNG equipment including turbines, compressors, pumps, etc., are typically stopped. The gas in the main and refrigeration loops will come to some equilibrium or settle-out pressure. This may take some time as liquid hydrocarbons in the loops can heat up and vaporize. Depending upon how long the plant will be shutdown, some amount of gas in the loops may be sent to flare to maintain safe pressure levels in the gas loops.

Maintenance Shutdown—As with a normal plant shutdown, the LNG equipment including turbines, compressors, pumps, etc., will be stopped. Once the equipment is stopped for a maintenance outage, each gas loop will be flared, removing hydrocarbons from the plant so maintenance activities can take place. Typically, each gas loop is flared in sequence and not all at once.

Startup—In some LNG plants, after a shutdown, the pressure in certain gas loops is too great to restart some equipment. For example, a full-pressure gas loop start may result in a gas turbine not having the required startup torque required to start a compressor under full gas suction pressure. This necessitates gas in the loop having to be flared to reduce the pressure in the loop to allow startup. Once the turbine is up to rated speed and has the required torque, gas must be added to the loop to obtain the correct operating pressures. As a result, this process is costly from a waste of gas inventory, from an environmental perspective, and from the time required to flare the gas and re-pressurize the loop. In modern LNG plants, steps are being taken to reduce this flaring and re-pressuring process by adding equipment such as helper motors to be used in the startup process.

In general, flaring of gas is looked upon poorly by environmental groups. So, in all cases, the amount of gas that is flared is minimized and only done when necessary.

Summary

A wide range of turbomachines are used in LNG plants, including centrifugal compressors, gas turbines, steam turbines, cryo-pumps, and expanders [59,60,61]. Typical power ranges for LNG turbomachines range from single digits in small-scale distributed LNG plants up to well above a hundred of thousand horsepower in large LNG export facilities. This chapter reviewed major turbomachinery in LNG plants with a special focus on how compressors, pumps, and gas turbines fit into the various LNG plant cycles, their integration, operation, and application limits [62]. Equipment covered included boil-off gas compressors, flash gas compressors, refrigeration compressors, LNG single- and two-phase expanders, gas turbine drivers, electric motors, and power generation gas turbines and steam turbines. The chapter also described and discussed how to best match turbomachinery for optimal service in LNG plants based on power, speed and utility needs of the most common LNG cycles. Operation and maintenance challenges that are unique for LNG plants were reviewed, including safety, sealing, materials, coating, and industry standards. The types of refrigeration compression trains, their application, sizing, speed match, and package features were also discussed.

References

[1] Shell LNG Outlook, 2023. https://www.shell.com/energy-and-innovation/natural-gas/liquefied-natural-gas-lng/lng-outlook-2023.html#iframe=L3dlYmFwcHMvTE5HX291dGxvb2tfMjAyMy8.

[2] GIIGNL – The International Group of Liquefied Natural Gas Importers, The LNG Process Chain, 2019. https://giignl.org/document/giignl-information-paper-n2-the-lng-process-chain.

[3] C. Steuer, Cost of liquefaction plant projects, in: Outlook for Competitive LNG Supply, Oxford Institute for Energy Studies, 2019, pp. 8–15. http://www.jstor.org/stable/resrep31040.11. (Accessed 10 February 2023).

[4] M.A. Qyyum, K. Qadeer, M. Lee, Comprehensive review of the design optimization of natural gas liquefaction processes: current status and perspectives, Ind. Eng. Chem. Res. 57 (17) (2018) 5819–5844, https://doi.org/10.1021/acs.iecr.7b03630.

[5] M. Roberts, F. Chen, O. Saygı-Arslan, Brayton refrigeration cycles for small-scale LNG, in: Gas Processing, 2015.
[6] L. Balascak, D. Healey, W. Miller, S. Trautmann, Air Products' Technologies for Small Scale LNG, Small Scale LNG, Oslo, Norway, 2011.
[7] J.C. Bronfenbrenner, Selecting a Suitable Process, LNG Industry, 2009.
[8] S. Mokhatab, D. Messersmith, Liquefaction technology selection for baseload LNG plants, Hydrocarbon Processing Magazine (2018).
[9] D.L. Andress, R.J. Watkins, Beauty of Simplicity: Phillips optimized cascade LNG liquefaction process, Am. Inst. Phys. Conf. Proc. 710 (1) (2004).
[10] M. Nouri, J. Rizopoulos, Condensate stabilization: how to get the most for your money, in: AIChE 2013 Spring Meeting, Houston, TX, 2013.
[11] A. Musardo, M. Pelella, P. Vinod, M. Weatherwax, G. Giovani, S. Cipriani, CO_2 compression at world's largest carbon dioxide injection project, in: Proceedings of the Second Middle East Turbomachinery Symposium, Doha, Qatar, 2013.
[12] K. Brun, S. Ross, B. Pettinato, T. Omatick, J. Thorp, Application of Hybrid Centrifugal Compressor and Pump Packages for Carbon Sequestration CO_2 Compression, Turbomachinery & Pump Symposia, Houston, TX, 2022.
[13] H. Fujieda, M. Iwamoto, Under the scenes of our lives high-pressure pump – CO_2 injection pump, Ebara Eng. Rev. 252 (2016).
[14] M. Taher, C.B. Meher-Homji, Design considerations for high pressure boil-off gas (BOG) centrifugal compressors with synchronous motor drives in LNG liquefaction plants, in: GT2019-90329, Proceeding of ASME Turbo Expo Phoenix, Arizona, USA, 2019.
[15] M. Taher, F. Evans, Centrifugal compressor Polytropic performance – improved rapid calculation results – cubic polynomial methods, Int. J. Turbomach. Propuls. Power 6 (2) (2021) 15, https://doi.org/10.3390/ijtpp6020015.
[16] E.W. Lemmon, I.H. Bell, M.L. Huber, M.O. McLinden, NIST Standard Reference Database 23: Reference Fluid Thermodynamic and Transport Properties-REFPROP, Version 10.0, National Institute of Standards and Technology, Standard Reference Data Program, Gaithersburg, MD, 2018.
[17] M. Taher, F. Evans, Centrifugal compressor polytropic performance evaluation: Taher-Evans methods, in: GT2023-104213, Proceeding of ASME Turbo Expo, Boston, Massachusetts, USA, 2023.
[18] Air Products, Large to mega-scale LNG plant processes and equipment, Liquefact. Nat. Gas (2021). https://www.airproducts.com/industries/lng.
[19] A. Brimm, S. Ghosh, D.J. Hawrysz, Operating Experience with the Split MR Machinery Configuration of the C3MR LNG Process, SPE Projects, Facilities & Construction, 2006, pp. 1–5.
[20] ConocoPhillips, Optimized Cascade Process, 2022. https://lnglicensing.conocophillips.com/what-we-do/lng-technology/optimized-cascade-process/.
[21] N.B. Khan, Process Efficiency Optimisation of Cascade LNG Process, Curtin University, 2018.
[22] C.B. Meher-Homji, D. Messersmith, H.P. Weyermann, G. Richardson, P. Pattrick, F.R. Biagi, F. Gravame, World's first aeroderivative based LNG liquefaction plant – design, operational experience and debottlenecking, in: First Middle East Turbomachinery Symposium, Doha, Qatar, 2011.
[23] J. Beard, M. Roberts, B. Kennington, US Equipment-Maker Air Products Employs Proven Technology for Larger Plants Now Being Proposed, LNG, 2018, pp. 48–50.
[24] S. Mokhatab, J.Y. Mak, J.V. Valappil, D.A. Wood, Natural Gas Liquefaction, Gulf Professional Publishing, 2014.
[25] Air Products, Mid-scale LNG capabilities for capacity from 0.25-2.0 MTPA designed for simplicity while delivering lower unit costs, Liquefact. Nat. Gas (2018). https://www.airproducts.com/industries/lng.
[26] S. Corbo, A. Guglielmo, R. Valente, G. Iurisci, New Challenges and Design for High Mach High Flow Coefficient Impeller for Large Size LNG Plant, Turbomachinery Laboratory, Texas A&M Engineering Experiment Station, 2018. https://hdl.handle.net/1969.1/175000.
[27] V. Jariwala, R. Chundru, B. Pettinato, Development and Application of Very High Flow Covered Stages for Process Centrifugal Compressors, Turbomachinery Laboratory, Texas A&M Engineering Experiment Station, Texas A & M University. Libraries, 2021. https://hdl.handle.net/1969.1/196728.
[28] M. Sandberg, C. Meher-Homji, J. Beard, Compressor selection for LNG liquefaction plants, in: Proceedings of the 38th Turbomachinery Symposium. Turbomachinery Laboratory, Texas A&M Engineering Experiment Station, 2019.
[29] M. Sandberg, Centrifugal compressors: matching the configuration to the application, Turbomach. Int. 58 (5) (2017) 20–24.
[30] J.A. Kocur, M.G. Muench, Impact of Electrical Noise on the Torsional Response of VFD Compressor Trains, Texas A&M University, Turbomachinery Laboratories, 2012. https://hdl.handle.net/1969.1/162983.
[31] L.S. Piergiovanni, E. Lerch, R. Zurowski, S.B. Kumar, R. Osman, B. Deo, B. Bahr, J. Sakaguchi, Y. Okazaki, T. Saito, Investigation of subsynchronous torsional interaction on LNG power plants, in: Proceedings of the 16th International Conference & Exhibition on Liquefied Natural Gas, Oran, Algeria, 2010.
[32] P. Rotondo, D. Andreo, S. Falomi, P. Jörg, A. Lenzi, T. Hattenbach, D. Fioravanti, S. De Franciscis, Combined Torsional and Electromechanical Analysis of an LNG Compression Train with Variable Speed Drive System, Texas A&M University, 2009. https://hdl.handle.net/1969.1/163110.
[33] B. Pettinato, J. Wilkes, Short Course 5: Lateral and torsional rotordynamics of machinery, in: Proceedings of the Fourth Biennial Asia Turbomachinery and Pump Symposium, Turbomachinery Laboratory, Texas A&M University, College Station, Texas, 2022.
[34] B.G. Heckel, F.W. Davis, Starter Motor Sizing For Large Gas Turbine (Single-Shaft) Driven LNG Strings, Texas A&M University, Turbomachinery Laboratories, 1998. https://hdl.handle.net/1969.1/163400.
[35] V. Patel, S. Rush, Submerged pumps and expanders with magnetic coupling for hazardous applications, in: Proceedings of the First Middle East Turbomachinery Symposium, Doha, Qatar, 2011.
[36] E.S. Karakas, R. Mollath, Cavitation Performance Improvement of an Industrial Cryogenic Centrifugal Pump by Implementing Variable Pitch Inducer, Turbomachinery Laboratory, Texas A&M Engineering Experiment Station; Texas A & M University. Libraries, 2021. https://hdl.handle.net/1969.1/196754.
[37] Bechtel Corporation, Corpus Christi Liquefaction Construction-Projects-Bechtel, Bechtel Corporation, Reston, VA, 2019. https://www.bechtel.com/projects/corpus-christi-liquefaction-project/. (Accessed 10 August 2021).
[38] D.A. Coyle, V. Patel, Processes and Pump Services in the LNG Industry, Texas A&M University, Turbomachinery Laboratories, 2005. https://hdl.handle.net/1969.1/163965.
[39] V.P. Patel, H.E. Kimmel, Fifteen years of field experience in LNG expander technology, in: Proceedings of the First Middle East Turbomachinery Symposium, Doha, Qatar, 2011.

[40] Qualls, W.R., Eaton, A.P., and Meher-Homji, C.B., n.d. "Liquid Expanders in the Phillips Optimized Cascade LNG Process," https://static.conocophillips.com/files/resources/smid_016_liquidexpanders.pdf.
[41] H.E. Kimmel, C. Chiu, H. Paradowski, Economic and environmental benefits of two-phase LNG expanders, in: Proceedings of the 14th International Conference and Exhibition of Liquefied Natural Gas, Doha, Qatar, 2004.
[42] E.S. Karakas, S. Ross, Refocus the energy, LNG Ind. Magaz. (2020).
[43] Y. Kikkawa, H.E. Kimmel, Interaction between liquefaction process and LNG expanders, in: Proceedings 2001 AICHE Spring National Meeting, Natural Gas Utilization Topical Conference, Houston, TX, 2001.
[44] J. Verkoehlen, Initial experience with LNG/MCR expanders in MLNG-Dua, in: Proceedings GASTECH 96, Vol. 2, Vienna, Austria, 1996.
[45] H.E. Kimmel, Variable speed turbine generators in LNG liquefaction plants, in: The 17th LNG/LPG/Natural Gas Conference & Exhibition, Vienna, Austria, 1996.
[46] L.G. Weisser, Hydraulic Turbine Power Generator Incorporating Axial Thrust Means, 1997. US Patent 5,656,205.
[47] K. Cholast, A. Kociembia, J. Heath, Two phase expanders replace Joule-Thomson valves at nitrogen rejection plants, in: 5th World LNG Summit, 2004.
[48] E.S. Karakas, D. Stasenko, Two-phase cryogenic expander for nitrogen rejection and cooling process in natural gas applications, in: Proceedings of ASME Turbo Expo, 2023. GT2023-101990.
[49] K. Cholast, K. Kociemba, H. Isalski, J. Madison, J. Heath, Two-phase LNG expanders, in: Gas Processors Association – GTL and LNG in Europe, Amsterdam, Netherlands, 2005.
[50] C. Houser, R.-J. Lee, N. Baudat, Efficiency Improvement of Open-cycle Cascaded Refrigeration Process for LNG Production, 2001. International Application Published Under the Patent Cooperation Treaty (PCT), Publication Number: WO 01/46634 A1.
[51] C. Meher-Homji, P. Pillai, R. Kurz, P. Rasmussen, LNG liquefaction plants-overview, design and operation, in: 45th Turbomachinery Symposium, Houston, TX, 2016.
[52] R. Kurz, J. Mistry, P. Davis, G.J. Cole, Application and Control of Variable Speed Centrifugal Compressors in the Oil and Gas Industry, 2020. PCIC2020-AT-119.
[53] R. Kurz, D. Reitz, D. Burnes, Performance of industrial gas turbines, in: 51st Turbomachinery Symposium, Houston, TX, 2022.
[54] R. Kurz, K. Brun, C. Meher-Homji, J. Moore, Gas turbine performance and maintenance, in: 41st Turbomachinery Symposium, Houston, TX, 2012.
[55] K. Brun, B. Pettinato, S. Ross, D. Bauer, A. Neil, R. Kurz, J. Thorp, Turbomachinery for Refinery Applications, Turbomachinery & Pump Symposia, Houston, TX, 2021.
[56] UNSW Sydney, 2020. http://www.materials.unsw.edu.au/tutorials/online-tutorials/2-temperature-impact. (Accessed 29 January 2020).
[57] R. Kurz, R.C. White, K. Brun, Surge control and dynamic behavior for centrifugal gas compressors, in: 3rd Middle East Turbomachinery Symposium, Doha, Qatar, 2015.
[58] R. Kurz, M. Ji, G. Beck, T. Allison, Small scale LNG concepts, in: ASME Paper GT2021-58989, 2021.
[59] K. Brun, R. Kurz, Compression Machinery for Oil and Gas, first edition, Gulf Professional Publishing, 2018.
[60] T. Omatick, K. Carpenter, The Heart of Liquefaction, LNG Industry, 2014.
[61] J. Wilkes, B. Pettinato, R. Kurz, J. Hollingsworth, D. Zhang, M. Taher, C. Kulhanek, F. Werdecker, D. Buche, G. Talabisco, Centrifugal compressors, in: Compression Machinery for Oil and Gas, 2019, pp. 31–133.
[62] K. Brun, R. Kurz, M. McBain, B.J. Setzenfand, E.S. Karakas, B. Pettinato, T. Omatick, D. Bauer, S. Ross, B. Hantz, J. Thorp, Turbomachinery in Liquefied Natural Gas Production Plants, Turbomachinery & Pump Symposia, Houston TX, 2023.

CHAPTER

6

Transport of natural gas

Avneet Singh[a], Justin Hollingsworth[b], Terry Kreuz[c], Karl Wygant[d], Rainer Kurz[a], and Kelsi Katcher[b]

[a]Solar Turbines, San Diego, CA, United States [b]Southwest Research Institute, San Antonio, TX, United States [c]National Fuel Gas Company, Williamsville, NY, United States [d]Ebara Elliott Energy, Jeannette, PA, United States

Value chain description—Gas gathering to use [1,2]

Overview

Natural gas is primarily methane (CH_4) with smaller quantities of other hydrocarbons. It was formed millions of years ago when dead organisms sunk to the bottom of the ocean and were buried under deposits of sedimentary rock. Subject to intense heat and pressure, these organisms underwent a transformation in which they were converted to gas over millions of years.

Natural gas is found in underground rock formations called reservoirs. The rock has tiny spaces called pores that allow them to hold water, natural gas, and sometimes oil. The natural gas is trapped underground by an impermeable rock called a cap rock and stays there until it is extracted.

Natural gas can be categorized as dry or wet. Dry gas is essentially gas that contains mostly methane. Wet gas, on the other hand, contains compounds such as ethane and butane in addition to methane. These natural gas liquids, or NGLs for short, can be separated and sold individually for various uses, such as in refrigerants and to produce products such as plastic.

Conventional natural gas can be extracted through drilling wells. Unconventional forms of natural gas such as shale gas, tight gas, sour gas, and coal bed methane have specific extraction techniques. Natural gas can also be found in reservoirs with oil and is sometimes extracted alongside oil. This type of natural gas is called associated gas. In the past, associated gas was commonly flared or burned as a waste product, but in most places today, it is captured and used.

Once extracted, natural gas is sent through small pipelines called gathering lines to processing plants, which separate the various hydrocarbons and fluids from the pure natural gas to produce what is known as pipeline quality dry natural gas before it can be transported. Processing involves four main steps to remove the various impurities: oil and condensate removal, water removal, separation of natural gas liquids, sulfur, and carbon dioxide removal. Gas is then transported through pipelines called feeders to distribution centers or stored in underground reservoirs for later use.

In some cases, gas is liquefied for shipping in large tankers across oceans. This type of gas is called liquefied natural gas, or LNG. Natural gas is mostly used for domestic or industrial heating and to generate electricity. It could also be compressed and used to fuel vehicles, and it is a feedstock for fertilizers, hydrogen fuel cells, and other chemical processes.

Natural gas development, especially in the United States, has increased as a result of technological advances in horizontal drilling and hydraulic fracturing.

When natural gas is burned, there are fewer greenhouse gas emissions and air pollutants when compared to other fossil fuels because methane has only one carbon atom for four hydrogen atoms. In fact, when used to produce electricity, natural gas emits approximately half the carbon emissions of coal. Despite fewer emissions, natural gas is still a source of CO_2.

In addition, methane is a potent greenhouse gas itself, having nearly 24 times the impact of CO_2. During the extraction and transportation process, natural gas can escape into the atmosphere and contribute to climate change. Natural gas leaks are also dangerous to nearby communities because they are colorless, odorless, toxic, and highly explosive.

Energy Transport Infrastructure for a Decarbonized Economy
https://doi.org/10.1016/B978-0-443-21893-4.00002-7

Copyright © 2025 Elsevier Inc. All rights are reserved, including those for text and data mining, AI training, and similar technologies.

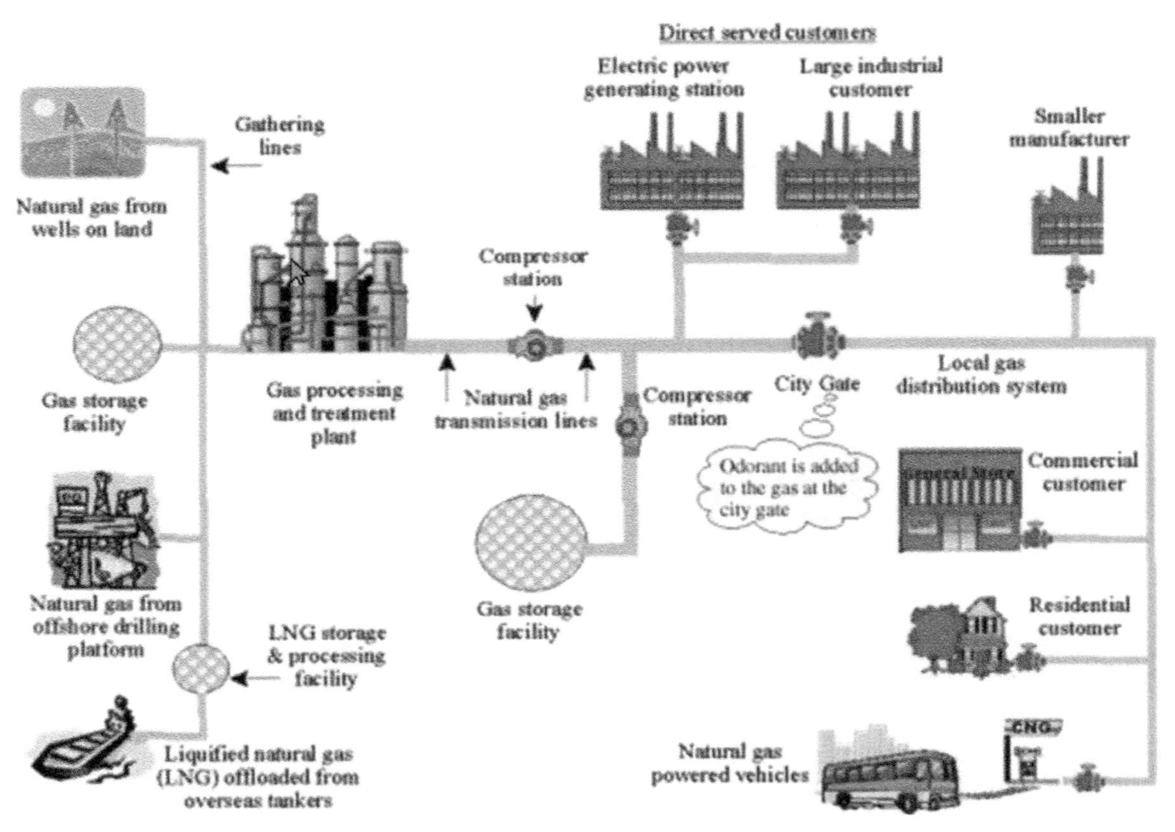

FIG. 6.1 Natural gas value chain [3].

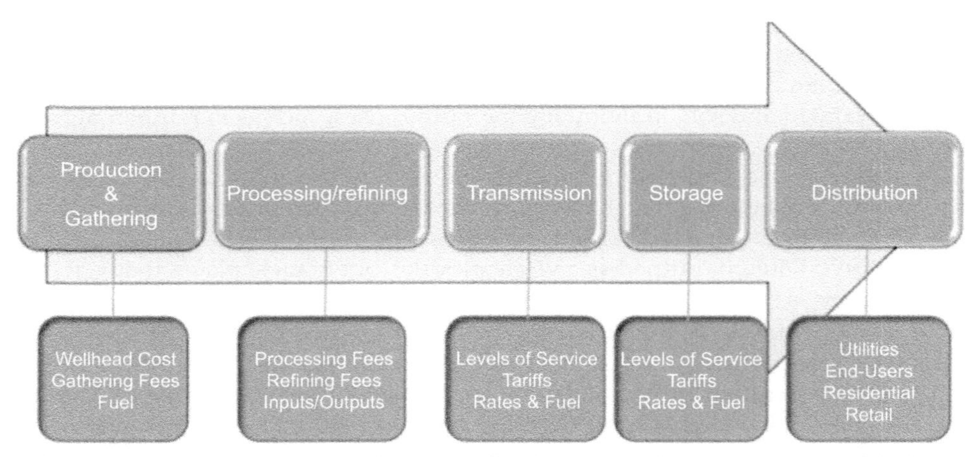

FIG. 6.2 Value chain for energy commodities [4]. *Used with permission from "Value Chain for Energy Commodities" by the John A. Dutton Institute for Teaching and Learning Excellence, The Pennsylvania State University. This image is licensed under CC BY-NC-SA 4.0.*

Figs. 6.1 and 6.2 show the value chain of natural gas and its uses.

Gas gathering

It would be impractical to connect each gas production well directly to the long-haul transportation grid. So, gas production is gathered at a collection point in the production field and processed into a marketable commodity before being delivered to a major pipeline. This is performed by an operation or set of providers often called "field services."

Gathering systems are a group of small pipelines that move gas (or oil) from wells and fields into a major processing facility or pipeline.

The gathering lines of the late 1800s and early 1900s were short, inefficient, and leaky. As onshore production expanded, gathering lines became much longer and more sophisticated, built with high-strength steel pipe, compression, measuring, and pressure regulation devices.

Specialized pipe-laying equipment was originally developed for gathering operations in the shallow offshore waters (15–60 m) of the US Gulf of Mexico. All over the world today, high-pressure gathering lines are laid at great water depths (over 1000 m), called subsea completions, to connect deep water production to onshore pipeline systems.

Gas processing

Natural gas at the wellhead or casing head gas has a widely varying composition depending on the production field and its associated reservoir characteristics. The safe, efficient operation of the natural gas market requires that all gas be processed (treated) to meet a consistent set of minimum quality standards, called *pipeline quality*.

Pipeline quality gas is free of liquid hydrocarbons, solids, water vapor, and contaminants and has a heat content within the pipeline-specified range, measured in Btu per cubic foot.

To meet the standards, the first step at the field processing facility is to remove any water and natural gas condensate. Typically, any wastewater is collected and treated before being sent back to the well or offsite for wastewater disposal. All contaminants, hydrogen sulfide, and carbon dioxide, mercury, and nitrogen are also removed.

At the next process step, the gas plant receives the gas condensate and separates it into marketable and valuable natural gas liquids (NGLs).

Finally, any NGLs—propane, butane, and pentanes—are separated from the natural gas stream, recovered, stored, and transported to liquefied petroleum gas (LPG) and petrochemical end-user markets.

Midstream activities

The midstream is the term for the function that provides the vital link between far-flung producing areas and the population centers where most consumers are located. Gathering, processing, fractionation, storage, and transportation pipeline companies are a major part of the midstream industry.

Midstream handles natural gas, natural gas liquids, and sulfur.

Fractionation plants, which remove NGLs from the produced oil and gas streams, are also key components of midstream activities. Here, the processed liquids are fractionated into products such as ethane, butane, propane, and LPG to be marketed as chemical feed stocks, fuel, or blend stocks for gasoline.

The pure natural gas is then compressed for transportation via pipeline.

LNG operations are often considered to be part of the midstream.

Gas transmission

There are various types of gas pipeline systems used throughout the world, as described below:

Production area (gathering) systems: When significant new reserves are discovered outside the established transportation system, gas producers often negotiate with pipeline companies to build pipeline facilities connecting the new production area to an existing long-haul pipeline.

Long-haul systems: Long-haul pipeline systems transport gas from major gas basins to major market areas. Pipeline diameters can range from 12 to 50 in. Compressor stations are needed every 70–100 miles for transmission.

Short-haul systems: Major urban or industrial areas that have built up in regions with significant gas supplies may be served by a grid of short-haul pipelines.

Reticulated systems: In some cases, these grids are reticulated, that is, the flow can easily be reversed to provide maximum flexibility to shippers.

Market area systems: Similarly, pipeline companies regularly look for opportunities to build extensions off their existing long-haul systems to new or rapidly growing market areas.

Gas storage

Natural gas can be stored underground in depleted reservoirs, aquifers, and salt caverns to balance seasonal demands. Depleted reservoirs are the most economically efficient storage facilities because their geology is known, and the extraction equipment is in place.

Aquifers naturally store water, but they can also be reconditioned to store natural gas. This is the most expensive storage option because:

- Determining their gas storage suitability requires time-consuming and expensive geologic studies, and
- Extensive extraction and gas retention infrastructure, including dehydration facilities, are required.

Through a process known as leaching, the salt in an underground dome can be extracted, leaving a strong structure for storing natural gas. Leaching is expensive, and salt domes are smaller than the other two types of storage facilities. However, gas can be injected and withdrawn from them much more quickly, providing more flexibility to the pipeline operator. Salt domes also retain gas much better than aquifers.

Storage strategy and services

Because the gas transmission business often has high seasonal fluctuations, well-placed gas storage helps pipelines maximize their ability to provide steady service to customers.

In North America, the highest demand for natural gas is in the winter as a result of home heating. In the past demand for natural gas was reduced in summer, so pipeline deliveries were lower. In recent years, because of increased demand from gas-fired power plants for air conditioning, demand has become somewhat less seasonal.

To maximize the use of pipeline capacity all year and create additional flexibility in a pipeline system, pipeline companies operate gas storage fields. Here, the operator uses gas from the pipeline to inject natural gas into the storage field when demand is low (summer) and withdraw it from the storage field during times of high demand (winter).

In the United States, natural gas storage is often contractually controlled by the pipeline's major customers.

Local distribution

In metropolitan areas, local distribution company (LDC) or local distribution zone (LDZ) companies are the last link in the natural gas value chain; they receive natural gas from long-haul pipelines and deliver it to thousands of homes, offices, stores, and industrial facilities.

This distribution system is another grid of small-diameter, low-pressure pipe. In contrast to the steady flow of gas in major pipelines, metropolitan gas is distributed on an as-needed basis. Power generators take delivery directly from a major pipeline because they can take large quantities of gas in a steady flow.

In addition, distributors add an odorizing agent, mercaptan, so that leaks are immediately recognizable.

The history of mercaptan addition started in 1937; in the small oil town of New London, Texas, a public school was heated with natural gas. At that time, delivered natural gas was odorless and colorless. On March 18, a gas leak was triggered by a spark, and the whole school exploded. Over 300 children and their teachers were killed.

As a direct result of this catastrophe, the US government passed a law requiring that the chemical "mercaptan" be put into natural gas to give it an identifying odor.

Liquified natural gas (LNG)

The newest aspect of the global natural gas business is LNG, and the chart shows a typical LNG value chain.

- Once received from the production facilities, the LNG processing starts with a liquefaction plant, at which the natural gas is cooled to a temperature of −160°C (−260°F) before storage and shipment.
- In this liquid state, LNG occupies only 1/600th of the volume it occupies as a gas. The result is a stable, high-BTU product that can be stored and shipped via specialized tankers to high-demand markets throughout the world.
- At the receiving location, a terminal called a regasification facility converts the LNG from its liquid state, ready to be transferred to a pipeline for transport to end-user customers.

Countries of origin for LNG liquefaction and exports include Algeria, Australia, Indonesia, Libya, Malaysia, Nigeria, Oman, Qatar, Trinidad, and Tobago. Qatar's LNG business has expanded the most rapidly of any of the producers.

LNG transport

After liquefaction, LNG is generally loaded onto specialized ocean-going tankers for shipment to global markets. These tankers carry double-hulled pressure insulated storage tanks that keep the LNG in its liquid form for safe transport.

Depending on the point of origin and final destination, these tankers can be in transit for several weeks before delivery is accomplished. Therefore, the carriers are insulated to limit the amount of LNG that "boils off" or evaporates during the voyage.

Current LNG carriers are typically up to 300 m long and require a minimum water depth of 15 m when fully loaded, with a load capacity of 125,000 cubic meters to 145,000 cubic meters.

More recently, there has been a very large increase in the sizes of vessels ordered, to as much as 250,000 cubic meters. Larger vessels have the advantage of reducing transportation and overall, LNG delivery costs. This size increase, however, will affect the design of the LNG plants and terminals.

By 2008, more than 140 LNG tankers were in service, delivering over 120 million metric tons per year.

LNG receipt and regasification

As Fig. 6.1 shows, regasification terminals can receive the LNG from the vessels into storage tanks, with enough storage tanks to both accommodate a vessel and provide uninterrupted flow into the natural gas pipeline.

Here, the imported LNG is then "regasified" through a series of pressure and temperature changes.

Regasification facilities tend to be very large and expensive. Newer terminals in the United States can deliver 1 BCF per day or more and can cost upward of US $1 billion per terminal.

From a commercial perspective, capacity at regasification terminals is governed by a terminal use agreement. Once regasified, the natural gas product is ready to be injected into a pipeline and moved to end-user markets.

Storage facilities—Surface

Surface storage facilities are the easiest part of the midstream to observe.

LNG above-ground tanks use an inner membrane or a steel-nickel inner tank lined with concrete to contain and help insulate the (very cold) LNG prior to regasification.

The common LPG sphere is found at NGL recovery facilities (often called gas plants) and refineries. It is round because that is the most structurally sound facility design to keep LPG under slight pressure.

Future importance of LNG

As crude oil reserves become increasingly difficult to produce and keep up with growing demand, natural gas will begin to replace fuel oil and diesel, especially in power generation.

Historically, natural gas needed to be tied to a pipeline to a regional market. The rapid pace of investment by gas producing countries in LNG capacity has significantly altered the value chain for natural gas. The importance of LNG to the supply chain is growing, especially in Europe, Japan, and China as indigenous production of gas in some countries (like the United Kingdom) declines, or pipeline supply networks are disrupted. Europe's support of the Kyoto Protocol and other carbon emission initiatives will also accelerate the use of gas.

This brings new opportunities for liquefied natural gas. LNG offers a global supply advantage and flexibility in the marketplace. Therefore, there are now investments around the world in liquefaction terminals, cryogenic ships, receiving terminals, and re-gasification plants.

Renewable natural gas

Renewable natural gas (RNG), also known as biomethane or green gas, is a sustainable alternative to traditional fossil-based natural gas. Derived from organic waste materials, RNG presents an opportunity to reduce greenhouse gas emissions, increase energy security, and promote a circular economy. This section explores the potential of RNG as a clean energy source, its production methods, applications, advantages, and challenges, as well as its future prospects. As organic waste decomposes, it releases biogas that is 40%–60% methane (CH_4). This biogas can be captured and refined to remove contaminants and increase its heat content. The resulting gas, RNG, can be used in place of or mingled with geologic or fossil natural gas (NG) in pipelines, fueling stations, and storage tanks, or as a "drop-in" fuel requiring no engine modifications in NG vehicles.

Biogas is the raw gas produced by the breakdown of organic materials in an oxygen-free (anaerobic) environment. After the removal of contaminants and other gases, biogas becomes RNG, which is typically 90%+methane. Biomethane is another name for RNG.

Not all biogas is converted to RNG. On farms, animal waste is often allowed to decompose in pits or ponds, where it produces methane. Methane, a powerful greenhouse gas (GHG) with 24 times the global warming potential of carbon dioxide (CO_2), is often released into the atmosphere.

At landfills and water resource recovery facilities (WRRFs), biogas is produced from the breakdown of organic waste and typically "flared" to convert its methane content to CO_2, which reduces (but does not eliminate) its global warming potential. Food waste from restaurants, institutions, and industrial food processors that is not delivered to biogas digesters or composted usually goes to landfills, where it, too, can release methane and CO_2 into the atmosphere.

The increasing demand for cleaner energy sources has spurred the development of RNG as an alternative to conventional natural gas. RNG is a versatile energy source produced from decomposing organic matter, such as agricultural, municipal, and industrial waste, as well as wastewater and landfills. By harnessing the biogas generated during the decomposition process, RNG can be converted into a valuable energy source, reducing greenhouse gas emissions and contributing to a more sustainable energy future.

There are two primary methods for producing RNG: anaerobic digestion and gasification.

- Anaerobic digestion: This biological process involves the decomposition of organic materials by microorganisms in an oxygen-free environment. The resulting biogas, primarily composed of methane and carbon dioxide, is captured and further purified to remove impurities and increase the methane content. The purified gas, referred to as RNG or biomethane, can then be injected into the natural gas grid or used for various energy applications.
- Gasification: In this thermochemical process, organic materials are exposed to high temperatures and controlled amounts of oxygen or steam, breaking down the waste into a synthesis gas (syngas) composed of carbon monoxide, hydrogen, and other trace components. The syngas is then further processed and purified to produce RNG.

RNG can be utilized in various applications, including:

- Power generation: RNG can be used as a fuel for electricity generation in combined heat and power (CHP) plants, contributing to a cleaner energy mix and reducing reliance on fossil fuels.
- Transportation: RNG can serve as a cleaner alternative fuel for natural gas vehicles, reducing greenhouse gas emissions and air pollution while also lowering fuel costs.
- Heating: RNG can be injected into the natural gas grid, replacing conventional natural gas for residential, commercial, and industrial heating purposes.
- Industrial processes: RNG can be used as a renewable feedstock in various industrial processes that require heat, steam, or chemical reactions.

RNG offers several advantages over conventional natural gas, including:

- Reduced greenhouse gas emissions: RNG production captures methane, a potent greenhouse gas, that would otherwise be released into the atmosphere. By using RNG in place of fossil-based natural gas, net greenhouse gas emissions can be significantly reduced.
- Energy security: RNG production relies on local waste resources, reducing dependence on imported fossil fuels and promoting energy security.
- Waste management: RNG production helps manage organic waste, diverting it from landfills and reducing the need for incineration, both of which have environmental impacts.
- Job creation: The RNG industry can create jobs in the areas of waste collection, processing, and transportation, as well as facility construction, operation, and maintenance.

Despite its potential, RNG faces several challenges that need to be addressed:

- Production costs: The cost of RNG production, especially in small-scale facilities, can be higher than that of conventional natural gas due to the need for specialized equipment and infrastructure.
- Infrastructure: RNG production and distribution require significant investments in infrastructure, including upgrading existing natural gas pipelines and building new ones.
- Feedstock availability: The availability of organic waste materials can be influenced by seasonal fluctuations and regional differences, posing challenges to consistent RNG (Figs 6.3 and 6.4).

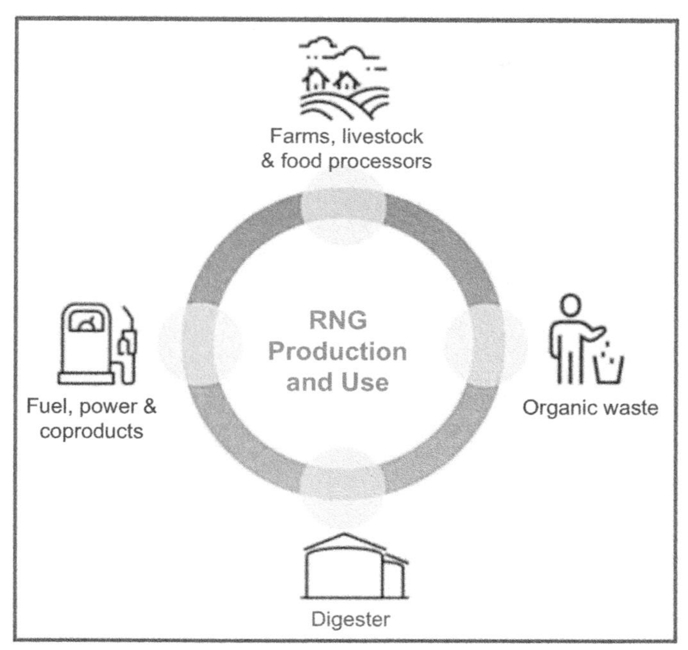

FIG. 6.3 RNG production and use [5].

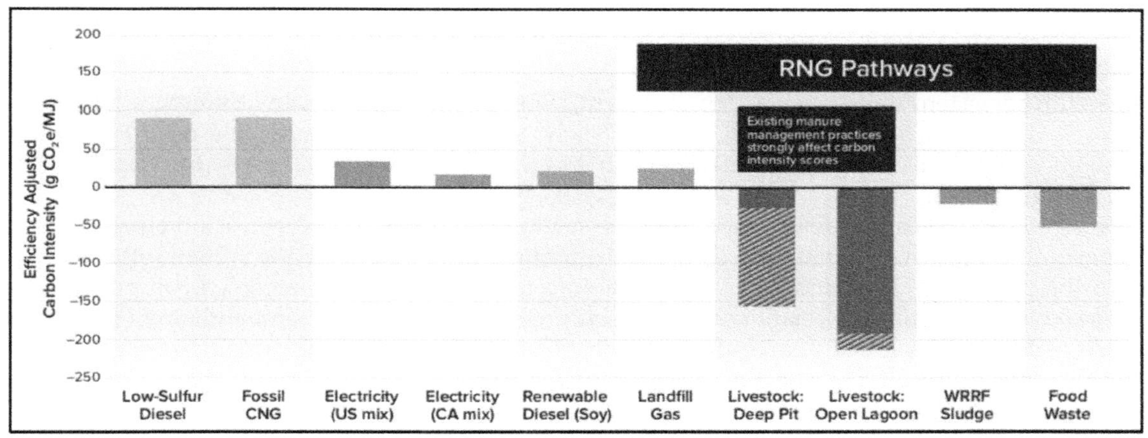

Some RNG pathways have very low carbon intensity (CI) scores because they capture emissions that would otherwise be released to the atmosphere. For farms with manure lagoons that currently emit high levels of methane, RNG production can yield negative CI scores. The diagonal-line overlays on bars represent the *range* of carbon intensity scores that can be achieved with corresponding RNG projects. (CA = California; CNG = compressed natural gas; CO_2e = carbon dioxide equivalent; g = gram; MJ = megajoule; RD = renewable diesel; WRRF = water resource recovery facility.)

FIG. 6.4 Carbon intensity (CI) of RNG pathways [5].

Compressors—Centrifugal compressors

The thermodynamics of gas compression apply to any type of compression, regardless of dynamic or positive displacement compressors. For a compressor receiving gas at a certain suction pressure and temperature, and delivering it at a certain discharge pressure, the isentropic head represents the energy input required by a reversible, adiabatic (thus isentropic) compression. The actual compressor will require a higher amount of energy input which is needed for ideal (isentropic) compression. This is best illustrated in a Mollier diagram (Fig. 6.5).

Head is the amount of work we have to apply to affect the change in enthalpy in the gas [6]. Physically, there is no difference between work, head, and enthalpy difference.

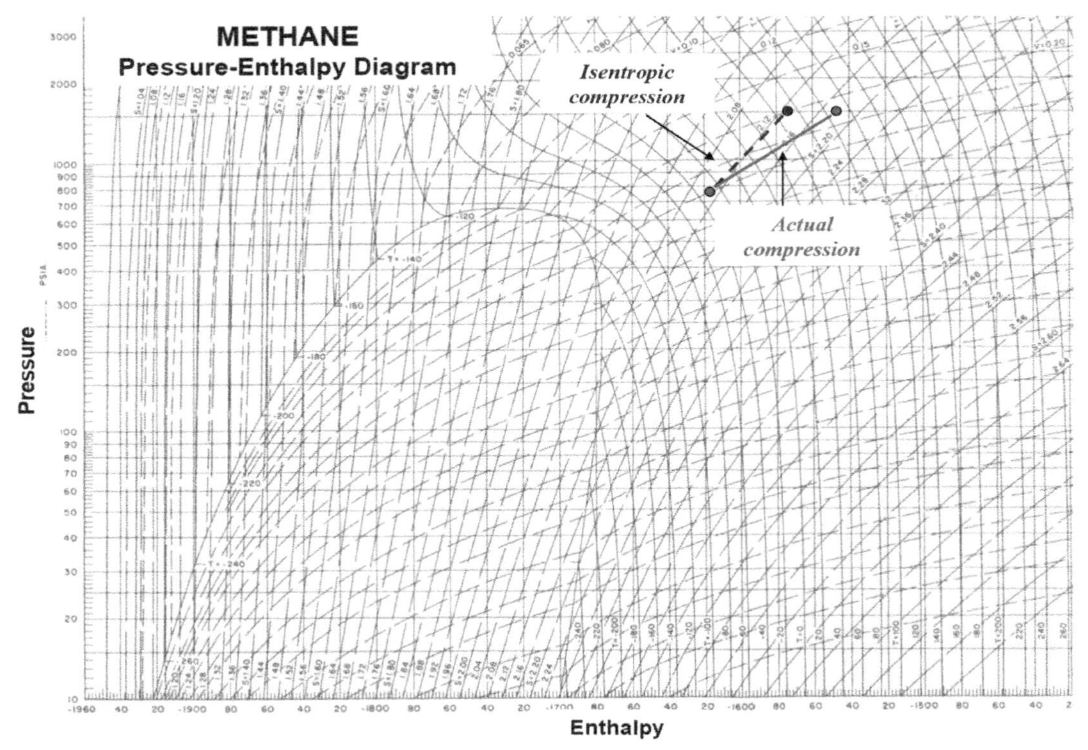

FIG. 6.5 Compression process in a Mollier diagram for methane. *Courtesy of Solar Turbines.*

In systems with consistent units (such as the SI system), work, head, and enthalpy difference have the same unit (e.g., kJ/kg in SI units). Only in inconsistent systems (such as US customary units), we need to consider that the enthalpy difference (e.g., in BTU/lbm) is related to head and work (e.g., in ft. lbf/lbm) by the mechanical equivalent of heat (e.g., in ft. lbf/BTU).

The compressor head (H) can be determined from the suction and discharge pressure and temperature, assuming the gas composition is known. The compressor is usually assumed to be an adiabatic system (otherwise, neither the isentropic nor the polytropic work and efficiency definitions are very useful). Intercooled compressors can only be considered piecewise adiabatic. The relationship between pressure, temperature, and enthalpy (h) is defined by the equations of state described below. By using the equations of state [7], the relevant enthalpies for the suction, the discharge, and the isentropic discharge states can be computed.

The isentropic head (H^*) is

$$H^* = h(p_d, s(p_s, T_s)) - h(p_s, T_s) \tag{6.1}$$

With the isentropic efficiency,

$$\eta_s = \frac{H^*}{H} \tag{6.2}$$

the actual head (H), which defines the power requirement as well as the discharge temperature, is

$$H = \frac{H^*}{\eta} = h(p_d, T_d) - h(p_s, T_s) \tag{6.3}$$

It should be noted that the polytropic efficiency is defined similarly, using the polytropic process instead of the isentropic process for comparison:

$$H = \frac{1}{\eta_p} \int_{p_1}^{p_2} v dp = \frac{H_p}{\eta_p} \tag{6.4}$$

and

$$H_p = \int_{p_1}^{p_2} v dp \tag{6.5}$$

The polytropic efficiency η_p is defined such that it is constant for any infinitesimally small compression step. For known work and polytropic work, the polytropic efficiency is

$$\eta_p = \frac{H_p}{H} \tag{6.6}$$

The actual head, which determines the absorbed power, is not affected by the choice of the polytropic or isentropic process. In order to fully define the isentropic compression process for a given gas, suction pressure, suction temperature, and discharge pressure have to be known. To define the polytropic process, in addition, either the polytropic compression efficiency or the discharge temperature has to be known. Both the isentropic and the polytropic processes are reversible. Both apply to adiabatic systems, but only the isentropic process is adiabatic. The polytropic process is the succession of an infinite number of isentropic compression steps, each followed by an isobaric heat addition. This (reversible) heat addition generates just the same temperature increase as the (irreversible) losses in the real process would generate. For centrifugal compressors, suction and discharge are defined at the suction and discharge flanges of the compressor. Since the majority of losses for reciprocating compressors occur at the valves, orifices, and pulsation bottles, these have to be included in the system when calculating work.

The actual flow (Q) can be calculated from standard flow[a] or mass flow once the density is known (from the equation of state), i.e.,

$$Q(p, T) = \frac{\rho_{std}}{\rho(p, T)} Q_{std} = \frac{W}{\rho(p, T))} \tag{6.7}$$

Finally, the aerodynamic or gas power of the compressor is determined to be:

$$P_g = \rho_1 Q_1 H = \frac{p_1}{Z_1 R T_1} Q_1 H \tag{6.8}$$

Mechanical losses occur in the gas compressor and the gearbox (if one is used). The adiabatic efficiency of a compressor does not include the mechanical losses, which typically amount to about 1%–2% of the absorbed power. The predicted absorbed power of a compressor should include all mechanical losses. By introducing a mechanical efficiency (η_m), typically 98%–99%, to account for bearing losses, the absorbed compressor power (P) becomes:

$$P = \frac{P_g}{\eta_m} \tag{6.9}$$

Understanding gas compression requires an understanding of the relationship between pressure, temperature, and density of a gas. An ideal gas exhibits the following behavior:

$$\frac{p}{\rho} = RT \tag{6.10}$$

where R is the gas constant, and as such, it is constant as long as the gas composition is not changed. Any gas at very low pressures ($p \rightarrow 0$) can be described by this equation.

For the elevated pressures we see in natural gas compression, this equation becomes inaccurate, and an additional variable, the compressibility factor Z, has to be added:

$$\frac{p}{\rho} = ZRT \tag{6.11}$$

Unfortunately, the compressibility factor itself is a function of pressure, temperature, and gas composition.

[a] Standard conditions are $p = 14.7$ psia, $T = 60$°F. Similarly, normal conditions are specified at $p = 101.325$ kPa, $T = 0$°C.

A similar situation arises when the enthalpy has to be calculated. For an ideal gas, we find.

$$\Delta h = c_p \cdot \Delta T = \int_{T_1}^{T_2} c_p dT \tag{6.12}$$

where c_p is only a function of temperature.

In real gas, we get additional terms for the deviation between real gas behavior and ideal gas behavior [8]:

$$\Delta h = \left(h^0 - h(p_1)\right)_{T_1} + \int_{T_1}^{T_2} c_p dT - \left(h^0 - h(p_2)\right)_{T_2} \tag{6.13}$$

The terms $(h^0 - h(p_1))_{T1}$ and $(h^0 - h(p_2))_{T2}$ are called departure functions because they describe the deviation of the real gas behavior from the ideal gas behavior. They relate the enthalpy at some pressure and temperature to a reference state at low pressure but at the same temperature. The departure functions can be calculated solely from an equation of state, while the term $\int c_p dT$ is evaluated in the ideal gas state.

Equations of state are semiempirical relationships that allow for the calculation of the compressibility factor as well as the departure functions. For gas compression applications, the most frequently used equations of state are Redlich-Kwong, Soave-Redlich-Kwong, Benedict-Webb-Rubin, Benedict-Webb-Rubin-Starling, and Lee-Kesler-Plocker [8].

In general, all of these equations provide accurate results for typical applications in pipelines, i.e., for gases with a high methane content and at pressures below about 3500 psia. Kumar [7], Beinecke [9], and Sandberg [10] have compared these equations of state regarding their accuracy for compression applications. For higher pressures and gases with high CO_2 and H_2S content, the Benedict-Webb-Rubin type EOS and, in particular, its derivative, the Lee-Kesler-Plocker EOS, provide more accurate results. REFPROP is currently considered the most accurate equation of state [10].

Centrifugal compressors

The typical centrifugal compressor is either a single-stage machine with an overhung rotor or a machine with multiple stages with a beam-style rotor. A stage consists of the inlet system (for the first stage) or a return channel (for subsequent stages), the impeller, the diffuser (either vaneless or with vanes), and after the last stage, a discharge collector or (in more modern machines) a discharge volute [6] (Fig. 6.6).

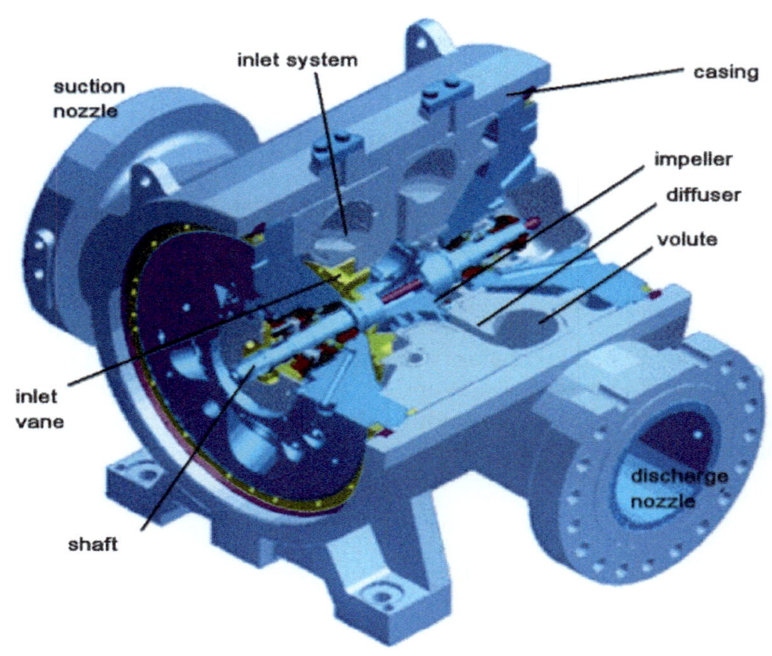

FIG. 6.6 Typical centrifugal compressor. *Courtesy of Solar Turbines.*

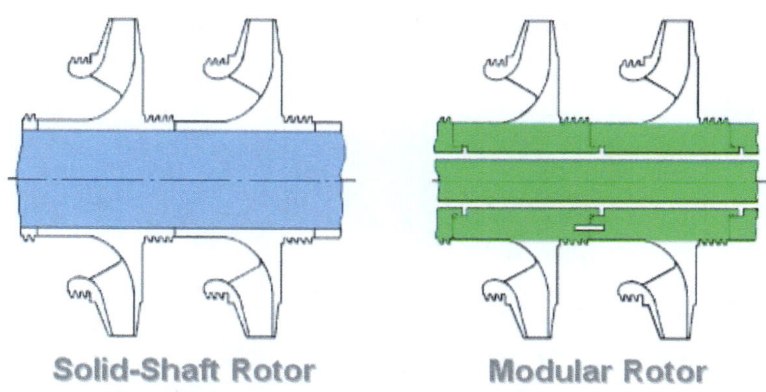

FIG. 6.7 Rotor designs for centrifugal compressors. *Courtesy of Solar Turbines.*

The gas enters the inlet nozzle of the compressor (Fig. 6.6) and is guided (often with the help of guide vanes) to the inlet of the first impeller. An impeller consists of a number of rotating vanes that impart mechanical energy to the gas. The gas will leave the impeller with increased velocity and increased static pressure. In the diffuser, part of the velocity is converted into static pressure. Diffusers can be vaneless or contain several vanes. If the compressor has more than one impeller, the gas will be brought in front of the next impeller through the return channel and the return vanes. After the diffuser of the last impeller in a compressor, the gas enters the discharge system. The discharge system can make use of either a volute, which can further convert velocity into static pressure, or a simple cavity that collects the gas before it exits the compressor through the discharge nozzle.

The rotating part of the compressor consists of all the impellers and the shaft. This rotor runs on two radial bearings (on all modern compressors, these are hydrodynamic tilt pad bearings), while the axial thrust generated by the impellers is balanced by a balance piston, and the resulting force is balanced by a hydrodynamic tilt pad thrust bearing. The compressor shaft can be either a solid shaft with the impellers shrunk or keyed on, or in modular rotors, the impellers can form part of the shaft (Fig. 6.7).

Rotordynamic stability of a compressor, that is, to operate without excessive vibrations within the desired range of speeds and pressures, is not only the key requirement for successful operation, but it also often limits the application of a compressor for certain applications. Particular importance relates to the capability to sufficiently dampen the various possible excitations the rotor system may be subject to, be it from seals, impellers, unbalance, and others [11].

To keep the gas from escaping at the shaft ends, dry gas seals are used on both shaft ends, except for overhung impellers, which only require one seal. Other seal types have been used in the past, but virtually all modern centrifugal compressors use dry gas seals, except for applications in air or nitrogen compression, where often carbon ring seals or labyrinth seals are used. For dry gas seals, the sealing is accomplished by a stationary and a rotating disk with a very small gap (about 5μm) between them. At standstill, springs press the movable seal disc onto the stationary disk. Once the compressor shaft starts to rotate, the groove pattern on one of the discs causes a separating force, making the seals run without mechanical contact with the sealing surfaces.

The pressure-containing casing is either horizontally or vertically split (Fig. 6.8). The casing as well as the compressor flanges have to be rated for the maximum discharge pressure the compressor will experience. Horizontally split casings are typically used for lower pressure applications (up to about 40bar (600psi) discharge pressure), while vertically split (barrel type) casings have successfully been used for discharge pressures up to 800bar (12,000psi).

The maximum number of impellers in a casing is usually limited by rotordynamic considerations. Therefore, the maximum amount of head that can be generated in one casing is limited. If more head is required, multiple casings, driven either by the same driver, or by separate drivers, have to be used. Another limitation for the head may be the temperature limits of the compressor (typically, the discharge temperatures are limited to about 175°C/350°F). If more head is required, the gas has to be cooled during the compression process.

Various design alternatives are available (Fig. 6.9).

Multibody tandems, that is, up to three compressor casings driven by the same driver, possibly with a gearbox either between the driver and the compressor train or between two of the compressors.

Compound compressors: This compressor contains multiple compartments, each with its own suction and discharge nozzle. All impellers are on the same shaft and face in the same direction.

Back-to-back compressors: This compressor contains two compartments, each with its own suction and discharge nozzle. The impellers are on the same shaft, but the impellers in the first compartment face in the opposite direction from the impellers in the second compartment.

FIG. 6.8 Casing designs for centrifugal compressors. Top: Barrel-type compressor during bundle removal, bottom horizontally split compressor.

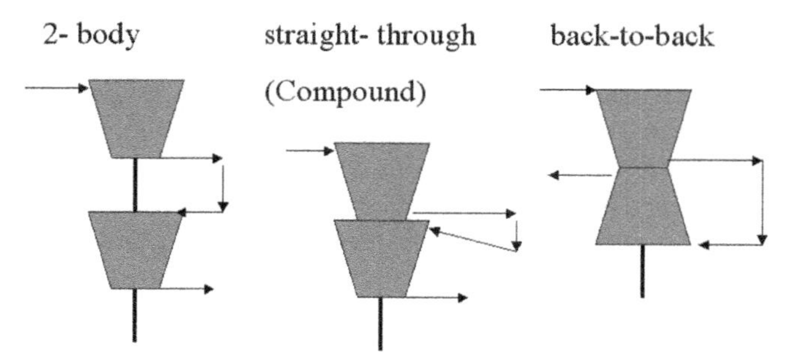

FIG. 6.9 Multibody, compound, and back-to-back compressors.

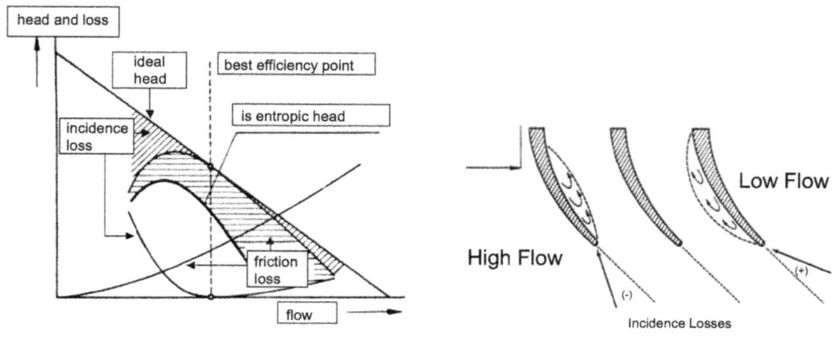

FIG. 6.10 Compressor operating at off design points. *Courtesy of Solar Turbines.*

Integral gear-type compressors: Overhung impellers are located at each end of multiple pinions, driven from a central bull gear.

Any of these configurations allows not only intercooling, but also sidestreams or gas off-takes.

The performance of centrifugal gas compressors is best displayed in a map showing isentropic efficiency and isentropic head as a function of the actual inlet flow. The means of control are added parameters. A compressor, operated at constant speed, is operated at its best efficiency point (Fig. 6.10). If we reduce the flow through the compressor (for example, because the discharge pressure that the compressor has to overcome is increased), then the compressor efficiency will be gradually reduced due to the increase in incidence losses. At a certain flow, a stall, probably in the form of a rotating stall, in one or more of the compressor components will occur. At further flow reduction, the compressor will eventually reach its stability limit and go into surge.

If the flow through a compressor at constant speed is reduced, the losses in all aerodynamic components will increase. Eventually, the flow in one of the aerodynamic components, usually in the diffuser but sometimes in the impeller inlet, will separate. It should be noted that stalls usually appear in one stage of a compressor first.

Flow separation in a vaneless diffuser means that all or parts of the flow will not exit the diffuser on its discharge end but will form areas where the flow stagnates or reverses its direction back to the inlet of the diffuser (i.e., the impeller exit).

Stall in the impeller inlet or vaned diffuser is due to the fact that the direction of the incoming flow relative to the rotating impeller changes with the flow rate through the compressor. Usually, vanes in the diffuser reduce the operating range of a stage compared to a vaneless diffuser. Therefore, a reduction in flow will lead to an increased mismatch between the direction of the incoming flow the impeller was designed for and the actual direction of the incoming flow. At one point, this mismatch becomes so significant that the flow through the impeller breaks down.

Flow separation can take on the characteristics of a rotating stall. When the flow through the compressor stage is reduced, parts of the diffuser experience flow separations. A rotating stall occurs if the regions of flow separation are not stationary but move in the direction of the rotating impeller (typically at 15%–30% of the impeller speed). A rotating stall can often be detected by increasing vibration signatures in the subsynchronous region. Onset of stall does not necessarily constitute an operating limit of the compressor. In fact, in many cases, the flow can be reduced further before the actual stability limit is reached.

If, again starting from the best efficiency point, the flow is increased, we also see a reduction in efficiency, accompanied by a reduction in head. Eventually, the head and efficiency will drop steeply, until the compressor will not produce any head at all. This operating scenario is called choke. For practical applications, the compressor is usually considered to be in choke when the head falls below a certain percentage of the head at the best efficiency point. Some compressor manufacturers do not allow the operation of their machines in deep choke. In these cases, the compressor map has a distinct high flow limit for each speed line.

The efficiency starts to drop off at higher flows because a higher flow causes higher internal velocities and thus, higher friction losses. The head reduction is a result of both the increased losses and the basic kinematic relationships in a centrifugal compressor. Even without any losses, a compressor with backward bent blades (as they are used in virtually every industrial centrifugal compressor) will experience a reduction in head with increased flow (Fig. 6.11). "Choke" and "Stonewall" are different terms for the same phenomenon.

Fig. 6.11 shows the map of a speed-controlled compressor. Such a map also needs to define the operating limits of the compressor.

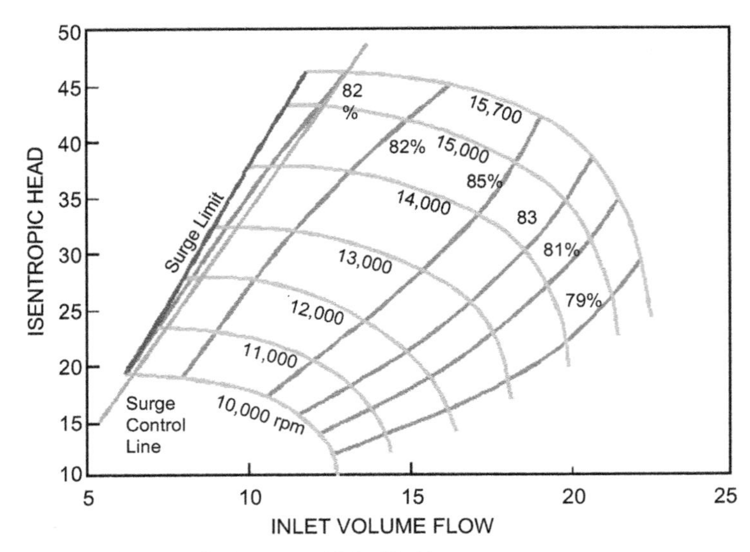

FIG. 6.11 Typical compressor map (variable speed). *Courtesy of Solar Turbines.*

The limitation for lower flow is the surge limit. Some manufacturers also limit the operation of their machines in the choke region, while others allow the operation of their machines anywhere in choke, as long as the head remains positive. Other limits include the maximum and minimum speed, the limits of vane settings, temperature limits, and others. The head-flow map does not automatically define the temperature limits of the compressor because the discharge temperature also depends on the gas composition and the suction temperature. With the information in the head-flow-efficiency map and known suction conditions, it can be calculated. The speed limits are either rotordynamic limitations or stress limits. It must be noted that a performance map as described will not change even if the inlet conditions are changed within limits.

Surge margin and turndown

Any operating point can be characterized by its distance from the onset of a surge. Two definitions are widely used: The surge margin:

$$SM(\%) = \frac{Q_A - Q_B}{Q_A} \cdot 100 \tag{6.14}$$

which is based on the flow difference between the operating point and the surge point at constant speed and the turndown:

$$\text{Turndown}\,(\%) = \frac{Q_A - Q_C}{Q_A} \cdot 100 \tag{6.15}$$

which is based on the flow difference between the operating point and the surge point at constant head.

Centrifugal compressor control

Previously, we discussed the operating characteristics of a compressor without any other control adjustments. In this case, the head the compressor has to generate uniquely determines the flow.

There are several ways to enhance the capability of the compressor to cover a wider range of different operating conditions (Fig. 6.12):

- variable speed
- inlet vanes
- diffusor vanes
- throttling (suction, discharge)
- recycling

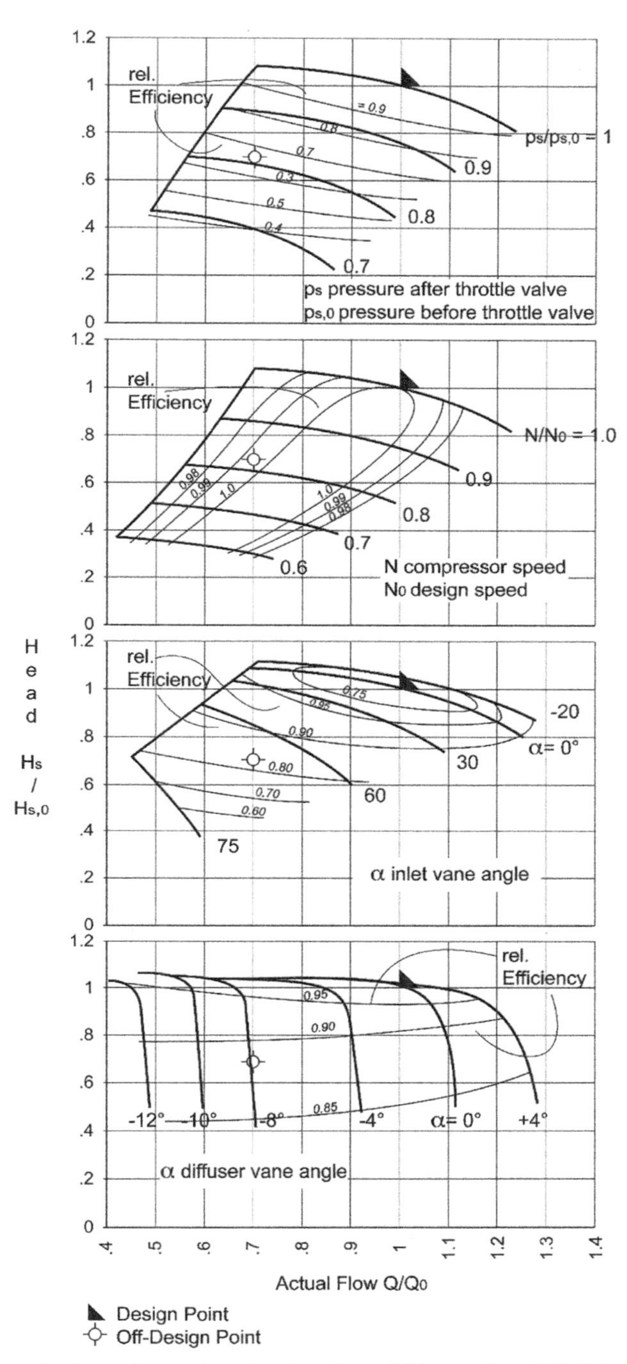

FIG. 6.12 Single-stage compressor maps for (top to bottom) suction throttle, variable speed, variable inlet vanes, and variable diffuser vanes. *From P.C. Rasmussen, R. Kurz, Centrifugal compressor applications, in: 38th Turbomachinery Symposium, Houston, TX, 2009.*

Variable speed

Compressor drivers that can operate at variable speeds (two shaft gas turbines, steam turbines, turbo expanders, electric motors with variable frequency drives, or variable speed gearboxes) allow the compressor to operate over a range of different speeds. The faster the compressor runs, the more head and flow it generates and the more power it consumes. The efficiency characteristics of the compressor are retained for different speeds, so this is a very efficient way of adjusting the compressor to a wide range of different operating conditions. Figs. 6.5 and 6.12 show the resulting map.

Adjustable inlet vanes

Modifying the swirl of the flow into the impeller allows for modification of the operating characteristics of the stage (Fig. 6.12). This can be accomplished by adding adjustable vanes upstream of the impeller. Increasing the swirl against the rotation of the impeller increases the head and flow through the stage. Increasing the swirl with the rotation of the impeller reduces the head and flow through the stage. This is very effective to increase the range in a single stage. In multistage compressors, the range increase is limited if only the first stage has adjustable vanes. The technical difficulty for high-pressure compressors lies in the fact that complicated mechanical linkage has to be actuated from outside the pressure-containing body.

Adjustable diffusor vanes

Vaned diffusers tend to limit the operating range of the compressor because the vanes are subject to increased incidence in off-design conditions, thus eventually causing stalls. Adjustable diffuser vanes allow for the adjustment of the changing flow conditions, thus effectively allowing for operation at much lower flows by delaying the onset of diffuser stall (Fig. 6.12). They will not increase the head or flow capability of the stage. In multistage compressors, the range increase is limited if only one stage has adjustable vanes. The technical difficulty for high-pressure compressors lies in the fact that complicated mechanical linkages have to be actuated from outside the pressure-containing body. Another issue is that for the vanes to operate, small gaps between the vanes and the diffuser walls have to exist. Ubiquitous leakage through these gaps causes efficiency and range penalties, in particular, in machines with narrow diffusers.

Throttling (suction, discharge)

A throttle valve on the suction or discharge side of the compressor increases the pressure ratio the compressor sees and therefore moves the operating point to lower flows on the constant-speed map. It is a very effective but inefficient way of controlling compressors (Fig. 6.12).

Recycling

A controlled recycle loop allows a certain amount of the process flow to go from the compressor discharge back to compressor suction. The compressor therefore sees a flow that is higher than the process flow. This is a very effective but inefficient way to allow the compressor system to operate at low flow.

Process control with centrifugal compressors driven by two shaft gas turbines

Centrifugal compressors, when driven by two-shaft gas turbines, are usually adapted to varying process conditions by means of speed control. This is a very elegant way of controlling a system because both the centrifugal compressor and the power turbine of a two-shaft gas turbine can operate over a wide range of speeds without any adverse effects. A typical configuration can operate at 50% of its maximum continuous speed and, in many cases, even lower. Reaction times are very fast, thus allowing a continuous load following using modern, PLC-based controllers.

A simple case is flow control: the flow into the machine is sensed by a flow metering element (such as a flow orifice, a venturi nozzle, or an ultrasonic device). A flow set point is selected by the operator. If the discharge pressure increases due to process changes, the controller will increase the fuel flow into the gas turbine. As a result, the power turbine will produce more power and cause the power turbine, together with the driven compressor, to accelerate. Thus, the compressor flow is kept constant. Both the power turbine speed and the power increase in that situation (Fig. 6.13).

If the discharge pressure is reduced or the suction pressure is increased due to process changes, the controller will reduce the fuel flow into the gas turbine. As a result, the power turbine will produce less power and cause the power turbine, together with the driven compressor, to decelerate. Thus, the compressor flow is kept constant.

Similar control mechanisms are available to keep the discharge pressure constant or to keep the suction pressure constant. Another possible control mode is to run the unit at maximum available driver power (or any other constant driver output). In this case, the operating points are all on a line of constant power, but the speed will vary.

The control scheme works for one or more compressors and can be set up for machines operating in series as well as in parallel.

If speed control is not available, the compressor can be equipped with a suction throttle or with variable guide vanes. The latter, if available in front of each impeller, is rather effective, but the mechanical complexity proves to be prohibitive in higher-pressure applications. The former is a mechanically simple means of control, but it has a detrimental effect on the overall efficiency.

Based on the requirements above, the compressor output must be controlled to match the system demand. This system demand is characterized by a relationship between system flow and system head, or pressure ratio. Given the large variations in operating conditions experienced by compressors, an important question is how to adjust

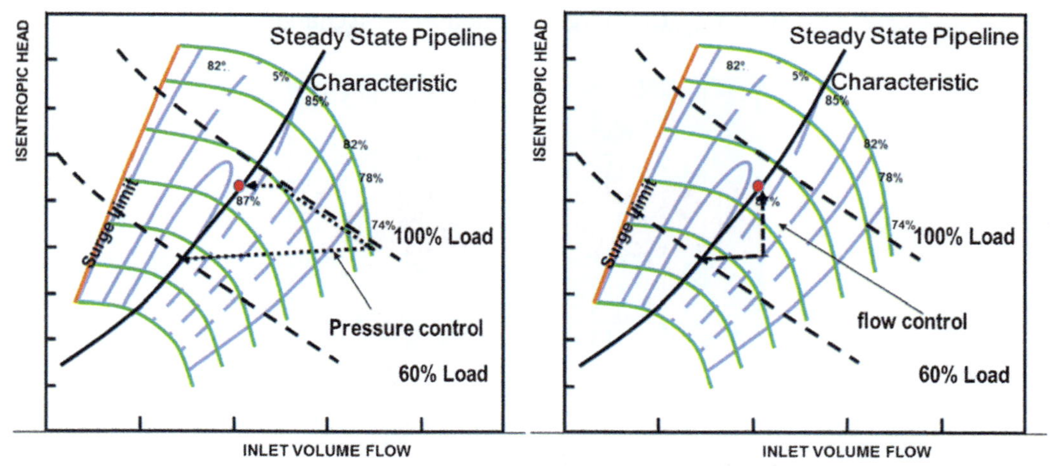

FIG. 6.13 Transient operation: load increase. (Left) Pressure control, (right) flow control [12].

the compressor to the varying conditions and, in particular, how this influences efficiency. Centrifugal compressors tend to have a rather flat head versus flow characteristic. This means that changes in pressure ratio have a significant effect on the actual flow through the machine. For a centrifugal compressor operating at a constant speed, the head or pressure ratio is reduced with increasing flow.

The resulting operating point of a compressor is determined by the head-flow characteristic of the system, the map of the compressor, and the available power to drive the compressor [13].

Surge control

Surge control systems are, by nature, surge avoidance systems. In general, the control system measures the gas flow through the compressor and the head it generates. The knowledge of head and flow allows us to compare the present operating point of the compressor with the predicted surge line (Fig. 6.10). If the process forces the compressor to approach the surge line, a recycle valve in a recycle line is opened. This allows the actual operating point of the compressor to move away from surge [14,15]. Well-designed surge control systems can reduce station flow to zero while keeping the compressor online. They will also make the transition from a fully closed recycle valve to an increasingly open recycle valve smooth and without disrupting the process.

Control for multiple process streams

Compressors or compressor trains often have to handle multiple process streams. To control N streams, the compressor train needs to have N control devices. For example, a two-body tandem with one sidestream can be controlled by a combination of speed control (for the entire train) in combination with a throttle valve for the sidestream. Another possibility for this case may be to use a unit recycle loop for control.

Start-up and shutdown

The start-up and shutdown processes require particular attention. Issues to be considered involve the speed-torque capability of the driver as well as the thermal balance in the recycle system. The former is in particular an issue with electric motor drives, where the available motor torque often depends on the capability of the electric grid or electric power generation. The latter involves the fact that in many instances, the compressor will pump gas into the recycle loop until the pressure it generates is high enough to go online. In a recycle loop that is not cooled, the temperature will increase (due to the work the compressors put into the gas), and if the temperature exceeds certain thresholds, the start has to be aborted. This issue is discussed in more detail by White [15].

Compressors—Reciprocating compressors

Reciprocating compressors are positive displacement compressors. In other words, compression is the result of trapping a gas in a space with a certain volume and subsequently reducing the volume. In a reciprocating compressor, the volume reduction is created by a piston making a reciprocating motion inside a cylinder. Gas change is affected by valves (Fig. 6.14).

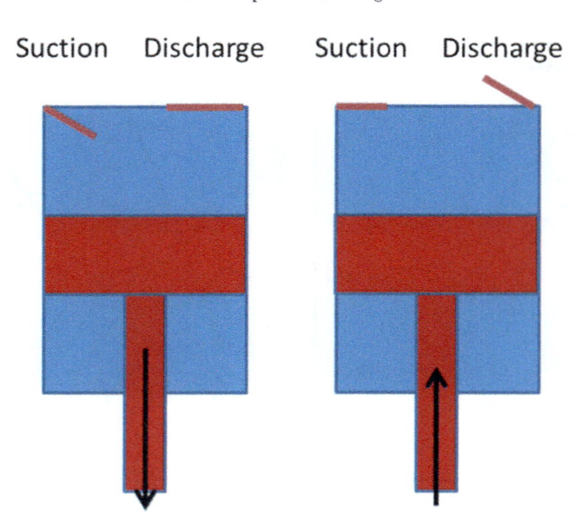

FIG. 6.14 Reciprocating motion.

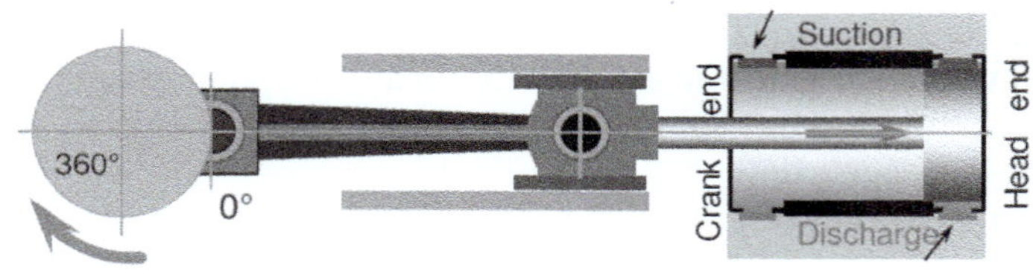

FIG. 6.15 Single-stage, double-acting compressor [16].

The piston is connected to the crankshaft via a piston rod (Fig. 6.15). The compressor takes gas in via the suction valve during the downward motion of the piston. Valves are essentially check valves and are operated by the pressure difference on either side of the valve. The suction valve opens when the pressure in the cylinder drops below the suction line pressure. Once the piston moves upward again, the pressure in the cylinder exceeds the suction line pressure, and the suction valve closes. The piston continues to travel upward, thus reducing the volume and increasing the pressure and temperature of the trapped gas. Once the gas pressure exceeds the discharge line pressure, the discharge valve opens, and gas is flowing to the discharge line. After passing the top dead center, the piston will move downward again, and the discharge valve will close once the cylinder pressure becomes lower than the discharge line pressure. After further travel, when the cylinder pressure becomes lower than the suction pressure, the suction valve will open again, and gas will flow in from the suction line. Then, a new compression cycle begins (Fig. 6.16).

Many compressor stages are designed as double-acting stages. Thus, both sides of the piston contribute to the compression effort (Fig. 6.15).

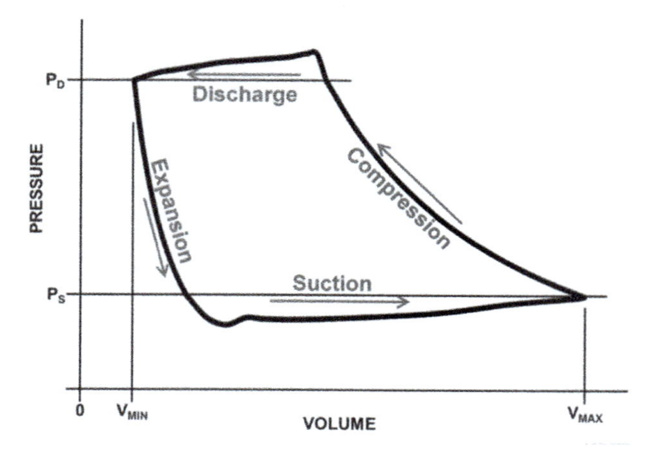

FIG. 6.16 Compression cycle for a reciprocating compressor in a p-V (pressure-volume) diagram [17].

Since the flow into and out of the cylinder is not continuous, pressure fluctuations are induced in the suction and discharge lines. These have to be dampened using a number of different methods, including orifices, damping bottles with choke tubes, or side branch absorbers. The pressure loss of these pulsation damping methods has to be included in the performance calculations of the compressor.

For higher pressure ratios, compressors are built with multiple stages in series, usually with intercoolers in between. This allows for high pressure ratios while limiting the discharge temperature. High-flow low-ratio compressors can also be designed with multiple stages in parallel to counter inherent flow limitations.

The major source of losses comes from the pressure loss in suction and discharge valves (Fig. 6.17). These losses depend on the pressure, gas density, and flow velocities, as well as the valve opening characteristics. Accordingly, the relative impact of valve losses tends to be reduced for higher pressure ratios and lighter gases (Fig. 6.18).

Commonly used control methods for reciprocating compressors are variable speed, variable clearance, timed valve closing, cylinder end deactivation, throttling, and recycling.

Increasing the speed simply increases the compression cycles per minute, thus impacting the flow. However, changing the speed will change pulsation frequencies, thus making the damping of these frequencies more difficult. Often, certain speeds have to be avoided. Also, higher speeds tend to reduce compressor efficiency due to dynamic effects. For gas engine drivers, changing the driver speed may impact efficiency and emissions characteristics.

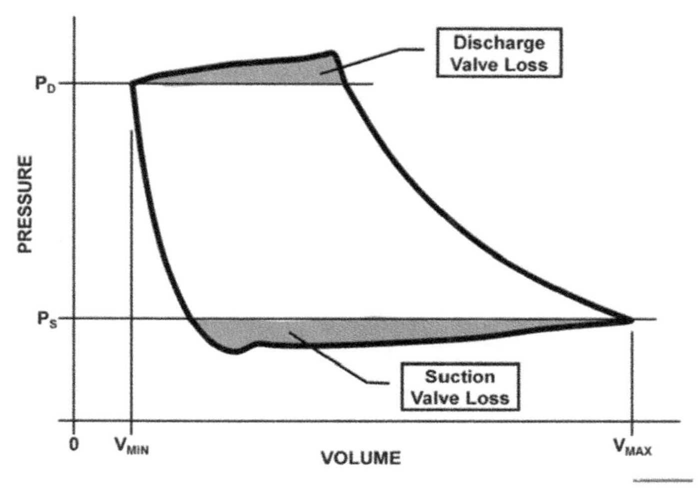

FIG. 6.17 Valve losses compressor [17].

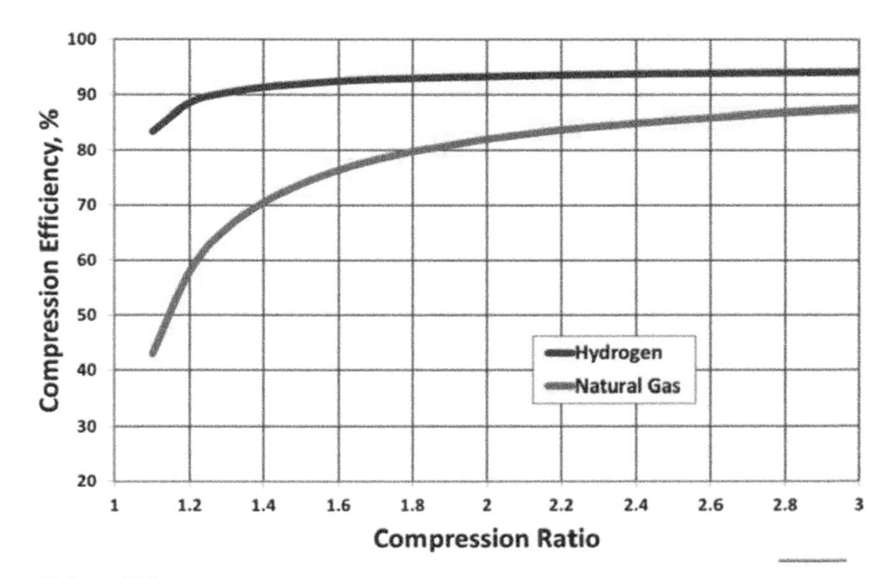

FIG. 6.18 Compression efficiency [17].

Variable clearance control (either with fixed pockets or variable pockets) allows for modification of the cylinder volume at the top dead center. The higher the clearance, the lower the volumetric efficiency of the cylinder. Volumetric efficiency denotes the ratio between piston displacement and actual capacity. Since different volumetric efficiencies barely affect the compression efficiency, modifying the clearance of a compressor allows for an elegant way to change the compressor flow capacity. The impact of a certain change in clearance on the compressor flow depends on the stage pressure ratio. Thus, clearance control is most effective for high pressure ratio stages.

Timed valve closing allows to delay the opening of valves, which can impact the pulsation frequencies of the system, but also allows to control the flow capacity.

Cylinder end deactivation involves keeping the valves for one or multiple cylinder ends open, thus essentially eliminating the flow contribution of that end. This will impact the pulsation characteristics of the compressors but allow for fast, significant changes in capacity.

Suction throttling impacts capacity but can actually increase the load of the compressor. It does not impact pulsations and allows for smooth changes in capacity. The method will reduce the system's efficiency.

Recycling, i.e., bringing all or part of the discharge flow back to suction via a controlled recycle valve, allows for capacity control but does not reduce load. Continuous recycling may require cooling the recycled gas.

Control concepts for reciprocating compressors have to take into account that the torque or power the compressor absorbs has to be lower than the torque or power from the driver to avoid driver stalling. In other words, control devices have to be used both for flow (capacity) control and for load control. Also, to be considered is the impact of the line pulsations that are induced by the compressor on the compressor's performance itself (Fig. 6.19).

The black broken lines show the situation for constant line pressures, and the red broken lines show the situation for fluctuating line pressures. The solid lines show the respective impact on the *p-V* diagram [17].

Without additional control measures, a reciprocation compressor stage shows very little change in actual suction flow, as opposed to a centrifugal compressor without any additional controls (as described in an earlier part of this chapter), where changes in pressure ratio led to a distinct change in flow (Fig. 6.20).

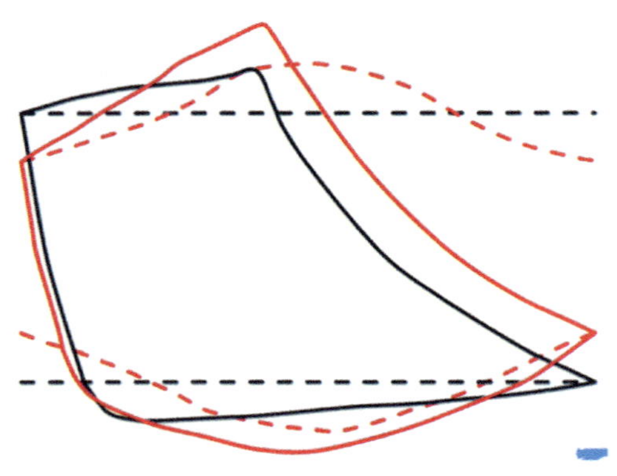

FIG. 6.19 Pressure volume diagram (see Figs. 6.16 and 6.17).

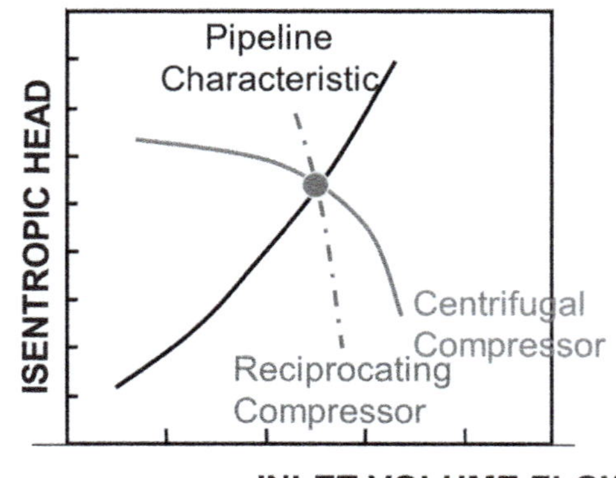

FIG. 6.20 Characteristic of a reciprocating compressor at constant speed and a centrifugal compressor at constant speed compared to a typical load characteristic for a pipeline.

The general arrangement of a multistage reciprocating compressor with pulsation bottles is shown in Fig. 6.21, and the layout of a four-throw compressor (i.e., four cylinders, double acting, and four piston rods from a common crankshaft) can be seen in Fig. 6.22.

Drivers—Gas turbines

A gas turbine converts the chemical energy of a fuel (for example, hydrogen or natural gas) into mechanical energy, which can be used to power a generator or other machinery, such as pumps or compressors. While gas turbines are extensively used for the propulsion of aircraft, we will focus in this text on their application in industrial uses. The gas turbine cycle is described as an (usually) open Brayton cycle and works as follows:

Air enters the compressor section of the gas turbine, where it is compressed to a higher pressure, and, as a result, to a higher temperature. It enters a combustor, where fuel is added and burned, creating a high-pressure, high-temperature

FIG. 6.21 Multistage reciprocating compressor. *Credit: SwRI Pulsation & Vibration Short Course.*

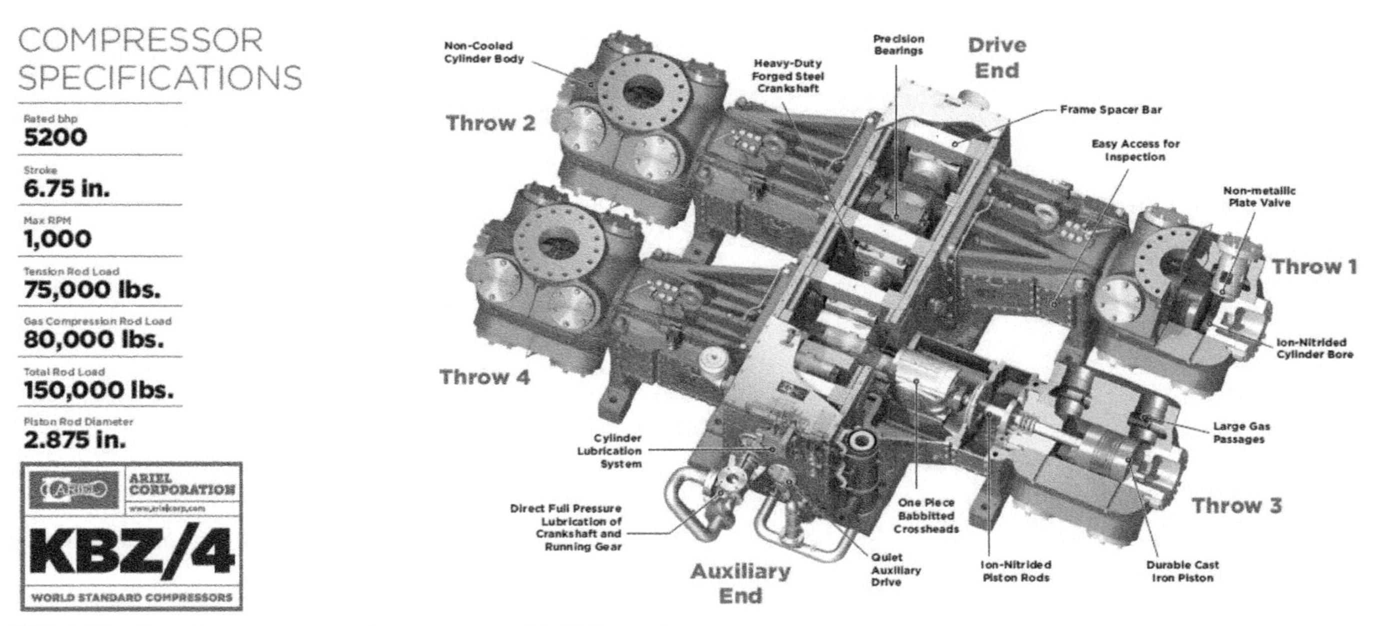

FIG. 6.22 Four throw compressor. *Image courteous of Ariel Corporation.*

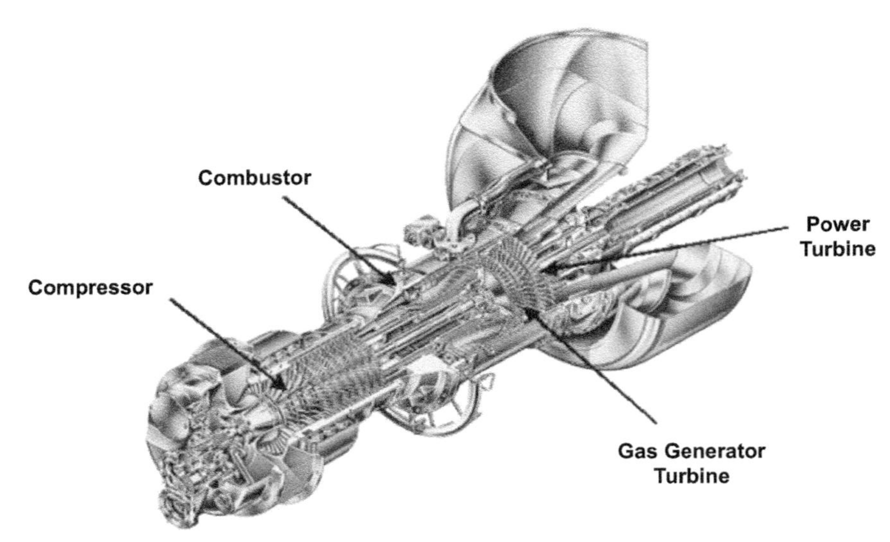

FIG. 6.23 Components of an industrial gas turbine [18]. *Courtesy of Solar Turbines.*

exhaust gas. This gas is expanded through a turbine, thus creating mechanical energy (Fig. 6.23). This energy is used to power the air compressor of the gas turbine, and the remaining energy can be used to drive a generator, pump, or compressor. The exhaust gas is then released to the atmosphere, but since it is still relatively hot, it can be used as a heat source for a steam, organic Rankine (ORC) or sCO_2 cycle [18].

The energy conversion from mechanical work into the gas (in the compressor) and from energy in the gas back to mechanical energy (in the turbine) is performed by means of aerodynamics, by appropriately manipulating gas flows. Leonard Euler (1754) equated the torque produced by a turbine wheel to the change in circumferential momentum of a working fluid passing through the wheel. Somewhat earlier (in 1738), Daniel Bernoulli stated the principle that (in inviscid, subsonic flow) an increase in flow velocity is always accompanied by a reduction in static pressure and vice versa, as long as no external energy is introduced. While Euler's equation applies Newton's principles of action and reaction, Bernoulli's law is an application of the conservation of energy [19]. These two principles explain the energy transfer in a turbomachinery stage (Fig. 6.24).

Besides being very efficient and extremely reliable, gas turbines are the fossil fuel-powered machines with the highest energy density. That means they give the highest output power per installed weight and installed footprint. They are also capable of operating in very harsh environments.

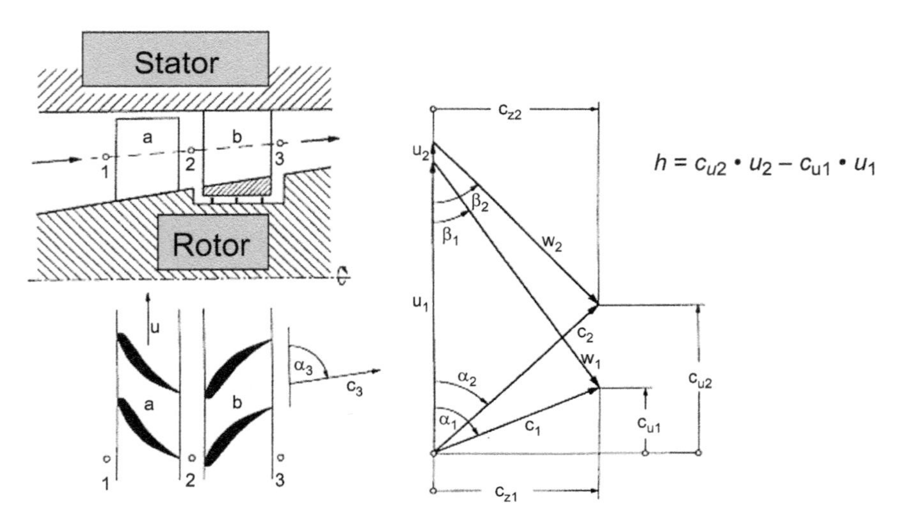

FIG. 6.24 Velocities in a typical compressor stage. Mechanical work h transferred to the air is determined by the change in circumferential momentum of the air [18]. *Courtesy of Solar Turbines.*

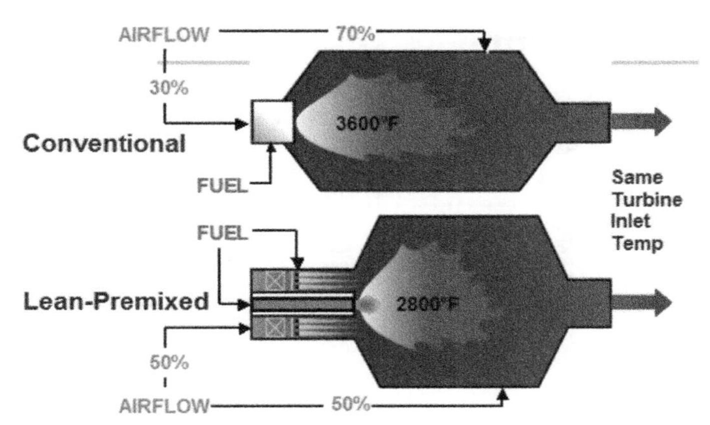

FIG. 6.25 Conventional and lean-premix combustion systems. *Courtesy of Solar Turbines.*

Another important topic is the combustion process and emissions control. Unlike in reciprocating engines, gas turbine combustion is continuous. This has the advantage that the combustion process can be made very efficient, with very low levels of products of incomplete combustion like carbon monoxide (CO) or unburned hydrocarbons (UHCs). The other major emissions component, oxides of nitrogen (NO_X), is not related to combustion efficiency, but strictly to the temperature levels in the flame (and the amount of nitrogen in the fuel). The solution to NO_x emissions, therefore, lies in lowering the flame temperature. Initially, this was accomplished by injecting massive amounts of steam or water into the flame zone, thus "cooling" the flame. This approach has significant drawbacks, not the least the requirement to provide large amounts (fuel-to-water ratios are approximately around 1) of extremely clean water. Since the 1990s, combustion technology has focused on systems often referred to as dry low NO_x combustion, or lean-premix combustion (Fig. 6.25). The idea behind these systems is to make sure that the mixture in the flame zone has a surplus of air, rather than allowing the flame to burn under stoichiometric conditions. This lean mixture, assuming the mixing has been done thoroughly, will burn at a lower flame temperature and thus produce less NO_x. One of the key requirements is the thorough mixing of fuel and air before the mixture enters the flame zone. Incomplete mixing will create zones where the mixture is stoichiometric (or at least less lean than intended), thus locally creating more NO_x. The flame temperature has to be carefully managed in a temperature window that minimizes both NO_x and CO. Lean-premix combustion systems allow to keep the NO_x, as well as CO and UHC emissions within prescribed limits for a wide range of loads, usually between full load and about 40% or 50% load. In order to accomplish this, the air flow into the combustion zone has to be manipulated over the load range [19].

Industrial gas turbines are designed either as single-shaft, two-shaft, or multispool engines. The latter are almost always derived from aircraft engine designs, where low weight is most important, even at the cost of higher complexity. Major components:

- Single-shaft gas turbines have all compressor and turbine stages running on the same shaft. They are usually used to drive generators (which require a constant speed), but rarely pumps or compressors.
- Two-shaft gas turbines consist of two sections: the gas producer (or gas generator) with the gas turbine compressor, the combustor, and the high-pressure portion of the turbine on one shaft, and a power turbine on a second shaft (Fig. 6.23). In this configuration, the high-pressure or gas producer turbine only drives the compressor, while the low-pressure or power turbine, working on a separate shaft at speeds independent of the gas producer, can drive mechanical equipment.
- Multiple spool engines: Industrial gas turbines derived from aircraft engines sometimes have two compressor sections (the HP and the LP compressor), each driven by a separate turbine section (the LP compressor is driven by an LP turbine by a shaft that rotates concentric within the shaft that is used for the HP turbine to drive the HP compressor), and running at different speeds. The energy left in the gas after this process is used to drive a power turbine (on a third, separate shaft), or the LP shaft is used as the output shaft.

The performance of gas turbines is impacted by ambient conditions such as temperature, site elevation, and load (Fig. 6.26). For two-shaft gas turbines, the speed of the power turbine, which determines the speed of the driven equipment, is also a factor.

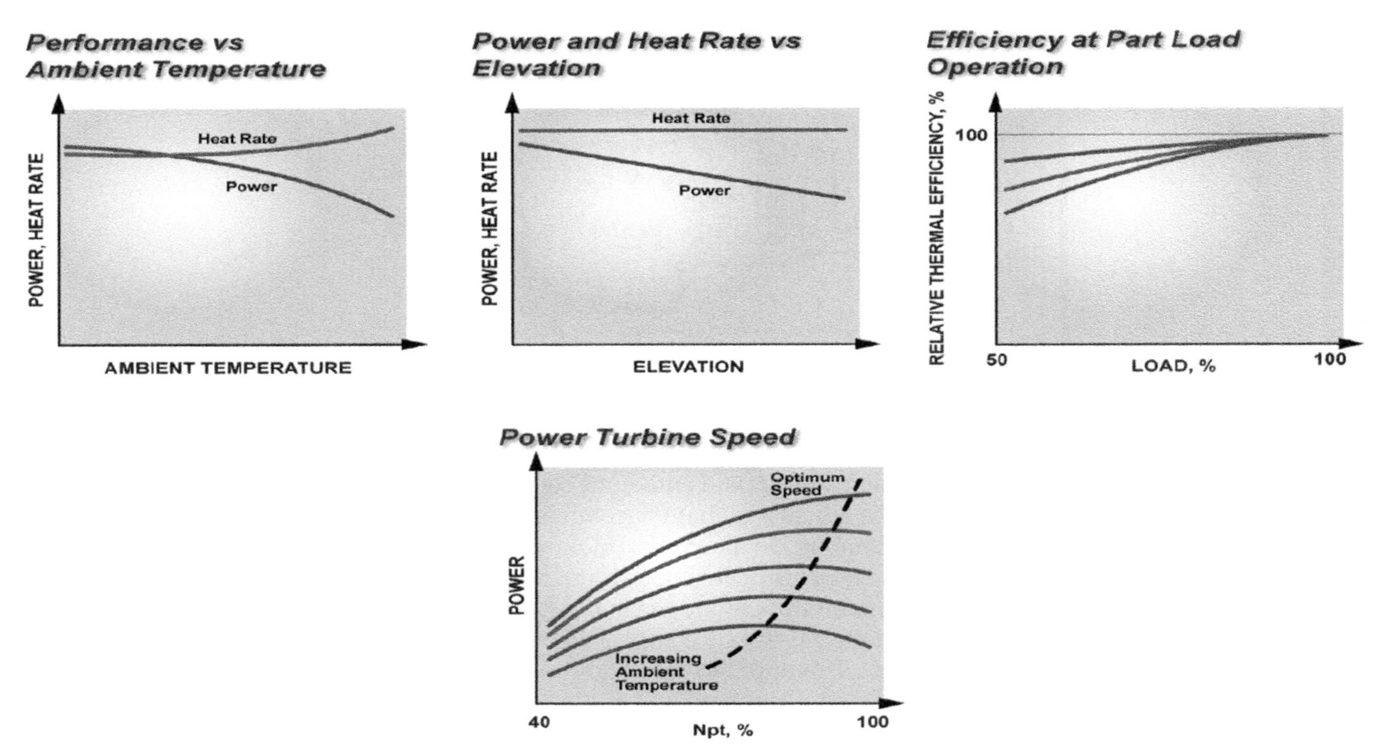

FIG. 6.26 Performance characteristics of a gas turbine. *Courtesy of Solar Turbines.*

Gas turbines ingest a large amount of ambient air during operation. Because of this, the quality of the air entering the turbine is a significant factor in the performance and life of the gas turbine. A filtration system is used to control the quality of the air by removing harmful contaminants that are present. The system is selected based on the operational philosophy and goals for the turbine, the contaminants present in the ambient air, and expected changes in the contaminants in the future due to temporary emission sources or seasonal changes [19].

Drivers—Electric motors

Electric motors are utilized to create mechanical work form electrical energy. Electric motors operate based on the interaction of electric current and electromagnetic fields (EMF). Lenz's Law states that the direction of an induced current in a conductor will always be such that it opposes the change that produced it. In simpler terms, when a magnetic field through a conductor changes, the induced current in the conductor will flow in a direction that creates a magnetic field opposing the change in the original magnetic field. By applying a changing magnetic field to a conductor, a force resisting the change is created, if the conductor is free to rotate, a motor is created.

The same principal in reverse is the basis of generators, where mechanical work is applied to the spinning rotor, creating a current in the generator.

Induction motors

Induction motors are simple AC machines. The induction motor is the most common type of large motor due to its simplicity, ruggedness, and low cost. It consists of two main parts: the stator and the rotor. The stator is the stationary part and consists of a laminated iron core with evenly spaced windings or coils. These windings when connected to an AC power supply creates a rotating magnetic field.

The rotor is the rotating part of the motor. It consists of a laminated iron core with conducting bars placed parallel to the motor's shaft. The bars are shorted at each end by conducting rings, forming a closed-loop or cage-like structure. The conducting bars are typically made of aluminum or copper and are typically permanently shorted to provide a closed circuit.

When the AC power is applied to the stator windings, it produces a rotating magnetic field. This rotating magnetic field induces currents in the "squirrel cage" rotor bars due to electromagnetic induction. As a result, magnetic fields are generated in the rotor bars, which interact with the magnetic field of the stator. This interaction causes the rotor to rotate, as the magnetic fields in the rotor bars try to align with the rotating magnetic field of the stator.

The squirrel cage rotor design offers several advantages, such as simplicity, durability, and self-starting capability. Since the rotor bars are permanently shorted, there is no need for additional brushes or slip rings to supply current to the rotor, making it maintenance-free. However, the speed of the squirrel cage induction motor is determined by the frequency of the AC power supply and the number of poles in the stator windings.

Squirrel cage induction motors are widely used in various applications, including pumps, fans, compressors, conveyors, and many industrial machinery applications. They are known for their robustness, reliability, and ability to provide high torque at low speeds, making them suitable for a wide range of tasks.

In theory, an induction motor could be designed for any industrial application and power, but in practice as the power level of the motor increases beyond around 20 MW, the improved efficiency of the synchronous motor overshadows the increase in complexity.

Synchronous motors

A synchronous AC motor is a type of alternating current (AC) motor that runs at a constant speed and maintains a fixed relationship, or synchronization, with the frequency of the power supply. Unlike induction motors, which rely on the principle of induction to generate a rotating magnetic field, synchronous motors have a rotor that rotates at the same speed as the rotating magnetic field produced by the stator.

The construction of a synchronous AC motor consists of a stator, which contains the stationary windings connected to the power supply, and a rotor, which is the rotating part of the motor. The stator windings create a rotating magnetic field when energized with AC power. The rotor of a synchronous motor can have different configurations, including salient pole and cylindrical rotor designs.

In a synchronous motor, the rotor is magnetized either by permanent magnets or by DC excitation through a separate DC power supply. The magnetic field of the rotor interacts with the rotating magnetic field of the stator, causing the rotor to rotate at the same speed as the rotating magnetic field. This synchronous speed is directly determined by the frequency of the power supply and the number of poles in the stator windings.

Synchronous motors have fixed stator windings electrically connected to the AC supply with a separate source of excitation connected to a field winding on the rotating shaft. A three-phase stator is similar to that of an induction motor. The rotating field has the same number of poles as the stator and is supplied by an external source of DC. Magnetic flux links the rotor and stator windings causing the motor to operate at synchronous speed. A synchronous motor starts as an induction motor, until the rotor speed is near synchronous speed where it is locked in step with the stator by application of a field excitation. When the synchronous motor is operating at synchronous speed, it is possible to alter the power factor by varying the excitation supplied to the motor field.

An important advantage of a synchronous motor is that the motor power factor can be controlled by adjusting the excitation of the rotating DC field. Unlike AC induction motors which run at a lagging power factor, a synchronous motor can run at unity or even at a leading power factor. This can help to improve the overall electrical system power factor, voltage drop, and also improve the voltage drop at the terminals of the motor (for DOL applications).

While synchronous motors can be made at any size, the additional complexity and cost means they are normally considered only at power levels above 20 MW.

The speed of the synchronous speed of AC motors is dependent on the line frequency (Hz) and the number of poles in the stator. Synchronous speed (in RPM) = (120 × frequency)/number of poles. Induction motors do not operate at a synchronous speed; due to the nature of the induced current in the rotor, a slip exists between the fields in the rotor and stator. Slip is typically 2%–5%; higher slip relates to higher torque generation but also relates to lower efficiencies.

No. of poles	Synchronous speed	
	50 Hz	60 Hz
2	3000	3600
4	1500	1800
6	1000	1200
8	750	900
12	500	600
16	375	450
18	333	400

Meeting the operational speed range of the compressor is important in gas compression systems because centrifugal compressors and most reciprocating compressors operate most efficiently in terms of capacity control by varying speed [20]. To vary the flow rate without speed control, centrifugal compressors involve suction or discharge throttling or recycling gas. Both of these capacity control options are significantly less efficient than changing the rotational speed of the centrifugal compressor. For reciprocating compressors, capacity may be varied by other means besides recycling flow and speed variation, such as opening volume pockets, deactivating the head-end of a cylinder, or delayed valve opening/closing.

However, speed variation provides substantially more rangeability and control of the reciprocating compressor throughput. For these reasons, a large majority of electric motor driven gas compression systems will require design for adjustable speed, typically accomplished through a variable frequency drive (VFD) controlling the motor or a variable speed hydraulic drive (VSHD) with a fixed speed motor. An alternative that is rarely used is to use a multispeed motor, available in two-speed, three-speed, or four-speed configurations.

Meeting the operational speed range of the gas compressor is the primary issue specific to this application in the selection of the electric motor drive train configuration. However, other common electric motor issues must also be considered for the gas compressor application as well. The cost, complexity, and reliability of the drive train will be impacted as more components are added.

Electric motor configurations include constant-speed motors driving the compressor via a variable speed gearbox (VSHD, Fig. 6.27) and variable frequency drive speed (VFD)-controlled motors driving the compressor either directly or via a gearbox (Fig. 6.28). The package for a constant-speed motor would look similar to Fig. 6.28.

Of importance for many applications are the performance characteristics of the driver, for example, the power as function of ambient conditions, or the power output at various output speeds. In general, a VFD-controlled motor is a constant torque machine, thus exhibiting a linear drop in power with speed (Fig. 6.29), implemented by maintaining a

FIG. 6.27 Electric motor drive package, constant-speed motor driving the compressor via a variable speed gearbox. *Courtesy of Solar Turbines.*

FIG. 6.28 Electric motor drive package, VFD-controlled motor driving the compressor via a speed increasing gearbox. *Courtesy of Solar Turbines.*

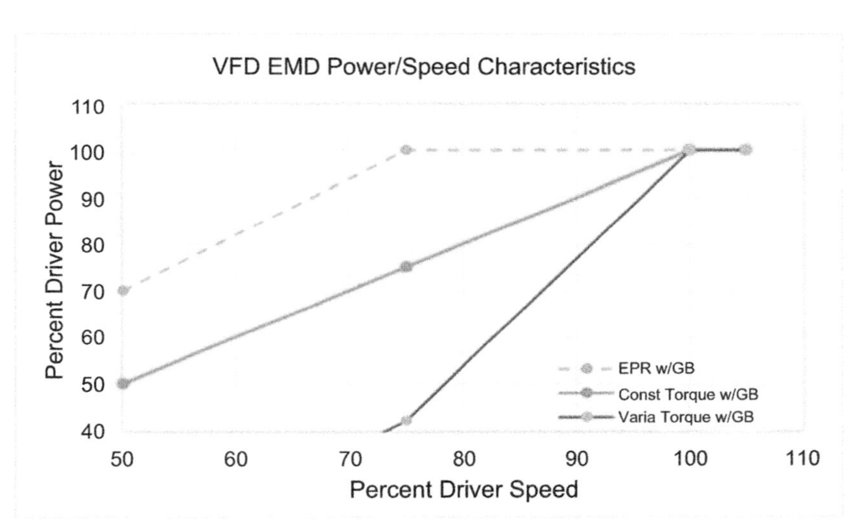

FIG. 6.29 Speed-power relationship for VFD-driven electric motors. *Courtesy of Solar Turbines.*

constant Volts/Hz ratio, until, above a certain corner frequency, the motor becomes power limited. There are exceptions to this behavior, where the motor is oversized to provide constant power over a wider range (expanded power range, EPR), or where, often for thermal reasons, the torque is reduced with speed [21]. The speed-power relationship has a significant impact on control concepts for variable speed drives. In particular, the linear reductions of power seen in most VFD-controlled electric motors impose a limit on the flexibility, compared to VSHD and two-shaft gas turbine drives.

The starting characteristics, including the amount of torque at low speeds, or, for constant-speed electric motors, the amount of additional current is required during starting must be considered (Fig. 6.30).

Whether the units are installed on shore, offshore or subsea, determines access to maintenance intervention, as well as the environmental conditions (for example, salt in the air) the equipment has to be designed for. For electric drivers, the question is also whether the electricity can be brought to site via transmission lines, or whether it has to be generated on site, usually with gas turbine driven generators.

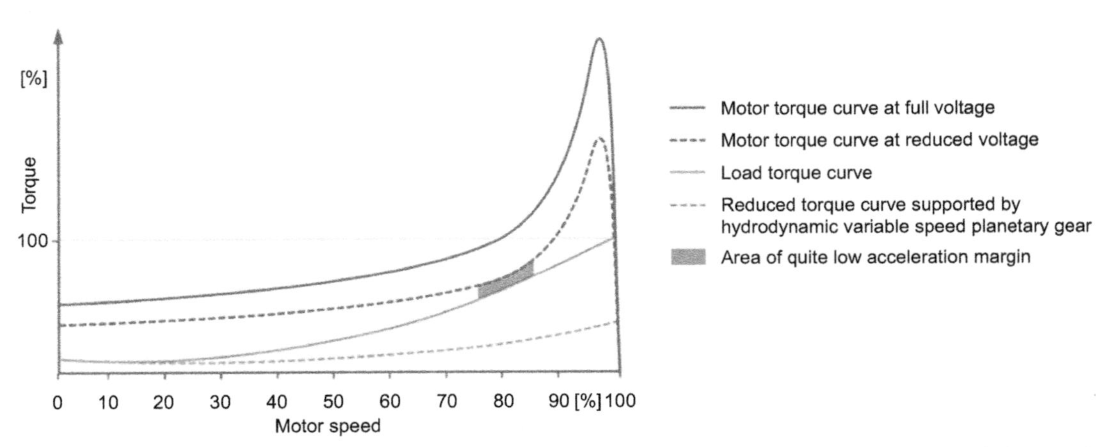

FIG. 6.30 Squirrel cage induction motor capability curve at different voltage levels and load torque curve. *From M. Glasbrenner, B. Venkataraman, R. Kurz, G.J. Cole, Electric-motor driven gas compressor packages: starting methods for large electrical motors and torsional integrity, in: Proc. 46th Turbosymposium, Hosuton, TX, 2017.*

Drivers—Reciprocating engines

The majority of reciprocating compressors used in the upstream and midstream industry segments today are driven by natural gas-fueled engines. The gas being compressed is also used to fuel the engine. The compressor is matched to the engine power and speed connected by a flexible coupling, commonly referred to as a separable compressor. Separables were built around 1970. Separables are available to match gas engines (less commonly to diesel engines) and are available in the power range of 200 KW up to around 7000 KW. Modern separable gas engines are almost exclusively four-cycle designs.

The primary market for reciprocating engines is for electric generators, and they are therefore designed to run at electric synchronous speeds such as 720, 900, 1200, and 1800 rpm (rpm) at 60 Hz frequency and 750, 1000, and 1500 rpm at 50 Hz. Natural gas-fueled engines are available from 40 to 6000 KW. As engines get larger, they run slower.

Integral engine-compressors

An engine/compressor configuration that is not prevalent today is the integral engine/compressor. This is a design where the compressor cylinders are mounted on the engine frame and driven by the engine crankshaft. Integral engine-compressor power levels were typically 4500 KW or less, running at 360 rpm or less, and very few have been built since the year 2000. Many thousands of integral engine-compressors are currently operating. Integral engine-compressors were built both as two- and four-cycle.

Compressor station design considerations

In natural gas transmission, compressor stations provide the necessary pressure boost to keep the natural gas flowing from the suppliers to the users. Compressor stations should be designed to operate efficiently to minimize parasitic losses on the pipeline and minimize any adverse effects on the pipeline (i.e., pulsations, capacity bottlenecks, etc.). This becomes challenging as pipeline conditions and demands change, as legacy compressor stations add new assets to meet demand, and as new stations are brought online. The ever-evolving operating conditions and requirements present a unique design and operational challenge.

This section provides a high-level summary of station design considerations that should be evaluated for new stations, new asset installs, and legacy assets with new operating conditions. These analyses evaluate piping code compliance, pulsation and vibration risks, and station operating efficiency.

Acoustic considerations

Piping acoustics are often top of mind when installing reciprocating compressors but should be considered for installations with both reciprocating and centrifugal compressor. The piping layout at the station will have inherent acoustic natural frequencies: piping lengths, volumes, and pipe stubs will each have an acoustic natural frequency dependent on the physical dimensions of the piping and the speed of sound in the gas. For an open-closed piping segment (pipe stub), the acoustic natural frequency is calculated as follows:

$$f_n = n\frac{c}{4L}, \quad n = 1,3,5,\ldots \tag{6.16}$$

where c is the speed of sound in the gas, and L is the length of the pipe stub. Here, a standing quarter wave develops (Fig. 6.31).

In an open-open piping segment, the acoustic natural frequency is a half-wave calculated as follows:

$$f_n = n\frac{c}{2L}, \quad n = 1,2,3,\ldots \tag{6.17}$$

The same half-wave calculation is used for a closed piping configuration, such as a vessel (Fig. 6.32).

Acoustic excitation can be supplied from numerous sources. Reciprocating compressors inherently produce significant acoustic excitation (or pulsations) at multiples of the compressor's running speed. If the acoustic excitation

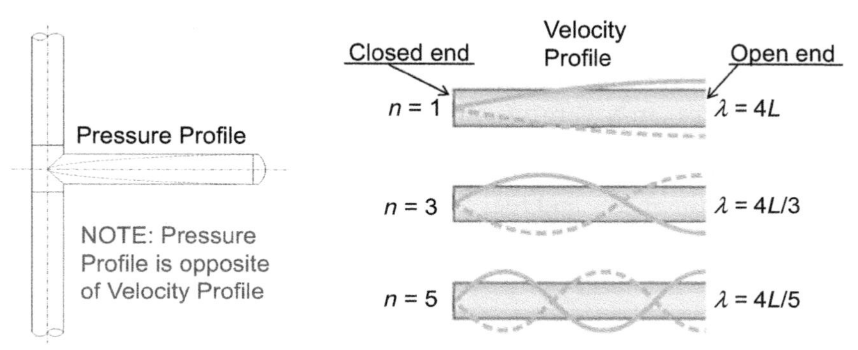

FIG. 6.31 Pipe quarter wave mode shape and frequency [22].

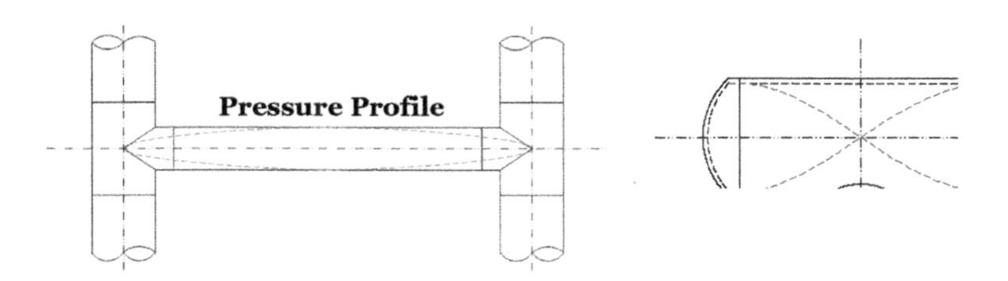

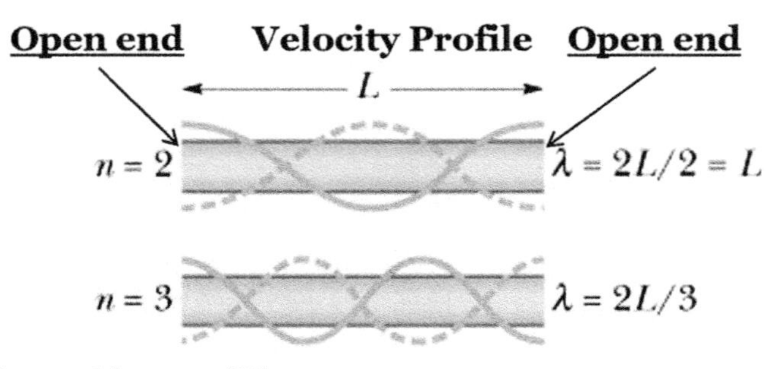

FIG. 6.32 Pipe half wave mode shape and frequency [22].

frequency aligns with the acoustic natural frequencies in the piping system, an acoustic resonance occurs, generating pulsations with significant amplitudes. Pulsations can become problematic when, for example,

- The acoustic/pulsation frequency couples with a mechanical natural frequency and causes vibrations,
- The pulsations cause significant errors through flow meters, and
- The pulsations feed back into the compressor cylinder and deter performance.

Acoustic and pulsation issues can also be present on centrifugal compressor installations. For example, acoustic excitation can be caused by blade pass or vortex shedding over pipe stubs. Acoustic resonances in these systems can also be problematic.

Pulsation control devices

Primarily in reciprocating compressor installations, pulsation control devices are designed and installed to attenuate and dampen the pulsation amplitudes generated by the compressor, reducing the magnitude of the acoustic excitation (reference API 618). Pulsation control devices often take the form of acoustic filter bottles (also called pulsation bottles or Helmholtz bottles), which attenuate acoustic excitation above the associated Helmholtz frequency (low-pass filter). These pulsation bottles are designed with a Helmholtz frequency below 1× running speed (if possible) to attenuate excitation energy across the full spectrum. However, in low-speed machines, this will result in a prohibitively large pulsation bottle; here, the bottles are typically sized for a Helmholtz frequency between 1× and 2× running speed (Fig. 6.33).

Where needed, orifice plates can be effective for broad-spectrum pulsation attenuation. Orifice plates can be installed at pipe flanges. Typical installation locations include the cylinder inlet and outlet, the inlet and outlet of the pulsation bottles, and the inlet/outlet of vessels (i.e., scrubbers, knockout drums, or attenuation volumes).

Pulsation bottles and orifice plates can be very effective at pulsation attenuation, but the associated process pressure drop must be considered. Any pressure loss must be compensated with additional head at the compressor to still meet the pipeline requirement. Excess pressure losses through pulsation bottles or orifice plates will result in a significant increase in compressor power requirements and fuel costs.

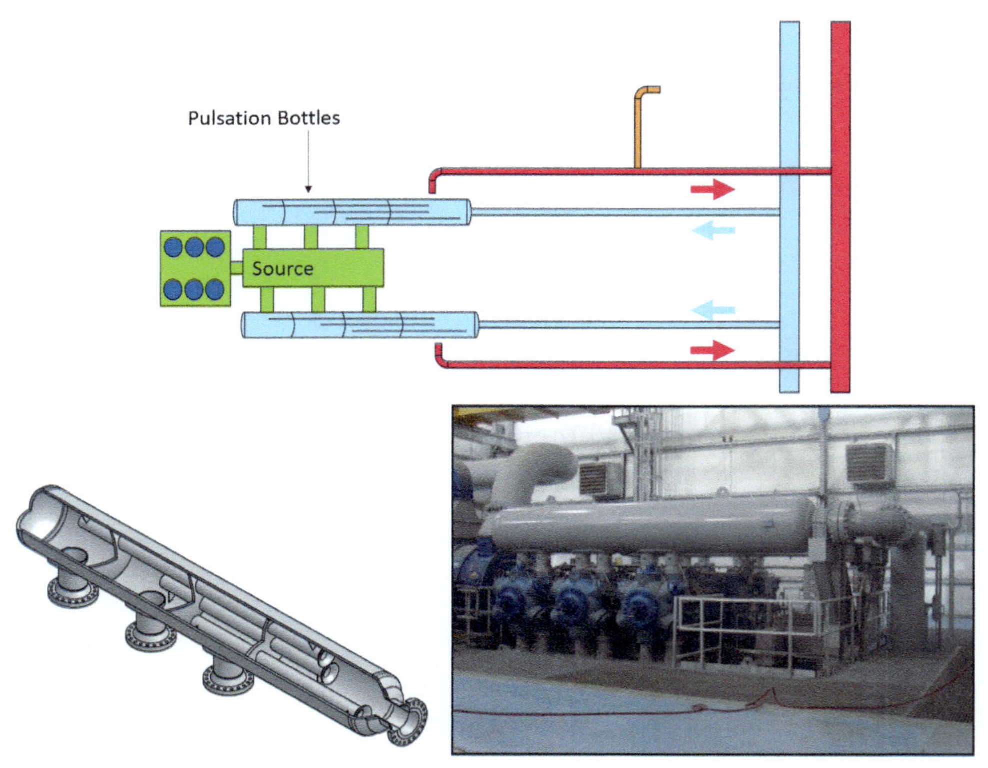

FIG. 6.33 Pulsation bottles on reciprocating compressors [22].

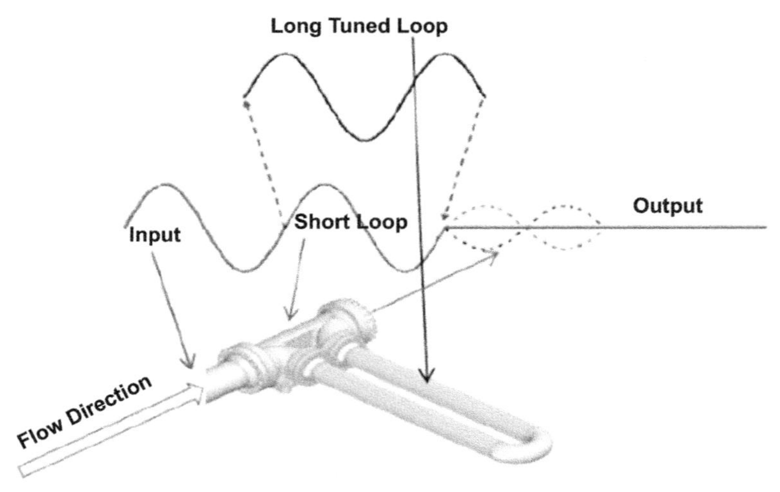

FIG. 6.34 Performance augmentation network (PAN) system for pulsation control [23]. *Credit: SwRI Pulsation & Vibration Short Course.*

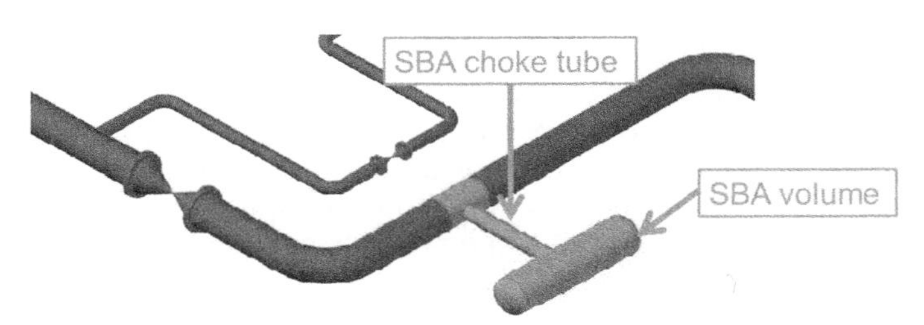

FIG. 6.35 Side-branch absorber for pulsation control [22].

Alternatively, low-pressure-drop pulsation control devices exist; examples include side-branch absorbers and performance augmentation network (PAN) systems that can be tuned to a specific acoustic frequency. These technologies can be useful to attenuate a problematic system-level resonance or tuned to 1× or 2× compressor running speed for fixed-speed machines (Figs. 6.34 and 6.35).

Flow-induced pulsations

In centrifugal compressor installations, flow-induced or vortex-shedding pulsations can occur. Vortices can be generated as the otherwise steady flow moves across a flow disturbance such as a tee, bend, or reducer; a restriction such as a valve or orifice plate; or an obstruction such as a strut, instrument location (e.g., thermowell), or tube. The most common source of low-frequency pulsations in a high-flow piping system is flow across a tee with a dead leg or pipe stub. Higher-frequency pulsations are typically associated with thermowells, valves, and orifices. The acoustic natural frequency associated with these obstructions can be calculated using the Strouhal number.

$$S_t = \frac{f_s d}{U} \tag{6.18}$$

The S_t is the Strouhal number, f_s is the vortex shedding frequency, d is the characteristic dimension of the obstruction, and U is the flow velocity in the unobstructed path. The Strouhal number ranges for various obstruction types are presented in Fig. 6.36.

In these cases, vortex shedding acts as the acoustic excitation source. If the acoustic excitation aligns with the acoustic natural frequency of the local piping system, an acoustic resonance can occur, generating pulsations with significant amplitudes. This is seen most often when the vortex shedding frequency off a tee is coincident with the quarter-wave response of the pipe stub (often a branch line with a normally closed valve). Here, pulsation amplitudes generated in the tee can become significant and impair the performance of the centrifugal compressor or flowmeter.

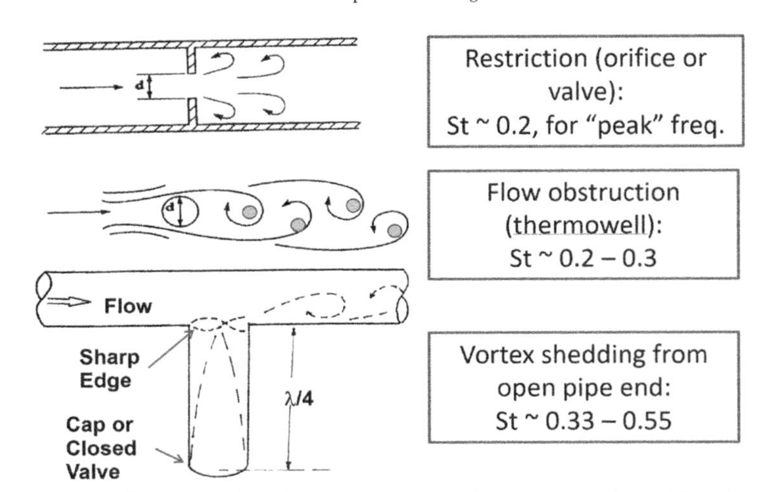

FIG. 6.36 Vortex shedding schematic and estimated Strouhal number ranges for excitation of an orifice, thermowell, and pipe stub [22].

Reciprocating compressor interaction with centrifugal compressors

At stations with both reciprocating and centrifugal compressors in operation, special care should be given to the operating conditions and the potential interaction between the two. Residual pulsations present at the centrifugal compressor will cause oscillations in the compressor's operating point. As the gas pressure fluctuates with the pressure waves (pulsations), the inlet pressure and inlet volume flow at the centrifugal compressor will shift, which could cause the compressor to surge. The residual pulsations at the centrifugal compressor can be used to generate an "operating map ellipse" of the uncertainty in the true operating point (Fig. 6.37). Based on research by SwRI [24], if more than 25% of the operating ellipse is across the surge line, it is expected that a surge could occur.

High-frequency noise and vibrations

High-frequency pulsations can also be problematic in terms of noise and vibration. Mechanical resonance and vibration are common on small-diameter and short branch lines (with higher MNF values), typically instrument lines. High turbulence levels in the discharge piping and blade-pass frequencies can also cause high external noise (110–130 dB). In some cases, the radial acoustic modes in the piping and pipe wall resonances can even be excited. This leads to concerns with pipe wall strain and acoustic fatigue, particularly at welded connections and other stress concentrators.

Mechanical considerations

All piping and compressor systems experience some vibration. The problem is when vibrations are high enough to cause failures, safety issues, reduce reliability, or are uncomfortable for the operators. Vibrations at compressor stations can be caused by an excessive excitation source, a mechanical resonance, an overflexible system, or some combination.

The most common excitation sources at reciprocating compressor stations are acoustic resonances (pulsations), mechanical unbalance at the compressor, gas compression loads (cylinder stretch), and excessive flow velocities or turbulence. Mechanical analyses should encompass the piping system as well as the compressor manifolds. API 618 offers multiple design approaches for reciprocating compressors:

- Design Approach 1: Empirical Acoustical Design (bottle sizing)
- Design Approach 2: Full Pulsation Analysis plus Mechanical Piping Review
- Design Approach 3: Design Approach 2 plus Mechanical Manifold Analysis

For centrifugal compressors, API 617 can be referenced for design guidelines, which also addresses high-speed rotordynamics, torsional behavior, and axial thrust loading. Nozzle loads from piping are addressed below.

Reciprocating compressor manifold

On reciprocating compressors, the mechanical design of the compressor manifold is often challenging. Many excitation sources exist here, and high excitation energy presents increased vibration potential. The "manifold" typically

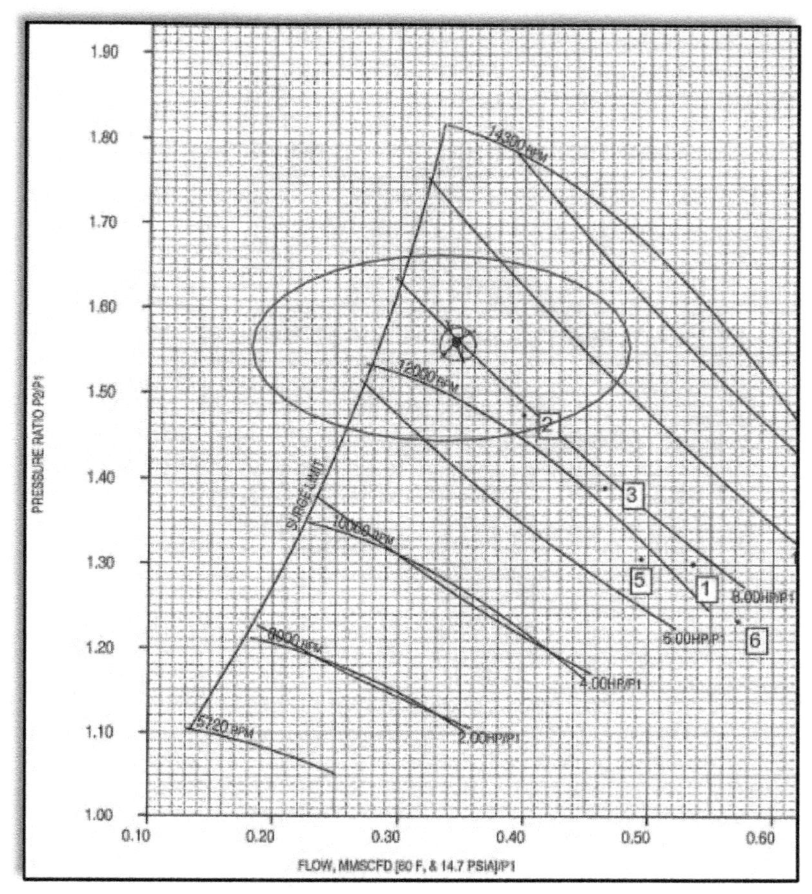

FIG. 6.37 Operating map ellipse showing the uncertainty in the centrifugal compressor operating point due to residual pulsations [22].

consists of the reciprocating compressor frame, distance piece, and cylinders, as well as the acoustic filter bottles and adjacent piping (Fig. 6.38).

A finite element model is often constructed to evaluate the mechanical natural frequencies (MNFs) of the compressor manifold, the forced response of the known excitation sources, and to predict the vibration amplitudes. The modal response frequencies of the various components in the manifold are compared to the compressor running orders to determine if and where mechanical resonances may occur. Resonance avoidance becomes very difficult for variable-speed machines, especially at higher orders.

When it is predicted that an interference will occur (between a modal response and a compressor running order), a forced response analysis is conducted to verify acceptable vibration levels and acceptable dynamic stress values. In the forced response analysis, the compressor unbalance, cylinder stretch, and acoustic shaking forces on the pulsation bottles are applied to the system and used to predict the vibration and stress amplitudes. See an example schematic below. Generally, 1.0 ips 0-pk and 3000 psi 0-pk are considered acceptable limits for the vibration and stress in the compressor manifold (Fig. 6.39).

Centrifugal compressor considerations

Piping restraints

When considering the station piping, pipe routing and restraints will be critical to the presence and magnitude of vibrations. A piping modal analysis can be completed to predict the MNF values and determine if the risk of mechanical resonance exists. Here, a simple beam model analysis can produce reliable results with a low computational cost. Alternatively, a piping restraint review can serve as a simple screening tool.

Most stations experience piping vibrations when the piping system is too flexible. The piping can be stiffened by adding clamp-type restraints with fabric liners for damping. For increasing stiffness, typically, clamps should be

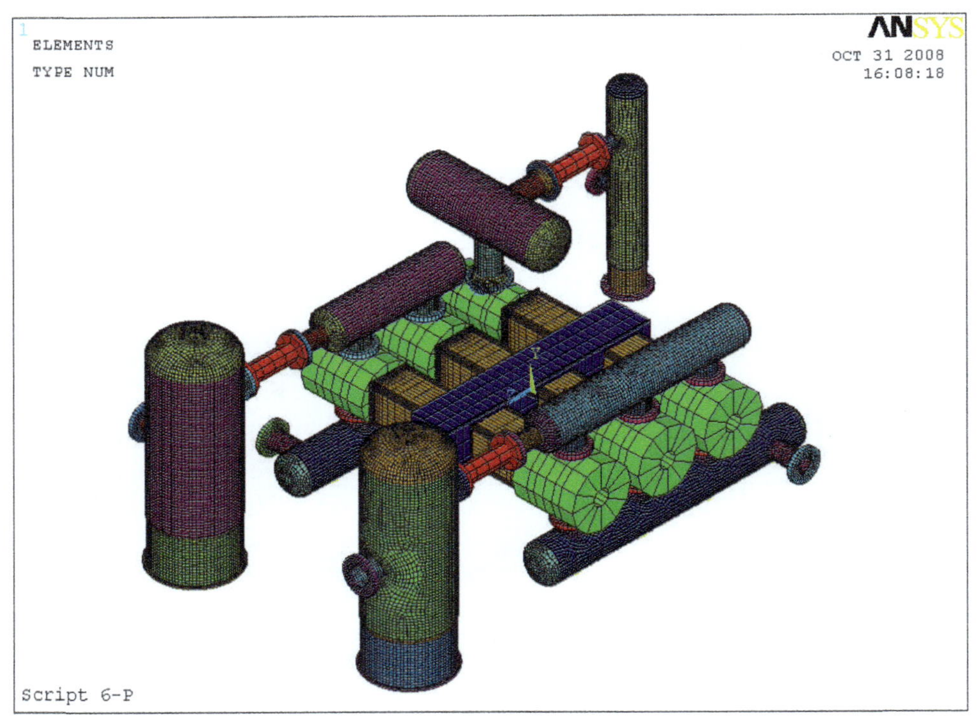

FIG. 6.38 Compressor manifold modal analysis and forced-response analysis [22].

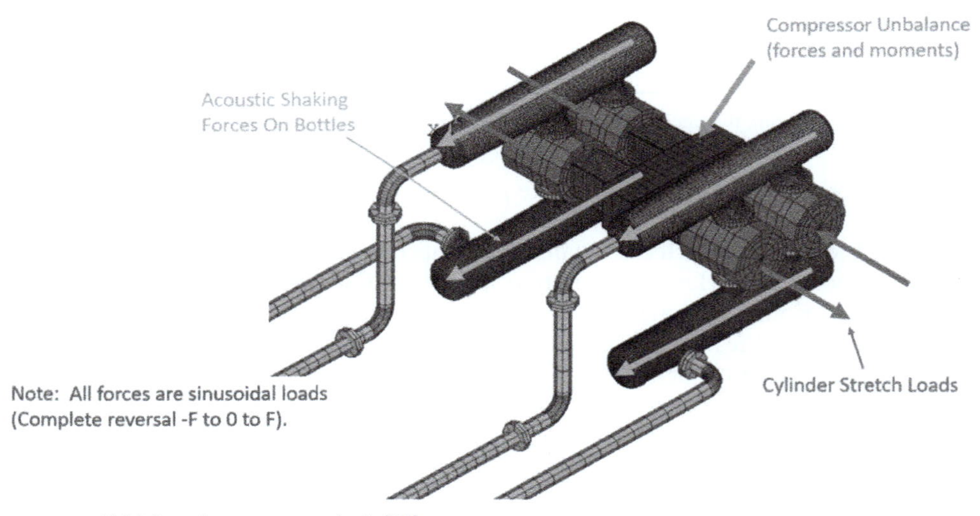

FIG. 6.39 Compressor manifold forced-response analysis [22].

located near elbows and tees, near concentrated masses such as valves, and periodically along long spans of piping.

In selecting pipe restraint types and locations, the mechanical piping modes should generally be above the excitation frequency. For reciprocating compressors, the lowest piping MNF should be 20% above 2× running speed. For centrifugal compressors, the piping MNFs should be above 10 Hz generally and above 15 Hz for lines with flow velocities above 100 ft./s. This generally means that compressors with higher running speeds require shorter spans between pipe clamps.

Note that mechanical and thermal considerations are competing when it comes to system stiffness: mechanically stiff systems reduce the risk of vibrations but will exacerbate thermal stresses in systems with large temperature changes.

Thermal stress considerations

Most compressors have elevated discharge temperatures, which cause thermal expansion in the discharge piping. This causes increased loads and stresses; in overconstrained piping systems, these stresses may become excessive. Compressor piping systems are evaluated per an ASME piping code (typically B31.1, B31.3, or B31.8) to calculate the static stresses due to thermal expansion, pressure, and weight loads as compared to the allowables. The resulting piping loads on equipment flanges are also evaluated compared to the OEM allowables, specifically at compressors and coolers. A thermal piping stress analysis can be used to select the appropriate pipe routing as well as the pipe restraint types and locations.

There are no absolute criteria for when a thermal pipe stress analyses should be conducted, but typically an analysis is done when:

- Operating temperatures above 130°F (54°C)
- Large-diameter piping (12″ and above)
- The design not similar to past installations
- Critical equipment load allowables (cooler, compressors, etc.)
- Critical applications (offshore, etc.)
- When required by the piping code

Piping stresses

Depending on the piping code applied, the piping stresses are evaluated over a range of load cases, including operating load cases (weight, pressure, temperature), sustained load cases (weight plus pressure), and expansion cases (considering the impact of temperature change alone). Analyses should also include wind, wave, and seismic loading, if applicable. Typically, the piping system is modeled as beam elements to reduce calculation time.

The thermal expansion is then calculated based on the pipe size, material, restraints, and routing. The thermal pipe stress model will then calculate the stresses throughout the piping network and evaluate the reaction loads at restraints. The stresses calculated in each load case are then compared to the piping code allowables for compliance (i.e., B31.1, B31.3, B31.8, Z662, etc.). Reaction loads at restraints should also be compared to the allowable loads that the clamp and pier can withstand.

In the case that piping stresses are in excess of the allowables, the piping restraints can be moved, removed, or changed to reduce stiffness. Thermal piping loops can also be added to increase piping system flexibility. Note that these edits are in direct contrast to the recommendations made in the mechanical design considerations section above.

Equipment nozzle loads

As the system piping and equipment grow thermally, the piping can impart large force and moment loads on the equipment nozzles or flange connections. Compressor and heat exchanger OEMs often specify nozzle load allowables to protect the integrity and operation of the equipment. In very stiff piping systems, the equipment nozzle loads can be significant.

For reference, allowable centrifugal compressor nozzle loads typically follow the methodology presented in API 617. Nozzle loads for coolers typically follow the methodology in API 661 or similar. Nozzle load limitations are often specified as forces in the X, Y, and Z directions as well as moments about the X, Y, and Z directions (Fig. 6.40). The limitations for each direction may or may not be the same, depending on the equipment configuration. The API 617 calculation also considers the combined loading on each nozzle in the form of $3 * F_r + M_r$ (using the resultant force and moment values) as well as the summation loads on the machine in the form of $2 * F_C + M_C$ (using the combined force and moment loads) (Figs. 6.41 and 6.42).

Pressure and capacity control

Challenges with off-design operation

Off-design operation is a challenge that most compressors and compressor stations face. Pipeline conditions change on every time scale, by the minute, hour, season, or year depending on the sources, supply, and demand. Specific challenges include, but are not limited to:

- Changes in gas composition over time
  - Causes changes in fluid properties such as density, heat capacity, and speed of sound.
  - Differences in density will change the volume flow requirement at the compressor inlet, moving the operating point on centrifugal compressor maps.

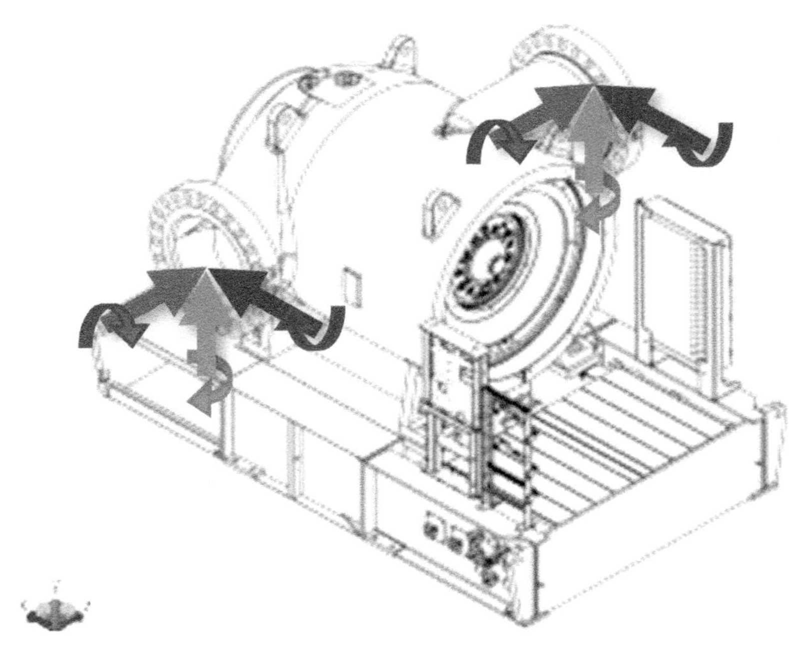

FIG. 6.40 Centrifugal compressor nozzle load directions shown [22] and solar turbines.

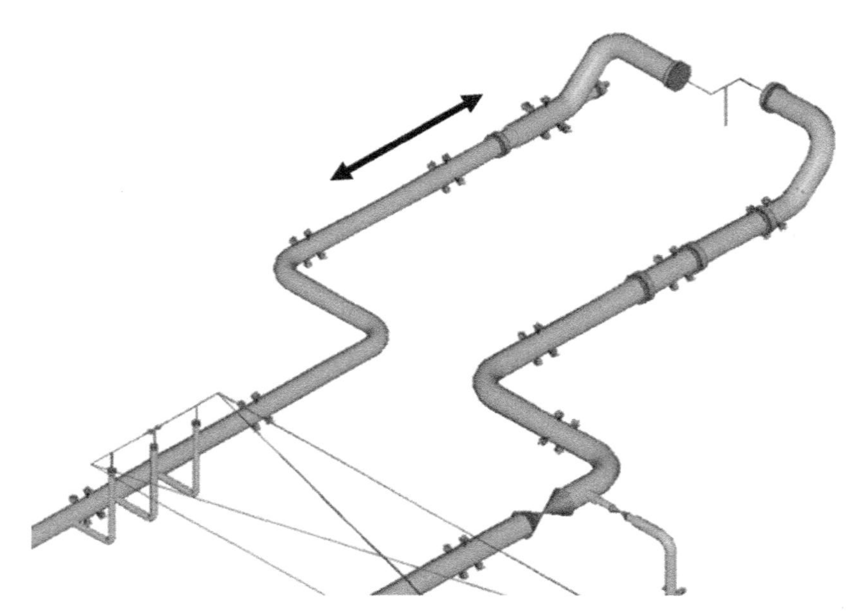

FIG. 6.41 Centrifugal compressor lateral piping, showing primary direction of thermal growth imparted on the compressor nozzle [22].

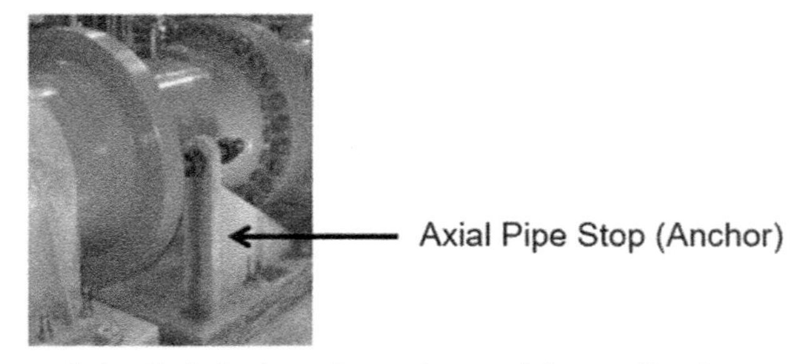

FIG. 6.42 Axial pipe stops are often needed to limit the thermal growth toward the centrifugal compressor [22].

  - Differences in the speed of sound will change the acoustic natural frequencies, potentially causing acoustic resonances where there previously were none.
  - Increased heavy hydrocarbons or water causing issues with liquid dropout, slugging, etc.
  - Difference in gas mole weight and heat capacity will impact the compression behavior and performance.
- Changes to the station inlet pressure, outlet pressure, and flow rate requirements by the pipeline
  - Increased station inlet pressure may be in excess of the pressure rating for the compressor or piping system, requiring suction pressure throttling.
  - Increased demand for pressure ratios or flow requirements may be in excess of what legacy compressors can provide.
  - Increased throughflow may cause excessive pressure losses through pulsation control devices, requiring more power input at the compressor.
  - Installation of new, larger units to support capacity can create issues with station pulsation and vibration control.
  - Installation of new centrifugal units in parallel with legacy reciprocating units.
- Station retrofit reverse flow direction in the pipeline
  - Significant changes in operating conditions and gas composition.
  - Addition of new station piping lines that must be re-evaluated for pulsation, vibration, and thermal.
- On new installations, economics favor 1–2 large compressors instead of many small compressors, which makes turndown challenging and forces off-design operation in many cases.

Changes in operating conditions can force compressors away from the design-point and best efficiency point, which increases the fuel requirement. In some cases, the change in pipeline requirements pushes the compressor outside the acceptable operating range. Two simple solutions employed at some compressor stations are pressure throttling and recycling.

Pressure throttling

If the station inlet pressure is in excess of the rated compressor inlet pressure, suction throttling is often used to cut the pressure to an acceptable value before it is fed into the compressor. This is a low-cost and simple solution, but it significantly increases the power requirement at the compressor and severely degrades station efficiency. If the suction pressure is cut by 200 psi, the compressor then has to generate an additional 200 psi of head to compensate. This is mostly an issue for reciprocating compressors.

Alternatively, if the new station operating conditions are outside of the acceptable operating range (e.g., a lower pressure ratio at a higher flowrate), the compressor speed may be increased to achieve the flowrate, but the discharge pressure is throttled to meet the pipeline required value. This may also be a solution for operating the compressor nearer to the design point. Improving the compressor's operating efficiency is desirable, but this is overall a very inefficient solution for pressure and capacity control.

Pressure throttling is rarely used for centrifugal compressors, as the change in speed allows for automatic adaptation.

Recycle

Alternatively, if the capacity requirement for the station is lower than what the compressor(s) can produce at the given pressure conditions, adding a recycle line is a simple, low-cost solution. Here, a portion of the gas from the compressor discharge is rerouted back to compressor suction through a throttle valve. Adjusting the recycle valve allows for easy control of the station throughput while allowing the compressors to continue operating near their design point. The bypass line can be added around each compressor or at the station level.

Again, this is a simple and effective solution, but inefficient. The recycled gas consumes power as it moves through the compressor(s), which is then lost across the recycle valve, so the station is consuming excess fuel. Also, for long-term operation in recycle mode, the recycled gas should be cooled (so excess energy is being consumed at the coolers as well). If a hot-gas recycle is used near the compressors, the suction gas will be heated; depending on the flow split, a hot gas recycle can cause the compressor to go down to a high-temperature in a matter of minutes. This is also a concern when using a hot gas recycle for surge control.

Options for efficiency improvement

Centrifugal compressors

Speed control on centrifugal compressors can be an effective option for expanding the operating range of the compressor, as opposed to recycling. Reducing the compressor speed will also reduce the compressor power in turndown conditions. However, speed control has limitations within the original operating map of the compressor as well.

For constant-speed machines, installing inlet guide vanes or variable inlet guide vanes can improve the aerodynamic efficiency under the new operating conditions, bringing compressor efficiency closer to the original design point. This is worth considering at the initial installation or as a retrofit.

On new installations, consider turndown and the most frequent operating conditions. Often, compressors are sized for the maximum operating conditions (pressure ratio or flow). This forces the compressor to operate well away from the design point the majority of the time. Alternatively, the compressor should be sized (if possible) so that the normal operating conditions are near the best efficiency point and the maximum operating conditions are at the edges of the compressor map but still achievable.

Reciprocating compressors

Speed control is also an option that should be considered for reciprocating compressors as well. Many legacy units have fixed speeds, but adding speed control allows for improved capacity control while saving compressor power.

Compressor valves on a reciprocating compressor can be a significant source of loss. If the valves are not properly sized, the valves can result in excessive pressure losses or limit the throughput capacity. The valve lift and spring stiffness should be selected for the most frequent operating conditions; poor selections on the lift and spring stiffness may result in excessive valve flutter, excessive valve open/close impact (leading to failures), and overconstriction of the flow path. Resizing and replacing compressor valves is a good retrofit option to improve performance.

For unloading and capacity control, cylinder unloaders should be considered; examples include valve finger unloaders and fixed or variable volume pocket unloaders. Unloaders can be used to deactivate one end of the cylinder to reduce capacity; this also has some reduction in compressor power. Pocket unloaders add volume to the cylinder to reduce the volume discharged by the cylinder end on each cycle, reducing compressor power. Both options present a better capacity control option compared to recycling. Note that, in these installations, the rod load and compressor unbalance should be reconsidered.

Instrumentation and data acquisition

Installing instrumentation throughout the station can be very helpful in quantifying performance and troubleshooting any operational issues. Pressure and temperature transducers should be installed in as many locations as possible. A mass flow meter on each unit is not typical, but it can be very useful in quantifying performance in real time.

Dynamic pressure transducers can be installed to measure pulsation amplitudes; it could be useful to monitor for residual pulsations at the inlet and outlet of a centrifugal compressor that is operating in parallel with reciprocating compressors.

Real-time vibration measurement on or near critical equipment can also be useful to identify increased vibrations over time. Most equipment (such as compressors and coolers) will come with some sort of vibration sensor included, but monitoring in additional areas could be useful.

Data acquisition setups can also vary greatly. Utilizing a data acquisition and control system that records and stores operating data can be very useful for troubleshooting poor performance. Also, historical data can be used to trend the compressor and station performance over a given time. Knowing the operating conditions that the station has produced can be very useful in making upgrade and retrofit decisions, such as when to restage.

Pipeline hydraulics

Natural gas can be transported over large distances in pipelines. Optimal pipeline pressures, depending on the length of the pipe as well as the cost of steel, are in the range of 40–160 bar (600–2500 psi) balancing the amount of power required to transport the gas with the investment in pipe. Most interstate or intercontinental pipeline systems operate at pressures between 60 and 100 bar (1000 and 1500 psi), although the pressures for older systems might be lower. The gas usually has to be compressed to pipeline pressure in a head station (usually coming from a gas plant). This head station often sees pressure ratios of 2–3. The pipeline compressors are arranged at regular distances along the pipeline, usually spaced for pressure ratios between 1.2 and 1.8. The distinction is sometimes made between mainline stations (that basically operate continuously) and booster stations that are only in operation sporadically to assist mainline compression.

In cross-country transmission of gases in pipelines, periodic booster compression is required, increasing pressure to overcome pressure drops caused by friction in the pipeline (Figs. 6.43 and 6.44).

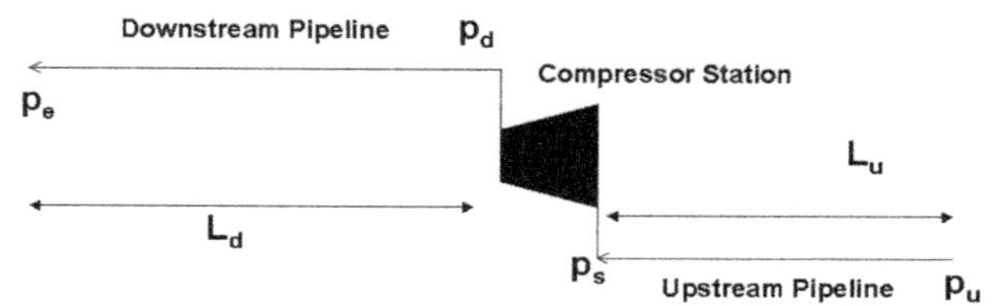

FIG. 6.43 Pipeline schematic. *Courtesy of Solar Turbines.*

FIG. 6.44 Pipeline compressor station with three compressors in parallel operation. *Courtesy of Solar Turbines.*

The more flow is sent through a given pipeline, the higher the station pressure ratio becomes (Fig. 6.45), as the friction losses increase with gas velocity.

Kurz [25] shows that for a given pipeline, under steady-state conditions, the head-flow relationship at the compressor station can be approximated by

$$H^* = C_p T_s \left[\left(\sqrt{\frac{1}{1 - \frac{C_3 + C_4' \cdot Q^2}{p_d^2}}} \right)^{\frac{k-1}{k}} - 1 \right] \tag{6.19}$$

where C_3 and C_4 are constants (for a given pipeline geometry) describing the pressure at the two ends of the pipeline, and the friction losses, respectively. This is displayed in Fig. 6.45. Eq. (6.19) shows a direct relationship between the flow transported in the pipeline and the required head, assuming the pipeline losses are known and the station discharge pressure is defined. This seems somewhat limiting, but one must consider that, to maximize flow in a pipeline system, the station discharge pressure will be at or close to the maximum operating pressure of the pipeline. This means that for a compressor station within a pipeline system, the head for the required flow is prescribed by the pipeline system.

Subsea pipelines often only have a head station but no stations along the line. They are either used to transport gas to shore from an offshore platform (see export compression) or to transport gas through large bodies of water. In either case, relatively high pressures (100–250 bar, 1500–3700 psi) are common [26].

A few onshore pipelines worldwide make use of the added supercompressibility of the natural gas at pressures above 140 bar (2000 psi, depending on gas composition) and operate as "dense-phase" pipelines at pressures between 125 and 180 bar (1800 and 2500 psi). Dense-phase operation at pressures above 140 bar (2000 psi) allows particular to avoid two-phase flows when ambient temperatures drop [6].

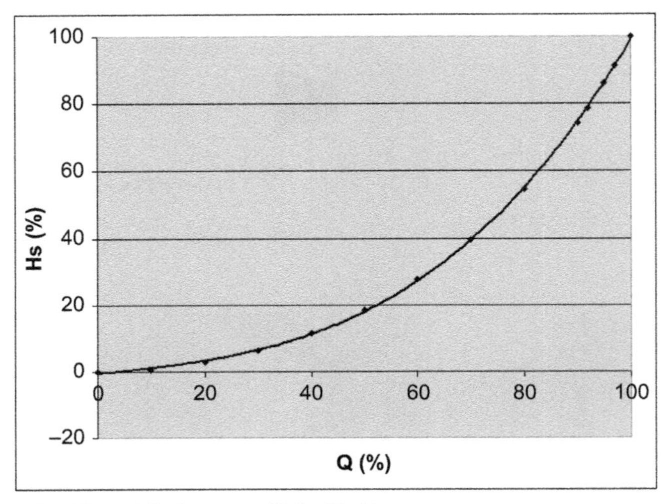

FIG. 6.45 Station head-flow relationship (Eq. 6.19). *Courtesy of Solar Turbines.*

If the throughput of a pipeline has to be increased, two possible concepts can be used: building a parallel pipe (looping), adding power to the compressor station (i.e., adding one or more compressors to the station), or a combination of both. If power is added to the station, the discharge pressure can be increased (assuming this is not already limited by the pipeline maximum operating pressure). The station will therefore operate at a higher-pressure ratio. The added compressors can either be installed in parallel or in series with the existing machines. If the pipeline is looped, the pressure ratio for the station is typically reduced, and the amount of gas that can be compressed with a given amount of power is increased. In either scenario, the existing machines may have to be restaged (for more pressure ratio and less flow per unit in the case of added power, for more flow and less pressure ratio in the other case).

In general, pipelines that have many take-offs, and interconnects (like in the United States) tend to operate more frequently in transient, non-steady-state conditions. Long, transcontinental pipelines that basically transport gas from point A to point B tend to operate closer to steady-state conditions. In most cases, compressors experience a significant range of operating conditions (Fig. 6.46).

Pipeline hydraulics are governed by the conservation of mass, momentum, and energy. To close the conservation equations, there is additionally a model for friction losses, a model for heat transfer with the environment, and an equation of state to describe the pressure-volume-temperature relationship for the gas. Simulations are typically performed numerically, where the pipeline is modeled piecewise; thus, methods of characteristics, finite volume, or finite element schemes are used.

Time dependent mass conservation requires:

$$\frac{\partial(\rho v)}{\partial x} + A\frac{\partial \rho}{\partial t} = 0 \tag{6.20}$$

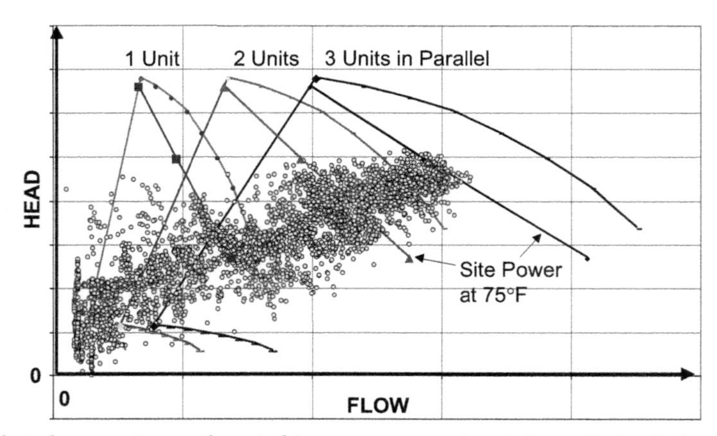

FIG. 6.46 Operating points collected over a six-month period in a gas compression station with 3 units in parallel [27].

where the through-flow area (A) is constant. The density is calculated from:

$$\frac{p}{\rho} = ZRT \tag{6.21}$$

The momentum balance is satisfied by:

$$\rho\left(\frac{\partial v}{\partial t}\right) + \rho v\left(\frac{\partial v}{\partial x}\right) = -\frac{\partial p}{\partial x} - \frac{\rho \cdot f_{DW} v|v|}{2D} \tag{6.22}$$

for a pipeline with no differences in elevation. The fundamental flow equation,

$$Q = C \cdot \frac{T_b}{P_b} \cdot e \cdot D_i^{2.5}\left[\frac{p_1^2 - p_2^2 - \frac{0.0375 \cdot SG \cdot (H_2 - H_1) \cdot p_a^2}{Z_a \cdot T_a}}{SG \cdot T_a \cdot L \cdot Z_a \cdot f_{DW}}\right]^{0.5} \tag{6.23}$$

also called the general flow equation, where $C = 47{,}880$ for flow (Q) in m^3/h and otherwise SI units [28], is widely used in the pipeline industry. It allows for the closure of the loss terms in the momentum equation. The equation is computationally efficient, requires fewer resources compared to other equations, and uses the Darcy Weisbach friction factor (f_{DW}) calculated as a function of the Reynolds number [29].

The compressor's operating point is always the result of the interaction between the compressor and its controls (speed, power, vane settings, etc.) and the characteristics of the system within which it operates. The system is usually characterized by the relationship between the compressor head and flow. The following are the typical cases:

(1) Systems requiring an increase in head with an increase in flow
(2) Systems requiring (more or less) constant head with changes in flow
(3) Systems with significant storage capacity, where the head requirement is a function of the amount of gas stored.

Examples for Type 1 are all pipeline systems, including transmission pipelines, but also gas gathering systems (Fig. 6.47). Type 2 systems often involve separators or other process equipment that need to operate at constant pressure. Another example is large reservoirs and situations where the gas has to be fed into a pipeline that operates at more or less constant pressure. Air compressors, which provide plant air at constant pressure, are another example. Refrigeration processes usually also require constant pressure, independent of flow.

Type 3 systems are often found in storage and withdrawal applications.

It should be noted that the steady-state characteristic of most systems is different from the transient characteristic. In transient operations, the gas inertia as well as mass storage effects have to be considered (see Fig. 6.13).

For any compressor characteristic (represented by its performance map), the compressor operating point is determined by the intersection of the system characteristic and the compressor characteristic.

The concept is explained for a pipeline application:

Depending on how steep the pipeline characteristic (Eq. 6.19, Fig. 6.45) becomes, additional compressors in a station may be arranged in series or in parallel (Fig. 6.48). Also, the characteristic requires the capability for the compressors to allow a reduction in head with reduced flow, and vice versa, in a prescribed fashion. The pipeline will therefore not require a change in flow at a constant head (or pressure ratio).

Letdown stations [31]

Letdown stations play a crucial role in natural gas transmission systems.

Efficient and safe delivery of natural gas from production fields to end-users relies on properly regulating gas pressures along the pipeline route. Pressure letdown stations are critical components that allow operators to reduce pipe pressures in a controlled manner.

The high-pressure gas leaving a compressor station or production field may be at pressures upward of 1000 psi. However, most distribution systems and end-users require pressures below 200 psi. Pressure letdown stations utilize valves and controls to gradually decrease transmission pressures to lower levels suitable for each stage of downstream transport and distribution. This stepped pressure reduction optimizes pipeline efficiency and capacity. It also prevents damage to facilities from excess pressure.

Several factors go into designing effective letdown stations. Key elements include the inlet and outlet pressure parameters, targeted flow rates, emergency shutdown needs, and noise control. Stations may utilize control valves,

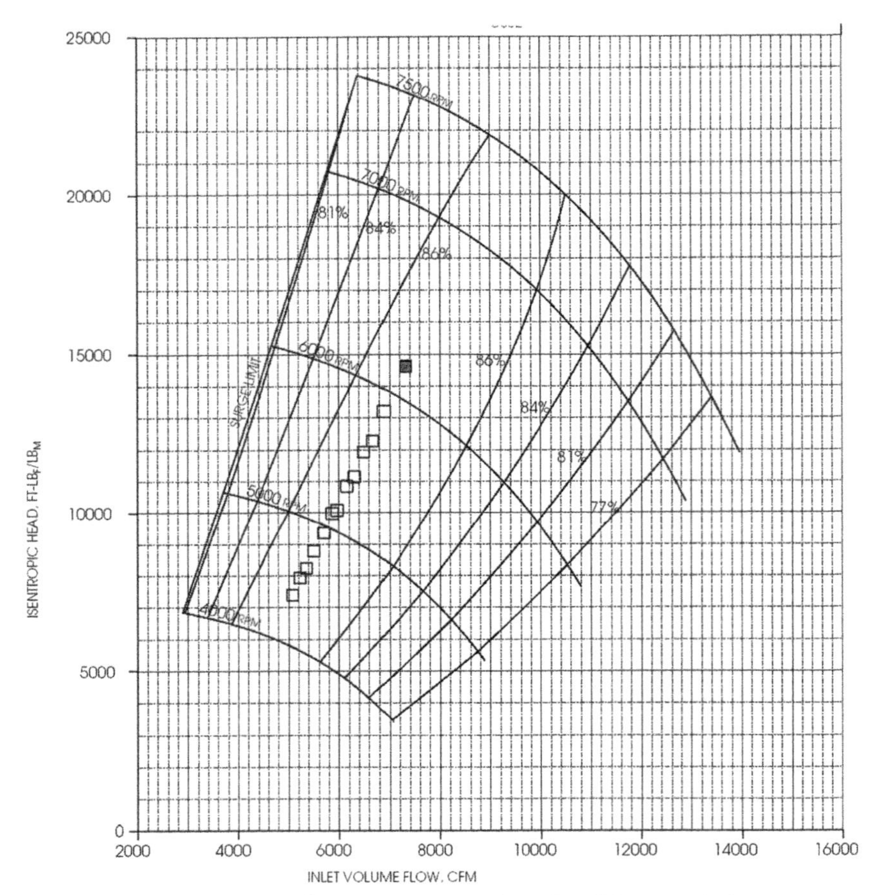

FIG. 6.47 System characteristics: Typical pipeline steady-state operating points plotted into a typical compressor performance map [30].

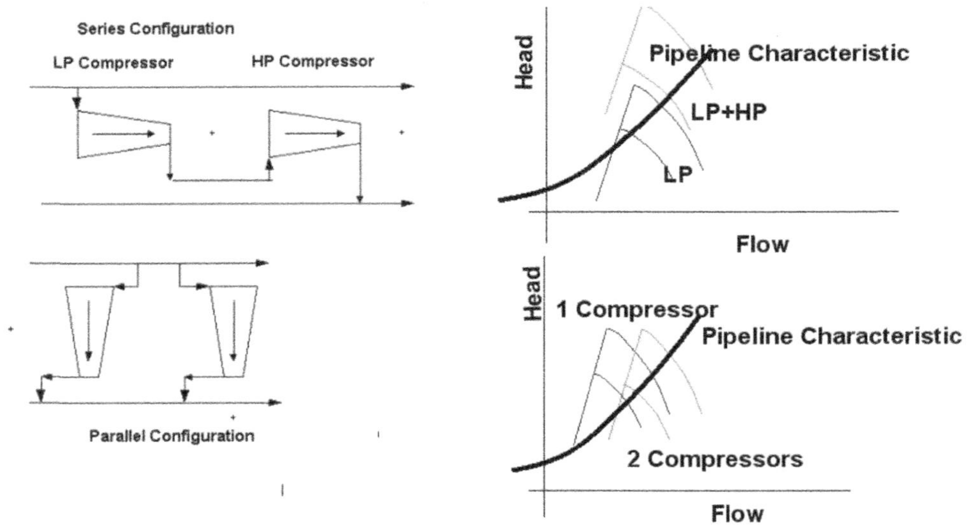

FIG. 6.48 Series and parallel compressors can cover the same duty.

monitoring instrumentation, emergency relief valves, filtration equipment, and surge control devices. Facilities must be designed to handle the maximum anticipated throughput without supply disruption or dangerous pressure spikes. Control systems and energy dissipation mechanisms are chosen to allow smooth, stable pressure regulation.

Proper pressure letdown station design and operation ensure gas flows efficiently at safe pressures along the entirety of a transmission pipeline. This prevents ruptures from excess pressures and allows maintaining contractual delivery pressures to downstream transfer points or distribution networks. Sufficient pressure reduction also serves to compensate for pressure drops due to friction within lengthy pipelines. Intelligently operated letdown stations are vital for economic and reliable natural gas transmission.

Each gas letdown and metering station branches off of the pipeline and is used to reduce pressure and meter the gas to the various users. For the pressure reduction and metering stations, the main equipment includes filters, heaters, pressure reducers and regulators, and flow metering skids. In addition, each station is generally equipped with drains for collection and disposal, an instrument gas system, and storage tanks.

Filter separators

Natural gas filter units are installed at each station to remove any entrained liquids and solids from the gas stream. The filters may comprise cyclonic elements to centrifuge particles and liquids to the sides of the enclosing pressure vessel. These particles and liquids will then drop down for collection in a sump, which can be drained periodically.

A station inlet filter-separator should be installed upstream of the flow meter. The filter separator is normally a horizontal unit with a full-size, quick-opening closure and access platform for element change out. The vessel should be equipped with level gauges, high liquid level switches, and a differential pressure transmitter across the filter elements. The filter-separator sumps should have automatic drain valves.

A condensate tank is installed for atmospheric storage of any liquids removed by the filter separator. Most tanks installed for this purpose are double-walled and installed on a concrete pad. The tank should contain a level gauge and a high liquid level switch.

Flow control

A control valve should be installed downstream of the meter run to control both the flow through the meter and the delivery pressure. This valve will primarily operate to limit the station throughput to prevent the incoming gas volume from exceeding the meter capacity or the nominated volume but will also be equipped with a pressure override.

The control valve is generally controlled by a gas flow computer (GFC) based on set points provided by the gas control center. The control valve will normally operate in the fully open position to minimize pressure losses through the station and should have a positioner, position indicator, and position transmitter.

The GFC would also monitor and control the facilities as well as perform custody transfer quality measurement. The GFC communicates all data to a central control console via the SCADA system.

At custody transfers, a gas chromatograph is generally used to determine the gas composition for purposes of calculating the gas gross heating value. These data are provided to the GFC for use in calculating the total gas heating value of the metered gas. A gas sample is taken from a continuously flowing location on the meter and regulator skid. The gas sample is secured at low pressure to minimize lag time utilizing a self-regulating sample probe and routed to the gas chromatograph and moisture analyzer. The moisture analyzer is provided to measure the water content of the gas. Depending on the sulfur content of the gas, a sulfur analyzer may be required.

Meter skid piping

The piping configuration on the meter skid should allow for bidirectional gas flow through the station through an appropriate piping and valve manifold. However, the gas flow through the meter and regulator should be in one direction only.

The control valve is installed between isolation ball valves to allow maintenance. It is prudent to install a manual bypass valve to allow continued operation during control valve maintenance activities.

Automatic shutdown valve

An automatic shutdown valve is normally installed at the pipeline connection. This valve should be remotely operated from the main operating system and equipped with local pneumatic controls, a hydraulic manual override, and open/close limit switches.

Blowdown of the meter station piping is accomplished by a vent stack located on the station inlet piping and vents on the meter skid located downstream of the meter and downstream of the flow control valve. Vent stacks may or may not include silencers, depending on the noise levels at the closest noise sensitive area (NSA).

Heaters

Natural gas heaters are installed to avoid the formation of hydrates, liquid hydrocarbons, and water as a result of pressure reduction. The gas heater is designed to raise the temperature of the gas so that after pressure reduction, the temperature of the gas will be above the dew point temperature at operating conditions and maximum flow. The heater is a water bath of natural circulation type maintained at a temperature between 158°F and 176°F. Where gas costs are high, an alternative is to use high-efficiency or condensing furnaces for the purpose of preheating the gas rather than the water bath heater.

Pressure reduction and regulation

The pressure-reduction system controls the supply pressure to the gas users at a regulated value. Each system consists of at least two trains of pressure reduction—one operating and the other standby. Each train will normally comprise two regulator valves in series.

Regulator valves should be sized for the maximum anticipated volumes at the minimum anticipated inlet pressure during those times of maximum volume. For stations serving multiple residential customers or other uninterruptible services, sufficient regulator capacity needs to be provided so the failure of one regulator valve run will not reduce the facility capacity below the required demand. Regulator valves in custody transfer stations are typically of the type that fail in the open position.

Sound pressure

Sound pressure levels at all service conditions should be considered. High noise levels (generally defined at greater than 110 dbA) can result in damage to regulators, control valves, control valve accessories, instrumentation, and downstream piping. The following are standard measures that can be taken to reduce sound pressure levels or to reduce the effects of the path:

- Install noise-attenuating trim or diffusers on the regulator,
- Install heavy wall pipe,
- Install insulation,
- Install silencers,
- Bury the regulators.

Overpressure protection

Stations do not require an overpressure relief device if a monitor regulator is installed in series in each regulator run or if a monitor regulator is installed in series with and common to all regulator runs.

Station relief device capacity should be the largest relief capacity requirement determined from the following criteria using flow and pressures:

- Failure of the single or largest capacity run that does not include a monitor regulator.
- Failure of all runs in which the regulators would fail to open upon failure of a single common instrument or instrument line.

Minimum relief device capacity for the failed regulator(s) should be the maximum total flow at the differential pressure between the inlet and outlet of the regulator(s), in the case where the inlet pressure is the upstream line MAOP or maximum source pressure, whichever is less, and the outlet pressure is the downstream MAOP plus allowed overpressure.

Metering system

The flow rate of the gas has to be measured at a number of locations for the purpose of monitoring the performance of the pipeline system and more particularly at places where custody transfer takes place. Depending on the purpose of metering, whether for performance monitoring or for sales, the measuring techniques used may vary according to the accuracy demanded.

Potential for future expansion should be considered when sizing the meter skid length.

Typically, a custody transfer metering station will comprise one or two runs of pipe with a calibrated metering orifice in each run. Should an ultrasonic meter be required, it should be designed to meet or exceed the requirements established for ultrasonic meters in AGA-9. Typically, the ultrasonic meter will be a multipath meter, and the meter tubes will be equipped with a flow conditioner. The fully assembled meter tubes should be calibrated at line pressure and full-flow conditions prior to use. Normally, the ultrasonic meter tubes will be designed for a minimum 10D upstream length from the flow conditioner to the meter and 5D lengths downstream of the meter. Further, the meter tube should be honed.

Pulsation

The elimination of pulsation through the use of pulsation control devices is an important step to take. Pulsation has a tendency to introduce meter errors. Computer analogs are used in the design of pulsation-control equipment. There are several methods of determining if levels of pulsation will cause meter errors, but assessment of the square-root error remains the best rule of thumb in determining if pulsation-control equipment is needed to improve absolute accuracy.

The square-root error is very predictable and is always positive. This error will always indicate a flow that is greater than the actual flow. Instability errors (which are pulsations that change the orifice coefficient) can vary in magnitude and can also be either positive or negative. A system with severe pulsation only needs a slight change in frequency (as little as a few Hz) to result in an error of several percent.

Cathodic protection

It is typical to separate the cathodic protection systems of the pipeline and meter station. This is normally done by installing insulation kits at the flange connections at the meter skid. Buried piping within the meter station, either upstream or downstream of the meter skid, should be cathodically protected from the associated pipeline's cathodic protection system.

Buildings

When conditions or regulations require a building to be utilized, typically a pre-engineered-type building designed in accordance with the International Building Code is used and is mounted on the meter skid to enclose the ultrasonic meter and flow control valve area only. Normally, the building will not be heated or insulated. Buildings are sized to allow the use of a meter-sensor removal tool.

The EFM and GC buildings are normally two separate buildings mounted on a common skid. As opposed to the meter building, the EFM building should be climate-controlled (heated and air-conditioned) and sized for the flow-control equipment and associated uninterrupted power supply (UPS). The GC building is not required to be climate-controlled but should have a hazardous gas detector with a warning strobe light.

Shipping

Gas gathering

Gas gathering systems are crucial components of the natural gas production and transportation processes. The subsea gas gathering process involves the collection of natural gas from offshore wells located on the ocean floor. Subsea gas gathering pipelines are then used to transport the gas to floating liquefied natural gas (FLNG) facilities for processing and transportation to onshore markets.

Subsea gas gathering pipelines are typically made of steel and are designed to withstand the harsh conditions of the ocean floor, including high pressure, low temperature, and corrosive seawater. The pipelines are laid on the ocean floor and connected to the subsea wellheads using specialized equipment and techniques. Optimizing the location of gathering floor sites was described by Arsenyev-Obraztsov [32].

- Determine a set of sections of the pay area where it is reasonable to locate bottom holes (the number of wells is predefined);
- Determine the geometrical configuration (chart) of the horizontal wellbore (the length of the horizontal wellbore is preset);

- Determine the allocation of the subsea production units (the number of units is prespecified);
- Determine central gathering unit allocation.

Once the natural gas is gathered and transported to the FLNG facility, it undergoes a series of processes to remove impurities and convert it into a liquid state for transportation. These processes include compression, liquefaction, and storage. The FLNG facility can then transport the liquefied natural gas (LNG) to onshore facilities using LNG carriers.

Subsea gas gathering systems and FLNG facilities play a critical role in meeting the world's growing demand for natural gas. These technologies enable offshore gas fields to be developed and produced economically and efficiently, thereby contributing to the energy security of many countries around the world.

Floating production storage and offloading (FPSO)

An FPSO is a type of offshore vessel used in the oil and gas industry to process, store, and offload crude oil, natural gas, and other hydrocarbons produced from subsea wells.

The FPSO typically receives the hydrocarbons from subsea wells through flexible pipelines or risers, and then processes and stores them on board. The processed hydrocarbons can then be offloaded to shuttle tankers, pipelines, or transported via a loading buoy to other vessels or facilities for further processing or transportation to market.

FPSOs are popular because they are able to operate in remote and deepwater offshore locations, where fixed structures or pipelines are not economically feasible. They can also be quickly deployed and redeployed, making them a flexible option for oil and gas companies. Additionally, the FPSO can be designed to withstand harsh weather conditions, ensuring safe and efficient operations in challenging offshore environments.

Floating liquefied natural gas (FLNG)

Floating liquefied natural gas (FLNG) describes an offshore facility tethered to the sea floor that recovers natural gas from a subsea field. The FLNG captures natural gas from below the sea floor, purifies, and liquefies the gas, and then transfers the resulting LNG to transport vessels/carriers. "The first FLNG cargo was shipped in 2017. Three FLNG vessels are now operational: Petronas' PFLNG Satu, Golar LNG's Cameroon FLNG, and Shell's Prelude. Of the seven major LNG projects sanctioned between 2015 and 2018, three were FLNG projects: Cameroon FLNG, Eni's Coral FLNG, and BP's Tortue FLNG." [33].

An FLNG is more economical than pumping the recovered gas to the shore. FLNGs do offer lower capital investment (CAPEX), manageable costs (OPEX), and flexibility to move to other sites. But there are limits on capacity; onshore sites offer larger capacity and are better suited to larger gas fields. An FLNG will have a faster build time than a comparable ground facility, due in part to lower permitting regulations. By using faster construction methods, projects can be completed in a shorter timeframe, which can lead to earlier cash flows. This was demonstrated by the final investment decisions (FIDs) made for the Coral FLNG project in Mozambique in 2017 and the Tortue FLNG project offshore Mauritania and Senegal in 2018. These projects were able to reach FID status due, in part, to their use of accelerated construction methods, which enabled them to begin generating revenue sooner than if they had used traditional construction approaches. This early cash flow helped support further development and investment in the projects. Both of these projects are the first LNG facilities in a frontier region. Expansions of the facilities in this area are likely to follow. Despite the success in the 2010s, the continued outlook for further FLNG FIDs is limited.

LNG carriers

LNG is a natural gas that has been cooled to a liquid state, which reduces its volume by about 600 times and makes it easier and more economical to transport over long distances. LNG carriers are typically large and have insulated cargo tanks to keep the LNG at a temperature of around −160°C, which is required to keep the gas in its liquid state.

The LNG is loaded onto the carrier at a liquefaction plant, which is typically located near the gas production site. Once loaded, the carrier sails to the receiving terminal, where the LNG is offloaded and re-gasified, usually through an onshore terminal or floating storage and regasification unit (FSRU), and then transported via pipeline to its final destination.

LNG carriers are designed to be highly specialized and safe, with features such as double hulls, redundant systems, and advanced navigation and communication equipment to ensure safe and reliable operations. They are also subject to strict international regulations and standards to minimize the risks associated with transporting large quantities of LNG.

Ship propulsion

It is important to understand the propulsion system of LNG transport vessels, as the natural gas being transported plays a role in that propulsion. Natural gas is transported in a liquid state, and during the course transport, some level of the gas "boils off" and becomes vapor. This boiled-off gas (BOG) is managed in several different ways: flaring, reliquification, or the gas can be transmitted to the drive system and used as fuel. The mechanism chosen for managing the BOG depends on the ship engine construction, amount of BOG generated, general ship construction, etc.

Electricity for the propulsion of LNG vessels can be provided by various types of generators. While the primary use of electricity is for the motor-driven propellers, many other ship systems also rely on electricity.

Traditionally, Marine Steam Turbine drives have been the most common propulsion system for LNG transport and similar vessels. However, with advances in technology, the dual-fuel diesel-electric (DFDE) system has become more prevalent. These engines are capable of burning both diesel oil and gas, providing greater efficiency than steam turbine propulsion vessels. In some cases, gas turbines have been used.

The efficiency gains of DFDE engines have made them a popular choice for LNG vessels, as they offer improved fuel economy and reduced emissions. While the main purpose of the electric system is to power the ship's propellers, the DFDE system's ability to generate electricity also supports other ship systems.

A tri-fuel diesel-electric engine is a type of propulsion system that can run on three different types of fuel: diesel, natural gas, and marine gas oil (MGO).

In this type of engine, a diesel generator produces electricity that powers the electric motors that turn the ship's propeller. The diesel generator can run on any of the three fuels, with the engine's control system automatically adjusting to optimize performance based on the fuel being used.

The ability to run on multiple types of fuel provides greater flexibility and resilience for ships that use tri-fuel diesel-electric engines. For example, if the price of diesel fuel rises or the supply is disrupted, the ship can switch to natural gas or MGO to continue operating.

Tri-fuel diesel-electric engines are commonly used in a variety of marine applications, including offshore vessels, ferries, and cruise ships, as well as in power generation for remote areas where access to natural gas is limited. They offer improved efficiency, reduced emissions, and greater flexibility compared to traditional diesel engines, making them an increasingly popular choice for shipbuilders and operators.

Marine steam turbine engine

A marine steam turbine is a type of propulsion system that uses steam to drive the ship's propellers. The steam is generated by heating water in a boiler, which creates high-pressure steam that is then directed to the turbine. The turbine converts the thermal energy of the steam into mechanical energy, which drives the propeller shafts and generates forward motion for the ship.

Marine steam turbines have a long history of use in naval and commercial vessels, with the technology dating back to the 19th century. They were particularly popular in the early 20th century, with many large ocean liners and battleships using steam turbine propulsion.

While marine steam turbines were once the most common propulsion system for large ships, they have largely been replaced by other types of engines, such as diesel and gas turbine engines. However, steam turbines are still used in some niche applications, such as in nuclear-powered ships.

Steam turbines offer advantages such as high efficiency, durability, and reliability, but they also require a significant amount of maintenance and can be expensive to operate. Modern propulsion systems, such as diesel-electric and gas turbine engines, offer improved fuel economy and reduced emissions, making them a more popular choice for many types of vessels.

Dual-fuel diesel electric (DFDE) propulsion systems

A dual-fuel engine is a type of internal combustion engine that is capable of running on both gaseous and liquid fuels. This innovative technology offers a flexible solution for power generation, as it allows operators to choose between different fuel sources based on availability, price, and environmental considerations.

One of the key advantages of a dual-fuel engine is its ability to operate in gas mode, which uses the Otto process. During this mode, the engine runs on a lean air-fuel mixture that is fed into the cylinders during the suction stroke. This results in high thermal efficiencies, with recorded figures exceeding 47%. This makes the dual-fuel engine an efficient and cost-effective option for power generation.

Alternatively, the engine can operate in diesel mode, which uses the diesel process. During this mode, diesel fuel is fed into the cylinders at the end of the compression stroke. The engine is optimized for running on gaseous fuels, and diesel fuel is typically used as a backup fuel in case of gas supply interruptions or emergencies.

The dual-fuel engine offers a seamless transition between different fuels during operation, without any loss of power or speed. This makes it an ideal choice for applications that require continuous and reliable power generation, such as in the marine, oil and gas, and power generation industries.

In addition to its flexibility and efficiency, the dual-fuel engine is designed to have the same output regardless of the fuel used. This means that operators can switch between different fuels without having to adjust the engine's performance settings. This makes the dual-fuel engine a highly versatile and user-friendly solution for power generation.

Natural boil-off gas burning

While the vessel is in service and "gassed-up," the boil-off from the tanks will either be burned as fuel in the main generating plant or disposed of in the gas combustion unit (GCU). The gas-burning process is initiated on the deck but monitored by the ship's engineers from the ECR. Normally, during sea voyages, one LD compressor and one fuel gas pump will supply fuel gas to the main generator engine to ensure stable gas supply to the diesel generators. If the boil-off gas (BOG) cannot be used or exceeds the requirements of the generating plant, the excess gas will be burned in the gas combustion unit.

Boil-off gas recovery system

A BOG recovery system or BOG re-liquefaction plant is used in LNG carriers to recover boil-off gas and return it to the cargo tanks. The BOG is taken from the cargo tanks, compressed to 4.5bar and −60°C, and passed through a nitrogen heat exchanger, where it is cooled to −160°C at between 2 and 4bar, converting it back into liquid. It then undergoes separation to remove incondensables before being returned to the tanks via loading or spray lines. Boil-off rate (BOR) is the amount of liquid evaporating from the cargo due to heat leakage, expressed as a percentage of the total liquid volume per unit time. Recent projected LNG carriers offer a BOR close to 0.1%, with the LNG carrier EKAPUTRA delivering a BOR of 0.1%/day as early as 1990 [34].

There are two types of gases in LNG carriers: natural boil-off gas (natural BOG) from cargo tanks and forcing boil-off gas (forcing BOG) generated by the forcing vaporizer. The natural BOG is controlled by the low-duty compressor (LD compressor) and sent to the main boilers through the low-duty heater. The flow rate is regulated by the LD compressor speed and inlet guide vane (IGV). If the natural BOG does not meet the boiler demand, the forcing vaporizer generates forcing BOG to supplement the natural BOG for the full speed range of the ship.

Boil-off gas combustion system

Combustion systems utilizing BOG systems are only used onboard LNG carriers. These burners and DFDE/TFDE are previously discussed in this section on propulsion. Described here is the process of capturing and moving the BOG. BOG is sent to the engine room via gas heaters using a low-capacity compressor and burned by the main boilers as fuel. The main boilers can operate in different modes, including exclusively BOG mode, combined BOG and fuel oil mode, and exclusively fuel oil mode. While steam turbine systems have been traditionally used for propulsion on LNG carriers, diesel engines that can use BOG as fuel are becoming a more popular solution due to their higher operating efficiencies.

Managing boil-off gas

Boil-off gas (BOG) and LNG carriers are specifically designed to carry natural gas in a cryogenic liquid state at a temperature of −163°C. The vaporization (or boil point) of LNG is −162°C. Chilling the LNG further would require additional energy. To minimize the energy required in chilling, the goal is to bring the LNG just below vapor temperature, use extensive insulation, and count on the thermal mass to maintain temperature in transport. Despite tank insulation designed to limit external heat from affecting the LNG, some residual heating is inevitable. Therefore, as a portion of the LNG moves above the boiling point, gas is formed. To limit the increasing pressure in the cargo hold, the BOG is moved away from the bulk of the LNG.

During transport, gas that boils off can be managed in part by a low-duty LNG compressor. The drive system of the compressor (electric motor) is separated from the compression system by means of a "bulkhead." The drive shaft for the compressor passes through the bulkhead, which is commonly sealed with either a positive pressure oil seal or an air seal. Due to the criticality of these compressors for propulsion, the units are commonly supplied in redundant pairs.

Low-duty (LD) compressor

The LD compressor (Fig. 6.49) is a multistage compressor that provides fuel gas to LNG carriers. It plays a critical role in maintaining the pressure of the cargo tank and supplying boil-off gas to the dual-fuel engine for use as fuel during the voyage. Shown is an integrally geared compressor, similar to a machine that is also used for onshore LNG applications compressing boil-off gas. The wide range of operating conditions associated with these applications can be realized relatively easily by an IGC compressor stage with overhung impellers since its diffuser area is accessible to actuate variable diffuser vanes from outside. In this way, a turn-down range of nearly 75% at constant discharge pressure can be achieved [35].

The LD compressor is designed for wide operation, with the first stage utilizing variable geometry diffuser (VGD) technology and two recycle loops for capacity control. This enables the compressor to adjust to varying operating conditions and maintain optimal performance.

High-duty (HD) compressor

The HD compressor is a key component used on LNG carriers and terminals to recirculate gas, transfer heated gas back to cargo tanks to warm up, and return vapor or gas generated during loading or initial cooling down to shore. It is typically a single-stage unit that is integrally geared for high performance.

During the on-loading or off-loading of LNG, the HD compressor applies compression to the gas as it is sent to shore from the terminals. This ensures that the gas is efficiently transferred and circulated throughout the system. The HD compressor is designed for flexible operation and utilizes reliable variable inlet guide vanes (VIGVs) to achieve this.

The VIGV is a simple yet effective technology that allows for precise control of the compressor's performance, enabling it to operate at peak efficiency across a wide range of conditions. This makes the HD compressor a versatile and reliable component that plays a critical role in the safe and efficient operation of LNG carriers and terminals.

Both the LD and HD compressors are critical components in the safe and efficient operation of LNG carriers and terminals, and their reliable performance is essential to ensuring the delivery of LNG to customers around the world.

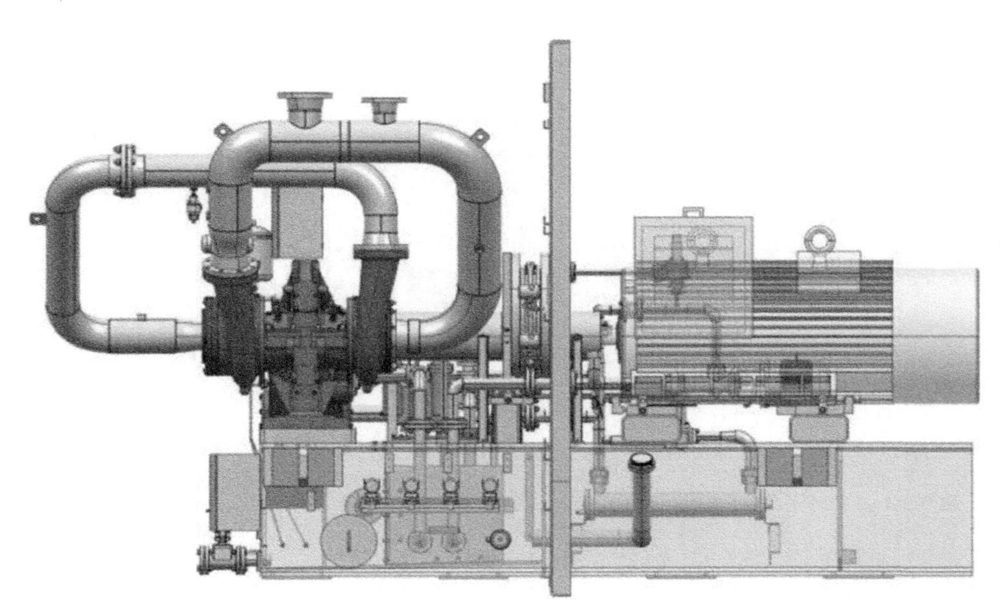

FIG. 6.49 Low-duty LNG compressor [35]. *Courtesy of Hanwha Power Systems.*

Loading and unloading

Terminal sites are equipped with articulated loading and unloading arms [36] for transferring LNG between ships and the terminal. To minimize the amount of LNG vaporization, both the loading and unloading arms and pipework are well insulated [37]. When tankers are loaded with LNG, the gas displaced from their tanks can be directed to boil-off or gas recovery storage tanks. The sections above provide more detail on boil off gas. This recovered gas can then be compressed and added to the local gas network or returned to the liquefaction plant as a liquid to refill the LNG storage tanks. Fig. 6.50 shows the marine loading and offloading process for transporting LNG from gas gathering and liquefaction plants to LNG terminals [38].

Loading of natural gas onto carrier

Loading LNG onto the carrier is a complicated process involving multiple safety and technical steps. The steps given below derive from [39]:

- *Line cool down*: In the process of cooling down the terminal, it is essential to start by instructing the terminal to initiate pumping at a slow rate for approximately 15 min. This gradual approach helps cool down both the terminal piping and the ship's headers. As the process continues, one must slowly increase the pumping rate until both the liquid main and spray headers have reached their desired cool temperatures, which typically takes around 15–20 min. Throughout this cooling process, it is crucial to closely monitor the pressure in the cargo tanks. If necessary, make adjustments to the HD compressor to ensure a constant vapor pressure. Additionally, pay special attention to prevent pipe sections from hogging by ensuring the liquid header and crossovers are cooled down and filled as quickly as possible.
- *LNG loading arms*: Before commencing the loading operation, it is imperative to cool down the cargo pipelines properly. There are several reasons for this. First, cooling the cargo lines helps minimize the risk of leaks occurring at joints, particularly where valves and other pipeline sections are involved. As these components contract when exposed to the cold LNG, cooling them in advance reduces the chances of leakage. Second, it helps avoid sudden shock loadings on bellows, which can occur when the pipes contract rapidly. Last, proper cooling is essential to prevent the formation of vapor locks in the pipelines. If LNG is introduced into a warm pipeline, the initial cargo can vaporize, creating a surge in pressure that may obstruct the loading process. Subsequently, this vapor could condense rapidly as the temperature decreases, potentially causing damage to the pipelines, valves, or connections.

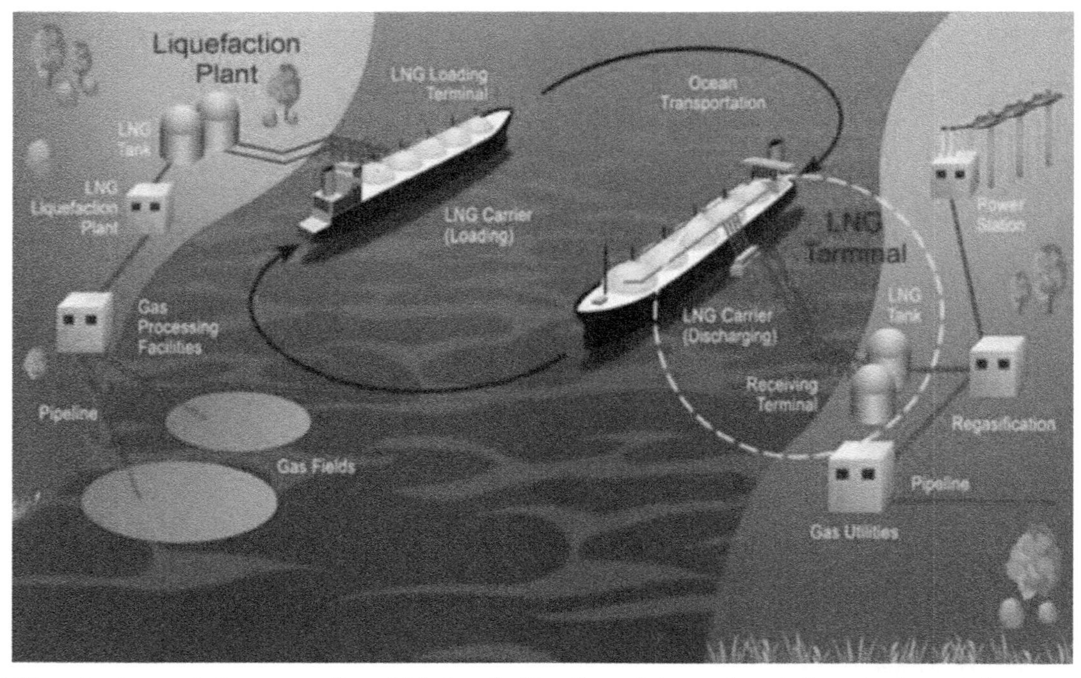

FIG. 6.50 LNG marine transportation system. *From I. Mustary, H. Chowdhury, B. Loganathan, F. Alam, Development of a computational model for optimal sourcing of LNG. Energy Procedia 110 (2017) 597–603. https://doi.org/10.1016/j.egypro.2017.03.191.*

- *Air purge of loading arms*: After connecting the loading arms, it is crucial to purge air from both the loading arms and the tips of the manifold pipes. To accomplish this, introduce nitrogen gas (N_2) into the loading arms through injection lines connected to them and pressurize it to approximately 4–6 kg/cm^{2}g. Once pressurized, open the ship's liquid manifold vent valve and vapor manifold vent/drain valve. This will release both air and N_2 gas into the atmosphere. This purging process should be repeated two or three times to ensure a thorough removal of air. Simultaneously, perform a leak test using a soap solution. The air purge operation should be considered complete when the oxygen content in the purged gas drops below 2%.
- *Loading arms cool down*: Cooling down the loading arms is a task performed from the shoreside using a small capacity pump. When discharging at the port, LNG is used to cool the arms by employing the ship's spray pump.
- *Loading operation*: The loading of LNG occurs through the loading manifolds into the liquid header and then into each tank's filling line. During this process, any boil-off and displaced vapor exit each tank via the vapor suction to the vapor header. Initially, the vapor flows freely to shore through the vapor crossover manifold. However, as the tank pressure increases, one compressor is activated to enhance gas flow to shore and control vapor main and cargo tank pressures. As the loading rate increases, it is crucial to closely monitor tank pressures and initiate one HD compressor as necessary. If the compressors are unable to handle the volume of boil-off and displaced gas, it may become necessary to reduce the loading rate accordingly.
- *Topping off*: As the LNG carrier nears the completion of cargo operations, it is important to stagger the filling of tanks in alignment with the cargo plan. This typically leaves a gap of 10–15 min between the completion of each tank. To ensure a smooth transition, notify the terminal well in advance and in accordance with the agreed procedure, indicating the intention to top off and reduce the loading rate. It is important to make this notification at least 30 min before reducing the rate. Note that membrane tanks typically fill to 98%, while Moss vessels fill to 99.5%, with independent alarms triggering at preset filling levels. The upper alarm will activate the ESD if prior alarms are ignored.
- *Deballasting*: The deballasting operation is conducted concurrently with the cargo loading operation. Prior to commencing deballasting, all ballast surfaces must undergo visual inspection to confirm they are free from oil or other pollutants. This inspection should be carried out through inspection hatches or tank lids, with particular attention paid to ballast tanks adjacent to fuel oil tanks. If equipped, gas detection and sampling systems may not always indicate the presence of hydrocarbons, especially in small quantities. Deballasting begins with gravity discharge and transitions to the use of ballast pumps as the ballast tanks' level approaches the vessel's waterline. It is crucial to adjust the ballast to maintain a slight stern trim, aiding in the stripping of ballast tanks. The flow rate of the ballast should be adjusted to keep the ship within 1 m of the arrival draft, or as specified by the terminal. Deballasting should typically be completed before the commencement of topping off the cargo tanks.
- *Filling rate of cargo tanks*: The IGC Code, which stands for the International Code for the Construction and Equipment of Ships Carrying Liquefied Gases in Bulk, came into effect on July 1, 1986. It is aligned with the International Convention on the Safety of Life at Sea, 1983, and the Regulations Relating to the Carriage and Storage of Dangerous Goods by Ship, which were revised in Japan. The IGC Code includes a chapter titled "Filling Limits for Cargo Tanks." LNG carriers registered in Japan are NK-class ships constructed in compliance with NK's "Rules and Guidance for the Survey and Construction of Steel Ships – Part N." These rules reflect the IGC Code requirements, even for LNG carriers built before the enforcement of the '83 SOLAS Convention. As a result, these ships meet the standards set for new vessels in the IGC Code.
- *Behavior of LNG in cargo tanks*: When LNG is loaded into the cargo tanks, the pressure of the vapor phase is kept substantially constant, slightly above atmospheric pressure. External heat passing through the tank insulation generates convection currents within the bulk cargo. This causes heated LNG to rise to the surface, where it vaporizes. The necessary heat for vaporization comes from the LNG itself, and as long as vapor is continuously removed while maintaining the pressure substantially constant, the LNG remains at its boiling temperature.

Unloading of natural gas from carrier

The unloading of the LNG from the carrier, conditioning the gas for the pipeline, and finally moving the gas through the pipeline involve a number of crucial steps.

- *Arrival of the natural gas carrier*: The initial phase involves the arrival of the LNG carrier at the terminal. The vessel, laden with liquefied natural gas (LNG), is maneuvered into position for the offloading process.

- *Connection to the receiving terminal*: The vessel is linked to the receiving terminal via an intricate network of hoses and loading arms. Precision is crucial in aligning and securing these components to avert any potential leaks, guaranteeing a secure and efficient transfer of the natural gas.
- *LNG transfer to the terminal*: The heart of the operation lies in the transfer of LNG from the carrier to the receiving terminal. Utilizing pumps and compressors, LNG is pumped into the terminal. Here, it undergoes a crucial transformation, transitioning from a liquid to a natural gas as it is warmed and vaporized, making it suitable for transportation through pipelines.
- *Compression and conditioning of natural gas*: To optimize its pressure, flow rate, and quality, natural gas often undergoes compression and conditioning. This phase involves the use of compressors, scrubbers, and other specialized equipment to eliminate impurities and moisture.
- *Measurement and monitoring*: Throughout the entire offloading process, a vigilant eye is kept on the natural gas. Key parameters such as flow rate, pressure, and composition are consistently measured and monitored. This meticulous oversight ensures that the gas aligns with regulatory standards and the precise requirements of the receiving facility.
- *Pipeline transfer*: In the final step, after compression, conditioning, and thorough measurement, the natural gas is channeled into the pipeline. An array of valves and fittings facilitates this transfer, usually under the meticulous control and supervision of advanced software and automation systems. These systems are integral to ensuring the process remains both safe and efficient.

Certifications

Centrifugal compressors associated with the BOG in LNG carriers frequently require independent design verification for maritime duty. These compressors have a direct impact on the safety and propulsion of the vessel. The two most common independent verification and certification entities are DNV and ABS. These entities provide design approval services, ensuring basic offshore design requirements are met. From an OEM perspective, timing is critical to the certification process. Starting the certification process too soon may mean that the design team can become stuck and not proceed as the project is not sufficiently defined for certification, or it may mean that the design is certified, but the application changes necessitating recertification. Having the certification too late means that purchase orders have already been issued prior to complete certification.

Det Norske Veritas (DNV)-GL

DNV GL was created in 2013 as the result of a merger between Det Norske Veritas (Norway) and Germanischer Lloyd (Germany). In 2021, DNV GL changed its name to DNV [40], while retaining its postmerger structure. DNV is the world's largest classification society, providing services for 13,175 vessels and mobile offshore units (MOUs) amounting to 265.4 million gross tonnes, which represents a global market share of 21%. DNV is also the largest technical consultancy and supervisory to the global renewable energy (particularly wind, wave, tidal, and solar) and oil and gas industries representing approximately 65% of the world's offshore pipelines that are designed and installed to DNV's technical standards.

ABS (American Bureau of Shipping)

ABS certification is awarded to ships that comply with industry standards and regulations, covering areas such as design, construction, and operation. It is not mandatory, but many ship owners choose to obtain it to demonstrate their commitment to safety, environmental protection, and quality management. In addition to certification, ABS provides technical advisory services and other support to the maritime industry.

References

[1] EKT Interactive. n.d. Natural Gas Value Chain - EKTInteractive.com [Internet]. [cited 2024 Jan 9]. Available from: https://ektinteractive.com/natural-gas/natural-gas-value-chain/.

[2] EKT Interactive. n.d. What is Natural Gas? [Internet]. [cited 2024 Jan 9]. Available from: https://ektinteractive.com/what-is-natural-gas/.

[3] H. Abudu, R. Sai, Examining prospects and challenges of Ghana's petroleum industry: a systematic review, Energy Rep. 6 (2020) 841–858, https://doi.org/10.1016/j.egyr.2020.04.009.

[4] Lesson 6 Introduction n.d. EBF 301: Global Finance for the Earth, Energy, and Materials Industries [Internet]. [cited 2024 Jan 9]. Available from: https://www.e-education.psu.edu/ebf301/node/700.

[5] Mintz M. n.d. Renewable Natural Gas (RNG) for Transportation [Internet]. 2021 [cited 2024 Jan 9]. Available from: https://www.anl.gov/sites/www/files/2021-03/RNG_FAQ_March_2021_FINAL_0.pdf.

[6] P.C. Rasmussen, R. Kurz, Centrifugal compressor applications, in: 38th Turbomachinery Symposium, Houston, TX, 2009.
[7] S. Kumar, R. Kurz, J.P. O'Connell, Equations of state for compressor design and testing, in: ASME Paper 99-GT-12, 1999.
[8] B. Poling, J.M. Prausnitz, J.P. O'Connell, The Properties of Gases and Liquids, fifth ed., McGraw-Hill, 2001.
[9] D. Beinecke, K. Luedtke, Die Auslegung von Turboverdichtern unter Beruecksichtigung des realen Gasverhaltens, in: VDI Bericht Nr. 487, Duesseldorf, Germany, 1983.
[10] M.R. Sandberg, Equation of state influences on compressor performance determination, in: 34th Turbomachinery Symposium, Houston, TX, 2005.
[11] K. Brun, R. Kurz, Compression Machinery for Oil and Gas, Elsevier Gulf Professional Publishing, Cambridge, MA, 2019.
[12] R. Kurz, K. Brun, Process control for compression systems, in: ASME PaperGT2017-63005, 2017.
[13] S. Ohanian, R. Kurz, Series or parallel arrangement in a two-unit compressor station, J. Eng. Gas Turbines Power 124 (4) (2002) 936–941.
[14] R. Kurz, R.C. White, Surge avoidance in gas compression systems, J. Turbomach. 126 (4) (2004) 501–506.
[15] R.C. White, R. Kurz, Surge avoidance for compressor systems, in: 35th Turbomachinery Symposium, Houston, TX, 2006.
[16] B. Boutin, P. Taylor, Basics of compressor valve design and troubleshooting, in: GMC 2015, Austin, TX, 2015.
[17] G. Phillipi, Basic thermodynamics of reciprocating compression, in: GMC 2015, Austin, TX, 2015.
[18] K. Brun, R. Kurz, Introduction to Gas Turbine Theory, third ed., Solar Turbines, San Diego, CA, 2019.
[19] R. Kurz, K. Brun, C. Mejer-Homji, J. Moore, F. Gonzalez, Gas turbine performance and maintenance, in: Proc. 42nd Turbomachinery Symposium, Houston, TX, 2013.
[20] M.G. Nored, J.R. Hollingsworth, K. Brun, Application Guideline for Electric Motor Drive Equipment for Natural Gas Compressors, Gas Machinery Research Council, Dallas, TX, 2009.
[21] R. Kurz, J. Mistry, P. Davis, G.-J. Cole, Application and control of variable speed centrifugal compressors in the oil and gas industry, in: IEEE/PCIC Conference, San Antonio, TX, 2021.
[22] Credit to SwRI Pulsation & Vibration Short Course.
[23] Borba, A.J.G.A. and Borges, F.O., "The Pros and Cons of Dry Gas Seals Installation in an Existing Synthesis Gas Compressor," Ammonia Technical Manual, 2007, pp: 127–138. http://www.iffcokandla.in/data/polopoly_fs/1.2466643.1437682608!/fileserver/file/507916/filename/010.pdf.
[24] K. Brun, S. Simons, R. Kurz, The impact of reciprocating compressor pulsations on the surge margin of centrifugal compressors, in: Gas Machinery Research Conference, Austin, TX, 2015.
[25] R. Kurz, M. Lubomirsky, Asymmetric solution for compressor station spare capacity, in: ASME paper GT2006-90069, 2006.
[26] M. Tobin, J. Labrujere, High pressure pipelines, A review of basic design parameters, in: GTS 2005, Moscow, 2005.
[27] R. Kurz, S. Ohanian, M. Lubomirsky, On compressor station layout, in: ASME Paper GT2003-38019, 2003.
[28] S.I. Hyman, M.A. Stoner, M.A. Karnitz, Gas flow formulas – an evaluation, Pipeline Gas J. 202 (14) (1975) 34–44.
[29] R. Kurz, M. Lubomirsky, R. Zamotorin, Transporting hydrogen-natural gas mixtures, in: PSIG 2110, Pipeline Simulation Interest Group Annual Meeting (Virtual), 2021.
[30] R. Kurz, K. Brun, Efficiency definition and load management for reciprocating and centrifugal compressors, in: ASME Paper GT2007-27081, 2007.
[31] https://www.pgjonline.com/magazine/2009/january-2009-vol-236-no-1/features/fundamentals-of-gas-pipeline-metering-stations.
[32] S.S. Arsenyev-Obraztsov, A.I. Ermolaev, A.M. Kuvichko, Optimal sea floor placement of the oil/gas production equipment, IOP Conf. Ser.: Mater. Sci. Eng. 700 (2019) 012011.
[33] Kelleher, L. Floating LNG Market Overview: Has FLNG Crested the Wave?, Gas Process & LNG. http://www.gasprocessingnews.com/articles/2019/08/floating-lng-market-overview-has-flng-crested-the-wave/.
[34] https://www.wartsila.com/encyclopedia/term/boil-off-rate-(bor).
[35] K. Wygant, et al., Tutorial on the application and design of integrally geared compressors, in: Asia Turbo-Pump Symposium, 2016.
[36] Flotech, Marine Loading Arms, 6 June 2020. https://www.flotechps.com/fluid-transfer/loading-arms/marine-loading-arms/.
[37] Herose, Your Guide to Cryogenic Insulation, 5 June 2020. https://www.herose.co.uk/wp-content/uploads/2015/09/Your_Guide_To_Cryogenic_Insulation.pdf.
[38] I. Mustary, H. Chowdhury, B. Loganathan, F. Alam, Development of a computational model for optimal sourcing of LNG. Energy Procedia 110 (2017) 597–603. https://doi.org/10.1016/j.egypro.2017.03.191.
[39] http://www.liquefiedgascarrier.com/loading-LNG.html.
[40] DNV, DNV GL changes name to DNV as it gears up for decade of transformation, 2021. Archived from the original on 2021-01-27. Retrieved 2021-03-06.

CHAPTER

7

Transport of hydrogen and carriers of hydrogen

Marybeth McBain[a], Justin Hollingsworth[b], Sebastian Freund[c], Tim Allison[b], Shane Harvey[a], Rainer Kurz[d], Gabe Glynn[e], Buddy Broerman[b], Terry Kreuz[f], Stephen Ross[a], Michael Müller[g], Subith Vasu Sumathi[h], Ramees K. Rahman[h], and Derrick Bauer[a]

[a]Ebara Elliott Energy, Jeannette, PA, United States [b]Southwest Research Institute, San Antonio, TX, United States [c]Energyfreund Consulting, Munich, Germany [d]Solar Turbines, San Diego, CA, United States [e]Atlas Copco, Stockholm, Sweden [f]National Fuel Gas Company, Williamsville, NY, United States [g]EagleBurgmann, Wolfratshausen, Bayern, Germany [h]University of Central Florida, Orlando, FL, United States

Introduction

The use of hydrogen in a decarbonized global energy market will unfold based on the deployment of commercially ready technology and the global population's accessibility to low carbon fuels. The extent to which hydrogen is produced, transported, and stored will depend on its ability to compete with other low-carbon fuels and energy storage forms, in terms of cost, storage duration, and availability of materials. The first two subsections of this chapter address the current mechanisms for transporting hydrogen, in both gas and liquid forms, and technologies, which support these transportation mechanisms. At the time of this publication, the contributors also felt it was pertinent to address where limits in current technology exist and what adaptations may be possible for new hydrogen machinery. Hydrogen transport through pipelines must confront additional risks of material corrosion and degradation due to the nature of hydrogen gas. Hydrogen embrittlement and variations of hydrogen degradation on steels are discussed as it is pertinent to reuse of previously installed steel pipelines that may be repurposed for hydrogen gas mixtures. However, other operators are considering the use of other materials, such as pipe within pipe, offering alternative materials and secondary containment methods such as coiled line pipe with an inner plastic lining on the wetted inner surface of the pipe.

Fundamentally, transporting hydrogen in pipelines must be viewed as a mechanism for energy storage when renewable power is curtailed or produced beyond the existing grid capacity (i.e., why produce hydrogen when renewable power will be more affordable?). Transportation of hydrogen can also be seen as the movement of curtailed energy (stored energy molecules in hydrogen) from the point of production to the point of use. In transporting hydrogen, pressurized gas pipelines are serving the same function of electrical lines as electrons could also be used to deliver the same stored energy. However, pipelines have been shown to be more efficient than electrical lines through a comprehensive, detailed study in 2021 [1]. It is also possible that the electrical grid capacity will become more constrained in the next two decades as added demands are introduced through more electrification (electric vehicles and electric motor-powered industrial uses). In addition, pipelines allow for the gas to be pressured up to the maximum allowable operating pressure (MAOP) of the system. In this regard, the operator can use linepack as stored energy, by operating

Energy Transport Infrastructure for a Decarbonized Economy
https://doi.org/10.1016/B978-0-443-21893-4.00012-X

Copyright © 2025 Elsevier Inc. All rights are reserved, including those for text and data mining, AI training, and similar technologies.

the compressor stations at the full design power to keep the system at the highest design pressure, which transforms a pipeline into a longer energy storage source (likely on the matter of days, depending on delivery pressures and flow rates in the system for the offtake).

This chapter also addresses other means of utilizing and requiring hydrogen transport: in liquid form for vehicular transport and fueling stations, in carrier forms (such as methanol and ammonia) for higher energy density and potentially lower cost and lower risk, and in terms of a larger energy storage network requiring storage of hydrogen to support transport networks. The debate on gas versus liquid transport is addressed as well, as two options for specifically different markets that lend themselves to distinctly one or the other forms of hydrogen, depending on volumes required, transport distance, and feasibility of each form.

Hydrogen gas inevitably finds its comparison with natural gas pipelines as natural gas pipelines are the widespread means of transport for natural gas and have been utilized reliably for the past 80–100 years. Liquefied natural gas (LNG) plants and the shipments of LNG by marine vessels have also evolved within the last 50 years to support LNG trading worldwide as a secondary transportation means through higher-density cryogenic temperatures for liquid methane. However, hydrogen is distinctly different from natural gas in many ways and must be viewed as an energy storage medium. Any comparisons will soon reveal the significant differences in future hydrogen pipelines or hydrogen liquefaction plants compared with natural gas, as hydrogen should be compared (in terms of cost and efficiency) with other viable decarbonized energy storage forms.

Hydrogen transport and storage will tend to look distinctly different technically and commercially compared with modern natural gas lines. It is also possible that the build-out of a hydrogen pipeline infrastructure will tend to be minimized due to the cost and risk involved. Instead, the hydrogen transport markets may emerge primarily through liquefaction and carrier forms of hydrogen that can be shipped at higher densities and with less risk—or alternatively, hydrogen may be produced closer to the point of use to minimize transport distances. For these reasons, liquefaction technologies and carrier forms of hydrogen are also addressed in this chapter on hydrogen transport.

Liquefied hydrogen has traditionally been used for space applications for rocket propulsion and in the semiconductor industry primarily for surface cleaning of wafers. The hydrogen industry has decades of experience generating liquid hydrogen, although only on a relatively small scale. The demand for hydrogen is expected to dramatically increase in the coming years as the mobility market shifts to carbon-free fuels such as hydrogen for use in fuel-cell electric vehicles (FCEVs) or smaller-distance mobility applications such as nonelectric, green-powered forklifts and golf carts.

The importance of hydrogen production and transportation in the energy transition and global net-zero commitments is more critical than ever before as renewables are becoming cheaper, and decarbonized fuels are becoming more of a necessity to curb global carbon emissions.

Hydrogen production—Via reforming (blue hydrogen)

Hydrogen may be produced from readily available natural gas through three primary processes, which are all methods of reforming the hydrocarbon molecules to produce hydrogen with the primary by-product being carbon monoxide and carbon dioxide. The carbon dioxide must be captured, compressed, and sequestered or reutilized (not emitted to atmosphere) to market the hydrogen as "blue hydrogen." The three main reformation processes for blue hydrogen are: (1) steam methane reforming, (2) auto-thermal reformation, and (3) partial oxidation of methane. Methane pyrolysis may also be used but is currently cost-prohibitive, due to the cost of disposal of carbon black.

Of these processes, steam methane reforming (SMR) is the most widespread and commercially utilized. SMR utilizes fired natural gas to produce the heat necessary for the endothermic reaction at a 3:1 ratio of carbon to hydrogen. The process is based on catalytic reforming to produce the CO shift when CH_4 is combined with oxygen, producing $CO+H_2+H_2O$ [2, chapter 5]. Two chemical reactions occur: the first combines methane with oxygen to produce hydrogen, water, and CO in this step: $CH_4+O_2 \rightarrow CO+H_2+H_2O$. This is typically followed by the second water gas shift reaction to produce additional hydrogen: $CO+H_2O \rightarrow CO_2+H_2O$. In the second step, the process is taken one step further to gain additional hydrogen from the water output in the first catalytic reforming step. Auto-thermal reforming (ATR) utilizes the same base chemical reactions but combines the methane with oxygen directly instead of methane with air. ATR does not have the nitrogen as a by-product in this reaction, but does require pure oxygen in the process, meaning an air separation unit must be added to the utility requirements of the plant to produce oxygen.

Additional heat is produced in the reaction, which may be incorporated into steam generation to elevate the energy efficiency of the process about 2.5 efficiency points higher. The energy efficiency is defined as the latent heat of vaporization (LHV) of hydrogen divided by the LHV of natural gas. In general, the SMR process energy efficiency ranges from 72% to 81% [2], assuming that natural gas content is entirely methane at its lower heating value of 47 MJ/kg. Often, natural gas will be 90%–96% methane for pipeline quality natural gas, and the remainder is made of up of heavier hydrocarbons, some nitrogen, and some CO_2, which will add inefficiencies to the SMR process to purify the gas and strip out the diluents.

The second method, auto-thermal reformation (ATR), is similar to SMR but differs by combining the natural gas with oxygen instead of air. This requires the addition of an air separation unit to produce oxygen. However, ATR is self-sustaining once the process is established at temperature, and it may be scaled easily to save capital costs of equipment. ATR processes tend to be much larger in volume of hydrogen produced compared with SMR, which typically has an optimized maximum size for a project of around 350 tons per day due to capital funding limits and equipment sizing. The largest SMR process in the United States was built in 2013 by Praxiar at Port Arthur, supplying 325 tpd or 135 mmscfd of hydrogen gas [3]. Another major advantage to ATR over SMR for the carbon capture and sequestration (CCS) side is that direct firing with oxygen eliminates by-products from the final production, namely nitrogen emissions. This means that the CO_2 is more easily filtered and refined (at lower cost) to produce CO_2 for sequestration without N_2 or NOX.

Finally, the third commercially available method of producing hydrogen from natural gas is partial oxidation. Partial oxidation utilizes a kinetic reaction (not catalytic) but also requires oxygen instead of air, similar to the ATR process. Unlike reforming, partial oxidation of natural gas to produce hydrogen has the advantage of a lower carbon per hydrogen production ratio at 2:1 instead of 3:1. As the process does not utilize a catalyst, the pretreatment steps are reduced significantly for partial oxidation, which helps to lower costs. However, due to the complexity of the process, partial oxidation is not utilized on as wide a scale as SMR or ATR for hydrogen.

Hydrogen production from water (green hydrogen)

Green hydrogen may be produced from water without the CO emissions of blue hydrogen, making this process appealing for excess, curtailed renewable energy to convert to stored energy in the form of hydrogen. Green hydrogen processes include alkaline electrolysis or exchange membranes. Alkaline electrolysis is commercially mature and gained widespread usage worldwide, although its scale is currently an order of magnitude smaller than SMR or ATR for hydrogen production (24 tons per day from electrolysis compared with 240 tons per day of hydrogen via SMR). In most typical electrolyzers, the anode and cathode of the water molecules are separated through diaphragms. Larger versions of the alkaline units are being developed, which do not require external circulation of the electrolytes, thus reducing the balance of plant requirements and energy loads to scale up to over 100 MW units.

Another option for electrolyzers is proton-exchange membrane electrolysis (PEM). PEM electrolyzers offer the ability to produce higher-purity hydrogen at elevated pressure, which helps to reduce the cost of transporting hydrogen through reducing filtration costs and later compression costs for transport. The main feature of PEMs is the membrane material used—which separates the cathode and anode reaction through a hydrogen ion conductive membrane. Alkaline membranes (similar to proton exchange) are also being developed that work off the membrane concept, doping the interface with the catalyst material. PEMs have the ability to produce hydrogen at higher discharge pressures, greatly reducing power requirements for hydrogen compressors to transport hydrogen through ship or pipeline. However, it must be cautioned that hydrogen production at a higher pressure does lead to larger gas carryover of the oxygen molecules and requires thicker membrane surfaces that can increase the amount of losses or lower efficiency of the process.

Hydrogen gas transport

If hydrogen gas through pipeline is considered a viable means of transporting energy, then several steps exist where compression is required in the transport chain. Depending on flows and pressure ratios, centrifugal compressors and reciprocating compressors can be utilized with various advantages and disadvantages to each. Diaphragm compressors may also be used for smaller flows with high head such as vehicle fueling stations. Another question is whether these compressors are driven by gas turbines, gas engines, or electric motors. Gas turbines and engines are being developed for higher hydrogen fuel content to work as mechanical drivers for the hydrogen compression.

Hydrogen transport in pipelines—Embrittlement issues

Among the most promising and developed technologies for lowering the carbon footprint on a worldwide scale is through the use of hydrogen. There is a great potential for low-carbon emission hydrogen production, and technological advancements for renewable hydrogen are possible for both near-term and long-term goals. For hydrogen to develop its full potential, it will be required that hydrogen can not only be economically generated, but it will have to be transmitted through pipelines, which make up the world's current energy infrastructure, unless other energy carrier forms develop cost-effectively.

Hydrogen transmission through the existing natural gas pipelines will be one of the most economical methods of transporting hydrogen from sites where it can be generated to all locations where it is needed (see Fig. 7.24 comparison of distance vs transport method). Hydrogen may be blended with methane for small volume transmission while dedicated lines will be used for higher volume hydrogen transmission in future applications. Utilizing the existing natural gas infrastructure is a far more cost-effective method compared with storing hydrogen in containers for transportation by rail or vehicle, if the material properties of existing natural gas lines permit such transport. The European Commission on the hydrogen economy has been making plans for hydrogen pipelines to transport hydrogen throughout Western Europe by 2050.

Hydrogen is known to have a detrimental influence on metals, including iron-based metals such as steels and other alloys, which have industrial significance. While the mechanical design strength of the material remains unchanged in terms of the yield and tensile strength, the ductility of the material is what can be reduced. From a fracture mechanics or from a defect tolerant design practice, defects within the metal that are acceptable without the presence of hydrogen may become detrimental in the presence of hydrogen because of the embrittlement of the material. An example of this loss of ductility is graphically illustrated in Fig. 7.1.

This testing shows the results of UNS S30403 stainless steel, commonly known as 304 L stainless steel, tensile tested in the presence of different hydrogen pressures. The results of the tensile test are compared against the same tensile test performed under standard atmospheric conditions. While the yield strength of the material did not change during any of the tensile tests, the ductility of the material continued to decrease with increasing partial pressure. The loss of the material's ductility translates to a loss in the toughness of the material. From a defect-tolerant design standpoint, a defect within a material will cause a fracture when the stress intensity of the defect exceeds the stress intensity threshold of the material. A reduction of the material stress intensity threshold in a hydrogen atmosphere means that a small defect size can become more critical in these applications where the toughness is reduced. This also translates into smaller imperfections or accidental exterior penetrations on the pipeline becoming a greater weakening effect in hydrogen lines compared with natural gas.

API Recommend Practice (RP) 941 provides guidelines for steels for hydrogen service at elevated temperatures and pressures in petroleum refineries and petrochemical plants. As steels have been utilized for hydrogen service for many years, this specification provides an excellent guideline for piping and pressure vessels that can be used for hydrogen transportation. API 941 states that carbon steel has been successfully utilized for hydrogen pressure vessels up to 69 MPa (10,000 psi) and temperatures up to 221°C (430°F), provided that weldments are stress relieved [5]. The heat-affected zone (HAZ) of a weldment can have a higher hardness at a localized area, so these welds should be stress relieved to ensure that they are less susceptible to hydrogen embrittlement.

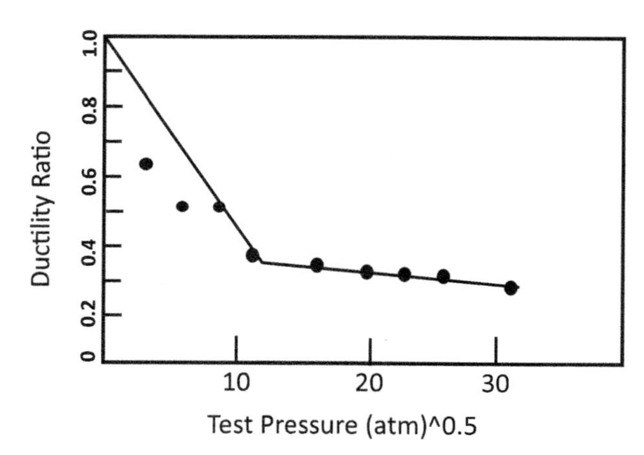

FIG. 7.1 Effect of hydrogen pressure on ductility of 304 L stainless steel tested to fracture in hydrogen gas at room temperature [4].

The most common form of carbon steel piping that is used commercially is ASTM A106, which it titled "*Standard Specification for Seamless Carbon Steel Pipe for High Temperature Service.*" The high temperature term in this title is a relative statement only used to indicate that this material is not rated for low-temperature service conditions. The ASTM A106 specification provides the processing requirements for common UNS G10180 or UNS G10200 (commonly referred to as 1018 or 1020) carbon steel with a minimum yield strength of 30 ksi [205 MPa]. These grades of carbon steel are referenced in API RP 941 and have been utilized for hydrogen applications at pressures and temperatures, which are referenced within the specification.

Along with API RP 941, ASME B31.12 titled "*Hydrogen Piping and Pipelines*" was adopted in 2009, which provides guidelines specifically for piping that will be used in hydrogen transportation. This specification provides design criteria for piping in hydrogen transportation. The wall thickness necessary for the piping is defined by the yield strength of the material along with the intended pipeline design pressure. Based upon these factors, a material performance factor (H_F) is applied as a derate factor for the design to account for the potential embrittlement effects of hydrogen. The piping weldments should be stress relieved to avoid localized areas with a high hardness that could lend themselves to hydrogen embrittlement; however, a stress relief operation is not always practical in pipeline applications. Hence, some operators are considering alternative materials to carbon steel for new hydrogen pipelines.

Through the utilization of the preheating the pipe to 300°F (149°C) at using a low carbon filler metal, slow-cooling the material from the weldment can help control the hardness of the weldment and the heat-affected zone. ASME B31.12 recommends keeping the heat-affected zone hardness below 241 BHN to avoid hydrogen embrittlement. It should also be mentioned that cold working of piping should be avoided in hydrogen pipeline applications. Cold working or bending of the piping can induce residual stresses within the material, and these higher residual stresses can make the material more susceptible to hydrogen embrittlement.

For the transportation of hydrogen in a mixture with hydrocarbons, hydrogen limits have been set to minimize the extent to which hydrogen may be introduced into existing natural gas pipelines, based on the compressor station components primarily—see "Hydrogen transport: Retrofits and pipeline station design for hydrogen mixtures" section. Ammonia has also been included in this discussion as it is a good carrier of hydrogen. For these conditions, standard carbon steel piping material can be successfully utilized. The general corrosion rate will be under 0.05 mm/year (0.002 in./year) if liquid water is kept out of the system.

Compressor embodiments for hydrogen gas transport

To support the various mechanisms of storage and hydrogen usage, the following specific hydrogen compressor applications are envisioned:

1. Compression from the electrolyzer (for green hydrogen) to pipeline pressure: This would require machines to compress from about 20 bar to pipeline pressures of approximately 80–100 bar. It may be possible in the future to build electrolyzers that produce hydrogen at pipeline pressure, thus avoiding the need for compression. Centrifugal, screw, or reciprocating compressors may be used typically for this application. *For green hydrogen production through electrolysis, the compressors to pipeline input for reciprocating compressors must be nonlubricated due to most of these applications' requirements for stringent gas purity standards in the pipeline.
2. Compression from a steam methane reforming process (blue hydrogen): The compression requirements are similar as for green hydrogen, but possibly higher ratios are required if the SMR generates hydrogen at 6–10 bar pressure (current technology limit). Centrifugal or reciprocating compressors may be used typically for this application. *For blue hydrogen production to pipeline entry, these reciprocating compressor installations may be lubricated or nonlubricated.
3. Boost compression in a pipeline: Pipelines will likely be constructed around similar MAOP limits based on carbon steel lines—which typically set the pipeline pressure at 1440 psi or 99.2 bar. If the pipeline has to cover a reasonable distance, a pipeline compressor station will be needed in regular intervals, with the intervals designed to maintain the minimum suction pressure for reasonable friction loss curves for the pipeline and lower velocity (10–12 m/s is typical for natural gas pipelines). Pipeline ratios are typically in the 1.2–1.5 range for a booster station. Centrifugal or reciprocating compressors are typically used for this application. The gas to be compressed in pipelines can be natural gas mixed with hydrogen from 0% to 100% hydrogen content. Given the large amount of existing and underutilized natural gas lines, it is likely that mixed natural gas and hydrogen lines will emerge. Most likely, these will be redesigned for a fluctuating hydrogen content due to the fluctuating availability of renewable electricity. Refs. [6–8] have provided calculations showing how varying hydrogen content affects power consumption and head requirements for the compressors, as well as the impact on transportation capacity of existing pipelines.

4. **Fuel gas compression:** These compressors would be installed when the hydrogen delivery pressure falls below fuel pressures for power generation or vehicle fuel gas transport. They are typically smaller (reciprocating, screw, or diaphragm compressors). For the turbine power generation side, the discharge gas pressure would be in the 20–30 bar range required for gas turbine fuel systems. Vehicles running on hydrogen gas typically operate at much higher pressures including heavy-duty trucks and commuter cars operate with fuel delivery at 700–900 bar.
5. **Hydrogen storage compression:** Given the rationale for hydrogen generation, it is likely that hydrogen or hydrogen natural gas mixtures will be subject to longer-term storage in aquifers, depleted gas fields, and salt caverns, assuming that they are able to contain hydrogen. Pressure ratios for the compressor will be from pipeline pressure (approximately 80–100 bar) to storage pressure, which could be as high as 200 or 300 bar. Centrifugal or reciprocating compressors may be used typically for this application.

If hydrogen is mixed with natural gas and transported in a natural gas pipeline, it is possible that some of the existing pipeline infrastructure is capable of accommodating a certain amount of hydrogen (less than 10% hydrogen). Material risks exist, however, due to the various compositions of steel in the ground and the effects of hydrogen embrittlement and degradation on these older carbon steel lines. Additional components at compressor stations may leak with even small amounts of hydrogen in the gas stream as many leakage points exist on flanges, regulators, and actuated valves. Additionally, many natural gas operators use instrument gas as the working pneumatic gas (not instrument air) to actuate any of the facility pneumatics. In summary, the capability of blending hydrogen into existing natural gas systems is limited by material concerns (for example, due to hydrogen embrittlement), by the capability of end users to handle certain amounts of hydrogen, and by the capability of the components and machinery involved at the compressor stations and metered delivery points. Kurz et al. [3,9], Allison et al. [1], Lubomirsky et al. [10], and Brun and Allison [2] have analyzed pipeline behavior if low amounts of hydrogen (up to 20% by volume) are added to an existing pipeline or if pipelines are operated with up to 100% hydrogen.

Adding hydrogen to an existing pipeline increases the compression work, as well as the total power consumption in the pipeline (Fig. 7.2), approximately 7× the compression power required for natural gas pipelines. Therefore, the amount of energy (either as fuel to a gas turbine or as electricity to a motor) to transport a certain amount of energy is increased (Fig. 7.3), and the transport capacity of the pipeline in terms of energy flow will be reduced (Fig. 7.4).

In Fig. 7.3, various levels of hydrogen mixed with pipeline quality gas are compared. The total energy content delivered in the pipeline is kept constant [3] to show the relative effect of the energy required to transport that same amount of energy delivered. The effect of the higher hydrogen content is pronounced, leading to approximately 60% more required energy at 20% hydrogen compared with pipeline gas with 0% hydrogen.

The transport capacity can be limited by either the maximum speed of the compressors or the available power of the drivers. For the example in Fig. 7.4, the power demand is increased up to about 7% hydrogen content and then starts to drop, because the driven compressors reach their speed limit. The pipeline operator then can decide to restage the

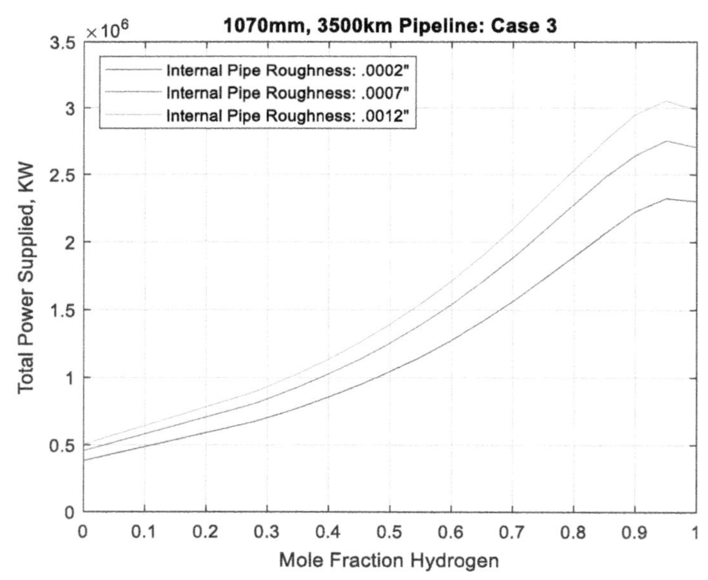

FIG. 7.2 Total power requirement for a 3500 km pipeline, 1070 mm diameter [1].

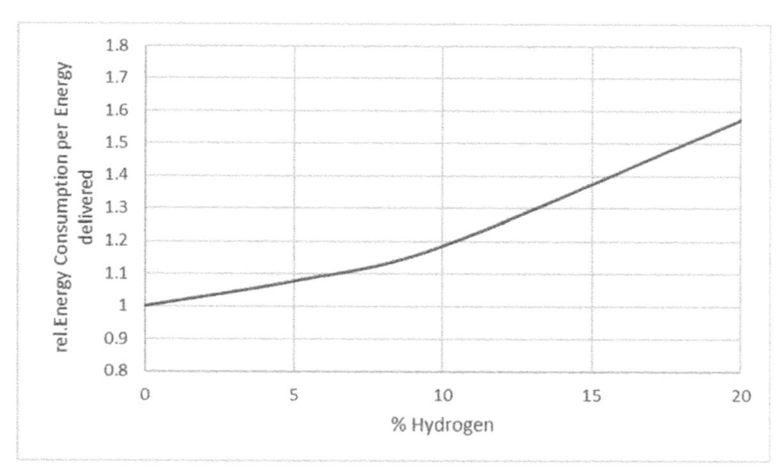

FIG. 7.3 Transportation efficiency of the pipeline as a function of hydrogen content in the pipeline gas, normalized with the transport efficiency for 0% hydrogen.

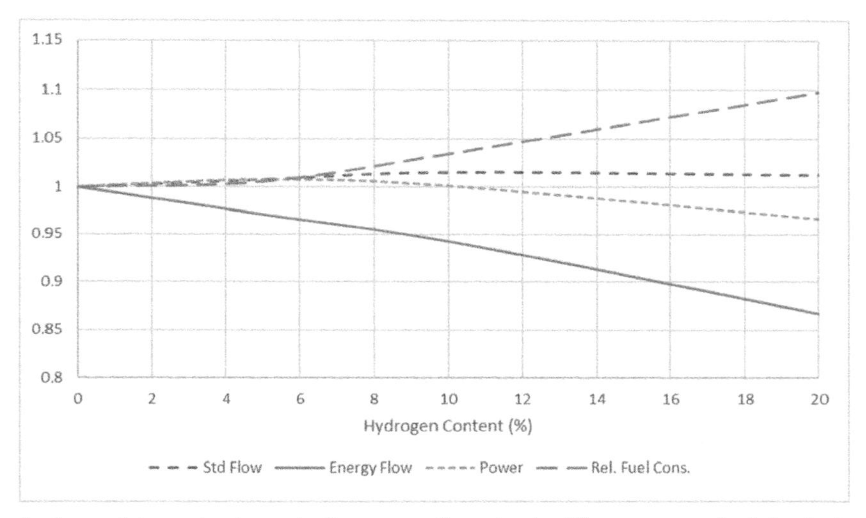

FIG. 7.4 Result of adding hydrogen into a natural gas pipeline: energy flow, standard flow, power and relative fuel consumption, normalized by the respective parameters at 0% hydrogen content [3].

compressors to provide more head and to add compression trains to increase the available power per compressor station. In the scenario described, if the hydrogen is produced from renewable energy, it would be highly likely that the hydrogen content in the natural gas would fluctuate as the hydrogen would tend to be produced during excess renewable power periods when the renewable power is curtailed, and hydrogen may then be produced from the excess. Kurz et al. [9] provide a detailed discussion on this topic.

It also has to be considered that even a small amount of composition impurity of the hydrogen, such as 4%–5% carbon dioxide, can substantially lower the compression work by a factor of 2. Therefore, it makes a significant difference at what pressure and composition the hydrogen is made available in various compression processes. Also, both the combustion characteristics and the compression work change if other components (such as carbon dioxide or methane) are part of the hydrogen produced. Lastly, the capability of hydrogen to cause material issues (such as hydrogen embrittlement) can be influenced by the presence of other substances in the gas composition.

While green hydrogen will likely be produced at a distance from the point of usage, blue hydrogen (for example, made from natural gas) would likely be produced near the point of usage. This follows from the fact that the transport of hydrogen is more energy-intensive than the transport of natural gas (as feedstock for blue hydrogen), and the transport of the CO_2 that would be generated back to a sequestration site. Kurz et al. [9] calculated this for a 1 GW power plant as shown in Fig. 7.5. They found that the required energy to transport hydrogen to the site of usage exceeds the

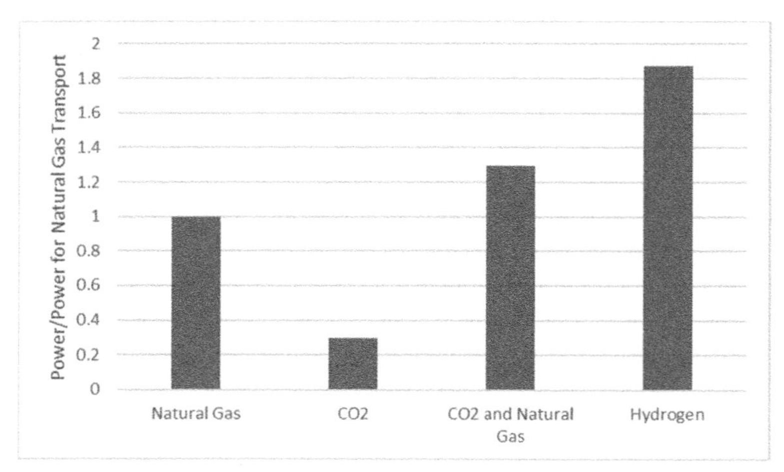

FIG. 7.5 Relative power requirement for transportation of natural gas and CO_2 (sequestered as by-product) vs hydrogen compared with transportation of natural gas to deliver fuel for production of a 1 GW power plant [9].

combined energy to transport natural gas to that site, and to transport CO_2 to a sequestration site, assuming that the distances in all three cases are identical.

Centrifugal compressors for hydrogen

Hydrogen is a very light gas with a high heat capacity and a very high speed of sound (Table 7.1). Given the thermodynamic properties of hydrogen, the amount of work required to create a certain pressure ratio and a certain amount of volume reduction is significantly more than for other gases such as natural gas, air, or CO_2 (Table 7.1). The higher work required can be explained by the isentropic relationship discussed in Eqs. (7.1)–(7.3). Additionally, because hydrogen has such a low density, when converted to a mass basis, hydrogen requires a larger volume flow, which additionally adds to the work required per molecular mass compared with natural gas or air. Hydrogen compression on a large scale can readily be accomplished with commercially available reciprocating and centrifugal compressors [9].

A key difference in the working principles of centrifugal and positive displacement machines is that centrifugal ("dynamic") compressors create head through Euler's equation (The head is directly related, through Euler's law, with the energy exchange and the changes in velocities in the compressor). The head of a compressor is largely a function of impeller tip velocity U and the number of impellers in a machine as shown in Eq. (7.1). Head translates into pressure ratio as a function of the gas composition, as shown by the gas properties given in Eq. (7.2) alongside the head term to equate to pressure ratio, P_2/P_1.

$$H = \Psi U^2 \tag{7.1}$$

Head is also related to the compressor pressure ratio for a centrifugal compressor by the isentropic relation shown in Eq. (7.2), noting the variable for the ratio of specific heats, gamma, shown in the exponent. Eq. (7.2) shows the approximate relationship between pressure ratio and head, assuming constant specific heat for T_1 and T_2:

$$\frac{P_2}{P_1} = \left(1 + \frac{\eta}{c_p \cdot T_1} \cdot H\right)^{\frac{\gamma}{\gamma-1}} \tag{7.2}$$

TABLE 7.1 Thermodynamic properties of hydrogen compared with natural gas and CO_2.

	Hydrogen	Natural gas	CO_2
Heat capacity (kJ/kg K)	14.3	2.3	0.839
Ratio of heat capacities	1.4	1.3	1.3
Speed of sound (m/s)	1320	450	280

A modified Euler's equation is shown in Eq. (7.3) for specific heat related to angular momentum (simplified for constant specific heat). This gives one the ability to relate specific heat, compressor speed, angular velocity, and the head produced directly. It is clear that the higher specific heat term in a hydrogen gas mixture will require higher rotational speeds/tip speeds to produce the same head required in a comparable (comparable pressure ratio) natural gas or air mixture. In simple terms, to maintain the same pressure ratio as shown in Eq. (7.2), the head must be increased to balance the higher specific heat term given in the denominator.

$$H_s = h_2 - h_1 = c_p*T_2 - c_p*T_1 = \omega*(r_2c_2 - r_1c_1) = H_{act}*n \tag{7.3}$$

Thus, for gases such as hydrogen with a large heat capacity (denominator c_p term in Eq. 7.2), the head translates into only a very little pressure ratio. To create a pressure ratio of 1.5, which is typical for pipeline applications, a pipeline compressor for natural gas would be a single- or two-stage compressor, while increased amounts of hydrogen mixed in natural gas would increase the number of stages, to ultimately, for 100% hydrogen requiring two compressor bodies.

For a given mass flow, the amount of work to get a certain pressure ratio with pure hydrogen is almost 10 times higher than that of natural gas (Fig. 7.6), which requires either multiple compressor bodies (more impellers) or compressors at higher speeds. Of course, hydrogen has a much higher energy density on a mass basis than natural gas (the lower heating value of hydrogen on a mass basis is 2.5 times that of natural gas), but even for the same energy flow, the compression work for hydrogen is four times higher.

Compressing pure hydrogen leads to a relatively small temperature rise with pressure ratio (Fig. 7.7) compared with other gases as its thermodynamic properties do not result in a higher temperature rise through the compression process compared with natural gas. Thus, intercooling capability is less critical. For centrifugal compressors, development by OEMs and research organizations is underway to find ways to increase the amount of work we can do per compressor stage or per compressor body. This can be accomplished by combinations of higher tip velocities, designs with less backswept blades, and more stages per compressor body (a stage is an impeller with its inlet system and its diffuser). Mach number limitations are generally not an issue due to the high speed of sound in hydrogen, which allows the gas velocity to be pushed upward without reaching Mach 1.

The compressor design choice of a closed- or open-face impeller will dictate present velocity limits. Shrouded impellers in centrifugal compressors utilize tip velocities of about 1000–1200 ft./s, while open-faced impellers have been built that can run at 2000 ft./s. Higher tip velocities increase stress levels in impellers, thus driving the selection to open-faced impellers.

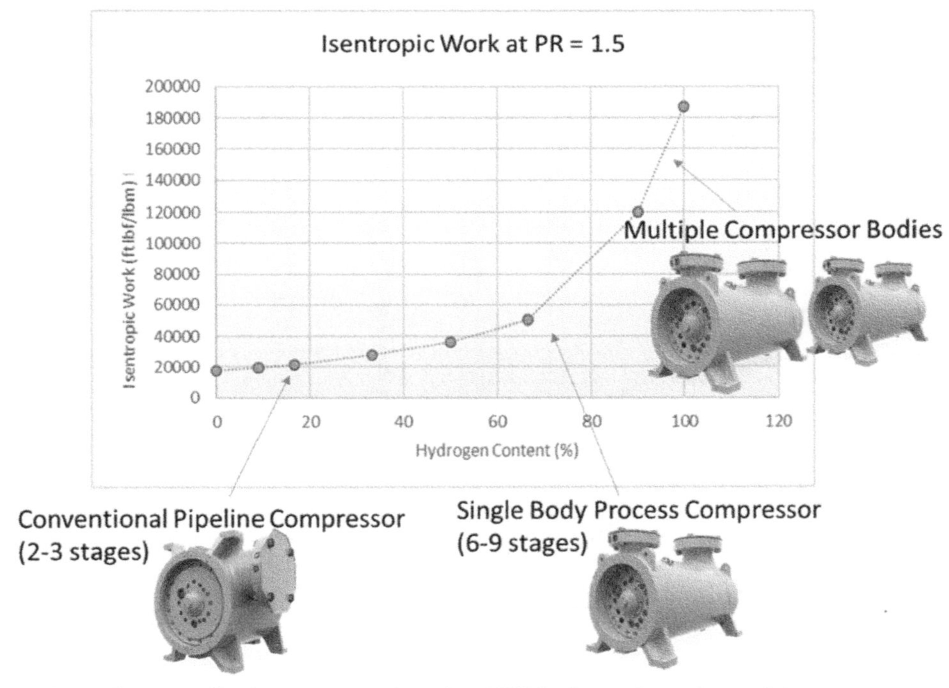

FIG. 7.6 Isentropic work as a function of hydrogen content from 0 to 100% hydrogen in methane mixtures.

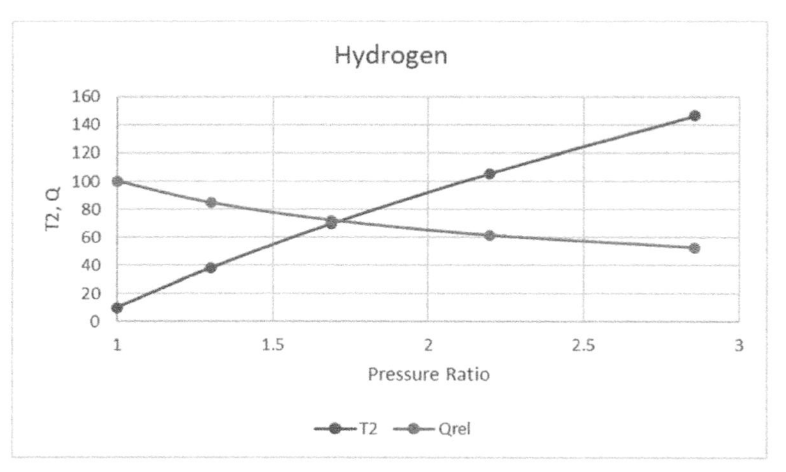

FIG. 7.7 Discharge temperature (°C) and volume reduction in hydrogen compression (no intercooling) [9].

Achieving higher tip velocities once the stress levels can be accomplished will help to increase the head in hydrogen machines. Doubling the tip velocity by itself would create four times the head due to the squared velocity term in Eq. (7.1). In other words, multiple conventional compressor bodies (multiwheel and multicasing compressors) can be replaced by a single body with higher tip velocities.

To accomplish lower stress at higher tip speeds, the choice of materials is driven to materials that have a high strength-to-density ratio, such as aluminum or composite materials. Of course, the materials used also have to be compatible with hydrogen. While open-faced impellers are routinely used in compressors with one or two impellers per shaft (for example, in integrally geared compressors), having multiple open-faced impellers on a shaft creates the challenge to adequately control the running clearances. In shrouded impellers, this is accomplished with labyrinth seals on the shroud, whereas in open-faced impellers, the clearance has to be maintained to avoid efficiency loss due to gas leaking across the impeller blades. Eisenlohr and Chladek [11] provide data on the influence of tip clearance, showing the impact on both maximum head and surge and choke for a highly loaded, open-faced impeller of a small gas turbine. These results are shown in Fig. 7.8, providing the critical impact of tip clearance control on compressor performance.

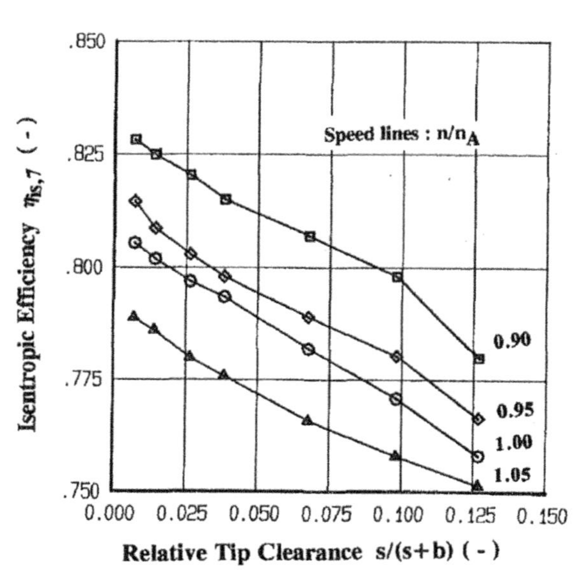

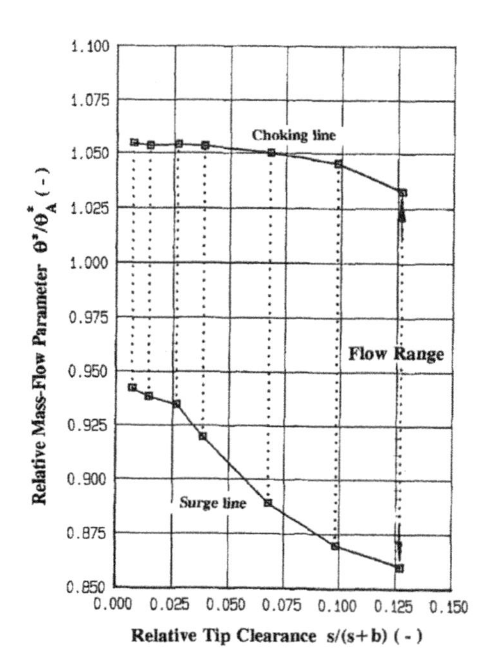

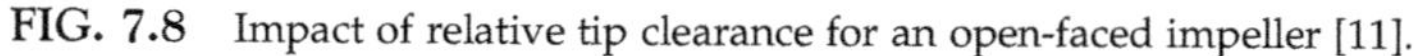

FIG. 7.8 Impact of relative tip clearance for an open-faced impeller [11].

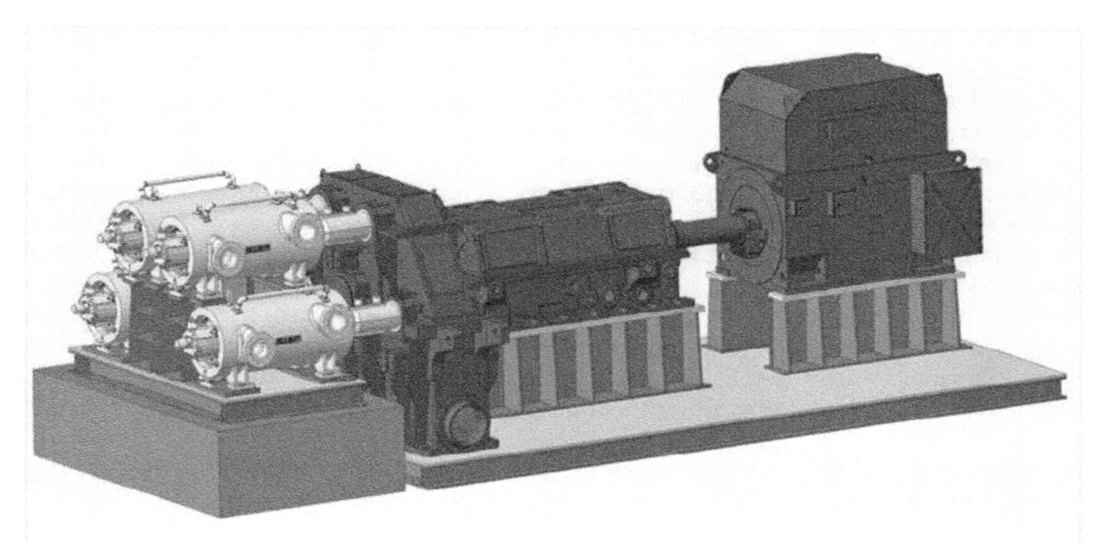

FIG. 7.9 Externally geared/compact arrangement of multiple compressor bodies for hydrogen compression [6].

For hydrogen at traditional centrifugal compressor speeds, the challenge is to maximize the impeller count per casing and the number of casings per unit depending upon the required delta pressure. In this effort, a multibody compressor is typically required. Multibody compressors may be in-line designs, externally geared or internally geared IGC for hydrogen service. To support the high head requirement, available hydrogen centrifugal compressors will utilize a higher number of casings for externally geared and in-line designs. Internally geared compressors (IGCs) may operate at higher rotational speeds up to 30,000–50,000 RPM, and also higher impeller tip speeds due to a smaller bore section, in an effort to accomplish the high head and limit the number of impellers and casings.

The external and internally geared options have different rotor dynamic-bearing concerns than each other. The inline compressor has to manage multiple stages on a single shaft or multiple bodies. The integrally geared compressors have to manage multiple pinions, typically mounting up to four compressor casings per each side of the multipinion gear. In Fig. 7.9 arrangement, the four, nearly identical, hydrogen compressors are connected to a multipinion gearbox of the design commonly used with integrally geared compressors. One of several design differences between the externally geared design and a traditional IGC is the relocation of the process piping nozzles from the top or bottom of the casings to the sides on the externally geared embodiment shown in Fig. 7.9, which necessitates also relocating the casing support feet from the centerline to the bottom of the casing. Both of these features are common to other centrifugal compressors, particularly for natural gas pipeline service.

Beyond the in-line design and multibody compressors, a final innovative arrangement may be utilized for some applications of limited pressure ratio with hydrogen—but with higher flow rate requirements. This option would utilize a double-ended motor with a compressor casing on each end and gearbox per casing to remove any radial constraints with the two body tandem. The double-ended motor configuration is easily adapted for higher flow rates and suits a limited pressure ratio range for hydrogen applications.

Material considerations in centrifugal compressors

Material selection and design choices are important topics for hydrogen compressors. In hydrogen machines, the impeller and shaft material must have sufficient strength while being light enough to minimize hoop stresses at high rotational speeds. Additionally, risk of hydrogen embrittlement poses additional limitations. Even at traditional speeds in a machine with a high number of impellers, certain material considerations must be factored in, given the risk of hydrogen embrittlement and potential corrosion concerns with wet hydrogen gas mixtures.

Hydrogen has been known to have a detrimental influence on metals, including iron-based metals such as steels and other metals, which have industrial significance. While the mechanical design strength of the material remains unchanged, the ductility of the material is what can be reduced. From a fracture mechanics or fit-for-service perspective, defects that are acceptable without the presence of hydrogen may become unstable in the presence of hydrogen. Hydrogen can also be responsible for cracking mechanisms within materials that are completely independent of an externally applied stress.

The following discussion, summarized from McBain et al. [12], comments on hydrogen's effects on compressor and pipeline material design. A notable, detrimental influence on the embrittlement of steels is atomic hydrogen. Atomic hydrogen is generated either by the dissociation of diatomic hydrogen (H_2) at elevated temperatures and pressures or by a cathodic reaction that occurs in the environment or with the surface of the steel itself. Atomic hydrogen can permeate the steel even more readily than diatomic hydrogen. The internal pressure can build up sufficiently high to create a fracture within the material itself. This mechanism is referred to as hydrogen-induced cracking (HIC). This type of cracking has many different nicknames including stepwise cracking, hydrogen pressure cracking, blister cracking, or hydrogen-induced stepwise cracking. The dangerous part of the hydrogen-induced cracking mechanism is that it can occur with no applied external stress.

HIC is most critical on components that are processed in a single direction, such as rolled plate and pipe, and is more of an issue for pipeline construction than for compressor design. When the hydrogen pressure builds up within these cracks, it can extend a crack, which then links up with an extending crack on a different plane of the material and creates a through-thickness crack that goes throughout the thickness of the material. This all occurs with no applied external stress.

For pipeline operators considering hydrogen blending with existing natural gas lines, HIC, and other hydrogen-associated influences, these effects can limit the use of hydrogen in older pipeline systems. See "Safety measures and hydrogen gas leakage detection" section for retrofits with hydrogen blending for more discussion.

Other components within the compressor need to be considered as well. At lower temperatures below 200°C (392°F), special considerations are not normally necessary for these components. Shaft seals can be an abradable seal, mica-filled PTFE material, or they can be a rub-tolerant material such as a carbon-filled PEEK or a PAI material. When following the guidelines within API 617 for the materials, there have not been any failures attributed to hydrogen embrittlement. Most failures in these compressors come from a carryover of ammonium chloride into the compressor, which is extremely corrosive. There is an effort by a number of compressor companies to find higher-strength materials that are not subjected to hydrogen embrittlement. There is no industry-recognized standard set as a fit-for-service testing. The development is of the materials, and testing conditions are reviewed on a case-by-case basis.

Provided that the temperature is kept below 200°C (392°F), relatively standard materials that are currently utilized in hydrogen compression and transportation can be utilized at pressures up to 10,000 psi (69 MPa). For compressors, there is an extensive service history at refineries for compressors under these conditions, and API 941 has recognized that standard carbon steel can be used in these conditions. Carbon steel is used for the pressure-containing components such as the process piping and the centrifugal compressor casing. Higher-strength steels can be used for the rotating components, which are kept under a maximum yield strength of 120 ksi (827 MPa), which provides a method to compress the hydrogen for transportation and storage.

Efforts are underway to find a material with a higher strength that is suitable for hydrogen compression. Higher-pressure pipelines will require materials with more stable carbides to avoid degradation during service. The requirements for high tip speeds on the one hand, and the impact of hydrogen embrittlement, may lead to the use of materials that are currently not generally used for compressors in oil and gas applications.

Reciprocating compressors for hydrogen

New hydrogen-reciprocating compressors are in contrast yet borrow some of the same reliable technology to some of the older designs of refineries from the past 50 years. The traditional refinery recycle hydrogen compressors used in hydrocracking and desulfurization are typically low-speed, lubricated units. Refineries have long utilized reciprocating compressors reliably for hydrogen recycle service either for hydrocracking, hydrotreating, or desulfurization. Future transportation compressor hydrogen units for pipeline and storage are evolving to meet the changing hydrogen market, but generally these are being designed in fairly high power ranges from 5 to 25 MW. These units are typically nonlubricated and use purged packing with block-mounted frames. Depending on the pressure ratio, these may also use water-cooled packing to keep discharge temperatures below 135°C. Some of these newer hydrogen-reciprocating compressors may utilize limited lubrication where the packing is still kept oil free. Nonlubricated compressors are required for most green hydrogen production and for compression at fuel stations for vehicles with fuel cells.

Due to the longer distance pieces and additional sealing, these units are larger and heavier than typical pipeline natural gas units. For the lower-pressure stages of compression, the larger cylinders can range from 20″ up to 48″ bore cylinders, with 4–6 cylinder valves per cylinder end. Some of the primary design features with reciprocating compressors for hydrogen that differ from traditional natural gas compressors are highlighted briefly below [2].

(1) *Typical package requirements*: No threaded connections; process gas piping requiring 100% x-ray; API 618 distance piece, often two compartments; purged packing; discharge temperature limits at 275°F; pulsation vessel changes due to acoustic property differences; water-cooled packing may be considered; welded heat exchangers required; gas detection suitable for hydrogen

(2) *Mechanical design*: Leakage can be a major issue through cylinder packing, distance piece and valves with hydrogen service, and effects losses and gas detection/fugitive losses.

(3) *Reciprocating compressor valves for hydrogen*: Cylinder valves tend to be the highest failure mechanism but tend to be worse in high-speed and high-density gas service. For hydrogen, accommodating greater flow areas for the higher volumes is necessary, but valve reliability in hydrogen is high especially for lubricated units. Flutter can be an issue. Valve design and development for hydrogen mixtures are an ongoing R&D effort by many OEMs due to the noted short run times (<6 months) for valves in nonlubricated cylinders on hydrogen [2, Fig. 9.19]. Ref. [2] also notes that nylon, PPS, and PEEK-based material are common for sealing elements in valves in hydrogen. PTFE-coated stainless valves are often used in many refinery applications for high hydrogen gas mixtures to counteract embrittlement effects on steel when used in valves.

(4) *Losses and efficiency*: Compressor losses increase with molecular weight. Increase is proportional to the square of piston speed and decreases proportionally with the square of valve area. The goal with hydrogen is to increase the valve effective area while keeping piston speeds as low as possible. This tends to be fairly easily accomplished with hydrogen because of hydrogen's low mole weight. Losses are low and efficiency will be higher in hydrogen compared with heavier gas mixtures.

(5) *Materials of construction*: Packing, seals, and material choices are based on pressure MAWP. API 618 5th edition, December 2007, Annex H does not currently require specialized construction materials for hydrogen. Materials listed by the API Annex are provided below for MAWP ranges in Table 7.2. Seal material tends to vary for hydrogen—preference to nylon, PPS, and PEEK-based material. Spring materials in valves also tend to be either derated for hydrogen or nickel or cobalt alloys due to hydrogen embrittlement concerns.

(6) *Pulsation and vibration acoustic control*: Pulsation bottles must be designed for the higher speed of sound in hydrogen, which tends to raise the natural acoustic frequencies in the system. The Helmholtz cutoff frequency will also be increased. The amplitudes of pulsation-induced forces can potentially be lower due to the lower gas density of hydrogen, but there are other variables that can impact pulsation levels. The larger wavelength of hydrogen will push acoustic natural frequencies upward, which could lead to an increase in vibration amplitudes or a decrease in vibration amplitudes, depending upon coincidences with compressor running speed.

(7) *Mechanical vibrations*: Skid-mounted equipment tends to be highly susceptible to higher vibrations with the faster running speeds. However, large powered hydrogen compressors over 5 MW will tend to be block-mounted with larger cylinders for the higher hydrogen volumetric flows. This block-mounted compressor tends to be heavier and easier for mechanical vibration control.

Reciprocating compressors create volume reduction and pressure ratio due to their geometry. This does not mean that their efficiency is automatically higher; it only means that one needs fewer stages to accomplish the required pressure ratio. As the power consumption is only determined by mass flow and head, there is no inherent advantage of reciprocating compressors for hydrogen, although often times space and operational needs dictate reciprocating units for pressure ratios greater than 3:1. Due to the valve losses being lower for the lower molecular weight of hydrogen, reciprocating compressors, from an efficiency standpoint, can maintain high efficiencies and low losses. However, due to valve wear with the low damping and less lubricated cylinder designs, reciprocating compressors can show lower reliability than other compressor technologies.

TABLE 7.2 Adapted from API 618 recommended materials of construction table for gas mixtures, including hydrogen, from Ref. [2, section 9.2.9, p. 375].

Material	Maximum allowable working pressure (bar)	Maximum allowable working pressure (psi)
Gray cast iron	70	1000
Nodular iron	100	1500
Cast steel	180	2500
Forged steel	No limit	No limit
Fabricated steel	85	1250

Nonlubricated compressors for hydrogen will have different surface finishes.

TABLE 7.3 Booster compressor application—Ps = 169 psia, Pd = 1440 psia, Q = 150 MMSCFD, Ts = 87°F.

	Recip compressor A	Recip compressor B	Recip compressor C
# of Units required	2 units	5 units	5 units
# of Throws per stage	3 cylinders for stage 1/3 cylinders for stage 2	2 cylinders for stage 1/2 cylinders for stage 2/2 cylinders for stage 3	3 cylinders for stage 1/2 cylinders for stage 2/1 cylinders for stage 3
Target speed range—RPM	360 RPM	713 RPM	1200 RPM
Cylinder bore size	13.5″–22.75″ bore	9.125″–17.875″ bore	10.0″–13.0″ bore
Stroke	14.0″ stroke	6.0″ stroke	6.0″ stroke
Rod diameter	4.0″ rod diameter	2.5″ rod diameter	2.875″ rod diameter
Lubricated or non-lubricated or both?	Lubricated or non-lubricated	Lubricated or non-lubricated	Lubricated

TABLE 7.4 Booster compressor application—Ps = 169 psia, Pd = 1440 psia, Q = 500 MMSCFD, Ts = 87°F.

	Recip compressor A	Recip compressor B
# of Units required	3 units	1 unit
# of Throws per stage	3 cylinders for stage 1/2 cylinders for stage 2	4 cylinders for stage 1/4 cylinders for stage 2
Target speed range—RPM	360 RPM	450 RPM
Cylinder bore size	19.75″ & 20. 5″ bore	8-cylinders—22.5″ & 30.5″ bore ×/35,000 hp motor
Stroke	14.0″ stroke No response	12.0″ stroke
Rod diameter	4.0″ rod diameter	5.5″ rod diameter
Lubricated or nonlubricated or both	Lubricated or nonlubricated	Lubricated or nonlubricated

For a pipeline booster application, a reciprocating compressor would typically be designed with two stages of compression to cover the 400–1440 psi range, with one intercooling step. Staging geometry is shown in Tables 7.3 and 7.4 for various low-speed and high-speed compressors designed for the pipeline booster application. The tables also compare low flow to high flow options with the contrasting flow rates in the two tables.

Pulsation bottles and orifice plate sizing for the above compressors will range in size based on cylinder volumes and flow rates. Bottles for hydrogen service can be empty bottles unless the pulsation analysis shows an excessive excitation of the bottle length response. If the bottle length response is a concern, a baffle with a pass-thru flow connection near the middle of the bottle or between each pair of cylinder nozzles is recommended to break up the bottle length response.

To handle the high-flow rate required, the cylinders would need to be fairly large, feeding into common suction and discharge bottles per stage. Cylinder valves would utilize low-lift, high-efficiency valves and compensate for higher discharge temperatures with specialized materials. In addition, pulsation bottles would need to be large (increased in proportion to the speed of sound change in hydrogen) to handle the surge pulse volume. For a compressor of this size, it is debatable whether the unit should be skid-mounted or concrete block-mounted and will depend on OEM recommendations as to the physical size of the cylinders, frame, and crankshaft required.

Diaphragm compressors for hydrogen refueling

The following subsection summarizes the larger discussion of diaphragm compressors available in Ref. [13]. Diaphragm compressors originated in the early 1900s. These hermetically sealed, positive displacement machines consist of two systems: a gas compression system and a hydraulic system. The gas compression system includes thin metal membranes, or diaphragms, which are clamped between an oil distribution and gas cavity plate and a process gas inlet and discharge check valves. The hydraulic system involves a motor-driven crankshaft connected to a reciprocating piston. The piston pressurizes a hydraulic fluid, which in turn causes the diaphragm group to sweep through a

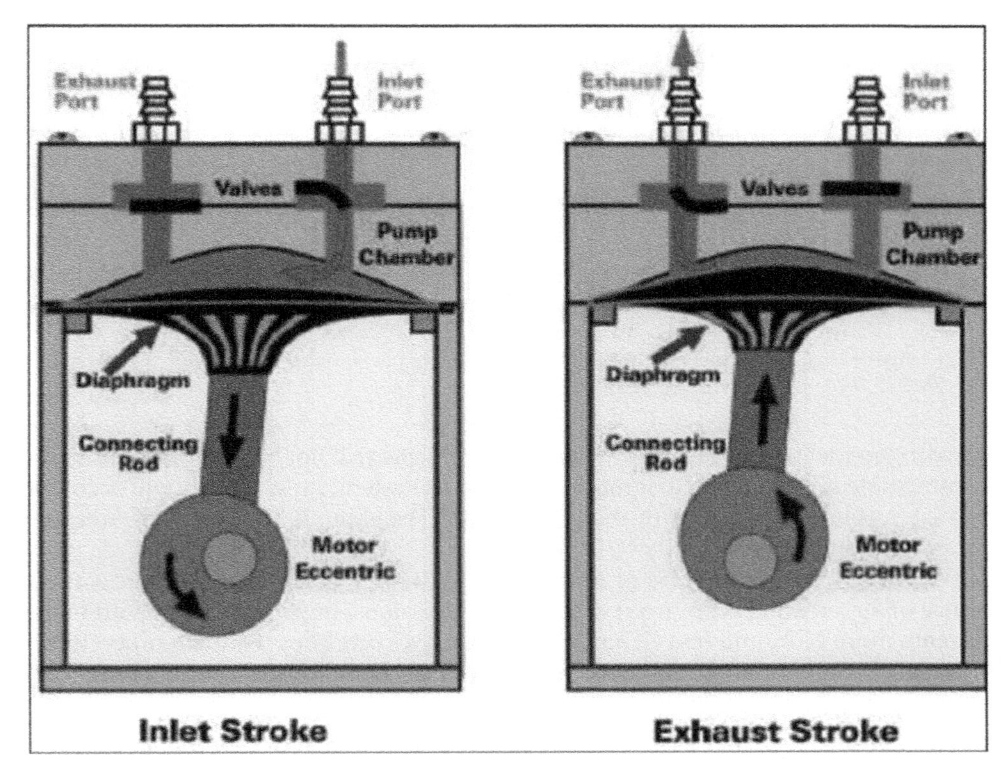

FIG. 7.10 Illustration of diaphragm compressor operation.

contoured cavity, thus moving the gas out of the compressor through the discharge check valve. The diaphragm group completely isolates the hydraulic fluid from the process gas.

A diaphragm compressor is a positive displacement machine, like a reciprocating compressor. The gas compression is achieved by use of a flexible membrane, as shown in Fig. 7.10, which is driven by a rod and crankshaft. Because no relative motion occurs between parts, no dynamic sealing is required. It is only the flexible membrane that comes into contact with the gas. Multistage compression can be applied and has been used to compress hydrogen up to 6000 psi.

Diaphragm compressors are being used for hydrogen fueling stations and are well suited to low-flow applications (approximately pressure near atmospheric up to 15,000 PSI and flows up to 500 SCFM). Due to the dynamics of the flexible membrane, this component is subject to fatigue fractures. The early-stage fueling stations in the California area are primarily being put in with diaphragm compressors. The reliability of these units is definitely lower, and challenges exist in designing a more robust membrane as the fueling stations are likely to require 24 h per day operation.

Diaphragm compressors utilize static seals at the outer circumference of the diaphragm group providing lubrication-free, leak-free, and contamination-free gas compression. This method of compression makes diaphragm compressors ideal for processing hazardous and high-purity gases. In addition, they are suitable for compression ratios up to 15:1 per stage, whereas conventional reciprocating compressors are limited to approximately 3:1 per stage. Diaphragm compressors can also produce discharge pressures up to 15,000 psig and are capable of compressing corrosive gases with minimal modifications to the materials of construction.

Safety measures and hydrogen gas leakage detection

Failed/delayed starts and risk of auto-ignition are clearly concerns if there is a gas leak near or in the gas turbine package, but there are also concerns for failed gas turbine ignition or flame-outs when unburned fuel enters the gas turbine exhaust system. This is also true for compressors not blown down, but not running, with hydrogen gas present. The amount of fuel that can enter the exhaust system between the time the control system detects the failure or flame-out and the fuel valve closes is long enough to completely fill the exhaust ducting of an engine. These hydrogen-specific safety measures are discussed by McBain et al. [14] and highlighted here.

The fuel-air ratio of this mixture in the exhaust is generally below the lower explosive limit when burning natural gas, so it will not burn. However, with increasing hydrogen, this mixture becomes flammable as shown in Fig. 7.11.

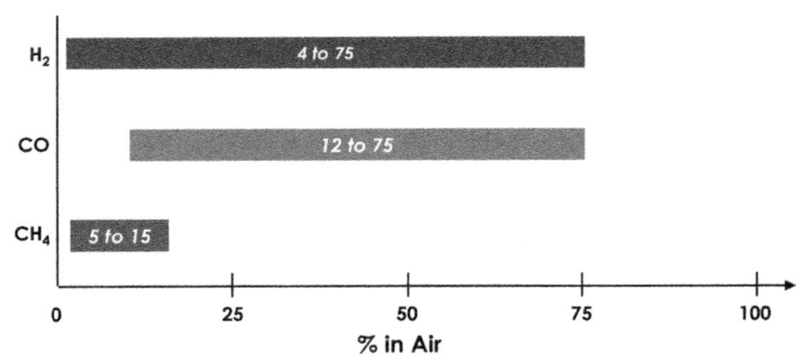

FIG. 7.11 Flammability range of H_2, CO, and methane in air at varying concentrations. *Source: Rich Dennis "Hydrogen storage: A brief overview of hydrogen storage options," TMCES 2020.*

Large amounts of hydrogen fuel will pose a risk, whereas natural gas will not be flammable until it drops to less than 15% in air. If this combustible mixture were to ignite in the exhaust system, pressure rise will occur, potentially causing damage to the exhaust system components of the gas turbine. The same is true of other areas on a compressor if exposed to air and an ignition source.

Emissions systems will not be entirely free of issues with hydrogen as the fuel as hydrogen will still need to be burned in the presence of air, creating NOx, even without CO emissions. In general, fuels with higher adiabatic flame temperatures will create more NOx and less CO and unburned hydrocarbons. Flame temperature increases with H_2 mixtures and will increase NOx emissions modestly for a conventional combustion system and very slightly for a lean premix combustor (assuming that the pilot fuel ratio is not changed).

For the compressor system, the components on skid should follow the appropriate hazard classification for the max percentage hydrogen of the design. The gas group does not change until hydrogen reaches over 20% for most of these components. For gas and fire detection, new devices will need to be designed to detect hydrogen in lower concentrations. Most importantly, special purge sequences should be added and used when there is a failed start or after a flameout before a subsequent attempt to restart.

Hydrogen transport: Retrofits and pipeline station design for hydrogen mixtures

Current industry trends suggest that the most probable future scenario for hydrogen in pipelines involves blending (up to 20% by volume). Many aspects of compressor stations need to be evaluated and/or analyzed when planning to convert an existing natural gas compressor station to a station that will compress a blend of hydrogen and natural gas. There are other industry indicators that suggest the eventual likelihood of 100% or near 100% hydrogen gas pipelines, which would result in additional evaluations and/or analyses, especially if existing natural gas pipeline stations get converted to hydrogen gas pipeline stations.

When evaluating the various components of a pipeline compressor station that would be exposed or impacted by the presence of hydrogen, different maximum threshold percentages of hydrogen are explored for each component. It was concluded by multiple research organizations and industry companies that some of the key aspects of the compressor station need to be evaluated and/or modified if 20% hydrogen will be introduced into the pipeline. Plans for blending hydrogen into existing natural gas pipelines are currently being discussed by pipeline governing organizations such as INGAA, PRCI, PHMSA, and EPRI in the United States. The current guideline authorities for the United States natural gas pipelines are evaluating a blend of up to 20% hydrogen (on a volumetric or molar basis) in existing natural gas stations, and likely leakage pathways that will exist at 20% hydrogen in blended natural gas with existing pipeline construction.

Compressing blends of hydrogen (up to 20%) and natural gas using reciprocating compressors would require a reevaluation of the compressor valves suitability (performance), cylinder materials, seals, lube oil, O-rings, and packing. In general, using reciprocating compressors for hydrogen blending applications (up to 20% hydrogen) should result in few required changes, and in some instances, no required changes. Centrifugal compressors can be modified by adding more staging (more wheels or casings) and possibly still utilize the same compressor with an upgraded power driver for up to 10% hydrogen.

Gas turbine and engine OEMs are working steadily to increase allowable hydrogen percentages up to 100% in the fuel stream. The primary issues related to the hydrogen firing are emissions controls (ability to retrofit NOx controls) and fuel handling. The actual combustion process and combustor nozzles in a turbine can generally handle the hydrogen fuel mixtures with very little design changes. Gas engine OEMs are primarily focused on the higher-speed flame front with hydrogen, and some redesign is required to accommodate greater than 50% hydrogen blends.

Fuel-handling systems must address filter usage for hydrogen, threaded fittings on the meter, GC, heater and filter packages, and measuring fuel quality with modified gas chromatographs (if this is required for the combustion controls).

If converting a natural gas pipeline station to a station that will compress a blend of hydrogen and natural gas, many aspects of the station redesign will need to be evaluated, including the following:

(1) Gas detectors with semiconductor technology.
(2) Gas turbine and engines with enclosures—safety monitoring systems.
(3) Gas chromatographs using helium as carrier fluids (which may not be able to detect hydrogen).
(4) Hot oil/condensate and steam boilers.
(5) Pulsation bottles and piping with threaded fittings.
(6) Blowdown systems.
(7) Nonpressure reducing valves.
(8) Gas hydraulic actuators on instrument gas-powered valves.
(9) All threaded process gas piping and PSVs with threaded connections.
(10) Gas flow meters especially with sensitivity in meter error to speed of sound or density changes.

Station design for pure hydrogen is an existing industry capability. There are currently approximately 1600 miles of hydrogen pipelines in operation in the United States of America [11]. There are many standards and guidelines for the design and development of pure hydrogen pipeline stations. A thorough list of applicable standards and safety guidelines for hydrogen pipeline stations and other applications has been compiled in Chapter 12 of the book titled *Machinery and Energy Systems for the Hydrogen Economy* [6].

Even though there are existing hydrogen pipelines, there are still design challenges associated with the installation, operation, and maintenance of hydrogen pipeline stations. Some of the key challenges for the installation of pure hydrogen pipeline stations are: the high capital costs and the impact of hydrogen leakage on operation and compressor reliability [11].

Regulatory framework for hydrogen pipelines

Hydrogen has gained significant interest as a clean energy carrier that can enable deep decarbonization across multiple sectors. However, infrastructure limitations have constrained widespread hydrogen deployment. Building out hydrogen pipeline networks can enable affordable and reliable hydrogen delivery, supporting expanded production and utilization. As hydrogen pipelines are still an emerging area, regulations continue to evolve to enable safe and sustainable development. This section summarizes the current regulatory landscape in 2023 for hydrogen-carrying pipelines in the United States, focusing on key federal and state requirements.

At the federal level, the Pipeline and Hazardous Materials Safety Administration (PHMSA) has primary jurisdiction over hydrogen pipelines. PHMSA currently allows hydrogen mixtures up to 20% by volume to be transported in existing natural gas pipelines without additional regulation. Pipeline operators must notify PHMSA if converting a natural gas pipeline to carry these hydrogen blends. For pipelines carrying pure hydrogen or blends above 20%, PHMSA's current regulatory framework provides operators flexibility in establishing safe operating parameters. PHMSA is working on more comprehensive regulations specific to hydrogen pipelines, expected to be proposed in 2025.

The National Fire Protection Association's NFPA 2 Hydrogen Technologies Code contains standards for hydrogen pipeline design, construction, operation, maintenance, and safety. While currently advisory, many states have adopted or are considering adopting NFPA 2, making compliance mandatory.

Key elements of NFPA 2 include:

- Materials requirements for hydrogen service
- Leak detection, surveillance, and alarms
- Setback distances from populated areas
- Emergency responder training requirements

The Department of Transportation's (DOT) Pipeline and Hazardous Materials Safety Administration (PHMSA) can grant special permits to allow variance from regulations on a case-by-case basis. Special permits have been leveraged by several hydrogen pipeline projects for design elements such as:

- Operating above 20% hydrogen blends
- Incorporating noncode materials such as fiberglass-reinforced plastic
- Reduced minimum yield strength requirements as a risk mitigation measure
- Permit conditions require additional safety measures such as enhanced inspection, surveillance, and control systems.

While federal regulations set minimum safety requirements, states can implement more stringent oversight through Public Utility Commissions (PUCs) or state fire marshal offices. Key state considerations include:

- Adopting NFPA 2 and International Fire Code standards
- Requiring PUC approval for siting, design, construction, and operation
- Establishing additional reporting and inspection requirements
- Some states such as California are developing hydrogen-specific regulations, whereas others are evaluating integrating hydrogen into existing gas pipeline oversight frameworks.

As hydrogen gains momentum worldwide as a clean energy carrier, many countries are working to develop regulations and codes to enable safe hydrogen pipeline deployment. Requirements vary across different geographies based on existing policy frameworks. This report summarizes current hydrogen pipeline regulations in key global markets outside the United States.

The European Union has developed overarching regulations for hydrogen market development and infrastructure buildout. These include technical standards for hydrogen pipelines. Key regulations for the EU include:

- Pressure Equipment Directive—technical standards for transport of industrial gases including hydrogen
- Gas Directive—framework for natural gas pipelines also applied to hydrogen
- Transmission System Operation Guideline—interoperability and data exchange rules
- Individual countries such as France, Germany, and the United Kingdom have also released national hydrogen strategies with pipeline safety codes.

As hydrogen scales globally, pipeline regulations will evolve from existing natural gas and industrial gas frameworks. Technical guidance is still limited, so demonstration projects rely heavily on industry best practices. Safety and public acceptance are key priorities guiding regulatory development.

Siting hydrogen pipelines requires assessment of population density, land use, environmentally sensitive areas, and underground utilities. Pipeline routing can be contentious without extensive public outreach. Pipeline construction standards address material selection, welding procedures, inspection and testing, pressure testing, burial depth, markers, and cathodic protection. Codes such as ASME B31.12 provide technical standards for design, construction, operation, and maintenance of hydrogen pipelines. Compliance helps address safety risks such as hydrogen's susceptibility to embrittlement and leakage.

Throughout ongoing operations, pipeline operators must establish and follow control room procedures, emergency response plans, preventative maintenance schedules, damage prevention programs, and public awareness initiatives. Key maintenance aspects include leak detection, corrosion control, identifying metal fatigue, and addressing thermal stresses from hydrogen's wide liquid temperature range. Cybersecurity measures are critical to prevent unauthorized access to pipeline operating systems. Real-time monitoring through supervisory control and data acquisition (SCADA) systems enables rapid response to anomalies.

Pipeline operators must report incidents such as fires, explosions, or major leaks to PHMSA and appropriate state agencies such as public utility commissions. PHMSA maintains a database of pipeline incidents for investigative and enforcement purposes. NTSB and other government entities such as the GAO periodically investigate major pipeline failures to determine the root causes and recommend additional safety measures to regulators. Findings help continuously improve the hydrogen pipeline regulatory framework.

As hydrogen infrastructure expands, pipeline safety regulations will increase in specificity and stringency. PHMSA is expected to propose the first comprehensive federal hydrogen pipeline regulations in 2025. States such as California and Texas are also working on updated codes. Industry standards will provide more detailed technical guidance as experience with hydrogen systems grows. Public education and emergency response training will become increasingly important to maintain stakeholder confidence.

Hydrogen pipelines are critical to enable scaled deployment across the energy system. Developing appropriate regulations that enable development while ensuring safety and reliability is an ongoing process. Collaboration between government, industry, academia, and the public is necessary for effective policy evolution. The current regulatory environment in 2023 provides a foundation while allowing flexibility for demonstration projects. Updated federal and state regulations expected over the next few years will provide more prescriptive guidance to support rapid scale-up of hydrogen infrastructure. Continued innovation and knowledge sharing will help advance hydrogen pipeline networks to achieve decarbonization goals.

Liquid hydrogen transport

Depending on geographical location and transportation opportunities, hydrogen liquefiers may either be localized smaller scale with production ranging from 10 to 15 tons per day (TPD) or large-scale industrial size with capacities of 30 TPD up to 200 TPD using multiple trains [15]. A Claude refrigeration cycle, a closed-loop Brayton cycle, or combination of these processes is used for hydrogen liquefaction. Traditionally, a Brayton cycle using helium as the primary refrigerant has been used for plant sizes up to 5 TPD, and a Claude cycle using hydrogen as the primary refrigerant has been used for plant sizes greater than 5 TPD.

An industrial-scale hydrogen liquefier cools a hydrogen feed gas stream in two main stages, a precooling stage and a liquefaction stage, to produce saturated liquid hydrogen (LH2) at −250°C (−418°F) at about 0.7 barg (10 psig) pressure. Fig. 7.12 shows a simplified block diagram of the hydrogen liquefaction process. The hydrogen feed gas stream is fed to the liquefier typically at 15–30 bar (217–435 psi) and cooled by available cooling utilities with a purity of approximately 99.998 mol-percent with less than 20 ppm of total impurities [8].

In the precooling step, the feed gas is cooled to an intermediate temperature using one or more refrigerants. Dividing the process into two or more steps allows less use of energy-intensive cycles with higher-boiling-point refrigerants before utilizing the more energy-intensive cryogenic refrigerants such as helium or hydrogen in the liquefaction step. Once the temperature is cold enough from precooling, the gas is sent to a cryogenic adsorption system to polish remaining impurities that could freeze up in liquefaction. The purity of feed gas flowing through the liquefaction heat exchangers is typically above 99.99999 mol-percent.

Before describing the catalytically conversion step in the liquefaction process, it is important to understand the ortho and para molecular hydrogen forms. These two different forms of molecular hydrogen (H_2) exist due to the different orientations of the hydrogen nuclei (protons) within the molecule. In ortho hydrogen, the two protons in the H_2 molecule have parallel spins, whereas in para hydrogen, the spins are antiparallel. The main difference in the thermodynamic properties between ortho and para hydrogen lies in their relative energies. Ortho hydrogen has a higher energy state than para hydrogen. This energy difference arises from the quantum mechanical effects associated with the nuclear spin states. The para state of molecular hydrogen (H_2) is desired for liquid hydrogen applications due to its lower heat capacity compared with ortho hydrogen. The lower heat capacity of para hydrogen allows it to reach lower temperatures and maintain them more efficiently. By utilizing para hydrogen, the heat capacity is reduced, meaning less energy is required to cool and maintain the liquid hydrogen at extremely low temperatures.

Additionally, the conversion from ortho to para hydrogen releases heat, which can lead to temperature increases and potential boiling issues in the storage and handling of liquid hydrogen. Furthermore, the lower heat capacity of para hydrogen enables better insulation efficiency in cryogenic systems, as there is less energy transfer required to

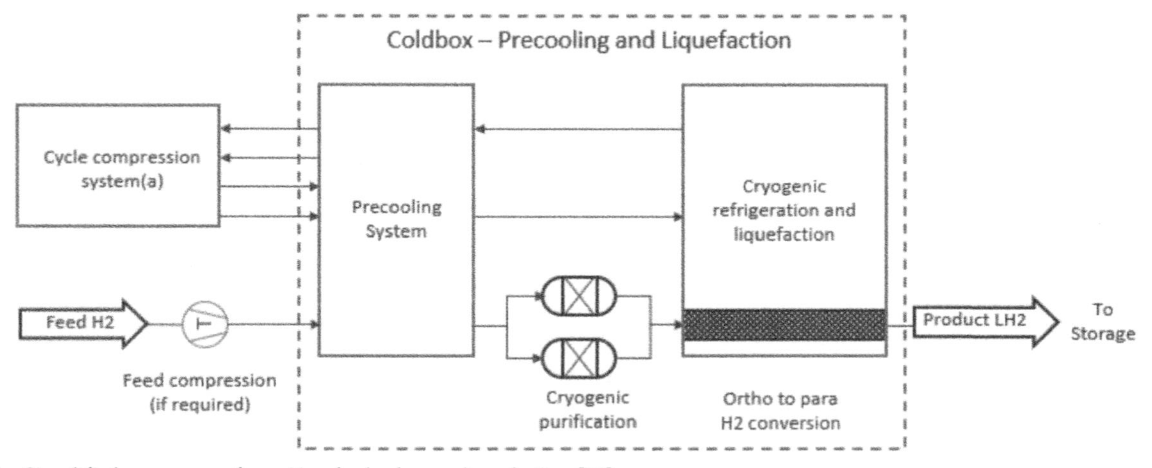

FIG. 7.12 Simplified process configuration for hydrogen liquefaction [16].

maintain the low temperatures. This leads to reduced heat losses and improved overall energy efficiency. Therefore, the preference for the para state in liquid hydrogen applications is primarily driven by its lower heat capacity, which allows for lower temperatures, better temperature stability, and improved energy efficiency in cryogenic systems.

The hydrogen gas is catalytically converted from normal hydrogen with 25% of para hydrogen to a higher final fraction of para hydrogen of around 95%–98% as it is cooled to a final liquefaction temperature. Catalytic conversion is an exothermic process that can be conducted inside the heat exchangers or in separate catalyst beds. Spherical or cylindrical storage tanks are commonly used for liquid hydrogen storage downstream of the liquefiers depending on the capacity of the plants.

Hydrogen liquefaction (H2L) processes are very energy-intensive. Therefore, a variety of process configurations have been built or conceptually proposed for large-scale applications using different refrigerants and refrigeration cycles. The main goal of H2L process optimization is to improve key performance indicators, SEC (specific energy consumption, KWh/KgH_2), and SLC (specific liquefaction cost accounting for capital, operation, and maintenance costs).

The SEC levels for current hydrogen liquefiers range from 11 to 13 kWh per kilogram of liquid hydrogen (kg LH_2). However, there is a potential benchmark of 6 kWh per kg LH_2 [16], which would require advanced technologies across all aspects of the liquefaction plant, including the recycle compression system and cryogenic refrigeration loops. The theoretical best possible SEC for hydrogen liquefaction is 3.9 kWh per kg LH_2. This value represents the minimum energy requirement to produce liquid hydrogen in a reversible Carnot process. It assumes an inlet pressure of 20 bar. It's important to note that achieving this theoretical minimum energy requirement is challenging and would require significant advancements in liquefaction technologies.

Commercially available technologies use liquid nitrogen, gaseous nitrogen, helium, or hydrogen as the refrigerants in both steps. There are also proposed conceptual processes that use a combination of two or more cycles for refrigeration or utilize a mixed refrigerant such as SMR and H_2/Ne in their cycles. Only a limited number of cryogenic refrigerants, such as helium, hydrogen, or their mixtures with neon, can be used in the final liquefaction cycle as freezing will occur with other refrigerants due to the very low temperatures required.

In general, hydrogen liquefaction processes can be categorized by the type of precooling and by the type of cryogenic refrigeration cycle used. Fig. 7.13 represents the main categories of hydrogen liquefaction cycles. The Joule-Thomson process uses a JT valve for expanding the refrigerant to its final cold temperature, while Brayton and Claude cycles respectively use turbines and a combination of turbines and JT valves in their cycle for expanding the cold refrigerant. Joule-Thomson cycles, however, are not used in industrial-scale applications due to their lower efficiency. Brayton and Claude cycles improve the overall energy efficiency of the process by using turbo-expanders as the near isentropic expansion provides lower temperatures than the Joule-Thomson isenthalpic process alone. Also, if the turbo-expander is paired with a compressor or generator, the expansion energy can be recovered. Turbo-expanders are a key piece of equipment in hydrogen liquefaction processes, as their application significantly minimizes the exergy loss.

Despite the variety of cycles that can be used for hydrogen liquefaction, the main equipment used in this process are the same: compressors, gas-phase turbo-expanders, and brazed aluminum or spiral-wound, high-efficiency heat exchangers. Liquid or flashing liquid hydrogen turbo-expanders have also been envisioned in several conceptual designs, but this equipment is still under development and not yet available in the industry.

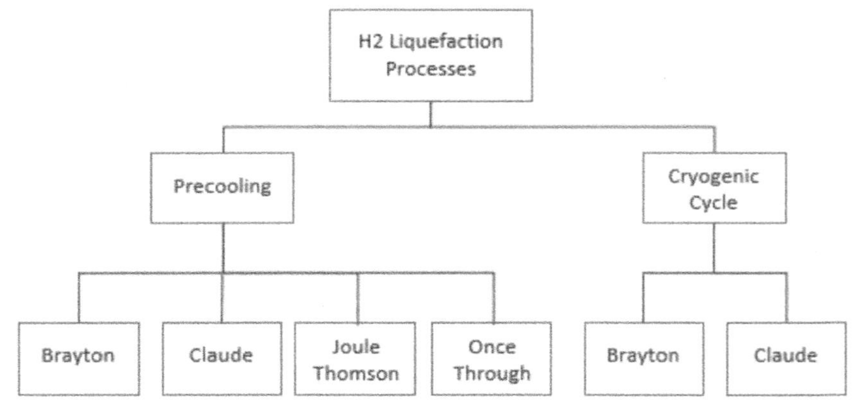

FIG. 7.13 Categories of refrigeration cycles Used in H_2 liquefaction process [16].

A once-through cycle utilizes a cold refrigerant, such as liquid nitrogen or LNG, for precooling without recirculation. Standard process and auxiliary equipment such as pumps, coolers, chillers, phase separator vessels, drums, piping, valves, instrumentation, analyzers, and measuring devices are common parts of the process.

Fig. 7.14 summarizes the various cycles used for either hydrogen precooling or liquefaction. Table 7.5 provides an overview of commercially available cycles using different refrigerants with their respective key performance indicators, SEC (specific energy consumption, KWh/KgH2), and capital cost.

A Brayton cycle design using helium as the cryogenic refrigerant has a lower capital cost mainly due to the possibility of utilizing oil-flooded screw compressors. Considering helium is inert and nontoxic, stringent requirements for design in hazardous service do not apply. Low capital investment is the main benefit of a Brayton helium cycle, but the energy efficiency is sacrificed. In contrast, Claude cycles that use pure hydrogen as the cryogenic refrigerant can achieve higher energy efficiency, but with higher capital costs due to a more complex and costly turbo-expander and compressor systems.

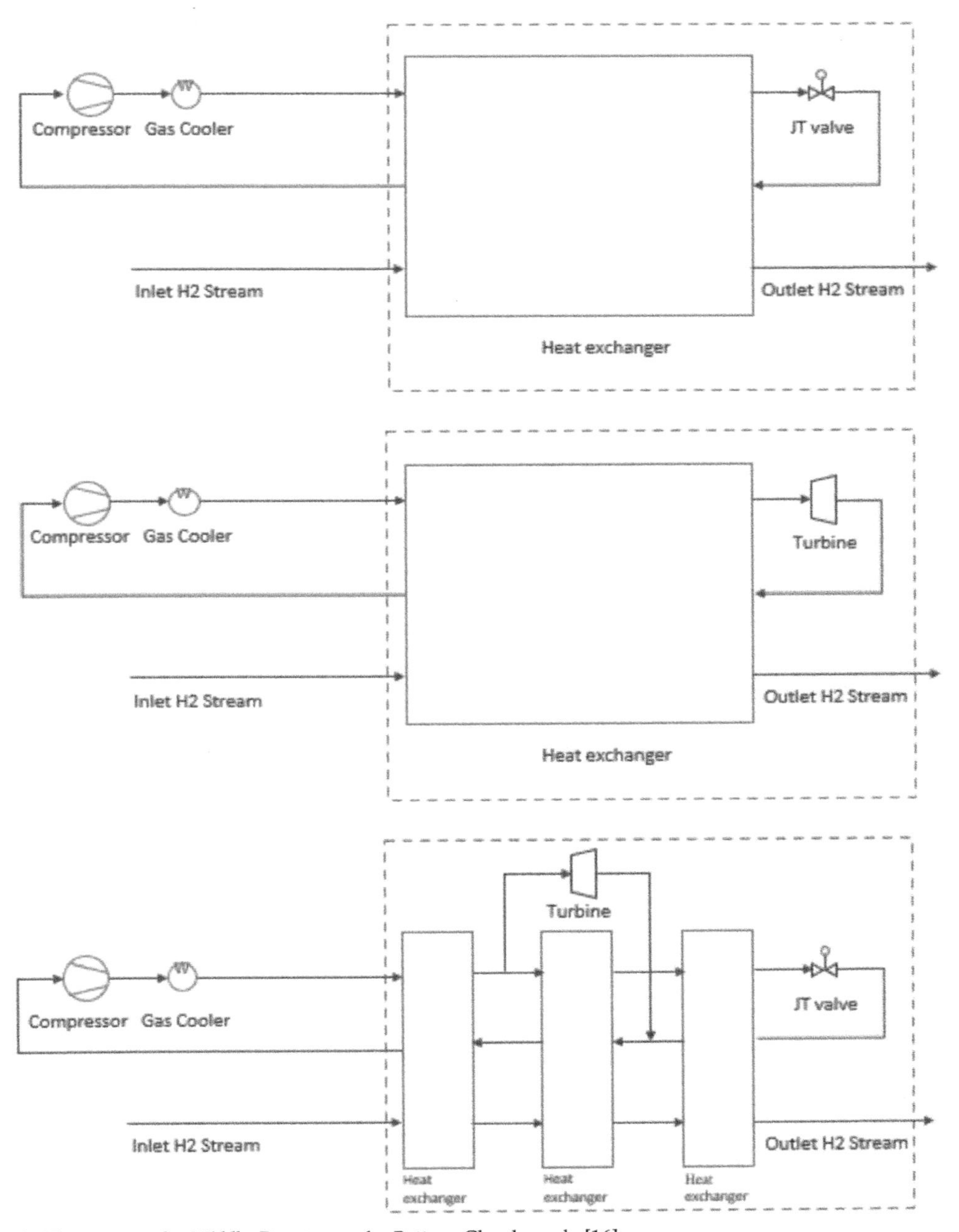

FIG. 7.14 *Top*: Joule-Thomson cycle, *Middle*: Brayton cycle, *Bottom*: Claude cycle [16].

TABLE 7.5 Qualitative overview of available H2L cycles in terms of SEC and CAPEX.

		Liquefaction cycle Pure H_2	He	Comments
Pre-cooling cycle	LN_2 (liquid nitrogen)	Claude cycle SEC = Medium to High Capex = Medium to High	Brayton cycle SEC = High Capex = Low	*Combination of different precooling technologies for precooling is possible to drive SEC down
	GN2 (gaseous nitrogen)	Claude Cycle SEC = Low to Medium Capex = Medium to High	Brayton Cycle SEC = High Capex = Low to Medium	Scaling up the process helps to bring SEC further down
	SMR (single mixed refrigerant)	Claude Cycle SEC = Low to Medium Capex = Medium to High	Brayton Cycle SEC = High Capex = Low to Medium	*Optimizing Claude cycle and utilizing mixed refrigerant (H2/Ne) can drive SEC down

Systems with low capital cost are mainly used for small-scale liquefiers (<10 TPD size) as the high capital cost becomes a prohibiting factor. Large-scale liquefiers (>10 TPD), however, can accommodate increased capital costs because achieving the lowest possible operating cost is a key factor for successful plant economics.

Large-scale liquefaction

As the hydrogen economy grows, both market demand and the need for more efficient transportation methods are driving the need to scale-up hydrogen liquefier capacities. Energy efficiency and required capital costs are considered the key factors for commercially viable hydrogen liquefiers at industrial scales. The key opportunities for optimization of liquefier performance for large-scale applications are:

(1) Optimization of the hydrogen Claude cycle for liquefaction utilizing mixed cryogenic refrigerants such as H_2/Ne with a higher compression ratio and more efficient turbomachinery
(2) Utilization of a more efficient precooling configuration including utilization of cascade or combined cycles
(3) Improvement in turbo-expander technologies, as the primary source of refrigeration for liquefaction technology, to achieve higher efficiency and higher energy recovery
(4) Development of flashing liquid/liquid expanders to replace liquid JT valve in the feed gas line
(5) Scale-up liquefaction plants' capacity to reduce capital investment per unit mass of produced liquid hydrogen, improving ROI of the investment

While cycle improvement using a cryogenic refrigerant with a higher compression ratio instead of pure hydrogen has the highest potential for efficiency improvement, it is more of a long-term solution as it requires investigation and pilot-scale experiments before it is qualified for industrial applications. For short-term and mid-term applications, improving Claude cycles using pure hydrogen is the most promising option for reducing SEC and SLC of industrial-scale liquefiers.

Unique machinery in liquid hydrogen—Hydrogen pumps

The working principle of pumps is different from that of compressors. Pumps increase pressure of the process fluid in liquid phase with negligible compression and minimal change in fluid density. Pressure predominantly rises as a result of restrictions within the system or pump, while maintaining a constant volume of the propelled fluid. These restrictions can stem from losses in the pump's internal passages and hydraulics or be associated with the broader system, such as restrictions in piping and external valves located outside the pump. The system restriction can be regulated by a discharge valve in a pumping application. In compressor applications, pressure increase is attained

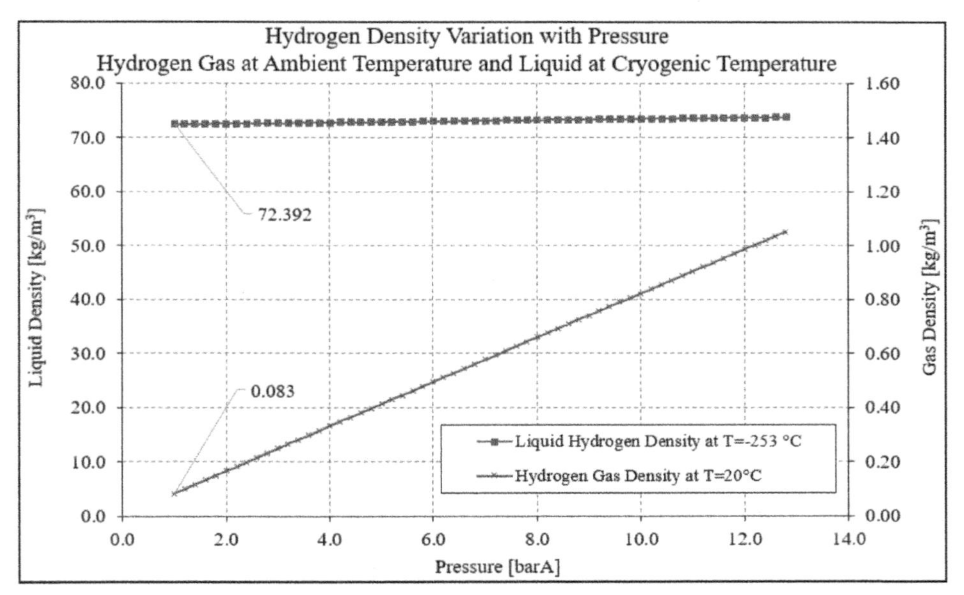

FIG. 7.15 Hydrogen density variation as liquid and gas at cryogenic temperature and ambient temperature, respectively [17].

through compression of the fluid by reducing its volume. Hydrogen in a liquid phase can be assumed incompressible since the variation of the density with pressure change is less than 3% for a given temperature.

Fig. 7.15 shows the variation in hydrogen density with pressure under gas and liquid phases. Hydrogen density is calculated according to published tables [18]. For a gas phase, pressurizing process fluid requires more energy due to the fact that fluid is compressible, and there is considerable amount of variation in density. With that, compressors require more energy to pressurize the fluid to the desirable condition. As the pumping applications do not result in change in density, and pressure increase is direct, pumps fundamentally use less energy and are more efficient for bringing a given volume to a desirable pressure. For hydrogen applications, this is an important consideration as density of the liquid hydrogen is around 850 times greater than hydrogen gas under atmospheric pressure. This is a considerable advantage of transportation and storage of hydrogen in liquid form.

However, there are many challenges to liquefy the hydrogen to be pumped in liquid form. Pump applications require hydrogen to be cooled to cryogenic temperatures of around −253°C (−424°F) under atmospheric pressure. The liquefaction process of hydrogen is very costly and requires cyclic compression and cooling of hydrogen in stages. Once hydrogen is liquefied, it must then be stored or transported in special cryogenic vessels or tankers to maintain cryogenic fluid temperatures. Due to the small molecular size and severe cryogenic conditions, boil-off gas and evaporative losses are extremely difficult to eliminate during transportation of hydrogen. Regardless, it is still considerably less expensive and more economical to liquefy hydrogen for transportation and storage purposes [Ref. 16].

Pumping applications

Hydrogen pumps are used in various industrial and some aerospace applications to pressurize and transport hydrogen in liquid phase. Depending on the process and application specifics, different types of hydrogen pumps are used to attain the most efficient and reliable operation.

Hydrogen pumps can be categorized into two primary types. For high-flow, relatively low-pressure applications, such as the transfer pumps used in displacing hydrogen liquid (LH_2) from cryogenic vessels to tankers, centrifugal pumps are preferred. The centrifugal pumps usually use a submerged motor design to eliminate leakage and boil-off through mechanical shaft seals. Due to a possibility of leakage and lack of reliability under extreme cold cryogenic temperatures, mechanical shaft seals are eliminated. These pumps use a common shaft for the motor and hydraulic components as shown in Fig. 7.16. These pumps are also vertically suspended inside a vacuum jacket, insulated pressure vessel with the necessary process piping. A picture of a vertically suspended cryogenic LH_2 pump is given in Fig. 7.16.

For high-pressure and relatively low-flow applications, reciprocating positive displacement pumps (PDPs) are the choice. These types of pumps are commonly used in hydrogen fueling applications. A reciprocating PDP uses a piston

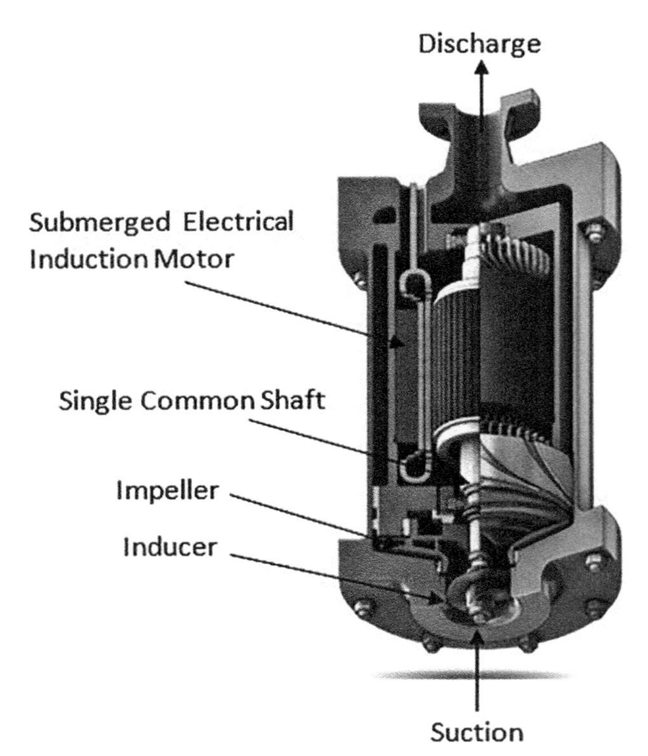

FIG. 7.16 Cryogenic pump internals—typical for hydrogen—integrated motor pump package for submerged tank operation. *Courtesy of Elliott Group - Ebara Corporation.*

and cylinder arrangement, connected to a crankshaft, and is driven with an electric motor or a hydraulic pump. The piston and cylinder section is vacuum-jacket insulated to maintain cryogenic temperatures. If the flow rate needs to be increased to meet the demand of the application, more than one set of piston and cylinder arrangements can be used in parallel. Fig. 7.17 illustrates the common area of application for both centrifugal pumps and reciprocating PDPs.

Hydrogen has been used as the primary fuel for rocket engines due to its light weight (low density with respect to other fuels) and its high combustible nature. Hydrogen turbopumps are used in rocket engine applications to supply fuel for combustion along with an oxidizer turbopump (Fig. 7.18). These pumps have been utilized to work in parallel with oxygen pumps to achieve the desired mixture for efficient combustion and hence propulsion. These pumps are driven by hot exhaust gas turbines to achieve high rotational speeds, which can be in excess of 50,000 RPM. Government space agencies and privately owned companies developed turbopumps in rocket engines, which greatly helped the advancements and technology improvements of industrial pumps for LH2 applications.

Challenges of hydrogen pumping and distinctions in pump design

The main challenges with LH_2 are the small molecular weight, relatively low density, and extremely cold pumping temperature. Special thermal insulations are utilized in pressure vessels of submerged motor pumps and the pump section of PDPs to maintain LH_2 temperature and minimize heat transfer and vaporization/boil off.

An extremely low pumping temperature of −253°C requires material selection with adequate impact strength. Thermal shrinkage properties of each different material used in conjunction with another must be considered in the design to ensure that the resulting stresses do not exceed the max allowable material stress. Hydraulic components are often made of 300-series stainless steel, and casings can be aerospace-grade aluminum alloy. Sand casting should be avoided, as there is a possibility of porosity due to sand inclusion and thermal effects during the manufacturing process, which can lead to hydrogen leakage. Hydrogen embrittlement (hydrogen-infused cracking) should be considered in the selection of materials for pump construction.

FIG. 7.17 Multistage LH_2 vertically suspended, submerged motor centrifugal pump. *Ebara Corporation, Japan.*

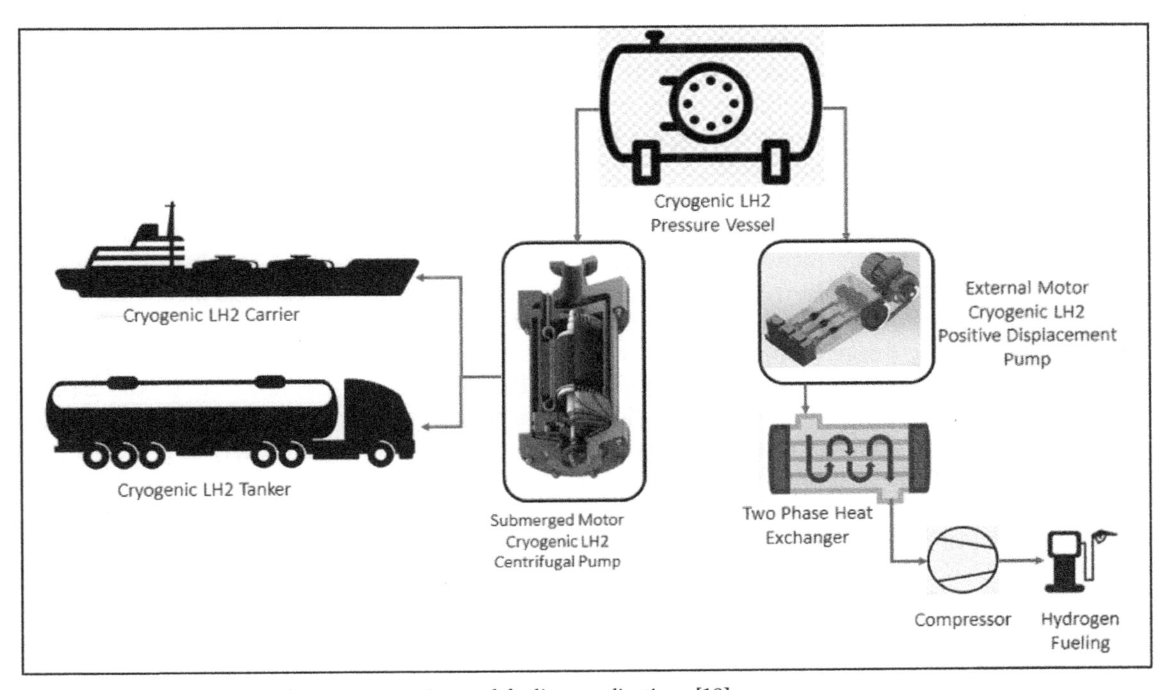

FIG. 7.18 Industrial LH_2 pumps used in transportation and fueling applications [19].

LH_2 pumps are often operated at relatively higher rotational speeds to achieve the desired pressure by the process. For a given process and with an identical pump hydraulic configuration, an LH_2 centrifugal pump must be operated 3.4 times faster in rotational speed with respect to a cryogenic liquid nitrogen (LN_2) pump to provide identical discharge pressure. An increase in rotational speed can be avoided by adding additional stages to the centrifugal pump. As for any pumping application, to attain an acceptable pump efficiency, rotational speed and stage count must be determined with the consideration of the pump specific speed (Ns), while for rotordynamics, reliability and cost are also taken into account.

Relatively low density also has a negative impact on a pump's rotordynamics, because the stiffness and damping decrease in pump hydrodynamic supports such as fluid film bearings and bushings. As mentioned before, to minimize the LH_2 boil off, mechanical seals are often not preferred for relatively high-flow pumping applications (centrifugal pumps). A submerged motor design approach eliminates the need of a mechanical shaft seal, but a motor can slightly increase the temperature of the pumped LH_2. Temperature increase is often not a concern; however, the ball bearings are lubricated with the pumped product in submerged motor pumps. There is no fully controlled lubrication and cooling system or loop in submerged motor pumps. Lubrication and cooling of the radial ball bearings are achieved by pumped products (fluid). LH_2 viscosity is estimated to be 0.014 cP at a cryogenic temperature of −253°C under 1 atm pressure. This requires ball bearings specifically designed and manufactured for LH2 applications to maintain sufficient lubrication. Another option is to support the rotating assembly by utilization of hydrostatic bearings, which do not rely on elastohydrodynamic film lubrication as much as a deep groove radial ball bearing. In addition, hydrostatic bearings can be operated at higher rotational speeds, if necessary, with respect to radial ball bearings.

For pumps with external motors utilizing mechanical shaft seals, the amount of leakage must be predicted along with the purity requirement for LH_2, as contamination of LH_2 with lube oil may not be allowed.

Unique machinery for liquid hydrogen—Turbo-expanders

Turbo-expanders, also known as expanders or expansion turbines, provide the cooling used for cryogenic processes such as liquefaction. Turbo-expanders are used to drop the liquid pressure of hydrogen to compatible tank and line pressure.

Process design optimization may require several iterations to reduce the number of turbo-expander stages while maintaining high machinery performance and low specific energy consumption. In addition, innovative aerodynamic and mechanical design features are required for pure hydrogen turbo-expanders. Liquefier cycle design considerations also include turndown capability, ease of startup, operation, and reliability.

Turbo-expander design for hydrogen liquefiers is challenging due to high isentropic enthalpy drop, low discharge volume, and deep cryogenic temperatures. Liquefier process requirements produce unique aerodynamic and mechanical designs, including low-flow turbo-expander wheels, high peripheral speeds, and robust thermal management. Due to its low mole weight, hydrogen has a relatively high enthalpy drop for a given pressure ratio, often forcing multistage solutions to satisfy blade tip speed, bearing speed, or rotordynamic limitations. Multistage expanders allow the speed of each stage to be customized for the gas properties and rotordynamic constraints, as the hydrogen gas expands with significantly different enthalpies.

Practical challenges arise from hydrogen's flammability and resistance to being contained by seals. Hermetic designs with no external shaft seals are often preferred to eliminate process loss and keep the hazardous gas contained. Hermetic solutions require the rotor-bearing systems be submerged in the process fluid, making material compatibility and bearing lubrication a challenge. Oil-free solutions, such as active magnetic or gas bearings, ensure that no lubrication oil can enter the process, where it could cause costly heat exchanger fouling.

Small turbo-expanders in hydrogen liquefaction service have historically utilized energy dissipating technologies, where any power absorbed by the turbo-expander is rejected in the form of heat and not recovered. As liquefiers scale up in size, energy recovery-based turbo-expander configurations become economically viable and improve plant specific energy consumption by adding free compression to the cycle or by generating electricity. Fig. 7.19 shows a typical magnetic bearing turbo-expander coupled with centrifugal compressor and also highlights design features required by liquefaction processes.

Alternative configurations with energy recovery through a generator are possible. Integrally geared expander-generator arrangements allow for one or more turbo-expander stages on a single train. This configuration is especially attractive when the cycle designs require several turbo-expander stages. Compressor stages may also be added to the integrally geared arrangement (referred to as a compander) where a motor drives the compressor stage(s) at reduced load due to the turbo-expander energy recovery. It is important to note that integral gearing requires external shaft sealing, which results in some loss of the process gas.

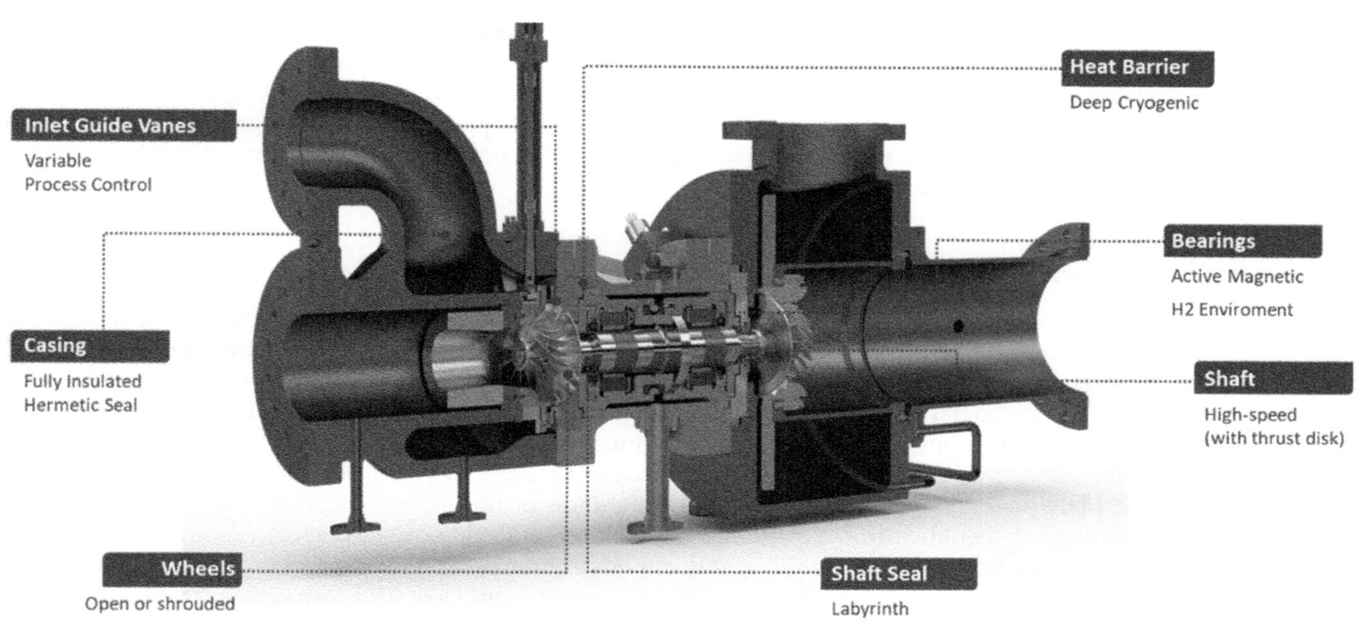

FIG. 7.19 Turbo-expander with power recovery via compressor load.

Cryogenic hydrogen turbo-expanders—Rotor-bearing system

Cryogenic turbo-expanders typically use active-magnetic, oil-lubricated, or gas bearings. Due to the low temperature and zero contamination requirement of the liquefaction process, turbo-expanders with oil-free bearings are often specified. Turbo-expanders with static or dynamic gas bearings have been successfully deployed in hydrogen liquefiers [17] at a smaller scale. However, these rotor-bearing systems have limitations in radial and axial load capacity at larger scales.

Active magnetic bearing (AMB) technology (Fig. 7.20) offers improved load capacity when compared with gas bearings and can typically handle process loads without external axial thrust balancing. Active magnetic bearings have no permanent magnets and are fully compatible with hydrogen service. Turbo-expanders with active magnetic bearings have been used in hydrogen-rich service for more than 30 years in the petrochemical industry.

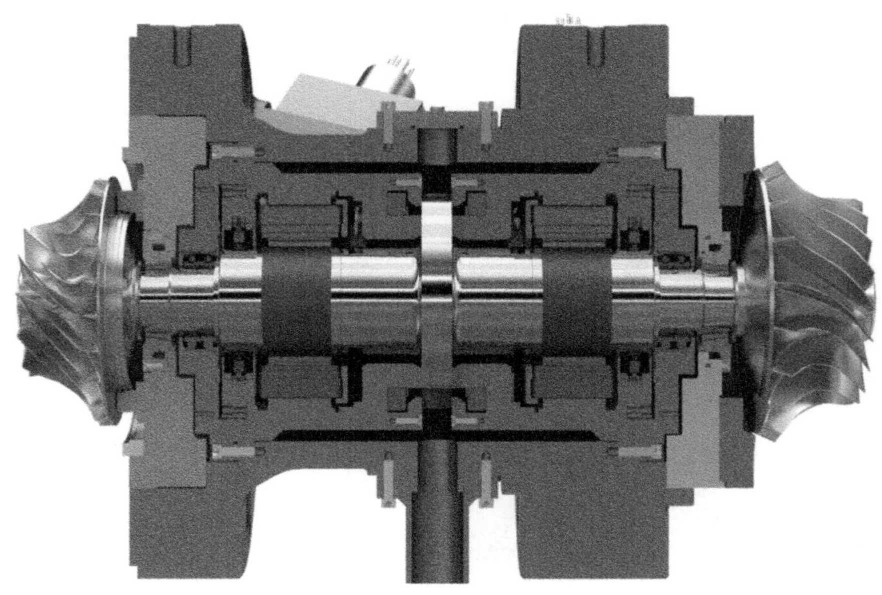

FIG. 7.20 Turbo-expander with active magnetic bearing (AMB) technology.

Minimizing turbo-expander stages requires high-rotating and high-peripheral speeds. Active magnetic bearings have limited speed capability when compared with gas or oil bearings, which may limit turbo-expander efficiency or require more stages at a smaller scale. As liquefiers grow, rotor speed decreases with larger diameter turbo-expander wheels and AMB speed limitations becoming less of a concern. Liquefiers 30 TPD and greater begin to utilize turbo-expander rotor-bearing designs similar in size to those used in the petrochemical industry, which have undergone decades of operating experience and optimization.

Cryo expanders—Thermal management

Net turbo-expander refrigeration depends both on aerodynamic performance and parasitic losses, such as thermal heat soak. Both heat conduction of the rotor and casing interface with the bearing carrier must be carefully analyzed when designing turbo-expanders for deep cryogenic service. Fig. 7.21 shows a thermal analysis performed on the cold-end turbo-expander in the hydrogen liquefaction cycle. As machinery size and duty increase, effects from the warm-rotor bearing system interface lessens and overall efficiency improves.

Insulation of turbomachinery in deep cryogenic service is important at all sizes. Turbo-expander casing flanges are designed in a single plane to interface with a cold box. Integration of the turbo-expander casing inside the cold box requires collaboration between the turbo-expander manufacturer and cold box supplier. A vacuum or other insulation can minimize heat gain from the surrounding ambient environment for both expander casing and interstage piping. Fig. 7.21 shows a typical warm- and cold-end turbo-expander interface with a vacuum-jacketed cold box.

Primary and secondary flow paths

For radial inflow turbo-expanders in hydrogen service, high enthalpy drop, low volume flow, and limitations in rotational speeds often shift designs toward nonoptimum wheel geometry. Turbines with a low degree of reaction may require backswept inlet blade angles and fully or partially shrouded impellers to improve performance. These challenges are further exaggerated at smaller scale where additional inefficiencies exist due to minimum blade thickness and tip clearances. Fig. 7.22 shows an example of the primary and leakage paths.

Given the low mole weight of both helium and hydrogen, other secondary flows in the process must be closely analyzed. To keep cold process gas away from the bearings, conventional oil-free seals require warm seal gas, which flows toward the process. This seal gas creates a significant performance loss when mixing with deep cryogenic hydrogen gas in the turbo-expander stage. Innovative shaft sealing for liquefiers is required to avoid this mixing. The parasitic heat loss from conventional shaft sealing on AMB can be as high as 4%–6% without additional mitigation.

Other losses to consider are turbo-expander wheel seals, where flow bypassing the turbo-expander wheel undergoes isenthalpic expansion, decreasing overall stage efficiency. Wheel seals are often required to aid axial thrust compensation. Robust axial thrust management, like that offered by an AMB system, allows for fewer wheel seals and therefore reduced parasitic losses.

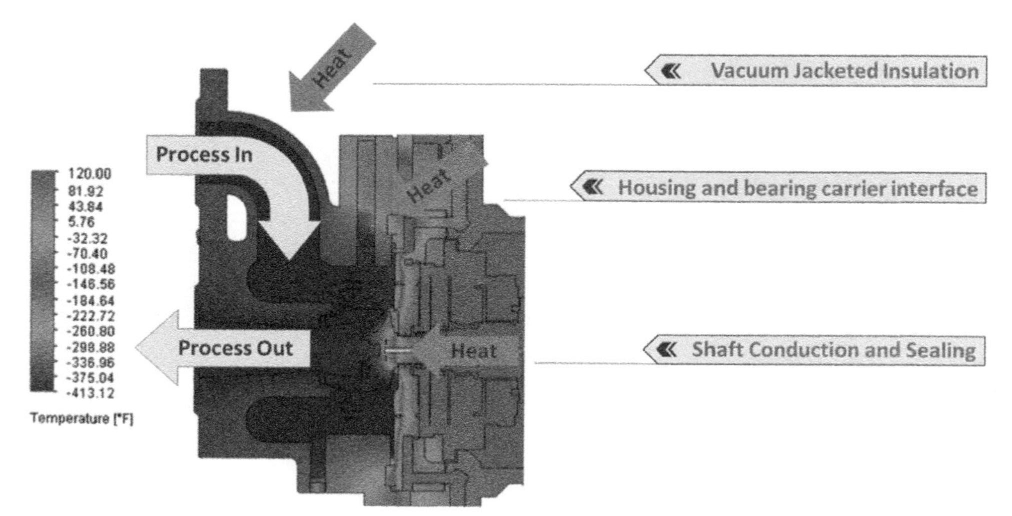

FIG. 7.21 Turbo-expander thermal analysis for cold-end hydrogen stage.

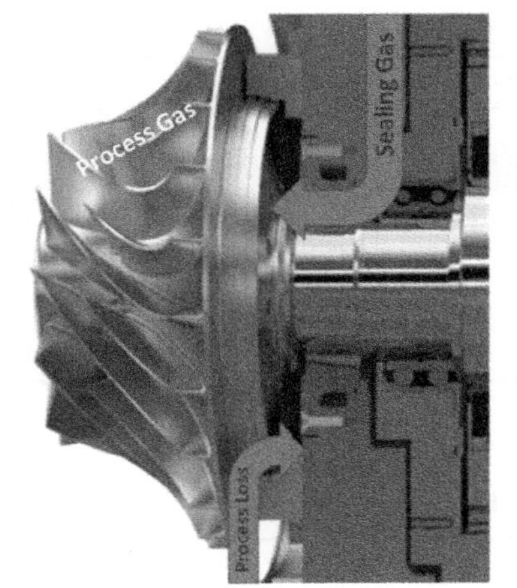

FIG. 7.22 Primary and secondary flow paths in a typical turbo-expander stage.

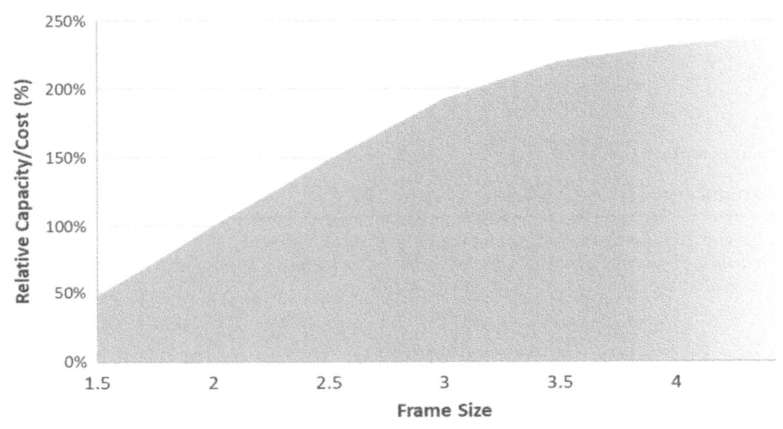

FIG. 7.23 Relative machinery capacity/cost for various turbo-expander frame sizes.

Machinery cost

Turbo-expanders are complex pieces of machinery that require many parts. When considering the entire package, part count remains relatively constant for different sizes of machinery unless features are removed. The machinery part count, instrumentation, and auxiliary support systems all have a large impact on cost at smaller sizes. As turbomachinery increases in size, material mass becomes a greater factor in the overall cost, resulting in cost scaling more proportional to machinery duty. Fig. 7.23 shows how this relationship impacts costs at different frame sizes.

Some recurring and nonrecurring costs in turbo-expander design and manufacturing can be improved with standardization. The projected demand for hydrogen liquefaction allows for modularization and/or built-to-order machinery, which significantly reduces engineering and test requirements per unit. Increased standardization and greater volumes of machinery can significantly impact the shape of the curve shown in Fig. 7.23.

At a certain threshold, improvements in economy of scale begin to diminish, and the decision between scaling machinery or simply adding stages has less of an impact on SEC and SLC. At that point, more favorable arrangements of machinery with more stages can be considered, which may further improve the overall process efficiency.

Gas vs. liquid hydrogen transport considerations

There are four main options for transportation of hydrogen: compressed gas tube trailers, liquid, gas pipeline, or chemical carriers. The economics behind the method in which you transport hydrogen varies depending on the application and location. Table 7.6 compares capital cost, operating cost, and feasibility of different hydrogen transportation options.

TABLE 7.6 Qualitative comparison of hydrogen transportation options.

	Compressed gas H_2 truck	Liq. H_2 truck	H_2 pipeline	Ammonia by ship	Liq. H_2 ship
Total capital costs	Low	Medium	High	High	Medium
Operating costs	High	Medium	Low	High	Medium
Transport cost per kg	High	Low	Low	Medium	Medium
Transport distances	Short Distance	Domestic	Continental	Intercontinental	Intercontinental
Consumer demand	1 to 10 TPD	10 to 500 TPD	100+ TPD	100+ TPD	100+ TPD

TABLE 7.7 Comparison of truck delivery to vehicle fueling stations.

Sample analysis	Gaseous H_2	Liquid H_2	Gasoline
Pressure (bar)	200	1	1
Temperature (°C)	Ambient	−253	Ambient
Mass/load (kg)	800	4200	26,000
HHV of fuel (MJ/kg)	141.9	141.9	48.1
HHV/load (GJ)	114	596	1252
Energy for mid-size fuel station (GJ/day)	876[a]	876[a]	1252
No. of trucks/same number of serviced vehicles	7.7	1.5	1

Original Analysis by Ulf and Eliasson (2006) Energy and the Hydrogen Economy.
[a] *Because of the superior tank-to-wheel efficiency of fuel cell vehicles, it is assumed that hydrogen-fueled vehicles need only 70% of the energy consumed by gasoline or diesel vehicles to travel the same distance.*

Compressing hydrogen in tube trailers is the simplest transportation method but is only suitable for short distances and relatively small volume applications because of its limited energy density per load. This method typically requires compression and storage at the end-use site, which makes the cost per kilogram expensive compared with alternative options.

Liquefying hydrogen significantly increases its density (it is 800 times denser than gaseous hydrogen at atmospheric pressure), which allows for more hydrogen to be transported in a single trip and makes the cost per kilogram more economical for further distances. The footprint at the end-use site is considerably smaller than gaseous H_2 storage. Table 7.7 compares trucked deliveries to vehicle fueling stations of gaseous H_2, liquid H_2, and traditional gasoline.

Transporting hydrogen by gas pipeline is (in some ways) a good fit for large industrial consumers, such as petrochemical or power plants that require a constant, high-volume flow, and natural gas pipelines provide a precedence of many decades of experience on how to operate high-pressure gas pipeline systems with many entries and offtakes. There are approximately 1600 miles (2575 km) of dedicated hydrogen pipelines installed in the United States for this purpose. Of course, pipelines require large capital expenditures and often encounter regulatory roadblocks due to their environmental impact. In the interim, hydrogen blending in existing natural gas pipelines is being investigated to reduce emissions.

Fig. 7.24 summarizes the optimal transport distance and project size in million tons of hydrogen per year. This figure highlights the various means of transporting hydrogen and why there is some overlap in the optimal choice, depending on distance and volume produced, as well as the ability to use repurposed natural gas lines. The use of repurposed pipelines obviously depends on the existence of gas infrastructure in the desired delivery area for hydrogen and the condition of those lines (corrosion, pressure rating or derating, steel susceptibility to hydrogen cracking, etc.), but it does open up the possibility of transporting longer distances more economically than converting to liquid organic carriers of ammonia.

Hydrogen carriers—Ammonia and methanol

In examining the carriers for hydrogen, ammonia and methanol are alternative liquid options for shipping hydrogen by marine vessel or rail. In Fig. 7.25, green hydrogen production as part of a renewable network is shown, without the use of a carrier gas. In this depiction, when excess renewable power is available, excess hydrogen could be

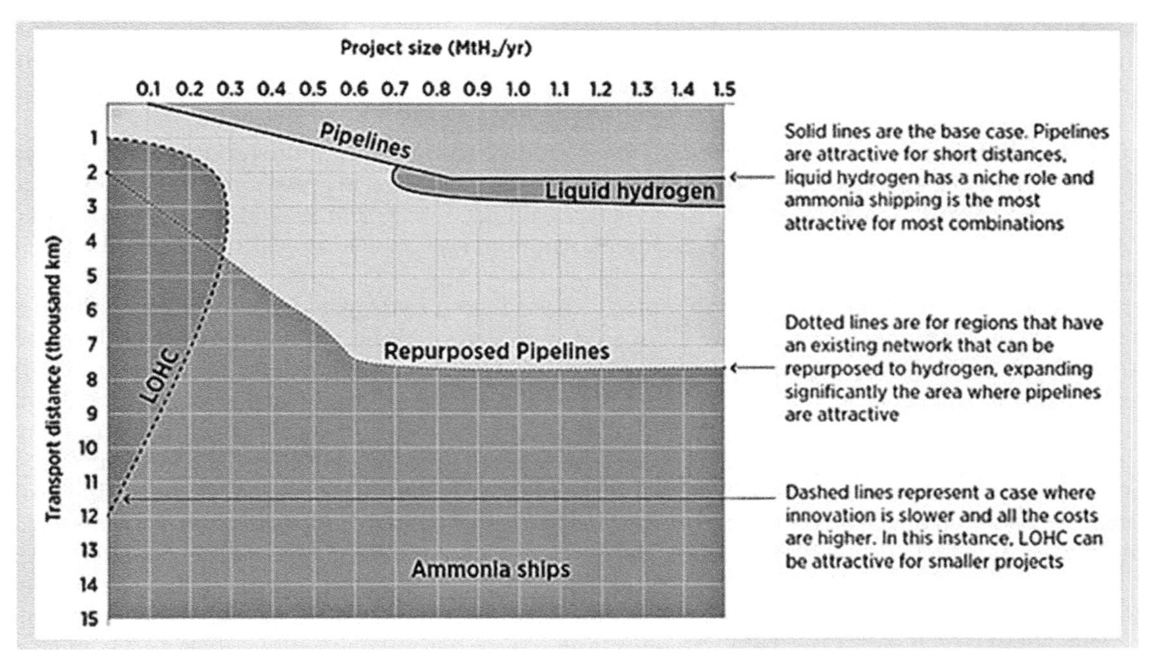

FIG. 7.24 Options for transporting hydrogen—Optimal selection for cost depends on distance transported and project size (in mass of hydrogen per year). *Source: "Global hydrogen trade to meet the 1.5°C climate goal: Part II – Technology review of hydrogen carriers"* © *IRENA 2022.*

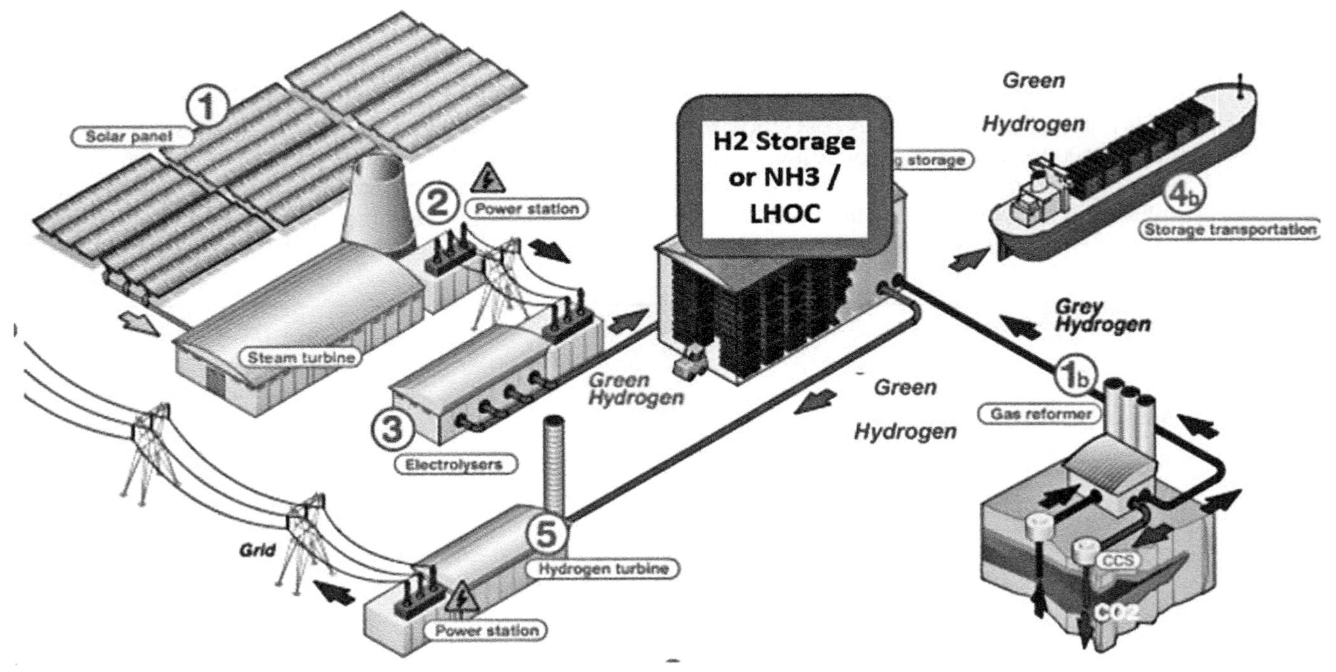

FIG. 7.25 Storage and green hydrogen production—Without carriers as part of a storage and power network.

produced and then shipped offsite or stored onsite, as high- pressure gas or cryogenic liquid (assuming no carriers are utilized). Hydrogen is then used to produce power when the renewables cannot provide the required grid power delivery through a gas turbine generator running on hydrogen gas. This concept can now be compared with the option of converting hydrogen to a carrier, such as ammonia or methanol, highlighted as follows.

Ammonia and methanol are two viable carriers for transporting hydrogen in a higher density, lower-risk state. Both of these molecules can be liquefied at higher temperatures and as such, vaporize at slower rates during the transport state (which can be an appreciable cost savings for long haul marine trips, for example). There are three essential steps to any hydrogen carrier that should be analyzed against one another to determine the most efficient transport mechanism: (1) conversion to carrier molecule, (2) transport vehicle for molecule and losses incurred due to loading/unloading/transport process, and (3) recovery of hydrogen in converting from molecule back to monetized form for hydrogen.

Ammonia is a well-developed industrial process, however, green ammonia requires adapting to a process utilizing green hydrogen (renewable powered electrolysis). As an interesting variation in green ammonia, process licensors and research studies are underway to develop electrochemical synthesis, which utilizes water directly instead of converting water to hydrogen, possibly saving capital equipment. Ammonia is liquefied at −33°C compared with hydrogen at −253°C—see Fig. 7.24.

Another major advantage is that the ammonia molecule has a higher hydrogen density. Additionally, the explosive or flammable range of Ammonia is fairly small compared with that of hydrogen when mixed with oxygen.

On the recovery side after transport, ammonia may be burned directly for fuel, sold for fertilizer, or undergo a cracking process to deliver hydrogen, leveraging all three possible commercial markets. These additional routes to sell ammonia in direct NH3 fuels or for direct fertilizer sales allows the losses in reconverting to be minimized if it is not necessary to convert back from ammonia to green hydrogen to monetize the hydrogen streams (Fig. 7.26).

Methanol is a basic industrial chemical with annual production volumes exceeding 100Mt. In addition to its use as a precursor in the production of various chemical products including plastics, paints, adhesives, solvents, etc., methanol is increasingly used as a fuel additive for gasoline and itself as a potentially renewable fuel for internal combustion engines with a lower heating volume of 19.9MJ/kg (about half the volumetric energy of gasoline or diesel). It is a potentially green, renewable fuel, because it can be synthesized from renewable feedstock with renewable energy.

Traditionally, "gray" methanol is produced by reforming coal or natural gas with unabated CO_2 emissions. Green methanol is produced in two distinct varieties: as "E-methanol" from hydrogen made by electrolysis with renewable electricity and CO_2 from biological or other climate neutral sources or as "biomethanol" from biogas or solid biomass through reforming or gasification into syngas. In both cases, a methanol synthesis reactor converts syngas into methanol that leaves the process as a distilled liquid ready to be sold.

Green methanol can be sold at a price premium over conventional gray methanol with buyers wishing to decarbonize their fuel source or chemical supply chain. The cost of biomethanol depends on the plant size and feedstock costs and may reach the level of gray methanol of several hundred USD per ton, while E-methanol, even with low electricity prices, is typically more costly. A cost study for green methanol published by Irena [20] shows for biomethanol a range from US$400 to 1000/ton, depending on Capex, Opex, and feedstock cost, with plant CAPEX accounting for US$200–300/ton. For E-methanol, the cost is driven mostly by hydrogen costs, and to a lesser degree, by the CO_2 cost. Irena estimates the 2020 costs of E-Methanol without incentive funding in a range from US$800 to 1600/ton but concludes significant future reduction potential with lower hydrogen costs.

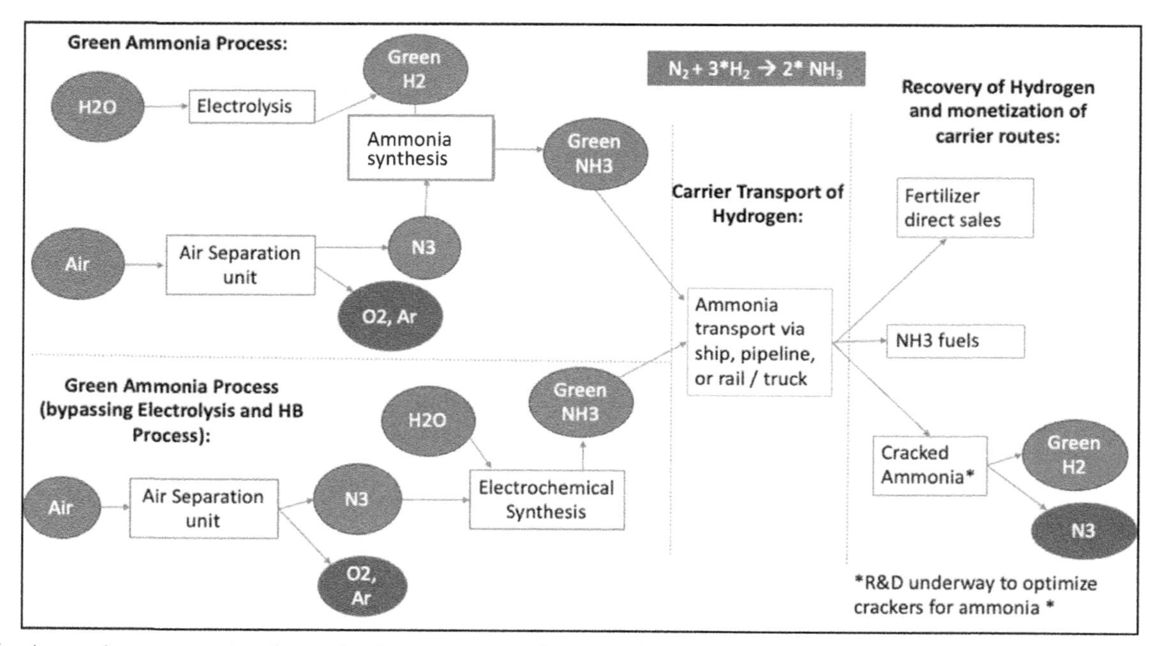

FIG. 7.26 Ammonia process options for production, transport, and recovery for both green ammonia via ammonia conversion process and direct electrochemical synthesis.

Biomethanol production starts by conversion of carbon-containing feedstock into syngas. For solid biomass, gasification produces a carbon-rich syngas. For methane from biogas, steam methane reforming is the preferred way to maximize H_2 production when sufficient CO_2 is available. An alternative pathway can be methane dry reforming with CO_2 and reduced steam for higher CO content or partial oxidation reforming, requiring less CO_2, or a combination thereof. Biogas with a high CO_2 content can be separated before subjecting it to SMR, providing an advantage to the overall process. This is despite the fact that enriching the syngas with a portion of the same CO_2 later is necessary. This is because the H_2 yield from SMR is lower, leading to a more complex removal of excess CO_2 from the product gases post-SMR or after the methanol reactor. This process is more complicated than the prior separation during biogas pretreatment when H_2S is also removed, and it involves techniques such as pressure swing absorption, liquid adsorption scrubbers, or membrane separation [21,22]. The mass of H_2 produced by an SMR per unit mass of methane (CH_4) input is typically in a range from 0.32 to 0.28, corresponding to a lower heating value efficiency of 77%–67%. Additionally, excess steam can be produced from the waste heat and used as process heat. Eventually, the syngas mixture (mainly H_2, CO, CO_2, and some CH_4 and H_2O) with sufficient CO_2 is fed into the methanol synthesis reactor.

E-methanol is made from hydrogen produced by water electrolysis with renewable electricity and carbon from a biological or climate neutral origin in the form of CO_2 that is captured from sources such as biomass combustion, biogas processing, bioethanol plants, or even direct air capture. Such sources where CO_2 is a by-product available in high concentration are naturally less expensive than, for instance, direct air capture. Water electrolysis typically requires about 50–60 kWh of electricity per kg of hydrogen. Depending on the source and type of electrolyzer, both hydrogen and CO_2 need to be compressed before being mixed in the desired ratio and fed into the methanol synthesis reactor.

There are two main chemical pathways that take place in a methanol synthesis reactor fed with H_2, CO_2, and CO. Both are exothermic and prefer elevated pressures and temperatures:

1. The CO-hydrogenation reaction $CO + 2H_2 \rightarrow CH_3OH$.
2. The direct CO_2 hydrogenation, producing water as by-product: $CO_2 + 3H_2 \rightarrow CH_3OH + H_2O$.
3. The indirect CO_2 hydrogenation with an endothermic reverse-water gas shift reaction ($CO_2 + H_2 \rightarrow CO + H_2O$) making CO to methanol.

Water is the by-product of 2 and 3 and needs to be separated. Overall, the stoichiometric ratio is 3 mol H_2 per mole of CO_2 for methanol reaction 2 or 3+1 and 2 mol of H_2 per mol of CO for reaction 1. The rate and equilibrium of each reaction depend on the concentration, catalyst, temperature, and pressure conditions. An optimal syngas composition would lead to a combination of a high conversion of carbon from CO and CO_2 to methanol, a high utilization of H_2, and subsequently, a high yield of methanol per mole of H_2. Chein et al. [23] have studied the influence of the mole ratios of CO_2/H_2 and CO/H_2 in the syngas systematically and found that a high ratio of CO/H_2 up to 0.36 and a lower ratio of CO_2/H_2 of less than 0.2 was beneficial for maximizing methanol yield, which means about two to three times as much CO as CO_2 in the synthesis reactor feed. A high CO_2 ratio leads to more excess water generation through reverse water gas shift (3) and through hydrogenation (2), which can compromise the catalyst [24]. In reality, the carbon source may be limited, and more CO_2 may be economically available than CO in the syngas per unit of available hydrogen.

A methanol synthesis reactor uses a catalyst with a large surface area over which the gas is passed in a pressurized reactor vessel. The common commercial catalyst is Cu/ZnO on Al_2O_3 [24], typically arranged as a packed bed reactor with coated pellets or spheres that are housed in reactor tubes, operating around 230–280°C and 50–120 bar. All three reactions are enabled by this catalyst and proceed to an equilibrium within this temperature and pressure window. The hydrogenation reactions are exothermic, and the reactor can be cooled (isothermal reactor), or the products increase in temperature and need to be cooled upon exit. As the reactions lead to equilibrium compositions, the process is either cascaded or has a recycling loop for the largest part of the products; length and residence time increase the yield. Cooler temperatures and higher pressures shift the equilibrium toward methanol, but the yield is limited by kinetics for temperatures less than 200°C. For long, tubular reactors, cooling for isothermal operation leads to 3× higher yields than for uncooled operation, which could be done in shorter reactors with a higher recycling rate [23]. An optimized process involves careful control of temperature, pressure, and flow rates including water/steam removal from the recycling stream to maximize methanol production efficiency. The reactor geometry has length and diameter as optimization parameters, which were explored through simulations by Chein et al. [23]. From the products leaving the reactor, methanol is condensed through cooling and subsequent flashing in a separator in one or two stages and finally distilled through rectification. The remaining gases are recycled; a gas fraction may be purged and burned for heating the SMR, preventing buildup of inert gases. The condensates, mostly water, can be fed back into the process for steam generation and reforming.

An illustrative example of the processes in an optimized, industrial-scale biogas-to-methanol plant can be found in a publication by Ghosh et al. [21]. This process involves one SMR and one synthesis reactor with direct CO_2 hydrogenation and has a low complexity yet high methanol yield. Chiefly, the main components are:

1. Biogas treatment with pressure swing absorption (PSA), membrane and/or absorption process for separation of CH_4 and CO_2 and removal of detrimental impurities from CH_4, foremost H_2S. Operates around atmospheric pressure but may require some compression for PSA or a blower at the inlet.
2. CH_4 compressor to SMR feed pressure of 15 bar
3. CO_2 compressor to synthesis reactor feed pressure of 50 bar
4. SMR operating at 15 bar, 900°C for optimum H_2/CO production with Ni catalyst. Externally fired by biogas in addition to process purge gas from the flash separators
5. Heat recovery steam generator (HRSG) for steam generation from SMR products operating at 15 bar and 900°C
6. Syngas cooler to precool syngas to ambient before compression to synthesis reactor feed
7. Syngas compressor to synthesis reactor feed pressure of 50 bar
8. Synthesis feed effluent heat exchanger (FEHE) in the synthesis reactor recycle loop for preheating feed and cooling product gas
9. Methanol synthesis reactor (Cu/Zn/Al/Zr catalyst), operating at 50 bar and 250°C
10. Product gas cooler to precool to ambient before high-pressure flash separator
11. HP flash separator operating at 48 bar, 38°C
12. Synthesis reactor recycle blower to recompress separator off-gas to reactor feed pressure of 50 bar
13. LP flash separator operating at 2 bar, 38°C
14. Distillation column for methanol-water separation as a compact, packed-bed column with head condenser (separating methanol and purge gas) and bottom reboiler (separating water). Excess steam from SMR can be used for heating.

A process flow diagram for this baseline biogas-to-methanol process described for large-scale plants by Ghosh et al. [21] is shown as follows (Fig. 7.27).

Through simulations, the authors compared different processes for their maximum methanol yield and found an optimum with a relatively low-complexity process with the following performance metrics. A molar conversion rate of carbon in biogas, SMR methane, and CO_2, to carbon in methanol of 70% was found. Based on SMR hydrogen output mass, 6.1 kg of methanol per kg of hydrogen could be synthesized with sufficient carbon. Based upon SMR input CH_4, the output mass of methanol is about 220%. Note that in these numbers, the biogas required for SMR heating (around 40%–45% of the total) is not included. When considering both CH_4 and methanol as a fuel defined by its lower heating value (LHV), an efficiency of conversion can be defined for the overall process based on LHV of methanol produced (19.9 MJ/kg) over LHV of total CH_4 (50 MJ/kg) including 40% heating gas for a 72% efficient SMR. For smaller systems, such as those used in decentralized production where methanol is an easily transported product, whereas biogas is not, the efficiency may be at 10%–15% lower, because small-scale SMR and the synthesis reactors both have lower yields of hydrogen and methanol, respectively. Such small scale, modular methanol plants have been conceptualized for applications using, e.g., biogas from agricultural waste or flare gas in the oil and gas industry at a size down to a few tons/day of methanol; even mini-scale plants packaged in containers are available, which produce several 100 kg methanol per day (although with a lower efficiency than small-scale plants).

Energy storage for hydrogen as part of transport network

To harness excess renewable power and bring sustained power with the inherent intermittent renewables, hydrogen can be stored as a gas, liquid, or carrier form. Hydrogen offers the ability to store large amounts of energy for longer amounts of time compared with other energy storage concepts. Underground storage (converted gas fields, salt caverns, or depleted reservoirs) is being considered for pressurized hydrogen gas, but testing is underway to determine leakage and how well reservoirs can transfer from natural gas to hydrogen storage.

Hydrogen compressors for storage will need to likely deliver at 3000 psi and above for injection at the top end of the reservoir. Initial boosters to take hydrogen from production (at atmospheric pressure up to 100 psi) to pipeline pressures will require hydrogen compressors capable of a high number of multiple stages due to significant head requirement.

Fig. 7.28 compares energy output vs. total discharge time for various energy storage concepts, including various batteries, flywheels, compressed air, and thermochemical (which would be in the class for hydrogen gas power).

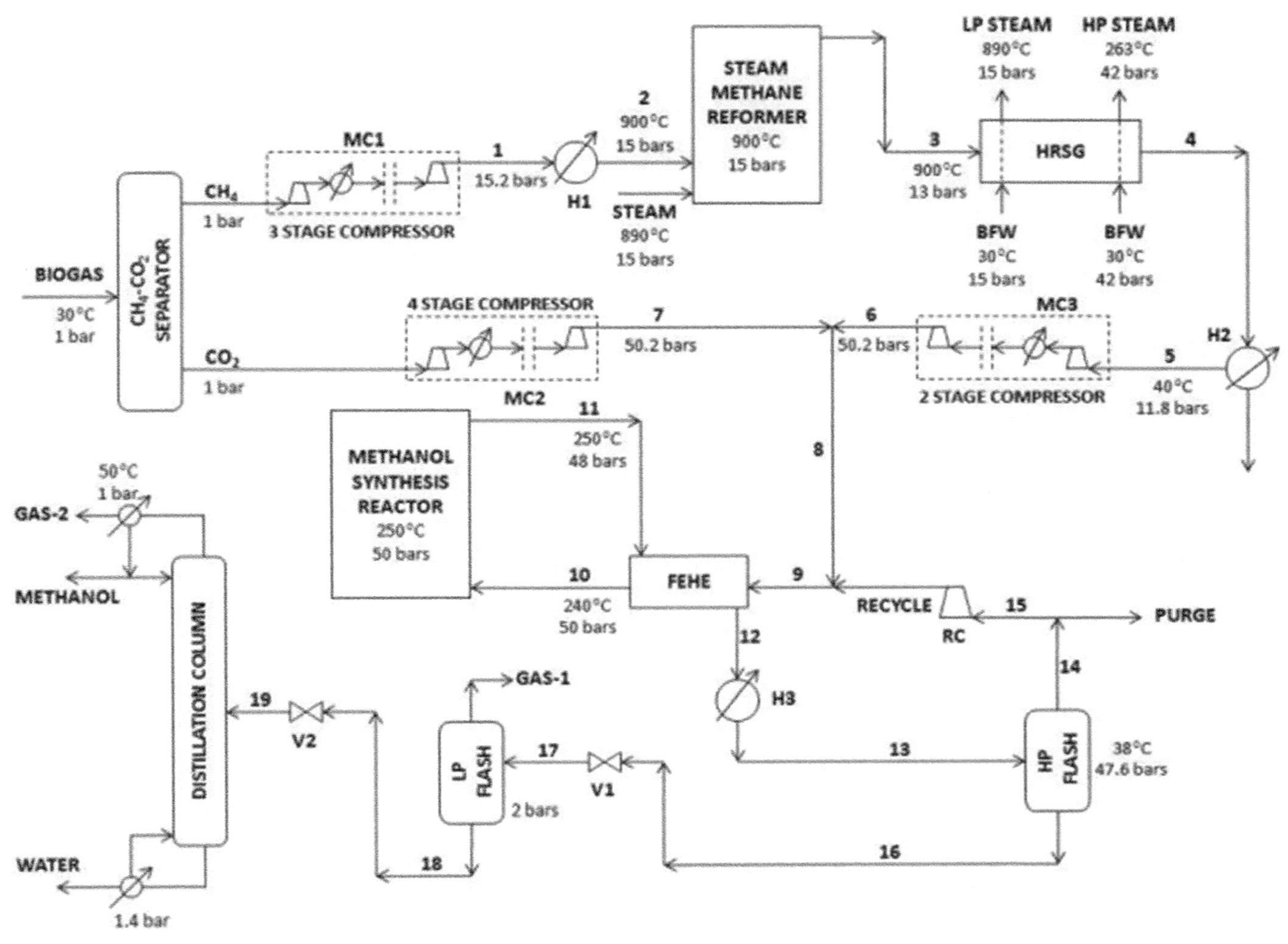

FIG. 7.27 Baseline biogas-to-methanol process with two-single reactors described [21].

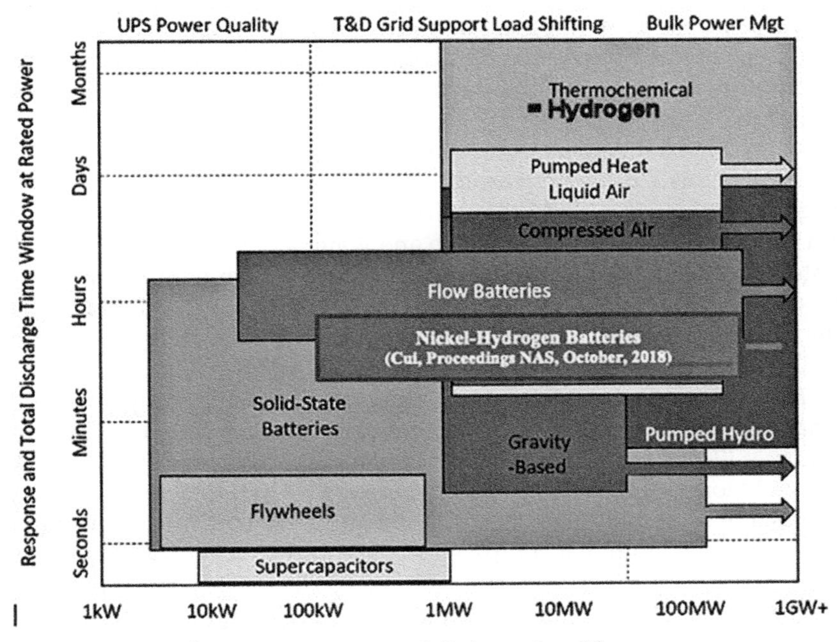

FIG. 7.28 Energy storage concepts compared—power output vs. total discharge time [1].

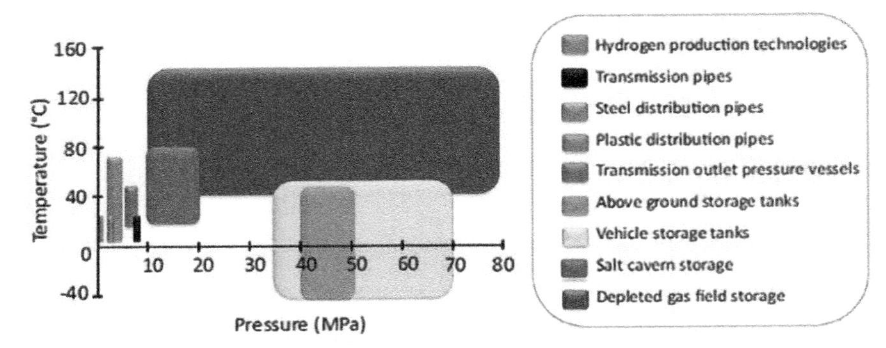

FIG. 7.29 Hydrogen storage pressures and temperatures [25].

Thermochemical storage is the only concept that provides storage on the order of months of time compared with hours or days and can provide 1 GW level energy output.

Finally, Fig. 7.29 provides a different comparison of energy storage concepts showing how different storage formats for hydrogen can be used over a pressure and temperature range that will be needed for larger volumes (higher duration and higher power). Depleted underground gas fields, vehicle storage tanks, and above-ground storage tanks can each provide the capability to store hydrogen at above 30 MPa and temperatures between 0°C and 120°C.

Ammonia for aviation fuels

The Federal Aviation Agency (FAA) estimates that ~2100 million gallons of fossil fuel is being used by the US aviation industry and is forecasted to grow to ~3000 million gallons by 2043. For every gallon of Jet fuel burned, approximately 21 gal of carbon dioxide are generated. The Intergovernmental Panel on Climate Change (IPCC)—a United Nations organization that assesses scientific, technical, and economic information on the effects of climate change—estimates that aircraft emissions correspond to ~2.5% of human-generated carbon dioxide emissions. To reduce the impact that carbon dioxide emissions have on the global temperature rise, governments and policymakers are looking to promote decarbonization technologies.

One of the alternative fuels that can push the decarbonization initiative is hydrogen. When hydrogen is burnt in pure oxygen to generate heat, the only by-product is water. However, when burnt in the air, some amount of NOx is also produced. One of the main challenges in implementing hydrogen combustion on a large scale is the complex infrastructure required to store hydrogen. To liquefy hydrogen, a temperature of ~33 K needs to be maintained. This is very energy-intensive, and construction of equipment that withstand this temperature is expensive. Hence, researchers are looking at different options to circumvent this challenge related to the storage and transport of hydrogen. One among them is to use ammonia as an energy source. Ammonia contains three atoms of hydrogen and has a latent heat of vaporization of ~18 MJ/Kg.

In the last few years, Japan started the quest to explore ammonia as a fuel source for power generation due to its excessive dependency on fossil fuels [26]. More recently, researchers from many countries such as the United States [27] and the United Kingdom [28] are exploring the utilization of ammonia for power generation. When ammonia burns in the air, the main products are molecular nitrogen and water along with the heat generated from combustion. However, ammonia contains nitrogen bonded to hydrogen, which at high temperatures in the presence of oxygen generates nitrogen oxide (NOx). Researchers are still working on strategies to minimize these NOx emissions, and several promising results have been published in the literature, which is close to 100 ppm of NOx without implementing selective catalytic reduction techniques [29].

However, the interest in the utilization of ammonia as a fuel for aviation came up only recently. Life cycle analysis on aircraft operated by ammonia shows there can be a significant reduction in greenhouse gases if ammonia is produced from renewable sources [30]. The National Aeronautics and Space Administration (NASA) has funded Boeing and the University of Central Florida under its University Leadership Initiative to explore the use of ammonia as an aviation fuel [31]. In Australia, Aviation H_2 plans to use liquid ammonia in its turbofan combustors [32] owing to the higher volumetric energy density of ammonia compared to gaseous hydrogen and ease of storage.

Despite this, the use of ammonia in aviation still poses several challenges. Some of these are (a) poor ignition characteristics of ammonia, (b) higher NOx formation, and (c) safety, environmental and health concerns. Poor ignition

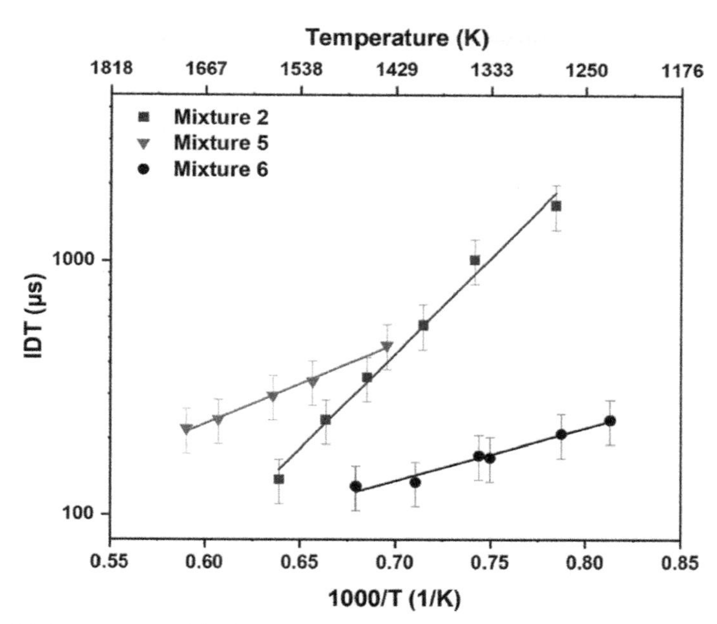

FIG. 7.30 Ignition delay times (IDT) of mixtures of hydrogen with ammonia and natural gas. Mixture 6 is H_2 in the air, mixture 2 is a 50/50 mixture of H_2/natural gas in the air, and mixture 5 is NH_3 in the air. All mixtures are at an equivalence ratio of 1.0. *Reproduced from J.B. Baker, et al., Experimental ignition delay time measurements and chemical kinetics modeling of hydrogen/ammonia/natural gas fuels. J. Eng. Gas Turbines Power 145 (4) (2022).*

characteristics of ammonia are evident from the ignition delay times of ammonia shown in Fig. 7.30. As evident, when mixed with hydrogen or natural gas, ammonia ignites faster. To circumvent this and to improve the combustion characteristics, ammonia blended with hydrogen is widely studied [33]. Another issue is related to NOx formation during the combustion of ammonia. One of the promising combustor designs that can reduce NOx during ammonia combustion is the rich-quench-lean (RQL) combustor design. In this, ammonia is burned in the first stage in fuel-rich conditions followed by injection of air in the second stage. The rich burn ensures that NOx formation from ammonia is minimal while ensuring maximum conversion of ammonia to hydrogen and nitrogen. In the second stage, air is injected, which ensures complete combustion of the hydrogen and leftover ammonia but in a fuel-lean condition.

This strategy has been modeled using a chemical reactor network (CRN) by several researchers [29]. However, the accuracy of CRN models is dependent on the accuracy of chemical kinetic models that are used, and none of the chemical kinetic models available in literature at present are validated for quantitative NOx kinetics in aviation engine conditions. Most of these works used chemical kinetic models that are validated with low-pressure and atmospheric-pressure data for NOx formation at larger timescales than in a typical aviation combustor [34,35]. Lastly, significant efforts are required in the safety, environmental and health aspects of ammonia leakage incident in storage facilities near airports and in aircrafts although many of these are already in place as ammonia is used in bulk quantities already in the fertilizer industry [36,37].

References

[1] T. Allison, J. Klaerner, S. Cich, R. Kurz, M. McBain, Power and Compression Analysis of Power-to-Gas Implementations in Natural Gas Pipelines with up to 100% Hydrogen Concentration, ASME Turbo Expo Paper GT2021-59398, 2021.

[2] K. Brun, T. Allison (Eds.), Chapters 3, Machinery basics, Chapter 5, Supply processes and machinery and Chapter 6, Transport and storage, in: Machinery and Energy Systems for the Hydrogen Economy, Elsevier, 2022, ISBN: 978-0323903943.

[3] R. Kurz, M. Lubomirsky, L. Cowell, K. Brun, F. Bainier, Hydrogen-combustion and compression, in: Proceedings of the 49th Turbomachinery Symposium, Houston, TX, 2020.

[4] J.M.R. Louthan, Hydrogen in metals and other materials: a comprehensive reference to books, bibliographies, workshops and conferences, Int. J. Hydrog. Energy 24 (10) (1999) 53–75.

[5] API Recommended Practice 941, Steels for Hydrogen Service at Elevated Temperatures and Pressures in Petroleum Refineries and Petrochemical Plants, API, 2016.

[6] K. Brun, S. Ross, B. Pettinato, T. Omatick, J. Thorp, Compression technology for a decarbonized energy economy, in: Proceedings of the 51st Turbomachinery Symposium, Houston, TX, 2022.

[7] Department of Energy, DOE Launches Bipartisan Infrastructure Law's $8 Billion Program for Clean Hydrogen Hubs Across US, 2022. https://www.energy.gov/articles/doe-launches-bipartisan-infrastructure-laws-8-billion-program-clean-hydrogen-hubs-across.
[8] P. Haeussinger, R. Lohmueller, A. Watson, Hydrogen, 2. Production, in: Ullmann's Encyclopedia of Industrial Chemistry, Wiley-VCH Verlag GmbH & Co. KGaA, 2000.
[9] R. Kurz, M. McBain, T. Allison, J. Moore, Compression turbomachinery for the decarbonizing world, in: Proceedings of the 51st Turbomachinery Symposium, Houston, TX, 2022.
[10] M. Lubomirsky, R. Zamotorin, A. Singh, R. Kurz, Compression requirements for carbon reduction, in: PSIG Conference, San Diego, CA, 2022.
[11] G. Eisenlohr, H. Chladek, Thermal tip clearance control for centrifugal compressor of an APU engine, ASME Turbomach. J. 116/626-634 (1994).
[12] M. Mcbain, D. Bauer, S. Ross, K. Brun, T. Allison, E. Broerman, Technology options for hydrogen compression, in: Gas Machinery Conference, Fort Worth, TX, 2022.
[13] K. Brun, R. Kurz, Chapter 5: Reciprocating compressors, in: Compression Machinery for Oil & Gas, Elsevier, 2019, ISBN: 978-0-12-814683-5.
[14] M. McBain, S.T. Omatick, K. Wygant, E. Karakas, K. Brun, J. Thorp, Applications and technology for hydrogen compression and pumping, in: Proceedings of the 52nd Turbomachinery Symposium, Houston TX, September, 2023.
[15] L. Mann, S. Mazdak, D. Patrick, B. Ershaghi, Economy of scaling hydrogen liquefiers and machineries, in: Gastech Conference, September, 2022.
[16] U. Cardella, Large-scale hydrogen liquefaction under the aspect of economic viability (Dissertation, TU München), mediaTUM, 2018. https://mediatum.ub.tum.de/doc/1442078/1442078.pdf.
[17] K. Ohlig, L. Decker, The latest developments and outlook for hydrogen liquefaction technology, AIP Conf. Proc. 1573 (1) (2014) 1311–1317.
[18] E.W. Lemmon, I.H. Bell, M.L. Huber, M.O. McLinden, NIST standard reference database 23: Reference fluid thermodynamic and transport properties, in: REFPROP, Version 10.0, National Institute of Standards and Technology, Standard Reference Data Program, Gaithersburg, 2018.
[19] N. Stetson, DOE hydrogen program perspectives [Conference presentation], in: Liquid Hydrogen Virtual Workshop, 2022. https://www.energy.gov/eere/fuelcells/liquid-hydrogen-technologies-workshop.
[20] IRENA and Methanol Institute, Innovation Outlook: Renewable Methanol, International Renewable Energy Agency, Abu Dhabi, 2021, ISBN: 978-92-9260-320-5. https://www.methanol.org/wp-content/uploads/2020/04/IRENA_Innovation_Renewable_Methanol_2021.pdf.
[21] S. Ghosh, V. Uday, A. Giri, S. Srinivas, Biogas to methanol: a comparison of conversion processes involving direct carbon dioxide hydrogenation and via reverse water gas shift reaction, J. Clean. Prod. 217 (2019) 615–626, https://doi.org/10.1016/j.jclepro.2019.01.171.
[22] R. Rinaldi, G. Lombardelli, M. Gatti, C. Visconti, M. Romano, Techno-economic analysis of a biogas-to-methanol process: study of different process configurations and conditions, J. Clean. Prod. 393 (2023) 1–3.
[23] R.-Y. Chein, W.-H. Chen, H.C. Ong, P.L. Show, Y. Singh, Analysis of methanol synthesis using CO_2 hydrogenation and syngas produced from biogas-based reforming processes, Chem. Eng. J. 426 (2021), https://doi.org/10.1016/j.cej.2021.130835.
[24] L.C. Grabow, M. Mavrikakis, Mechanism of methanol synthesis on cu through CO_2 and CO hydrogenation, ACS Catal. 1 (4) (2011) 365–384, https://doi.org/10.1021/cs200055d.
[25] A. Hassanpouryouzband, E. Joonaki, K. Edlmann, N. Heinemann, J. Yang, Thermodynamic and transport properties of hydrogen containing streams, Sci. Data 7 (2020) 222. https://www.nature.com/articles/s41597-020-0568-6.
[26] Independent Commodity Intelligence Services, Japan's government embracing ammonia as fuel of the future in zero-carbon emissions drive, 2020, Available from: https://www.icis.com/explore/resources/news/2020/10/28/10568460/japan-s-government-embracing-ammonia-as-fuel-of-the-future-in-zero-carbon-emissions-drive/.
[27] M. Pierro, et al., High-Fuel Loading Ignition Delay Time Characterization of Hydrogen/Natural Gas/Ammonia at Gas Turbine-Relevant Conditions inside a High-Pressure Shock Tube, 2022.
[28] A. Valera-Medina, et al., Ammonia for power, Prog. Energy Combust. Sci. 69 (2018) 63–102.
[29] S. Mashruk, H. Xiao, A. Valera-Medina, Rich-quench-lean model comparison for the clean use of humidified ammonia/hydrogen combustion systems, Int. J. Hydrog. Energy 46 (5) (2021) 4472–4484.
[30] Y. Bicer, I. Dincer, Life cycle evaluation of hydrogen and other potential fuels for aircrafts, Int. J. Hydrog. Energy 42 (16) (2017) 10722–10738.
[31] NASA, Boeing, UCF to study zero-carbon ammonia jet fuel, Retrieved from https://www.ammoniaenergy.org/articles/nasa-boeing-ucf-to-study-zero-carbon-ammonia-jet-fuel/.
[32] Ammonia Energy, Aviation H_2 chooses ammonia to develop carbon-free flight in Australia, Retrieved from https://www.ammoniaenergy.org/articles/aviation-h2-chooses-ammonia-to-develop-carbon-free-flight-in-australia/.
[33] J. Baker, et al., Ammonia hydrogen ignition measurements for clean aircraft propulsion, in: AIAA SCITECH 2022 Forum, American Institute of Aeronautics and Astronautics, 2021.
[34] P. Glarborg, et al., Modeling nitrogen chemistry in combustion, Prog. Energy Combust. Sci. 67 (2018) 31–68.
[35] B. Mei, et al., Experimental and kinetic modeling investigation on the laminar flame propagation of ammonia under oxygen enrichment and elevated pressure conditions, Combust. Flame 210 (2019) 236–246.
[36] Y.H. Chung, et al., Fire safety evaluation of high-pressure ammonia storage systems, Energies 15 (2022), https://doi.org/10.3390/en15020520.
[37] A. Roy, P. Srivastava, S. Sinha, Dynamic failure assessment of an ammonia storage unit: a case study, Process Saf. Environ. Prot. 94 (2015) 385–401.

CHAPTER

8

Transport of carbon dioxide

Robert Pelton[a], *Peter Renzi*[b], *Kevin Supak*[c], *Rainer Kurz*[d], *Kelsi Katcher*[c], *Rahul Iyer*[e], *Jon Bygrave*[f], *Karl Wygant*[a], *Jason Wilkes*[c], *and Klaus Brun*[a]

[a]Ebara Elliott Energy, Jeannette, PA, United States [b]Engineering Design Group, LLC, Buffalo, WY, United States [c]Southwest Research Institute, San Antonio, TX, United States [d]Solar Turbines, San Diego, CA, United States [e]KCK Group, Cupertino, CA, United States [f]Hanwha Power Systems, Houston, TX, United States

Sources of carbon dioxide

CO_2 originates from a number of places; these are generally separated as anthropogenic and nonanthropogenic. Anthropogenic sources refer to human activities that release carbon dioxide into the atmosphere. These sources are major contributors to the increasing concentration of CO_2, driving climate change and global warming. The main anthropogenic sources of CO_2 include the following [1]:

1. Burning of Fossil Fuels: The combustion of fossil fuels, such as coal, oil, and natural gas, for energy production, transportation, and industrial processes is the largest anthropogenic source of CO_2 emissions. It accounts for approximately 75%–80% of total CO_2 emissions.
2. Industrial Processes: Certain industrial activities, such as cement production, iron and steel manufacturing, and chemical production, contribute to CO_2 emissions. Together, these industrial processes account for around 15%–23% of total CO_2 emissions.
3. Land-Use Changes: Deforestation and land-use changes result in the release of CO_2 stored in forests and other vegetation, making up about 5%–10% of total CO_2 emissions. While this changes the balance of CO_2 generated or released, this source cannot be captured and transported.
4. Waste Management: Decomposing organic waste in landfills produces methane, which, when oxidized in the atmosphere, contributes to CO_2 emissions. Additionally, waste incineration releases CO_2 directly. This source accounts for approximately 3%–5% of total CO_2 emissions.

In addition to anthropogenic CO_2, nonanthropogenic CO_2 sources account for a substantially lower portion of the total CO_2 generated and released into the atmosphere. Nonanthropogenic sources refer to natural processes that release carbon dioxide into the atmosphere. These sources have been part of the Earth's carbon cycle for millions of years and include the following:

1. Geological Activities: Volcanic eruptions release CO_2 and other greenhouse gases, contributing approximately 0.2%–0.5% of total CO_2 emissions on an annual average basis. While volcanic emissions can be significant over short periods, they are relatively small compared with anthropogenic sources on a global scale.
2. Ocean-Atmosphere Exchange: The oceans act as a carbon sink, absorbing and releasing CO_2 through natural processes. While the oceans currently absorb more CO_2 than they release, human activities are disrupting this balance, leading to ocean acidification and potential CO_2 release in the future.
3. Biosphere Fluxes: Natural processes in the biosphere, such as respiration of plants and animals, also lead to CO_2 emissions and uptake. Overall, the biosphere acts as a carbon sink, absorbing CO_2 through photosynthesis.
4. CO_2 Dome and Other Natural Geologic Sources: Certain geological formations can trap CO_2, leading to localized accumulations known as CO_2 domes. While not a significant contributor on a global scale, they may release CO_2 in specific areas.

Energy Transport Infrastructure for a Decarbonized Economy
https://doi.org/10.1016/B978-0-443-21893-4.00007-6

Copyright © 2025 Elsevier Inc. All rights are reserved, including those for text and data mining, AI training, and similar technologies.

While nonanthropogenic sources of CO_2 have been part of the Earth's carbon cycle for millennia, the rapid increase in anthropogenic emissions over the past century has disrupted the natural balance, leading to higher concentrations of CO_2 in the atmosphere and driving climate change. Understanding and addressing both anthropogenic and nonanthropogenic sources are essential in developing effective strategies to mitigate CO_2 emissions and combat climate change.

It's important to reiterate that these percentages are rough estimates and can vary depending on various factors. The burning of fossil fuels, including both precombustion and postcombustion emissions, remains the most substantial contributor to CO_2 emissions. Industrial processes, especially cement production and iron/steel manufacturing, also make significant contributions. Natural sources contribute only a small fraction of total CO_2 emissions on an annual average basis.

In this chapter, we will focus on the sources of CO_2 related to the energy sector (burning of fossil fuels) and large-scale industrial processes, which produce large quantities of CO_2. Those sources related to land-use changes and agriculture will not be addressed.

Burning of fossil fuels

The burning of fossil fuels has played a significant role in driving the global economy since the beginning of the industrial revolution. Fossil fuels are used for various purposes including energy production, chemical manufacturing, transportation, heating, manufacturing, etc. The energy sector is the largest producer of CO_2 emissions by a wide margin. The bulk of these emissions comes from the burning of a range of fossil fuels including: coal, oil, and natural gas. The energy from combustion is used primarily for small- and large-scale propulsion, heating, and power generation applications. Fossil fuels are burned in boilers for creating heat and steam, gas turbine engines for power and propulsion, and gas and diesel-fueled internal combustion engines for transportation. In most of these applications, the resulting CO_2 that is generated is exhausted to the atmosphere. Fossil fuels have high energy density and release significant amounts of heat when burned. This high energy density makes combustion engines well suited to transportation applications, where the fuel must be carried with the cargo. It also improves the economies of supplying large amounts of fuel to power plants. Fossil fuels are also used as raw material and heat sources for many industrial processes including the manufacture of steel, cement, fertilizers, and other chemicals.

The burning of fossil fuels results in the release of significant quantities of exhaust gases into the environment. The primary exhaust gases of concern are classified as greenhouse gases. Fig. 8.1 shows that CO_2 comprises about 76% of the global greenhouse gas emissions, with the bulk of those emissions coming from the combustion of fossil fuels and other industrial processes. These emissions can impact the environment in many ways including contributing to global climate change, air and water pollution, and resource depletion.

Precombustion sources of CO_2

Before some fuels can be burned, additional processing is necessary, which may release some CO_2 into the environment. Crude oil is also processed through a refining process to separate the feedstock into various components including gasoline, propane, diesel, etc. Typical distillation processes involve heating the crude oil through direct fired heaters, burning a fossil fuel, or indirectly through steam or hot oil. In most cases, the heating of the feedstock involves combustion and release of additional CO_2. Natural gas typically undergoes processing before export to market to prepare the fuel for combustion.

Natural gas reservoirs contain significant quantities of impurities, principally CO_2 and H_2S, and additional treatments must be applied to remove these, which are usually amine or membrane separations. The processes produce high-purity CO_2 streams, which can be stored. There are two operating natural gas plants that capture and store CO_2 from natural gas processing: the Sleipner plant in the North Sea and the In Salah plant in Algeria. CO_2 captured from natural gas processing is also used in several enhanced oil recovery (EOR) projects in the USA. For the separation of CO_2 in natural gas processing, the coabsorption of hydrocarbons and H_2S and carryover to the CO_2 product stream may be an issue. In the Sleipner project, the CO_2 stream is 98% pure with the main contaminant being methane [3]. There is also the Gorgon LNG and carbon capture and sequestration (CCS) project offshore Western Australia where the raw natural gas contains 15% CO_2. The CO_2 is stripped away at a processing plant before it is geologically sequestered at a depth of about 2 km. As of May 2023, over 8 million tons of CO_2 had been captured and stored [4].

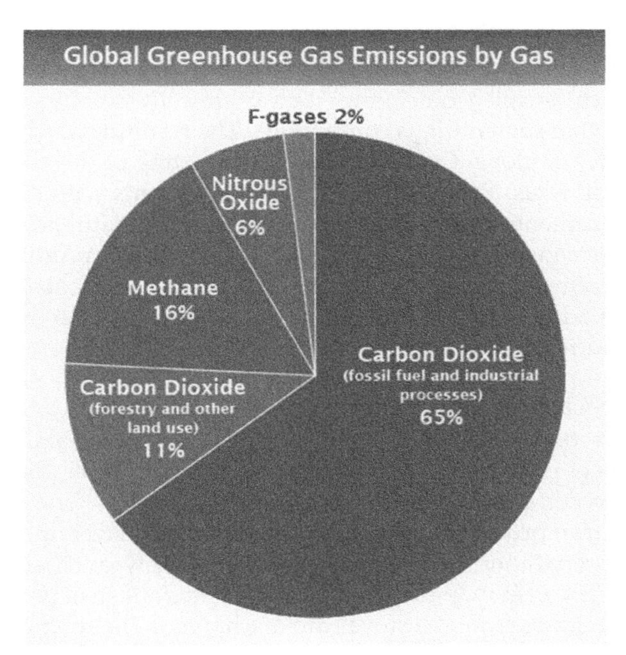

FIG. 8.1 Type of global greenhouse gas emissions [2]. *United States Environmental Protection Agency.*

Postcombustion sources of CO_2

During the combustion of a fossil fuel, hydrocarbon is mixed with oxygen. The result of this reaction is the generation of heat and combustion by-products. The main constituents of the by-products in the exhaust from combustion of a hydrocarbon are nitrogen (65%–75%), CO_2 (8%–18%), water (1%–10%), and oxygen (1%–8%). The flue gas will also contain varying amounts of other gases and particulate matter depending on the type of fuel and combustion process. It is estimated that global CO_2 emissions from the combustion of fossil fuels reached 36.6 billion tons in 2022. The bulk of these emissions produces CO_2 in near-ambient conditions in a mixture with the other flue gases. Capture efforts focus on the separation, compression, transport, and eventual utilization or sequestration of the CO_2.

Industrial processes

CO_2 emissions arising from nonpower industrial processes are substantial, although much less significant than from the power generation sector. In many of these processes, there is the potential to capture and transport the CO_2. The main industrial processes with significant CO_2 emissions include:

- Cement Production: 7%–8% of global CO_2 emissions
- Iron and Steel Production: 7%–8%
- Lime Production: 2%–3%
- Hydrogen and Ammonia Production
- Air Separation
- Chemical Production

Cement production

Production of cement is the largest industrial source of CO_2 emissions after the power sector, accounting for over 1 billion tons of CO_2 per year or about 8% of global CO_2 emissions. Cement production involves the calcination of limestone and has large process emissions of CO_2. Large quantities of heat energy are needed to drive the process, which is usually derived from the combustion of fossil fuels. NOx, SO_2, CO, and CO_2 are the primary gaseous emissions in the manufacture of cement. Smaller quantities of VOCs, NH3, chlorine, and HCl may also be emitted [5]. Emissions may also include partial combustion products. The concentration of CO_2 in the flue gases from cement production is 15%–30% by volume, considerably higher than from fossil fuel power plants. Postcombustion capture technologies can therefore be applied to cement production plants.

Portland cement, which is manufactured in the highest volumes, for example, requires limestone (calcium carbonate) to be heated to 1450°C (2640°F) in a kiln to remove its carbon, producing lime (calcium oxide) in a calcination reaction. The calcium oxide, which is called quicklime, then chemically combines with the other materials in the mix to form calcium silicates and other cementitious compounds. The resulting hard substance is called "clinker." This chemical reaction is a major emitter of global CO_2 emissions. The heating of the kiln requires significant amounts of energy, usually from burning fossil fuels. There are other types of cements with somewhat different raw materials, such as in calcium sulfoaluminate cements that allow for lower firing temperatures and produce less CO_2 but generate more sulfur oxides. A cement plant consumes 3–6 GJ of fuel per ton of clinker produced, depending on the raw materials and the process used. Most cement kilns today use coal and petroleum coke as primary fuels and, to a lesser extent, natural gas and fuel oil. In addition to improving energy efficiency of the process, the use of alternative fuels (biomass, waste, hydrogen) would reduce the carbon footprint.

Iron, steel, and metallurgical coke production

The iron and steel industry is an energy-intensive activity and a major industrial CO_2 emitting sector, accounting for about 650 Mt of CO_2 per year. Steel production at an integrated iron and steel plant is accomplished using several related processes, and emissions occur at each step of the production process. These processes include: (1) coke production; (2) sinter production; (3) iron production; (4) iron preparation; (5) steel production; (6) semifinished product preparation; (7) finished product preparation; (8) heat and electricity supply; and handling and transport of raw, intermediate, and waste materials [6]. The vast majority of CO_2 emissions from steel production come from blast furnace stove stacks where the combustion gases from the stoves are discharged. The relative composition of the exhaust of a typical blast furnace gas has been estimated to be 60% N_2, 28% CO, and 12% CO_2 [7]. Postcombustion carbon capture applied to this dilute CO_2 exhaust stream is likely to produce similar impurity estimates to those from the power sector. Metallurgical coke is used in blast furnaces to reduce iron ore to iron. Coke is produced by destructive distillation of coal in the oxygen-free atmosphere of coke ovens until the most volatile components are removed, and these stack gases are a source of CO_2.

Lime production

Lime is produced through the calcinations of limestone, dolomite, or other mineral materials, and rotary kilns are the most prevalent type of kilns used in the process. CO_2, CO, SO_2, and NOx are all produced in lime kilns, and emissions are affected by the properties of the fuel used to heat the kiln, the properties of the mineral feed material, the quality of the lime produced, the type of kiln used, and the type of pollution control equipment used. Toxic species in the exhaust gases from lime kilns are metals such as arsenic, cadmium, chromium, nickel, and HCl [8]. The exhaust gas from a lime kiln contains around 50% CO_2 [6].

Air separation

Carbon dioxide (CO_2) is a gas that is present in the Earth's atmosphere. Air separation is the process of separating air into its individual components including: nitrogen, oxygen, and argon (see for constituent gases of the atmosphere [9]). During this process, a significant amount of CO_2 is also generated as a by-product (Table 8.1).

Air separation is commonly achieved through fractional distillation. Fractional distillation [10] involves separating the different components of air based on their boiling points. Cryogenic air separation units are typically used for this process and are designed to produce nitrogen or oxygen, with argon often being coproduced. Distillation of air using at least two distillation columns is the only viable source of rare gases such as neon, krypton, and xenon. Advanced air separation processes are also used to recover helium.

In addition to fractional distillation, various other noncryogenic methods can be used for air separation such as membrane, pressure swing adsorption (PSA), and vacuum pressure swing adsorption (VPSA). They are commercially used to separate a single component from ordinary air. High-purity oxygen, nitrogen, and argon are most commonly used in the semiconductor device fabrication process and require cryogenic distillation.

The CO_2 produced during air separation is usually vented into the atmosphere [11]. However, there are various methods to capture and store CO_2, such as CCUS technologies that can help reduce the emissions of this potent greenhouse gas into the atmosphere. Therefore, it is important to consider the environmental impact of CO_2 emissions during air separation processes and to explore mitigation strategies.

Hydrogen and ammonia production

The production of hydrogen is the first step in the manufacture of ammonia using the Haber-Bosch process. Around half of all globally produced hydrogen is used to produce ammonia, and 80% of ammonia manufactured worldwide is

TABLE 8.1 Makeup of atmosphere [9] (excluding water vapor).

Gas	Symbol	Atmosphere content
Nitrogen	N_2	78.08%
Oxygen	O_2	20.95%
Argon	Ar	0.93%
Carbon dioxide	CO_2	0.04%
Neon	Ne	18.182 ppm
Helium	He	5.24 ppm
Methane	CH_4	1.70 ppm
Krypton	Kr	1.14 ppm
Hydrogen	H_2	0.53 ppm
Nitrous oxide	N_2O	0.31 ppm
Carbon monoxide	CO	0.10 ppm
Xenon	Xe	0.09 ppm
Ozone	O_3	0.07 ppm
Nitrogen dioxide	NO_2	0.02 ppm
Iodine	I_2	0.01 ppm
Ammonia	NH_3	trace

used to produce inorganic nitrogen-based fertilizers. There are several processes for producing hydrogen from fossil fuel or biomass feedstocks including: steam reforming, auto-thermal reforming, partial oxidation, and gasification. Technology selection depends on economics, feedstock source, and plant flexibility. All involve the application of solid-fuel gasification or natural-gas reforming technologies to produce a syngas, which is purified by a gas cleanup step to produce a reformed syngas mix of H_2. The water-gas shift reaction process converts syngas to a mixture of CO_2 and hydrogen in varying proportions. In the case of H_2 production, the CO_2 must be removed to produce a purified stream [12]. As the process is quite similar to precombustion capture, parallels may be drawn in terms of the composition of the produced CO_2 stream.

Other processes

Other petrochemical processes, such as the production of ethylene and methanol, are also amenable to CO_2 capture and storage. CO_2 can also be captured from processes involving biomass, such as the fermentation of sugar to produce bioethanol.

Characteristics of CO_2

Carbon dioxide is a covalently bonded, nonpolar molecule consisting of two oxygen atoms and one carbon atom. CO_2 is found naturally on Earth and is essential to sustaining life as it is the primary component of the photosynthesis process. The phase diagram of CO_2 is shown in Fig. 8.2. CO_2 has similar behavior to synthetic refrigerants as it can change phase with relatively small differences in pressure and temperature, and its use as a working fluid in refrigeration cycles is becoming more common due to its nontoxic (in low concentrations), nonflammable, and non-ozone-depleting properties. CO_2 is also being considered in power generation cycles due to its high fluid density, low critical temperature, and chemical stability.

Gas

CO_2 exists as a gas at atmospheric pressure, and its density is 1.84 kg/m^3 at 298 K, which is about 50% more than air. CO_2 is highly soluble in liquids, odorless in low concentrations, and transparent to visible light. CO_2 gas absorbs

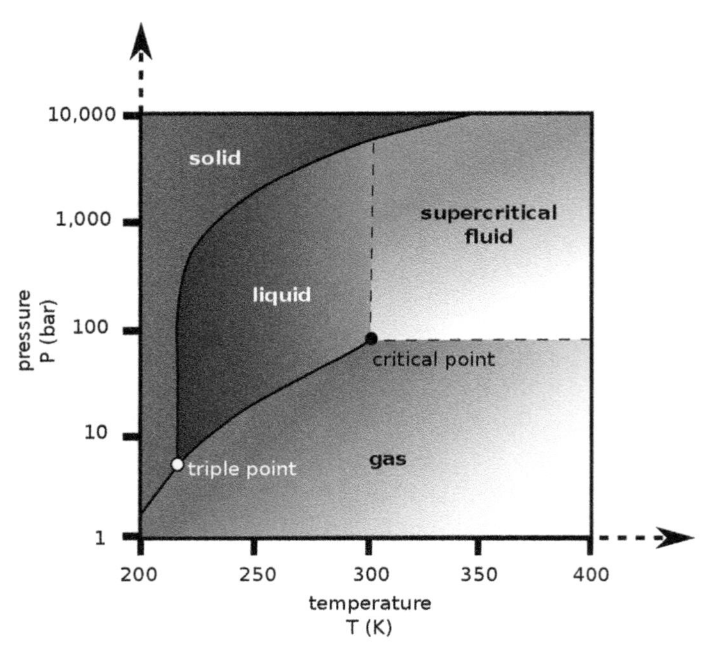

FIG. 8.2 Pure CO_2 phase diagram [13].

infrared radiation making it a greenhouse gas that traps heat within Earth's atmosphere. While it is not the strongest absorber of infrared energy when compared with other greenhouse gases, it is overall increasing in concentration in the atmosphere due to human activity and is long-lived. CO_2 gas is an asphyxiant at local atmospheric concentrations of 10% or more and can produce unconsciousness or death. Given its high density, it tends to remain near the ground due to buoyancy before it disperses in the atmosphere due to diffusion. The high Joule-Thompson coefficient for CO_2 gas makes it an effective fluid for rapidly chilling processes in several industries including, but not limited to: HVAC, medical, food and beverage, and chemical production.

Supercritical

Dense-phase or supercritical CO_2 (sCO_2) exists at temperatures and pressures above 304.1 K and 73.7 bar, respectively. Like other supercritical fluids, sCO_2 fills volumes like a gas but has a high density like liquids. The density of sCO_2 near the critical point is 467.6 kg/m^3 and increases rapidly to 703.7 kg/m3 with a 10 bar increase in pressure. Supercritical CO_2 is gaining attention as a low-temperature, environmentally friendly solvent in industrial and food and beverage processes for chemical extraction and cleaning. Government and commercial interest in sCO_2 as a solvent and its use in power cycles, carbon sequestration, and enhanced oil recovery has resulted in recent technology developments in machinery, seals, and characterizing the thermosphysical properties near the critical point including the effects of impurities and materials compatibility.

Liquid

Liquid CO_2 only forms at pressures above the triple-point conditions of 5.17 bar and 216.55 K and temperatures below 304.1 K. Liquid CO_2 is transparent and odorless, and the density is 1101 kg/m3 near the triple point at saturation conditions. Liquid CO_2 has more compressibility when compared with water, and its viscosity is at least 10 times lower than that of water. Liquid CO_2 is commonly transported in high pressure, and the US Department of Transportation (DOT) approved cylinders for use in the food and beverage industry to carbonate liquids and in fire extinguishers for displacing oxygen needed for combustion. Liquid CO_2 is also currently transported in pipelines for the oil and gas industry for enhanced oil recovery.

Solid

At atmospheric pressure and temperatures below 194.7 K, CO_2 exists as a solid and is commonly referred to as dry ice. Dry ice sublimes at temperatures above 194.7 K making it an effective cooling agent in the food and other industries when refrigeration is not possible. Dry ice particles are also used in blast cleaning of industrial surfaces, because it does not leave residue behind. The density of dry ice increases with decreasing temperature and can range from about 1.55 g/cm3 to 1.7 g/cm^3. Dry ice has low thermal and electrical conductivity, and the enthalpy of sublimation is 571 kJ/kg.

Multiphase

Under saturated, two-phase conditions, the density ratio of liquid to gaseous CO_2 is quite low when compared with typical fluids transported in pipelines. Therefore, the flow patterns that develop in piping transporting two-phase CO_2 will be different when compared with the behavior of traditional oil and gas production systems. However, most pipelines that transport CO_2 operate above the critical pressure, thus reducing the risk of developing two-phase flow. It is important to note that with the increasing demand of transporting CO_2 for subsurface sequestration, there is higher potential to cross the phase boundaries, and more research is needed to understand two-phase CO_2 behavior in large-diameter piping to quantify flow patterns at various gas-to-liquid fractions and velocities. Two-phase CO_2 also exists during rapid decompression, such as a pipeline rupture event.

Additionally, condensable fluids, like water, could be present in future CO_2 transport pipelines, thus presenting additional pressure drop and corrosion concerns from liquid dropout and the formation of carbonic acid. At the time of this publication, there are no public performance data available for separators or technology that can remove free water from high-pressure CO_2-dominant systems.

Impact of impurities

Impurities contained in the CO_2 stream have an impact on the design and operation of the pipeline system. Therefore, knowledge of the thermodynamic properties with regard to the relationship between pressure, volume, temperature, and their combined effects is important. At the triple point (5.2 bar, −56°C), CO_2 can exist as a solid, liquid, or gas. However, at temperatures and pressures beyond the critical point (74 bar, 31°C), CO_2 is in the supercritical phase. More importantly, the presence of impurities in the CO_2 stream alters the critical point, which is the highest pressure on the phase diagram. This affects the operating pressure range and increases the possibility of two-phase flow in the CO_2 transport pipeline. Experimental data on binary mixtures of CO_2 with other impurities are widely available. However, most of the experiments were focused on CO_2/H_2O, CO_2/CH4, CO_2/N_2, and CO_2/H_2S, while only a few involved effects of O_2, SO_2 and that may be present in the CO_2 stream captured from the fossil fuel power plants. The presence of impurities alters the critical pressure of the CO_2 stream due to the differences in the vapor pressure of various constituent species and thus, affects the repressurization distance (the distance between compression/pumping stations to overcome the frictional pressure drops) along the CO_2 transport pipeline.

To alleviate the impact of impurities on the possibility of two-phase flow, the operating pressure of the CO_2 transport pipeline needs to be increased, and suitable points of repressurization need to be identified. A small alteration in the working conditions close to the CO_2 critical point can result in a significant change in CO_2 density. For example, the density will double for a decrease of about 10°C from the critical temperature. This has both technical and cost implications on the hydraulic system of CCUS pipeline systems. To keep the CO_2 stream at the supercritical phase throughout the CO_2 transport pipeline, a pump-based system is recommended for flow repressurization. Furthermore, the variation in the pipeline depth can be expected to induce changes in the temperature and pressure of the CO_2 stream as a result of differences in the surrounding pressure, especially in a marine environment. The design and establishment of CO_2 transport pipelines are dependent on several factors such as viscosity and thermal conductivity, and these influence the calculation of its hydraulic properties, as well as its ability to transfer heat. The viscosity of pure CO_2 decreases with an increase in temperature and reduces further with the presence of impurities. The reduction in CO_2 viscosity increases the efficiency of transport along the pipeline, as the pressure losses throughout the pipeline are reduced.

Impurities in CO_2 streams from CCUS flue gas are a product of fossil fuel combustion, mainly containing N_2, CO_2, H_2O, and O_2 due to excess air in the combustion process. Nitrogen-containing impurities primarily include oxides, such as NO and NO_2, which are collectively known as NOx. Other potential impurities are oxides of sulfur (SO_2, SO_3), commonly referred to as SOx, and hydrogen sulfide (H_2S). Thus, the likely impurities in the CO_2 stream separated from coal-fired power plant flue gas are NOx, SOx, H_2O, O_2, and H_2S, and various gaseous impurities have been identified that exist in the CO_2 stream, such as N_2, O_2, SO_2, CH4, H_2O, CO, and H_2S. Operating conditions of CO_2 transport pipelines, such as pressure, differ depending on whether the pipeline is located within the onshore or offshore environment. For this reason, these pipelines need to be managed under stringent control of contaminants; for example, the Dynamis project [14] recommended levels of impurities for CO_2 transport through pipeline. Notably, the key influencing factors are the differences in CO_2 capture and separation technology, as well as fuel used at the CO_2 source. Potential impurities in CO_2 streams captured from a coal-fired power plant using the monoethanolamine (MEA) process were widely examined. These studies concluded that to give a complete account of impurities in the CCUS processes, there is a need to consider various technologies used for CO_2 separation and likely impurities to be expected from those technologies. There is a requirement to limit the free-water content to <600 ppm for certain operations. In the same vein, it has been reported by the Norwegian-based company Statoil in the North Sea that the water content for the first compression state is 3.9%mol and 0.3%mol at the third stage. It has been reported that the presence of other impurities, such as CH_4, N_2, H_2O, and amines in the CO_2 stream, affects the solubility of H_2O. The presence of free water is significant in CO_2 transport, because free water may result in a phase split that, in turn, could trigger hydrate formation and pipe blockage, as well as pipeline corrosion due to carbonic acid formation. Water solubility in CO_2 drops sharply as pressure increases between 50 and 60 bar and then shows a rapid increase with stabilization at 60–80 bar. It can be observed that the CO_2 solubility in water increases considerably after the change of CO_2 phase from gaseous to liquid. However, it is essential to understand the difference in the impurities content among different phases during pressure drop, especially when free water is readily available. Unfortunately, the impurities in the CO_2 stream are a vital subject with regard to supercritical CO_2 transport that is not totally understood at present.

It is important to avoid hydrate formation in the CO_2 transport pipeline. Operating away from the hydrate formation zone is essential to prevent the pipeline from blockage that will lead to a forced shutdown of the system and will increase the energy consumption required for subsequent startup of the system. Following the results from the Dynamis project, at a temperature of approximately 10°C lower than the system operating condition, stringent, free-water content specification is required to prevent hydrate formation. There is a possibility of hydrate formation when free water is present in a significant amount, and both temperature and pressure are in the hydrate formation zone. Nevertheless, hydrates may still be formed at a very low temperature, even though the amount of free water in the CO_2 stream is negligible. In this sense, transport of CO_2 at a low temperature and a high pressure along a pipeline located on the sea bed increases the risk of hydrate formation. Efforts are being made, especially at the demonstration stage, to compress and transport water-free CO_2, but this may be difficult at the project implementation stage where the mixing of CO_2 streams from different sources is expected. Therefore, in terms of operational parameters, the specification of the drying condition of CO_2 is important. Work is required to identify the free-water content that is allowable under particular operating conditions and that would pose minimal corrosion issues in the CO_2 transport pipeline.

The impact of typical impurities in high-pressure gas or dense-phase transported CO_2 streams from fossil fuel power plants and other industries fitted with carbon capture and storage technologies must also be investigated. The ultimate composition of the CO_2 stream captured from fossil fuel power plants or other CO_2-intensive industries and transported to a storage site using high-pressure pipelines will be governed by safety, environmental and economic considerations. The project, therefore, aims to elucidate optimum levels of CO_2 purification for carbon capture processes in consideration of downstream impurity impacts on pipeline transport, geological storage, and the purification costs. To complement this, key gaps in knowledge relating to the impact of impurities on the chemical, physical, and transport properties of the CO_2 stream under different operating conditions will be addressed. Impurities in CO_2 captured from combustion-based power generation with CCUS can arise in a number of ways. Water is a major combustion product and is considered an impurity in the CO_2 stream. The elements inherently present in a fuel, such as coal, include sulfur, chlorine, and mercury, are released upon complete or incomplete combustion and form compounds in the gas phase, which may remain to some extent as impurities in the CO_2 after it is captured and compressed. The oxidizing agent used for combustion (such as air) may result in residual impurities of N_2, O_2, and Ar. These same impurities may also result from air ingress into the process. The materials and chemicals used for the CO_2 separation process, such as monoethanolamine in the case of postcombustion capture or selexol in precombustion capture, and their degradation products can also be carried over into the CO_2 stream constituting a further class of impurity.

Due to its toxicity, limits have been suggested for carbon monoxide, but these vary widely in the literature. The removal of particulates from CO_2 streams is driven by the need to prevent damage or fouling of equipment. Design parameters for particulates have been reported as 0–1 ppmv [15]; however, it may be possible to specify limits for certain particle size ranges. For other components that may be present in CO_2 streams (e.g., HCl, HF, NH3, MEA,

TABLE 8.2 Coal/biomass oxidation products [16].

Coal/biomass oxidation products	
Complete	**Partial**
H_2O, SO_x, NO_x, HCl, HF	CO, H_2S, COS, NH_3, HCN
Volatiles	**Biomass alkali metals**
H_2, CH_4, C_2H_6, C_3+	KCl, NaCl, K_2SO_4, KOH etc.
Trace metals	**Particulates**
Hg ($HgCl_2$), Pb, Se, As etc.	Ash, PAH/soot
Oxidant/air ingress	**Process fluids**
O_2, N_2, Ar	Glycol, MEA, Selexol, NH_3 etc.

Selexol), little or no information is available to understand their downstream impacts on transport and storage and determine maximum allowable amounts. Further work is therefore required to understand the impacts of these species in transport and storage applications and to elucidate potential crossover effects.

Impurities contained in the CO_2 streams from different carbon capture technologies may be classified broadly by origin into three main categories arising from fuel oxidation, excess oxidant/air ingress, and process fluids. It is possible that impurity species arise from different sources. For example, NH_3 may arise as an oxidation product or as a process fluid. The typical oxidation products that derive from coal and/or biomass combustion are given in Table 8.2. These fuels are considered for use with CCUS and produce a larger range and higher level of CO_2 impurities in comparison to those of CO_2 derived from natural-gas combustion with CCUS.

Major and minor complete oxidation products of coal and biomass form the common impurities of water, SOx, NOx, and halogens. Partial oxidation products such as carbon monoxide (CO) and hydrogen sulfide (H_2S) may arise from fuel-rich conditions encountered in gasifiers as employed for integrated gasification combined cycles (IGCC). Volatiles comprising hydrogen and light hydrocarbons are formed from fuel devolatilization with heating. Biomass fuels contain higher levels of alkali metals in comparison to coal and could form a class of CO_2 impurities, the main species being chlorides, sulfates, and hydroxides of potassium and sodium. Trace metals contained in fuel may be released to the gas phase on combustion and propagate into the CO_2 stream. These metals may exist in the CO_2 stream in elemental or oxidized forms, such as mercury dichloride $HgCl_2$, and may require removal due to operational and environmental health reasons. Particulates in the form of ash and soot with polycyclic aromatic hydrocarbon (PAH) precursors are another type of oxidation impurity. Oxygen, nitrogen, and argon are CO_2 impurities that can arise from excess oxidant used for combustion or air ingress into the boiler. The oxidant/air ingress species are referred to as "inerts" by some authors. A final class of CO_2 impurities are the process fluids used for CO_2 separation such as monoethanolamine and selexol. Other contaminants may arise from the power plant or CCUS process such as machinery lubricants or metals, but they are not discussed further in this analysis, because no estimates or measurements have been made, and they are not expected to be present in CO_2 streams in levels that would cause concern.

The dense-phase scenarios tend to produce the least desirable qualities for pipeline transportation. These CO_2 streams have the lowest proportions of CO_2 and have the highest bubble-point curves, compressibility, and Joule-Thomson coefficient and the lowest densities, speed of sound, and thermal conductivities. However, it is worth noting that these scenarios have the lowest viscosities. The impurities influence a wide range of thermodynamic properties, including the density of the stream, the specific pressure drop, and the critical point. As a consequence, the pipeline design parameters such as diameter, wall thickness, inlet pressure, minimum allowable operational pressure (MAOP), and the distance between booster stations are potentially subject to change.

To enable containment of CO_2 mixtures in liquid form, the containment system must keep the product (above the triple-point pressure) at pressures and temperatures that are above the bubble and melting point lines. This zone varies for different impurity scenarios. Some of the impurity scenarios can be transported in the liquid phase at around −57°C and 1–1.5 MPa, with others requiring higher pressures.

Temperature has a more significant effect on density than pressure for both pure CO_2 and CO_2-containing impurities. To increase the density, and therefore capacity, of the pipeline, the inlet temperature should be as low as possible. For fracture control, the saturation pressure of the CO_2 stream is a critical variable that will determine the required pipeline dimensions and toughness to prevent a long running ductile fracture. The saturation pressure is dependent on the composition of the CO_2 mixture and also on the operating conditions of the pipeline. Impurities with lower critical temperatures than CO_2 will raise the saturation pressure, and those with higher critical temperatures than CO_2 will lower the saturation pressure. It has been shown that H_2 in particular has the most potent effect, in terms of %mol addition, in raising the saturation pressure, and therefore, has the most detrimental effect on fracture control for CO_2 pipelines.

Further works by the National Energy Technology Laboratory (NETL) [15] and the Dynamis project [14] have provided recommended impurity limits for CO_2 stream components in studies of CO_2 capture utilization and storage systems. Limits are suggested based upon a number of different factors, and these quality guidelines may serve as a basis for conceptual studies. The presence of impurities in CO_2 can shift the boundaries in the CO_2 phase diagram to high pressures, meaning that higher operating pressures are needed to keep CO_2 in the dense phase. For pipeline and storage applications, the total concentration of the air-derived, noncondensable species (N_2, O_2, and Ar) should not exceed 4% due to the impact on compression and transport costs. In addition, these species can reduce the CO_2 structural trapping capacity in geological formations by a greater degree than their molar fractions [17].

Hydrogen may be present in precombustion capture-derived CO_2 streams and is believed to impact required pipeline inlet pressures significantly [18]. Enhanced Oil Recovery (EOR) applications require stricter limits, particularly O_2, which should be kept below 100 ppm due to it promoting microbial growth and a reaction with hydrocarbons. CCUS specifications for water are set to limit corrosion due to the formation of in situ carbonic acid [19], clathrate formation, and condensation at given operating conditions. Reported guidelines for water vary widely and can be dependent on the concentration of other species present in the stream such as acid gases. Sulfur species (H_2S, COS, SO_2, and SO_3) pose a corrosion risk in the presence of water and should be removed to a certain level, and there are additional toxicity concerns for H_2S.

There are, however, a number of technology approaches for removing sulfur species including newer developments for carbon capture applications. SO_3 can form in pulverized fuel plants that utilize postcombustion or oxy-fuel combustion capture techniques. As this species reacts quickly with water, so it can be removed by water contacting units. NETL [15] recommends that the target for SO_2 be 100 ppmv on the basis of its IDLH (immediately dangerous to life or health) level. NOx species may be present in CO_2 streams as combustion by-products and also pose a corrosion risk due to nitric acid formation [20]. There are a number of traditional and novel CCUS approaches for NOx limitation. The IDLH limits of NO and NO_2 are 100 and 200 ppmv, respectively, and a limit of 100 ppmv has therefore been proposed for CCUS-derived CO_2 streams. Among the numerous trace metal species that could be present in CO_2 streams, mercury receives attention due to its toxicity and corrosion effects on a number of metals.

Safety concerns

Reliability is the capability of an engineering system or a component to operate under a set of operating conditions for a specified period to produce a desired result. Based on this definition, a system or component can be described as unreliable when it can no longer maintain or operate under a specific set of operating conditions over time to produce a desired result. Therefore, measures need to be taken at the design stage to ensure that systems are made reliable over their useful life cycle. A necessary consideration of reliability, availability, maintainability, and operability (RAMO) characteristics of the CO_2 transport pipeline makes a significant, positive contribution to achieving reasonable economic life cycle costs. Importantly, it has been claimed that there is little experience to date on the actual behavior of anthropogenic CO_2 in the supercritical phase, and this poses a number of challenges for the integrity, reliability, safety, and cost efficiency of the pipeline. It is a common understanding within the industry that the CO_2 transport pipeline network should be designed and developed within the remits of that of the oil and gas industry. The reliability and maintenance challenges should be considered at the design phase of the CO_2 transport pipeline. At this stage, it becomes imperative to resolve the challenges related to impurities content in the CO_2 stream, material selection, corrosion and fracture prevention, and operation and maintenance of the entire system. Reliable pipelines for CO_2 transport will require a well-organized maintenance culture. Furthermore, the current literature has emphasized the importance of reliable means of corrosion prediction that are necessary for the prevention of leakage, accidental discharge, and loss of CO_2 resulting from corrosion. Finally, for effective control of CO_2 pipeline integrity, a management regime is required. This incorporates, among a number of other aspects, selection of material, inspection and monitoring, maintenance, operation, corrosion mitigation, evaluation of risks together with the concept of communicating these risks.

Transport of CO_2 takes place under a high pressure and in a supercritical phase. Depressurization of the system may occur as a result of pipeline failure or planned maintenance. Loss of pressure can also occur due to the length and geometry of the CO_2 transport pipeline. It has been estimated that the maximum CO_2 release rate from a faulty pipeline is estimated at a range of 0.001–22 tons/s. More importantly, these figures depend on the pipe diameter, puncture size and level of impurities that may affect the CO_2 stream phase, operating temperature and pressure, and whether the CO_2 release and dispersion are planned or accidental. Furthermore, a change in the CO_2 phase gives rise to dry ice formation in the pipeline surroundings, which has an indirect effect on the concentration and impurities around the faulty pipeline. An industrial-scale experiment on the release and dispersal of CO_2, known as CO2PIPETRANS [21], was conducted by BP and Shell. The data gathered from this experiment were used to validate simulations of CO_2 release and dispersion. From the material integrity viewpoint, it is necessary to have control of the rate of depressurization, as depressurization that occurs too fast can accelerate the temperature drop rate within the pipeline causing the steel wall to become brittle.

Economics do not favor transportation of a large amount of CO_2 at a low pressure over a long distance. Therefore, transport of CO_2 should be carried out at a high pressure, and as a consequence, this may pose some health and safety risks. In the assessment of environmental risks for the CO_2 transport pipeline, ensuring the safe operation of the high-pressure pipeline has been identified as a major risk. It has been indicated that an emergency planning zone (EPZ) around the pipeline, which requires detailed emergency response planning, needs to be considered at the design and planning stages.

CO_2 is known to be neither toxic when released in small quantities nor explosive. However, if the CO_2 transport pipeline is accidentally ruptured, it can release a considerable amount of CO_2 into the air that could pose harm to humans under particular circumstances. Considering the fact that certain regions of the earth, such as the European Union, are characterized by a high population density and that some of the CO_2 capture sites are located near cities, existing regulations should be strengthened to route high-pressure pipelines away from buildings and dwellings. Moreover, care must be taken to significantly reduce the impurities content in the CO_2 stream that can pose injury or harm to humans, such as H_2S. In this sense, CO_2 transport pipelines must be buried deep enough to prevent digging equipment from reaching them. Furthermore, crack arrestors should be fitted in CO_2 pipelines, and for urban transit pipelines, a pressure release mechanism, such as a supervisory control and data acquisition (SCADA) system, should be fitted. Based on the experiences of the natural gas pipelines industry, failure rates associated with leaks for CO_2 transport pipelines are estimated to range between 0.7 and 6.1×10^{-4} yr/km. Most of the recorded failures to date were caused largely by third-party interference, pipeline corrosion, material, and construction defects, such as welds, movement of ground, or operator errors. Leakage could also be a result of existing or induced defects, fractures, or along a spill position. Currently, there are not enough empirical data and experience to accurately determine the likelihood of failure of CO_2 transport pipelines compared with natural gas pipelines. This is further complicated due to the presence of impurities in the CO_2 stream. When considering pressures for offshore and onshore pipelines, several authors maintained that the offshore CO_2 transport pipeline route can be designed for higher pressures than the onshore route (up to 300 bars). This is because of reduced risks associated with the human population onshore. Nevertheless, protection from or resistance to rupture by simple unintended acts, such as a dragging boat anchor, needs to be considered in the risk profile.

Purification

The required purity of CO_2 is dependent on its end application. For EOR or CCUS applications, NETL recommends the following limits on impurities (Table 8.3). This table shows the impurities from a combustion process and what limits are required for purification by the CPU (CO_2 processing unit), as shown in Fig. 8.3. The CPU is divided into two sections, the warm part and the cold part. The warm part includes compression, a scrubber (wash) for removal of condensable gases (NOx, SOx) and an absorber containing molecular sieves for water removal (down to less than 10 ppm), and activated carbon to adsorb any residual NOx or SOx. The cold part, sometimes referred to as the cold box, is a cryogenic distillation column for removal of noncondensable gases such as O_2, N_2, and argon. There are several technology vendors providing different solutions for the CPU, and the details of their designs will be discussed later in this section.

Within the "warm part" of the CPU, the technology development involving the removal of the NOx and SOx is considered an emerging technology where several equipment providers have proposed options (Air Products, Linde, Praxair and Air Liquide). Key technologies provided are scrubbing and knockout drums by all four technology licensors. Knockout is typically performed at each stage of compression, whereas scrubbing is performed as a separate unit operation from compression to remove condensable gases.

Air Products proposes the use of the sour compression process (based on lead chamber reaction) to knock out 99% of the SOx as H_2SO_4 and remove at least 95% of NOx as HNO_3 and HNO_2 during the compression of the CO_2-rich flue gas. While the Linde proposes the use of the LICONOX process, whereby 99% of the SOx is removed in the scrubbers. The

TABLE 8.3 Recommended CO_2 composition limits for EOR and CCUS [22].

Component	Unit (max unless otherwise noted)	Carbon steel pipeline		Enhanced oil recovery		Saline reservoir sequestration		Saline reservoir CO_2 & H_2S Co-sequestration		Venting concerns (see Section 3.0)
		Conceptual design	Range in literature	Conceptual design	Range in literature	Conceptual design	Range in literature	Conceptual design	Range in literature	
CO_2	vol% (min)	95	90–99.8	95	90–99.8	95	90–99.8	95	20–99.8	Yes-IDLH 40,000 ppmv
H_2O	ppm_v	500	20–650	500	20–650	500	20–650	500	20–650	
N_2	vol%	4	0.01–7	1	0.01–2	4	0.01–7	4	0.01–7	
O_2	vol%	0.001	0.001–4	0.001	0.001–1.3	0.001	0.001–4	0.001	0.001–4	
Ar	vol%	4	0.01–4	1	0.01–1	4	0.01–4	4	0.01–4	
CH_4	vol%	4	0.01–4	1	0.01–2	4	0.01–4	4	0.01–4	Yes-Asphyxiate, Explosive
H_2	vol%	4	0.01–4	1	0.01–1	4	0.01–4	4	0.02–4	Yes-Asphyxiate, Explosive
CO	ppm_v	35	10–5000	35	10–5000	35	10–5000	35	10–5000	Yes-IDLH 1200 ppmv
H_2S	vol%	001	0.002–1.3	0.01	0.002–1.3	0.01	0.002–1.3	75	10–77	Yes-IDLH 100 ppmv
SO_2	ppm_v	100	10–50,000	100	10–50,000	100	10–50,000	50	10–100	Yes-IDLH 100 ppmv
NO_x	ppm_v	100	20–2500	100	20–2500	100	20–2500	100	20–2500	Yes-IDLH NO-100 ppmv, NO_2-200 ppmv
NH_3	ppm_v	50	0–50	50	0–50	50	0–50	50	0–50	Yes-IDLH 300 ppmv
COS	ppm_v	Trace	Trace	5	0–5	Trace	Trace	Trace	Trace	Lethal @ high concentrations (>1000 ppmv)
C_2H_6	vol%	1	0–1	1	0–1	1	0–1	1	0–1	Yes-Asphyxiant, Explosive
C_3+	vol%	<1	0–1	<1	0–1	<1	0–1	<1	0–1	
Part.	ppm_v	1	0–1	1	0–1	1	0–1	1	0–1	
HCl	ppm_v	N.I.[a]	N.I.[a]	N.I.[a]	N.I.[a]	N.I.[a]	N.I.[a]	N.I.[a]	N.I.[a]	Yes-IDLH 50 ppmv
HF	ppm_v	N.I.[a]	N.I.[a]	N.I.[a]	N.I.[a]	N.I.[a]	N.I.[a]	N.I.[a]	N.I.[a]	Yes-IDLH 30 ppmv
HCN	ppm_v	trace	trace	trace	trace	trace	trace	trace	trace	Yes-IDLH 50 ppmv
Hg	ppm_v	N.I.[a]	N.I.[a]	N.I.[a]	N.I.[a]	N.I.[a]	N.I.[a]	N.I.[a]	N.I.[a]	Yes-IDLH 2 mg/m^3 (organo)
Glycol	ppb_v	46	0–174	46	0–174	46	0–174	46	0–174	
MEA	ppm_v	N.I.[a]	N.I.[a]	N.I.[a]	N.I.[a]	N.I.[a]	N.I.[a]	N.I.[a]	N.I.[a]	MSDS Exp Limits 3 ppmv, 6 mg/m^3
Selexol	ppm_v.	N.I.[a]	N.I.[a]	N.I.[a]	N.I.[a]	N.I.[a]	N.I.[a]	N.I.[a]	N.I.[a]	

[a] *Not enough information is available to determine the maximum allowable amount.*

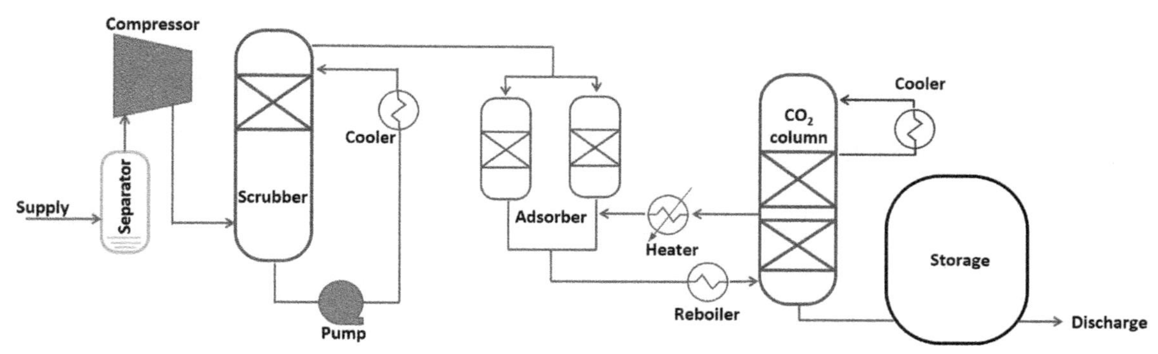

FIG. 8.3 Flow diagram of a CO_2 purification and liquefaction plant.

cleaned gas is compressed to 15bar to convert NO to NO_2; then, NO_2 is removed using alkali wash in the scrubber (based on NH_3 water or NaOH). This would result in a removal of at least 95% of NOx as spent salt of nitrite and nitrate.

Praxair presented two possible options for pretreatment of the flue gas. The first option uses sulfuric acid wash to recover nitric acid. This would result in a clean gas containing 50–100ppm SOx and less than 50ppm NOx. The second option uses activated carbon to adsorb any SOx and NOx resulting in dilute acid during regeneration. And Air Liquide proposes the use of $NaCO_3$ to quench and scrub the flue gas to reduce the SOx down to less than 10ppm and the removal of the NOx at the knockout drum of the compressor and final removal at a separate distillation column of the cold box.

Among the four options presented above, only the process of Air Liquide has been tested in a large-scale pilot plant. The other approaches have been tested only at smaller-scale pilot facilities (i.e., <10t/d CO_2). Future work in this area will primarily focus on the area of scaling-up, process intensification, and process integration.

Inert removal through a cryogenic process (cold part) is also referred to as cryogenic distillation (shown in Fig. 8.3 as the refrigeration unit) and uses an auto-refrigeration cycle using impure CO_2 as a refrigerant. The cold part of the CPU primarily functions to separate the other noncondensable CO_2 components, mainly consisting of O_2, N_2, and Ar. This part of the CPU technology is an evolutionary development of the liquefaction plant used in the current fleet of industrial and food-grade CO_2 production. The newer innovation of this technology is the use of the auto-refrigeration cycle using impure CO_2 as a refrigerant. The cold part of the CPU could consist of a cycle having partial condensation and/or a cryogenic distillation column. The primary driver in the design of the cycle is dependent on the limits of the CO_2's oxygen content. Future development in this area would only require engineering data of impure CO_2 to validate refrigeration performance. Newer technologies would focus on improving the refrigeration cycle and could also include other possible refrigerants other than the impure CO_2. It should be noted that the best practices developed in technologies used by the LNG industry could also be adapted to this process. Linde has investigated a range of configurations for the cryogenic part of the CPU.

Corrosion

The importance of corrosion in the CO_2 transport pipeline cannot be underestimated as it would affect the integrity of the pipeline infrastructure. A number of studies have been conducted on the subject of impurities and their corrosivity in the transport of CO_2. It has been highlighted that the presence of free water in the CO_2 stream transported through the pipeline should be avoided. Some of those studies evaluated the effects of H_2O and other impurities on corrosion in different pipeline materials. There is a correlation between the moisture content in the CO_2 stream and the rate at which the interior wall of the CO_2 transport pipeline corrodes. However, the research on the allowable level of free water in the CO_2 stream that will not cause pipeline corrosion is limited. There are two views: one saying that the free water content should be limited to as low as 50ppm, while the other indicates that, in the worst-case scenario, it should not exceed 600ppm, as above this level, corrosion of the pipeline material may occur. In practice, some of these sources recommended that in the presence of a large quantity of SO_2, lower levels of moisture must be considered. SO_2 naturally was noted to be more acidic when dissolved in water and could intensify the corrosion of the pipeline.

The damage resulting from corrosion of the pipeline material increases with a decrease in temperature. There is therefore an increased risk for corrosion when a reduction in the temperature occurs along with a drop in the operating

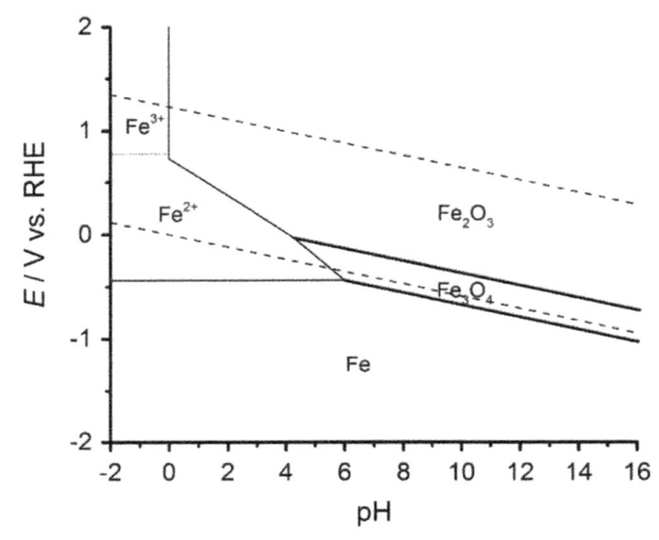

FIG. 8.4 Pourbaix diagram for the iron—Water system [23].

pressure or at a time of total depressurization in parts of the pipeline. Apart from the formation of carbonic acid in the aqueous phase, which reduces the pH and increases the risk of corrosion, a key challenge of CO_2 transport through pipeline is the presence of impurities such as NOx, SOx, and H_2S that segregate to the aqueous phase. The segregated aqueous phase forms in situ sulfuric and nitric acids, which cause a further drop in the pH of the solution.

When analyzing the effect of impurities on corrosion, it was estimated that in a worst-case scenario, the fluid pH could be as low as 3.2, attributed to carbonic acid alone. Similarly, in the event of formation of an isolated, water-rich aqueous phase, CO_2 saturates it, producing a pH of approximately 3. The explanation (both theoretical and experimental) is of the mutual solubility of H_2O in CO_2 as well as CO_2 in H_2O. The manner in which low pH impacts the pipeline material can be predicted to a degree by the Pourbaix diagram for iron (Fig. 8.4). The Pourbaix diagram is an illustration of a phase diagram outlining electrochemical stability for different redox states of an element. The water redox line is important in the Pourbaix diagram for elements such as Fe. Water in liquid form is stable within the redox lines. However, below the H_2 line and above the O_2 line, liquid water is unstable relative to H_2 and O_2, respectively. An active metal such as Fe can only show stability below the H_2 line. Therefore, metallic Fe displays instability when it contacts water and undergoes some reactions. Under such conditions, these reactions occur irrespectively of the potential (V) and pH.

In the design and operation of CO_2 transport pipelines, corrosion and material selection are of significant consideration. Before material selection is carried out, it is necessary to identify the full stream composition together with the whole range of operating conditions that all system equipment will be exposed to. Again, consideration should be given to the steady state as well as the dynamic excursion situations such as shutdown, startup, and upsets. In CO_2 pipeline transport, corrosion and corrosion mechanism considerations take into account: free-water phase, CO_2 corrosion and O_2 corrosion of carbon steel, corrosion-resistant alloys, stress corrosion, hydrogen damage, liquid metal embrittlement, and degradation of nonmetallic parts. There are factors militating against CO_2 pipeline corrosion prevention procedures, and these include: lack of selective protection of low-grade carbon steel materials, absence of knowledge of correct metallurgy inhibitor test application, inadequate correlation of surface monitoring procedures with internal rate of corrosion, and negligence on the significance of complementing laboratory tests with field trials.

Infrastructure and value chain

Capture

CO_2 capture is an integral part of several industrial processes, and accordingly, technologies to separate or capture CO_2 from flue gas streams have been commercially available for many decades. In practice, the most appropriate capture technology for a given application depends on a number of factors, including the initial and final desired CO_2 concentration, operating pressure and temperature, composition and flow rate of the gas stream, integration with the original facility, and cost considerations. The cost of capturing CO_2 can vary significantly, mainly according to the concentration of CO_2 in the gas stream from which it is being captured, the plant's location, energy and steam supply, and integration with the original facility. For some processes, such as ethanol production or natural gas processing or after oxy-fuel combustion in applications such as power generation or cement, CO_2 can be already highly concentrated. This CO_2 can be simply pretreated if necessary (e.g., dehydration) and then compressed for transport and storage or use at a relatively low cost. For example, the cost of separating out the CO_2 contained in natural gas—which is often required for technical reasons before the gas can be sold or liquefied—can be as low as USD 15/t-USD 25/t. For more diluted CO_2 streams, including the flue gas from power plants (where the CO_2 concentration is typically 3%–14%) or a blast furnace in a steel plant (20%–27%), the cost of CO_2 capture is much higher (over USD 40/t of CO_2 and sometimes more than USD 100/t), accounting on average for around 75% of the total cost of carbon capture, utilization, and storage (CCUS).

CO_2 capture costs for hydrogen refer to production through steam methane reforming (SMR) of natural gas; the broad cost range reflects varying levels of CO_2 concentration. The lower end of the cost range applies to CO_2 capture from the concentrated "process" stream, while the higher end applies to CO_2 capture from the more diluted stream coming out of the SMR furnace. As mentioned earlier, the cost of CO_2 capture is much lower for concentrated sources such as hydrogen production, coal to chemicals and natural gas processing than for power generation, cement and steel production, and direct air capture (DAC). The CO_2 concentration of raw natural gas varies considerably by reservoir, ranging from CO_2-free natural gas in Siberian fields to exceptionally high shares of 90% CO_2 content in some fields in South-East Asia. Raw natural gas produced from the Norwegian Sleipner field has a CO_2 concentration of 9%, which is considered to be high compared with many other fields. The low capture cost is also due to the high pressure of the captured CO_2 stream, which reduces cost for CO_2 compression. Most CO_2 capture systems have been designed to capture around 85%–90% of the CO_2 from the point source, which results in the lowest cost per ton of CO_2 captured. However, in a net-zero energy system, higher capture rates—approaching 100%—will be needed. This is technically and economically achievable according to recent studies.

Apart from natural-gas sweetening, capture from these processes has not been tested on a large scale. Different processes must be considered individually for their suitability to CO_2 capture. Many of the major industrial emissions sources are suitable for CO_2 capture in terms of emissions per source and the concentration of CO_2 in waste gas streams. Implementation of CCUS technologies for most industrial activities (e.g., boilers, iron and steel furnaces, and cement) requires a capture step applied to low concentration CO_2 streams. The technical feasibility of this in each case will depend upon the layout of the industrial plant. In some instances, industrial activities already apply some form of CO_2 removal or capture as an inherent part of the process and therefore emit a relatively pure CO_2 stream. These types of activities include natural gas processing, hydrogen production for ammonia and subsequent fertilizer production, and synthetic fuel production such as coal-to-liquids and gas-to-liquids. Estimates and measurements of the ranges and levels of impurities in the CO_2 capture streams from industrial processes have limited availability. The following subsections provide brief descriptions of some of the main industries considered for CCUS and present data of the derived CO_2 stream composition where possible.

Liquefaction

Liquefaction of CO_2 is an intermediate step for storage or ship transport. Two processes are suggested. The traditional method is based on external refrigeration, and the other is an integrated refrigeration process. In the external refrigeration process, traditional refrigeration based on ammonia was selected. In the internal refrigeration process, liquefaction is achieved by compression, cooling, and expansion of the CO_2. A diagram of this basic process is shown below in Fig. 8.5.

Simulation models have been developed for cycle options. A process based on ammonia refrigeration was calculated to be most cost optimum. There are, however, still possibilities for improvements especially for the internal refrigeration process. The life cycle cost was calculated for the two product pressures 600 kPa and 1500 kPa. The difference in liquefaction cost between an external- and internal-based process was larger for low pressure (600 kPa) than for a medium pressure (1500 kPa).

In the external cooling process, the CO_2 is compressed to a high pressure and then cooled and expanded to the delivery pressure. The cooling is called "external," because the refrigerant is not in contact with the main (CO_2) gas. The first separator separates condensed components such as water before the compression. After compression and cooling, liquid (which is mainly water) is removed. The CO_2 is condensed in the heat exchanger with evaporating ammonia as the cooling medium. The evaporated ammonia is then compressed, cooled, and expanded in a traditional refrigeration circuit.

The simple refrigeration process has a much lower energy consumption compared with the simple base case internal refrigeration process. The processes based on traditional ammonia refrigeration have been calculated to be both

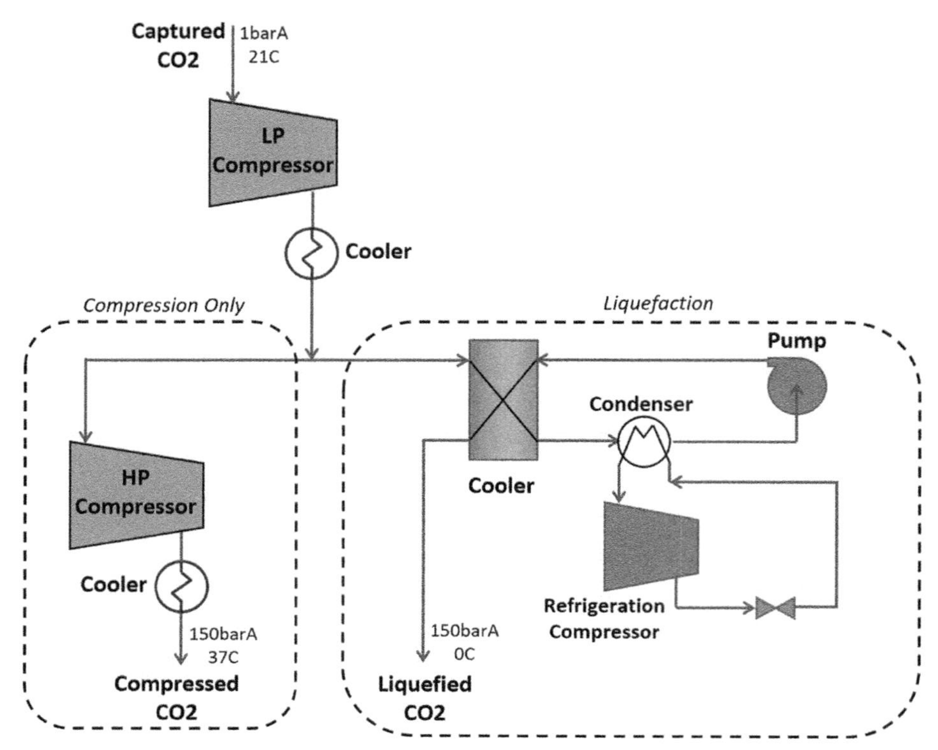

FIG. 8.5 CO_2 liquefaction process.

energy and cost optimum. However, the improved process with internal refrigeration was not far from being energy optimum. The simple process based on internal refrigeration had a very high energy consumption. The process was, however, not very realistic since it was assumed that there is no intercooling in the compressor. The calculations indicate that it is the process with internal refrigeration, which has the largest potential for improvements.

The most expensive process units from both a capital and an operating cost point of view were the compressors. Because of that, the energy optimum process was also the cost optimum process. The improved alternatives had energy consumptions close to each other. Among the evaluated cases, the most cost optimum process was a process based on external refrigeration. The investment was estimated to 23 million Euro, and the operating cost was estimated to 4 Euro/ton. The investment was estimated to be 4 million Euro higher, and the operating cost was estimated to be only 0.13 Euro/ton higher for the internal refrigeration process.

Transport

Captured CO_2 needs to be transported to a storage site for injection into a geological formation. CO_2 is generally transported in the dense phase at temperature and pressure ranges between 12°C and 44°C and 85 barA and 200 barA. The lower pressure limit is set by the phase behavior of CO_2 and should be sufficient to maintain single conditions, while the upper pressure limit is mostly due to economic and material concerns. There are two ways by which large amounts of CO_2 may be transported:

1. Compression of CO_2 to dense phase (>74 bar) for pipeline transport
2. Refrigeration of CO_2 to liquid phase for transport by ship, truck, or other vehicles

Captured CO_2 usually contains water. Water must be removed before transport to prevent CO_2 and water from forming acids that can corrode pipelines and other equipment. Dehydration is typically done in conjunction with compression or refrigeration. The indicative costs of pipeline and shipping transport modes vary significantly with scale and transport distance. The CCUS value chain consists of various components, each with a range of costs that vary with different drivers. CO_2 must be captured, compressed and dehydrated, transported to the injection site, and finally, injected and monitored.

Project characteristics determine project costs in any location. For example, the Northern Lights project [24], which plans to transport CO_2 by ship from various ports to a storage site under the seabed of the North Sea, is targeting storage costs of €35–50/tCO_2. Pipeline costs are strongly affected by economies of scale. This is the case for dense-phase pipelines (>74 bar) or gas-phase pipelines. All else being equal, gas-phase pipelines are larger in diameter than dense-phase pipelines, which tends to make them more expensive. As such, bulk transport of CO_2 is usually done under dense-phase conditions. Although specific pipeline costs vary from country to country, the general pattern of these cost curves will be observed in all locations. Pipeline costs are remarkably high at small flowrates, falling rapidly with increasing flow before effectively leveling off once flows reach the megatonne range. The strong influence of pipeline economies of scale is a key driver of the development of CCUS hubs. Megatonne CO_2 sources such as power stations, gas processing plants, or other large industrial sources should be able to support an economical CO_2 pipeline on their own. These can then serve as anchor customers for a hub, enabling smaller CO_2 sources to also use the pipeline without incurring the much higher pipeline costs observed at small flowrates. For very long transport distances and mid-range tonnages, shipping can become more economical than pipelines. Shipping does not have the same economies of scale as pipelines, but has the advantage that it can be deployed in a modular fashion—starting with one ship and scaling up over time as needed. It can also be directed to different storage sites, which may be useful if price competition between storage sites emerges.

The availability of infrastructure to transport CO_2 safely and reliably is an essential factor in enabling the deployment of CCUS. The two main options for the large-scale transport of CO_2 are through pipeline and ship, although for short distances and small volumes, CO_2 can also be transported by truck or rail, albeit at a higher cost per ton of CO_2. Transport by pipeline has been practiced for many years and is already deployed at large scale. Large-scale transportation of CO_2 by ship has not yet been demonstrated (Technology Readiness Level TRL 4–7) but would have similarities to the shipping of liquefied petroleum gas (LPG) and LNG. Nonetheless, considerable possibilities for innovation remain, in particular for offshore unloading of CO_2, and spillovers from the general shipping industry, including automation and new propulsion technologies. Economic factors and regulatory frameworks are the main considerations in the choice of a CO_2 transport mode. Pipelines are currently the cheapest way of transporting CO_2 in large quantities onshore and, depending on the distance and volumes, offshore as well. There is already an extensive onshore CO_2 pipeline network in North America (Fig. 8.6) with a combined length of more than 8000 km—mostly in the United States. These onshore pipelines currently transport more than 70 Mt/year of CO_2, mainly for EOR.

In June 2020, the Alberta Carbon Trunk Line (ACTL) in Canada came online with a pipeline capacity of 14.6 Mt CO_2, with significant excess capacity (some 90%) to accommodate CO_2 from future CCUS facilities. The ACTL received CAD 560 million (USD 430 million) in capital funding from the Canadian and Albertan governments, slightly below half of the CAD 1.2 billion (USD 920 million) estimated project cost. There are also two CO_2 pipeline systems in Europe and two in the Middle East. Proximity to pipeline infrastructure has been a key driver for recent CCUS project announcements in the United States. The share of pipeline transportation in the total cost of a CCUS project varies according to the quantity transported as well as the diameter, length, and materials used in building the pipeline. Other factors include labor cost and the planned lifetime of the system. Location and geography are significant factors that affect the total cost as well. In most cases, transport represents well under one-quarter of the total cost of CCUS projects. Pipelines located in remote and sparsely populated regions cost about 50%–80% less than in highly populated areas. Offshore pipelines can be 40%–70% more expensive than onshore pipelines. There are strong economies of scale based on pipeline capacity, with unit costs decreasing significantly with rising CO_2 capacity. Pipeline costs are also likely to differ substantially among regions as new projects are developed. The cost of new pipelines is estimated to be generally 30% lower in Asia than in Europe.

While the properties of CO_2 lead to different design specifications compared with natural gas, CO_2 transport by pipeline bears many similarities to high-pressure transport of natural gas. Repurposing existing natural gas or oil pipelines, where feasible, would normally be much cheaper than building a new line. Design pressure and remaining service life are the two main considerations to be taken into account to evaluate the repurposing of existing oil and gas pipelines. Repurposed pipelines typically operate at lower pressure, which leads to a reduction in CO_2 transport capacity compared with higher-pressure, purpose-built CO_2 pipelines. Furthermore, many existing oil and gas pipelines have been in operation for more than 20 years. A case-by-case analysis is necessary to evaluate their remaining life, taking into account internal corrosion and the remaining fatigue life in particular.

Pipeline costs are highly sensitive to scale and location. CO_2 transportation by ship to an offshore storage facility offers greater flexibility, particularly where there is more than one offshore storage facility available to accept CO_2. The flexibility of shipping can also facilitate the initial development of CO_2 capture hubs, which could later be connected or converted into a more permanent pipeline network as CO_2 volumes grow. Today, only around 1000 tons of food quality CO_2, which is 99.9% pure, is shipped in Europe every year from large point sources to coastal distribution terminals.

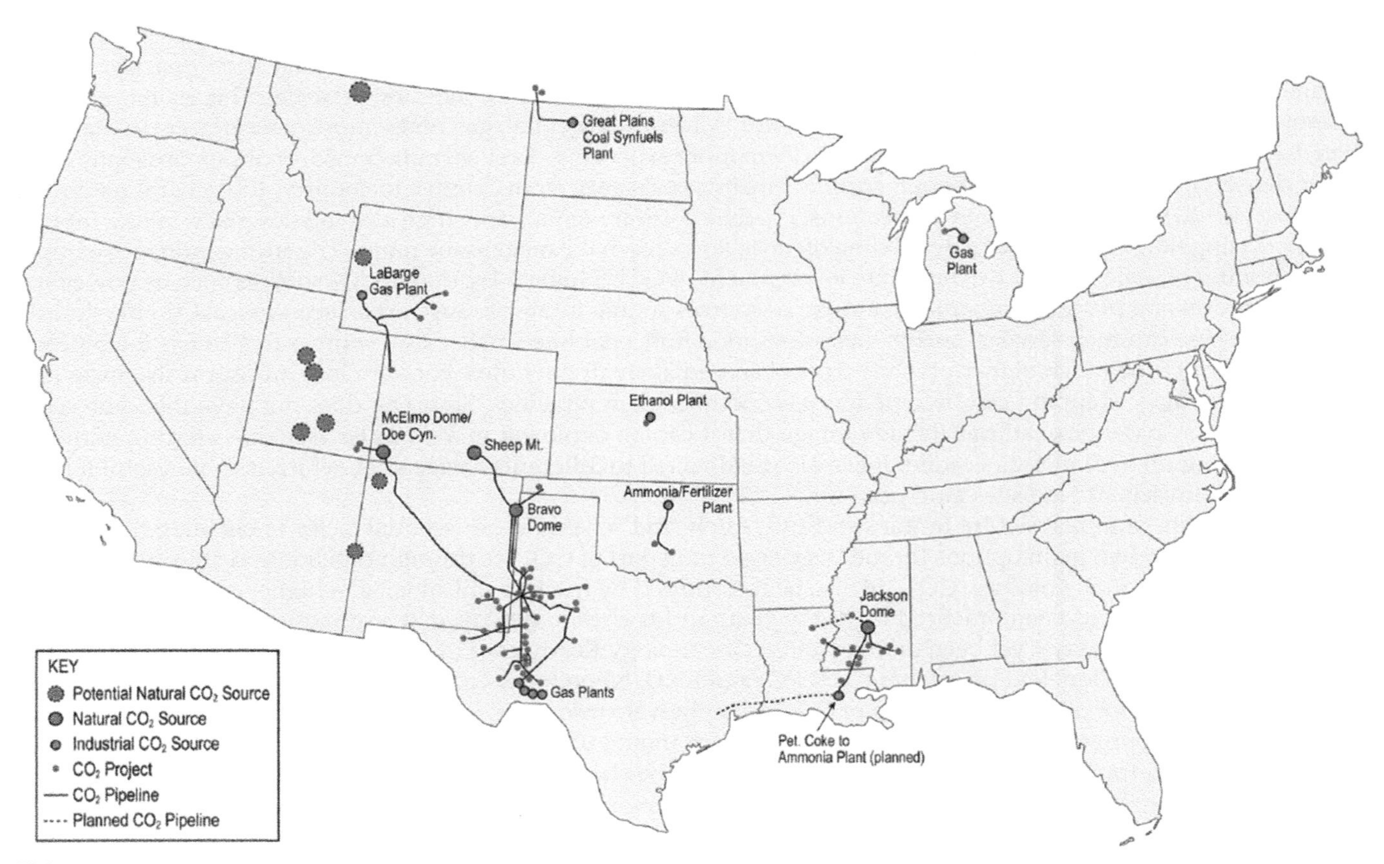

FIG. 8.6 US CO_2 pipelines [25].

In recent years, interest in CO_2 shipping has increased in several regions and countries where offshore storage has been proposed, including in Europe, Japan, and Korea. Large-scale transportation of CO_2 by ship has not yet been demonstrated but would have similarities to shipping of LPG and LNG. The supply chain would consist of several steps. CO_2 would first have to be liquefied and stored in tanks before being loaded onto ships for transport. Destinations may be other ports or offshore storage sites. Unloading onshore would be relatively straightforward, based on experience with current CO_2 shipping operations and from large-scale shipping of other gases, such as LPG and LNG. Several studies have investigated the potential to repurpose existing LPG and ethylene carriers for CO_2 transport, but this would face significant challenges for ships that were not originally designed for CO_2 transport. Conditioning and injection, or direct injection to the storage site after conditioning on a ship, are not yet proven and the processes are less well understood. Shipping CO_2 by sea may be viable for regional CCUS clusters. In some instances, shipping can compete with pipelines on cost, especially for long-distance transport, which might be needed for countries with limited domestic storage resources. The share of capital in total costs is higher for pipelines than for ships, so shipping can be the cheapest option for long-distance transport of small volumes of CO_2 (up to around 2 Mt/year). This would be the case with several early industrial CCUS clusters across Europe.

Regarding the temperatures, the upper temperature limit is determined by the compressor station discharge temperature and the temperature limits of the external pipeline coating material, while the lower limit is determined by the winter ground temperature of the surrounding soil. The specification of water in currently operating pipelines ranges between 640 ppmv and 20 ppmv to avoid the formation of free water in the pipeline at the operating conditions. However, while it is known that the presence of impurities will affect the solubility of water in CO_2, there has been little research into the absolute effects of these impurities, and the published data are limited. Generally, lower discharge temperatures are desirable for transporting CO_2, particularly for supercritical CO_2, as this can greatly increase the density and reduce the size of transmission pipelines. Cooling the CO_2 at the compressor discharge to near or below ambient temperatures, using ambient air or cooling water to cool the suction temperature, results in large power savings in a CO_2 compression train. Ambient cooling, whenever it is available, is always preferable and more efficient than mechanical refrigeration.

Scenario compositions containing CO and H_2S are at risk from forms of stress corrosion cracking in the presence of water, and therefore, material specification would have to be carefully considered for these scenarios. To transport the scenario compositions studied in this report by ship, high-pressure and low-temperature conditions are required to maintain the fluid in its liquid phase. This renders these compositions uneconomical for transportation in the cryogenic liquid phase. It is feasible to transport high-purity CO_2 streams by ship where the bubble point line remains sufficiently close to that of pure CO_2. A suitable pressure and temperature combination for CO_2 streams with a very high purity is 0.6 MPa and −57°C. An equivalence to smaller class, Type C LNG ships, shows that the tank operating pressures and temperatures are within existing ship design scope. Increasing the tank pressure moves the scenarios into the liquid phase. However, most of the scenarios require an unfeasibly large pressure for ship transportation in the liquid phase.

The saturation pressure is dependent on the composition of the CO_2 mixture and also on the operating conditions of the pipeline. In general, impurities with lower critical temperatures than CO_2 will raise the saturation pressure, and those with higher critical temperatures than CO_2 will lower the saturation pressure. It has been shown that H_2 in particular has the most potent effect, in terms of % mol addition, in raising the saturation pressure, and therefore has the most detrimental effect on fracture control for CO_2 pipelines.

The amount of CO_2 transported through pipeline is highest in the supercritical phase as a result of its high density in this phase in comparison with other phases. Furthermore, transport of CO_2 in the supercritical phase is regarded as the most cost-effective method of transport from the CO_2 capture point to the point of its utilization or storage through pipeline. The amount of CO_2 transported per unit volume is maximized in this phase, because the supercritical fluid possesses the density of a liquid and the viscosity of a gas. However, for the captured CO_2 to be transported in the supercritical phase, it has to be compressed to a pressure that is higher than the critical pressure to prevent two-phase flow in the CO_2 transport pipeline. The condition under which CO_2 is transported to the storage site is primarily dependent upon the availability of the means of CO_2 transport, such as a ship, truck, or pipeline. Some are of the opinion that the amount of CO_2 to be transported along with the distance between the CO_2 capture facility and storage site should be considered to determine the most economically feasible mode of transport. The presence and type of impurities influence the properties of the CO_2 fluid. The power requirement for compression of a CO_2 stream with impurities is higher than that for pure CO_2. This is a result of an increase in the critical pressure of the mixture with an increase in the impurities content. In the same vein, it is believed that if the CO_2 stream with impurities reaches a two-phase situation along the pipeline, there will be a larger drop in pressure compared with the pure CO_2 stream. Finally, it has been reported that the CO_2 transport pressure ranges between 50 and 100 bar. At the pressures of 50–100 bar, the water solubility limit is restricted from 0.3×10^{-2} to 0.4×10^{-2} (mole basis). Commenting on the issue of free water condensation, some have stated that before the transportation of CO_2 through a pipeline, efforts should be made to purify, dehydrate, and compress it to a supercritical pressure of 145 bar. In summary, there has been considerable work carried out on the effect of each impurity on both critical point and pipeline repressurization distances. Most research on the effect of impurities on the thermodynamics of transported CO_2 is largely based on mono and binary considerations. For this reason, it is essential to quantify the holistic impacts of CO_2 impurities on transport line performance. This should be conducted at different impurity contents, for example, up to 20%.

The presence of other impurities, such as CH4, N_2, H_2O, and amines in the CO_2 stream, affects the solubility of H_2O [26,27]. Similarly, Yang [28] noted a considerable reduction in water solubility in the liquid phase when 5% CH_4 was added. The presence of free water is significant in CO_2 transport, because free water may result in a phase split that, in turn, could trigger hydrate formation and pipe blockage, as well as pipeline corrosion. Moreover, water solubility in CO_2 drops sharply as pressure increases between 50 and 60 bar [29] and then shows a rapid increase with stabilization at 60–80 bar. However, it can be observed that the CO_2 solubility in water increases considerably after the change of CO_2 phase from gaseous to liquid ingress (Fig. 8.7). It is essential to understand the difference in the impurities content among different phases during pressure drop, especially when free water is readily available.

In summary, the future CO_2 transport pipeline will require intermingling of CO_2 fluids from different sources; monitoring levels of impurity, which may inadvertently lead to corrosion, is important. The following questions need to be addressed:

- What is the best effective procedure to abate most avoidable corrosion costs (should be addressed at the conceptual phase)?
- There is a need for further research to determine the effect of elevation on the fluid properties as a result of pressure drop, how the supercritical nature of the fluid is lost temporarily and how quickly this can recover?

Again, at what height could the supercritical/dense nature fail to converge? Moreover, it is expected that as the CCUS industry grows, more power plants and industrial operators will connect to an already installed trunk pipeline. This has an obvious economic advantage over point-to-point operations as shown in some of the demonstration

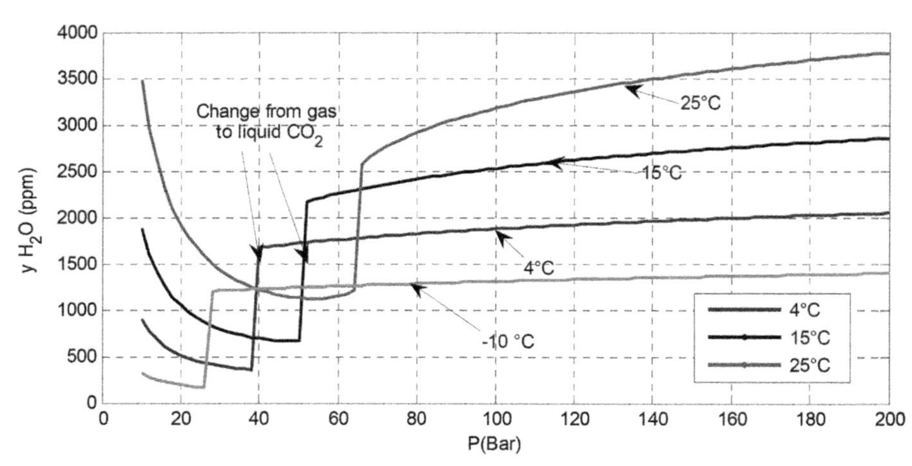

FIG. 8.7 Solubility of water in pure CO_2 as a function of temperature and pressure [14].

projects. However, work is needed to develop a method of determining the optimum pipe diameter to avoid over-specification of pipe size in anticipation of future growth in a region.

Energy losses result from the existence of impurities, which affect the thermodynamics of the CO_2 phase. In an event following transport, depressurization, or fracture formation, involving rapid cooling, understanding the heat transfer characteristics of the CO_2 transport pipeline is crucial. It is essential to accurately understand and represent the correlation between the physical properties of the CO_2 stream, such as temperature and pressure, expressed in terms of other physical-dependent properties including density, viscosity, and thermal conductivity. This is because there is a considerable phase difference between CO_2 and other similar fluids such as natural gas transported through the pipeline. A direct link exists between the energy requirements and the operating pressure when considering supercritical fluid flow in CO_2 pipeline transport. It has been shown that the four major components of pressure drop, which include friction, acceleration, local, and gravitational, can be distinguished. In the pipeline transport of CO_2 in the supercritical phase, it is essential that the operating temperature is maintained at a desired level. If necessary, heaters and insulation need to be applied at some locations of the CO_2 transport pipeline to prevent hydrate formation. Loss of energy in the CO_2 transport pipeline can be analyzed by estimating the amount of heat transferred to the environment that is proportional to the heat transfer coefficient and the temperature difference between the pipe wall and the surrounding environment. Furthermore, in the CO_2 transport pipeline, energy analysis should involve heat loss to the pipeline surroundings, depressurization resulting from an accidental discharge, and planned maintenance. The energy drop along the pipeline is proportional to the length of the pipeline, though other factors such as the nature of the pipeline material, ambient temperature, and insulation, where applicable, need to be taken into account. Importantly, on an increase in the ambient temperature, the density of CO_2 reduces, causing an increase in velocity of the fluid flow. As a result, a pressure drop occurs. The implication of this is that further pressure drop results in higher operating costs. Determination of the maximum safe CO_2 pipeline distances to subsequent booster stations as a function of inlet pressure, environmental temperature, and ground heat transfer rate can be carried out by commercially available energy analyses.

Available CO_2 storage or utilization is a prerequisite for the production of emissions-free blue H_2. It is also a prerequisite for CCUS from power plants. CO_2 can be stored onshore or offshore. The Texas Gulf area anchors the world's leading H_2 production systems (1/3 of US requirements per year). Thus, coupling current gray H_2 production with CCUS, the CO_2 footprint in this area can be reduced. The eastern portion of the US Gulf Coast H_2 system overlays existing CCUS infrastructure—the Denbury system, which was developed to bring CO_2 to recover oil reserves though EOR. Several large SMR facilities are proximate to the Denbury system and could be linked to it through pipeline to initiate the move from gray to blue H_2. Over time, a CCUS system could be expanded, tapping into the additional active and depleted reservoirs throughout the US Gulf Coast both onshore and offshore. Adding new blue H_2 capacity also makes a lot of sense for the Gulf Coast area as companies such as Air Products have already announced blue H_2 projects with CCUS.

Compression

CO_2 compression to the dense phase (>74 bar) is required for storage, thus making the best use of the available void space in the storage formation. For pipeline transport, compressing to dense phase also reduces the cost of the

CO_2 pipeline compared with gas-phase transport. CO_2 compression cost has two main components: capital cost of the equipment and energy cost to drive the compressor. Compression energy costs scale linearly with flow—doubling the flow will double the compression energy cost, all else being equal. As such, there is no cost advantage to increasing the scale for energy cost. Compression capital costs do experience economies of scale, up to a point. Commercially available CO_2 compression systems are available up to a maximum power rating in the order of 40 MW. For CO_2 flows requiring more power than this (over around 3 Mtpa of CO_2), multiple compression trains will be required. At this point, the economies of scale will have been exploited. As compression systems rated over 40 MW become available, it may be possible to extend the economies of scale to somewhat higher flowrates. Although there continue to be incremental improvements in compression technology—mainly aimed at increasing efficiency and reliability—significant improvements in CO_2 compression costs are not anticipated due to the maturity of compressor technology.

Incentive

The most widely known regulatory support of underground CO_2 injection is Section 45Q of the United States Internal Revenue Code [30] established in 2008 and updated following the Inflation Reduction Action of 2022 [31]. The policy is designed to incentivize carbon capture, utilization, and storage (CCUS), and a variety of project types are eligible. The scope/design, performance, and monitoring requirements for qualifying for these and other credits can be technically complex. As of the writing of this text, the 2022 expansion of the 45Q credits are currently established as:

- \$85/metric ton of CO_2 permanently stored
- \$60/metric ton of CO_2 used for enhanced oil recovery (EOR) or other industrial uses of CO_2, provided that net emissions reductions can be clearly demonstrated
- \$180/metric ton of CO_2 generated by direct air capture (DAC) permanently stored
- \$130/metric ton of CO_2 generated by direct air capture (DAC) permanently stored

Financial supports such as tax credits and other subsidies generally require a minimum scale of project to be eligible. In the case of 45Q credits, the following minimum scales apply:

- 18,750 metric tons per year for thermal power plants, provided that at least 75% of the CO_2 produced is designed to be captured
- 12,000 t per year for other emitting facilities such as a steel plant
- 1000 t per year for DAC facilities

Specific to the 45Q program, the 2022 changes in US Federal Law required that projects have until January 2033 to begin construction to qualify.

In addition, 45Q more than triples incentive for direct air capture facilities. Credits were the same as point-source capture prior to Inflation Reduction ACT (IRA), but the bill increases them to US\$130/ton if utilized and US\$180/ton if stored. The amount of capture to be eligible also decreases from 100,000 tons to just 1000 of CO_2 per year. While DAC projects are still costing north of US\$400/ton today, leading developers are setting the long-term cost of capture goals for DAC at ~US\$100–130/ton. Coupled with other revenue channels such as the California Low Carbon Fuel Standard [32], carbon removal credits in the voluntary carbon market, as well as a rapidly growing sustainable aviation fuel space for CO_2 utilization, we believe the DAC industry in the United States is particularly well positioned to scale quickly in the coming years. The Global CCS Institute [33] had developed a country readiness index based on its assessment of supportive policy, legal and regulatory framework, and storage resources. These factors are plotted against a nation's potential storage indicator. Based on that analysis, the United States is among the countries most ready for commercial carbon capture deployment and is by far the most advantaged given its storage potential. According to the International Energy Agency (IEA) [34], the majority of stationary emission sources in the United States are located close to potential geological storage sites; 85% of emissions come from plants located within 100 km of a site and 80% within 50 km. To put these distances into context, the average distance over which CO_2 is currently transported by pipeline between existing CCUS facilities is around 180 km, and the maximum is around 375 km.

The upgraded 45Q tax incentives could substantially accelerate carbon capture investments in the United States due to (1) higher price of incentive; and (2) lower volume threshold under which a facility can be eligible for the credit. We note that the higher credit level would make carbon capture and storage a viable decarbonization solution for a large number of industries with close to 350 million tons of annual CO_2 emissions in the United States. The comprehensive climate incentives included in the IRA will likely catalyze development of carbon and hydrogen hubs in the United States that could potentially leapfrog those in progress in Europe. While the concept of a hydrogen hub or a carbon capture hub is already gaining momentum in the United States, a large-scale, low-carbon industrial hub will likely include three key attributes: zero-carbon electricity, clean hydrogen production, and carbon capture, storage, and utilization. To be competitive, hubs should be situated near low-cost, clean electricity resources, advantageous geologic storage (such as deep saline formation for CO_2 storage and salt caverns for hydrogen storage), and expandable infrastructure (such as pipelines, docks, distribution systems).

From a transition perspective, development of low-carbon industrial hubs will be the most effective pathway to achieve large-scale and rapid decarbonization, particularly for hard-to-abate sectors. Industrial clusters of companies/facilities (aka a hub) are poised to take the most advantage of government incentives as the group can also share the investment costs associated with the necessary transport infrastructure, thereby supporting economies of scale and further reducing unit costs/risks. Companies that are early champions and/or anchors in such hub development would also likely get first-mover advantage in reducing their climate risk exposure and capturing new market opportunities. There are already 22 hydrogen hub proposals under development plus several carbon capture coalitions being formed in the United States. This was in response to the passage in 2022 of the Infrastructure Investment and Jobs Act [31], which allocates an unprecedented >US$9 billion and >US$12 billion of investment toward CCUS and hydrogen projects, respectively. These investments will essentially work in conjunction with various incentives in the IRA—such as the US$5.8 billion Advanced Industrial Facilities Deployment Program, the credit for production of clean hydrogen, and the 45Q credit for CCUS—to accelerate adoption of CCUS and production of clean hydrogen in the United States. Given the attractiveness of economics discussed earlier, we believe projects could move forward rather quickly on the back of the IRA incentives.

- Carbon capture allocated spending: US$3.5bn
- Direct air capture hubs (aim four regional hubs): US$3.5bn
- Carbon capture pilot and demonstration program: US$2.5bn
- CO_2 storage commercialization program: US$2.1bn
- Carbon capture transportation infrastructure program: US$0.3bn
- Carbon utilization and procurement grant program: US$0.12bn
- Direct air capture prize: US$0.1bn
- Carbon capture tech program: US$0.05bn
- Funding for Class VI well permits at EPA and States 45Q CCUS tax credit: CBO cost estimate US$3.2bn
- Point source capture: US$60/ton if utilized; US$85/ton if stored
- Direct air capture: US$130/ton if utilized; US$180/ton if stored; available for 12 years of operation

The gulf coast and midwest regions of North America will likely dominate initial carbon capture developments. The Great Plains Institute identified eight regions where highly concentrated industrial and power generating facilities coincide with opportunities for permanent geologic carbon storage. The midwest, the Illinois Basin, and the Gulf Coast (mostly Houston and southern Louisiana) have the highest concentration of industrial and power facilities and thus aggregate emissions from these facilities. Moreover, the majority of CCUS projects in planning are located along the Gulf Coast. There are also two major pipeline projects in discussion in the midwest by developers Summit Carbon Solutions and Navigator in partnership with landowners, local, and industry partners. While more specific details on the requisite criteria for eligible projects to receive government funding should emerge with the application opening process in 4Q22, the Department of Energy (DOE) will likely favor those that have the potential to form a regional hub. In fact, respondents to the DOE's Request for Information (RFI) proposal identified potential host site regions for the point-source carbon capture demonstrations that were almost spot on with potential carbon and hydrogen hub sites identified by the GPI.

Usage

CO_2 utilization or "carbon recycling" CO_2 can be used as an input to a range of products and services. The potential applications for CO_2 include direct use, where the CO_2 is not chemically altered (nonconversion) and the transformation of CO_2 to a useful product (conversion). Today, around 230 Mt of CO_2 is used globally each year. The largest consumer is the fertilizer industry, which uses 125 Mt/year of CO_2 as a raw material in urea manufacturing, followed by the oil and gas industry, which consumes around 70–80 Mt per year for EOR. Other commercial uses of CO_2 include food and beverage production, cooling, water treatment, and greenhouses, where it is used to stimulate plant growth. New CO_2 use pathways, involving chemical and biological technologies, offer opportunities for future CO_2 use. Many of these pathways are still in an early stage of development, but early opportunities are already being realized. There are three main categories of CO_2-based products: fuels, chemicals, and building products.

Fuels

The carbon in CO_2 can be used to convert hydrogen into a synthetic hydrocarbon fuel that is as easy to handle and use as a gaseous or liquid fossil fuel. The production of such fuels is highly energy-intensive and is most economically viable where both low-cost renewable energy and CO_2 are available. The largest plant currently in operation is the George Olah facility in Iceland, which converts around 5500 tons of CO_2 per year into methanol using hydrogen produced from renewable electricity [35].

Chemicals

The carbon in CO_2 can be used as an alternative to fossil fuels in the production of chemicals that require carbon to provide their structure and properties. These include polymers and primary chemicals such as ethylene and methanol, which are building blocks to produce an array of end-use chemicals. The need for hydrogen and energy varies significantly according to the chemical and production pathway. An example of a company active in the field is Covestro, which is operating a facility to produce around 5500 tons of polymers per year in Dormagen, Germany. CO_2 substitutes up to 20% of the fossil feedstock normally used in the process.

Building materials

CO_2 can be used in the production of building materials to replace water in concrete, called CO_2 curing, or as a raw material in its constituents (cement and construction aggregates). The CO_2 is reacted with minerals or waste streams, such as iron slag, to form carbonates, the form of carbon that makes up concrete. This conversion pathway is typically less energy-intensive than for fuels and chemicals and involves permanent storage of CO_2 in the materials. Some CO_2-based building materials can have superior performance compared with their conventional counterparts. A few applications, such as the use of CO_2 in concrete mixing, are already commercially available in some markets today. Two North American companies, CarbonCure and Solidia, are leading the commercialization of CO_2-curing technology, with CarbonCure now operating some 175 facilities in the United States and Canada. The British company Carbon8 is among the companies using CO_2 to convert waste materials into aggregates as a component of building materials. It is currently operating two commercial plants and aims to have five to six plants in operation eventually.

The prospects for CO_2-based products are very difficult to assess, as the technologies are generally at an early stage of development for many applications. Policy support will be crucial since they are likely to cost a lot more than conventional and alternative low-carbon products, mainly because of their high-energy intensity. The market for CO_2-based products is expected to remain small in the short term, but could grow rapidly in the longer term. A high-level screening of the theoretical potential for CO_2 use shows that it could reach as much as 5 $GtCO_2$/year for chemicals and building materials and even more for synthetic hydrocarbon fuels. However, in practice, these levels are unlikely to be attainable, mostly for economic reasons. Synthetic hydrocarbon fuels are unlikely to be able to compete with direct use of low-carbon hydrogen or electricity in most applications, but could become important in sectors that continue to need hydrocarbon fuels as the energy sector approaches net-zero emissions and where other fuel alternatives are limited, such as aviation.

For example, it is projected that synthetic kerosene could meet around 40% (250 Mton) of aviation energy demand in 2070, requiring around 830 Mt of CO_2. Large-scale deployment of CO_2-based chemicals and fuels would involve large amounts of renewable electricity for their production, in particular for the generation of low-carbon hydrogen. In the Sustainable Development Scenario, the production of synthetic aviation fuels alone in 2070 requires around 5500 TWh, which is around 8% of all the electricity produced worldwide in 2070. The extent to which the capture and utilization of CO_2 contribute to reducing emissions varies considerably depending on the origin of the CO_2 and the way the CO_2 is used.

EOR

Enhanced oil recovery (EOR) is a process for extracting oil that has not already been retrieved through conventional oil recovery techniques. One common technique is when compressed gases are forcefully injected into the well in a way that both forces the oil to the surface and reduces its viscosity. The less viscous the oil, the easier it flows and the more cheaply it can be extracted. Although various gases can be used in this process, CO_2 is used most often. With the newer practices of CO_2 capture and transport, the use of CO_2 for enhanced oil recovery could be a growth market. As regulations may force many industrial CO_2 producers to capture CO_2, the oil producers have a use for CO_2 to enhance oil recovery, whereby the oil producers may profit or charge industrial producers to dispose of the CO_2 in the oil fields or in active or abandoned oil reservoirs. Centrifugal compressors are used throughout the CO_2 capture, transportation, or injection processes [36].

Methods of transport

Pipeline

CO_2 is a relatively heavy gas with a low speed of sound, a low specific heat capacity, and a critical pressure and a critical temperature in a range of typical compression applications. The combination of these features leads to the fact that turbo compressors achieve high pressure ratios and a high volume reduction per stage, but will often operate close to the speed of sound. For high overall pressure ratios, intercooling is essential to reduce power consumption and to keep operating temperatures within customary limits. The high volume reduction per stage leads to challenges with matching impellers of subsequent stages and makes it advantageous to use designs with gearboxes to increase the speed for higher-pressure stages.

Operating conditions for pipelines and for sequestration can be in the range of supercritical or dense-phase regions for CO_2. These regions are characterized by high fluid density, while CO_2 still has the characteristics of a gas (for example, it is compressible, has a low viscosity, and fills available space), it has a density that resembles liquids. Supercritical fluids do not show any liquid-vapor phase change when the temperature is reduced. These regions are characterized by very strong sensitivity of the density-to-temperature changes. This is a challenge for any compressor [37].

The thermodynamic concept of compressing or pumping CO_2 has to be considered for the different phases of CO_2 as a gas, CO_2 as a liquid, CO_2 in the two-phase region, supercritical CO_2, and dense-phase CO_2 [38]. Because the critical pressure (73.9 bar) and temperature (31.1°C) of CO_2 are in a range that is accessible for compression duties and achievable for pipeline transport, the operation in supercritical and dense phases must be discussed. Furthermore, CO_2 in the supercritical or dense phase has a density similar to natural gas liquids (600–900 kg/m^3), while still behaving like a gas (that it is compressible and fills available space). It also has a low viscosity under these conditions, which makes the pipeline transport in the supercritical phase attractive.

IPCC [38] indicates that prospective areas for CO_2 sequestration do not line up with large sources of CO_2 production. Further, in many areas, the sequestration sites may be offshore. This indicates the need for CO_2 transportation solutions. While CO_2 can be transported in pipelines over a wide range of pressures, transport in the dense-phase region has a number of advantages. While CO_2 is still compressible and behaves like a gas in terms of viscosity, its density is close to that of a liquid. Thus, transport in dense phase is very energy-efficient and certainly the preferred mode for longer distances [37].

While in natural gas pipelines, elevation changes are only responsible for secondary effects, but in dense-phase CO_2, elevation changes lead to significant changes in gas pressure. Therefore, the pipeline has to be sized such that under all operating conditions and operating temperatures, the CO_2 stays about the critical pressure and does not cross into the two-phase region. Also, rapid density changes, as a result of temperature changes, have to be considered when sizing equipment. The properties of CO_2 in dense phase raises the question whether pumps or compressors can be used in this application (Fig. 8.8). CO_2 in dense phase is compressible, and there are no phase changes. The duty can thus be handled efficiently by both compressors and pumps [39,40].

Calculations, using commercially available pipeline simulation software, show that for a 30 in (760 mm) diameter pipeline flowing 1300 MMSCFD of CO_2 modeled after an existing CO_2 pipeline [41], there is a pressure drop from 152 bar to 90 bar after a distance between stations of 210 km (130 miles). No elevation changes were simulated. The power consumed for recompression from 90 bar to 152 bar is 8550 kW (11,500 hp) with a single-stage centrifugal compressor running at low tip speed (machine Mach number 0.52) at an assumed but realistic polytropic efficiency of 86%. The pressure ratio is higher than what would be used in an optimized natural gas pipeline.

Transporting CO_2 in dense-phase conditions, Fig. 8.8, has a number of advantages since it combines a high density with a viscosity that is in the same range as a gas, rather than that of a liquid. Supercritical CO_2 is compressible, so increased pressure leads to increased density, and like a gas, it fills any available volume. Unlike gas in the subcritical range, there is no phase change when it is cooled. Existing pipelines operate at pressures between the critical point and 200 bar, typically between 85 and 150 bar to maintain single-phase operations in the dense-phase region. The upper pressure limit is based on the mean allowable operating pressure and maximum allowable pressure as determined by the physical pipeline design. Pressure will vary along the length of the pipeline due to viscous, or pipe friction losses, change in elevation, and thermodynamic effects associated with changes in temperature. Pressure changes due to environmental thermal input and elevation profile can be significant for CO_2 pipelines. In addition to thermal flux from the environment, isenthalpic effects associated with a sudden pressure drop, such as across a throttling valve, can cause localized temperature changes. Care must be taken to open valves slowly to minimize local thermal gradients, which can result in phase changes and transients within the piping system. If a phase change were to occur at a valve, significant damage could result from the high gas flow velocities associated with the liquid-to-gas phase

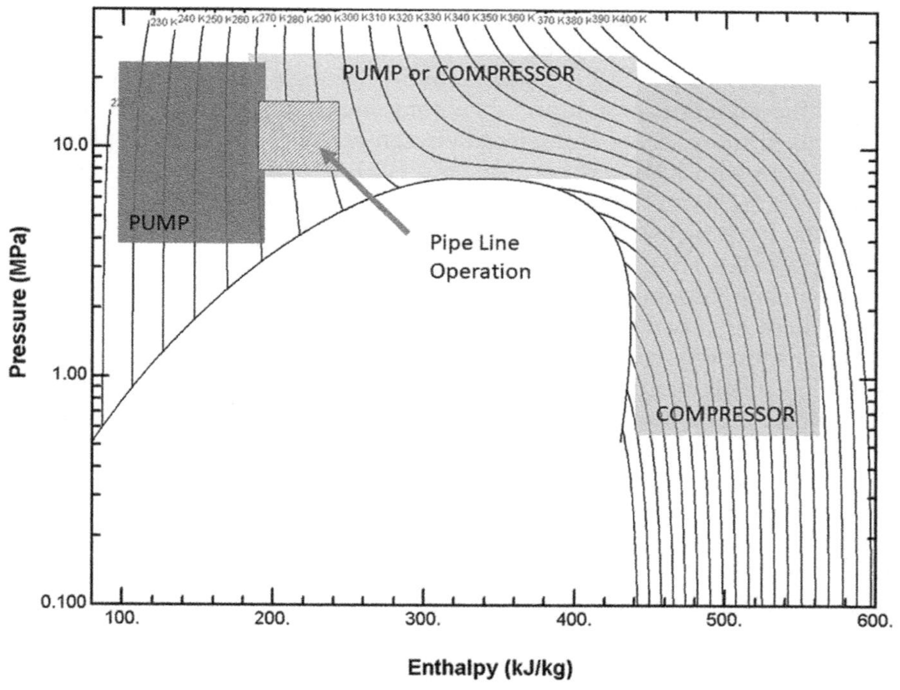

FIG. 8.8 Pipeline transport of CO_2 in dense phase [39].

transition [42]. Due to these features, both pumps and compressors can be used as boosters for pipelines. Compressors would run relatively slow, but significantly faster than a pump. The pressure ratios are low; booster stations will be placed about 150–250 km apart, thus requiring a pressure ratio of about 1.5–1.7.

At elevated pressures above the critical pressure, CO_2 can be transported in a dense-phase state (Fig. 8.8) using pumps or compressors that maintain pressure and temperature above its critical point. This strategy of pumping the CO_2 versus compressing CO_2 in gas-phase compressors is often dictated by the length of the pipeline. For shorter pipelines, compression in the gaseous phase is preferable, because the overall power consumption will be less important, and pressurization above 2200 psi is not required. Thus, the gas-phase compressors operate around typical natural gas pipeline pressures, but the maximum operating pressure is limited by dew point considerations. However, for greater volumes of CO_2, as shown in Fig. 8.9, it is worth the cost of initially compressing the CO_2 to the dense phase/supercritical state as it can then be pumped at overall lower power costs than compression. Fig. 8.9 illustrates this

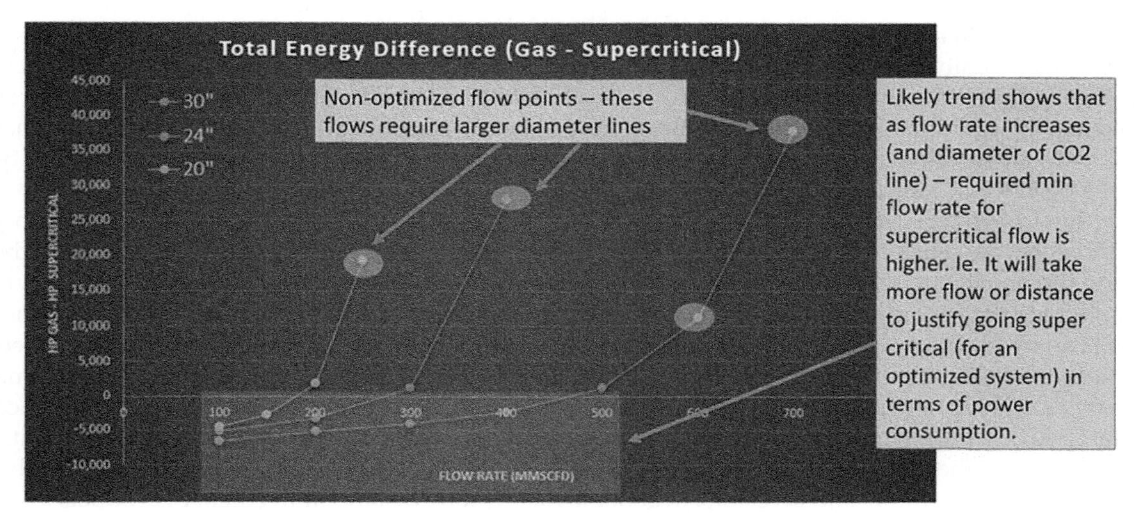

FIG. 8.9 CO_2 transport in gas or supercritical state [41].

breakover point, where the overall power consumption between the supercritical state and gas phase (*y*-axis) will be less for the supercritical pumps, at three diameters of 20″, 24″, and 30″ lines and for flow rates between 100 and 700 mmscfd (*x*-axis). For example, for a 24″ CO_2 pipeline (middle orange line; dim gray in print version), the power spent in compressing up to the supercritical state and then pumping CO_2 is less than the gaseous compression power for flow rates of 300 mmscfd or more. A key difference between dense-phase and gas-phase pipelines is that in gas pipelines, the maximum gas pressure is limited, because liquid CO_2 dropout has to be avoided, while in dense-phase applications, a minimum pressure has to be maintained to avoid liquids (Fig. 8.8).

It should also be noted that the pipeline flows should be optimized to a diameter such that a larger diameter choice will incur less pressure drop and keep the power consumption lower. Once the power consumption difference starts to increase exponentially, a pipeline designer would choose to increase the pipeline diameter to incur a lower power demand in either the gas phase or supercritical state. Finally, temperature effects should also be taken into account as the dense-phase CO_2 transport will be more sensitive to temperature changes, especially for the pipeline route ground temperatures. A CO_2 pipeline optimization study should consider these temperature effects to assure that the pumps can maintain pipeline pressures given the expected fluid temperatures and mixture composition.

Another consideration is that transporting CO_2 in dense phase will not allow the reuse of existing natural gas pipelines, which are typically rated for approximately 1500 psi (105 bar) MAOP. While 1500 psi (105 bar) is still above the critical pressure, operability concerns (for example, the large change in volume for relatively small changes in temperature, Fig. 8.8) may prevent operation in that range. The option then becomes to reduce the operating pressure even further, i.e., to 400–600 psi (28–41 bar) MAOP, because at higher pressures, changes in ambient temperature may lead to the formation of liquid CO_2 (Fig. 8.8). Operating existing pipelines at reduced pressures greatly reduces the flow capacity for a given pipe diameter compared with supercritical transport by a factor of 4 or 5. On the other hand, the compression can be accomplished with conventional pipeline compressors.

The specific energy requirement for CO_2 pipeline transport depends on a number of factors, such as the inlet pressure, impurity content in the CO_2 stream, pipe diameter and length, and heat transfer coefficient. Due to the pressure loss along the pipeline, the compression or pumping stations are required to maintain the CO_2 stream in the supercritical phase. Therefore, both the cost and the energy requirement of the CO_2 transport pipeline are expected to increase for the routes located in a difficult terrain of variable altitude. The total energy requirement for the CO_2 transport pipeline comprises the power requirement to compress the CO_2 stream to the pipeline inlet pressure and the power requirement for recompression of the CO_2 stream to compensate for the pressure losses along the pipeline. The latter is not only influenced by the efficiency of the compressor but also primarily by the temperature of the pipeline environment and the thermal insulation layer, both of which affect the operating conditions of the CO_2 transport pipeline. It has been shown that for a postcombustion CO_2 capture, a 20% reduction in the compression power requirement can be achieved when the CO_2 stream is only compressed to the critical pressure, under which it becomes a supercritical fluid, and then is pumped, as opposed to being further compressed to the desired pipeline inlet pressure. In the same vein, there are different power requirements for refrigerated and nonrefrigerated compression strategies in comparison to isothermal compression, which is assessed to be 30%–40% higher.

Generally, flow assurance is dependent on many factors including the allowable level of impurities in the CO_2 stream, the operating conditions of the CO_2 transport pipeline (pressure and temperature), and the potential for hydrate formation. In a flow assurance assessment, the dynamic or non-steady state is important. This is because by their nature, it is usually difficult to determine the frequency of occurrence of various operating states, such as shutdown and startup. Several sources have described these phenomena including an initial startup, planned shutdown, and planned startup after planned shutdown, and planned startup after nonplanned shutdown emergencies. These sources have developed some understanding on several conditions including temperature, pressure, density, and viscosity, among others that affect the flow assurance of the CO_2 transport pipeline.

Operating the CO_2 transport pipeline under a two-phase condition is not desirable, as this presents a particular difficulty during startup. However, to overcome this difficulty, the CO_2 stream is initially compressed and then recompressed along the pipeline to a higher pressure than the nominal operating pressure. This not only affects the energy requirement but also has an impact on the nominal operation pressure design for the CO_2 transport pipeline. Of equal significance is an operation under a long-lasting shutdown and cool-down scenario, for example, after weeks of low mass flow rate, increasing the flow rate becomes essential for a subsequent startup procedure. As mentioned earlier, recompression distance is dependent on the impurity content, as well as the pipeline diameter. If the presence of impurities is large, the CO_2 transport pipeline will need to be operated at a higher pressure to sustain the supercritical phase.

A cost estimation of the CO_2 transport pipeline projects is important, because this determines the feasibility of the project for the potential operators and investors. In general, for any long distance movement of products to occur, there must be an overwhelming economic incentive based on the demand, similar to hydrocarbon production and the transport chain. This can also be applied to the transport of CO_2 through pipelines. However, the value of CO_2 is given on the basis of both environmental and societal needs for it to be stored, rather than the monetary value of CO_2 itself. Furthermore, several sources claimed that the economics of scale are required to reduce the cost of CO_2 transport through single, large-capacity pipelines. This is important as it has been estimated that the CO_2 transport pipeline constitutes about 21% of the overall cost of a full-chain CCUS project. The cost of a CO_2 transport pipeline varies from one project to another and depends on the amount of CO_2 to be transported, the diameter and length, and the material of the pipeline. Other important factors that affect the cost of CO_2 transport are labor cost and expected system lifetime.

While CO_2 is considered to be a nontoxic—and therefore relatively safe—gas, there are certain hazards and risks that need to be considered in the transportation of CO_2 by pipeline, as well as through other forms of transport. The whole idea of limiting CO_2 emissions is to transport it in a fashion that eliminates leaks as much as possible. Therefore, attention should be paid to the appropriate design of the joints between the pipes, valves, compressors/pumps, and other equipment being used in the transportation. Leaks do not only add to the greenhouse gas load of the operation, but they can also be hazardous to personnel and other equipment in the vicinity due to hot or cold impingement. As CO_2 is heavier than air, it can also collect in enclosed areas whether they have a roof or not, displacing the air in the area and potentially causing suffocation.

Attention should also be paid to pressure drops in the system due to the proximity to the saturation line of CO_2 when operating supercritically. An excessive pressure drop can reduce the temperature very quickly, dropping the operating condition into the vapor dome and producing liquids. These could be detrimental to the performance of the equipment at best, but could be the cause of erosion, shock loading of components, and other damage-inducing effects. In extreme cases, the temperature could reduce so much that the CO_2 turns solid, causing blockage of the flow or severe damage to components due to impact. Refer to the following figures showing the temperature margin between the isenthalpic expansion path for CO_2, air, and methane relative to their saturation lines when starting from 150 bar and 30°C (Fig. 8.10) and 60°C (Fig. 8.11). CO_2 clearly drops into the vapor dome in both cases, while the other two gases stay safely above their respective saturation lines.

Marine transport

Most major CO_2 generation sites and potential storage locations are not colocated. This will result in a growing need for all types of CO_2 transport, including marine shipping. The current fleet of CO_2 carriers is small and uses medium-pressure tanks (15 bar at −28°C), used primarily for food-grade CO_2 transport. Some of these smaller vessels, usually with capacities between 800 and 1000 m^3, are able to carry both CO_2 and LPG. As the demand for CCUS increases, marine transport infrastructure will need to migrate to larger vessels to transport large volumes of CO_2 to permanent storage sites, such as depleted oil fields. One option to increase shipping capacity is by utilizing high-pressure tanks (35–45 bar at 0–10°C). It is anticipated that transporting CO_2 at a higher pressure and temperature will reduce the overall shipping costs by reducing the energy required to cool the gas for loading and reheat the gas at unloading. At the same time, low-pressure transport systems are also being explored, which would utilize tanks at 7 bar and −49°C. The lower tank pressure allows for much larger tank volumes at lower cost relative to the medium- and high-pressure

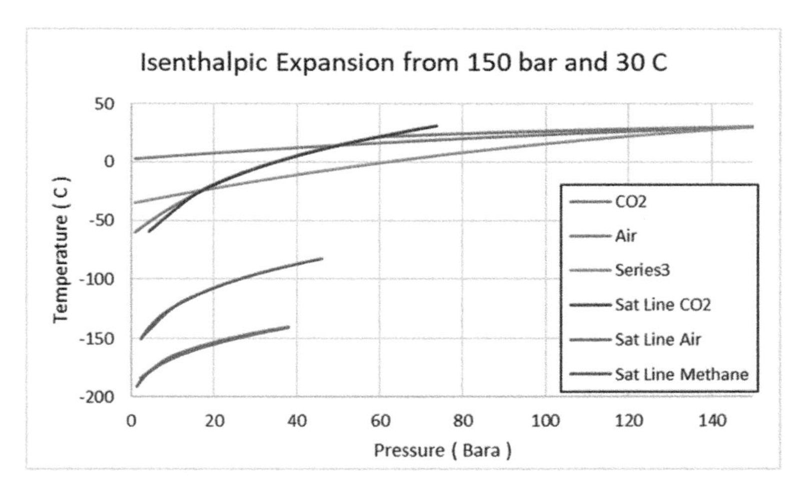

FIG. 8.10 CO_2 expansion from 150 bar and 30°C.

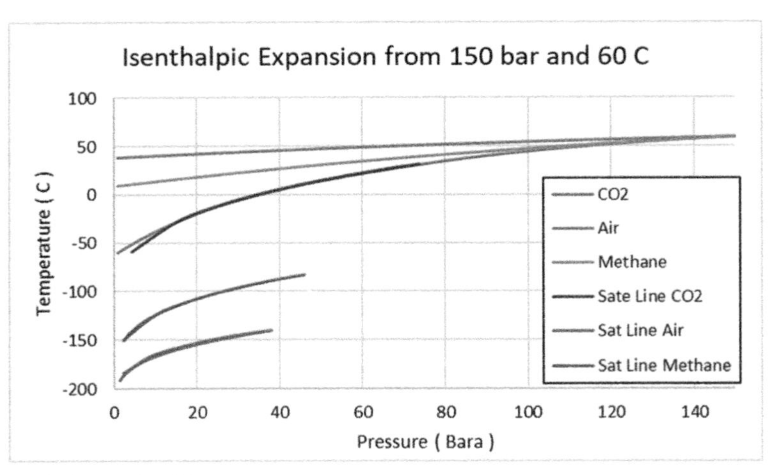

FIG. 8.11 CO_2 expansion from 150 bar and 60°C.

options. However, low-pressure shipping would require more energy in the liquefaction and cooling of the CO_2 for this application. The final selection of the optimal shipping pressure will depend on the capabilities of the tanks. Conventional gas carriers use large cylindrical tanks, which is an option for CO_2 tankers as long as the pressure stays low. Another option (which just received DNV approval) is a ship design with vertically stacked, small-diameter cylinders [43]. Smaller tanks allow for higher-pressure transport. This configuration has the added advantage, over larger tanks, of reduced liquid CO_2 sloshing.

It is expected that the ship designs and technology will evolve to meet the demands of the growing market. The codes and standards of safety and design are also evolving to match the changes in the industry. DNV Rules Part 5, Chapter 7 have been updated to include a class for "Tankers for CO_2." Additionally, CO_2 carriers must also comply with the existing regulation of the International Code of the Construction and Equipment of Ships Carrying Liquefied Gases in Bulk (IGC Code) [44].

Marine transport of CO_2 requires a similar infrastructure to what is currently used for natural gas transport. This would include:

1. CO_2 liquefaction
2. Intermediate storage
3. CO_2 loading facilities
4. CO_2 transport ship
5. CO_2 receiving facility

The CO_2 shipping chain is shown in Fig. 8.12. The first step is liquefaction of the CO_2 supply to increase the density for loading. The liquefaction process is energy-intensive and should be located near adequate power resources required to drive the compressors in the liquefaction process. Depending on the type of ship, the CO_2 will be liquefied and cooled to a temperature of between −50°C and 0°C. Next, it will be transferred to local storage tanks until it can be loaded. To load the vessel, the shipping facility will need liquid CO_2 pumps and a boil-off gas management system. Additional on-shore facilities/infrastructure typically includes controls, safety systems, and environmental protection

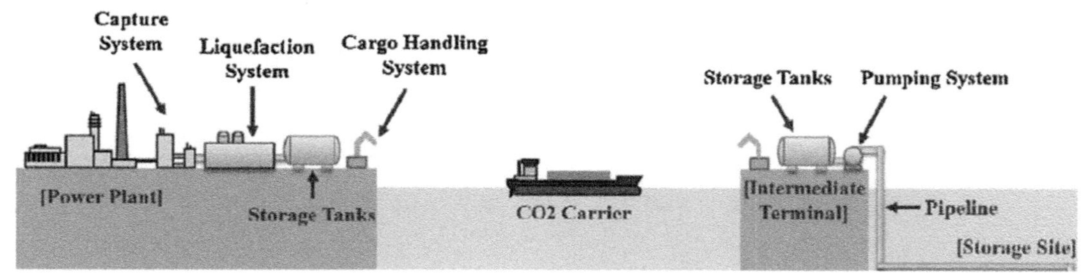

FIG. 8.12 Carbon dioxide shipping chain [45].

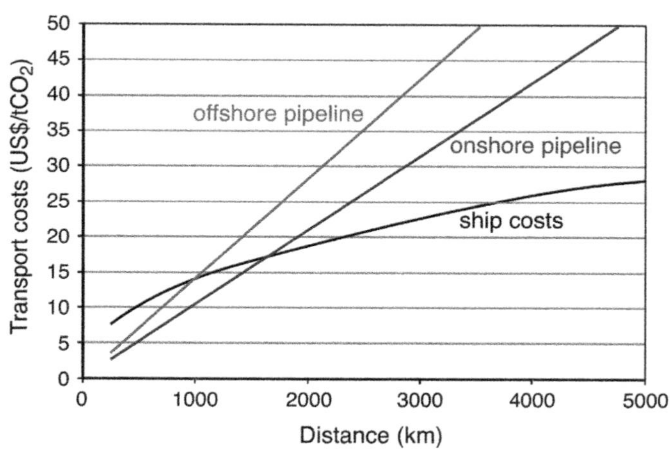

FIG. 8.13 Costs of marine shipping of CO_2 vs. pipelines [38].

measures, loading arms, or hoses. After transport, the carriers are unloaded at a receiving terminal. A receiving terminal requires a similar infrastructure as at the loading terminal to ensure safe handling of the fluid.

The choice of transportation method is determined by the geography between shipping points, the amount of fluid to be moved, and the expected future demand. For large volumes, either marine shipping or pipelines are the most cost-effective method of transport. Fig. 8.13 shows that for distances under ~1000 km, offshore pipeline transport is more cost-effective than marine shipping. For greater distances, the shipping is expected to have lower cost.

Rail and road

Transporting liquefied CO_2 by truck or rail can be an option when a pipeline infrastructure does not exist or is too costly to construct. Trucks are able to transport up to 25 tons of CO_2, and rail cars can each transport up to 80 tons at low pressure [46]. The basic process for truck and rail transport is similar to marine and involves the following basic steps:

1. Liquefaction
2. Buffer Storage
3. Transport
4. Transfer to Receiving Site Storage

The energy penalty of transporting through road and rail comes from the liquefaction stage and pressurization to supercritical pressures at the permanent storage location. This is in contrast to pipelines, which typically already operate at supercritical conditions. Trade studies are currently being conducted for vessels or tube trailers that are rated to higher pressures for truck or rail to transport CO_2 under supercritical conditions to eliminate boil-off, which can be up to 10% of the mass depending on the distance and environmental conditions. Additionally, the use of lightweight, fiber-reinforced polymer storage tanks is being explored to replace metallic materials in these transportation modes. It is currently estimated that the cost of transporting CO_2 through truck exceeds the tax credit of $85 per ton of permanently stored CO_2 when it is transported further than 150 miles and if you transport more than 100,000 tons per year. Transporting through rail is currently estimated to cost at least $125 per ton (baseline) and increases moderately to $150 per ton for distances up to 1000 miles for transporting more than 100,000 tons per year, Fig. 8.14.

Fig. 8.15 compares the primary methods of transporting CO_2. Pipelines and ships provide the greatest capacity and are best suited to support moving large volumes of CO_2 for CCUS. Rail capacities are much smaller, and road transport provides the smallest capacity and is best suited for meeting small-scale industrial needs and local food and beverage demands.

Compression/pumping

Compression

CO_2 is a relatively heavy gas with a low speed of sound, a low specific heat capacity, and a critical pressure and critical temperature in a range of typical compression applications. The combination of these features leads to the fact

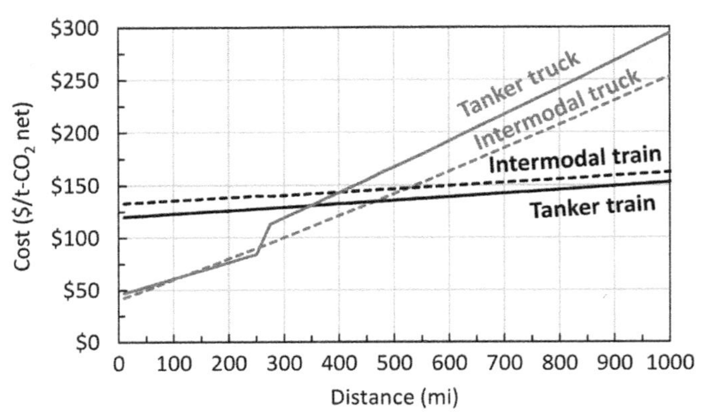

FIG. 8.14 Cost for shipping 100 kt-CO_2/Year [46].

Method	Pressure	Temperature	Phase	Capacity	Remarks
Pipelines	4.8–20 Mpa	283–307 K	Vapour, dense phase	~100 Mt CO_2/year	• Higher capital costs, lower operating costs • Low-pressure pipeline system is 20% more expensive than dense phase transmission. • Well-established for EOR USE.
Ships	0.65–4.5 Mpa	221–283 K	Liquid	>70 Mt CO_2/year	• Higher operating costs, lower capital costs • Currently applied in food and brewery industry for smaller quantities and different conditions. • Enhanced sink-source matching
Motor Carriers	1.7–2 Mpa	243–253 K	Liquid	>1 Mt CO_2/year	• 2–30 tonnes per batch • Not economical for large-scale CCUS projects • Boil-off gas emitted 10% of the load
Railway	0.65–2.6 Mpa	223–253 K	Liquid	>3 Mt CO_2/year	• No large-scale systems in place • Loading/unloading and storage infrastructure required • Only feasible with existing rail line • More advantageous over medium and long distances

FIG. 8.15 CO_2 transportation alternatives. *From H. Al Baroudi, A. Awoyomi, K. Patchigolla, K. Jonnalagadda, E.J. Anthony, A review of large-scale CO_2 shipping and marine emissions management for carbon capture, utilisation and storage, Appl. Energy 287 (2021).*

that turbo compressors achieve high-pressure ratios and a high volume reduction per stage, but will often operate close to the speed of sound. For high overall pressure ratios, intercooling is essential to reduce power consumption and to keep operating temperatures within customary limits. The high volume reduction per stage leads to challenges with matching impellers of subsequent stages and makes it advantageous to use designs with gearboxes to increase the speed for higher-pressure stages, thus keeping flow coefficients within optimal ranges for maximized efficiency.

Operating conditions for pipelines and for sequestration can be in the range of supercritical or dense-phase regions for CO_2. These regions are characterized by high fluid density; while CO_2 still has the characteristics of a gas (for example, it is compressible, has a low viscosity, and fills available space), it has a density that resembles that of liquids. Supercritical fluids do not show any liquid-vapor phase change when the temperature is reduced. These regions are characterized by very strong sensitivity of the density to temperature changes. This is a challenge for any compressor [37].

The thermodynamic concept of compressing or pumping CO_2 has to be considered for the different phases of CO_2 as a gas, CO_2 as a liquid, CO_2 in the two-phase region, supercritical CO_2, and dense-phase CO_2 [38]. Because the critical pressure (73.9 bar) and temperature (31.1°C) of CO_2 are in a range that is accessible for compression duties and achievable for pipeline transport, the operation in supercritical and dense phases must be discussed. Furthermore, CO_2 in the supercritical or dense phase has a density similar to natural gas liquids (600 to 900 kg/m^3), while still behaving like a gas (it is compressible and fills available space). It also has a low viscosity under these conditions, which makes the pipeline transport in the supercritical phase attractive.

Pumping

Liquid CO_2 is generally stored at low temperatures (typically −20°C or so) and modest pressures (~20 barg) in heavily insulated containers. Even then, a small amount of CO_2 vaporizes continuously (this is known as boil-off gas) and causes the pressure in the container to increase. The pressure is regulated by blowing off small amounts of CO_2, or for larger containers, the contents are chilled to maintain an equilibrium pressure inside. Pumping the CO_2 in liquid form is unlikely to be a viable option except over very short distances due to the CO_2 picking up temperature from the medium surrounding the pipe, and boil-off gas will be generated, leading to a progressively greater and unwanted mixed-phase flow. An option to compress flow of the transported CO_2 through a pipeline is to pump it as a dense-phase fluid. In reality, the pump will be quite similar to the supercritical compressor in that there will still be a slight volume change as the pressure increases, and additional caution will be required to ensure off-design conditions remain within the pump's performance characteristic and the CO_2's pressure and temperature vary in operation.

Drivers

As with all machinery that consumes power, the transportation method chosen for the CO_2 will require something to produce the motive power. The choice of driver will come down to the availability of the energy source: combustible fuel for internal combustion engines (ICEs) and electricity for motors. The choice will also depend on the cost of the energy source required and to a certain extent, the power required by the driven equipment being used to push the CO_2 along the pipeline. Electric motors are typically the driver of choice for lower power applications, because the efficiency of combustion engines, particularly of gas turbines, reduces considerably as the power requirement goes below 10 MW. A 10 MW CO_2 pumping application will mean a fairly significant pipeline capacity unless the pumping stations are significantly spaced out leading to higher pressure drops between them. Therefore, it is likely that many CO_2 pipeline boosting station applications will be motor-driven. Electric motors can be single-speed direct drives for typical pumps or with a gearbox to match the required speed of a compressor. A single-speed compressor would likely require variable inlet guide vanes to provide some performance flexibility as flow rates and pressure requirements may vary during the operation of the pipeline. Electric motors can also be variable speed to provide more flexibility of operation and better match the capacity and pressure ratio required at the boosting station.

Sequestration

Purpose

The business case for capturing CO_2 for the purpose of either utilization or storage is often driven or supported by various regulatory and incentive programs depending on geography or jurisdiction. Further, the injection of CO_2 and associated compounds can in most jurisdictions be an environmentally regulated practice. Given that the chief purpose of underground CO_2 injection is the quantifiable and permanent avoidance of atmospheric release of CO_2, incentive and regulatory structures generally focus on these two dimensions of the practice.

Overview

Storing CO_2 involves the injection of captured CO_2 into a deep, underground geological reservoir of porous rock overlaid by an impermeable layer of rocks, which seals the reservoir and prevents the upward migration of CO_2 and escape into the atmosphere. There are several types of reservoirs suitable for CO_2 storage, with deep saline formations and depleted oil and gas reservoirs having the largest capacity. Deep saline formations are layers of porous and permeable rocks saturated with salty water (brine), which are widespread in both onshore and offshore sedimentary basins. Depleted oil and gas reservoirs are porous rock formations that have trapped crude oil or gas for millions of years before being extracted and can similarly trap injected CO_2. When CO_2 is injected into a reservoir, it flows through it filling the pore space. The gas is usually compressed first to increase its density, turning it into a liquid. The reservoir must be at depths greater than 800 m to retain the CO_2 in a dense liquid state. The CO_2 is permanently trapped in the reservoir through several mechanisms: structural trapping by the seal, solubility trapping in pore space water, residual trapping in individual or groups of pores, and mineral trapping by reacting with the reservoir rocks to form carbonate minerals. The nature and the type of the trapping mechanisms for reliable and effective CO_2 storage, which vary within and across the life of a site depending on geological conditions, are well understood thanks to decades of experience in injecting CO_2 for EOR and dedicated storage. CO_2 storage in rock formations (basalts) that have high concentrations of reactive chemicals is also possible, but it is in an early stage of development (TRL 3). The injected CO_2 reacts with the chemical components to form stable minerals, trapping the CO_2. However, further testing and research are required to develop the technology, notably to determine water requirements, which can be considerable. There are large basaltic formations in several regions around the world, and both onshore and offshore sites have been considered for storage. Such formations also exist in places such as India, where there may be limited conventional storage capacity, potentially opening up new opportunities for CCUS.

The overall technical storage capacity for storing CO_2 underground worldwide is uncertain, particularly for saline aquifers where more site characterization and exploration are still needed, but potentially very large. The availability of storage differs considerably across regions with Russia, North America, and Africa holding the largest capacities. Substantial capacity is also thought to exist in Australia. The overall technical storage capacity for storing CO_2 underground worldwide is uncertain, but potentially very large. These global estimates are based on an estimated average CO_2 storage capacity per cubic kilometer of sedimentary rock. While this methodology has limitations, it offers a consistent approach to obtaining global CO_2 storage capacity estimates. The vast majority of the estimated CO_2 storage capacity is onshore in deep saline aquifers and depleted oil and gas fields. Storage capacity is estimated to range from 6000 Gt to 42,000 Gt for onshore sites. There is also significant offshore capacity ranging from 2000 Gt to 13,000 Gt (taking into account only sites within 300 km of the shore, at water depths of less than 300 m, and outside the Arctic and Antarctic).

Even the lowest estimates of global storage capacity of around 8000 Gt far exceed the 220 Gt of CO_2 that is stored over the period 2020–2070 in the Sustainable Development Scenario [47]. Despite the stark regional variation in storage capacity, only a few countries might face a shortfall in domestic storage capacity over that time frame. Theoretical storage capacity far exceeds that needed in the Sustainable Development Scenario to 2070. While notional storage volumes are considerable, a smaller fraction will most likely prove to be technically or commercially feasible. CO_2 storage capacity is analogous to oil or gas insofar as it is a natural resource requiring exploration and appraisal involving extensive data gathering. While success rates might prove to be higher than in the oil and gas exploration sector, failure rates, costs and delays in the exploration, and the appraisal phase are likely to be significant. The process of moving along the scale from undiscovered resource status to subcommercial and then commercial status can take between 5 and 12 years for petroleum assets and even longer for undiscovered saline formation.

A valuable first step in characterizing the progress of storage sites has been made by the OGCI's CO_2 Storage Resource Catalogue [48], which classifies storage sites in 13 countries following closely the definitions of the storage resources management system (SRMS). The majority of resource assessments are not project-based and are automatically categorized as noncommercial on the SRMS scale. The possibility that CO_2 stored underground could leak out has raised questions about the effectiveness of CCUS as a climate mitigation measure and public concerns about safety risks. Decades of experience with large-scale CO_2 storage has demonstrated that the risk of seepage of CO_2 to the atmosphere or the contamination of groundwater can be managed effectively. The probability and potential impact of such events have been studied comprehensively and have been found to be generally low, with risks declining over time. Nonetheless, careful storage site selection and thorough assessment are critical to ensure the safe and permanent storage of CO_2 and to reduce risks to acceptable levels. Thorough assessment includes detailed modeling of the anticipated behavior of the CO_2 over time, together with ongoing monitoring, measurement, and verification. A robust legal and regulatory framework is important to ensure appropriate site selection and safe operation of geological CO_2 storage sites. This already exists in many countries. Project developers and the public authorities have to address public concerns through effective stakeholder engagement.

Costs

The cost of developing CO_2 storage sites will be an important factor in how quickly CCUS is deployed in the coming decades in some regions, though costs are generally expected to be low relative to CO_2 capture. Current and estimated CO_2 storage costs vary significantly depending on the rate of CO_2 injection, the characteristics of the storage reservoirs, and the location of CO_2 storage sites. The cost of developing new sites, especially where CO_2 storage has not been carried out before, is very uncertain, particularly with regard to the effect of reservoir properties and characteristics. In some cases, storage costs can be quite low. Indeed, when the CO_2 is stored as a consequence of CO_2-EOR operations, the cost of storage can effectively be negative net of the incremental revenues from oil production. More than half of onshore storage in the United States is estimated to be below USD 10/tCO_2, which would typically represent only a minor part of the overall cost of a CCUS project. Depleted oil and gas fields using existing wells are expected to be the cheapest. About half of offshore storage is estimated to be available at costs below USD 35/tCO_2. Similar cost curves are expected to apply in other regions, but further research is needed to confirm this. Injecting, storing, and monitoring CO_2 within the subsurface are well established. The drivers for cost and future cost reductions are found in three key areas: site selection, deployment, and technology advancement.

Site selection

The maturity of the technology adopted from the oil and gas industry and environmental services provides higher confidence in cost estimates. There is a broad range of geological storage costs. For example, the National Petroleum Council estimated storage in the United States at $1–$18 per ton of CO_2 [49]. The factors that contribute to this range in costs are attributed to the site and include: access (offshore is significantly more expensive than onshore storage), existing land uses and access. A well-characterized site (previous oil and gas, CO_2 exploration or development) has lower development costs than unexplored sites. For example, depleted oil and gas fields have a significant amount of data attained during production of hydrocarbon, requiring less additional data to prove the suitability of the storage formation.

Surface facilities, offshore platforms, pipes, and wells can be reused reducing the capital investment required. For storage capacity/injectivity, large storage formations with higher injection rates require fewer injection wells per ton of CO_2 injected. As detailed earlier, the analysis found an onshore site with existing data, and infrastructure reuse is the lowest cost. The most expensive is an offshore site with little data, and no existing infrastructure to be repurposed for CO_2 storage. Ease of deployment, CO_2 footprint, and postclosure requirements all impact the ongoing costs of monitoring an operation.

The majority of operating projects to date have targeted onshore, deep saline formations. However, offshore saline formations are increasingly being developed, especially in the North Sea of the United Kingdom, EU, and Norway. Depleted oil and gas fields have only been used for pilot and demonstration projects to date, undoubtedly reflecting the lowest cost of research and development. Despite the lower overall costs, commercial CCUS facilities in the development pipeline are not solely pursuing oil and gas fields. Future CCUS operations comprise a mix of deep saline formations and oil and gas fields. Access to depleted fields is not the primary reason for this mix. The primary driver for developing deep saline formations with large capacity and high injection rates appears to be increasing CO_2 storage rates and improving economies of scale. This increased CO_2 storage rate is evident in that the majority of CCUS hubs in development are pursuing these options, including CarbonNet (Australia), Northern Lights (Norway), and PORTHOS (The Netherlands).

Deployment

A high injection rate per well and CO_2 stored per site are not the only ways to achieve economies of scale. Increasing the rate of deployment of CCUS overall will also reduce the costs for CO_2 storage operations. To date, the manufacturing of CO_2-specific materials and experience in CO_2 operations, although mature, is still small scale compared with the oil and gas industry. In 2018, around 80 Mtpa of natural and anthropogenic CO_2 was injected [50]. To meet climate targets, over 5000 Mtpa of anthropogenic CO_2 must be injected by 2050. As exploration and appraisal for CO_2 storage sites become routine, a 20% reduction in appraisal costs is expected due primarily to the development of CO_2-specific seismic and well drilling processes. The IEA Greenhouse Gas R&D Programme (GHG) estimates that 30–60 storage sites must be developed each year to meet the IEA SDS. In terms of new wells, this equals 300–1200 wells annually. This roll-out of infrastructure (rigs, platforms, wells, piping) may result in a material reduction in costs overall as CO_2 corrosion-resistant steel, cement, and other components are manufactured at greater scale. The Quest CCUS Facility in Canada has cited improved future economies of scale of infrastructure, and refinement of the CO_2 storage process will reduce future operational costs [51].

Technology advancement is expected to deliver modest reductions in the cost of storage. Future savings are seen in the refinement of existing equipment, digital innovation, and automation. Cost reductions of over $45 million in CAPEX and $60 million in OPEX are estimated for a theoretical future CCUS facility storing in an offshore saline formation according to the IEAGHG [52]. These cost reductions are mainly attributed to digital innovations (automation and predictive maintenance).

Gas purity requirements

Noncondensable impurities in a CO_2 stream reduce the density of the gas stream, which leads to a drop in the total storage capacity of a reservoir. Less pure CO_2 would fill storage sites up more quickly, incurring higher injection and storage costs. A geochemical reaction between the CO_2 stream and in situ brine and minerals in the storage formation can reduce permeability and increase pore pressures in a geologic storage site. Fractures and pore spaces can be blocked by mineral precipitation. The geometry of the hydraulic capillaries can be altered by the growth in secondary minerals in the brine that result from the precipitation process. The formation of stronger acids due to the presence of water and SO_2 or H_2S can reduce the pH of the formation water, forming a highly acidified zone in which rapid mineral dissolution of the carbonate and silicate minerals may actually increase the porosity. However, at the edge of the injection zone, the increase in pH actually results in the precipitation of secondary minerals, which can reduce the porosity and, potentially, the formation permeability.

A requirement on purity of the effluent CO_2 stream will greatly increase the cost of capture and may eliminate from choice some of the more cost-effective capture processes that are not capable of achieving high purity. However, impurities in the injected CO_2 may affect the efficiency and safety of CO_2 storage in underground formations in various ways such as decreasing storage capacity and CO_2 injectivity and reducing storage integrity through chemical reactions. The measure of storage capacity for impure CO_2 and the impact of the impurities have been established. Normalized CO_2 storage capacity can be calculated for any CO_2 mixtures, and the impact of the impurities can be clearly seen. Noncondensable impurities, which cannot be liquefied at ambient temperature, such as N_2, O_2, and Ar, result in a significant decrease of CO_2 storage capacity. The degree of decrease in the capacity is a function of pressure, temperature, and composition. For all mixtures of supercritical CO_2 (sCO_2) and noncondensable gases, there is a maximum decrease of the storage capacity at a given temperature. The pressure corresponding to this maximum decrease in storage capacity will change with temperature.

In contrast to noncondensable gases, impurities that are more easily condensable than CO_2, such as SO_2, could increase the storage capacity, and there is a maximum increase at a certain temperature. However, for reference, SO_2 will react with residual H_2O in the gas stream and form H_2SO4, which is corrosive. This can be rationalized from the consideration that SO_2 decreases the average distance between molecules of the mixture—an opposite effect of that of the noncondensable gases.

The change of density caused by noncondensable gas impurities results in lower injectivity of impure CO_2 into geological formations. Above a threshold pressure range, the injectivity could reach the level of pure CO_2 due to lowered viscosity. However, more condensable impurities such as SO_2 may have an effect of increasing the injectivity through increasing density of the CO_2 stream. Noncondensable gas impurities increase the buoyancy of the CO_2 plume. This would decrease the sweep efficiency of injected CO_2. As a result, the efficiency of solubility trapping and residual trapping of CO_2 would decrease.

Among reactive impurity species, SOx, H_2S, and NOx would have the greatest chemical effects on the rocks. Based on previous studies, the effect of SOx on reduction of rock porosity and injectivity would be much smaller than commonly thought. However, if H_2S and SOx are coinjected, such as in the case where CO_2 streams from precombustion capture and oxy-fuel combustion are merged, deposition of elemental sulfur in the pore space could be a concern for pore plugging and, hence, injectivity reduction. After termination of injection, the acidic impurities would promote the corrosion of cements in the presence of water. CO_2 containing SOx, NOx, and O_2 impurities would be more corrosive to cements and steels than pure CO_2 or CO_2/H_2S mixtures.

Boundary effects

Measuring, permanence, and monitoring underground carbon capture in the form of CCUS and carbon capture in the form of CO_2 enhanced oil recovery (EOR) are different practices with fundamentally different mechanisms of CO_2 accumulation. This technical distinction informs the means by which CO_2 sequestration is monitored in these differing cases.

For CCS injection wells:

- Injecting CO_2 in a well in the presence of impermeable boundaries results in rising pressure as a function of quantities injected.
- With pressure increasing over time, the ability to inject declines over time.
- The actual CO_2 capacity of an injection reservoir is significantly less than what would initially be estimated based on volumetric information of the reservoir.
- As a result, the importance of accurate measurements of rates and pressures at wellhead and surface facilities become more important for CCS and remain important over a longer period of time.
- CCS regulations are more concerned with postinjection duration that can be considered to be many decades or even centuries in length. Project developers and operators generally expend more effort on the postinjection than on the injection period, with significant monitoring and verification activity required to evaluate the status of the stored CO_2.

For CO_2 EOR:

- The presence of multiple producer wells in CO_2 EOR means that the average pressure in the reservoir will ultimately attain a nearly steady state.
- Ascertaining the amount of CO_2 retained in the subsurface during CO_2 EOR requires subsurface monitoring.
- After CO_2 EOR, there is a blowdown phase as a reservoir moves onto another recovery process or is abandoned altogether.

Permanence and timescales

Both CCS and CO_2 EOR injection can last for decades at a single site. Permanence requirements vary by jurisdiction and program, but in general terms, they require significant monitoring and careful accounting. There are certain criteria and standards that geologic carbon sequestration projects must adhere to acquire appropriate certification. The accounting requirements cover emissions associated with CCS projects, including:

- Emissions from CCS operations
- CO_2 surface leakage
- Above-ground fugitive emissions
- Post-well closure emissions

With regards to 45Q, the U.S. Environmental Protection Agency (EPA) is empowered to establishing and enforcing any CCUS regulations as part of the Underground Injection Control (UIC) Program. At the heart of these regulations is the establishment of the Class VI wells [53] which are used to inject CO_2 into deep rock formations for long-term storage. The U.S. Department of Energy (DOE) has no responsibility in developing regulations for underground CO_2 storage. However, DOE does support the continued development and field testing of technologies that can be used by operators to verify that the regulations relating to the safe storage of CO_2 underground are met. In December 2010, the EPA finalized the minimum federal requirements under the Safe Drinking Water Act (SDWA) for underground CO_2 injection for the purpose of geologic storage. This final rule applies to owners or operators of wells that will be used to inject CO_2 into the subsurface for the purpose of long-term storage. The UIC Class VI regulations set minimum technical criteria for the permitting, geologic site characterization, corrective action (if necessary), financial responsibility, well construction, operation, monitoring, well plugging, postinjection site care (PISC), and site closure of Class VI wells for the purposes of protecting underground sources of drinking water (USDW).

Monitoring, verification, accounting (MVA), and assessment are important parts of making the storage of carbon dioxide (CO_2) safe, effective, and permanent in all types of geologic formations. Monitoring occurs before, during, and after the injection phase of a CO_2 storage project. The MVA plan for storage projects can have broad scopes covering CO_2 storage conformance and containment, monitoring techniques for internal quality control, and verification and accounting for regulators and monetizing benefits of geologic storage. The location of the injected CO_2 plume in underground formations can also be determined, through monitoring, to satisfy operating regulatory requirements to ensure that potable groundwater and ecosystems are protected throughout the project life cycle.

Monitoring technologies can be deployed for atmospheric, near-surface, and subsurface applications to ensure that injected CO_2 remains in the targeted storage formation, as well as to check for indicators of possible CO_2 migration out of a storage complex. There is a large portfolio of technologies available for monitoring storage of projects, many of which are highly developed due to decades of use and experience gained in the oil and gas industry, as well as through advancements through targeted research and development (R&D).

- Atmospheric monitoring tools are used to measure CO_2 density and flux in the atmosphere above underground storage sites. Three types of tools that are used for identifying and quantifying CO_2 in the atmosphere are optical CO_2 sensors, atmospheric CO_2 tracers, and eddy covariance (EC) flux measurement techniques.
- Near-surface monitoring techniques are used to measure CO_2 in regions that extend from the top of the soil zone down to the shallow groundwater zone. Tools that measure CO_2 effects in the near-surface region include geochemical monitoring tools, surface displacement monitoring tools, and ecosystem stress monitoring tools.
- Subsurface monitoring tools are used to detect and monitor the migration of CO_2 that has been injected into a geologic storage reservoir, characterize faults and fractures, and monitor for any potential microseismic activity that may be present in the storage complex.
- Subsurface monitoring tools include well logging tools, downhole monitoring tools, subsurface fluid sampling and tracer analysis, seismic-imaging methods, high-precision gravity methods, and electrical techniques.

Through the National Energy Technology Laboratory's (NETL) Advanced Storage R&D area, research has allowed for a portfolio of available monitoring technologies for all types of CO_2 storage situations. NETL continues to develop and field test advanced monitoring technologies, as well as support protocols, to decrease the cost and uncertainty in measurements needed to satisfy regulations for tracking the fate of subsurface CO_2 and quantify any emissions to the atmosphere.

Carbon capture

At present, it appears that the carbon-capture portion of future power plants will be based on one of three general technological approaches:

1. Precombustion decarbonization (hydrogen generation)
2. Postcombustion CO_2 capture
3. Oxy-fuel combustion

Each technology has advantages and disadvantages. Some have been proven in the chemicals production industry; others, though holding much future promise, are still in the laboratory development stage.

Capture technologies for precombustion CO_2

Most existing H_2 production plants for refining, chemical, and agricultural use employ a steam methane reformer (SMR) to convert hydrocarbon feeds, such as natural gas and steam, into synthesis gas, which comprises H_2, CO, CO_2, unconverted methane, and a small amount of inerts. To maximize H_2, the synthesis gas is cooled and shifted in a water-gas shift reactor to convert CO and water to H_2 and CO_2. In a gray H_2 scheme, the shifted syngas is separated in an H_2 pressure swing adsorption (PSA) unit to generate a high-purity H_2 stream and a low-pressure tail gas stream that is sent to the reformer furnace as fuel, with additional natural gas to supply heat for the endothermic SMR reaction. Accordingly (and why this process for hydrogen generation is termed "gray"), all of the carbon dioxide from the natural gas combustion eventually exits the system as CO_2 in the furnace stack. To reduce the carbon emissions of an existing gray H_2 asset, CO_2 can be captured from three locations (shifted syngas, PSA tail gas, or flue gas). The best place economically for CO_2 capture is at the shifted syngas because of its high purity and concentration.

Similar to what was mentioned for post- and oxy-combustion, the pressure and concentration will determine the type of separation and the compressor power (once the CO_2 is separated) to get the CO_2 to pipeline specification. The most effective way to produce high-pressure CO_2 is with a POX H_2 process for producing blue H_2. Gas POX technology is an O_2-based system with direct firing in a refractory lined reactor, but it is a noncatalytic process that does not consume steam and has no direct CO_2 emissions. All of the CO_2 is in the product stream and, therefore, at a high pressure and purity. Essentially, all of the CO_2 is captured as high pressure and purity-shifted syngas.

The Rectisol process with intermediate water-gas shift (WGS) is one of the most effective procedures for precombustion CO_2 capture from IGCC plants firing a heavy fuel. It offers multiple benefits such as desulfurization, additional H_2 generation through WGS, H_2 separation, and CO_2 capture all in a single, integrated train. This process configuration has been applied in the Pernis refinery/127-MW IGCC project. The Pernis plant is the first IGCC facility equipped with CO_2 separation, although the separated CO_2 is currently vented. Thanks to this arrangement, Pernis could be considered the only sequestration-ready IGCC plant in the world. Single-circuit Rectisol processes also are running at the 350-MW Vresova IGCC project in the Czech Republic and at a Schwarze Pumpe station in Germany.

Selexol is another physical solvent competitive with Rectisol. There are 55 Selexol operating units in syngas and natural gas service in the world. In operating IGCC plants, Selexol isn't as popular; however, if H_2 production or CO_2 capture is the priority, Selexol moderately outperforms Rectisol.

Capture technologies for post-combustion CO_2

The first approach (e.g., the simple addition of a separate, postcombustion CO_2-capture system to a power plant) is the most straightforward technique. End-of-pipe treatment of flue gases produced by conventional fossil fuel-fired plants belongs to this category. However, the technique's economic efficiency is rather low. The huge volumes of flue gas containing relatively little CO_2 must be handled by conventional absorption processes requiring very large and expensive equipment. What's more is that the efficiency penalty that the technique imposes on the power plant is huge (on the order of 25%–35%). However, postcombustion capture seems eminently suitable for the retrofitting of existing facilities, because it does not affect the upstream (fuel) part of the plant.

Many commercial technologies being proposed for CO_2 capture are not new and have proven effective as components of industrial processes. Many of those processes are technologically mature and available. For example, chemical absorption and physical absorption are ready to CO_2 capture in bulk quantities today, but at a prohibitive cost. R&D studies suggest that chemical absorption may be more suitable than physical absorption for postcombustion decarbonization. Physical absorption may be a better fit with precombustion decarbonization.

Alkanolamines are an example of chemical absorption that are considered by many as the best candidates for post-combustion decarbonization of flue gases. They have been well proven as decarbonization solvents in the gas processing, chemical, and petroleum industries for more than 50 years. Fig. 8.16 is a flow diagram of a typical process of this sort. The upstream absorption stage cools the CO_2 stream and removes particulates from it. Next, the cooled and cleaned stream enters the absorption tower, where it makes contact with the alkanolamine solvent in countercurrent flow. The gas to be absorbed enters the absorber at its bottom, flows up, and leaves at the top. The solvent enters the top of the absorber, flows down, and emerges at the bottom. CO_2 is chemically bound to the solvent by the exothermic reaction of the gas with the amine in the solvent.

The liquid amine, CO_2-rich solvent then leaves the bottom of the absorber and passes into the stripping tower through a cross heat exchanger. In the CO_2 stripper, the mixture is heated with steam to liberate the CO_2 from the solvent as the acid gas. This step is carried out at a lower pressure than the previous absorption step to enhance desorption of CO_2 from the liquid. The CO_2 is now ready for further steps of compression, transport from the power plant to a storage site, and sequestration. The hot, lean amine solution then flows through the cross heat exchanger, where it contacts the rich amine solution from the absorber. The lean amine solution from the cross heat exchanger is then returned to the top of the absorption tower.

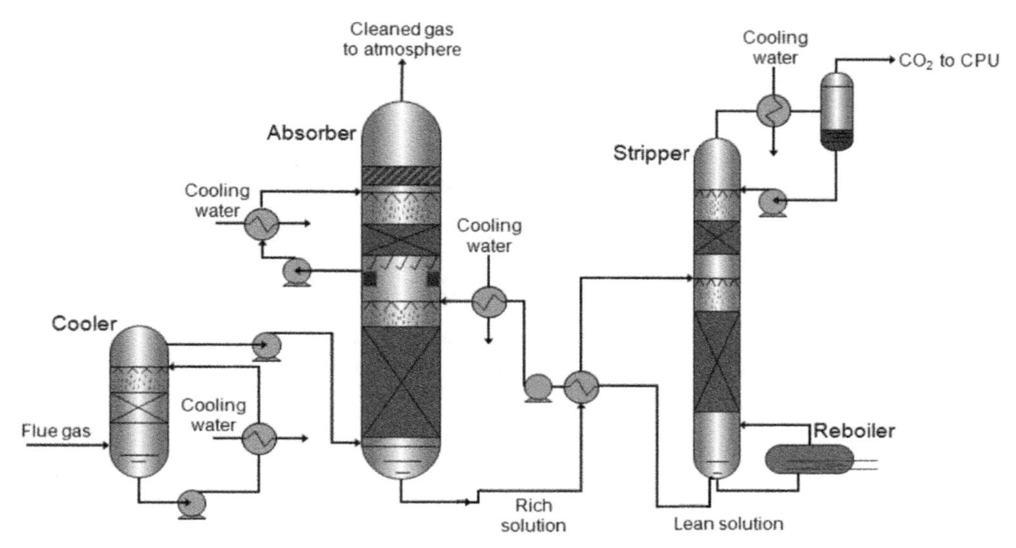

FIG. 8.16 After-the-fact approach. The process flow of a typical flue gas decarbonization system [54].

Amine absorption has been practiced at a large scale in the natural gas processing industry to remove hydrogen sulfide (H_2S) and CO_2 from the fuel. Adapting the technique to flue gas decarbonization is problematic for two reasons. First, CO_2 is present in large quantities in flue gas, but H_2S is only an impurity to be removed from natural gas. Second, decarbonization of natural gas must address the presence of H_2S, but there is no H_2S in flue gas. The greatest obstacle to postcombustion decarbonization is the low pressure (atmospheric) of the flue gas. Only chemical solvents with high reaction energies such as alkanoamines can economically scrub CO_2 under such low partial pressures. The term "amine" refers to a group of organic compounds that can be derived from ammonia (NH_3) by replacing one or more H_2 molecules by organic radicals. Amines are classified according to the number of hydrogen atoms replaced.

Primary amines (RNH_2) include monoethanol amine (MEA) and diglycolamine (DGA). There is considerable industrial experience with primary amine chemical absorption solvents, especially with MEA. MEA, one of the most frequently used solvents for CO_2 capture, has been the traditional solvent of choice for CO_2 absorption and acid gas removal in general. It is the cheapest technique, but it generates the most reaction heat (1.9 MJ/kg). Because MEA's molecular weight is the lowest of the primary amines, it has the highest theoretical absorption capacity, but it also has the lowest boiling point, so there may be solvent carryover in the CO_2 removal and regeneration steps. Another drawback of MEA is its high reactivity with carbon oxysulfide (COS) and carbon disulfide (CS_2), which degrades the solvent. In addition, the CO_2 itself is a strong corrosive agent. Techniques based on primary amines have been used in the industry. An example is the Fluor Daniel Econamine FG process, which uses MEA concentrations of around 30% by weight to successfully remove 80%–90% of the CO_2 from the flue of an ABB Lummus process. In the latter process, the MEA concentration is around 20% by weight.

Secondary amines (R2NH) include diethanolamine (DEA) and di-isopropylamine (DIPA). Secondary amines have a lower capture reaction heat and some advantages over primary amines. For example, the reaction heat of CO_2 with DEA is only 1.5 MJ/kg, compared with 1.9 MJ/kg for primary amines. This makes the use of secondary amines more economical in the regeneration step than using MEA. However, secondary amines share the other downsides of primary amines.

Tertiary amines (R3N) amines, including triethanolamine (TEA) and methyl-diethanolamine (MDEA), are even less reactive. They require the least heat to liberate the CO_2 from the solvent. For example, MDEA's capture reaction heat is just 1.3 MJ/kg. Because tertiary amines react more slowly with CO_2, they must be circulated more quickly than primary and secondary amines. On the upside, tertiary amines degrade and corrode more slowly than primary and secondary amines.

Oxy-fuel combustion

The second approach to carbon capture is oxy-fuel combustion, which is also called oxy-fuel decarbonization or O_2/CO_2 firing. It is a much more elegant technique than postcombustion CO_2 capture, because pure oxygen is used as the oxidant in the combustion process instead of air. This approach completely eliminates nitrogen and its by-products from the process. Instead of nitrogen, wet CO_2 is the primary by-product of combustion. The CO_2 is recycled in a semi-closed cycle serves as the working fluid for the power cycle. Oxy-fuel combustion cycles operate at high pressure and with high concentrations of CO_2 making CCUS much simpler than in conventional combustion-driven power cycles.

For processes with a high CO_2 concentration, such as oxy-fuel combustion or in gas processing facilities, physical solvents may be used to further increase the CO_2 stream for storage. With this increase in concentration and pressure, the CO_2 capture equipment is smaller, and physical solvents can be used with lower energy penalties for regeneration. Using physical solvents, CO_2 concentration can be three times higher, while pressure upstream of the gas turbine is typically 20 times higher. Volume concentration of the CO_2 is therefore 60 times higher compared with typical flue gas from a coal plant. The advantage in this case is lower heat consumption in the solvent regeneration step. No additional heat is necessary, and the stripping is driven mainly by the pressure release (flash distillation).

Power cycles

For many of the reasons mentioned earlier, CO_2 is a desirable working fluid for refrigeration cycles, heat pumps, and power generation cycles. A few select cycles are presented here for reference.

Indirect heated power cycles

Supercritical CO_2 has also been gaining traction in the power generation space, as it can provide higher thermal efficiencies than a traditional steam cycle at turbine inlet temperatures above 500°C with substantially smaller turbomachinery (approximately one-tenth the size). The high density of CO_2 in the supercritical phase lends to high power

FIG. 8.17 Turbine rotor for a 10 MW Net sCO_2 recompression brayton cycle.

density allowing flow paths and overall fluid component sizing to be substantially smaller throughout the system, including piping, heat exchangers, valves, and turbomachines. See the image of a turbine rotor for a 10 MW net demonstration power plant presented in Fig. 8.17.

Several indirectly fired, closed-loop power cycles have been studied to understand the performance capabilities of these sCO_2 cycles. See three example cycles presented in Fig. 8.18. An indirect cycle refers to a power cycle where the heat is added to the working fluid (CO_2) through a primary heat exchanger. In these indirect cycles, the heat transfer fluid can take many forms, meaning the sCO_2 power cycle can support power generation using a variety of conventional and renewable fuels. In an indirectly fired cycle, the flue gas from burned natural gas could be used to heat the CO_2 through the primary heater. Alternatively, concentrated solar energy or nuclear can be used as the primary heat input for the power cycle; here, a heat transfer fluid such as molten salt or solid particles can be used to collect heat from the solar collector or reactor, store the heat (making for a dispatchable energy system), and then heat the CO_2 for power generation.

Considering the cycles presented in Fig. 8.18, the simplest form is the Brayton cycle (Fig. 8.18A). In this cycle, the CO_2 is circulated in a closed loop through a compressor, heater, turbine, and then cooler. Enhancements to this cycle have been considered for improving thermal efficiencies. Fig. 8.18B presents a recuperated Brayton cycle. In a recuperated Brayton cycle, a CO_2-to-CO_2 heat exchanger (recuperator) is used in between the cold side and hot side of the cycle to preheat the working fluid before it enters the heater and precool the fluid before it enters the cooler. This reduces the heat input and heat rejection duties to maintain the target operating temperature, which increases the overall cycle efficiency. The third cycle presented here (Fig. 8.18C) is the recompression Brayton cycle. This configuration is considered to be the most efficient cycle for high-temperature power generation applications. A recompressor and two recuperators are added to the basic Brayton cycle. The CO_2 flow is split between the two compressors; the flow is cooled before entering the main compressor, then runs through the low-temperature recuperator where it remixes with the recompressor flow. The flow split between the compressors can be varied in off-design conditions to maintain high performance. This cycle is currently being demonstrated at the Supercritical Transformational Electric Power (STEP) demo pilot plant at Southwest Research Institute in San Antonio, Texas.

It should be noted that the sCO_2 Brayton cycle and recuperated Brayton cycle are also applicable and desirable for low-enthalpy power generation systems such as waste heat recovery and geothermal. As expected, following the

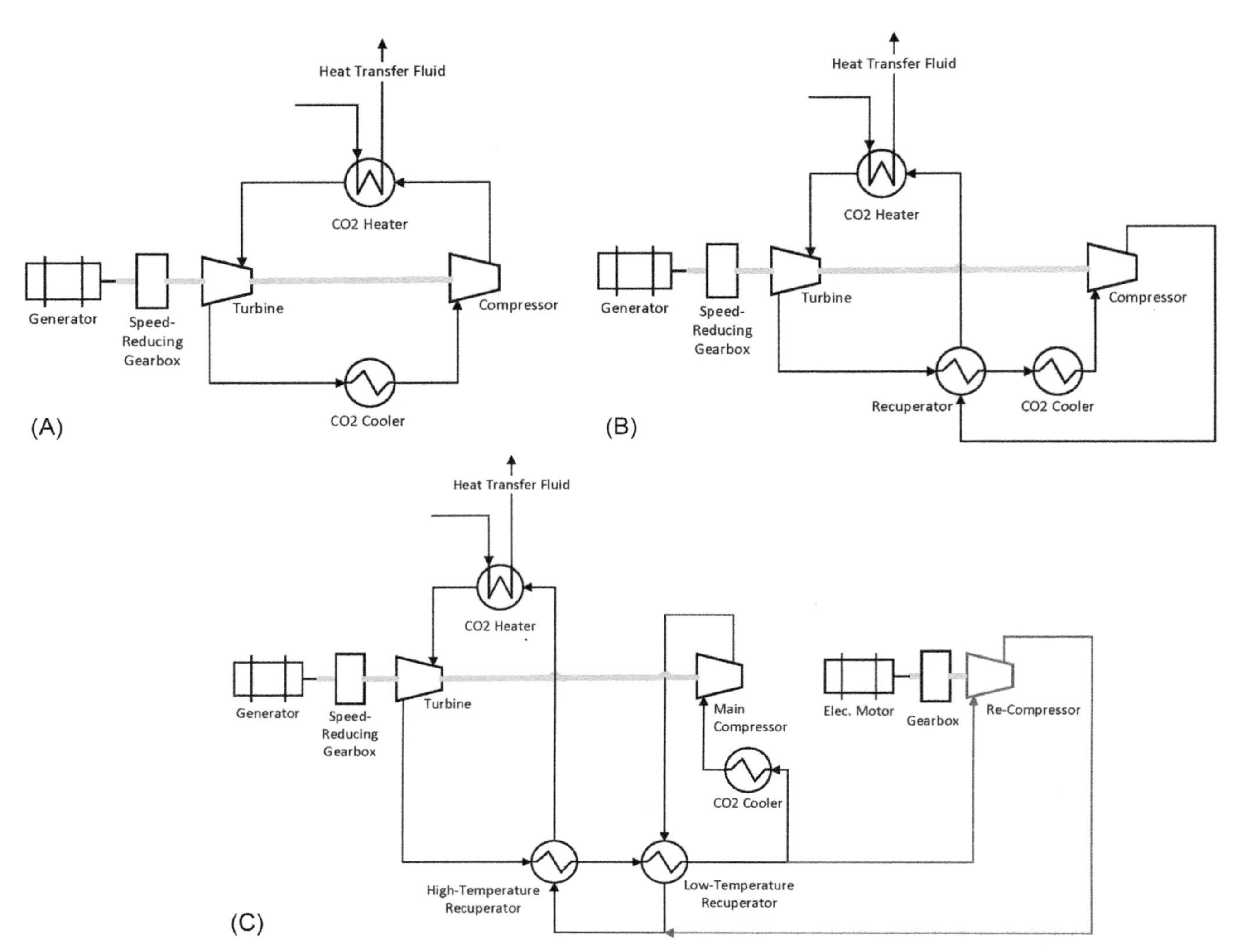

FIG. 8.18 Cycle schematic for three example indirect power cycles: (A) Brayton cycle, (B) recuperated Brayton cycle, and (C) recuperated recompression Brayton cycle.

Carnot efficiency trend, the thermal efficiency of these low-temperature cycles is much less than the power generation cycles with 700°C turbine inlet temperature. However, these systems still hold merit as industries strive to decarbonize and improve plant-level efficiencies.

Direct fired power cycles

Alternatively, direct-fired sCO_2 oxy-combustion power cycles are also under consideration, namely the Allam cycle or Allam-Fetvedt cycle (see Fig. 8.19). This cycle is similar to the recuperated Brayton cycle presented earlier, but it operates transcritical meaning the low-temperature side of the cycle drops into liquid phase so the CO_2 can be pumped. Then, upstream of the turbine, pure oxygen and hydrocarbon fuel (e.g., methane) are injected and burned directly in the sCO_2 stream, providing the thermal energy input to the cycle. Downstream of the turbine and recuperator, the postcombustion species (primarily water and CO_2 with a small portion of CO and other by-products) are removed from the stream to maintain mass balance in the system. Merits of this cycle include high-thermal efficiencies due to the high turbine inlet temperatures, inherent carbon capture, and the cycle producing water and high-pressure, useable CO_2.

The Allam-Fetvedt cycle is currently being researched and demonstrated at a 50 MWth test facility in La Porte, Texas. This facility is owned and operated by NET Power LLC, supported by Constellation Energy Corporation, Occidental Petroleum Corporation (Oxy) Low Carbon Ventures, Baker Hughes Company, and 8 Rivers Capital.

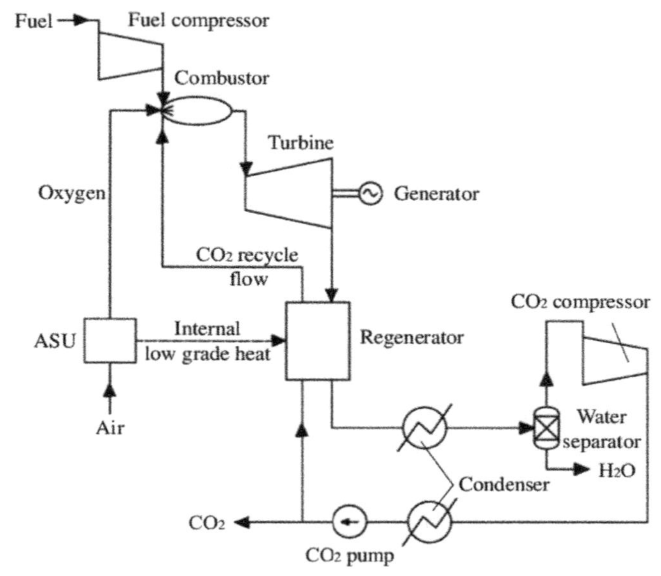

FIG. 8.19 Direct-fired oxy-combustion sCO_2 power cycle (Allam cycle) [55].

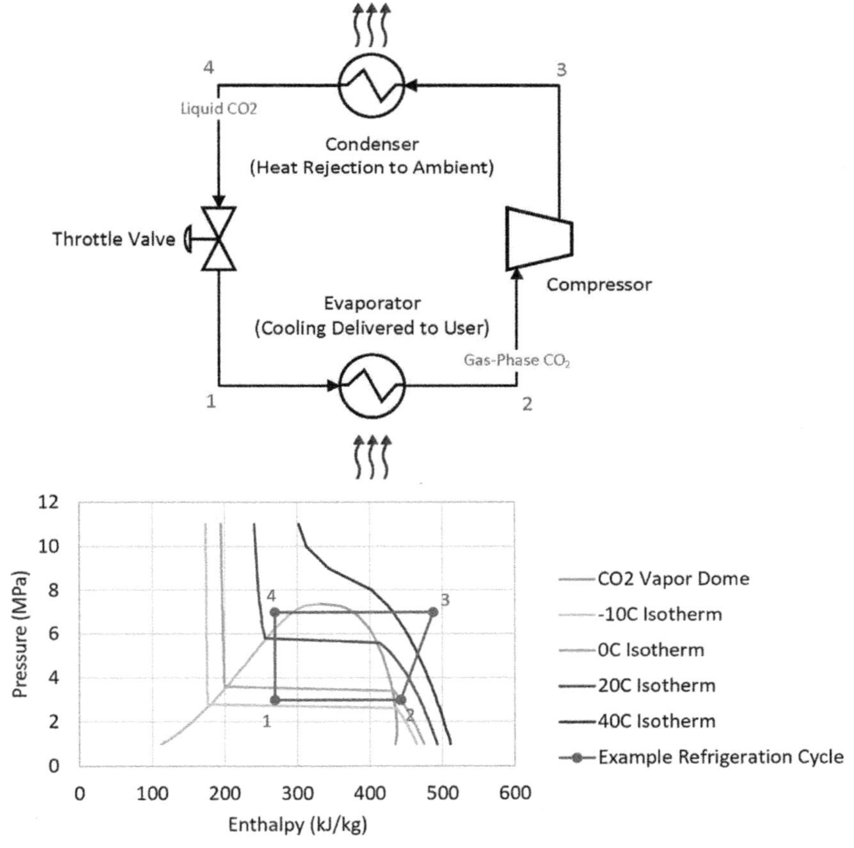

FIG. 8.20 Example refrigeration cycle, fluid properties using NIST REFPROP.

Refrigeration and heat pumps

CO_2 is designated as Refrigerant R744 and is a popular working fluid in both refrigeration cycles and heat pumps. The working principle of a refrigeration cycle and heat pump is the same; in its simplest form, the cycle moves heat from a lower-temperature source to a higher-temperature sink using a compressor, heat exchangers, and a throttle valve (or an expander). See a schematic and example cycle diagram presented for a refrigeration cycle in Fig. 8.20.

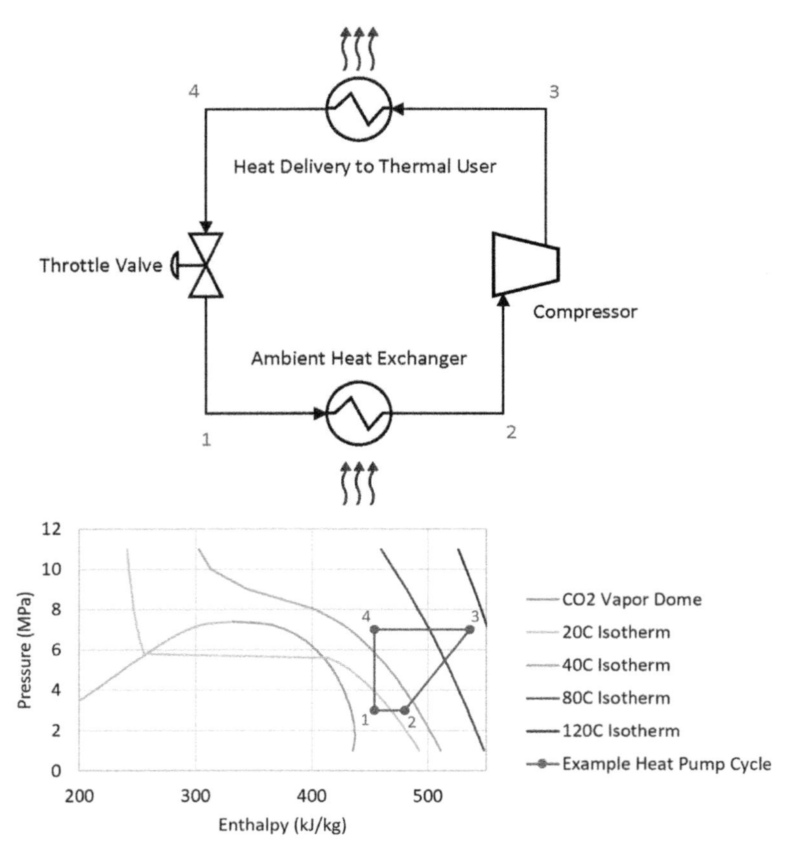

FIG. 8.21 Example heat pump cycle, fluid properties using NIST REFPROP.

Heat is added to the cycle from the low-temperature source (i.e., providing cooling to the thermal user). The pressure and temperature of the CO_2 are then increased across the compressor. Heat is then rejected from the cycle at the condenser at the higher-temperature sink (often atmosphere through an air-cooled heat exchanger). The pressure is then reduced across the throttle valve (also reducing temperature through Joule-Thompson cooling).

The operating principle of a heat pump is the same. However, the cold side of the cycle is often tied to atmosphere. Here, the near-ambient temperature is boosted across the compressor to provide higher temperature heat to a thermal user. The CO_2 is then throttled across a valve to reduce the pressure and temperature, allowing more heat to be collected from atmosphere. See a schematic and example heat pump cycle presented in Fig. 8.21.

References

[1] C2ES, C2ES.org. [Online], 2023, Available from: https://www.c2es.org/content/international-emissions/#:~:text=Globally%2C%20the%20primary%20sources%20of,72%20percent%20of%20all%20emissions.
[2] O.E. Edenhofer, AR5 Climate Change 2014: Mitigation of Climate Change, 2014.
[3] European CCS Demonstration Project Network, Thematic Report: Characterization Session (Storage), 2012. http://cdn.globalccsinstitute.com/sites/default/files/publications/92451/thematic-reportcharacterisation-session-storage.pdf.
[4] Chevron, Fact Sheet: Gorgon Carbon Capture. [Online], 2023, Available from: https://australia.chevron.com/-/media/australia/publications/documents/gorgon-CCS- -fact-sheet-dec-2022.pdf.
[5] EPA, Portland Cement Manufacturin. AP 42, Fifth Edition, Vol. 1, 1995. Ch. 11, Sec 11. 6 Washington DC.
[6] G. Last, M. Schmick, Identification and Selection of Major Carbon Dioxide Stream Compositions, Pacific Northwest National Laboratory, Richland, WA, 2011.
[7] EPA, Inventory of US Greenhouse Gas Emissions and Sinks, 430-R-10-006, 2010.
[8] US EPA, Fact Sheet Final Rule to Reduce Toxic Air Emissions from Lime Manufacturing Plants, U.S. Environmental Protection Agency, Washington, DC, 2003.
[9] US Department of Commerce, Atmosphere, Introduction to the Atmosphere, National Oceanic and Atmospheric Administration, 2023. Available from: https://www.noaa.gov/jetstream/atmosphere.
[10] J.R. Couper, W.R. Penney, J.R. Fair, Chemical Process Equipment: Selection and Design, Elsevier, 2011.

[11] H. Wei, S. Da, Life cycle assessment of carbon capture and utilization technologies for CO_2 emissions reduction in air separation industry, J. Energy Procedia (2017).
[12] P. Zakkour, G. Cook, CCS Roadmap for Industry: High-purity CO_2 Sources: Sectoral Assesment, 2010.
[13] B. Finney, M. Jacobs, Pressure–temperature phase diagram of carbon dioxide, 2005. https://en.wikipedia.org/wiki/Carbon_dioxide#/media/File:Carbon_dioxide_pressure-temperature_phase_diagram.svg.
[14] E. de Visser, Dynamis CO_2 quality recommendations, Int. J. Greenh. Gas Control (2008).
[15] M. Woods, M. Matuszewski, Quality Guideline for Energy System Studies: CO_2 Impurity Design Parameters, National Energy Technology Laboratory (NETL), Pittsburgh, PA, 2012.
[16] Porter RFMPMea., The range and level of impurities in CO_2 streams from different carbon capture sources, Int. J. Greenh. Gas Control (2015).
[17] J. Wang, D. Ryan, E.J. Anthony, A. Wigston, L. Basava-Reddi, N. Wildgust, The effect of impurities in oxyfuel flue gas on CO_2 storage capacity, In Int. J. Greenh. Gas Control 11 (2012) 158–162.
[18] B. Wetenhall, J.M. Race, M.J. Downie, The effect of CO_2 purity on the development of pipeline networks for carbon capture and storage schemes, Int. J. Greenh. Gas Control 30 (2014) 197–211.
[19] I. Cole, P. Corrigan, S. Sim, N. Birbilis, Corrosion of pipelines used for CO_2 transport in CCS: is it a real problem? Int. J. Greenh. Gas Control 5 (2011) 749–756.
[20] S. Sea, Investigating the effect of salt and acid impurities in supercritical CO_2 as relevant to the corrosion of carbon capture and storage pipelines, Int. J. Greenh. Gas Control (2013).
[21] H. Holt, Discharge and dispersion for CO_2 releases from a long pipe: experimental data and data review, in: DNV Symposium Series, 2015.
[22] (NETL) NETL, CO_2 Impurity Design Parameters, 2013.
[23] S. Perry, Pourbaix diagrams as a root for the simulation of polarization curves for corroding metal surfaces, in: 232nd ECS Meeting, Fort Washington, MD, USA, 2017.
[24] Northern Lights, Accelerating decarbonisation. [Online], 2023, Available from: https://norlights.com/.
[25] National Energy Technology Labratory, Carbon Dioxide Enhanced Oil Recovery, 2010.
[26] Y. Xiang, Z. Wang, X. Yang, Z. Li, W. Ni, The upper limit of moisture content for supercritical CO_2 pipeline transport, J. Supercrit. Fluids 67 (2012) 14–21.
[27] D.C. Thomas, Carbon Dioxide Capture for Storage in Deep Geologic Formations-Results from the CO_2 Capture Project: Vol. 1-Capture and Separation of Carbon Dioxide from Combustion, Elsevier, 2005.
[28] S. Yang, D. Akhiyarov, D. Erickson, G. Winning, Challenges Associated with Flow Assurance Modeling of CO_2-Rich Pipelines, NACE International, 2012.
[29] Y.-S. Choi, S. Nesic, D. Young, Effect of Impurities on the Corrosion Behavior of CO_2 Transmission Pipeline Steel in Supercritical CO_2—Water Environments, Environmental Science & Technology, 2010.
[30] United States Internal Revenue Service, IRS, Internal Revenue Code, IRS, Washington, DC, 2023.
[31] US Internal Revenue Service, Inflation Reduction Act of 2022. [Online], 2023, Available from: https://www.irs.gov/inflation-reduction-act-of-2022.
[32] California Air Resouces Board, Low Carbon Fuel Standard. [Online], 2023, Available from: https://ww2.arb.ca.gov/our-work/programs/low-carbon-fuel-standard.
[33] The Global CCS Institute, 2023. Available from: https://www.globalccsinstitute.com/.
[34] United States Environmental Protection Agency, Greenhouse Gas Inventory Guidance: Direct Emissions from Stationary Combustion Sources, 2020.
[35] Carbon Recycling International, George Olah Renewable Methanol Plant. [Online], 2023, Available from: https://www.carbonrecycling.is/project-goplant.
[36] B. Kea, Application of hybrid centrifugal compressor and pump packages for carbon sequestration CO_2 compression, in: Turbomachinery and Pump Symposium, Houston, 2022.
[37] R. Kurz, T. Allison, J. Moore, M. Mcbain, Compression turbomachinery for the decarbonizing world, in: Proc. Turbomachinery Symposium, Houston, TX, 2022.
[38] IPCC, Carbon Capture and Storage, Cambridge University Press, 2005.
[39] R. Kurz, J. Mistry, M. McBain, M. Lubomirsky, Compression applications in the decarbonizing discussion, IEEE Trans. Ind. Appl. (2023).
[40] M. Mohitpour, P. Seevam, K.K. Botros, B. Rothwell, C. Ennis, Pipeline Transportation of Carbon Dioxide Containing Impurities, ASME Press, New York, 2012.
[41] R.P. Doctor, Transport of CO_2, IPCC Special Report on Carbon Dioxide Capture and Storage, 2005.
[42] A.M. McClung, J. Moore, A. Lerche, Pipeline transport of supercritical CO_2, in: Gas Machinery Conference, Austin, TX, 2012.
[43] DNV, DNV supports innovations in CO2 carrier design. [Online], 2022, Available from: https://www.dnv.com/expert-story/maritime-impact/DNV-supports-innovations-in-CO2-carrier-design.html.
[44] International Maritime Organization, IGC Code. [Online], 2023, Available from: https://www.imo.org/en/OurWork/Safety/Pages/IGC-Code.aspx.
[45] Noh, H., et al., Conceptualization of CO_2 terminal for offshore CCS using system engineering process, Energies 12 (22) (2019) 4350.
[46] W. Li, C. Myers, Multi-modal modelling for decarbonization scenarios and industrial decarbonization, CDR, and CO_2 conversion, in: Roadmap for CO_2 Transport Fundamental Research Workshop, NETL, Dublin, Ohio, 2023.
[47] IEA, 2023. https://www.iea.org/data-and-statistics/charts/world-captured-co2-by-source-in-the-sustainable-development-scenario-2020-2070. Available from: https://www.iea.org/data-and-statistics/charts/world-captured-co2-by-source-in-the-sustainable-development-scenario-2020-2070.
[48] Oil and Gas Climate Initiative, CO_2 Storage Resource Catalogue. [Online], 2023, Available from: https://www.ogci.com/ccus/co2-storage-catalogue.
[49] National Petroleum Council Report, Meeting the Dual Challenge: A Roadmap to At-Scale Deployment of Carbon Capture, Use, and Storage. Washington, DC, 2019, https://dualchallenge.npc.org/downloads.php.

[50] Global CCS Institute, Global Status of CCS Report: 2020, 2020.
[51] Shell, www.shell.ca. [Online], 2023, Available from: https://www.shell.ca/en_ca/about-us/projects-and-sites/quest-carbon-capture-and-storage-project.html.
[52] IEAGHG, Beyond LCOE: Value of Technologies in Different Generation and Grid Scenarios, 2020.
[53] EPA, US EPA. [Online], 2023, Available from: https://www.epa.gov/uic/class-vi-wells-used-geologic-sequestration-carbon-dioxide.
[54] S. Hasan, A. Abbas, G. Nasar, Improving the carbon capture efficiency for gas power plants through amine-based absorbents, Sustainability 13(1) (2021) 72.
[55] A.N. Rogalev, N.D. Rogalev, V.O. Kindra, E.Y. Grigoriev, B.A. Makhmutov, The flow path characteristics analysis for suoercritical carbon dioxide gas turbines, in: E3S Web of Conferences, Vol. 124, 2019, p. 124.

CHAPTER

9

District heating and cooling

Kelsi Katcher[a], *Terry Kreuz*[b], *Adam Neil*[c], *and Jordan Nielson*[a]

[a]Southwest Research Institute, San Antonio, TX, United States [b]National Fuel Gas Company, Williamsville, NY, United States [c]Ebara Elliott Energy, Jeannette, PA, United States

Abbreviations

CHP combined heat and power
EGS enhanced geothermal system
HVAC heating, ventilation, and cooling
TES thermal energy storage

Understanding district heating/cooling

District heating systems transport thermal energy, in the form of heated or chilled fluids (most commonly water or steam), from a centralized source to support the heating and cooling needs of multiple users or buildings. Typically, the circulating fluid is heated or chilled by an external thermal source, which is then distributed to the thermal users via insulated pipes. District heating systems can be configured to provide heating services or a combination of heating and cooling. See a schematic of a district heating system shown in Fig. 9.1.

District heating systems provide many potential benefits depending on the system setup and the thermal user's needs. Instead of utilizing individual heat pumps or boilers in each building, for example, a larger central heat pump can supply multiple homes or buildings, removing duplicate equipment, improving efficiencies, reducing overall energy consumption and greenhouse gas emissions, and potentially reducing the utility cost for users. Connecting the thermal energy systems for multiple buildings or facilities provides economies of scale that allow for the deployment of a more efficient and resilient energy source. Operation at this scale also allows for alternate energy or fuel sources to be used, including waste heat, biomass, and geothermal, as well as integrated systems with renewable energies such as wind and solar. Furthermore, at this scale, industrial-grade equipment such as condensing economizers can be used to boost thermal efficiency.

District heating systems are well suited to support highly populated areas (such as city downtown areas or business districts), residential communities, campuses, military bases, healthcare facilities or districts, airports, and industrial districts. By combining the thermal loads of multiple buildings, the overall input energy can be reduced. District heating is particularly beneficial in districts where both heating and cooling needs exist. The central system can be configured to provide both the required heating and cooling duties, but the net thermal duty produced at the central station to balance the cycle will be less than producing the thermal duty individually at each facility. Also, the thermal duty produced at the central energy station will be more stable, with gradual changes in the net duty required by the cycle, as compared to the steep ramping in heating and cooling demand that would be required locally at each facility. See a summary of the benefits of district heating systems in Fig. 9.2, as compiled by the DOE Office of Energy Efficiency and Renewable Energy (EERE).

Potential applications

In the United States, existing district heating systems are operating in each of the applications mentioned previously. Major US cities with downtown district heating networks include New York, Boston, Philadelphia, San

Energy Transport Infrastructure for a Decarbonized Economy
https://doi.org/10.1016/B978-0-443-21893-4.00017-9

Copyright © 2025 Elsevier Inc. All rights are reserved, including those for text and data mining, AI training, and similar technologies.

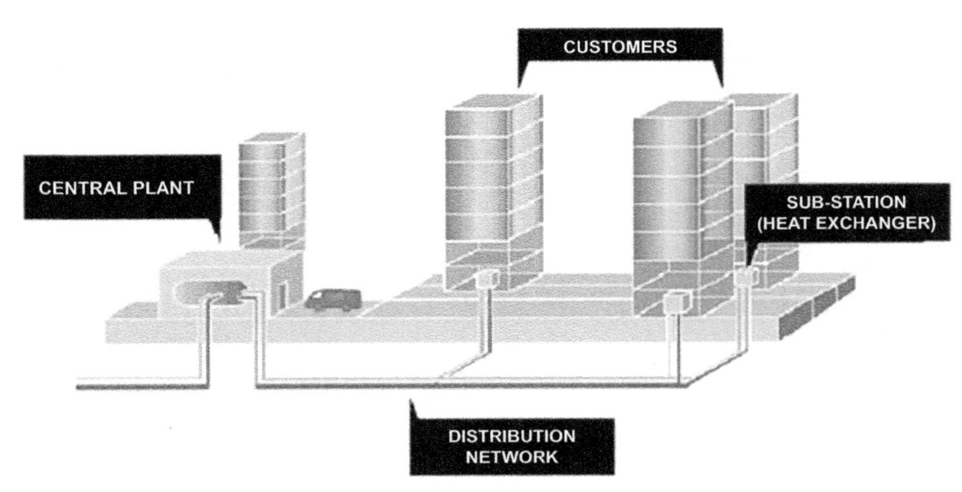

FIG. 9.1 Schematic of a typical district heating system [1].

Benefits to Customers	• Higher energy efficiency • Lower building costs (no separate boilers, chillers, or other related hardware) • Easier building operation and maintenance • Enhanced building aesthetics and comfort (reduced noise and vibration) • Improved reliability (industrial-grade district energy equipment is more robust than commercial equipment installed at building level)
Benefits to Cities and Communities	• Reduced first cost for new development • Flexibility in use of fuel sources, including local or regional fuel sources (wood waste, biomass, waste heat, etc.) that keep energy dollars recirculating in local economy • Architectural and aesthetic advantages, with roofs free of mechanical equipment • Grey water/treated sewage effluent usable for condenser water (owing to central plant scale), conserving potable water for consumption • Capacity to provide baseload power and heat for microgrids, enhancing resilience and reducing regional greenhouse gas emissions
Benefits to Grid Infrastructure	• Reduced peak demand (enabled by aggregating loads and shifting peak demand with thermal energy storage) • Fewer natural gas peaking stations • Lower transmission and distribution costs
Benefits to the Environment	• Reduced air emissions, including greenhouse gases, as a result of greater fuel efficiency of district energy systems that include CHP • Increased adoption of renewable energy sources at scale, replacing higher-emitting central station generation with low- and zero-emitting technologies • Improved stormwater management owing to free roof space, which can be used for low-impact storm water management strategies and mitigation of excessive runoff

FIG. 9.2 Benefits of district heating systems as compiled by the DOE Office of Energy Efficiency and Renewable Energy [2].

Francisco, Denver, Minneapolis, and more [2]. In 2012, there were more than 660 district heating systems operating in the United States, providing heating to 5.5 billion square feet of floor space and cooling to 1.9 billion square feet [3]. These installations reportedly delivered 5386 MMBtu/h. of hot water, 52,857 MMBTu/h. of chilled water, and 188 million lb/h. of steam. Since 2012, the district heating capacity in the United States has steadily increased and is predicted to continue increasing into the foreseeable future, particularly the capacity of hot water and cooling systems (see Fig. 9.3). In 2022, district heating supplied approximately 9% of the global heating demand in buildings and industry [4].

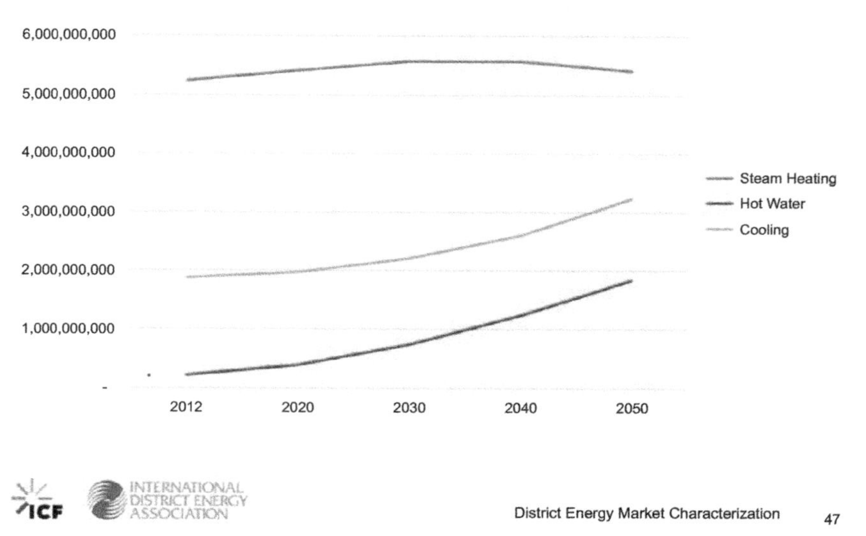

FIG. 9.3 Historical and predicted district heating capacity in the United States as reported by the U.S. Energy Information Administration (EIA) [3].

Heating applications

Most existing district heating systems use steam as the circulating fluid. However, newer systems are trending toward hot water (liquid-phase) distribution. The heat capacity can be supplied from many different sources including fossil fuels, geothermal, waste heat from industrial processes, and renewables. The generated heat is then delivered to support a variety of needs, primarily comfort or space heating, but applications can also include electrical and mechanical power generation, process heating, and hot water production, among other things. Two examples of existing district heating networks are given below.

New York City has 105 miles of underground piping that supplies steam to over 1500 buildings to provide heating and power [5]. Fig. 9.4 shows a cogeneration plant that delivers steam at 450–475 °F to the network of underground piping. This steam is used to heat buildings and produce hot water. During the summer months, the steam delivered to many buildings is also used for air conditioning; the steam powers the compressor on large chiller systems used for comfort cooling. Steam is also utilized for sterilization in hospitals, humidification in museums, and to generate power for dry cleaners.

Germany is also home to an extensive district heating infrastructure. More than 80% of the district heating in Germany is currently produced through combined heat and power (CHP) generation, with heat often delivered in the form of hot water [6]. This involves utilizing the waste heat from an electric power generation plant for heating applications. Leveraging CHP systems reduces the total energy consumption by 30%–50% compared to generating the heat and power separately. Considering Berlin alone, the district heating network is comprised of 2000 km of piping; the installed capacity of heat generation assets totals 5.6 GWth; and the district heating demand totals approximately 10 TWh per year, supplying approximately 1.3 million households.

Looking forward, cities in Germany are striving to connect 100,000 buildings annually to district heating networks as part of their decarbonization plan [7]. In 2022, Germany also launched a program, supported by the European Commission, to promote green district heating systems based on renewable energy and waste heat [8]. In this initiative, they are striving to integrate renewable and waste heat into 75% of new district heating system construction, upgrade existing networks to incorporate renewables, and focus on installing solar thermal generation facilities, heat pumps, and heat reservoirs.

Fourth generation district heating networks

So-called fourth generation district heating networks are systems that operate at much lower temperatures [9]. The comfort heating temperature for residential and commercial spaces is typically 21°C, and the hot water temperature for residential use is typically 55°C. For applications where comfort heating and water heating are the primary goals, low-

FIG. 9.4 Cogeneration plant in New York City delivering steam at 450–475 °F to district heating network [5]. *Source: Photo by Charles Parker, https://www.pexels.com/photo/steam-pipe-on-road-among-skyscrapers-on-sunny-day-5847394/.*

grade heating systems can be used. Operating the distribution network at lower heat transfer fluid temperatures improves system efficiency because the lower temperature differential between the working fluid and the ground means lower thermal losses. These systems can also be designed with energy efficiency improvements to reduce demand and operate with lower mass flow rates, resulting in smaller piping. Delivery of the heat to users in these systems may require larger radiators or underfloor heating to effectively transfer the heat. But operating with low-grade heat also allows for more heat source options, including lower-temperature waste heat recovery.

Cooling applications

Most existing district energy systems supply heating services. However, in some cases, it is beneficial to deploy a district cooling system. These systems typically take the form of an industrial-scale chiller plant that produces chilled water (around 45 °F). The chilled water is then distributed through the piping network to provide air conditioning to buildings; the chilled water flows through heat exchangers at each building, absorbing heat, and then returns to the central plant for heat rejection. During the summer months, air conditioning demand typically creates 50%–70% of the peak electricity demand [10]. By combining the cooling duties of many buildings or facilities and applying economies of scale at the central plant (including system efficiency improvements), the total energy consumption can be reduced, which can help reduce strain on the electric grid. An example district cooling system is illustrated in Fig. 9.5.

As mentioned in the previous section, cooling can also be provided to buildings by (1) utilizing the delivered steam to drive steam turbine chillers or (2) utilizing delivered steam or hot water to drive absorption chillers.

Residential and campus systems

In this application, a central heating and cooling system will supply heated and/or chilled fluid to the various buildings through a network of pipes. Piping is often buried under roadways from the central station, and branch lines run to and from each building. The buried pipes are sometimes installed in underground tunnels to allow maintenance access to the lines as well as any valves or pumps. The heat transfer fluid from the district supply line can exchange with the individual heating/cooling system fluid in each building or may directly provide the heating/cooling.

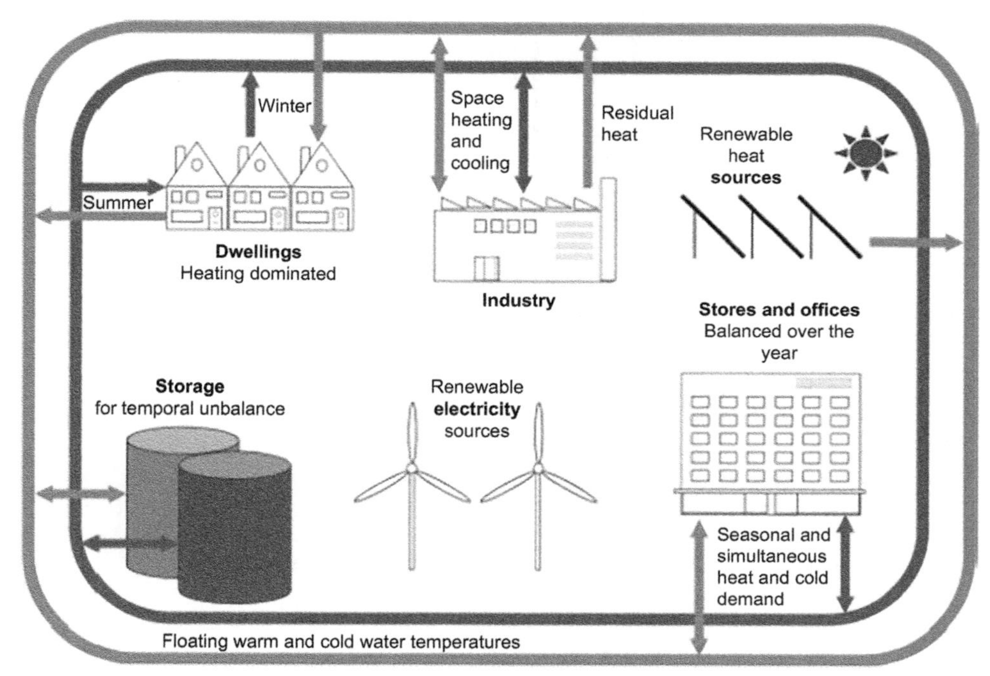

FIG. 9.5 Schematic of a district heating and cooling system for supporting seasonal conditioning demands [11]. *Source: Wikipedia, https://commons.wikimedia.org/wiki/File:Cold_District_heating,_schematic_function.png.*

A four-pipe system can be installed to include a heating supply and return line as well as a cooling supply and return line, allowing individual facilities to switch between heating and cooling. In the case that some facilities are heating, and some are cooling, the central pumping station adjusts the operation and the total thermal duty to maintain a steady supply. This reduces the overall energy (electrical or fuel) consumption to provide the thermal duty at each building because some of the cooling and heating loads are offset at the central system. These systems commonly run heated/chilled water or steam for applications including comfort heating/cooling, radiant heating, and snow and ice melting.

A 48 MW district energy system currently supplies the Texas Medical Center in Houston, Texas. This installation employs a combined heat and power (CHP) system to provide electricity, chilled water, and steam to support 18 healthcare facilities. Having dedicated electricity, heating, and cooling provides stability and resiliency to these critical infrastructure sites. During Hurricane Harvey and its aftermath in 2017, the Texas Medical Center was able to continue operating as usual without any interruption in the thermal or electrical supply [2].

Data center applications

The large thermal and electrical demands at data centers are also attractive for district energy and microgrid concepts. In the United States, data centers account for approximately 2% of the electricity used annually [12]. Many new data centers are being constructed with dedicated power plants to provide stability and resiliency, separated from the local utility grid. The large cooling demand for the servers is also critical to the data center's operation.

Considering district energy systems, data centers could leverage a CHP system to meet both electrical and thermal demands. Here, the waste heat from the electricity production could be used along with an absorption chiller to deliver cooling. Two examples of such systems are (1) a 780 kW microturbine CHP system installed to support the university data center at Syracuse University and (2) a 4.6 MW CHP unit installed at BP's Helios Plaza in Houston, TX, allowing the 24/7 trading operations to continue through outage events.

Alternatively, the heat produced by data centers could be used as heat input to a district heating network. As an example, Amazon data centers are taking this approach; one installation in Dublin is supplying heating to numerous public, commercial, and residential buildings [13]. The cooling water exiting the air handling units (absorbing excess heat from the data center) is run through several heat pumps, where the water temperature is boosted from 25°C to 85°C. This hot water is then delivered to the various thermal users as illustrated in Fig. 9.6. It is expected that this system reduces CO_2 emissions by 11,400 tons per year (60% reduction compared to having individual boilers on site). Similar installations have been proposed or commissioned by Meta, Microsoft, and Fortum.

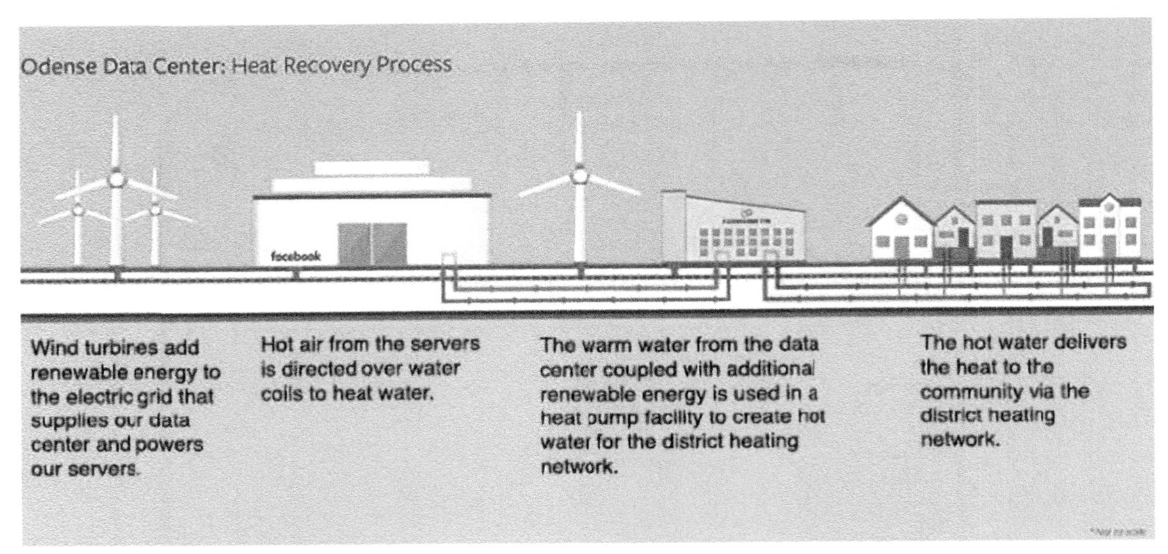

FIG. 9.6 District energy system supplied from data center server cooling augmented with heat pumps [14].

Challenges

District heating and cooling systems face many challenges that may pose barriers to widespread installation and deployment. The key challenges to this technology are:

1. Co-location of producer and users: For district energy systems to be viable, the available source/sink and a district of sufficient energy demand density must be co-located. For sources such as geothermal or industrial waste heat, the users must be within reasonable proximity to the heat source. Alternatively, if combustible fuels or a central heat pump are used at the central pumping station for producing the thermal energy, the location of the district is not limited geographically, but there must be sufficient space available for the central station to be commissioned, and the permitting for emissions, noise, etc. must be approved. This may be particularly challenging in urban areas where population density and thermal demand density are high.
2. Infrastructure and capital investment: Large-scale geothermal district heating and cooling systems require substantial infrastructure and investment. The complexity of designing and implementing such systems, including constructing the distribution network and integrating with existing infrastructure, poses a significant challenge. For this reason, there must be a sufficiently high thermal demand density in order for the implementation of a district system to be economically viable. This also presents a market penetration risk and barrier if it has to compete with energy systems that can produce at a lower cost per thermal duty. This may be particularly true with 'renewable' district heating systems based on geothermal, solar, wind, or waste heat.
3. Regulatory and institutional barriers: These systems often require significant cooperation between different stakeholders, including homeowners, local businesses, and local, regional, and national authorities. Navigating through different regulatory jurisdictions can be a complex and time-consuming process.
4. Heat distribution: The efficiency of a district system depends on its ability to distribute heat effectively and minimize heat loss. Older buildings or those not designed with energy efficiency in mind may need significant retrofitting to benefit from district heating.

Addressing these challenges requires a multi-pronged approach, including public education campaigns, training programs for installers, financial incentives, and streamlined regulatory processes. For district heating and cooling systems, fostering collaboration among all stakeholders is crucial.

Trends in policy

District heating and cooling systems have gained more interest over the past several years, and more district heating markets are emerging. This is particularly true in Europe, where they have received increased policy support since the energy crisis in 2022 [4]. Example of programs ramping up over the past 2 years include:

- The Czech green district heating scheme received €401 million in support from the European Union in 2023 [4].

- Programs in the United Kingdom are introducing heat zoning regulations, targeting that 18% of the heat consumption in the UK can be supplied through heat networks by 2050 [4].
- Regulations in Denmark were relaxed to allow geothermal heat projects to be exempt from the existing price regulations in order to spur more development programs [4].
- The district heating network capacity in Vancouver, Canada is being expanded to include 6.6 MW of sewage and waste water heat recovery equipment, capturing latent heat and boosting temperature through heat pumps [4].
- The first project in China was launched in 2023 to utilize waste heat recovery from nuclear power plants for district heating [4].

Looking forward, policy and financial support will be crucial for market penetration, scale-up, and decarbonization of district heating systems. Without incentives, utility providers have little motivation to deviate from what has been working, namely relying on natural gas or other fossil fuels for heat. Many countries are beginning to include district heating and cooling efforts as part of their decarbonization pathway. The increased support by way of regulatory, policy, grants, and other financial incentives provides a more economically viable path for district energy networks. Alternatively, increasing taxes or financial penalties for fossil fuel consumption or emissions may also drive the adoption of district heating systems due to their potential for increased efficiency or electrification.

Apart from economic considerations, integration challenges, and public acceptance can pose a significant barrier. The establishment of consumer protection rights in new district heating markets, comparable to those previously in place with other utilities, will be critical to securing public trust and acceptance to allow the deployment of new systems [4]. Performing heat mapping studies at the municipal or regional level, involving impacted communities, will be critical to gain support and confidence; such studies are being performed in Ireland and Germany [4]. Furthermore, the implementation of policies to require district or motivate heating connections as part of the zoning requirements can help overcome the resistance to retrofits or connecting facilities to the district network; such policies are being proposed under the Energy Security Bill in the UK [4].

Scale: How do we define a "district"?

In district energy systems, the heated or chilled water must be piped from the central pumping station to the thermal users without significant temperature change. As noted earlier, transporting thermal energy over long distances is a key challenge. Therefore, district heating is best suited for areas of high thermal-demand density. A suggested thermal demand density of at least 120 TJ/km^2 has been proposed by O'Shea [13]. Research by Stratego has proposed that districts with thermal demand densities of 30–100 TJ/km^2 will require fourth generation advanced district heating systems, densities of 100–300 TJ/km^2 are doable with current district heating technology, and over 300 TJ/km^2 are highly desirable districts [15].

Additional critical factors for consideration are the connectedness of the thermal users, the magnitude of the thermal users, the size of the thermal producer, and the proximity to the thermal producer.

Heat sources and sinks

Combustible fuels

In the United States, most district heating systems are based on fossil fuels, with natural gas accounting for 74% (see Fig. 9.7) and coal and oil also fueling a smaller portion of the systems. Other fuels included renewable natural gas (from sources such as agriculture and landfills), biomass (forest product waste, agricultural waste), and incineration of solid waste.

Considering district heating globally, in 2022, district heating networks supplied approximately 17 EJ of heat, with China, Russia, and Europe accounting for more than 90% of the district heat production. Of the heat supplied, nearly 90% of systems are based on fossil fuels. Coal was the leading supplier at 48% of the total district heat demand, primarily in China; coal was followed by natural gas, supplying 38% of the heat demand (see Fig. 9.8 from the IEA). The remaining 10% of district heat was supplied from municipal waste, waste heat recovery, nuclear, and renewable sources [4].

Fossil fuel-based district heating systems typically employ a boiler or furnace. Here, water is directly heated or boiled by burning fossil fuels. The hot water or steam is then supplied through the distribution networks to the users.

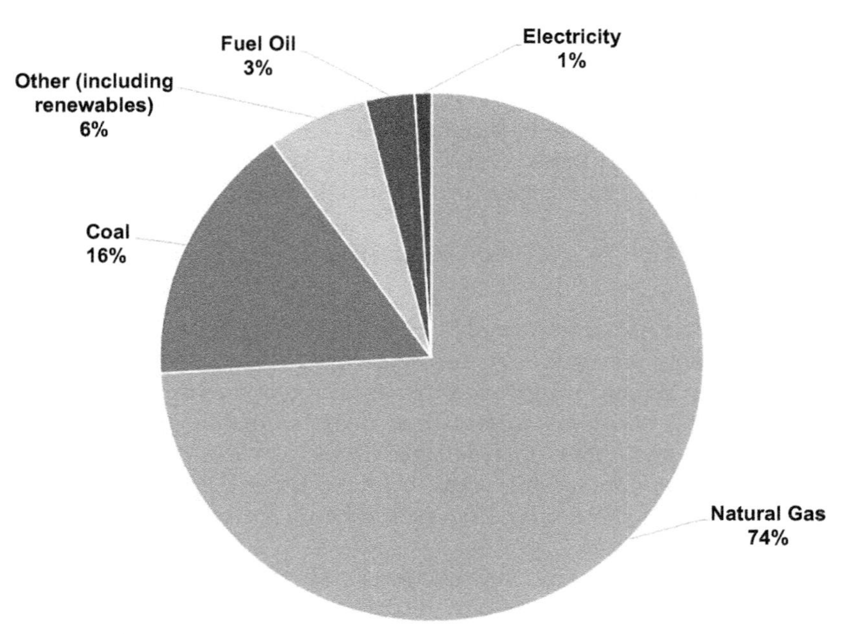

FIG. 9.7 Heat sources used to supply district heating systems in the United States, as compiled by the DOE Office of Energy Efficiency and Renewable Energy [2].

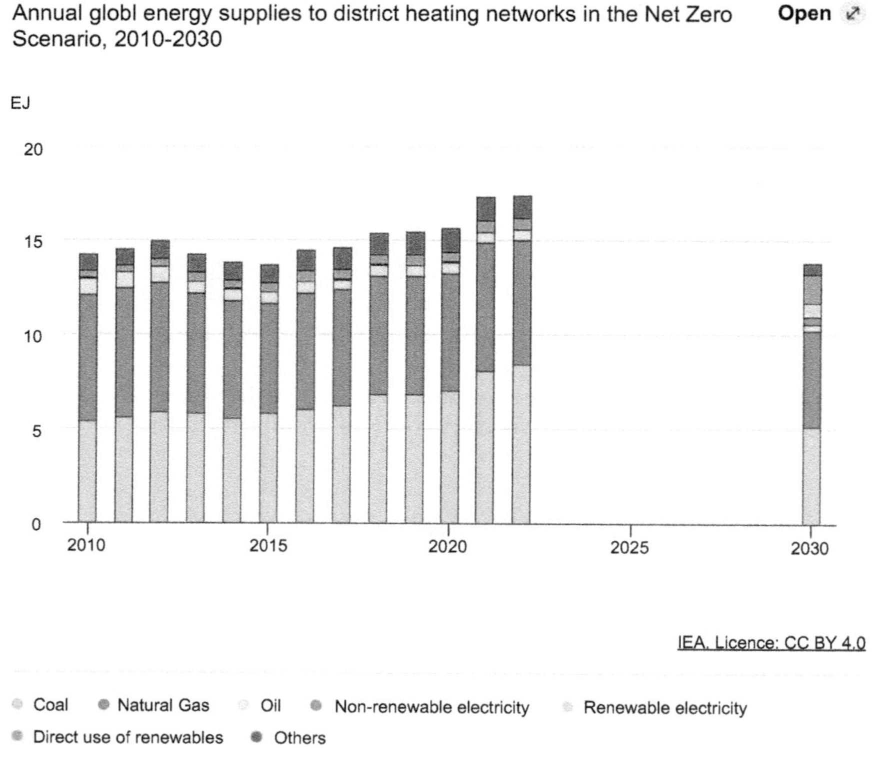

FIG. 9.8 Heat sources used to supply district heating systems globally, as reported by the International Energy Agency [4].

See schematics of a boiler and furnace system shown in Fig. 9.9. Note that reheat and economizer stages can be added to these systems to improve the thermal efficiency and increase the supplied heat duty.

Supplying existing natural gas systems with renewable natural gas or biogas has the potential to decrease overall emissions and global warming potential in the region, especially in cases where methane would have been emitted to the atmosphere if uncaptured.

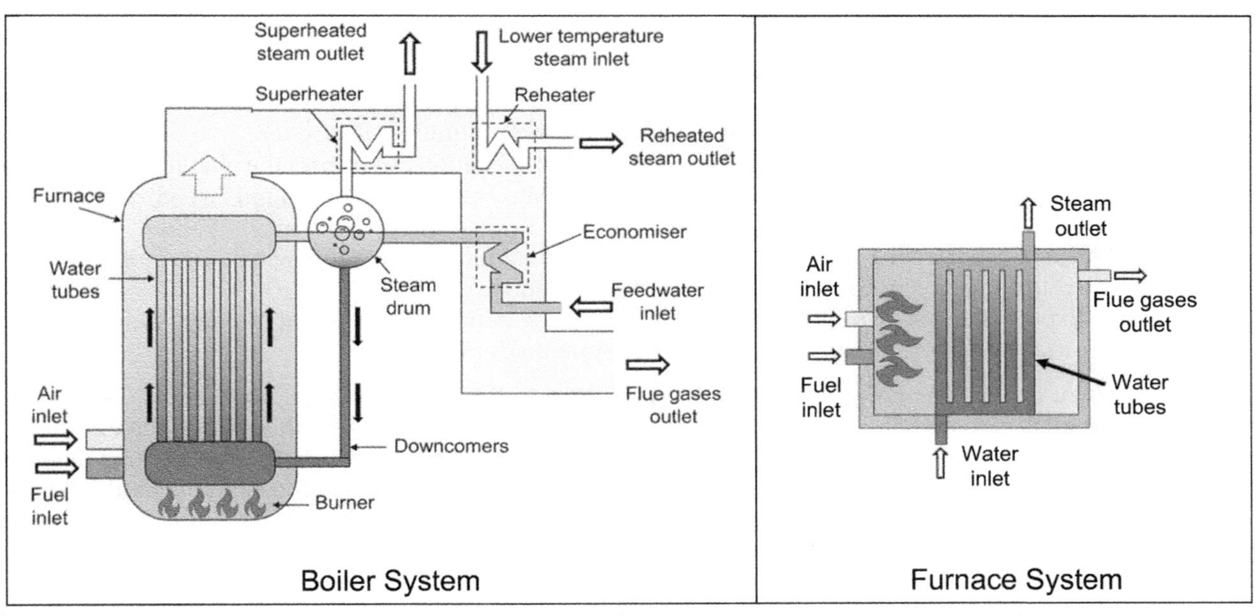

FIG. 9.9 Schematic of a boiler system (left) for supplying hot steam and a furnace system (right) for supplying hot water to district heating networks [16].

Geothermal

Geothermal energy refers to the thermal energy stored in the Earth's crust. This is a type of renewable energy that comes from the heat generated by the Earth's core, produced primarily by the decay of radioactive materials. Geothermal energy has several advantages as a renewable energy source. It is reliable, available 24/7, and does not depend on weather conditions like wind and solar power. Geothermal plants also have a relatively small land footprint compared to other renewable energy sources, and they emit very low levels of greenhouse gases and other air pollutants.

Geothermal district heating and cooling systems leverage the Earth's consistent underground temperatures to provide climate control for clusters of buildings or whole districts. Utilizing the natural thermal energy reservoir beneath the Earth's surface, these systems can offer a sustainable and energy-efficient alternative to traditional heating, ventilation, and air conditioning (HVAC) systems.

There are two main types of geothermal energy systems: direct-use systems and geothermal power plants. Direct-use systems involve using geothermal water or steam directly for heating, cooling, and other thermal applications, such as greenhouses, aquaculture, and hot springs resorts. Geothermal power plants, on the other hand, use geothermal fluids to generate electricity using power turbines (see illustration in Fig. 9.10). The waste heat from the power cycle can then be used for district heating, similar to the CHP concept presented earlier.

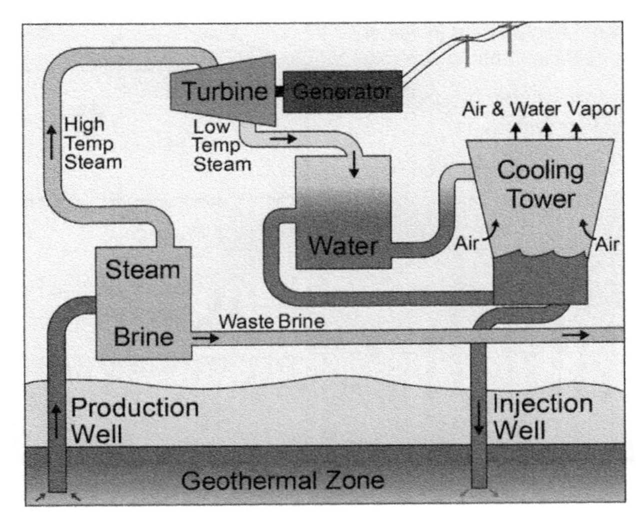

FIG. 9.10 Example schematic of a geothermal power generation system [17].

The major challenges of geothermal district heating and cooling systems include:

- High Upfront Costs: The primary challenge in implementing geothermal systems is the high initial cost, including site preparation, drilling, and equipment installation. For residential applications, the cost can range from $10,000 to $30,000, while for commercial and industrial applications it can be significantly higher.
- Site Suitability: The geology, hydrology, and thermal gradient of a location can impact the viability and efficiency of geothermal systems. Not all areas are suitable, and a detailed site assessment is required, which adds to the overall cost.
- Space Requirements: Geothermal systems require space for ground loop installation, which may not be available in densely populated urban areas or locations with challenging landscapes.
- Awareness and Expertise: Lack of public awareness about the benefits of geothermal systems and a shortage of skilled installers and service providers can impede widespread adoption.

To limit the cost of drilling, shallow geothermal wells can be installed with a heat pump to boost the temperature above or below the ground temperature. The temperature a few meters beneath the Earth's surface remains fairly constant year-round, around 10–15°C (50–60°F), providing a reliable source of heat in the winter and cool air in the summer. The heat pump, illustrated in Fig. 9.11, uses a refrigerant to transfer heat between the building and the Earth via a ground heat exchanger, typically consisting of pipes buried underground. In the heating mode, the heat pump extracts heat from the ground, elevates its temperature, and distributes it within the building. During cooling mode, the process is reversed; the heat pump extracts heat from the building and releases it into the ground, thereby cooling the indoor air.

While the challenges of implementing geothermal systems are significant, they are not insurmountable. With targeted strategies, policy support, and investment in research and development, the potential of geothermal energy can be harnessed to provide sustainable and efficient heating and cooling solutions across diverse sectors and communities. In many regions, incentives such as tax credits, grants, and low-interest loans are available to offset the high initial costs of geothermal heating and cooling systems. Furthermore, these systems tend to have lower long-term operating costs than traditional systems due to their high efficiency and lower maintenance needs. Geothermal heating and cooling systems also have longer lifespans, typically around 20 years for the heat pump and up to 50 years for the ground loop.

Residential applications

Geothermal heating and cooling systems in residential buildings can provide heating, cooling, and hot water, replacing traditional HVAC systems. These systems can be utilized in various residential settings, from individual homes to apartment buildings.

The average cost for installing a geothermal heating and cooling system in a residential property ranges from $10,000 to $30,000. However, homeowners can recover this investment over time through reduced energy costs—often between 30% and 60% savings on heating and 20%–50% on cooling per year.

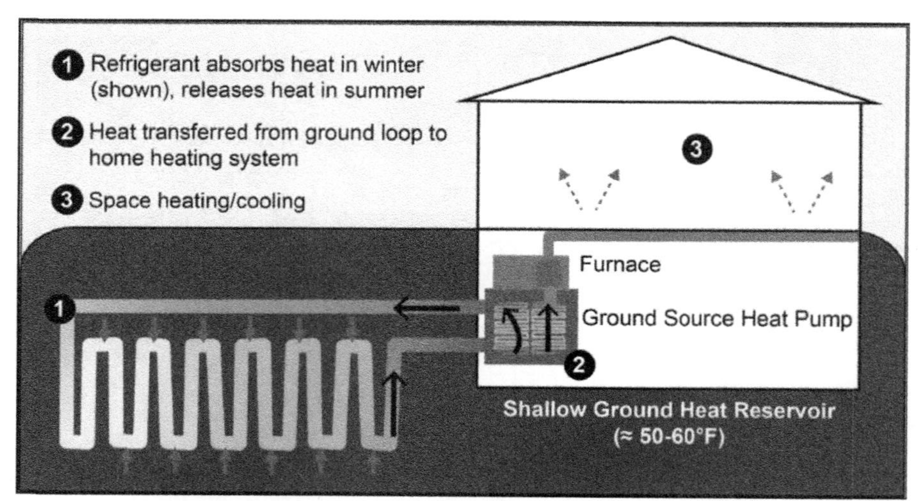

FIG. 9.11 Schematic of a geothermal based heat pump for supplying heat [17].

Commercial applications

In the commercial sector, geothermal heating and cooling systems are used in facilities such as offices, hotels, hospitals, schools, and retail complexes. These systems provide comfortable indoor environments and a reliable hot water supply, vital for various commercial operations.

The capital cost for commercial geothermal heating and cooling installations is site-specific and depends on factors such as system size, building requirements, and ground conditions. However, the energy savings (approximately 25%–50% compared to conventional systems) can offset the higher upfront costs in the long run.

Industrial applications

In the industrial sector, geothermal heating and cooling systems can be used for space heating and cooling, process heating, and cooling loads, often providing substantial energy savings. For example, these systems can maintain precise temperatures necessary in industries like food and beverage, pharmaceuticals, and high-tech manufacturing.

Industrial geothermal heating and cooling installation costs are highly variable, given the diverse and specific needs of different industries. However, similar to other sectors, the significant energy savings and reduced maintenance costs can recoup the initial investment over time.

Waste heat

Waste heat sources can be used in a very similar way to geothermal. Here, the waste heat stream can be used to directly heat water (or another heat transfer fluid), which is then distributed to users. If the source temperature of the waste heat stream is low, a heat pump can be employed to further increase the temperature supplied to the thermal users.

The magnitude of waste heat is immense, totaling nearly 2000 PJ in the United States alone [18]. Examples of waste heat sources that could be utilized for district heating systems are quantified in Fig. 9.12 and include:

- Exhaust or flue gas from thermal power generation plants (CHP)
- Exhaust or heat rejection from chemical or industrial processes (i.e., fuel refining, cement production, steel making, glass making, forest and paper products, textiles, food, and beverage, etc.)
- Wastewater or sewage

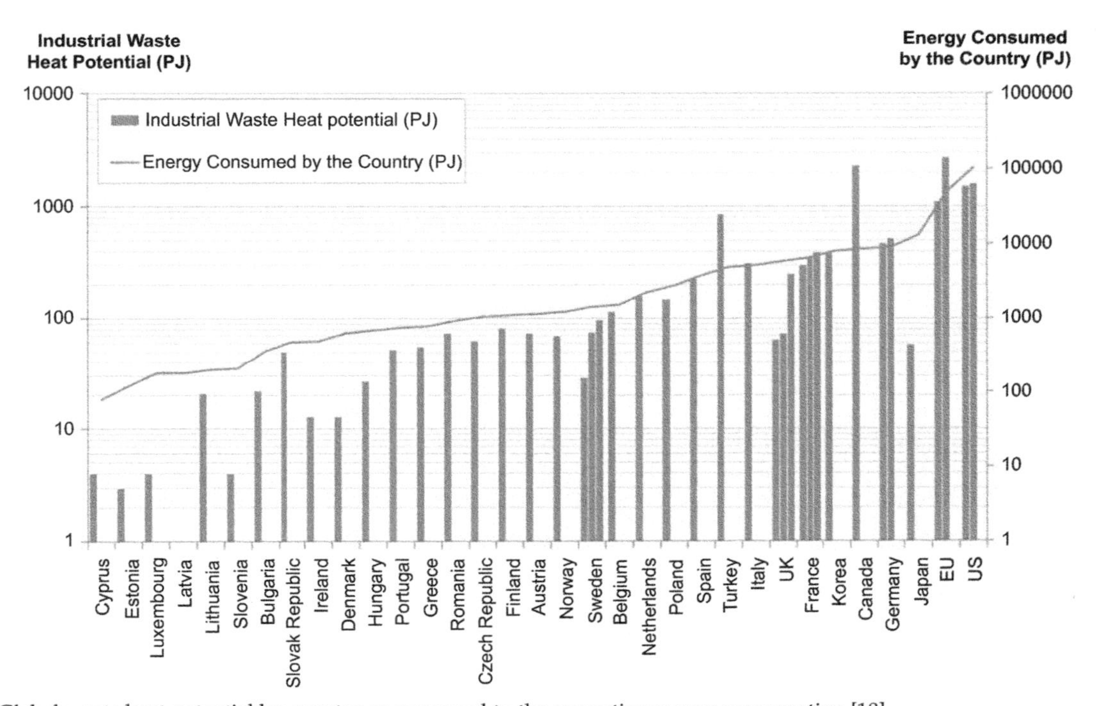

FIG. 9.12 Global waste heat potential by country as compared to the respective energy consumption [18].

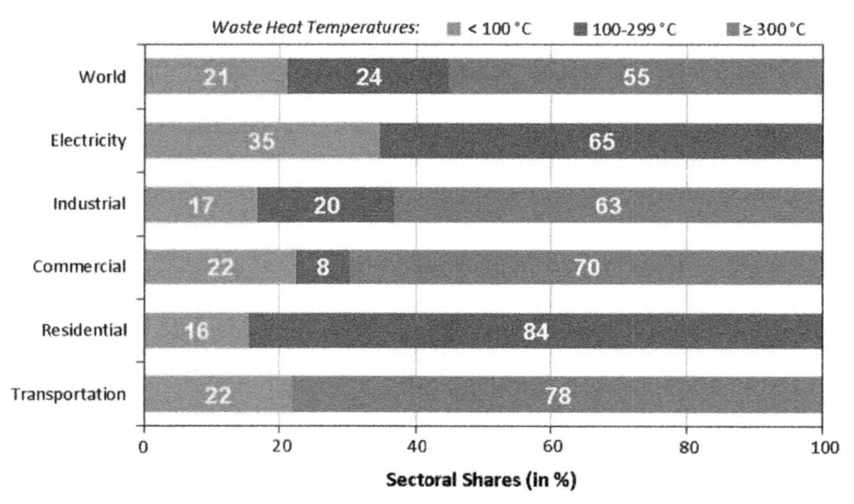

FIG. 9.13 Waste heat quantities categorized by sector and source temperature [18].

Considering the industrial sector, 83% of the waste heat is emitted at temperatures above 100°C [18] as shown in Fig. 9.13. This source temperature would be more than sufficient to directly heat water or generate steam for a district heating network.

Thermal power station (heat extraction)

Combining district heating and power generation is a technique that involves using waste heat from power generation to provide heating to nearby buildings and homes through a district heating system. This approach is also known as CHP or cogeneration. According to 2012 data from the International District Energy Association (IDEA), 281 existing CHP installations are providing district heating service; these systems reportedly supply over 6700 MWth of heating duty along with 30 million MWh of electricity [2].

In a typical power generation plant, a significant amount of heat is rejected after the process fluid runs through the turbine for electricity generation. This waste heat can be captured and used to heat water or other fluids, which can then be circulated through a network of pipes to provide heat to nearby buildings. See an example installation in Fig. 9.14. Here, a steam generator is installed in the exhaust duct downstream of a gas turbine. The high-grade steam is then directed through a steam turbine bottoming cycle to produce additional electricity. The low-grade steam exiting the

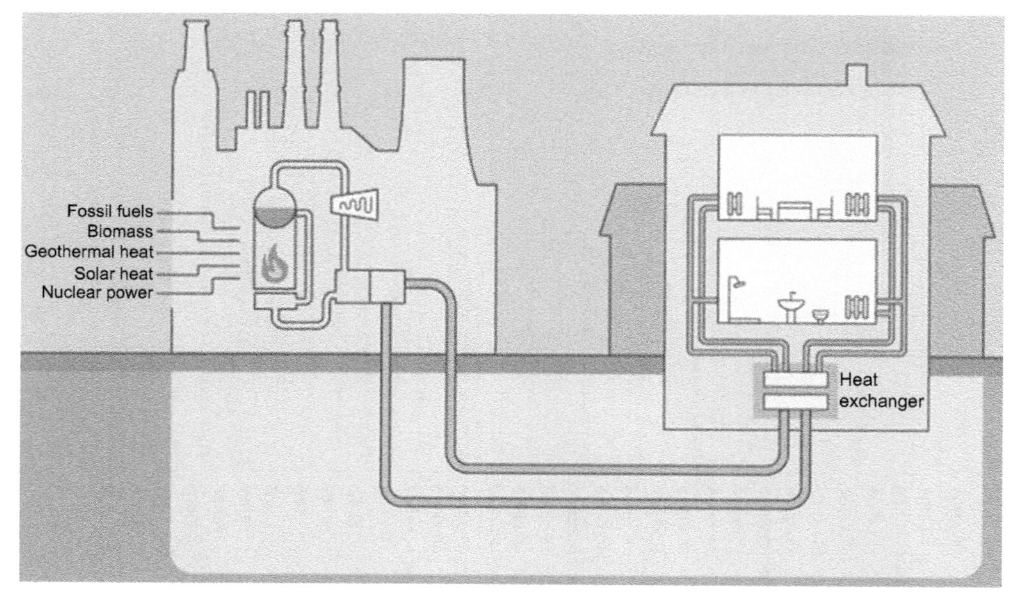

FIG. 9.14 Schematic of a steam power plant using waste heat from a turbine discharge to supply a district heating network [19]. *Source: Wikipedia, https://commons.wikimedia.org/wiki/File:District_heating.gif.*

steam turbine is then routed to a district heating network. Similarly, economizer heat exchangers could be installed in the gas turbine exhaust duct downstream of the steam generator to heat water for a liquid water district heating loop.

Utilizing the waste heat through a district heating system, the overall thermal efficiency of the process can be increased. This is because the waste heat that would otherwise be lost is instead used to provide a useful service, reducing the need for separate heating systems and reducing overall energy consumption.

Combining district heating and power generation has several advantages. It can reduce energy costs and greenhouse gas emissions by increasing the overall efficiency of the energy system. It can also provide a reliable source of heating for nearby buildings, which can be particularly useful in colder climates. As shown in the figure above, CHP plants could be especially useful in microgrid installations where a campus or community is supplied both electrical power and heat from a single, local system.

Solar and wind

Integrating more renewable resources into utility networks is a common theme among all decarbonization roadmaps. Solar and wind energy could be utilized to support district heating and cooling systems as well. Examples of renewable district heating and cooling systems include:

- Direct solar thermal systems
- Utilizing electricity for heat generation
- Utilizing electricity to operate heat pumps

In a direct solar thermal system, solar collectors are used to directly heat water, which is then supplied to thermal users. The schematic below presents this concept. Thermal fluid storage can also be incorporated to improve the year-round consistency in the heated water supply as illustrated in Fig. 9.15. See two examples for solar collectors also shown in Fig. 9.16 (e.g., parabolic troughs and flat plate collectors) [21]. In both cases, the solar energy is concentrated in order to heat the water passing through the tubes.

Alternatively, the electricity produced from solar PV cells or wind turbines can be used to generate heat through electric heating elements and/or heat pumps. Installing dedicated PV cells or wind turbines to support district heating networks is an option for reducing greenhouse gas emissions and replacing the use of fossil fuels. Green electricity can be used in multiple ways.

- Dedicated PV cells or wind turbines could generate electricity to power electric heating elements that directly heat the water or steam. The heat duty is then balanced by bringing more or fewer assets online or through the use of energy storage.
- Dedicated green electricity could be used to power the pump or compressor in a heat pump cycle. The heat pump could be referenced to ambient air, a body of water, a waste heat stream, or a geothermal reservoir. The heat pump would then boost the supplied district temperature above or below the reference point to provide the required heating or cooling duty. Reference the figure presented in the section "Data Center Applications" earlier.

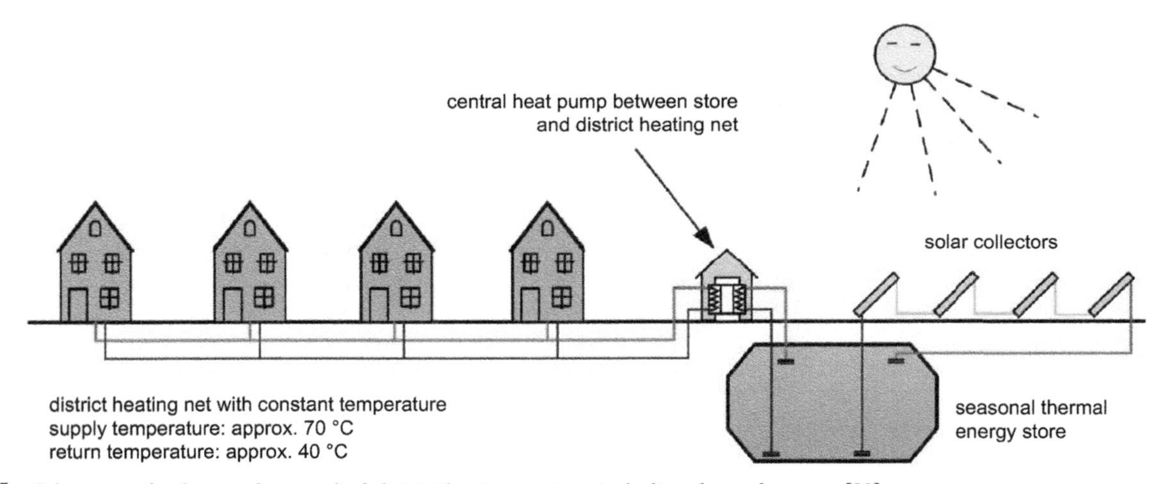

FIG. 9.15 Schematic of a direct solar-supplied district heating system, including thermal storage [20].

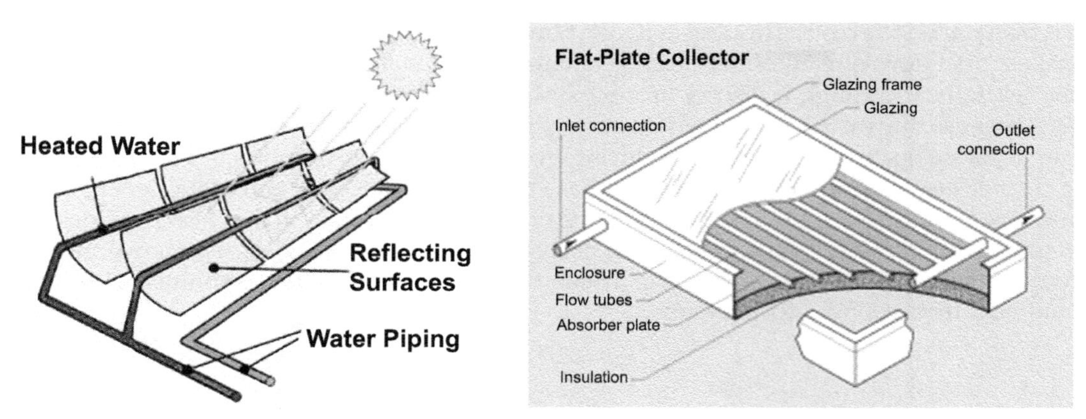

FIG. 9.16 Schematics for two types of solar collectors [21]. *Wikimedia Commons. (August 10, 2015). Flat Plate Glazed Collector [Online]. Available: https://upload.wikimedia.org/wikipedia/commons/4/40/Flat_plate_glazed_collector.gif; Wikimedia Commons. (August 10, 2015). Line Focus Collector [Online]. Available: https://upload.wikimedia.org/wikipedia/commons/thumb/a/ad/Solarpipe-scheme.svg/2000px-Solarpipe-scheme.svg.png.*

- Alternatively, district heating and cooling systems could be built integral to the solar and wind assets installed to support the electric grid. Here, the district heating/cooling system could be used to manage curtailment. When the electricity being produced (from renewable assets or others) is in excess of the current demand, this electricity could be used to charge a thermal store. The stored thermal energy could then be deployed as needed to support the district's thermal demand.

Installing district heating and cooling systems with renewable energy sources and the electricity grid could provide an integrated approach to help balance the grid and improve overall energy efficiency and effectiveness.

Heat pumps

In its simplest form, a heat pump is a cycle that moves thermal energy from a low-temperature source to a high-temperature sink. This is achieved through the use of a process fluid with favorable fluid-thermal properties (i.e., refrigerants) and work addition through a compressor or pump. See a simple schematic shown in Fig. 9.17.

In a heat pump, heat is added to the cycle from a low-temperature source (at the evaporator). The refrigerant typically enters the evaporator in the liquid phase at a very low temperature. From the evaporator, the refrigerant is run through a compressor. The pressure rise through the compressor is selected to give a substantial temperature rise in the refrigerant. Heat is then rejected at the condenser, adding heat to a higher-temperature source. The refrigerant is then expanded across a throttle valve, orifice, or similar (this could also be a turbine for energy recovery) to reduce the refrigerant temperature before returning to the evaporator.

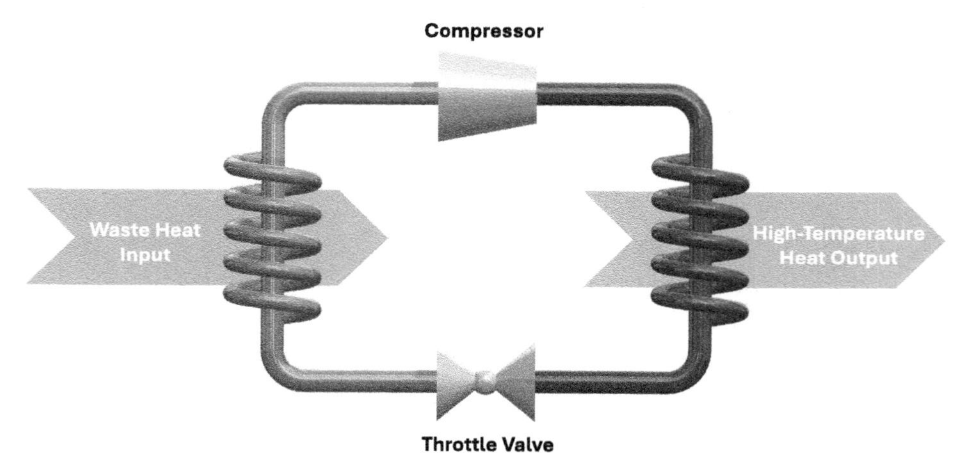

FIG. 9.17 Conceptual schematic of a heat pump [22].

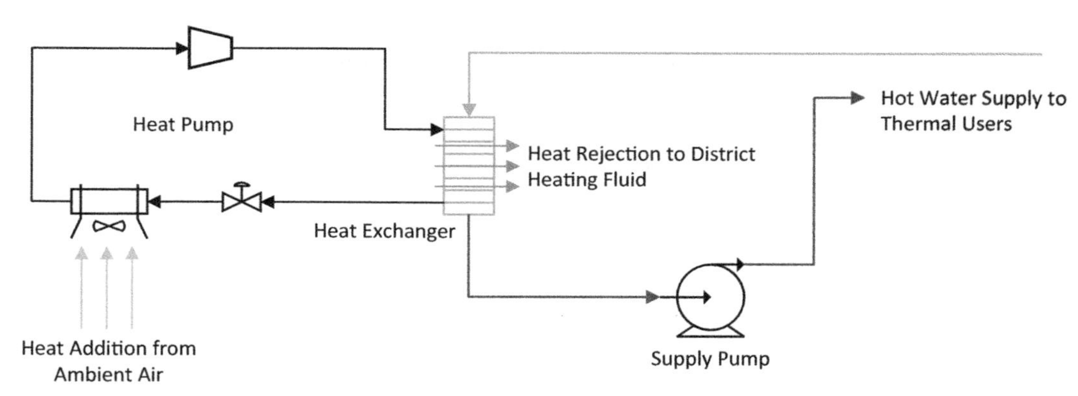

FIG. 9.18 Cycle schematic for a central heat pump supplying a district heating system.

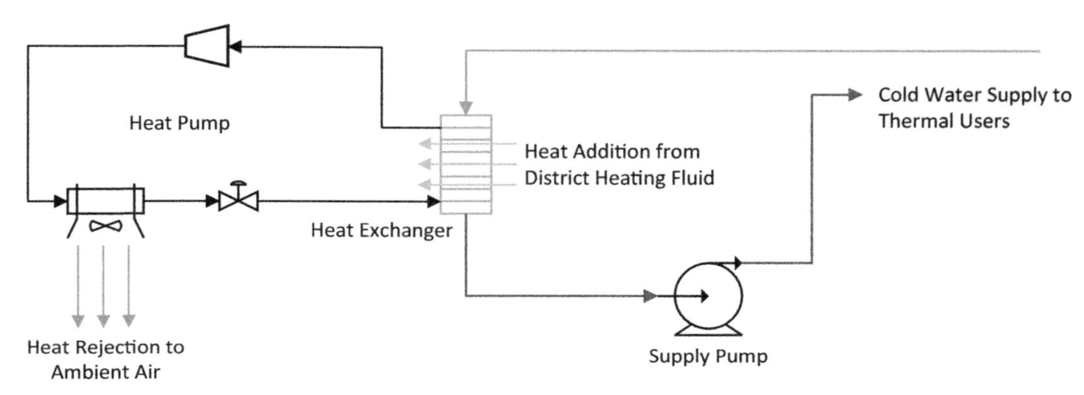

FIG. 9.19 Cycle schematic for central heat pump supplying a district cooling system.

Heat pumps can be very useful for (1) boosting the temperature of a warm stream to a more useful level or (2) reducing the temperature of a cold stream to a useful level. In this way, heat pumps can be used to produce hot streams for district heating or cold streams for district cooling. See the cycle schematics in Figs. 9.18 and 9.19 for a heat pump referenced to ambient air: one supplying heat and one supplying cooling.

Heat pumps are useful because they can be used to make usable hot or cold streams from moderate-temperature sources, such as ambient air, large bodies of water, waste heat streams, or geothermal. The amount of temperature lift required across the compressor (required head) drives the amount of power input to the heat pump cycle; a larger temperature lift from the referenced source temperature will require a larger power input.

Ambient

Ambient-referenced heat pumps are very common because the heat pump can be deployed anywhere, and there are no geographical limitations. Heat is usually exchanged with air through a large array of finned tubes; air is typically forced across the tube bundle using large axial fans. A key benefit of this type of system is the simplicity; heat pumps can be packaged with commercial components over a wide range of thermal scales.

In these installations, the power to operate the fans should be considered an additional parasitic loss on the heat pump. Also, the year-round changes in ambient temperature will greatly impact the performance of the heat pump; for example, in cold weather, a much larger temperature lift (and higher power consumption) will be required to produce hot water with the same temperature as compared to warm weather.

Large bodies of water

Large bodies of water can also be used as a reference heat source for heat pumps. Examples can include lakes, rivers, seawater, wastewater streams, and groundwater. Using a water source as the anchor for the heat pump can be desirable because the specific heat capacity of water is larger than that of air, so the heat exchangers do not need to be as large. Also, large bodies of water may offer a more consistent source/sink temperature throughout the seasons.

Downsides to using a water-referenced heat pump include geographical limitations and limitations on how much heat can be consumed by the body of water. In the case of a flowing water source (river, wastewater stream, etc.), the flow velocity and temperature of the water stream will impact the heat pump performance. Additionally, in drought years, river- and lake-referenced systems may be severely limited. Additionally, applications such as deep lake and seawater heat rejection may gather resistance or face regulatory limitations. The ecosystem impact of continuous heat addition must be considered.

Geothermal and waste heat

As noted in the above sections, heat pumps can also utilize geothermal reservoirs or waste heat sources. Geothermal-referenced systems are advantageous because the sub-surface rock temperature is very constant throughout the year. Also, the thermal mass of the formation is immense, meaning that heat consumption and rejection to the formation will result in relatively small changes in rock temperature. Depending on the thermal duty vs surface area of the sub-surface heat exchanger, it is possible that the region local to the district heating piping network will become cooler (or hotter) than the surrounding rock with time because the thermal conductivity of rock is low. However, if a geothermal system is referenced for cooling during the summer months and heating during the winter months, a well-designed sub-surface heat exchanger can minimize the impact on formation temperature.

Similarly, low-grade waste heat streams can be used as the heat source for a heat pump. The heat pump can then achieve sufficient temperature lift above the waste heat stream temperature to produce a more useful hot water stream. In comparing to an ambient-referenced heat pump, a waste heat stream supplying even 50–70°C will require significantly less temperature lift (and significantly less power consumption).

Systems and cycles for district heating/cooling

Common cycles and working fluids

District heating and cooling systems are designed to provide heating and cooling to multiple buildings or structures from a central source. These systems typically use a working fluid to transport thermal energy between the central source and the buildings being served. The working fluids used in district heating and cooling cycles can vary depending on the specific design of the system, but some common examples include:

1. Water: Water is the most common working fluid used in district heating and cooling systems. In these systems, hot water is typically used to provide heating to buildings, while chilled water is used for cooling. Water is an excellent heat transfer medium, and it is readily available and relatively inexpensive.

   Hot water provides some benefits over steam, as it typically has less distribution losses for the same thermal duty [2]. Hot water distribution is also well suited for applications where solar thermal, industrial process waste heat, or data center heat rejection are being used as the heat source.
2. Steam: Steam can also be used as a working fluid in district heating systems, particularly in areas where there is an abundance of steam available from industrial processes. Steam has a higher heat capacity than water and can transport more thermal energy per unit of mass.
3. Refrigerants: Some district cooling systems use refrigerants such as R134a, R410a, R1234ze, or CO2 as a working fluid. These refrigerants absorb heat from the buildings being cooled and transport it to a central cooling plant, where the heat is dissipated.
4. Thermal oils: In some district heating systems, thermal oils such as Dowtherm or Therminol are used as the working fluid. These oils have a high boiling point and are capable of transferring heat at higher temperatures than water.

Overall, the choice of working fluid will depend on a variety of factors, including the specific design of the district heating or cooling system, the thermal loads of the buildings being served, and the availability and cost of the working fluid.

Considering the cycle structure, the distribution network is relatively simple. The heat transfer fluid passes through a primary heat exchanger at the central plant to increase or decrease the temperature. The heat transfer fluid is then pumped through the distribution network to the individual users. Each user will have a dedicated heat exchanger or receiver equipment to extract the thermal duty. The heat transfer fluid is then returned to the central pumping station. See the cycle schematic below.

The cycle structure at the central pumping station can take many forms, depending on the heat source/sink. In the schematic shown in Fig. 9.20, a heat pump referenced to ambient is shown.

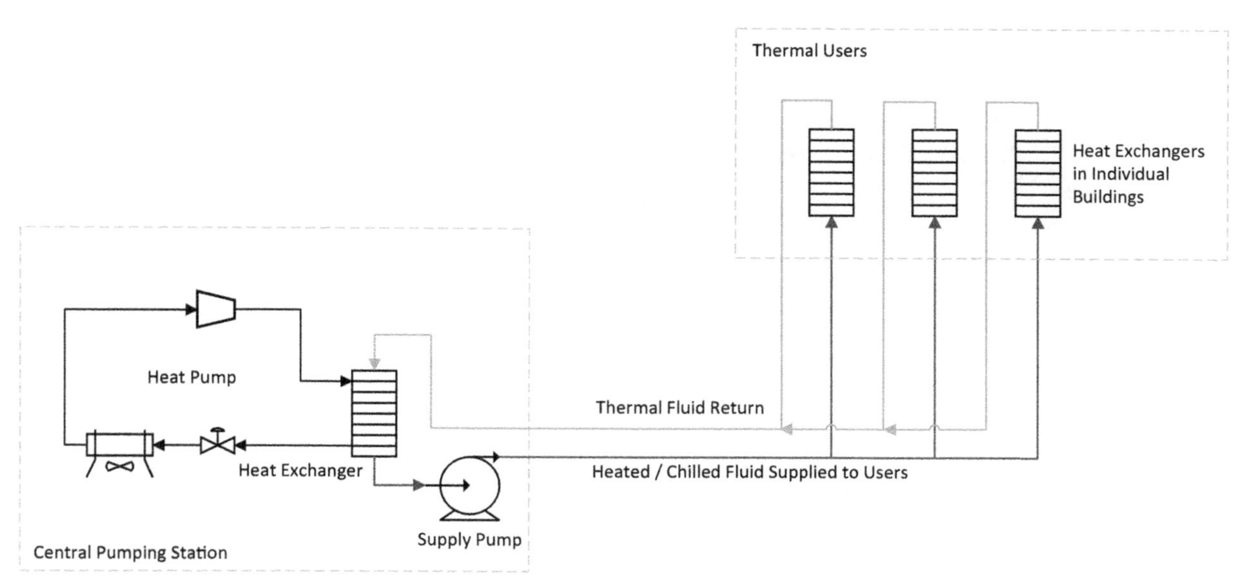

FIG. 9.20 Schematic of a heat pump referenced to ambient air, supplying heat to the district of users.

Central pumping station

The central pumping station is typically co-located with the primary thermal input for the district heating network. Here, the heat transfer fluid (typically water) interfaces with the heat source/sink through the primary heat exchanger(s), and the prime mover of the heat transfer fluid produces head and flow to supply the distribution network.

The pumping station is required to manage the temperature of the fluid as well as the flow rate in order to properly meet the thermal demand and balance the thermal duty of the system. This often means that the station has to operate at turndown conditions. The station is often sized to meet the peak demand for thermal capacity but only operates at this point for a small percentage of the year. Operation at part load (turndown) is then accomplished by flow control or temperature control. Depending on the heat source/sink, temperature control is not always a viable option as it may be tied to another process whose operating point cannot be altered. Therefore, turndown is often controlled using flow control on the heat transfer fluid; this may take the form of a recycle line, throttling, or pump speed control. Flow control through recycle or throttling are simple to install and operate, but these control methods are inefficient as they do not reduce the pump power consumption at turndown conditions. Here, variable speed control is the best choice if possible.

Distribution network

From the central pumping station, the heated or chilled fluid is distributed to the thermal users through the distribution network. This often takes the form of an extensive piping network buried underground. Factors key to the distribution network are line size, insulation, and controllability. Properly selecting line size is critical to minimizing frictional losses while managing system costs. Pipe material and insulation selection are also critical. Distribution pipes with high thermal energy losses will degrade the system's efficiency and effectiveness. Many legacy systems were built with cast iron piping networks. However, many new installations utilize high-temperature plastic piping for chilled water systems or hot water systems up to 180 °F at relatively low pressures [23]. Plastic piping has the advantages of being flexible as well as being delivered in long continuous runs, sometimes hundreds of feet, which eliminates many joints and connections in the network. Also, plastic piping can be delivered pre-insulated with jacketing to significantly reduce the installation time. See the example installations shown in Figs. 9.21 and 9.22.

Distribution headers are often installed under roadways, with smaller feed lines supplying individual users. The buried pipes can be installed in underground tunnels to allow maintenance or access to any valves and pumps. In a large distribution network, valving may be required to isolate portions of the network if certain users do not require continuous input. Also, in distribution networks spanning long distances, booster pumps may be required to periodically increase the pressure in the heat transfer fluid to overcome frictional losses.

FIG. 9.21 Example of buried distribution network piping [23]. *From REHAU. "INSULPEX® pre-insulated PEXa piping." INSULPEX | Mechanical & plumbing | REHAU.*

FIG. 9.22 Example of fourth generation district heat network insulated piping [24]. *Source: Wikipedia, https://commons.wikimedia.org/wiki/File:District_heating_pipelines_V%C3%A4ster%C3%A5s_1.JPG.*

Distribution networks may take the form of a two-pipe or four-pipe system. Depending on the heat transfer fluid, the heat source/sink at the central pumping station, and the variation in heating and cooling demand by the users, it may be necessary to supply heated and chilled fluid to the distribution network in separate flow loops. This would be accomplished with a four-pipe system. If the total thermal duty (heating and cooling demand) can be sufficiently balanced, then a two-pipe system can be installed. Most existing systems are considered two-pipe systems (meaning one supply line and one return line) because, in these systems, all thermal users are demanding the same deliverable (i.e., all users are requiring heat, all users are requiring steam, etc.).

Equipment

Pumps

Pumps or compressors are a critical component in district heating networks as they are required to actually transport the heat from the central pumping station or the heat source/sink to the users. For water-based systems, several manufacturers currently supply large centrifugal pumps that support this application well. See an example in Fig. 9.23. For larger districts, multiple pumps may be installed in parallel to meet the heat demand and flow capacity as shown in Fig. 9.24.

Heat exchangers

Heat exchangers are also a critical component for district heating systems. One or multiple primary heat exchangers will be installed at the central pumping station to heat or cool the heat transfer fluid. For thermal users, heat exchangers will be required in each building as well (as shown in the cycle schematic above).

FIG. 9.23 Example hot water pump for the distribution network, image courtesy of DESMI [25].

FIG. 9.24 Example distribution network with an array of pumps, image courtesy of DESMI [26].

District heating applications often use plate type heat exchangers (see Fig. 9.25). Plate heat exchangers offer advantages over shell and tube type heat exchangers in that they are more compact. Brazed / diffusion bonded or gasketed plate heat exchangers can be used. Brazed or diffusion bonded plate heat exchangers offer a hermetic solution to minimize the risk of leaks; the interface seals in gasketed plate heat exchangers can wear and eventually begin to leak over time, especially in applications with high thermal cycling. Based on this consideration, brazed or diffusion bonded plate heat exchangers are superior; however, these units cannot be disassembled to be inspected, cleaned, or maintenance. For this reason, the type of heat exchanger seal should be selected carefully based on the application and working fluids. Note that plate heat exchangers can be used as the primary heat exchanger at the central plant as well as in individual buildings.

Heat exchangers may also take different forms depending on the heat source or sink. Heat exchanger manufacturers offer some source-specific solutions. This heat exchanger would be well-suited to a CHP application. This heat exchanger would be installed in the exhaust gas duct from the power generation process. Water would then be routed into this heat exchanger to generate steam, which could then be supplied to the distribution network.

FIG. 9.25 Plate-style heat exchangers common in district heating systems [27, 28]. *Source: Wikipedia, https://commons.wikimedia.org/wiki/File:Plate_heat_exchanger_-_dismantled_pic02.jpg.*

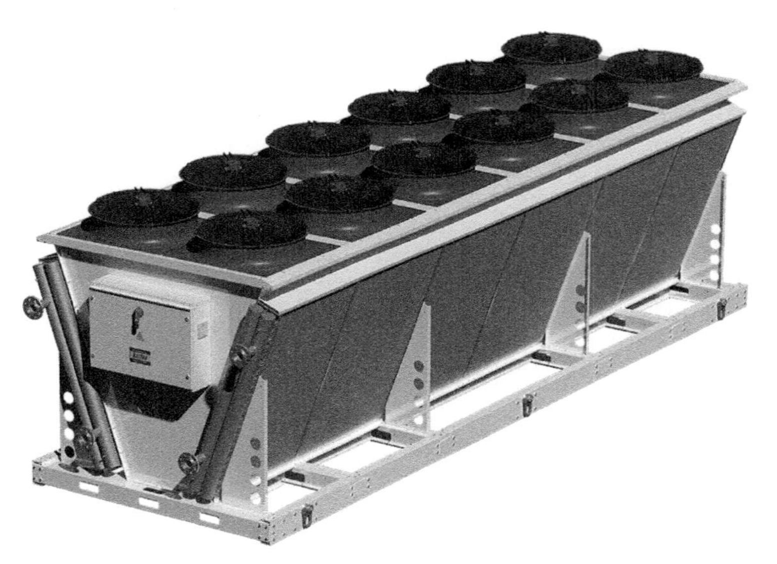

FIG. 9.26 Example heat pump skid. *Credit: Used with permission from Kaltra GmbH, https://www.kaltra.com/products/air-cooled-condensers.*

Considering solar input, solar heating arrays could be used at the primary heat exchanger at the central pumping station.

For heat pump applications referenced to ambient, large process to air heat exchangers would be utilized. See an example heat pump skid shown in Fig. 9.26. The V-shape bays are comprised of finned tube bundles; the heat pump fluid would pass through these tubes. Large axial fans pull ambient air across the tube bundles to increase the temperature in the heat pump fluid.

Heat pumps to boost temperature

In some district heating networks, it may be beneficial to install heat pumps for individual users. This would allow the delivered temperature to be significantly higher than the heat transfer fluid temperature (see the schematic shown in Fig. 9.27). In these applications, the heat pump would be referenced to the distributed fluid (likely hot water), which

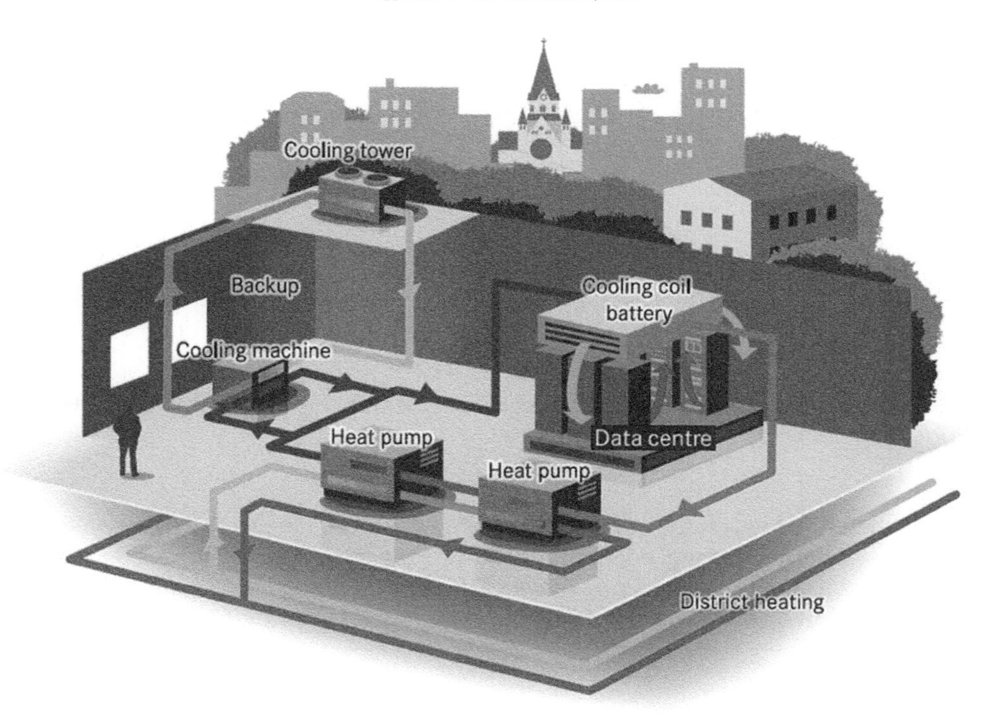

FIG. 9.27 Schematic of a hybrid district energy system with individual heat pumps located at specific users to boost the supply temperature [29]. *Source: Heat pumps in district heating and cooling systems, Part of Today in the Lab – Tomorrow in Energy? Technology Report, 17 November 2020; https://www.iea.org/articles/heat-pumps-in-district-heating-and-cooling-systems.*

still reduces the on-site power consumption compared to operating an air-referenced heat pump. Operating the local heat pump referenced to the distributed fluid also allows for some stability, as compared to air-referenced heat pumps that must deal with ambient temperature swings throughout the year.

Installing local heat pumps could be used to generate on-site steam from a hot-water district heating network. Additionally, local heat pumps could be useful in implementing low-grade fourth generation district heating networks, making district heating viable in low density areas.

Other applications and combined systems

Enhanced geothermal systems

Enhanced geothermal systems (EGS) are a type of geothermal energy technology that seeks to extract heat from deep underground rock formations that do not have sufficient permeability or water content to support conventional geothermal systems. EGS technology involves drilling deep wells into hot rock formations, fracturing the rock, and then injecting water or other fluids into the fractures to create an artificial geothermal reservoir.

EGS systems work by circulating a working fluid, such as water or a refrigerant, through the hot rock formations to extract heat, which can then be used to generate electricity or provide heating and cooling. Residual heat is transferred from the working fluid through a heat exchanger, and the fluid is then re-injected into the well, and the cycle starts again.

EGS has the potential to significantly expand the availability of geothermal energy resources, as it can be deployed in areas where conventional geothermal systems are not economically feasible. However, EGS technology is still in the early stages of development and faces several technical challenges, including managing the high pressures and temperatures in the deep rock formations and minimizing the risk of inducing seismic activity through the fracturing process. Despite these challenges, EGS is seen as a promising technology for expanding the use of geothermal energy and reducing greenhouse gas emissions.

When considering geothermal for power generation, district heating should still be considered as a CHP solution. Geothermal reservoirs generally produce temperatures in the range of 100–250°C. In this range, thermal efficiencies for

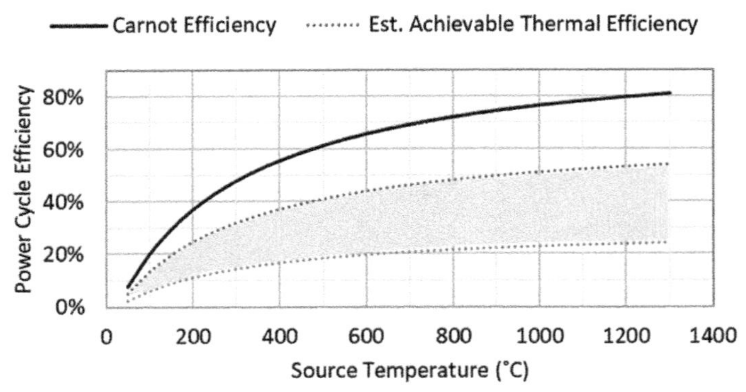

FIG. 9.28 Estimated achievable power cycle efficiencies and Carnot efficiency as a function of source temperature, referencing ambient as the sink temperature.

power generation are inherently low, on the order of 10%–25%; see the plot in Fig. 9.28. This means that (in a Brayton or Rankine cycle), 75%–90% of the geothermal energy input to the power cycle will be rejected at the cooler before returning to the process compressor. In utility scale, power generation installations, the amount of rejected heat will be vast. Utilization of this excess low-grade heat (directly or with a heat pump) for district heat applications holds significant merit.

Thermal energy storage

Thermal energy storage (TES) is a technology that allows excess thermal energy to be stored for later use, providing a way to manage energy demand and reduce energy costs. There are several types of thermal energy storage systems, including:

1. Sensible heat storage: This is the most common type of TES system and involves storing thermal energy by raising the temperature of a material, such as water, concrete, or gravel. When the stored heat is needed, the material is circulated through a heat exchanger to transfer the heat to a working fluid, such as air or water, which can then be used for space heating, domestic hot water, or industrial processes.
2. Latent heat storage: This type of TES system stores thermal energy by changing the phase of a material, such as melting or solidifying a phase-change material (PCM). When the stored heat is needed, the PCM is heated or cooled to release or absorb the stored energy, which can then be used for space heating, cooling, or other thermal applications.
3. Thermochemical storage: Thermochemical storage involves storing thermal energy by using a chemical reaction to absorb or release heat. For example, a salt hydrate can be heated to release water vapor and store the heat, and then cooled to reabsorb the water vapor and release the stored heat.
4. Stratified thermal storage: This type of TES system involves storing hot and cold fluids in separate layers within a storage tank. When the stored heat is needed, the hot fluid is drawn from the top of the tank, and when cooling is needed, the cold fluid is drawn from the bottom of the tank.
5. Sorption storage: Sorption storage involves storing thermal energy by adsorbing or absorbing a gas or liquid onto a porous material, such as activated carbon. When the stored heat is needed, the gas or liquid is released and the stored energy is transferred to a working fluid.

These different types of TES systems can be used for various applications, including building heating and cooling, industrial processes, and solar thermal energy storage. The choice of TES system depends on the specific application, as well as factors such as the required storage capacity, thermal efficiency, and cost.

References

[1] U.S. Energy Information Administration (EIA), U.S. District Energy Market Characterization, Prepared by ICF and IDEA, 2018, p. 1. Available at https://www.eia.gov/analysis/studies/buildings/districtservices/pdf/districtservices.pdf.

[2] U.S. Department of Energy, Office of Energy Efficiency and Renewable Energy, Combined Heat and Power Technology Fact Sheet Series: District Energy Systems Overview, 2020, DOE/EE-2125. September. Available at: https://www.energy.gov/eere/amo/articles/combined-heat-and-power-technology-fact-sheet-series-district-energy.

[3] U.S. Energy Information Administration (EIA), U.S. District Energy Market Characterization, Prepared by ICF and IDEA, 2018. Available from https://www.eia.gov/analysis/studies/buildings/districtservices/pdf/districtservices.pdf.
[4] International Energy Agency (IEA), District Heating, Available at: https://www.iea.org/energy-system/buildings/district-heating.
[5] Young M., How the New York City Steam System Works, Untapped New York, Available at: https://untappedcities.com/2021/07/09/new-york-city-steam-system/.
[6] EnBW Company, District Heating – Energy with a Future, Available at: https://www.enbw.com/company/the-group/energy-production/district-heating/.
[7] C. Kyllmann, Germany to Connect 100,000 Buildings to District Heating Annually, Clean Wire Energy, 2023. Available at: https://www.cleanenergywire.org/news/germany-connect-100000-buildings-district-heating-annually.
[8] European Commission, State Aid: Commission Approves €2.98 billion German Scheme to Promote Green District Heating, Press Release, 2022. 2 Aug. Brussels.
[9] CAX, District Heating Design: Fourth Generation District Heat Networks Employ Lower Temperatures and Recycle Heat, Available at: https://www.icax.co.uk/District_Heating_Design.html.
[10] International District Energy Association, District Cooling, Available at: https://www.districtenergy.org/topics/district-cooling.
[11] Austin Energy, Commercial Services: District Cooling, Available at: https://austinenergy.com/commercial/commercial-services/district-energy-cooling/district-cooling.
[12] Combined Heat and Power Alliance, Combined Heat and Power Potential in Data Centers, 2021, 30 Jun. Available at: https://chpalliance.org/resources/publications/combined-heat-and-power-potential-in-data-centers/.
[13] C. Metcalfe, Heat from an Amazon Data Center is Warming Dublin's Buildings, Reasons to be Cheerful, 2023. 20 Feb. Available at: https://reasonstobecheerful.world/data-center-heat-green-energy/#:~:text=When%20servers%20in%20the%20Amazon,unusual%20form%20of%20green%20energy.
[14] B. Schweber, Harvesting Data Center Heat: Opportunity or Obstacle, EE Times, 2023. 31 Jan. Available at: https://www.eetimes.com/harvesting-data-center-heat-opportunity-or-obstacle/.
[15] Stratego Enhanced Heating & Cooling Plans, Quantifying the Potential for District Heating and Cooling in EU Member States, Project No. IEE/13/650, Work Package 2, Background Report 6. Co-funded by the Intelligent Energy Europe Programme of the European Union.
[16] I. De La Cruz, C.E. Ugalde-Loo, District heating and cooling systems, in: Microgrids and Local Energy Systems, 2021, https://doi.org/10.5772/intechopen.99740. Published: December 15th. Available from: https://www.intechopen.com/chapters/78400.
[17] Center for Sustainable Systems, University of Michigan, Geothermal Energy Factsheet. Pub. No. CSS10-10, 2022, Available from: https://css.umich.edu/publications/factsheets/energy/geothermal-energy-factsheet.
[18] W. Chen, Z. Huang, K.J. Chua, Sustainable energy recovery from thermal processes: a review, Energ. Sustain. Soc. 12 (2022) 46, https://doi.org/10.1186/s13705-022-00372-2.
[19] Environmental and Energy Study Institute, The Role of District Energy/Combined Heat and Power in Energy and Climate Policy Solutions, 2009, 21 Apr. Available from https://www.eesi.org/briefings/view/the-role-of-district-energy-combined-heat-and-power-in-energy-and-climate-p.
[20] D. Bauer, R. Marx, H. Drück, Solar District heating for the built environment-technology and future trends within the European project EINSTEIN, Energy Procedia (2014) 57, https://doi.org/10.1016/j.egypro.2014.10.303.
[21] Energy Education. Solar collector. Available from: https://energyeducation.ca/encyclopedia/Solar_collector.
[22] ARANER, Large Heat Pumps in District Heating Systems, Available from: https://www.araner.com/blog/large-heat-pumps-in-district-heating-systems.
[23] Plastic Pipe Institute (PPI), District Energy Heating & Cooling, Available from: https://www.plasticpipe.org/BuildingConstruction/BuildingConstruction/-Applications-/DistrictEnergy-Heating-Cooling.aspx.
[24] ICAX, District Heating Design: Fourth generation district heat networks employ lower temperatures and recycle heat, Available from: https://www.icax.co.uk/District_Heating_Design.html.
[25] DESMI, Pumps for District Heating, Available from: https://www.desmi.com/segments/utility/applications/district-heating/.
[26] DESMI, High Efficiency Pumps for Aalborg District Heating, Available from: https://www.desmi.com/customer-stories/high-efficiency-pumps-for-aalborg-district-heating/.
[27] University of Strathclyde Engineering, District Heating From Wind: Kirkwall, Available from: https://www.esru.strath.ac.uk/EandE/Web_sites/11-12/District_heating_from_wind/otherequipment.html.
[28] Alfa Laval. AlfaNova Fusion Bonded Plate Heat Exchangers. Available from: https://www.alfalaval.us/products/heat-transfer/plate-heat-exchangers/fusion-bonded-plate-heat-exchangers/alfanova/.
[29] G. Rueter, Can Heat Pumps Replace Fossil Fuels for Heat?, NATURE AND ENVIRONMENT, GERMANY, 2022 Sep 16. Available from: https://www.dw.com/en/heat-pumps-district-heating-decarbonize-energy-crisis-russian-oil-and-gas/a-63053664.

CHAPTER

10

Gas-to-liquids and other decarbonized energy carriers

Subith Vasu Sumathi[a], *Ramees K. Rahman*[a], *Rahul Iyer*[b], *Sebastian Freund*[c], *and Karl Wygant*[d]

[a]University of Central Florida, Orlando, FL, United States [b]KCK Group, Cupertino, CA, United States [c]Energyfreund Consulting, Munich, Germany [d]Ebara Elliott Energy, Jeannette, PA, United States

Gas-to-liquids

Gas-to-liquids (GTL) is a technology that converts natural gas or other gaseous hydrocarbons into liquid fuels and other valuable products [1]. The process typically involves two main steps: gasification and Fischer-Tropsch synthesis. In the gasification stage, the gas feedstock is transformed into a synthesis gas (syngas) through a combination of high temperature and pressure, along with steam, oxygen, or air. The syngas, consisting primarily of carbon monoxide and hydrogen, is then subjected to the Fischer-Tropsch (FT) synthesis, which is converted into liquid hydrocarbons using catalysts. GTL technology offers several advantages, including the ability to monetize stranded gas reserves and the production of high-quality, clean-burning fuels. GTL fuels have low sulfur and aromatic content, reducing emissions of pollutants, as FT catalyst systems are generally intolerant of sulfur compounds, and therefore, a desulfurizing step is required during gasification. GTL coproducts, such as synthetic lubricants and waxes, have applications in various industries. Despite its potential, GTL faces challenges related to high capital costs, energy efficiency, and environmental concerns. However, ongoing research and development efforts continue to improve the efficiency and viability of GTL technology as a means to harness natural gas resources and diversify the liquid fuel market.

Natural gas and other gaseous hydrocarbons are typical raw materials for gas-to-liquids (GTL) technology. Natural gas, which primarily consists of methane, is the most common feedstock for GTL processes. It is often sourced from conventional natural gas reserves, shale gas deposits, or associated gas produced during oil extraction. However, GTL technology can also utilize other gaseous hydrocarbons, such as stranded or flared gases, or coal seam gas. It can also use solid fuels such as coal [2] and biomass as feedstocks [3]. Overall, the flexibility of GTL technology allows for the utilization of various gaseous hydrocarbons, enabling the conversion of stranded or unconventional gas resources into liquid fuels.

GTL technology has the potential to contribute to greenhouse gas reduction efforts. On a life-cycle basis, GTL fuels can have lower emissions of greenhouse gases compared to conventional petroleum-based fuels [4]. While tailpipe GHG emissions are essentially identical, upstream emissions associated with hydrocarbon production can be lower if renewable sources are used. Shell, which is the largest operator of GTL FT (Pearl GTL, Ras Laffan, Qatar) production globally, provided a life-cycle analysis owing most GHG benefits to upstream GHG reduction such as flare mitigation or the use of renewable or biomethane resources [5]. Additionally, during the GTL process, the carbon monoxide and hydrogen present in the syngas are converted into liquid hydrocarbons, resulting in reduced levels of sulfur, nitrogen, and aromatic compounds in the final product. As a result, GTL fuels exhibit lower levels of pollutants and emissions that contribute to air pollution and climate change. GTL fuels also have a higher cetane rating, which improves combustion efficiency and reduces particulate matter emissions. Furthermore, GTL can utilize stranded or flared natural gas reserves, which would otherwise be wasted, reducing methane emissions, a potent greenhouse gas [6]. While the overall greenhouse gas reduction potential of GTL depends on factors such as feedstock selection, process efficiency,

Energy Transport Infrastructure for a Decarbonized Economy
https://doi.org/10.1016/B978-0-443-21893-4.00003-9

Copyright © 2025 Elsevier Inc. All rights are reserved, including those for text and data mining, AI training, and similar technologies.

and energy source, it has the potential to contribute to the transition toward lower-carbon and cleaner energy systems, supporting efforts to mitigate climate change.

Some of the products that can be produced using the GTL technology are listed as follows:

- Gasoline: GTL technology can produce high-quality gasoline with excellent combustion properties and a lower sulfur and aromatic content compared to conventional gasoline.
- Diesel: GTL diesel is a cleaner-burning alternative to conventional diesel fuel, with lower levels of sulfur, particulate matter, and aromatics. It typically has a higher cetane rating, leading to improved combustion efficiency and reduced emissions.
- Jet fuel: GTL can produce jet fuel that meets the stringent specifications required for aviation use, offering reduced emissions and improved fuel efficiency compared to traditional jet fuels.
- Naphtha: GTL naphtha is a feedstock for the production of various petrochemicals, including plastics, solvents, and chemicals used in the manufacturing industry.

In addition to liquid fuels, GTL technology can also produce other valuable products as follows:

- Lubricants: GTL technology can produce lubricants that offer similar or better performance and stability compared to lubricants produced from conventional crude oil. Since the properties are essentially the same, they can be used in various applications, including automotive, industrial, and aviation sectors.
- Waxes: GTL waxes are used in applications such as candles, coatings, packaging, and personal care products.
- Chemicals: GTL processes can produce a variety of chemicals, including alcohols, olefins, and solvents, which find applications in the production of plastics, detergents, and other chemical products.

These GTL products offer advantages such as improved environmental performance, reduced emissions, and enhanced product quality compared to their conventional counterparts, contributing to sustainable and cleaner energy systems.

Gasoline

In 1859, Edwin Drake accomplished the remarkable feat of drilling a crude oil well in Pennsylvania. As part of his efforts, Drake distilled the oil to obtain kerosene, which served as a valuable source of lighting during that time. Despite the distillation process also producing gasoline and other petroleum products, Drake did not have a use for them and disposed of them [7]. The true value of gasoline as a fuel only became apparent in 1892, when the automobile was invented. Subsequently, gasoline gained recognition as a valuable fuel source. By the year 1920, the roads witnessed the presence of approximately 9 million gasoline-powered vehicles, and service stations offering gasoline began to emerge throughout the country. Presently, gasoline stands as the primary fuel for nearly all light-duty vehicles in the United States.

Gasoline is a mixture of several hydrocarbons. This includes aromatics, olefins, and other straight-chain hydrocarbons [8]. Often, gasoline is blended with oxygenates such as ethanol. In the United States, gasoline used in light vehicles comprises nearly 10%–15% ethanol by volume. Several researchers have shown that blending gasoline with ethanol reduces its sooting tendency and reduces particulate matter emissions [9,10]. In some other parts of the world, gasoline is blended with methyl tertiary butyl ether (MTBE) to improve its combustion properties [11].

The vapor pressure of gasoline is higher than that of fuel oils and other heavier products obtained during crude oil distillation. Additionally, the flash point of gasoline is less than −23°C. Due to this, particular attention is required during the storage and handling of gasoline. This includes the use of a floating roof or pressurized tank for storage to reduce vaporization loss and the use of antistatic additives in gasoline during transportation to reduce the risk of fire hazards.

Gasoline is a highly versatile fuel that can exhibit a wide variety of compositions, depending on various factors. The most common form of gasoline comprises a mixture of hydrocarbons derived from crude oil, typically consisting of carbon atoms ranging from 7 to 12 in length. However, gasoline can also contain additives such as ethanol, which is often blended in varying proportions to enhance octane ratings and reduce harmful emissions. Additionally, different regions or countries may have specific regulations governing gasoline composition, leading to variations in the presence of oxygenates, aromatics, and other compounds. Furthermore, advancements in fuel technology have led to the development of specialized gasoline formulations, such as high-performance fuels with increased octane levels or biofuels derived from renewable sources. Overall, the composition of gasoline is a complex blend that continues to evolve, driven by factors such as environmental considerations, engine efficiency, and regulatory standards.

In addition to hydrocarbons and additives, gasoline can also contain trace amounts of impurities and contaminants that may vary depending on the refining and distribution processes. These impurities can include sulfur, benzene, and other particulate matter. However, stringent regulations and advancements in refining techniques have significantly reduced the presence of these pollutants in modern gasoline. Furthermore, as the demand for cleaner and more sustainable energy sources grows, there is a rising interest in alternative fuels that can replace or supplement gasoline. These alternatives include electric vehicles powered by batteries or hydrogen fuel cells, as well as biofuels derived from plant-based sources.

When storing gasoline in a storage facility, the following things need to be considered:

- Storage containers: Use approved storage containers or tanks specifically designed (e.g., API 650 or 620) for gasoline, which are capable of handling low vapor pressure. These containers should have proper seals and closures to minimize evaporation and leakage.
- Vapor recovery systems: Install vapor recovery systems in storage tanks to capture and control gasoline vapors. These systems help reduce emissions and improve safety by preventing the release of flammable vapors into the environment.
- Ventilation: Ensure proper ventilation in storage areas to disperse any accumulated gasoline vapors. Good ventilation reduces the concentration of flammable vapors, minimizing the risk of ignition.
- Fire safety: Implement robust fire safety measures, including fire suppression systems, fire extinguishers, and emergency response protocols, according to National Fire Protection Association (NFPA) standards. Proper grounding and bonding are also critical to preventing static electricity buildup and potential ignition.
- Transport vessels: Use specialized tanks or containers designed for gasoline transport. These vessels should have proper venting and pressure relief systems to prevent over-pressurization during transportation.
- Avoid overfilling: Do not overfill storage containers or transport tanks. This can be accomplished by equipping storage containers with level sensors with auto-shutoff or alarms. Allow sufficient headspace to accommodate thermal expansion and prevent excess pressure buildup.
- Compliance with regulations: Adhere to local regulations and safety codes governing the storage and transportation of gasoline, including requirements for vapor recovery, containment, labeling, and safety equipment.
- Training and awareness: Provide training to personnel involved in gasoline storage and transport on safe handling procedures, vapor control, emergency response, and the importance of following safety protocols.
- Regular inspection and maintenance: Regularly inspect storage containers, transport vessels, and associated equipment for any signs of damage, leaks, or malfunction. Perform routine maintenance to ensure the integrity of the storage and transport systems.
- Spill prevention and response: Implement spill prevention measures, such as secondary containment systems and proper transfer procedures, to minimize the risk of spills during storage and transport. Establish spill response protocols and have appropriate spill cleanup materials readily available.

By considering these specific factors and implementing appropriate safety measures, the storage and transport of gasoline can be conducted in a manner that minimizes hazards, ensures regulatory compliance, and promotes the safety of personnel and the environment.

Fuel oils

Fuel oils, also known as heating oils or residual fuels, are a type of petroleum-based fuel commonly used for heating, power generation, and industrial processes. Fuel oils are derived from the heavier fractions of crude oil, which are left behind after the refining process. They consist of a complex mixture of hydrocarbons with higher molecular weights, making them thicker and less volatile than gasoline or diesel fuels. The viscosity of fuel oils can vary depending on their intended use, with lighter grades typically used for heating and heavier grades for industrial applications. Fuel oils are commonly categorized by their sulfur content, with lower sulfur fuels being preferred due to environmental concerns. However, high-sulfur fuel oils are still used in certain industries that employ pollution control technologies. Despite efforts to transition to cleaner energy sources, fuel oils continue to be a reliable and cost-effective option for many heating and industrial needs, although their usage is gradually being replaced by cleaner alternatives in some regions.

Fuel oils have distinct properties that make them suitable for specific applications. Their high energy density and slow combustion rate make them ideal for use in boilers, furnaces, and power plants, where a steady and controlled release of heat is required. Fuel oils are often used in maritime shipping as well, powering large vessels and providing the necessary propulsion for long-haul journeys. In certain industries, fuel oils serve as feedstocks for various processes, including asphalt production and the manufacturing of lubricants and waxes. While fuel oils offer advantages in terms of energy content and affordability, they also present challenges related to their environmental impact. The combustion of fuel oils releases pollutants such as sulfur dioxide, nitrogen oxides, and particulate matter, contributing to air pollution and potential health hazards. As a result, there is increasing pressure to adopt cleaner alternatives and improve the efficiency of fuel oil combustion to minimize emissions and promote a more sustainable energy future.

Fuel oil transportation involves the movement of fuel oil from refineries or storage facilities to end users or distribution points. Due to its relatively high viscosity and density, fuel oil is typically transported via tanker trucks or barges. Tanker trucks provide a flexible and efficient means of delivering fuel oil to residential and commercial customers. These trucks are equipped with specialized compartments to store and transport the fuel oil safely. Some are equipped with heating systems to keep fuel oil viscosity within limits to enable efficient pumping. Barges, on the other hand, are used for larger-scale transportation, especially for delivering fuel oil to coastal regions or remote areas with limited access to pipelines. Barges can transport large volumes of fuel oil, reducing the need for multiple trips and optimizing the supply chain. The transportation of fuel oil requires adherence to safety regulations and precautions to prevent spills, leaks, or accidents that could harm the environment or public safety. Regular maintenance and inspection of transportation vessels and equipment, along with effective spill response plans, are critical to ensuring the safe and efficient transportation of fuel oil.

Fuel oil storage involves the containment and safekeeping of fuel oil in designated facilities or tanks until it is ready for use. These storage facilities are crucial for maintaining a steady supply of fuel oil to meet demand, especially during peak usage periods. Fuel oil storage tanks come in various sizes and configurations, ranging from small aboveground tanks for residential use to large underground or aboveground tanks used in commercial or industrial settings. Proper storage practices include considerations such as tank material, location, capacity, and safety measures. Tanks should be constructed from materials that are resistant to corrosion and leakage, and they should be situated in areas that comply with local regulations and safety codes. Adequate ventilation and fire prevention measures, such as installing flame arrestors or fire suppression systems, are essential to mitigate the risks associated with fuel oil storage. Regular inspections, maintenance, and monitoring of storage tanks are vital to detect any leaks or damage early on and ensure the integrity of the fuel oil storage system. Effective fuel oil storage practices help ensure a reliable fuel supply, minimize environmental risks, and promote the safe handling and distribution of fuel oil.

When storing fuel oil, there are several specific factors to consider to ensure safety and maintain the integrity of the stored fuel. These include the following:

- Tank material: Choose a storage tank made of suitable materials that are resistant to corrosion and leakage. The common tank material of construction is steel. The tanks are often insulated to prevent heat loss, especially when designed for storing high-viscosity fuel oils.
- Heating system: Fuel oil tanks are equipped with heating coils to keep fuel oil viscosity within limits to ensure pumps are not overloaded.
- Pipelines: If high-viscosity fuel oils are transported, pipelines should be equipped with heating and insulation to maintain the fuel oil viscosity within pump design limits. Large pipelines are required to have thermal relief valves to prevent rupture due to the increase in pressure from thermal expansion.
- Tank size and capacity: Tank size depends on factors such as anticipated consumption, delivery frequency, and available space. It is important to ensure that the tank's capacity aligns with local regulations and safety guidelines.
- Location: Place the tank in a well-ventilated area away from potential ignition sources, such as open flames, electrical equipment, or heat sources. Consider environmental factors such as soil stability, flood risks, and proximity to water bodies.
- Spill containment: Implement measures to contain potential spills, such as installing secondary containment systems or using impermeable barriers around the tank area. Regulations related to this can be found in NFPA 30. Typically, the bund wall (secondary containment) needs to be tall enough to contain spillage from the largest tank in the area. This helps prevent fuel from reaching the soil or nearby water sources in case of a leak or spill.
- Ventilation: Ensure adequate ventilation around the tank to prevent the accumulation of flammable vapors. Vent pipes and openings should be properly designed and regularly maintained.
- Maintenance and inspection: Regularly inspect the tank for signs of damage, corrosion, or leaks. Perform routine maintenance, including cleaning the tank, checking valves and fittings, and testing fuel quality.
- Fire safety: Implement fire prevention measures, such as installing fire suppression systems, proper grounding and bonding, and keeping firefighting equipment nearby. Educate personnel on fire safety protocols and ensure the availability of fire extinguishers.

- Compliance with regulations: Adhere to local regulations and safety codes related to fuel oil storage. Obtain any necessary permits or certifications.
- Monitoring and alarms: Install fuel level monitoring systems and leak detection alarms to promptly identify any abnormal conditions or fuel loss.
- Training and emergency response: Provide training to personnel on proper handling, spill response procedures, and emergency protocols. Establish clear communication channels and contact information for reporting incidents or seeking assistance.

By considering these factors, individuals and organizations can ensure the safe and effective storage of fuel oil while minimizing potential risks to people, property, and the environment.

Biofuels

Biofuels have a long and rich history as a transportation fuel. The industry, as a set of feedstocks, technologies, and infrastructure has undergone several stages of development over the past nearly 200 years. This section considers ethanol, biodiesel, and renewable diesel/jet fuel in this discussion.

A simple study of biofuels, like most other forms of transportable/storable infrastructure fuels, can be broken down into three basic elements: Feedstock, conversion process, and final product (Fig. 10.1). We will use this framework to discuss each of the three fuels named above:

In general, the history of industrial biofuels has been driven by a search for a suitable final product that can be used with conventional or close-to-conventional internal combustion engine cycles such as spark-ignition gasoline, diesel, and more recently, jet engines. It is also important to note that biofuels have been subject to a wide range of regulatory factors across multiple geographies, which have shaped the dynamics of these industries. The regulatory supports of various biofuels (most notably ethanol) include reducing reliance on gasoline and therefore crude oil importation, reducing smog-forming pollutants, and more recently, reducing life-cycle greenhouse gas emissions.

A federal initiative in the United States promoting the adoption of renewable diesel and biodiesel is the Renewable Fuel Standard (RFS). This nationwide policy mandates a minimum volume of renewable fuels in transportation fuel sold in the country. Renewable diesel and biodiesel are two fuels falling under this standard. According to the Environmental Protection Agency (EPA), drop-in fuels, such as renewable diesel and biodiesel, closely resembling conventional gasoline or diesel, can seamlessly "drop in" to replace or augment existing petroleum-based fuels.

At the state level, an illustration of a program advocating for low-carbon and renewable fuels is the California Low-Carbon Fuel Standard (LCFS). This initiative aims to reduce the carbon intensity of California's transportation fuel pool by introducing more low-carbon alternatives, guided by a life-cycle assessment. This approach results in the certification of state-approved fuel pathways [12].

According to the U.S. Energy Information Administration, renewable diesel gets some of the most favorable GHG reduction scores in these programs, which means participants in them are increasingly opting for renewable diesel to meet rising renewable fuel targets. In addition to establishing renewable fuel targets, state and federal tax incentives also increase the demand for renewable diesel. Ethanol. The foundational history of the modern internal combustion engine is linked to ethanol, which was used by Nicolaus Otto in the development of the modern four-cycle engine in 1826 [13]. Ethanol, or ethyl alcohol, is a simple compound of carbon, hydrogen, and oxygen commonly written as CH_3CH_2OH, C_3H_5OH, C_2H_6O, or EtOH, where the abbreviation Et stands for ethyl. Known for thousands of years, ethanol's history is that of a grain alcohol, first produced as early as 5400 BCE [14].

As a fuel, ethanol has a higher octane number and lower evaporation pressure (and higher evaporation heat) compared to gasoline. Ethanol also contains 34.7% oxygen by weight, making it an oxygen additive in many gasoline fuel blends. Ethanol has a lower LHV and HHV than gasoline, meaning that the heating value of a blend decreases as more ethanol is blended. These fundamental physical characteristics impact blending ratios with gasoline and the performance of engines operating in different environments. The operating points of an engine can be optimized for specific ethanol-gasoline blends, which are sometimes balanced against the need for certain material compatibility of engine components. Vehicle manufacturers, fuel producers, and regulators in the United States have adopted specific blend ratios for the general use of 10% ethanol, 15% ethanol, and 85% ethanol. In Brazil, 100% ethanol is frequently used as a motor fuel. Today, essentially all on-road gasoline sold in the United States contains 10% ethanol by volume.

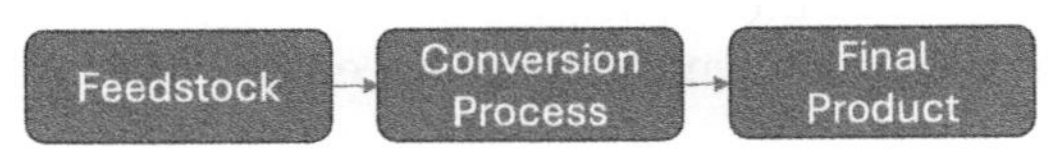

FIG. 10.1 Simplified framework showing elements of biofuels.

The performance of engines operating on these various blends has been extensively studied and has resulted in a mature set of fuel blend specifications in the United States, Brazil, and China, which are the three largest ethanol-consuming markets in the world. In addition, Canada, India, Thailand, Germany, Argentina, France, Japan, and the United Kingdom all blend ethanol to varying degrees in their gasoline pools [15]. Before understanding the advantages of ethanol blends, it is important to understand ethanol's problematic shortcomings.

First and foremost, ethanol is corrosive. When reacting with ambient oxygen, acidic compounds form, which can lead to corrosion of the fuel system and ultimately faster engine wear. As fuel systems in vehicles have advanced to become "airtight" to reduce evaporative emissions, this issue has been largely resolved by vehicle advances. Second, ethanol is hydrophilic, and water and ethanol are miscible. This causes any water that may be present in gasoline to dissolve with the ethanol and can cause the combination to drop out of the solution. Third, low ethanol blends, in some cases, can accelerate microbial growth, which can also accelerate corrosion. These last two challenges are only relevant in situations where ethanol-containing gasoline is stored for long periods. Antimicrobial fuel additives for these applications are widely available but can have follow-on environmental concerns [16].

From a regulatory standpoint, the United States and Brazil, the two leading producers and consumers of ethanol, have actively endorsed domestic ethanol production as a direct response to the 1973 oil crisis. In 1975, the Brazilian government initiated the National Alcohol Program (Portuguese: Programa Nacional do Álcool), a comprehensive nationwide initiative financially supported by the government. This program aimed to replace fossil fuel-based automobile fuels, such as gasoline, with ethanol derived from sugarcane, thereby promoting bioethanol as a sustainable fuel source [17].

The US federal government and various US states have implemented a nuanced and phased set of supports for ethanol use over the years, starting in 1978 by exempting E10 (including the gasoline portion) from the federal excise tax on motor fuels (Energy Tax Act 1978). The Energy Policy Act of 2005 established the first volumetric mandate for ethanol inclusion in the form of the Renewable Fuel Standard (RFS). The RFS requires motor fuels to contain an increasing portion of renewable fuels (primarily met by biofuels), which lowers the dependence of the United States on foreign oil and reduces greenhouse gas (GHG) emissions. At that point, the RFS (and the subsequent RFS2 of 2007) became the primary driving force behind increasing ethanol use [18]. Additionally, the Clean Air Act requires the use of oxygenated gasoline in areas where wintertime carbon monoxide levels exceed federal air quality standards. Without oxygenated gasoline, carbon monoxide emissions from gasoline-fueled vehicles tend to increase in cold weather. Winter oxygenated gasoline programs are implemented by the states.

Enabled by the reduction of CO, and unburned HC emissions, higher compression ratios of engines have become the norm. The reduction of CO emissions is caused by the wide flammability and oxygenated characteristics of ethanol. Therefore, improvements in power output and energy efficiency (not to be confused with miles-per-gallon) are achieved [19].

Ethanol production

Today, ethanol is produced on an industrial scale as an energy product primarily via the fermentation of sugars and starches contained within industrially farmed corn (generally in the United States) or sugar cane (generally in Brazil). The vast majority of fuel ethanol is produced in the United States and Brazil, making use of the dominance of the abovementioned crops, respectively (Fig. 10.2).

In the United States, corn (specifically #2 yellow dent corn) is the primary feedstock for ethanol production. Corn is typically processed in one of two process technologies known as wet-milling or dry-grinding processes (Figs. 10.3 and 10.4).

The majority of US ethanol production is from dry-grind technology. The traditional dry-grind process grinds the whole corn kernel and mixes it with water and enzymes. The mash is then cooked to liquefy the starch further. The mash is then cooled and mixed with more enzymes to convert the remaining sugar polymers to glucose before fermenting into ethanol [20]. The components of the kernel not fermented include the germ, fiber, and protein, which are concentrated in the distiller's dried grains that are produced as coproducts. While dry milling is less capital-intensive, it also yields less ethanol per bushel of corn than wet milling [20].

Wet milling involves steeping the corn for up to 48 h to assist in separating the parts of the corn kernel. Processing the slurry separates the germ from the rest of the kernel, which is processed further to separate the fiber, starch, and gluten. The fiber and corn gluten become components of animal feed, while the starch is fermented to become ethanol, corn starch, or corn syrup [21].

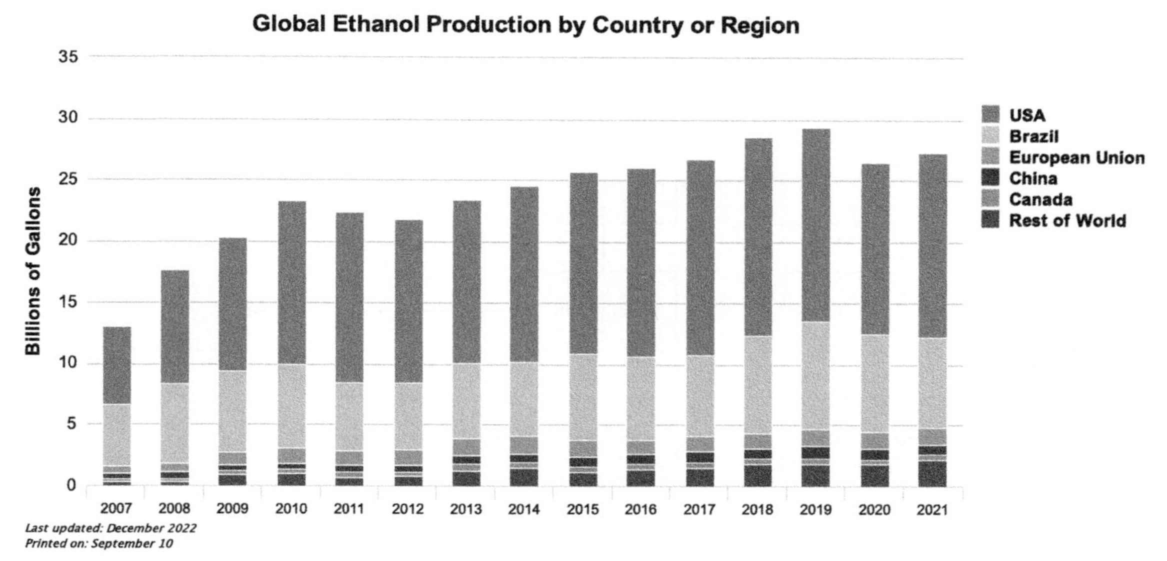

FIG. 10.2 Global ethanol production. *Courtesy Renewable Fuels Association and U.S. Energy Information Administration.*

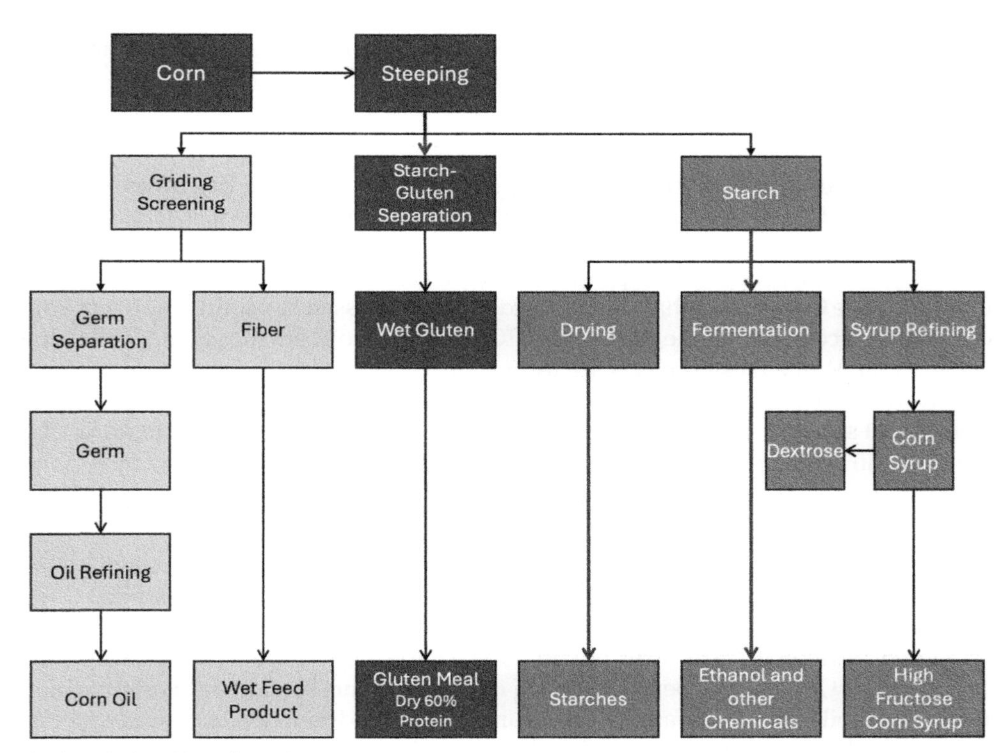

FIG. 10.3 Wet mill ethanol plant block flow diagram.

Ethanol feedstock and GHG reductions

The environmental impacts and benefits of biofuels are a well-studied topic, encompassing everything from soil science to water eutrophication, process emissions, and fuel use. Standards for conducting these analyses have been developed and include ISO 14040 and methodologies developed by US national labs such as the GREET (Greenhouse Gases, Regulated Emissions, and Energy Use in Transportation) model. Both major regulatory drivers of biofuel use in the United States (RFS and LCFS) make use of life cycle analysis to determine qualification for mandates and tax credits, but only the LCFS provides incentives for incremental improvements in GHG reductions.

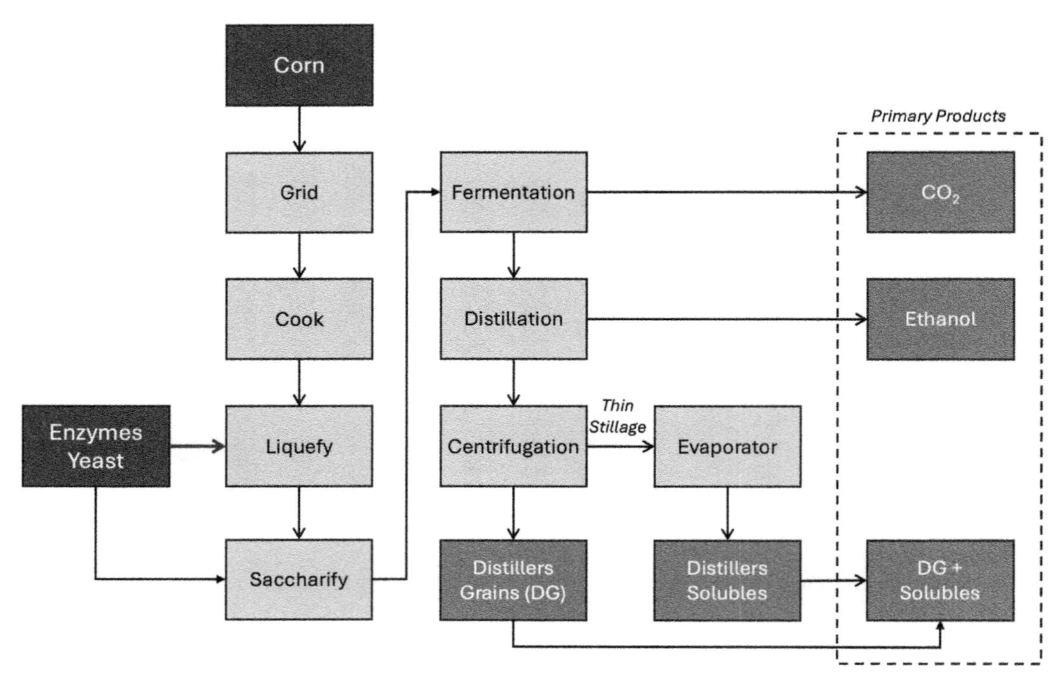

FIG. 10.4 Dry mill ethanol plant block flow diagram. *Used with permission from "Dry grind ethanol process" by Caroline Clifford, The Pennsylvania State University. This image is licensed under CC BY-NC-SA 4.0.*

These policies have the net result of incentivizing the use of feedstocks that have the lowest agricultural inputs (such as synthetic fertilizers) and the lowest land-use change associated with initiating industrial farming. Other federal policies include 45Q (discussed in the CCS portion of this book.)

Ethanol production by either wet or dry mill processes is well suited to CCS. This is because the fermentation process produces a relatively pure CO_2 stream that can be dried and compressed to produce a sequestration-ready stream. Sequestering this CO_2 stream could bring the life-cycle GHG impact of corn ethanol down well below 25% of conventional gasoline's impact [22]. While these projects are technically feasible, the geology required for CCS and the high capital costs associated with collecting and transporting CO_2 make these projects economically challenging. Many projects are in the development stages, and many more could become feasible with newly expanded CCS incentives. CCS is discussed in further detail in Chapter 8 of this publication.

Biodiesel

Rudolf Diesel invented the diesel engine in the 1890s. Similar to ethanol being a foundational fuel in the development of the modern four-cycle gasoline engine, the diesel engine could run on a variety of fuels, including direct vegetable oils. Diesel engine technology was featured in 1900 at the Paris Exposition powered by peanut oil, but eventually, the scalability, cost/price, and performance of fossil diesel fuel rendered this simple biofuel obsolete [23].

While "straight vegetable oil" referred to as SVO in the industry, could function technically, performance was poor compared to much more sophisticated refined diesel fuel. This reality drove researchers to develop a chemical process to improve vegetable oils as fuel by converting the oil to an ethyl ester. In 1937, G. Chavanne was granted a Belgian patent for an ethyl ester of palm oil, which was followed by a public demonstration in 1938 wherein a passenger bus fueled with palm oil ethyl ester drove the route from between Brussels and Louvain, a distance of approximately 30km [24]. During the next decade (from 1939 to 1945), global petroleum fuel supply chains were heavily strained by World War II. During this period, vegetable oil and biodiesel blends were used more commonly but soon lost relevance as petroleum supplies grew and refined product value chains stabilized.

Like ethanol with gasoline, biodiesel is generally used in specific blend amounts with diesel fuel. Biodiesel can be blended and used in many different concentrations. The most common are B5 (up to 5% biodiesel) and B20 (6%–20% biodiesel). B100 (pure biodiesel) is typically used as a blendstock to produce lower percentage blends and is rarely used as a transportation fuel. The molecular differences between biodiesel and fossil diesel result in complex changes to engine and fuel performance.

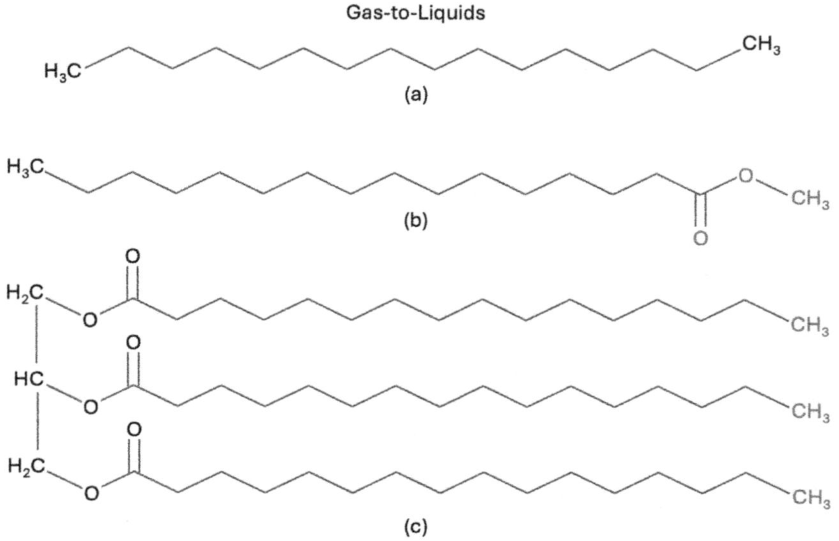

Gas-to-Liquids

(a)

(b)

(c)

FIG. 10.5 Structure of (A) petroleum diesel, (B) biodiesel, and (C) vegetable oil or triglyceride [25].

Biodiesel, like other liquid fuels, is a combination of long-chain carbon molecules with hydrogen atoms but with an additional ester functional group (–COOR). Fig. 10.5 provides a cartoon of a typical biodiesel molecule and a typical petroleum diesel molecule.

As depicted above, a biodiesel molecule can typically be comprised of a 16- or 17-carbon chain along with an ester group. In general, a longer molecule will increase the heat of combustion and cetane number, which in a diesel engine can decrease NOx emissions. Indeed, biodiesel blends do provide for a measurable increase in NOx emissions compared to conventional petroleum diesel [25]. The longer molecules also increase viscosity, which can be problematic for fuel systems. A typical biodiesel molecule will also have less branching, which also contributes to a higher gel point. Biodiesel tends to thicken and "gel up" at low temperatures more readily than petroleum diesel. Some types of oil are more of a problem than others [26]. However, less branching also increases the cetane number of the fuel, which has been favorable to biodiesel's use as a blendstock for lower-grade diesel fuels. Finally, saturation is a factor that drives reduced NOx emissions and increases lubricity, which is also a valuable attribute for biodiesel as a blendstock, which was particularly relevant during the desulphurization transition of diesel fuels. Biodiesel has a higher oxygen content (typically 10%–12%) than petroleum diesel. Generally, this results in lower emissions but reduces LHV by volume. Typically, as compared to petroleum diesel, biodiesel delivers lower peak engine power by approximately 4% [27] (Fig. 10.6).

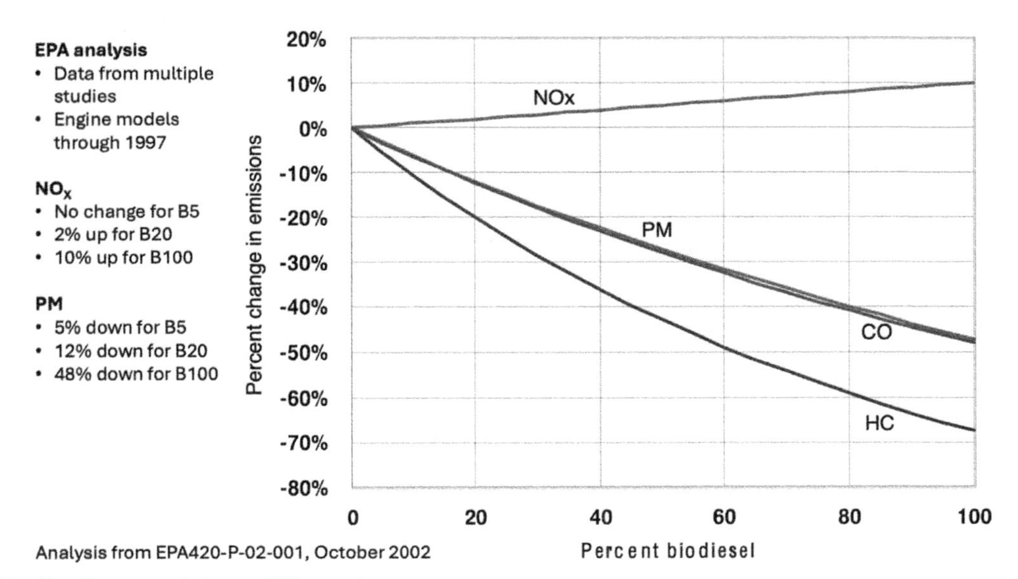

FIG. 10.6 Biodiesel's effect on emissions—Older engines.

Like ethanol, there is a range of fuel additives available to compensate for most, if not all, of the performance challenges that biodiesel may have when compared to petroleum diesel. The most common among these are cold-flow improvers. These additives improve the cold-weather performance of biodiesel by limiting its ability to gel, which has been a challenge in more extreme environments for diesel fuel as well.

Biodiesel manufacturing process—Transesterification

In contrast to ethanol production, which is generally a fermentation process, biodiesel production is a chemical process that converts vegetable oils or animal fats plus an alcohol plus a catalyst to produce a mixture of long-chain monoalkylic esters. Commercially, methanol is used most commonly as the alcohol and a strong base such as KOH or NaOH is used as the catalyst.

Methanol and the base catalyst (such as NaOH) are combined, wherein the NAOH separates into ions. Once the catalyst is prepared, 1 mol of feedstock triglyceride will react with 3 mols of methanol, so excess methanol has to be used in the reaction to ensure a complete reaction. The three attached carbons with hydrogen react with OH-ions and form glycerin, while the CH_3 group reacts with the free fatty acid to form the fatty acid methyl ester. Water as a byproduct of the reaction is generally unwanted, as it drives the production of soaps (Fig. 10.7).

Fig. 10.8 shows a schematic of the process for making biodiesel. Glycerol is formed and has to be separated from the biodiesel. Both glycerol and biodiesel need to be removed and recycled in the process. Water is added to both the biodiesel and glycerol to remove unwanted side products, particularly glycerol, that may remain in the biodiesel. The wash water is separated similarly to solvent extraction (it contains some glycerol), and the trace water is evaporated out of the biodiesel. Acid is added to the glycerol to provide neutralized glycerol.

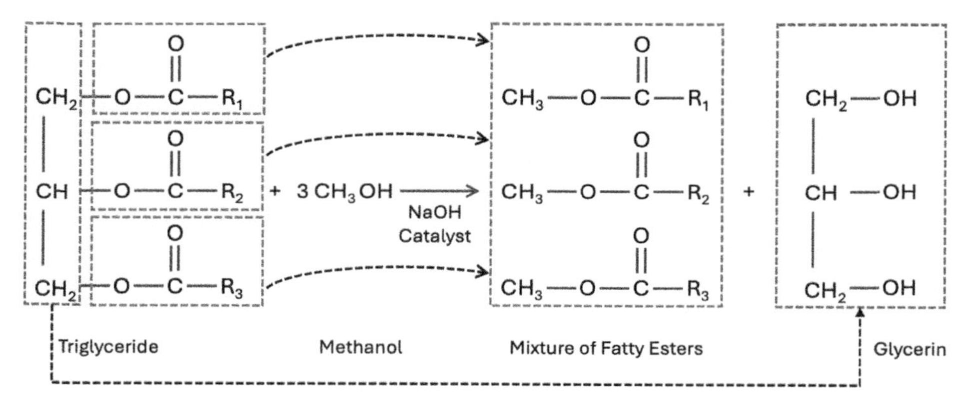

FIG. 10.7 Chemistry of biodiesel production. *Source: Biobased Energy Education Materials Exchange System (BEEMS), Ohio State University and USDA.*

	Stoichiometric	Typical
Fat or Oil	100 lbs.	100 lbs.
+		
Alcohol (methanol)	10 lbs.	16-20 lbs.
+		
Catalyst (NaOH; 1% w/w oil)	1 lbs.	1 lbs.
↓		
Biodiesel (Methyl Ester)	100 lbs.	100 lbs.
+		
Glycerin	10 lbs.	10 lbs.

FIG. 10.8 Conversion of fatty acid into biodiesel. Note excess alcohol. *Source: Biobased Energy Education Materials Exchange System (BEEMS), Ohio State University and USDA.*

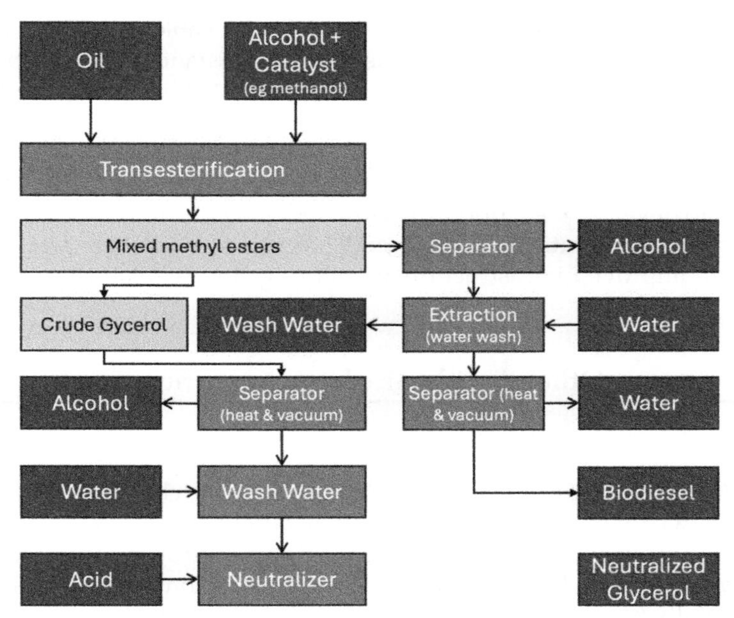

FIG. 10.9 Simply biodiesel process schematic using transesterification. *Source: Biobased Energy Education Materials Exchange System (BEEMS), Ohio State University and USDA.*

Fig. 10.9 shows process challenges that arise most frequently from water inclusion; therefore, the initial reactants, such as the feedstock oils, should be as water-free as possible. Water reacts with the triglyceride to make free fatty acids and a diglyceride. Water also affects the catalyst by dissociating the sodium or potassium from the hydroxide, and the ions Na^+ and K^+ then react with free fatty acids to form soap. In such a scenario, the sodium that was intended to be employed as a catalyst is now bound with the fatty acid and is unusable and unrecoverable. All natural oils contain some amount of free fatty acids, which is a critical aspect of feedstock selection and preparation, to be discussed in the next section.

Methanol

Methanol, also known as methyl alcohol or wood alcohol, is a colorless, volatile liquid with a distinct odor. It is the simplest alcohol compound, consisting of a single carbon atom bonded to three hydrogen atoms and one hydroxyl group (CH_3OH). Methanol has a wide range of applications across various industries. It is commonly used as a solvent in chemical processes, as a fuel in racing and high-performance vehicles, and as a feedstock in the production of formaldehyde, acetic acid, and other chemicals. Methanol can also be utilized as an alternative fuel source, either by itself or as a blend with gasoline or diesel, owing to its high octane rating and clean-burning properties. Furthermore, methanol is being explored as a potential energy carrier in fuel cell technologies and as a means to store and release renewable energy. While methanol has numerous industrial uses, it is important to handle it with care as it is highly toxic and flammable. Safety precautions and proper handling procedures are essential to ensure the safe utilization of this versatile chemical compound.

In recent years, there has been growing interest in methanol as a sustainable and renewable fuel option. Methanol can be produced from various sources, including natural gas, coal, biomass, and carbon dioxide. Biomethanol, derived from biomass feedstocks such as agricultural waste or dedicated energy crops, offers a potential pathway to reducing greenhouse gas emissions and reliance on fossil fuels. Methanol can also be produced using carbon dioxide captured from industrial processes or directly from the atmosphere, making it a promising candidate for carbon capture and utilization (CCU) technologies. Additionally, methanol can be easily transported and stored, making it a flexible energy carrier for remote or off-grid applications. However, further research and investment are needed to optimize methanol production methods, improve its energy efficiency, and ensure the development of sustainable feedstock options. With continued advancements and support, methanol has the potential to play a significant role in the transition to a more sustainable and low-carbon energy future.

Methane transportation refers to the process of moving methane gas from its source to various destinations for various applications. Methane, the primary component of natural gas, is a versatile and abundant fuel that is used for heating, electricity generation, and as a feedstock in various industrial processes. Methane can be transported through pipelines, which form an extensive network across regions and countries, allowing for the efficient distribution of natural gas to homes, businesses, and power plants. Pipelines offer a reliable and cost-effective means of transportation, with the ability to transport large volumes of methane over long distances. In addition to pipelines, methane can also be transported via liquefied natural gas (LNG) carriers. LNG is created by cooling methane to extremely low temperatures, which reduces its volume and allows for easier storage and transport. LNG is particularly useful for transporting natural gas to areas without pipeline infrastructure or for international export. Methane transportation plays a vital role in ensuring the availability of natural gas for energy consumption and industrial use, contributing to the global energy supply chain. However, as the world focuses on reducing greenhouse gas emissions, the environmental impact of methane leaks during transportation and the overall lifecycle of natural gas are important considerations that need to be addressed through improved infrastructure, monitoring, and regulatory measures.

Other decarbonized energy carriers

Ammonia

Ammonia is a compound composed of nitrogen and hydrogen, with the chemical formula NH3. It is a colorless gas with a pungent odor and is known for its versatile applications in various industries. One of the primary uses of ammonia is as a key component in the production of fertilizers. Due to its high nitrogen content, ammonia serves as an essential source of nutrients for plant growth. It is also employed in refrigeration systems, as it has excellent heat transfer properties and is environmentally friendly, with no ozone depletion potential. Ammonia is utilized as a precursor for the production of various chemicals, including plastics, fibers, explosives, and pharmaceuticals. It can also be used as a cleaning agent in household and industrial settings. While ammonia has several valuable applications, it is essential to handle it with caution as it is highly toxic and can pose health risks if not handled properly.

In recent years, ammonia has gained attention as a potential alternative fuel source with low carbon emissions. It has the potential to be used as a clean and sustainable energy carrier, particularly in the field of transportation. Ammonia can be easily liquefied, enabling efficient storage and transportation. When used in fuel cells, it can generate electricity with only water and nitrogen as byproducts, avoiding greenhouse gas emissions. Research and development efforts are underway to improve ammonia fuel production, storage, and utilization technologies to make it a viable and environmentally friendly energy option. However, challenges remain in terms of handling and distributing ammonia safely, as it is corrosive and can be hazardous if mishandled. With further advancements and infrastructure developments, ammonia has the potential to contribute to a more sustainable energy landscape, reducing reliance on fossil fuels and mitigating the impact of climate change.

Ammonia transportation involves the movement of ammonia from production facilities or storage sites to various destinations for industrial or commercial use. Ammonia is typically transported in liquid form under pressure or refrigeration to maintain its stability and prevent evaporation. The most common method of ammonia transportation is through dedicated tanker vessels, rail tank cars, or tanker trucks. These transportation modes are equipped with specialized containment systems, including tanks or containers made of materials compatible with ammonia. Safety measures are paramount during ammonia transportation to mitigate the risk of leaks or spills, as ammonia is highly toxic and can pose health hazards. Proper labeling, hazard communication, and adherence to regulatory requirements are crucial to ensuring the safe handling and transport of ammonia. Additionally, due to ammonia's corrosive nature, regular inspections, maintenance, and cleaning of transportation equipment are necessary to prevent deterioration and maintain integrity. Overall, ammonia transportation requires strict adherence to safety protocols and regulations to protect both human health and the environment.

Ammonia transportation involves specific considerations to ensure this volatile compound's safe and efficient movement. The design and construction of ammonia transport vessels prioritize containment and safety. Tanks and containers are typically made of materials such as stainless steel or specialized alloys that can withstand the corrosive properties of ammonia. In addition, ammonia transportation often requires refrigeration or pressure systems to maintain the substance in its liquid state. Proper loading and unloading procedures and secure sealing of containers are essential to prevent leakage and minimize the risk of exposure. Ammonia transport operators follow strict regulations regarding labeling, placarding, and documentation to ensure proper identification and handling. Emergency response plans and spill containment measures are in place to address any unforeseen incidents. By adhering to these stringent protocols, ammonia transportation can be conducted safely and reliably, supporting its utilization in various industrial sectors.

Ammonia storage involves the careful handling and containment of ammonia to ensure its stability and prevent leaks or releases into the environment. Ammonia is typically stored in tanks or vessels specifically designed for ammonia storage, constructed from materials resistant to corrosion and compatible with ammonia. These storage facilities are equipped with safety features such as pressure relief valves, emergency vents, and leak detection systems to mitigate the risks associated with overpressure or potential leaks. Adequate ventilation and temperature control systems are in place to prevent the buildup of ammonia vapors. Proper labeling and signage are used to identify ammonia storage areas and communicate potential hazards. Regular inspections and maintenance activities are conducted to monitor the condition of storage tanks, valves, and associated equipment, ensuring their integrity and minimizing the risk of leaks or failures. Training programs and safety procedures are implemented to educate personnel on safe handling practices, emergency response protocols, and the use of personal protective equipment. By adhering to these strict safety measures, ammonia storage can be conducted securely and controlled, minimizing the risks to both human health and the environment.

Liquid organic hydrogen carriers

A fairly new method for transporting hydrogen is chemical bonding to an organic solvent, liquid organic hydrogen carriers (LOHC). Certain organic solvents can chemically absorb hydrogen by hydrogenation in an exothermic reaction under pressure. The solvent thereby stores significant amounts of hydrogen as a liquid at ambient pressure and temperature and can be transported easily. For use, it is dehydrated in an endothermic reaction to release the hydrogen by heating at low pressure. Examples of solvents potentially used as LOHC include toluene, naphthalene, benzyltoluene, and N-ethylcarbazole. An overview of promising LOHC and current research and development has been written by Chandra and Yoon [28].

The advantage of LOHC is the ease of transporting the liquid by any means of liquid trucking or shipping in tanks. This compares well to the difficulties of transporting the same mass of hydrogen either as cryogenic liquid or as a compressed gas. The amount of hydrogen that can be stored in LOHC ranges from about 5% to 7% by weight.

The issue with LOHC is the requirement for the two catalytic reactors; especially the dehydrogenation requires significant heat input to liberate the hydrogen The enthalpies of desorption of various LOHC range from about 50 to 70 kJ/mol-H_2 [29]. For an exemplary system in which LOHC provide H_2 for a fuel cell (LHV 242 kJ/mol), some 30% of the H_2 may have to be combusted to provide the dehydrogenation heat, with a corresponding system efficiency hit. A reduction of this efficiency hit would be possible with a lower enthalpy of desorption and when the dehydrogenation reactor catalysts were sufficiently active at lower temperatures. In these cases less heat would be needed or the reaction temperatures would be low enough to utilize the fuel cell's waste heat (around 80°C). When using LOHC for supplying a combustion engine instead of a fuel cell, more waste heat and at a higher temperature can be captured to provide the reaction heat, making LOHC more promising for such an application. Likewise, in industrial applications supplied with hydrogen via an LOHC where a waste heat source is available, the system efficiency may not be compromised as much.

A commercial LOHC system with reactors, tanks, equipment, and services is offered, for example, by the company Hydrogenious LOHC Technologies GmbH. They used the liquid hydrocarbon benzyltoluene [30], which is hydrogenated in an exothermic process at 25–50 bar and 250°C to perhydro-benzyltoluene, releasing about 9 kWh/kg-H_2 of heat while absorbing about 6% of the initial liquid mass of hydrogen. After transport and storage, the hydrogen can be released again at 1–3 bar at about 300°C with about 11 kWh/kg-H_2 heat input. When not using waste heat, and not accounting for any use of heat released, this is an energy input of approximately one-third of the hydrogen's lower heating value. Additionally, about 5% of the lower heating value will be used in the form of electricity for a compression of, e.g., 2 bar during dehydrogenation back to a hydrogenation pressure of 35 bar. Depending on the heat source, LOHC can be a quite energy-intensive way to store hydrogen, comparable with liquefaction. At an energy cost of $50/MWh, the energy for dehydrogenation and for compression amounts to about $0.7/kg-$H_2$. The transport of LOHC in a truck trailer equates to about 1.5 t of hydrogen, which is about 50% more than compressed hydrogen but only half of the amount of liquid hydrogen per truck. The cost of benzyltoluene is about $3–$6/l, the cost of the reaction, heat exchange, storage, and compression equipment is not known, but at least the tanks are comparatively inexpensive as they are neither insulated nor pressure vessels. The total LOHC storage cost as the annuity of the equipment and solvent divided by the amount of hydrogen passed through plus the energy and operating costs can be estimated just above $1/kg-$H_2$ for large-scale, mature processes.

Formic acid

Formic acid, also known as methanoic acid, is a colorless and sharp-smelling liquid with the chemical formula HCOOH. Among the simplest organic acids, it occurs naturally in various sources, including certain plants and animal secretions. With diverse applications across industries, formic acid serves as a preservative and antibacterial agent in animal feed and food preservation. Additionally, it functions as a coagulant in rubber production, a deicing agent at airports, and a pH regulator in various industrial processes.

Playing a crucial role in the textile and leather industries, formic acid is integral to dyeing and tanning processes. Moreover, it acts as a fundamental raw material for producing various chemicals such as formate salts, esters, and pharmaceuticals. Its utility extends to serving as a reducing agent in specific chemical reactions. Despite its versatility, it's essential to note that formic acid is highly corrosive and poses hazards if mishandled. Adhering to safety measures such as proper ventilation, protective equipment, and careful storage is imperative for ensuring secure handling and usage.

Formic acid has gained attention in recent years as a potential sustainable energy carrier and a source of hydrogen. It can be used in fuel cells as an alternative to more conventional hydrogen storage methods. By selectively decomposing formic acid, hydrogen gas can be released, which can then be used as a clean fuel for various applications, including fuel cell-powered vehicles. This process offers advantages such as high hydrogen content, ease of storage and transport, and the potential for renewable production of formic acid from carbon dioxide capture and utilization technologies. Additionally, formic acid has garnered interest in the field of carbon dioxide (CO_2) capture and storage. It can react with CO_2 to form formate salts, which can be utilized in CO_2 sequestration or as a precursor to produce valuable chemicals. Ongoing research and development efforts continue to explore the potential of formic acid as a versatile and sustainable compound in energy and environmental applications.

The storage of Formic acid requires careful handling and appropriate containment to ensure safety and maintain the integrity of the stored substance. Formic acid is highly corrosive and can pose health risks if mishandled. When storing formic acid, it is crucial to use containers made of materials that are resistant to corrosion, such as high-density polyethylene (HDPE) or stainless steel. These containers should have proper seals and closures to prevent leakage or evaporation. Storing formic acid in a well-ventilated area is also important, away from heat sources, direct sunlight, and incompatible substances. Proper labeling and clear signage should be used to identify the storage area and communicate the hazards associated with formic acid. Adequate personal protective equipment (PPE) should be provided to those handling the substance. Regular inspections of the storage containers and monitoring for leaks or damage are essential to ensuring the integrity of the storage system. In the event of spills or leaks, appropriate spill response protocols should be in place to safely contain and clean up the formic acid. By following these storage guidelines and safety measures, the risks associated with formic acid storage can be minimized, ensuring the well-being of personnel and preventing harm to the environment.

Formic acid transportation involves careful handling and adherence to safety measures to ensure this corrosive substance's secure and efficient movement. During formic acid transportation, it is crucial to use specialized containers or tanks made of materials resistant to corrosion, such as stainless steel or high-density polyethylene (HDPE), that can withstand the acid's corrosive nature. These containers should be securely sealed to prevent leakage or spills. Transportation vehicles, such as tankers or trucks, should be equipped with appropriate safety features, including pressure relief valves and leak detection systems, to mitigate the risks associated with over-pressurization or potential leaks. Proper labeling and placarding of the vehicles are necessary to communicate the hazards and precautions associated with formic acid. Transport personnel should be adequately trained on handling procedures, emergency response protocols, and the use of personal protective equipment (PPE). Regular inspections and maintenance of transportation equipment are essential to ensuring their integrity and minimizing the risk of leaks or failures during transit. By adhering to these stringent safety protocols and regulations, formic acid transportation can be conducted with a strong focus on minimizing risks to both human health and the environment.

Solid-state storage of hydrogen

Chemical absorption in metal hydrides

Metal hydrides are light metal alloys that can chemically absorb and later desorb hydrogen under varying temperature and pressure conditions and be used for storage and transport. Exemplary materials include MgH_2, $LaNi_5H_6$ and $NaAlH_4$. The alloys are pulverized creating a large surface area, additives for catalyzing the reaction and stabilizing a porous microstructure are added, and the whole is contained in pressure tanks with heat exchangers. The absorption reaction is exothermic, but it requires elevated temperatures. MgH_2, for instance, needs to be heated up to 425°C at 20bar of pressure to store about 7% of hydrogen on a magnesium powder mass base [31]. Other researchers reported storing 7.6% H_2 at about 300°C–350°C, and 20bar in alloyed MgH_2 with very fine compressed powder containing graphite for improved thermal conductivity [32]. For desorption, an endothermic reaction, the material is heated until the desired pressure is reached; the pressures are much lower than for absorption, e.g., 1.5bar, at a similar temperature.

The heat of the reversible reaction is 74 kJ/mol H_2; this equates to about 30% of the lower heating value (the molar heat of combustion of H_2 without condensation is 242 kJ/mol). This heat would typically be lost in the process. First, to start the hydrogenation reaction, some warm-up is needed, and then the heat of the reaction needs to be dissipated by cooling. Later, the metal hydrate powder needs to be heated again and kept at a similar temperature for desorption by adding heat during the reverse process. However, this significant energy consumption can be mitigated by using heat storage for the heat transfer fluid between absorption and desorption. One interesting solution for an "adiabatic" tank uses a phase-change material, a metal alloy, with a melting temperature of 340°C [32]. This keeps the insulated tank heated at the reaction temperature once heat losses are compensated, and the direction of the reversible reaction is determined by the pressure. With an increasing pressure of around 20 bar, the storage is filled with H_2, while at a decreasing pressure of around 1.5 bar, H_2 is released. Besides the loss in pressure, this system may offer good storage efficiency.

The advantages of metal hydride storage include relatively modest pressures and a solid material that can be easily stored in light-pressure cylinders, i.e., safety. However, reaction kinetics and thermal conductivity may be limiting the flow rates. The storage system, besides the tanks, requires heat exchangers and transfer fluids, and depending on the alloy, rather high temperatures are required for operation. Recycling the heat may be difficult, resulting in reduced storage efficiency, and the output pressure may be too low for some applications, while the input pressure may require additional compression.

Generally, metal hydride storage concepts are still subject to research for both stationary and mobile applications, with an exemplary mobile application for fuel cell hydrogen storage in the U212 and U214 classes of submarines and a stationary application storing 120 kg of hydrogen for an electrolyzer-fuel cell system at Griffith University in Brisbane [33].

Physical adsorption and desorption

Adsorption-based storage methods for hydrogen use porous materials with a large internal surface area and fall somewhere between gaseous and liquid storage in terms of storage density and temperature ranges. For physical adsorption, still low or even cryogenic temperatures and increased pressures may be required; for desorption, heat or decreasing pressures are needed. The material groups that have been found promising for adsorption-based hydrogen storage include but are not limited to, metal-organic frameworks, activated coal, carbon nanotubes, and aerogels. As of 2023, research publications and lab prototypes can be found, but commercial applications of adsorption storage have not been reported [34].

One class of adsorbents, metal-organic frameworks, has been found to offer an internal surface area of up to 4000 m^2/g and can store up to 6% of hydrogen at pressures around 50 bar [35]. The material is enclosed in a pressure vessel with internal gas diffusion and heat exchange provisions. Cryogenic temperatures and increased pressures are required for adsorbing hydrogen, and a significant heat of adsorption needs to be removed in the process, in addition to any heat of compression. The low temperatures, e.g., around −200°C, required for high uptake of hydrogen, make the system expensive because of the need for a low-temperature chiller or the use of liquid nitrogen [34].

One example of a practical adsorbed hydrogen storage system is the "Cryogenic Flux Capacitor" described by Swanger et al. [36], which has been developed and tested in a laboratory. The cryogenic flux capacitor uses a nanoporous aerogel with an internal surface area of about 1000 m^2/g in a pressure tank with an integrated heat exchanger, which permits heating the unit at a controlled rate for pressure rise and discharge and cooling the unit with, e.g., liquid nitrogen to facilitate charging. The prototype being developed is expected to reach storage densities of approximately 70 kg/m^3 or 8.5 GJ/m^3 and will be sized for storing 3 kg of hydrogen [34].

References

[1] A.P. Steynberg, Chapter 1 - Introduction to Fischer-Tropsch technology, in: A. Steynberg, M. Dry (Eds.), Studies in Surface Science and Catalysis, 152, Elsevier, 2004, pp. 1–63.
[2] A.P. Steynberg, H.G. Nel, Clean coal conversion options using Fischer–Tropsch technology, Fuel 83 (6) (2004) 765–770.
[3] R. Rauch, A. Kiennemann, A. Sauciuc, Chapter 12 - Fischer-Tropsch synthesis to biofuels (BtL process), in: K.S. Triantafyllidis, A.A. Lappas, M. Stöcker (Eds.), The Role of Catalysis for the Sustainable Production of Bio-fuels and Bio-chemicals, Elsevier, Amsterdam, 2013, pp. 397–443.
[4] G.S. Forman, T.E. Hahn, S.D. Jensen, Greenhouse gas emission evaluation of the GTL pathway, Environ. Sci. Technol. 45 (20) (2011) 9084–9092.
[5] . https://www.energy.gov/eere/vehicles/articles/assessment-environmental-impacts-shell-gtl-fuel.
[6] A.J. Kidnay, W.R. Parrish, D.G. McCartney, Fundamentals of Natural Gas Processing, CRC Press, 2019.
[7] D.J. Soeder, D.J. Soeder, The history of oil & gas development in the U.S, in: Fracking and the Environment: A Scientific Assessment of the Environmental Risks From Hydraulic Fracturing and Fossil Fuels, Springer International Publishing, Cham, 2021, pp. 37–61.
[8] M. Mehl, W.J. Pitz, C.K. Westbrook, H.J. Curran, Kinetic modeling of gasoline surrogate components and mixtures under engine conditions, Proc. Combust. Inst. 33 (1) (2011) 193–200.

[9] M.Z. Jacobson, Effects of ethanol (E85) versus gasoline vehicles on cancer and mortality in the United States, Environ. Sci. Technol. 41 (11) (2007) 4150–4157.
[10] S. Barak, R.K. Rahman, S. Neupane, E. Ninnemann, F. Arafin, A. Laich, et al., Measuring the effectiveness of high-performance co-optima biofuels on suppressing soot formation at high temperature, Proc. Natl. Acad. Sci. USA 117 (7) (2020) 3451–3460.
[11] J. Badra, F. Alowaid, A. Alhussaini, A. Alnakhli, A.S. AlRamadan, Understanding of the octane response of gasoline/MTBE blends, Fuel 318 (2022) 123647.
[12] A.E. Farrell, D. Sperling, S. Arons, A. Brandt, M. Delucchi, A. Eggert, et al. (2007). A low-carbon fuel standard for California Part 1: Technical analysis, Transportation Sustainability Research Center, UC Berkeley. Retrieved from https://escholarship.org/uc/item/8zm8d3wj.
[13] I. Glassman, R.A. Yetter, N.G. Glumac, Combustion, Academic Press, 2014.
[14] M. Battcock, S. Azam-Ali, Fermented Fruits and Vegetables: A Global Perspective, Food & Agriculture Org, 1998.
[15] https://www.ers.usda.gov/amber-waves/2023/june/fuel-ethanol-use-expanding-globally-but-still-concentrated-in-few-markets.
[16] G.V.S. Luz, B.A.S.M. Sousa, A.V. Guedes, C.C. Barreto, L.M. Brasil, Biocides used as additives to biodiesels and their risks to the environment and public health: a review, Molecules [Internet] 23 (10) (2018).
[17] M.B. Bastos, Brazil's Ethanol Program—An Insider's View, Energy Tribune, 2007. Archived from the original on 2011-07-10. Retrieved 2008-08-14.
[18] C. Johnson, K. Moriarty, T. Alleman, D. Santini, History of Ethanol Fuel Adoption in the United States: Policy, Economics, and Logistics, United States, 2021.
[19] M. Koç, Y. Sekmen, T. Topgül, H.S. Yücesu, The effects of ethanol–unleaded gasoline blends on engine performance and exhaust emissions in a spark-ignition engine, Renew. Energy 34 (10) (2009) 2101–2106.
[20] G.S. Murthy, V. Singh, D.B. Johnston, K.D. Rausch, M.E. Tumbleson, Evaluation and strategies to improve fermentation characteristics of modified dry-grind corn processes, Cereal Chem. 83 (5) (2006) 455–459.
[21] S. Rajagopalan, E. Ponnampalam, D. McCalla, M. Stowers, Enhancing profitability of dry mill ethanol plants, Appl. Biochem. Biotechnol. 120 (1) (2005) 37–50.
[22] J. Dees, K. Oke, H. Goldstein, S.T. McCoy, D.L. Sanchez, A.J. Simon, W. Li, Cost and life cycle emissions of ethanol produced with an oxyfuel boiler and carbon capture and storage, Environ. Sci. Technol. 57 (13) (2023) 5391–5403.
[23] G. Pahl, Biodiesel: Growing a New Energy Economy, Chelsea Green Publishing, 2008.
[24] G. Knothe, History of Vegetable Oil-Based Diesel Fuels. The Biodiesel Handbook, Elsevier, 2010, pp. 5–19.
[25] R. McCormick, Effects of Biodiesel on NOx Emissions, National Renewable Energy Lab.(NREL), Golden, CO, 2005.
[26] P.V. Bhale, N.V. Deshpande, S.B. Thombre, Improving the low temperature properties of biodiesel fuel, Renew. Energy 34 (3) (2009) 794–800.
[27] M. Zheng, M.C. Mulenga, G.T. Reader, M. Wang, D.S.K. Ting, J. Tjong, Biodiesel engine performance and emissions in low temperature combustion, Fuel 87 (6) (2008) 714–722.
[28] P.R. Chandra, M. Yoon, Potential liquid-organic hydrogen carrier (LOHC) systems: a review on recent progress, Energies 13 (22) (2020) 6040. https://doi.org/10.3390/en13226040.
[29] P. Preuster, C. Papp, P. Wasserscheid, Liquid organic hydrogen carriers (LOHCs): toward a hydrogen-free hydrogen economy, Acc. Chem. Res. (2016). https://doi.org/10.1021/acs.accounts.6b00474.
[30] M.M. Distel, J.M. Margutti, J. Obermeier, A. Nuβ, I. Baumeister, M. Hritsyshyna, A. Weiβ, M. Neubert Large-scale H_2 storage and transport with liquid organic hydrogen carrier technology: insights into current project developments and the future outlook, Energy Technol. (2024) 2301042. https://doi.org/10.1002/ente.202301042.
[31] E. Rivard, M. Trudeau, K. Zaghib, Hydrogen storage for mobility: A review, Materials 12 (12) (2019) 1973.
[32] D. Fruchart, M. Jehan, N. Skryabina, P. de Rango, Hydrogen solid state storage on MgH2 compacts for mass applications, Metals 13 (5) (2023) 992.
[33] J.B. Von Colbe, J.-R. Ares, J. Barale, M. Baricco, C. Buckley, G. Capurso, et al., Application of hydrides in hydrogen storage and compression: Achievements, outlook and perspectives, Int. J. Hydrog. Energy 44 (15) (2019) 7780–7808.
[34] R. Kurz, et al., Chapter 6 - Transport and storage, in: K. Brun, T. Allison (Eds.), Machinery and Energy Systems for the Hydrogen Economy, Elsevier, 2022, pp. 215–249. https://doi.org/10.1016/B978-0-323-90394-3.00003-5.
[35] U. Bünger, E. Naess, M. Schlichtenmayer, M. Hirscher, I. Senkovska and S. Kaskel, Analysis of hydrogen storage in porous adsorption materials, in: 18th World Hydrogen Energy Conference 2010 – WHEC 2010, Proceedings, 2010, p. 291.
[36] A.M. Swanger, J.E. Fesmire, Cryogenic flux capacitor for solid-state storage and on-demand supply of fluid commodities, Google Patents, 2019.

CHAPTER

11

Future trends

Stephen Ross[a], *Brian Pettinato*[a], *Sterling Scavo-Fulk*[a], *Brian Hantz*[a], *and Kevin Supak*[b]

[a]Ebara Elliott Energy, Jeannette, PA, United States [b]Southwest Research Institute, San Antonio, TX, United States

Introduction

The existing global energy infrastructure provides various forms of liquid and gaseous fuels that are delivered through a vast transportation network. This transportation network utilizes pipelines, marine tankers, barges, rail cars, tanker trucks, and intermodal canisters to deliver fuel. Rules for the transportation of fuels continue to evolve and are governed by domestic and international regulations formed by regulatory bodies such as the U.S. Department of Transportation Pipeline and Hazardous Materials Safety Administration (PHMSA) as well as the UN Sub-Committee of Experts on the Transport of Dangerous Goods (TDG).

Pipelines

Pipelines are generally the most economical method for transporting liquid and gaseous fuels, though this is dependent on the distance and other factors. The global daily capacity of operational oil and gas pipelines is in excess of 200 million barrels of oil equivalent. The combined length of these pipelines is more than 1.18 million km with over half that length located in the Americas and about a quarter in Europe. As of 2021, there were over 200,000 km of planned expansions in the world [1]. Within the United States alone, the American Petroleum Institute estimates that there are more than 305,000 km of liquid petroleum pipelines, 480,000 km of natural gas transmission pipelines, and 3,200,000 km of gas distribution pipeline.

Tanker ships

Tanker ships are generally considered to be the second most efficient method for transporting liquid and gaseous fuels next to pipelines, though for very long distance shipments, they can actually be the most efficient means of transportation. There are several types of tankers for transporting oil, refined products, liquefied petroleum gas (LPG), and liquefied natural gas (LNG). Shipping sizes are constrained by harbor capability and shipping routes, including canal capability. The Seawise Giant is the largest supertanker ever built, having 564,763 deadweight tonnage (DWT) and a capacity for 4 million barrels of oil. The ship was incapable of navigating the Suez Canal, the Panama Canal, or the English Channel.

Deadweight tonnage is one means of classifying ships (Fig. 11.1). Several additional and more meaningful terms have been developed to describe the maximum-sized ships that can navigate specific locations in a laden condition based on ship length, width, and draft [2]. Q-Max defines the largest LNG carriers that can dock in Qatar. Malaccamax defines the largest ships that can navigate the straits of Malacca: Suezmax, the Suez Canal; Baltimax, the Baltic Sea; Chinamax, Chinese ports; Panamax, the original Panama Canal opened in 1914; Neopanamax, the current Panama Canal opened in 2016 and expanding tonnage from 52,500 DWT to 120,000 DWT;. Seawaymax, the St. Lawrence Seaway. Today's global petroleum tanker fleet has a capacity of around 629 million tons.

Energy Transport Infrastructure for a Decarbonized Economy
https://doi.org/10.1016/B978-0-443-21893-4.00016-7

Copyright © 2025 Elsevier Inc. All rights are reserved, including those for text and data mining, AI training, and similar technologies.

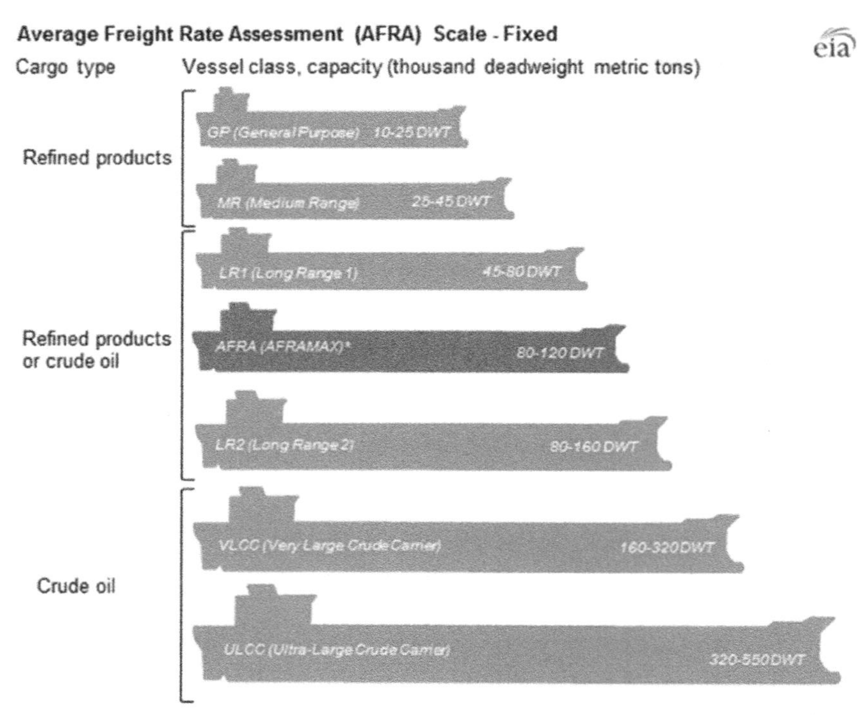

FIG. 11.1 Average freight rate assessment [2].

Tanker ships have evolved from single-hull to double-hull construction. The tanker ship Exxon Valdez, which spilled 260,000 barrels of crude after hitting a reef, was a single-hull tanker ship. Single-hull tanker ships were phased out of oil transportation in 2015 as part of the International Maritime Organization (IMO rulings). Greater regulation of shipping emissions is expected to take place in the future. The Suez Canal is undergoing an expansion to keep the canal from getting blocked, but the Suezmax size will remain the same.

The latest evolution of tanker shipping is not so much related to the ships themselves, but to the global infrastructure. Since the 2016 Panama Canal upgrade, ports throughout the world are being upgraded to handle Panamax ships.

Containerized shipping

Other means for transporting liquid and gaseous fuels are by railroad tank cars, tanker trucks, and intermodal containers (Fig. 11.2). Rules for shipping containers continue to evolve, governed by national and international regulations. Table 11.1 describes North American tanker rail cars, tanker trucks, and intermodal ISO tanks.

In 2015, the United States DOT-111 tank car, which was the work horse for railway petroleum, was slated for phase-out and replacement by the DOT-117 tank car. Older tank cars can be retrofitted to the DOT-117 standard and are described as DOT-117R. Retrofits include an insulating jacket, pressure relief devices, and a head shield. The phase-out schedule runs from January 1, 2018 to May 1, 2029, depending on the car's configuration and the cargo being carried. In 2023, a derailment in East Palestine, Ohio, involved the derailment and subsequent head and/or shell

FIG. 11.2 DOT-112 rail car [3].

TABLE 11.1 Containerized shipping.

Cargo	Examples of fuel cargo	Rail car	Tanker truck	ISO tank
Gases liquefied by pressure only	LPG, butane, anhydrous ammonia	DOT-112 129,847 ℒ	MC-331 Capacity: 43,900 ℒ	UN T50 Capacity: 51,000 ℒ
Cargos at temperatures <−90°C (<−130F)	LNG, LH2	DOT-113 113,562 ℒ	MC-338 Capacity: 55,000 ℒ	UN T75 Capacity: 47,000 ℒ
Petroleum liquids	Crude oil, distillates, ethanol, ethanol blends	DOT-117 TC-117 ℒ	MC-306 (DOT-406) Capacity: 34,000 ℒ	UN T3 Capacity: 44,000 ℒ

breaches of 7 out of 16 of the DOT-111 cars that derailed. Three DOT-117 cars were also involved in the derailment, but performed much better [4].

Tanker trucks for highway transport are manufactured in a variety of shapes and capacities depending on the product transported and required service. Larger capacity tanker trucks are typically used for longer distances while smaller trucks can provide local deliveries. The smaller product tanks can be mounted on the truck chassis while larger tanks are hauled as semitrailers.

ISO tank containers are also called ISOtainers. The tank containers can be transported by a variety of modes the same as a box container; on a ship, on a rail car, or on a truck trailer.

Future trends for existing networks

Future trends for these existing networks are toward (1) sustainability, (2) modernization (automation), (3) diversification of fuel types (integration of renewables), and (4) diversification of fuel sourcing (security of supply).

Sustainability initiatives in the global energy transportation infrastructure are taking on many forms, including the direct elimination of leakage emissions, electrification, improvements in efficiency, carbon sequestration, and pivoting away from coal. These initiatives seem to have always been around in some form, especially politically, starting with the United Nations Framework Convention on Climate Change adopted in 1992 at the Earth Summit in Rio de Janeiro. However, the recent spate of global regulatory requirements has been a fairly recent phenomenon born from this movement.

Elimination of leakage emissions is a particularly strong current and future trend. The Global Methane Pledge (GMP) is an international initiative with the goal of cutting anthropogenic methane emissions by at least 30% from 2020 levels by 2030. The United States, European Union, and 11 other countries launched the GMP energy pathway to accelerate methane emissions reduction in the fossil energy sector and working toward the minimization of methane flaring and both methane and CO_2 emissions across the value chain to the fullest extent possible [5]. The World Bank lays out the financial case for flaring and emission reduction (FMR) projects and sets a goal for eliminating routine flaring. In 2015, the World Bank introduced the Zero Routine Flaring by 2030 initiative [6].

To accomplish flaring and emission reduction, further infrastructure and technology are required. Conventional alternatives to flaring include (i) gas reinjection back into the ground, or (ii) large to mid-scale capturing, processing, and transporting the gas for consumption. In the case of remote and small production wells, capturing the gas isn't particularly economical and the infrastructure often doesn't exist for processing and transportation. This stranded gas would typically be flared or vented on site. Significant efforts by suppliers are being made to develop solutions for the economic capture and utilization of this small scale associated natural gas in amounts below 15 MMSCFD (or 0.15 BCM/y) from upstream production facilities [7]. These efforts include miniature gas-to-liquids (Mini-GTL); power generation using a small-scale gas engine; scaled gas processing unit with NGL recovery; power generation using a microturbine or small-scale gas turbine; mini-LNG; mini CNG; small-scale gas-to-chemical (GTC) plants; water purification or desalination; cogeneration; adsorbed natural gas storage and transport. Many of these solutions are in a pilot phase and not yet widely adopted.

Further emission reduction also requires advanced leak detection technology. The traditional method of leak detection involves personal visual inspection. Several technologies are being brought forth and developed that offer high accuracy and real-time monitoring capabilities. These technologies include laser-based sensors, airborne detection by aircraft or drone, distributed fiber optic sensing, data analytics and artificial intelligence, and satellite detection. These technologies provide precise geolocation, proactive maintenance, and the ability to detect leaks over long distances [8].

EU goals were primarily driven by climate concerns toward diversification away from fossil fuels, but there has remained a dependence on natural gas, not only as a fuel but also as a chemical feedstock. The sudden shutdown of the Nordstream pipeline in 2022 shifted the energy focus from emissions reductions toward security of supply (SOS). It is perhaps a strange irony that a mild winter, which could be attributed in part to global warming, enabled Europe to get through its winter without completely using up all its gas reserves. The loss of Nordstream supply required natural gas imports by other means. The world stepped up to deliver natural gas to Europe in a relatively short period of time.

Increasingly, security of supply is an energy focus. When pipelines cannot be constructed, other means of energy transportation, such as rail, try to step into the void. Liquefied natural gas has been shipped by rail using ISO modal shipping containers (also known as ISOtainers) particularly in Japan to reach rural areas where pipeline construction is not economical. More recently, the US PHMSA authorized transportation of LNG in bulk railcars effective August 2020 [9], but then suspended that authorization effective October 2023 [10] after a change in administration.

Solid-state hydrogen storage

The US Department of Energy developed some technical goals and objectives for hydrogen storage by 2020. These objectives focus on onboard automotive hydrogen storage achieving 1.8 kWh/kg at $10/kWh, novel conversion processes to reduce the need and cost for carbon fiber tanks by 25% from $13/lb. to around $9/lb., the development of a rechargeable storage system for portable power applications with targets of 1.0 kWh/kg and 1.3 kWh/L at $0.4/Wh$_{net}$, and the development of storing system for material handling equipment targeting 1.7 kWh/L at $15/kWh$_{net}$ [11]. There are two basic categories of hydrogen storage: physical and material where physical is the storage of hydrogen by changing its physical phase to increase its volumetric density (compressed or liquefied) and material is storing hydrogen using other materials as carriers [12].

According to Aire Liquide, a hydrogen fuel cell vehicle would require 1 kg of hydrogen to drive 100 km; stored at atmospheric pressure; this would take up 11 m^3 (11,000 L) or roughly the volume a large utility vehicle trunk [13]. This is orders of magnitude more than the required volume of unleaded gasoline needed to travel 100 km. This is understandable since the density of a liquid is higher than that of a gas. While the other assumptions for why 1 kg of hydrogen is required for 100 km of travel are not included in this assertion, this still demonstrates, at minimum, that there is at least one major problem with just switching cars from running on gasoline to pure hydrogen, fuel storage. The best way to address this is to increase the density of hydrogen stored. The ways to increase the density of hydrogen immediately make the hydrogen fuel storage more challenging. There are three main storage options to address this increased hydrogen density requirement: high-pressure gaseous hydrogen, cryogenic liquid hydrogen, and hydride-based solid. Since hydrogen is in a gaseous state at normal atmospheric pressure and room temperature conditions, a lot of energy is required to increase its density to a liquid form. Solid metallic hydrides are different though. Hydrogen does not undergo a phase change to achieve a solid state. Rather, it combines with something else already in a solid state, and similar to charging a battery with electrons, the reaction is reversible.

Solid-state hydrogen storage is being development for transporting hydrogen with relative ease and safety. This format is being pursued for two main reasons. The first is that high-pressure hydrogen tanks need to be cylinders, which are an awkward shape for fitting into a standard car. The second is the energy penalty of having to compress the hydrogen high enough to have a reasonable gas density within the tank for travel. Solid storage would allow for the storage tank to be in any shape necessary to easily fit within the vehicle in question and does not have the energy loss from highly compressed hydrogen.

There are two types of solid, material-based storage: sorbents and metal hydrides.

Metal hydride solid storage is achieved by bonding hydrogen with another material on an atomic level. One such material is lithium nitride. This can combine with hydrogen to form lithium amide and lithium hydride releasing hydrogen as a gas in the reverse process. Hydrogen Storage Materials-Advanced Research Consortium (HyMARC) is a multiinstitutional project working on developing this technology including Lawrence Livermore, Lawrence Berkeley, and Sandia national Laboratories and several other universities and institutions. Some studies with this material led the scientists to hypothesis that the process is too long due to the need to form intermediate species before the lithium nitride completely absorbs the hydrogen to form lithium amide and lithium hydride. Research was performed "to understand the chemistry of a nanoconfined lithium-nitrogen-hydrogen system wherein each pore within the confined space is measured in nanometers" [14]. During this research, the Sandia team developed a method whereby a porous carbon matrix confined the lithium nitride and speeds up the reaction. Numerous experiments and molecular dynamic simulations were performed to demonstrate how reducing the particle size increased cycling time. They found that the favorable thermodynamics of the intermediate phase formations became unfavorable as the particle size decreased due to the increasing energy penalty in decreasing particle size. Thus, as long as the particles are smaller than the threshold, no intermediate phases form and the reaction speed is maximized.

The discovery of this particle size threshold for storing hydrogen as lithium hydride is a great step toward mobilizing hydrogen use and guiding the design of more efficient hydrogen storage. The chemical understanding gained by the researchers has aided in identifying obstacles in using other materials. A promising material for solid storage is magnesium diboride, which becomes magnesium borohydride when hydrogenated and "has one of the highest known theoretical capacities of any material to store hydrogen," but its "hydrogenation reaction…can take a week" [14]. Further research by HyMARC is being conducted to model and evaluate different materials on a macroscale in actual use case conditions [14]. This modeling at a macroscale proves to be quite computationally expensive, and even high-performance computers are not yet powerful enough. Research characterizing this reaction uncovered that hydrogen gas reacts at dislocations within the MgB2, defect sites. This fundamental understanding of the chemical reaction is guiding researchers in developing a way to engineer the material for faster hydrogen storage including exploring if trace amounts of other materials might act as catalysts in the reaction. More technical information on this can be found in published work from Livermore Research Laboratory.

Sorbent material hydrogen storage stores hydrogen in its molecular dihydrogen state. This has advantages over metal-hydride atomic hydrogen storage when paired with systems requiring dihydrogen fuel as opposed to atomic hydrogen fuel. There are two main disadvantages to sorbent hydrogen storage. One is hydrogen's relatively weak absorption enthalpy, typical of a gas-to-solid reaction, as compared with chemical hydrogen bond formation or interstitial atomic hydrogen in metal hydrides [15]. The other disadvantage is that hydrogen has a relatively large van der Waals dimension compared with atomic hydrogen [15]. Dihydrogen absorption into a substrate can still have a higher density than compressed hydrogen. Hydrogen absorption into a substrate uses a framework generated by London dispersion forces above the temperature and pressure triple point. Additionally, the measurement of how much hydrogen has been absorbed is actually the Gibbs surface excess measured either gravimetrically or volumetrically [15]. Hydrogen absorbent enthalpy, while not a single-valued quantity, ranges from 5 to 10 kJ/mol and is reported as a differential following Henry's Law value. Some of the sorbent materials under investigation "range from coordination polymers to activated carbons" due to the high gravimetric density requirements [15]. High surface area paired with high micropore volume is the geometric jackpot for high hydrogen absorbents.

Storing hydrogen in a solid would be especially helpful for hydrogen-powered vehicles where fitting a canister of compressed hydrogen would be very difficult due to the bulky nature of the requirements for such a pressure vessel. This also removes the risk that the compressed hydrogen explodes in the event of an accident. Storing hydrogen onboard as a solid allows the designer the possibility to fit the matrix material in whatever shape and location is most efficient for the vehicle. However, current technology is limited to 2%–3% hydrogen to total weight of the tank, and there are other aspects at play such as temperature and pressure of the usage cycle [13]. These aspects require current solid hydride storage technology to be further developed before they will surpass liquid or high-pressure hydrogen storage.

Hydrogen blend with methane and transported by pipeline

Methane is currently transported through pipeline and has a vast network of pipelines to get it from source to consumer. Hydrogen pipelines exist already, with the majority of it in the United States located along the Gulf Coast. However, there is a lot of research being conducted into repurposing the extensive natural gas network for transport of hydrogen with minimal redesign with the hydrogen power future vision in mind. There are a number of technical difficulties with just replacing the gas within a methane pipeline with hydrogen. Among these are hydrogen embrittlement, mole weight, flashpoint, and flammability range. These difficulties will require different pipeline station compression equipment than what is standard for methane. This is discussed in much greater detail in Chapter 7 on hydrogen transport.

Retrofitting the existing large network of methane pipelines is possible but will require different monitoring systems, regulatory practices, and equipment updated as pipelines switch from natural gas to blended methane/hydrogen to pure hydrogen if existing pipeline networks are to be used. Some of this development is already in process since hydrogen is already being transported in pipelines. However, these pipelines were designed and built for pure hydrogen, whereas converting natural gas pipelines to pure hydrogen pipelines will require modifications to include different monitoring systems and abide by guidelines specific to pure hydrogen transport.

Hydrogen storage projects

Some current projects, demonstrations, and future research include methanol to hydrogen fuel cells, a pure hydrogen pipeline test, and geologic hydrogen sources.

The first methanol-to-hydrogen fuel cell system was successfully tested on June 27, 2023, by Powercell Group, e1 Marine, and RIZ Industries [16]. The 200 kW propulsion chain string test was performed in Gothenburg, Sweden, on land, and they expect to be able to scale up to megawatt-level propulsion chains for other types of vessels. An inland push boat, using only a 1.4 MW hydrogen fuel cell system, was successfully tested.

On June 25, 2023, PipeChina succeeded in their first higher-pressure pure-hydrogen pipeline test [17]. Their goal is to build the country's first long-distance pipeline at over 400 km with the first-phase pipeline being capable of handling 100,000 metric tons of hydrogen each year. They plan to have the potential to increase capacity up to 500,000 tons. Their success was with 6.3 Mpa (913.7 psi) hydrogen charging and 9.45 Mpa (1371 psi) pipeline blast test [18].

A June 26, 2023, article by Alan Ohnsman discusses current research into geologic hydrogen. This geologic hydrogen is hypothesized to be in wells similar to oil and could have as much as 10 million megatons underground [19]. Being in wells, this could be a productive shift for drilling companies from oil to hydrogen. The technology would be a little different, but the premise of drilling down to a hydrogen well would be very similar to existing drilling practices now. This shift could also alleviate some of the pushback from oil companies since they would be able to just add hydrogen to their existing portfolio as the transition away from oil based fuels is underway.

Hydrogen liquefaction

Introduction

Liquefied hydrogen is garnering significant attention as an alternative energy source and energy storage mode in recent years. Hydrogen is the lightest element, and liquefying this light gaseous hydrogen requires extremely low temperatures (~20 K) to achieve and maintain liquefaction. While the technology to achieve and maintain these temperatures is challenging, liquefied hydrogen is already used for many important applications, and its application is expected to continue to grow in the near future for many industries. One of the key drivers for the growth of the liquid hydrogen market is the growing demand for reduced dependence on fossil fuels and concerns over carbon emission levels.

While many applications utilize liquid hydrogen directly, hydrogen liquefaction will still be an important part of the value chain for compressed hydrogen applications due to the involvement of tanker truck, freight, and ship transport of liquid hydrogen. Liquefied hydrogen has a much higher density at 71 kg/m^3, compared with compressed gaseous hydrogen (700 bar) at 40 kg/m^3. Transporting a liquid at a higher density means tankers can carry larger payloads, refueling is faster, and highly pressurized storage containers are not needed. While steam reforming is currently expected to hold the largest market share of hydrogen production, electrolysis for green hydrogen production is forecasted to experience significant growth as demand for green hydrogen from renewable energy sources increases.

As of 2021, there are dozens of hydrogen liquefiers worldwide currently in operation, with a total capacity of approximately 470 tons per day (tpd). North American facilities account for more than 85% of this total, with the majority of remaining capacity being produced in Europe and Asia (primarily Japan) [20]. The current state of the art typically includes plant production capacities in the relatively small-scale (1–10 tpd) to medium-scale (10–50 tpd). In 2022, a HySTRA pilot project successfully demonstrated the liquid hydrogen supply chain by delivering hydrogen produced and liquefied in Australia through tanker ship to Japan [21].

Liquid hydrogen is currently used as a common liquid rocket fuel in the aerospace industry, due to its highest specific impulse of any known rocket propellant. The recent growth in space programs is expected to lead to a significant increase in the demand for liquid hydrogen. In 2022 alone, the rapid growth and demand for advanced satellites resulted in 186 rocket launches to orbit, an increase of 40 compared with 2021, and a trend expected to grow in the near future [22].

Liquid hydrogen can also be used as a fuel for internal combustion engines or fuel cells in the automotive and transport industries. For example, many consumer car manufacturers such as Toyota, Hyundai, and Honda have invested heavily in the idea of hydrogen-fueled consumer vehicles, with Toyota largely considered to be the most active and invested. While much of this activity has involved hydrogen fuel cells utilizing compressed gaseous hydrogen, Toyota has more recently begun experimenting with a liquid hydrogen combustion engine in a liquid hydrogen Toyota Corolla as part of a 24-h race to learn more about this potential [23]. Liquid hydrogen provides a higher energy density for a given fuel tank volume and is therefore an active area of research for the automotive industry, where the low volumetric energy density can be challenging.

The aviation industry is also beginning to see active research in liquid hydrogen as a fuel source for clean aviation fuel. Airbus, for example, is actively working to develop a zero-emission commercial aircraft by 2035 utilizing liquid hydrogen tanks [24]. In the near term, Airbus expects metallic storage tanks to be used; however, this introduces the potential for long-term improvements with composite tanks to reduce cost and improve efficiency. Similar projects are underway by other manufacturers to deliver hydrogen-powered aircraft. ZeroAvia, for example, aims to certify multiple hydrogen-electric engines for regional turboprops by 2027 [25].

While hydrogen liquefaction technologies have existed for many decades (largely in support of the NASA space programs), this technology has not significantly evolved much since the 1960s. That is, until recently, when hydrogen was identified as an important energy vector for the carbon-neutral economy. Significant research is now underway to improve liquefaction process efficiencies, storage technologies, and scaling the process.

Thermodynamic modeling and liquefaction plants/processes

One of the primary interests in designing better optimized processes for the liquefaction of hydrogen is improved thermodynamic data and models of cryogenic hydrogen. Zhang et al. [26] reiterated recommendations to publish more experimental data on the properties (density, sonic velocity, heat capacity, etc.) for hydrogen in both the gaseous and liquid phases at temperatures below 50 K. Additional measurements on para-hydrogen above 100 K would also help to resolve contributions due to ortho-hydrogen in many existing measurements. Improving the accuracy and reducing uncertainty in these values will serve to aid simulations of the hydrogen liquefaction processes and provide critical insight into potential improvements to reduce the cost of hydrogen liquefaction in the future.

This gap in understanding of hydrogen's cryogenic properties is related to the current gap in energy requirements in current hydrogen liquefaction plants. The current state of the art for hydrogen liquefaction plants involves relatively small-scale (1–10 tpd) to medium-scale (10–50 tpd) capacities, with a typical plant energy efficiency of >13.8 kWh/kg. In 2022, the Department of Energy (DOE) hosted a workshop of industry leaders in the field of liquid hydrogen, where the expectation was stated that larger plants (>100 tpd) would be capable of reaching as low as 6–7 kWh/kg [27].

Small-scale liquefaction plants are considered CAPEX-favorable, while large-scale liquefaction plants are OPEX-favorable. Scaling the liquefaction process down for smaller plants is typically not a concern and involves relatively well-developed methods; however, much of the ongoing research and development are focused on the challenge of scaling the liquefaction process up. A common bottleneck in these plants is the turbomachinery, including the compressors; however, improvements to insulation and energy efficiency/recovery are also important [27].

According to the DOE, targets for future large-scale hydrogen liquefaction plants include a 300 ton/day production capacity with an energy consumption of 6 kWh/kg_{LH2} [28]. As stated by Zhang et al., a wealth of research into liquefaction processes is ongoing to find energy consumption and efficiency improvements. Presently, modified Claude cycles, where hydrogen is both the working fluid and final product, is most commonly employed. Compared with the basic Linde-Hampson and commonly used Claude cycles, more advanced expander cycles (Brayton, Collins, and mixed refrigerant) show significant promise for improved efficiency. Ansarinasab et al. [29], for example, demonstrated that a mixed refrigerant cycle with isentropic expansion may be able to produce significantly lower specific energy consumption and increased efficiency compared with commonly used Claude or Brayton cycles.

While many liquefaction plants today employ some version of a Brayton or Claude main refrigeration cycle, the long-term focus will likely involve high-pressure Claude cycles or mixed refrigerant cycles to achieve greater than 100 tons per day of production with improved specific energy consumption. The liquid nitrogen precooling cycles are also targeting novel mixed refrigerants to improve performance, while the commonly used reciprocating compressors may ideally be replaced with centrifugal compressors to meet the demand for increased capacity while minimizing energy consumption [26].

There is still active discussion, however, that greater commercial value may exist in novel developments with small-scale plants. Dr. Jacob Leachman [30] discusses that a significant source of system losses in the liquefaction of hydrogen involves the compression of the hydrogen and liquid nitrogen precooling. Leachman notes that the electrolysis process can be used to pressurize hydrogen through electrochemical compression, which may allow smaller liquefiers to forego the mechanical complexity of mechanical compressors. Additionally, research into utilizing a Stirling cycle in lieu of liquid nitrogen precooling may allow a more efficient cooling cycle for small-scale plants. The combination of efficiency improvements and a large network of small, efficient liquefiers would open the possibility of a more distributed liquid hydrogen network with lower global losses due to a reduced need for on-site storage. This may prove valuable in the future as more renewable energy sources emerge. With distributed energy sources, a distributed network of liquefiers may be more commercially viable. However, a highly distributed network of smaller liquefiers may necessitate the need for many cryo-compressors for pipeline transport, which require more development.

To meet DOE targets, the hydrogen liquefaction industry will require a number of these discussed optimizations, such as reconfigured liquefaction processes with high efficiency and lower energy costs, optimized scale of capacity to minimize cost per kilogram of produced liquid hydrogen, strategic placement of liquefaction terminals, and improved recovery of "waste cold" from other terminals (e.g., LNG terminals, boil-off). Liquefaction processes will also benefit from component-level improvements such as more efficient turbomachinery, heat exchangers, and refrigerants, and active research and development will be required in all of these fields to improve the commercialization of liquid hydrogen.

In 2021, Air Products opened a new hydrogen liquefaction plant in La Porte, Texas, to produce 30 tpd liquid hydrogen for the western United States, with another plant announced in March 2022 to produce 10 tpd of green liquid hydrogen in Arizona [31]. In 2022, Air Liquide opened its largest hydrogen liquefaction facility in North Las Vegas, Nevada, to produce 30 tpd of liquid hydrogen as well. More and larger capacity facilities are expected to open as demand increases and the liquefaction process becomes more refined; however, none so far have developed plans to achieve the DOE target of 300 tpd.

Al Ghafri et al. [32] provide a summary of the challenges facing various components of the hydrogen liquefaction processes (Fig. 11.3), highlighting crucial areas of research in the future.

Overall
- Increase efficiency, decrease energy consumption and cost
- Modular limitations
- Integration of renewable energy
- Translation of conceptual models to industry
- Thermodynamic model accuracy for process design and simulation
- Hydrogen embrittlement and leakage

Compression
- High energy demand
- Large number of compression stages required
- Flow rate limits of piston compressors
- No evidence of turbo-compressors use in industrial LH_2 plants

Pre-Cooling
- High energy demand of producing and recycling LN_2
- Limited industry examples of LNG or MR used for H_2 cooling
- MR thermodynamics not well understood
- Lack of knowledge on costs and environmental consequences for N_2, LNG and MR cycles

Cryogenic Cooling
- Freeze out of impurities from H_2
- Optimizing heat exchanger sizes
- Accuracy of thermodynamic models

Catalysts and OP conversion
- Accurate OP ratio measurement
- Optimizing residence time of H_2 in the heat exchanger
- Limited conversion rate data for different catalysts
- Catalyst longevity
- Pressure drop in heat exchangers

Expansion
- Maximizing work recovery
- J–T expansion inefficiency
- Large-scale oil-free and efficient turbine expanders
- Optimizing flow rates for different expansion technologies

FIG. 11.3 Challenges facing different parts of the liquefaction process. *Taken from S.Z. Al Ghafri, S. Munro, U. Cardella, T. Funke, W. Notardonato, J. M. Trusler, E.F. May, Hydrogen liquefaction: a review of the fundamental physics, engineering practice and future opportunities. Energy Environ. Sci. 15 (7) (2022): 2690–2731.*

Compressors and liquid hydrogen pumps are often cited as one of the major areas of focus for plant efficiency improvement for the future [27]. These components are a major contribution to total cost not only in the liquefaction processes, but they also dominate the overall cost for refueling transport. Low-suction pressure compressors are still in development and may offer significant benefits to system costs, as well as a good liquid hydrogen pump.

Ortho/para-hydrogen catalysts

Diatomic hydrogen molecules consist of two covalently bonded hydrogen atoms. With these two atoms, molecular hydrogen can exist in two spin states, referring to the associated magnetic moment of each proton. These two spin states are known as para-hydrogen and ortho-hydrogen. In para-hydrogen, the spins of the hydrogen atoms are opposed, while in ortho-hydrogen, the spin states are aligned. This results in a small difference of 1.455 kJ/mol in the energy levels of the two spin states, which can become significant hurdles for the liquefaction process.

The lowest energy state of diatomic hydrogen is reached in the para-hydrogen state with opposed molecular spins. The concentration of each spin isomer is temperature dependent—at room temperature, hydrogen is roughly 75% ortho-hydrogen and 25% para-hydrogen. As temperature is decreased in cryogenic liquefaction, ortho-hydrogen is thermodynamically unstable as molecules become close enough to interact and convert to para-hydrogen in an exothermic conversion process. The heat released is a critical factor in boil-off losses of liquefied hydrogen. To combat this inefficiency, catalysts are introduced prior to the liquefaction to convert ortho-hydrogen to para-hydrogen, preventing the majority of exothermic conversions leading to boil-off losses after the liquefaction.

Current catalysts involved in speeding up this conversion process are often commercially available iron hydroxides and chromium oxides. While research on novel catalytic conversion exists (for example, Dr. Jacob Leachman discusses research opportunities to improve ortho/para conversion through quantum effects [27]), much of the focus for future improvements is primarily on the implementation of existing catalysts. Particularly, how the catalyst is introduced to the process and its effect on conversion efficiency and the overall liquefaction process is a major research area. Dr. Raja Amirthalignam [27] explains that the catalyst is typically either in the heat exchanger or in a separate vessel, with current research efforts indicating that there is a larger advantage to introducing the catalyst in the heat exchanger. Additionally, Hartl [33] mentions that catalyst dilution during its introduction is one factor that has been shown to impact efficiency depending on temperature, in some cases demonstrating an increase in efficiency with lower dilution. Hence, there are many variables that may impact catalyst effectiveness that need to be studied in detail.

It should be noted, however, that the thermodynamic modeling involved in the liquefier remains a challenge due to the model deterioration with the catalyst introduction. Park et al. [34] state that the pressure drop in a cylinder filled with ortho-para catalyst can differ by a factor of 5 with typical predictions with Ergun's equation, and further research is ongoing to best model and implement the catalytic conversion process. More research will nevertheless be needed to determine the optimal application and analysis of these catalysts for future improvements.

Storage and transport

Bulk storage of liquid hydrogen remains a challenge due to the extremely low temperature that must be maintained to reduce boil-off losses. Similar to LNG, ambient heat will ultimately leak into the tank and produce some temperature rise in the liquid hydrogen, resulting in some amount of boil-off. This boil-off will increase the pressure in the tank, eventually requiring the excess gas to be removed and recompressed for reliquefaction or compressed storage elsewhere. However, due to the much lower temperature required for liquid hydrogen than LNG, these boil-off losses can be quite significant. While the amount of boil-off and heat leak will depend on the surface-area-to-volume ratio of the storage tank and tank design and insulation, the boil-off can in some cases reach 1%–5% per day [26]. Even sloshing of the liquid inside the tank during transport can be problematic, sometimes requiring baffle designs to minimize kinetic energy transfer to the liquid hydrogen. This presents a significant challenge to the overall efficiency of liquid hydrogen as a fuel, and many efforts have been focused on improving this aspect of hydrogen liquefaction.

NASA currently owns the largest cryogenic storage tank in the world, with a usable capacity of 4732 m^3 (1.25 million gallons) at its Kennedy Space Center in Florida. However, future storage tanks are expected to be significantly larger. In 2021, the DOE selected a consortium of experts led by Shell International Exploration and Production Inc. to demonstrate the feasibility of a large-scale liquid hydrogen tank with a capacity of 20,000–100,000 m^3 [35]. In line with these ambitions, McDermott International Ltd. announced in 2021 a new record-breaking tank designed by CB&I Storage Solutions capable of storing 40,000 m^3 (10.5 million gallons), with development underway for a 100,000 m^3 tank under DOE funding [36]. Kawasaki Heavy Industries is similarly working on the design of a liquid hydrogen tanker ship with up to four 40,000 m^3 storage tanks for liquid hydrogen transport and a 160,000 m^3 transport capability. The tanker ship would notably utilize the natural boil-off gas as a fuel to power the ship and further reduce CO_2 emissions [37]. Similarly, a press release by Alfa Laval stated Shell International Trading and Shipping Company Ltd. entered into an agreement with Alfa Laval regarding the development of a gas combustion unit for use on liquid hydrogen carrier ships [38].

Larger storage tanks benefit from basic scaling properties: for a spherical tank, the volume increases with the cube of radius, while the surface area increases with the square. Thus, larger tanks utilized for higher liquid hydrogen demand will benefit from reduced heat leaks (and subsequent boil-off losses) over its surface area for a given volume of liquid hydrogen. However, while the improved economy of scale exists for larger tanks, this is not the only aspect of improving the efficiency of liquid hydrogen storage. The tank design and insulation are another critical factor. The developments by CB&I Storage Solutions involve evacuated glass bubble thermal insulation used for NASA's 1.25-million-gallon liquid hydrogen tank, and an active insulation involving an internal heat exchange to actively maintain low temperatures and reduce boil-off. The evacuated glass bubble insulation is used in lieu of the more commonly used perlite powder system and is cited to reduce boil-off losses by as much as 46% based on studies conducted on a retrofitted tank at NASA Stennis Space Center in Mississippi [39].

Active insulation is another method used to reduce boil-off to near zero. Active insulation refers to the inclusion of an internal heat exchanger as part of a refrigeration cycle to actively remove heat from the stored liquid hydrogen. New research indicates the NASA method, referred to as Integrated Refrigeration and Storage (IRAS), demonstrates the capability to maintain zero or near-zero boil-off losses for long time periods [40]. With this system, ambient heat leak into the tank is removed by an internal heat exchanger as part of a Brayton cycle cryogenic refrigerator. The economic analysis of the "zero" boil-off indicated that every dollar spent on electricity to power the active internal heat exchanger may save roughly $7 of liquid hydrogen (based on typical electricity and hydrogen production costs at the time) [41].

It should be noted, however, that advancements in storage tank insulation to reduce boil-off are primarily aimed toward tanks requiring longer-term storage. For tanks experiencing continuous use, such as fueling applications, the transfer losses dominate the system since the boil-off losses in the tank do not have the time to become relatively significant. For longer-term storage, the combination of insulation materials, passive cooling (by redirecting cold boil-off gases to intercept incoming heat leaks), and active cooling may prove to be a valuable approach in the future.

As mentioned previously, while global market demand and transport are pushing for larger tanks, lightweight liquid hydrogen storage tanks are also being actively developed for aviation use, where size and weight are important factors. Airbus has targeted the first hydrogen-powered commercial aircraft by 2035 [24]. The ZeroE concepts under development would utilize cryogenic liquid hydrogen storage tanks to fuel modified gas turbine engines, with additional hydrogen fuel cells for complementing electrical power. In the short term, these tanks are expected to be metallic; longer-term development goals for Airbus involve composite materials to further reduce weight. This goal of commercial hydrogen-powered aircraft has been followed in late 2022 by an announcement that Rolls-Royce and EasyJet successfully ran a modern aero engine on hydrogen [42], an active area of research that is expected to continue in the future for hydrogen-powered flight.

Future outlook

While many applications of liquid hydrogen continue to grow in the near future, research into improved liquefaction and storage/transport processes of liquefied hydrogen will continue to play a critical role in the success and widespread adoption of each end use. Currently, the energy input to liquefy hydrogen can be prohibitive, so large efforts across all aspects of the liquefaction and distribution process are required to improve economic feasibility. These areas include the better understanding and modeling of hydrogen thermodynamics, the use of improved models to develop better optimized liquefaction processes, the improved use of ortho/para-hydrogen catalysts, greater development into higher-efficiency turbomachinery for hydrogen liquefaction plants, larger-scale bulk storage with improved insulation capabilities, as well as reduced transport losses through truck, freight, or ship.

Regarding timelines, liquid hydrogen is already being used as a common rocket fuel and is in development for liquid hydrogen-fueled trucks, with maritime applications expected within the mid-2020s. Applications in the aviation sector are expected to take more time and may not be fully developed until 2040 or later. Each sector presents unique challenges that remain subjects of active research to improve the commercial viability of the hydrogen economy.

CO_2 research

Traditional uses of CO_2 are mostly in the production of fertilizer, enhanced oil recovery, and the food and beverage industry. However, with carbon capture utilization becoming a global priority, new uses for large quantities of CO_2 are being investigated by several industries and research groups to avoid the transport and storage costs for sequestration. It is important to note that the overall carbon footprint and energy requirements of these uses must be evaluated to ensure that they are not net carbon producers when compared with existing processes or production methods. Some of these uses include:

- Fuel and intermediate chemical production, including methane, olefins, methanol, gasoline, and aviation fuels. The chemical conversion process involves combining CO_2 with H_2 in a direct reaction using hydrogenation or in an indirect reaction using reverse water-gas shift to produce the syngas CO.
- Chemical production includes polymers, foams, and resins made by using the chemical reaction of CO_2 with heat and fossil-based raw materials.
- Concrete cured with CO_2.
- Construction aggregate production by combining CO_2, heat, and waste materials such as iron slag or coal fly ash.
- Enhancing biological processes such as food and algae production.
- Advanced carbon material production includes composites, graphene, and nanotubes.

There are several research areas that are currently being investigated by government organizations and pipeline companies related to CO_2 transportation. The highest priority topic is understanding the effects of impurities (H_2O, CH_4, H_2, O_2, SOx, NOx, etc.) related to material compatibility and thermodynamic behavior. The source of these impurities is associated with carbon capture from fossil-fueled power plants and other industrial processes. Concentration limits for individuals and combinations of impurities are being established by conducting material compatibility tests and pressure-volume-temperature (PVT) tests to understand the impacts on material integrity and the fluid mixture phase behavior. Another active area of research is understanding pipeline decompression behavior with CO_2 dominate mixtures, as rupture events for CO_2 pipelines behave differently and can be more catastrophic than in traditional fluids. A European joint research group managed by Climit has conducted research in all of these areas and has compared the results with an active CO_2 sequestration project in Norway called Snohvit. However, more research is needed to understand phase behavior and the validity of transport models, including pressure drop predictions and transient behavior such as shutdowns and startups. Another active area of research is determining if existing pipelines can be utilized to transport CO_2, given the challenges previously described.

Several projects related to the energy transition are active at the time of this publication. The US Department of Energy (DOE) has invested heavily in distribution and gathering hubs for both CO_2 and hydrogen through various front-end engineering design (FEED) studies and pilot-scale demonstrations.

CarbonSAFE began in 2016 by addressing the key gaps on the critical path toward carbon capture and storage (CCS) deployment to reduce the technical risk and cost of carbon transport and storage. CarbonSAFE has identified at least 20 sites to perform pilot-scale demonstrations. Europe's Northern Lights project is funding similar studies and demonstrations to connect carbon producers with transport and storage networks.

H2Hubs is another program initiated by the DOE to establish 6–10 regional clean hydrogen hubs across the United States as part of an $8 billion-dollar hydrogen program. H2Hubs will connect hydrogen producers, consumers, and local infrastructure to accelerate hydrogen energy use and storage. The goal of this program is to provide capital funding to demonstrate the production, processing, delivery, storage, and end use of clean hydrogen.

DOE's Supercritical Transformational Electric Power (STEP) program has the goal to operate a 10 MWe pilot-scale facility using a supercritical CO_2 power cycle. Supercritical CO_2 power cycles have the potential to lower the cost of electricity production as compared with traditional steam power generation cycles due to the compact nature of the machinery associated with the high density of CO_2. This project will validate component and cycle performance over a range of operating conditions and is targeting a 2%–5% improvement in overall thermal efficiency with an associated reduction in emissions, fuel, and water usage.

New energy projects announced online

News announcements regarding new projects in the energy field are often related to green hydrogen or electrolyzer technologies. However, research and projects regarding methane or natural gas are continuing on several fronts.

Macaw Energies Ltd. has signed a collaboration agreement with GTUIT LLC to use GTUIT's technology to treat natural gas to produce liquefied natural gas through its mobile liquefaction units. The agreement provides a platform for constructing the first 10 of Macaw's US liquefaction units and integrating GTUIT's gas treatment with the liquefaction technology. The LNG would be created at the wellhead from potentially flared gas and could be sold or used as a diesel fuel alternative for use in the oilfield. Macaw CEO Slim Hbaieb discussed the company's plans in an email interview with a Midland, TX, newspaper [43]. Operations could begin as early as 2024 and expand as dictated by the marketplace.

French oil major TotalEnergies has invested in Ductor, a Finnish startup company that has developed a new technology to process organic waste. The high nitrogen content of some organic waste usually makes it unattractive as a source for biomethane. Using the new process will allow TotalEnergies to market biomethane production, while Ductor will oversee the production of sustainable biofertilizers. Currently, Ductor has a portfolio of around 15–20 projects, including some in advanced stages. Applications would be in both Europe and the United States [44].

Linden Cogeneration (Linden Cogen) has successfully completed the commissioning process for its hydrogen blending initiative at the Phillips 66 (Houston) Bayway Refinery. The initiative takes refinery gas containing hydrogen by-products and blends it with natural gas to fuel the Unit 6 gas turbine. This will curb CO_2 emissions by reducing the amount of natural gas used for power and steam generation. Additionally, the project is improving the efficiencies of the overall refinery, the flare system, and Linden Cogen operations [45].

Fossil fuels cannot quickly be replaced with hydrogen, renewable energy sources, or nuclear fusion. This can mean that in the short term, carbon capture and sequestration projects will be promoted in the near term.

Carbon capture and storage (CCS) and carbon dioxide removal (CDR) will "play a key role in decarbonization" as companies execute plans on their net-zero emission targets, according to a recent report by S&P Global Ratings. In a sample of 25 of the highest-revenue oil and gas companies, all of them plan to use at least one among the options of CCS, CDR, or carbon credits to meet their decarbonization goals, the rating firm said in a report authored by its Sustainability Research team. Of the companies in the sample, CCS capacity in 2022 represented 7% of their scopes 1 and 2 emissions, with most activity coming from oil and gas majors in the United States and Europe. According to the World Economic Forum, scope 1 emissions are direct emissions that a company causes by operating the things that it owns or controls, while scope 2 emissions are indirect emissions created by the production of the energy that an organization buys. Plans for the deployment of CCS and carbon capture, utilization, and storage (CCUS) would see capacity rise from 50 million tons currently to 325 million tons by 2030, which includes targets for enhanced oil recovery and solutions to capture emissions from other companies. Of the firms in the sample, only 60% revealed their expected future capacity, only 56% identified the specific investment costs required, and 24% said they would use the captured carbon for enhanced oil recovery, but often these goals are expressed in vague terms. CDR is a group of both nature-based and technological solutions that remove carbon dioxide from the atmosphere and permanently store it in geological or ocean reservoirs. An example of CDR is planting trees, as in reforestation. Technological CDR solutions such as direct air carbon capture and storage, which removes and stores carbon from ambient air, are at much earlier stages of development, with "technical and economic challenges still to be overcome," the report said [46].

For example, a large direct air capture proposal, Project Bison, was running well behind schedule as of late 2023. Announced in September 2022 by the startup CarbonCapture Inc., the initial phase of the project was set to come online at the end of 2023. It would scale up to remove 5 million metric tons of carbon from the air annually by 2030 (the largest DAC plant in operation today, Climeworks' Orca facility in Iceland, is only capable of removing 4000 tons per year). However, as of June 2023, the California-based company still had not determined where to site its plant or how to power it, according to officials involved with the project. In addition, Frontier Carbon Solutions LLC, the firm CarbonCapture will rely on to transport and store the carbon, was only granted the permits it needs to inject CO_2 deep underground by the state of Wyoming in December 2023. CarbonCapture plans to build Project Bison somewhere in Sweetwater County, in the southwestern corner of Wyoming. The exact site would be within about 15 miles of three Class VI carbon storage wells that Frontier is planning to drill. The artist's conception shows more than 100 house-sized modules on an arid plain. The 55-person DAC startup had not begun hiring in Wyoming. However, CEO Adrian Corless said he expected to double the company's headcount over 12 months and open a new manufacturing facility. The company had applied for matching funds from the DOE's $3.5 billion DAC hub program. He estimated that the facility would initially require about 5 MW of power, with demand increasing to as much as 1 gigawatt when Project Bison is completed [47].

Other plans are progressing at a faster rate. Incentivized by subsidies included in the 2023 Inflation Reduction Act, oil and gas companies including Occidental Petroleum, Chevron, ConocoPhillips, and ExxonMobil are covertly buying up subsurface rights along the Gulf Coast to establish carbon dioxide storage sites. An estimated 480,000 acres of land have been targeted for CO_2 storage in Texas and Louisiana, with offshore leasing in the Gulf of Mexico amounting to twice that size. So far, much of the attention has focused on a roughly 600-mile stretch of the Gulf Coast between Corpus Christi in Texas and New Orleans. That area is filled with petroleum refineries, chemical plants, and other facilities that make it one of the biggest concentrated sources of industrial CO_2 emissions in the United States. The area also has the right kind of geology for keeping CO_2 in place underground, as well as an existing network of pipes that can be used to move the gas between the facilities that emit it and the wells where it will be injected. Sequestration is a new business; only four CO_2 injection wells have been issued permits as of 2023 in the United States, although nearly 100 are under review. There are many legal and regulatory details that are still unsettled, from who exactly owns the rights to the "pore space" in the rocks that trap the carbon dioxide to who's responsible if the compressed gas seeps out of the leased area into someone else's property underground. Some environmentalists and geologists warn that we may not fully understand what will happen when pressurized gas is pumped in such large quantities into the earth and that unexpected leaks could be dangerous or even fatal if enough CO_2 comes out [48].

As noted earlier, green hydrogen and electrolyzers seem to dominate new project announcements. A majority of green hydrogen projects are located in Europe.

Cepsa (Madrid, Spain) and GETEC, one of Europe's leading energy service providers, have reached a cooperation agreement by which, at 2026, the Spanish energy company will supply green hydrogen and its derivatives to GETEC for use by its industry clients as part of their decarbonization goals. GETEC provides heating, cooling, and electricity to industry clients in various sectors, such as the chemical industry, automotive, food industry, pharma industry, biopolymer industry, and paper industry. It has regional platforms in Germany, the Benelux countries, Switzerland, and Italy. Cepsa is developing 2GW of green hydrogen at its two energy parks in Andalusia, southern Spain, as part of its 2030 Positive Motion strategy to become a leader in the production of renewable hydrogen and advanced biofuels. The two hydrogen plants, with a €3 billion investment, will form part of the Andalusian Green Hydrogen Valley, for which Cepsa has recently signed a number of partnership agreements across the hydrogen value chain [49].

ABB, Inc. (Zug, Switzerland) is collaborating with Lhyfe (Paris), a world pioneer in the production of renewable hydrogen, and Skyborn, a global leader in renewable energy, to jointly realize and optimize one of Europe's most ambitious renewable hydrogen projects ever, SoutH2Port. The project is to be located in close proximity to Skyborn's 1 GW offshore wind farm, Storgrundet in Söderhamn, Sweden, where Skyborn and Lhyfe recently entered into a sales purchase agreement with Stora Enso for an industry property of around 40 ha. Once fully operational, SoutH2port is expected to produce about 240 tons of hydrogen per day with an installed capacity of 600 MW, making it one of the largest suppliers of renewable hydrogen in Europe [50].

Stargate Hydrogen Solutions OÜ (Tallinn, Estonia) has signed a grant agreement with the European Commission under the Horizon Europe program. The project, coordinated by the Oceanic Platform of the Canary Islands (PLOCAN), will receive €10.7 million for demonstrating the full value chain of green hydrogen. The project will utilize renewable energy from a 6 MW offshore wind facility to produce green hydrogen using a marinized high-efficiency electrolysis unit. A 1 MW Gateway series electrolyzer from Stargate will be installed in the onshore hydrogen production plant at the PLOCAN site on Gran Canaria. The generated green hydrogen and oxygen will be used at the local hospital complex. The production of green hydrogen in maritime environments is still in its technological infancy, with only a handful of demonstration projects ongoing in Europe [51].

Djewels B.V., 100% owned by HyCC, has contracted McPhy Energy S.A. (La Motte Fanjas, France) for the electrolyzer supply and with Technip Energies (Paris) for the design and construction of the planned 20 MW green hydrogen plant. Djewels will be a state-of-the-art electrolysis facility located in Delfzijl, the Netherlands. The plant will be operated by HyCC using electrolyzers from McPhy to produce up to 3000 tons of green hydrogen per year from renewable power and water. The green hydrogen can be used by OCI Methanol Europe for the production of renewable methanol to reduce CO_2 emissions by up to 27,000 tons per year. Other companies supporting the project include Gasunie, DeNora, and Hinicio [52].

Electrolyzer production is taking place in Europe and the United States. Another contract for McPhy Energy S.A. (La Motte Fanjas, France) will be for the supply, assembly, and commissioning of one McLyzer 800-30 with a capacity of 4 MW for Plansee Group. The electrolyzer will operate on the Plansee Group manufacturing site, based in Reutte, Austria, where green hydrogen will be used to produce carbon-neutral high-performance metals for various high-tech applications such as semiconductors, electronics, or medical. Today, hydrogen on the Plansee Group production site is provided through steam reforming as an integral part of metal production and is used by the company Plansee High Performance Materials and tool manufacturer Ceratizit. The equipment provided will enable the site to gradually replace the production of green hydrogen produced with renewable energy. To ensure operational reliability and availability, McPhy has also signed a long-term maintenance agreement in the framework of this contract [53].

Nel Hydrogen ASA (Oslo, Norway) announced its plans to build a $400 million new automated gigawatt electrolyzer manufacturing facility in Michigan. When fully developed, the facility will employ more than 500 people and be among the largest electrolyzer manufacturing plants in the world. The Michigan facility will have a production capacity of up to 4GW of alkaline and PEM electrolyzers. Going forward, Nel will build on its fully automated alkaline manufacturing concept invented at Herøya in Norway. Similarly, the company's expansion of the facility in Wallingford, CT, will play a critical role in creating a blueprint for scaling up the production of PEM electrolyzers. Nel's PEM electrolyzers have been developed with decades of support from the US Department of Energy [54].

Plug Power Inc. (Latham, NY) is part of a consortium of companies that received a $21.8 million grant from the European Commission to build an offshore hydrogen production plant. As a member of the nine-company consortium HOPE (Hydrogen Offshore Production Europe), Plug will design and deliver a 10MW proton exchange membrane (PEM) electrolyzer system to the site in the North Sea, off the port of Ostend, Belgium. The HOPE project will produce up to 4 tons a day of green hydrogen at sea, which will be transported to shore by pipeline, compressed, and delivered to customers for mobility needs and small industries in Belgium, northern France, and the southern Netherlands within less than a 200-mile radius. HOPE is the first offshore project of this size in the world to begin actual implementation, with the production unit and export and distribution infrastructure due to come online in mid-2026. The HOPE project is being coordinated by Lhyfe (France) and implemented by eight European partners. In 2022, Plug and Lhyfe pioneered the proof of concept for the world's first floating offshore hydrogen production plant, Sealhyfe [55].

thyssenkrupp nucera AG & Co. KGaA (Dortmund, Germany) has signed an agreement for a North American project under which a company has contractually secured the supply of the standardized 20MW "scalum" electrolysis modules with a total installed capacity in the high multihundred megawatt range. The two companies have agreed not to disclose further details of the contract. Thyssenkrupp nucera has already contracted more than 3GW of the capacity of alkaline water electrolysis. These include a more than 2GW electrolysis plant for Air Products in Saudi Arabia, making it one of the world's largest green hydrogen projects; the delivery of Shell's new 200MW hydrogen plant in the port of Rotterdam; and H2 Green Steel's green steel plant. These reference projects prove that thyssenkrupp nucera is a world-leading technology provider for the industry, ranging from multi-100MW up to the gigawatt power range [56].

References

[1] M. Hussein, World's Oil and Gas Pipelines, Al Jazeera, 2021.

[2] U.S. Energy Information Administration, Oil tanker sizes range from general purpose to ultra-large crude carriers on AFRA scale, Today in Energy (2014). September 16. Available from: https://www.eia.gov/todayinenergy/detail.cfm?id=17991

[3] BCP 2.1, Figure available from: https://www.gbrx.com/railcars/33700-gallon-lpg-pressure-tank-car/.

[4] U.S. Department of Transportation Pipeline and Hazardous Materials Safety Administration Office of Hazardous Materials Safety, Safety Advisory Notice for DOT-111 Tank Cars in Flammable Liquid Service, 2023, March 22. Available from: https://www.phmsa.dot.gov/news/phmsa-safety-advisory-notice-dot-111-tank-cars-flammable-liquid-service.

[5] Directorate-General for Energy, Global Methane Pledge: From Moment to Momentum, 2022, November 17. Available from: https://energy.ec.europa.eu/publications/gmp-year-1-factsheet_en#files.

[6] G. Lorenzato, S. Tordo, B. van den Berg, H.M. Howells, S. Sarmiento-Saher, Financing Solutions to Reduce Natural Gas Flaring and Methane Emissions. International Development in Focus, World Bank, Washington, DC, 2022. Available from: http://hdl.handle.net/10986/37177. License: CC BY 3.0 IGO.

[7] Report on Small-scale Technologies for Utilization of Associated Gas (English), 2021. World Bank Group, Washington, DC. Available from: http://documents.worldbank.org/curated/en/305891644478108245/Report-on-Small-scale-Technologies-for-Utilization-of-Associated-Gas.

8] Energy5, The Future of Natural Gas Leak Detection Emerging Technologies, 2024, January 10. Available from: https://energy5.com/the-future-of-natural-gas-leak-detection-emerging-technologies.

[9] Pipeline and Hazardous Materials Safety Administration (PHMSA), Hazardous Materials: Liquefied Natural Gas by Rail, Federal Register, 2020. 85 FR 44994. Available from: https://www.federalregister.gov/documents/2020/07/24/2020-13604/hazardous-materials-liquefied-natural-gas-by-rail.

[10] Pipeline and Hazardous Materials Safety Administration (PHMSA), Hazardous Materials: Suspension of HMR Amendments Authorizing Transportation of Liquefied Natural Gas by Rail, Federal Register, 2023. 88 FR 60356. Available from: https://www.federalregister.gov/documents/2023/09/01/2023-18569/hazardous-materials-suspension-of-hmr-amendments-authorizing-transportation-of-liquefied-natural-gas.

[11] United States Department of Energy, Hydrogen Storage (Chapter 3.3 from Multi-Year Research, Development, and Demonstration Plan), United States Department of Energy, 2025. [Online] Available: https://www.energy.gov/sites/default/files/2015/05/f22/fcto_myrdd_storage.pdf. Accessed 14 July 2023.

[12] M. Yang, R. Hunger, S. Berrettoni, B. Sprecher, B. Wang, A review of hydrogen storage and transport technologies, Clean Energy (2023) 190–216, https://doi.org/10.1093/ce/zkad021.

[13] Air Liquide, n.d. Storing Hydrogen. Air Liquide Energies. [Online] Available: https://energies.airliquide.com/resources-planet-hydrogen/how-hydrogen-stored.

[14] A. Chen, A Solid Hydrogen-Storage Solution, Lawrence Livermore National Laboratory, 2018. [Online] Available: https://str.llnl.gov/2018-01/wood.

[15] United States Department of Energy. Sorbent Storage Materials. n.d.. Energy.Gov. [Online] Available: https://www.energy.gov/eere/fuelcells/sorbent-storage-materials.
[16] J. Max, Successfully Conducted Tests on New Methanol to Hydrogen Fuel Cell System, Hydrogen Fuel News, June 27, 2023. [Online] Available: https://www.hydrogenfuelnews.com/hydrogen-fuel-cell-menthol/8559344/#:~:text=A%20unique%20hydrogen%20fuel%20cell%20concept%20using%20methanol-to-H2,between%20Powercell%20Group%2C%20e1%20Marine%20and%20RIX%20Industries.
[17] Z. Xin, PipeChina Successfully Test Hydrogen Pipeline, China Daily, June 27, 2023. [Online] Available: https://global.chinadaily.com.cn/a/202306/27/WS649a2ffda310bf8a75d6bc52.html.
[18] PipeChina West Pipeline Company, China Succeeds in First High-Pressure Pure-Hydrogen Pipeline Test, CCTV, June 25, 2023. [Online] Available: https://www.cctvplus.com/news/20230626/8330911.shtml#!language=1.
[19] A. Ohnsman, Forget Oil. New Wildcatters Are Drilling for Limitless 'Geologic' Hydrogen, Forbes, June 26, 2023.
[20] Y. Qiu, H. Yang, L. Tong, L. Wang, Research progress of cryogenic materials for storage and transportation of liquid hydrogen, Metals 11 (7) (2021) 1101.
[21] CO2-Free Hydrogen Energy Supply-Chain Technology Research Association, HySTRA, 2022. https://www.hystra.or.jp/en/project/. (Accessed 24 July 2023).
[22] S. Dowling, What Are the Odds of a Successful Space Launch?, BBC Future, 2023. https://www.bbc.com/future/article/20230518-what-are-the-odds-of-a-successful-space-launch. (Accessed 24 July 2023).
[23] F. Wallace, Unleashing the Power: Liquid Hydrogen Fuels Debut Race Car, Hydrogen Fuel News, 2023. http://www.hydrogenfuelnews.com/liquid-hydrogen-first-race-car/8558968/. (Accessed 24 July 2023).
[24] Airbus.com, How to Store Liquid Hydrogen for Zero-Emission Flight, 2021. https://www.airbus.com/en/newsroom/news/2021-12-how-to-store-liquid-hydrogen-for-zero-emission-flight. (Accessed 24 July 2023).
[25] PR Newswire, GENH2 and ZeroAvia Sign MOU to Develop Liquid Hydrogen Technologies for Airports, 2023. https://www.prnewswire.com/news-releases/genh2-and-zeroavia-sign-mou-to-develop-liquid-hydrogen-technologies-for-airports-301832145.html. (Accessed 24 July 2023).
[26] T. Zhang, J. Uratani, Y. Huang, L. Xu, S. Griffiths, Y. Ding, Hydrogen liquefaction and storage: recent progress and perspectives, Renew. Sust. Energ. Rev. 176 (2023) 113204.
[27] Department of Energy and Hydrogen and Fuel Cell Technologies Office, Liquid Hydrogen Technologies Workshop, 2022. https://www.energy.gov/eere/fuelcells/liquid-hydrogen-technologies-workshop. (Accessed 24 July 2023).
[28] Energy.gov, DOE Technical Targets for Hydrogen Delivery, 2023. https://www.energy.gov/eere/fuelcells/doe-technical-targets-hydrogen-delivery. (Accessed 24 July 2023).
[29] H. Ansarinasab, M. Mehrpooya, M. Sadeghzadeh, An exergy-based investigation on hydrogen liquefaction plant-exergy, exergoeconomic, and exergoenvironmental analyses, J. Clean. Prod. 210 (2019) 530–541.
[30] J. Leachman, Could Smaller Hydrogen Liquefiers Be Better?, Washington State University, 2020. https://hydrogen.wsu.edu/2020/10/06/could-smaller-hydrogen-liquefiers-be-better/. (Accessed 24 July 2023).
[31] Air Products, Air Products to Build Green Liquid Hydrogen Production Facility in Arizona, 2022. https://www.airproducts.com/company/news-center/2022/03/0308-air-products-green-liquid-hydrogen-production-facility-in-arizona. (Accessed 24 July 2023).
[32] S.Z. Al Ghafri, S. Munro, U. Cardella, T. Funke, W. Notardonato, J.M. Trusler, E.F. May, Hydrogen liquefaction: a review of the fundamental physics, engineering practice and future opportunities, Energy Environ. Sci. 15 (7) (2022) 2690–2731.
[33] M. Hartl, R.C. Gillis, L. Daemen, D.P. Olds, K. Page, S. Carlson, G. Muhrer, Hydrogen adsorption on two catalysts for the ortho-to parahydrogen conversion: Cr-doped silica and ferric oxide gel, Phys. Chem. Chem. Phys. 18 (26) (2016) 17281–17293.
[34] J. Park, H. Lim, G.H. Rhee, S.W. Karng, Catalyst filled heat exchanger for hydrogen liquefaction, Int. J. Heat Mass Transf. 170 (2021) 121007.
[35] R. Soni, U.S. Energy Department Picks Shell-led Consortium for Hydrogen Storage Project, Reuters, 2021. https://www.reuters.com/business/energy/us-energy-department-picks-shell-led-consortium-hydrogen-storage-project-2021-10-13/. (Accessed 24 July 2023).
[36] McDermott, McDermott's CB& I Storage Solutions Completes Conceptual Design for World's Largest Liquid Hydrogen Sphere, 2021. https://www.mcdermott-investors.com/news/press-release-details/2021/McDermotts-CBI-Storage-Solutions-Completes-Conceptual-Design-for-Worlds-Largest-Liquid-Hydrogen-Sphere/default.aspx. (Accessed 24 July 2023).
[37] Kawasaki Heavy Industries, Ltd, Kawasaki Obtains AIP for Large, 160,000m3 Liquefied Hydrogen Carrier, 2022. https://global.kawasaki.com/en/corp/newsroom/news/detail/?f=20220422_3378. (Accessed 24 July 2023).
[38] Alfa Laval, Shell Signs Agreement with Alfa Laval to Develop a Gas Combustion Unit (GCU) for Hydrogen Boil-Off Gas, 2022. https://www.alfalaval.com/media/news/2022/shell-signs-agreement-with-alfa-laval-to-develop-a-gas-combustion-unit-gcu-for-hydrogen-boil-off-gas/. (Accessed 24 July 2023).
[39] J.P. Sass, W.S. Cyr, T.M. Barrett, R.G. Baumgartner, J.W. Lott, J.E. Fesmire, Glass bubbles insulation for liquid hydrogen storage tanks, in: AIP Conference Proceedings, Vol. 1218, No. 1, American Institute of Physics, 2010, pp. 772–779.
[40] W.U. Notardonato, A.M. Swanger, J.E. Fesmire, K.M. Jumper, W.L. Johnson, T.M. Tomsik, Zero boil-off methods for large-scale liquid hydrogen tanks using integrated refrigeration and storage, in: IOP Conference Series: Materials Science and Engineering, Vol. 278, No. 1, IOP Publishing, 2017, p. 012012.
[41] A. Swanger, World's Largest Liquid Hydrogen Tank Nears Completion, Cryogenic Society of America, Inc, 2023. https://www.cryogenicsociety.org/index.php?option=com_dailyplanetblog&view=entry&year=2022&month=05&day=05&id=48%3Aworld-s-largest-liquid-hydrogen-tank-nears-completion. (Accessed 24 July 2023).
[42] Rolls-Royce, Rolls-royce and EasyJet set New World First, 2022. https://www.rolls-royce.com/media/press-releases/2022/28-11-2022-rr-and-easyjet-set-new-aviation-world-first-with-successful-hydrogen-engine-run.aspx?ref=sparkofgenius.org. (Accessed 24 July 2023).
[43] M. McEwen, Midland Could Host First Flare-to-LNG Pilot for Macaw Energies, Midland Reporter-Telegram, 2023. [Internet] [cited 2023 October 22]; [about 2 p]. Available from: https://www.mrt.com/business/oil/article/midland-host-first-flare-to-lng-pilot-macaw-18142367.php.
[44] TotalEnergies invests in Ductor to facilitate biomethane production, 2023. Offshore Technology [internet]. May 25 [cited 2023 October 22]; [about 2 p]. Available from: https://www.offshore-technology.com/news/totalenergies-invests-in-ductor/?utm_campaign=GD%20Oil%20%26%20Gas%20Prospect%20Newsletters&utm_medium=email&_hsmi=72201820&_hsenc=p2ANqtz-94ERuW77s_

WaiKtnDwCPpOxacbjNEYImG8cV2xHBBkSHmNPhoPZLFOO_j8DtpRGeQ3v4XQl1Ek0s3FZiBtf-0ZAB6y4w&utm_content=72201820&utm_source=hs_email&cf-view&cf-closed.
[45] M. Bailey, Phillips 66 Supplies Hydrogen for Blending Project with Linden Cogen, Chemical Engineering, 2023. [internet]. [cited 2023 October 22]; [about 1 p]. Available from: https://www.chemengonline.com/phillips-66-supplies-hydrogen-for-blending-project-with-linden-cogen/?oly_enc_id=3658C9935723E8X.
[46] R. Teodoro, Carbon Capture, CO2 Removal to Play Key Decarbonization Role: S&P Global, Rigzone, 2023. [internet]. [cited 2023 October 24]; [about 2 p]. Available from: https://www.rigzone.com/news/carbon_capture_co2_removal_to_play_key_decarbonization_role_sp_global-12-jun-2023-173029-article/.
[47] C. Hiar, Project Bison, a Large Carbon Removal Proposal, Faces Delays, E&E News, 2023. [internet]. [cited 2023 October 24]; [about 2 p]. Available from: https://www.eenews.net/articles/project-bison-a-large-carbon-removal-proposal-faces-delays/.
[48] P. Dvorak, New land grab by oil giants is deep underground, Wall Street J. (2023). [internet]. [cited 2023 October 24]; [about 7 p]. Available from: https://www.wsj.com/articles/new-land-grab-by-oil-giants-is-deep-underground-34cd5e97.
[49] M. Bailey, Cepsa and GETEC Reach Agreement to Supply Green Hydrogen to Industry Clients in Europe, Chemical Engineering, 2023. [internet]. [cited 2023 October 25]; [about 1 p]. Available from: https://www.chemengonline.com/cepsa-and-getec-reach-agreement-to-supply-green-hydrogen-to-industry-clients-in-europe/?oly_enc_id=3658C9935723E8X.
[50] M. Bailey, ABB Collaborates with Lhyfe and Skyborn on One of Europe's Largest Renewable Hydrogen Projects, Chemical Engineering, 2023. [internet]. [cited 2023 October 25]; [about 1 p]. Available from: https://www.chemengonline.com/abb-collaborates-with-lhyfe-and-skyborn-on-one-of-europes-largest-renewable-hydrogen-projects/?oly_enc_id=3658C9935723E8X.
[51] M. Bailey, New Offshore Green-Hydrogen Demonstration Project Planned for the Canary Islands, Chemical Engineering, 2023. [internet]. [cited 2023 October 25]; [about 1 p]. Available from: https://www.chemengonline.com/new-offshore-green-hydrogen-demonstration-project-planned-for-the-canary-islands/?oly_enc_id=3658C9935723E8X.
[52] M. Bailey, McPhy and Technip Energies Contracted for HYCC Green-Hydrogen Plant, Chemical Engineering, 2023. [internet]. [cited 2023 October 25]; [about 1 p]. Available from: https://www.chemengonline.com/mcphy-and-technip-energies-contracted-for-hycc-green-hydrogen-plant/?oly_enc_id=3658C9935723E8X.
[53] M. Bailey, McPhy Energy Awarded an Electrolyzer Contract for "Green Metal" Project in Austria, Chemical Engineering, 2023. [internet]. [cited 2023 October 25]; [about 1 p]. Available from: https://www.chemengonline.com/mcphy-energy-awarded-an-electrolyzer-contract-for-green-metal-project-in-austria/?oly_enc_id=3658C9935723E8X.
[54] M. Bailey, Nel Plans Large-Scale Electrolyzer Manufacturing Facility in Michigan, Chemical Engineering, 2023. [internet]. [cited 2023 October 25]; [about 1 p]. Available from: https://www.chemengonline.com/nel-plans-large-scale-electrolyzer-manufacturing-facility-in-michigan/?oly_enc_id=3658C9935723E8X.
[55] M. Bailey, Plug Power to Supply Electrolyzer for European HOPE Project, Chemical Engineering, 2023. [internet]. [cited 2023 October 25]; [about 1 p]. Available from: https://www.chemengonline.com/plug-power-to-supply-electrolyzer-for-european-hope-project/?oly_enc_id=3658C9935723E8X.
[56] M. Bailey, Thyssenkrupp Nucera to Supply Electrolyzers for Green Hydrogen Project in North America, Chemical Engineering, 2023. [internet]. [cited 2023 October 25]; [about 1 p]. Available from: https://www.chemengonline.com/thyssenkrupp-nucera-to-supply-electrolyzers-for-green-hydrogen-project-in-north-america/?oly_enc_id=3658C9935723E8X.

CHAPTER

12

R&D facilities for pipeline research

Jason Wilkes[a], *Kevin Supak*[a], *Buddy Broerman*[a], *and Guillermo Paniagua-Perez*[b]

[a]Southwest Research Institute, San Antonio, TX, United States [b]Purdue University, West Lafayette, IN, United States

Introduction

Research and development (R&D) facilities play a crucial role in advancing pipeline research, particularly when it comes to the transport of carbon dioxide (CO_2), methane, and hydrogen. These pipelines are at the forefront of the global energy transition, driving efforts to mitigate greenhouse gas emissions, enhance energy security, and foster a sustainable future. The significance of R&D facilities in this context cannot be overstated, as they form the backbone for innovation, knowledge expansion, and the successful deployment of these critical pipeline systems.

R&D facilities are instrumental in driving technological advancements in pipeline research. For CO_2 transport, efficient and secure pipeline systems are essential for carbon capture, utilization, and storage (CCUS) projects, which hold the potential to significantly reduce CO_2 emissions from industrial sources and power plants. By investing in R&D, researchers can develop novel materials, coatings, and technologies that address challenges like corrosion, pressure, and temperature fluctuations associated with CO_2 pipelines.

Similarly, for methane transport, R&D facilities are pivotal in advancing leak detection, pipeline integrity, and safety measures. Methane, the primary component of natural gas, is a potent greenhouse gas, and minimizing its release during transportation is crucial for climate change mitigation. This is the most mature pipeline in operation today, and generally is well understood; however, R&D facilities still exist that have been improving the safety, reliability, and performance of natural gas pipelines for half a decade.

For hydrogen pipelines, which are still emerging technologies, R&D facilities play a key role in developing and testing materials capable of withstanding hydrogen embrittlement and optimizing hydrogen blending with existing natural gas infrastructure. As hydrogen is blended into a natural gas stream, the impact of that hydrogen on the compression and power generation systems transporting the natural gas blends and hydrogen can lead to unexpected results. As hydrogen gains traction as a clean energy carrier, the role of R&D facilities are critical in ensuring its safe and efficient transport through dedicated pipelines.

Safety and environmental concerns are paramount when it comes to pipeline transportation. R&D facilities provide controlled environments for testing and simulating various operational scenarios, thereby identifying potential hazards and designing measures to mitigate risks effectively. By fostering research in pipeline materials, corrosion resistance, and leak detection systems, R&D facilities contribute to ensuring the integrity and reliability of CO_2, methane, and hydrogen pipelines, reducing the chances of accidents or environmental impacts.

Finally, in the absence of comprehensive and well-defined guidelines for CO_2, methane, and hydrogen pipeline transport, R&D facilities play a key role in establishing best practices and industry standards. Through collaborative research efforts, these facilities can contribute to the formulation of guidelines that address safety, efficiency, and environmental considerations. This, in turn, fosters regulatory confidence and facilitates the seamless integration of these pipelines into existing energy infrastructure.

These R&D facilities act as hubs for knowledge exchange and collaboration among researchers, industry experts, and policymakers. This network encourages the sharing of expertise, data, and insights, accelerating progress and

Energy Transport Infrastructure for a Decarbonized Economy
https://doi.org/10.1016/B978-0-443-21893-4.00011-8

Copyright © 2025 Elsevier Inc. All rights are reserved, including those for text and data mining, AI training, and similar technologies.

fostering innovation in pipeline research. The collaborative nature of R&D facilities ensures that stakeholders can collectively address challenges, pool resources, and explore new avenues for sustainable energy transport.

CO_2 pipeline research facilities

In today's dynamic technological landscape, the rapid adoption of multidisciplinary optimization strategies and additive manufacturing techniques has opened the door to the accelerated introduction of novel designs into the market. These advancements promise to transform industries and revolutionize pipeline research and development. However, this transformative potential is currently constrained by the imperative need for experimental validation at various stages of technological readiness. To bring new pipeline designs to fruition, rigorous experimental testing is indispensable. Moreover, the challenge extends beyond component testing; there is an ongoing need to continually assess new instrumentation against established calibrated techniques, all while maintaining optical access under elevated pressures and temperatures. This multifaceted landscape necessitates the formulation and implementation of refined experimental procedures.

Without these refined procedures, the prospect of drastically reducing the time-to-market for emergent aerothermal concepts may remain elusive. The promise of groundbreaking innovations that could reshape the pipeline industry might remain confined to the drawing board. Specialized facilities have been conceived to bridge this gap between innovation and practical application. These facilities are designed to facilitate short-duration tests, which focus on heat transfer measurements while streamlining operational costs, and continuous-operation facilities, offering measurement techniques of unparalleled precision, albeit requiring higher operational costs. Furthermore, carbon dioxide (CO_2), methane, and hydrogen gases introduce a wide range of Reynolds and Mach numbers, pushing the boundaries of experimentation and exploration in pipeline design.

Experimental work in sCO_2 is marked by high costs, intensive labor, and risks associated with unforeseen practical challenges. Sandia National Laboratories (SNL) has shared its experiences in constructing and operating turbocompressor testing loops alongside guidelines for the design and operation of sCO_2 R&D systems [1,2]. Similarly, BMPC's publications shed light on practical aspects of sCO_2 Brayton system testing [3]. The landscape of supercritical CO_2 has undergone significant developments during the past half a decade, which reflect a global effort encompassing a range of specialized topics and practical challenges. Sharing experiences and lessons learned is crucial in navigating the complexities of sCO_2 technology. The global landscape shows a diverse range of contributions. Yu et al.'s analysis shows that the United States, China, South Korea, Australia, and India are the key players in sCO_2 Brayton cycles [4].

TABLE 12.1 Global sCO_2 test facilities.

Organization	Name test facility	Location	Timeframe/status
SNL	More than one test rig	Albuquerque, NM, USA	Research ongoing with several test rigs
SwRI	SwRi SunShot facility, 1 MWe sCO_2 test loop	San Antonio, TX, USA	SunShot finished (2011–2018), last paper published in 2018
Echogen Power Systems	*EPS100*	Akron, OH, USA (Olean, NY)	Test program in Olean, NY, USA completed in 2014; research ongoing at Akron test facility
Naval Nuclear Laboratory op. by BMPC	*Integrated System Test (IST)*	West Mifflin, PA, USA	Finished
GTI Energy, SwRI, GE Global Research, DOE many others through Joint Industry Program	*STEP Demo*, 10 MWe sCO_2 Pilot Plant Test Facility	San Antonio, TX, USA	Commissioning and start-up expected late 22 or early 23, testing in 23
Net Power, Baker Hughes	*50-MWth test facility, Serial #1 Utility Scale Plant*	La Porte, TX, USA	La Porte demonstration site connected the grid in fall 21 (2012–2021), first utility-scale power plant in Permian West, TX expected to go online in 26
KAERI, KAIST, POSTECH	*SCIEL*	Daejeon, KR	Finished
KAIST, KAERI	*SCO_2PE*	Daejeon, KR	Finished
KIER	5 cycles so far	Daejeon, KR	Full 5th cycle commissioning expected in 2020, not yet published
TIT, IAE	Bench scale test facility	Tokyo, JP	Finished

TABLE 12.1 Global sCO_2 test facilities—cont'd

Organization	Name test facility	Location	Timeframe/status
Shouhang, EDF	10 MWe supercritical cycle + CSP demonstration	Shouhang, CN	2018–2023, commissioning of retrofit cycle to industrial CSP plant planned end of 2021, no news since end of 2019
CHNG, TPRI	5 MW fossil-fired supercritical CO_2 power cycle pilot loop	Xi'an, CN	Experiments were planned in 2020, no news since19
Indian Institute of Science, SNL	Test facility for supercritical CO_2 Brayton cycle	Bangalore, IN	Research seems ongoing
CSIRO	Solar-Driven Supercritical Brayton Cycle	Newcastle, AUS	Project completed (2012–2017), now collaboration with US DOE $\geq$ joined the STEP Demo project
The University of Queensland	Refrigerant and Supercritical CO_2 Test Loop, *PHIPL*	Queensland, AUS	No news on test loop since 2016; ASTRI project ongoing, demonstration planned end of 22
Baker Hughes	Prototype compressor test rig	Florence, IT	2018–2021, results paper published in 2022
CVR	*SUSEN* test loop	Prague, CZ	Since ~2007, research on several experimental projects (e.g., COMPASsCO_2) is ongoing
IKE	*SCARLETT*	Stuttgart, DE	Finished, team involved in other sCo$_2$ projects
Project Consortium (University Duisburg-Essen, CVR, University of Stuttgart)	*sCO_2-HeRo* loop	Duisburg, DE	Projects sCO_2-HeRo (2015–2018), sCO_2-Flex (2018–2021) and sCO_2-4-NPP (2019–2022) finished
Cranfield University	*Rolls-Royce sCO_2 Test Rig*	Bedfordshire, UK	Research ongoing, in operation
Brunel University, Engoia	*HT2C* facility	London, UK	Project I-ThERM finished (2015–2021)
LUT University	*LUTsCO_2* facility	Lappeenranta, FIN	Only design published so far, research ongoing
TU Wien	sCO_2 test facility	Vienna, AUT	Project SCARABEUS (2019–2023) ongoing, experiments expected in 23

European test facilities are typically small-scale test facilities and focused on turbomachinery, and several studies have honed in on specific aspects of sCO_2 applications such as integrating Brayton cycles in nuclear applications [5]. A comprehensive list of experimental facilities is provided in Table 12.1 for reference.

TU Wien commissioned a supercritical carbon dioxide (sCO_2) test facility 2018 to test both supercritical and transcritical operation modes and enable heat transfer measurements. The facility's development revealed Inadequate Commercial CO_2 Pumps for cycle experiments. The test facility, as shown in Fig. 12.1, comprises five major components: High- and Low-Pressure Sides, Piston Pump, Shell-and-Tube Heat Exchanger, Expansion Valve, and Water Cooler. The facility utilizes an electrically heated thermal oil for heat input, and CO_2 of food-grade quality (Biogon C E 290) is used as the working fluid. Comprehensive safety measures, including a Hazard and Operability Study and the implementation of safety valves and burst discs, were critical in the facility's design. The facility underwent rigorous testing, including pressure and a leak test with nitrogen.

It is important to recognize that pure CO_2 transport is not novel and that there are several facilities around the world that have been separating CO_2 from production wells prior to injecting it into saline aquifers for permanent storage or supplying it for enhanced oil recovery. Examples of these facilities include the Gorgon LNG production facility in Western Australia, the LaBarge gas processing plant in Wyoming, and the Sleipner gas field in the North Sea. However, these pipelines that transport CO_2 have been designed to operate above the critical pressure, thus transporting the fluid in single-phase. These facilities have high-purity requirements, particularly for water, oxygen, and H_2S content, to ensure that the risk of corrosion is reduced. More research is needed to understand how the critical pressure can be affected by impurities and to manage subsequent multiphase flow patterns in the event that free gas is present in the system.

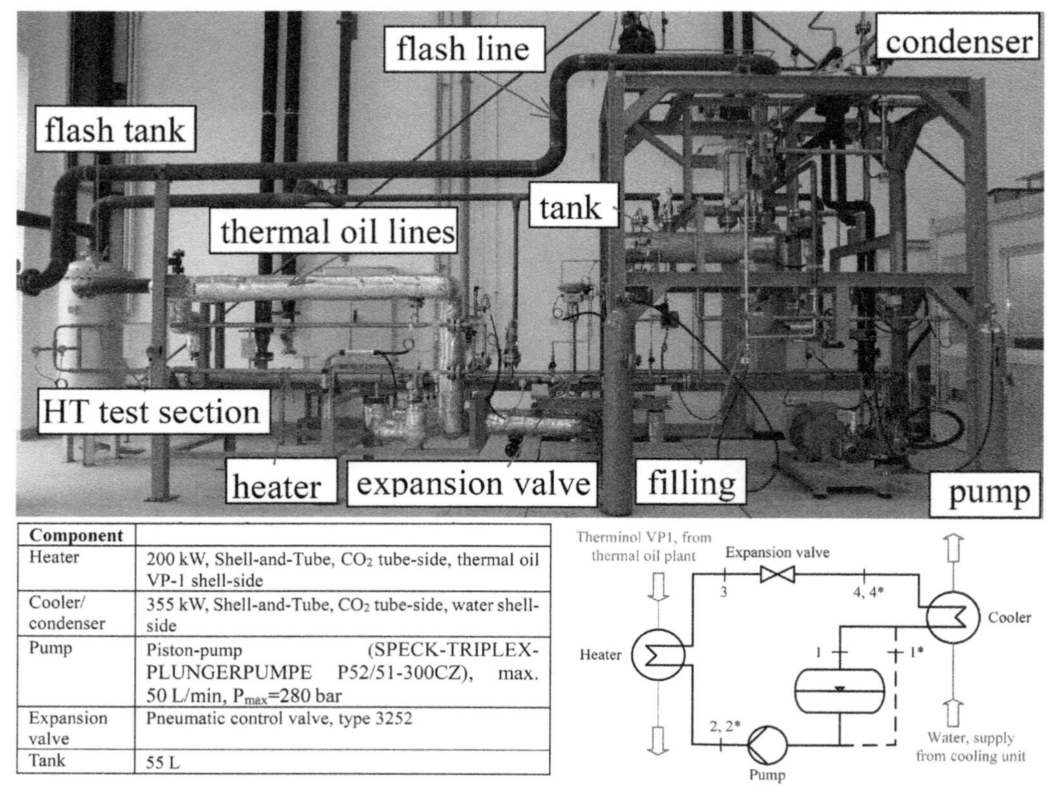

Component	
Heater	200 kW, Shell-and-Tube, CO_2 tube-side, thermal oil VP-1 shell-side
Cooler/ condenser	355 kW, Shell-and-Tube, CO_2 tube-side, water shell-side
Pump	Piston-pump (SPECK-TRIPLEX-PLUNGERPUMPE P52/51-300CZ), max. 50 L/min, P_{max}=280 bar
Expansion valve	Pneumatic control valve, type 3252
Tank	55 L

FIG. 12.1 Components of the test facility at TU Wein [6].

There are only a few test facilities worldwide that have the capability to conduct supercritical and multiphase CO_2 flow experiments at field-relevant scales. These include SINTEF and the Institute for Energy Technology (IFE) in Norway, as well as the Southwest Research Institute (SwRI) in the United States. The primary purpose of these test facilities is to acquire transport-related data and validate field equipment. The transport-related data include instruments for measuring pressure drop and temperatures, observations of flow patterns in multiphase flow, conditions that lead to hydrate formation, and effects of impurities in both horizontal and vertical orientations. These data are used to design pipelines and injection systems in a safe and efficient way. Equipment such as pumps, compressors, valves, flow meters, and materials are often tested prior to their deployment in the field to ensure that they perform as designed.

IFE has recently published some pure CO_2 two-phase flow data in vertically and horizontally oriented 2-in. piping [7,8]. At the time of this publication, these are the only known data sets related to two-phase CO_2 flow and both observed flow patterns that are different from traditional oil and gas fluids at similar superficial velocities. This is in large part due to the lower density ratio between liquid and gaseous CO_2 vs crude oil and natural gas. SINTEF has announced future experiments that will study CO_2 flow with impurities in their facilities that were originally designed for oil and gas pipeline research [9]. Southwest Research Institute has several facilities that study CO_2 flow across the gas, liquid, and supercritical phase boundaries. These facilities were constructed to validate machinery design for power cycles that utilize CO_2 as the working fluid and are currently being adapted to study two-phase CO_2 flow in horizontal piping in addition to testing flow components such as subsurface safety valves and flow meters.

Though there are many variants of a pipeline CO_2 research facility, these facilities primarily focus on the crucial aspects of CO_2 transport, which involve one or more of the following crucial functions of a pipeline:

1. Compression
2. Pumping
3. Heat removal
4. Control
5. Metering
6. Separation
7. Safety

CO_2 compression research

Most of the research on CO_2 compression has studied the performance of the compressors near-critical point behavior for power applications. This selection of compressor inlet state is based on reducing compressor power for indirectly fired recompression cycles. These facilities operate in a state where the fluid is slightly above the critical point at design condition where the fluid is dense, has low viscosity, and the power is comparable to transport compressor stations; thus, they can be adapted to study relevant pipeline compression topics. Fig. 12.2 shows the piping layout for Southwest Research Institute's Integrally Geared Compressor/Expander Test Facility. This facility was originally constructed as part of a United States Department of Energy DOE cooperative agreement (EE0007114).

Fig. 12.3 shows a P&ID for the compressor loop. Discussing some of the main features, the nominal power limit of the test loop is 3 MW based on motor power. This enables the facility to transport between 60 and 120 kg/s of CO_2 at a maximum pressure of approximately 4000 psi. The facility is water-cooled with a process cooler outside of the lab building (see Fig. 12.2) for ease of access to the water supply and to minimize the potential risk associated with a failure on the water supply lines. The piping straight lengths upstream and downstream of the compressors and turbine were sized based on the ASME PTC-10 standard for instrumentation locations. Allowing sufficient straight lengths upstream and downstream of pressure and temperature measurement locations is necessary to ensure process measurements are not disturbed by flow instabilities generated in the compressors, turbines, valves, etc.

Each line in the main process loop was initially sized based on the required pressure/temperature rating, the expected flow velocity through the line, and the associated pressure drop through the line. In general, the piping was sized to limit the flow velocities in the pipe to 100 ft./s or less (30 m/s). SwRI has experience with operating sCO_2 test loops with flow velocities over 200 ft./s (61 m/s). However, the elevated piping in this test loop is highly susceptible to vibration. Limiting the flow velocities in the piping reduces the amount of excitation energy present in these elevated lines. Initial pipe sizes (based on flow velocities) were further validated for the necessary pressure and temperature rating of each section of the cycle. The piping material, pipe size, and pipe schedule were chosen using ASME B31.3 piping code (2008 version) for carbon steel and stainless steel and ASME B31.1 (2012 version) for Inconel 625.

The estimated piping lengths shown in Fig. 12.2 were used to estimate the frictional pressure losses through each section of the loop. The pressure losses were allowed to exceed 5 psi in lines where a large pressure drop would be taken across a control valve. In this case, the pressure losses through the piping are insignificant compared to the expected process pressure drop and were not used for line sizing. See the maximum-flow operating conditions,

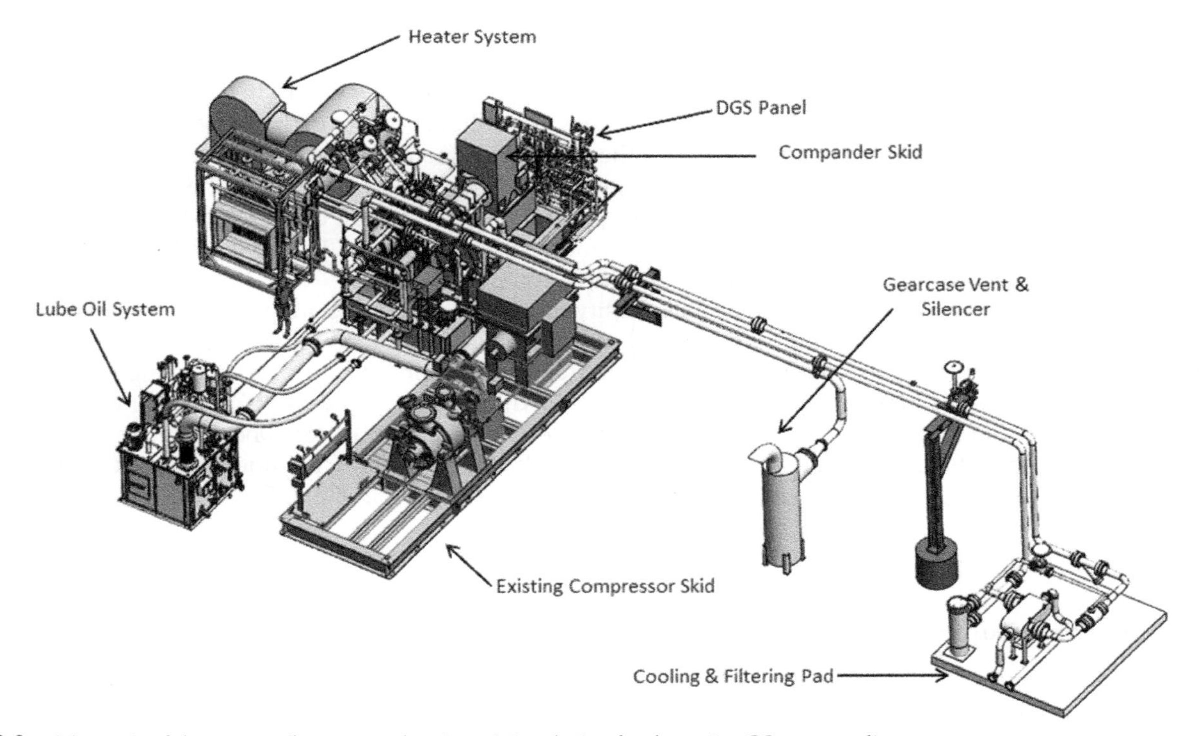

FIG. 12.2 Schematic of the compander system showing piping design for the main sCO_2 process lines.

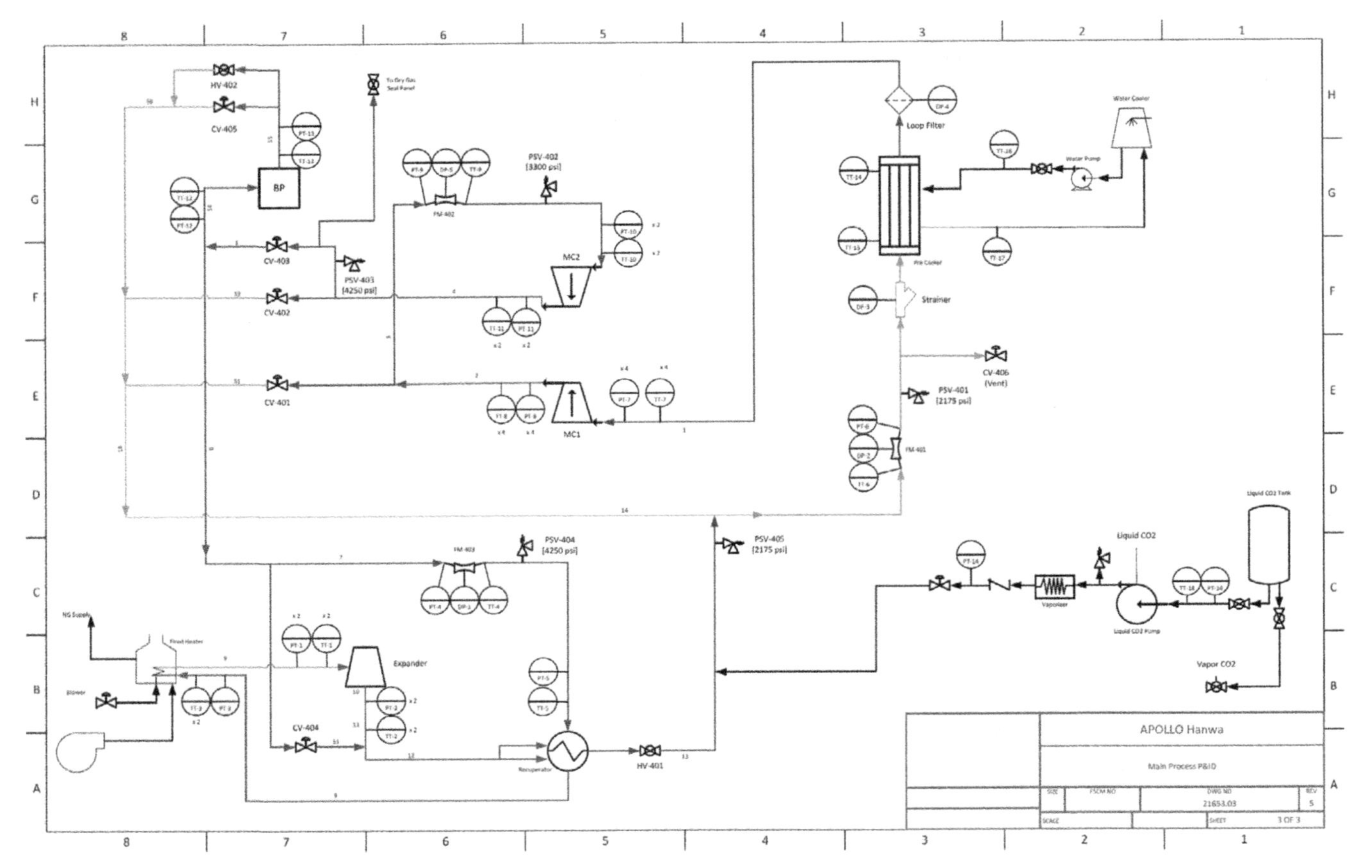

FIG. 12.3 P&ID for reduced flow test loop.

preliminary pipe sizing, and pressure drop predictions summarized in. It should be noted here that stainless steel piping was chosen for the majority of the test loop as opposed to carbon steel to protect the various loop components. Supercritical CO_2 is highly corrosive in nature, and utilizing stainless allows for this loop to be utilized to investigate the behavior of different gas mixtures with combinations of water vapor or other gasses that may cause corrosion concerns. When using carbon steel piping, the sCO_2 will dislodge small particles (e.g., rust) from the inner wall of the piping. These small particles are then carried through the loop components, where they can cause surface damage, fouling, and blockage (a big problem in a closed-loop facility). This is of highest concern for the compressor/expander impellers and the microchannel heat exchangers (the recuperator and the process cooler). Additionally, using all carbon steel piping would require more frequent change-outs for the filter elements. Overall, installing stainless steel piping ensures the longevity of the test loop and process-critical components.

It should be noted that, for its original intent, this facility also supported the testing of a high-temperature, high-pressure expander stage. While this configuration is not particularly applicable to pipeline research. It has since been reconfigured to allow for heated gas through the heater to rejoin the main flow prior to traveling to the main loop cooler. This enables the user of the facility to study the mixing of different gas streams of CO_2 along with testing components such as coolers and air-cooled CO_2 heat exchangers at elevated temperatures. One additional feature that has been added to this loop includes the addition of a separator. While this is currently intended to knock out liquid water from the CO_2 post-oxy-fueled combustion, it also enables the testing of liquid CO_2 separation and reinjection technologies.

CO_2 pumping and multiphase flow research

While much of the gas coming into CO_2 compressor stations is likely to be gaseous, the temperature separating dense phase supercritical CO_2 that should be pumped (60 °F) from supercritical fluid that should be compressed (90 °F) is relatively small, where fluid between these two extremes may require a combination of pumping and

FIG. 12.4 Nuovo pignone dense phase CO_2 pump (Southwest Research Institute).

compression. Fig. 12.4 shows a dense phase pump that enables the characterization of dense phase fluid behavior. This pump takes in dense phase CO_2 at approximately 8°C at 85bar and pumps it up to a pressure of 250bar. It has since been repurposed to study liquid CO_2 flow behavior across inclined or declined flow sections, as well as studying the flow across expansion devices and turbines. Unlike the previous facility, this pump is significantly lower in power, allowing for only 700kW of pumping power at a flowrate of approximately 16kg/s.

Fig. 12.5 shows a long horizontal run connected to the discharge of the liquid CO_2 pump. This section can be reconfigured to study a variety of interesting phenomena relating to dense phase CO_2 behavior for pipelines, injection wells, and transport phenomena. For injection wells, there is a substantial concern regarding erosion of orifices and chokes as CO_2 is expanded from high-pressure pipeline pressure to the reservoir pressure. This facility will enable testing with a

FIG. 12.5 High-pressure CO_2 multiphase pressure-drop test facility at Southwest Research Institute.

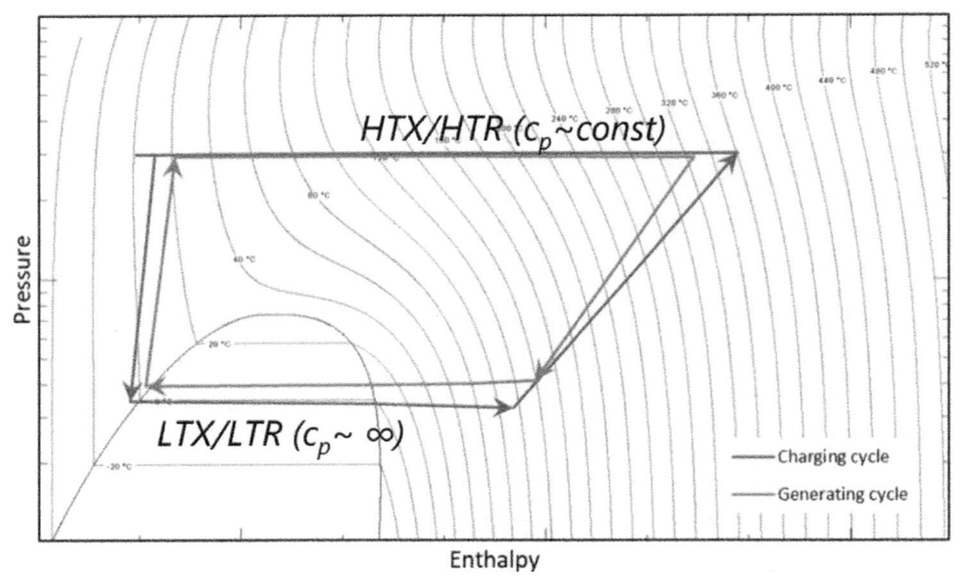

FIG. 12.6 CO_2-based energy storage cycle.

pressure ratio of approximately 3 across such orifices in the liquid regime. Additionally, noise- and vibration-related phenomena with these features can also be studied. Fig. 12.6 shows an example PH diagram for an energy storage cycle that researchers are considering using sCO_2. This cycle can increase efficiency as the turbine discharge pressure drops further into the dome. While this is not particularly relevant to CO_2 transport, the phenomenon impacting the performance of the expander will still yield valuable information regarding the performance and operation of expansion devices in pipeline applications.

CO_2 sequestration compression

In discussing post-combustion carbon capture, there are few research facilities that have been designed to study CO_2 compression from atmospheric pressure up to pipeline pressures. While many of these compression processes are considered commercial, the method used to compress this gas has a large impact on plant performance. One example of this is a facility described by Moore et al. [10] designed to investigate the performance of an isothermal barrel compressor (Fig. 12.7) in comparison with a typical straight through compressor. This research paper addresses the pressing issue of mitigating carbon dioxide (CO_2) emissions from power plants and other significant sources of greenhouse gases by optimizing the compression process, which can negatively impact both the availability of power plants and their operational costs. Preliminary analysis presented by Moore et al. suggests that the compression process alone can reduce the typical efficiency of a power plant by 8% to 12%.

The primary goal of the study was to determine the most efficient means to transport CO_2 from various point sources including power generation plants, elevate the pressure of CO_2 to pipeline levels using the minimum amount of energy necessary. Researchers have explored various thermodynamic processes for pressure increases in both liquid and gaseous states. Additionally, they have considered alternative methods such as the liquefaction of CO_2 and liquid pumping.

One of the key findings is the potential of isothermal compression, which can be achieved using integrally geared machines with multiple pinions driven off a common bull gear. This arrangement allows for the implementation of up to 10 stages of compression. However, it is noted that the reliability of integrally geared compressors does not match that of in-line centrifugal machines widely used in the oil and gas industry, and additionally these machines are often operated at lower pressure than barrel style machines; therefore, the researchers designed an internally cooled, in-line centrifugal compressor diaphragm capable of dissipating the heat generated during compression without the need for external intercoolers, relying instead on internal heat exchangers and liquid cooling.

Significant challenges exist in cooling a high-velocity gas within the compressor due to the limited surface area and the necessity of minimizing pressure drop in the gas stream. Researchers employed 3D Computational Fluid Dynamics analysis to arrive at an optimal design that provides effective heat transfer while adding minimal pressure drop. This research builds upon a previous single-stage prototype diaphragm and focuses on a multi-stage cooled diaphragm

FIG. 12.7 Internally cooled CO_2 barrel compressor (Dresser Rand Datum).

design, enhancing mechanical strength and manufacturing processes. Testing involved a full-scale 3 MW, 6-stage back-to-back centrifugal compressor operating in a closed-loop test facility under a range of conditions.

In the context of reducing CO_2 emissions, this work aims to significantly reduce the power penalty associated with carbon capture. Traditional power plants, including Pulverized Coal (PC) and Integrated Gasification Combined Cycle (IGCC) plants, can incur substantial power penalties—up to 27%–37% for PC plants and 13%–17% for IGCC plants—due to the compression required for carbon capture. This compression represents a substantial portion of the overall penalty. The goal is to minimize this penalty through innovative compression concepts, thereby achieving the desired pressure for CO_2 transport with minimal energy consumption.

Power savings from using the internally cooled diaphragm are calculated and compared to conventional back-to-back compressors, showing potential savings of 10.4%–11.7%. The compressor in this work is designed for a mass flow equivalent to the CO_2 produced by a 35 MW coal-fired power plant, and it incorporates features to enhance efficiency.

In conclusion, this research addresses a critical aspect of carbon capture and storage by seeking innovative ways to reduce the energy requirements for CO_2 compression in power production. Such advancements are pivotal in combating climate change and achieving sustainable energy practices. The facility remaining can be utilized to test further equipment of its kind.

CO_2 heat transfer research

Modern electronics cooling thermal management necessitates compact, efficient, and environmentally friendly solutions. Supercritical carbon dioxide (sCO_2) is a promising alternative to conventional two-phase cooling systems. Utilizing additive manufacturing (AM), an aluminum cold plate was developed for sCO_2 heat transfer experiments. A neutron radiography image validated the adequate tolerance of AM, revealing manufacturing imperfections.

This research showcases the potential of sCO_2 in electronics cooling, particularly in optimizing cold plate design and operational parameters. Optical Backscattering Reflectometry enabled the measurement of heat transfer coefficients and detailed cold plate wall temperature maps.

The use of supercritical CO_2 in gas turbine cooling remains underexplored due to the operational challenges posed by high-temperature and high-pressure conditions. The University of Central Florida's Center for Advanced Turbomachinery and Energy Research Lab has pioneered an initiative to provide experimental heat transfer data pertinent to gas turbine applications. The facility detailed in Fig. 12.8 outlines the flow path. Key components are the preheaters for heating, a booster pump for CO_2 injection, the large shell and tube recuperator, and the test section that can include either 1. jet impingement experiment segment where additional heat is added or 2. a pressure vessel with a pin fin test section. The facility has a ventilation system for CO_2 leakage and CO_2 concentration sensors to maintain levels below

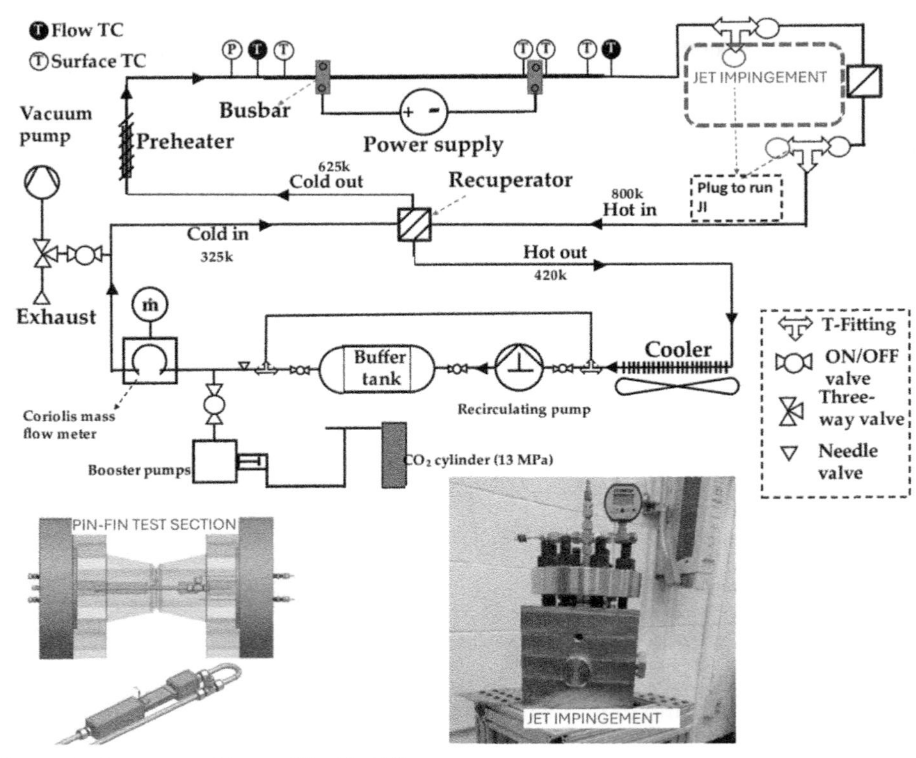

FIG. 12.8 sCO_2 facility at the University of Central Florida [11].

1000 ppm. Fifteen rope heaters along yellow (dark gray in print version) and red (light gray in print version) lines, regulated by eight variacs, add heat to the flow. Concurrently, a recuperator shell is filled and pressurized with nitrogen to a safe limit of around 120 bar.

The National Solar Thermal Test Facility (NSTTF) comprises heated particle and sCO_2 flow loops for testing heat exchangers for thermosolar Brayton power plants with thermal energy storage operating at temperatures over 700C. Fig. 12.9 Left shows a 40-kWth heat exchanger with 12 parallel fluidized bed channels, working on counterflow with sCO_2 microchannels, enclosed by diffusion-bonded stainless-steel plates, is designed for testing. Fig. 12.9 Right

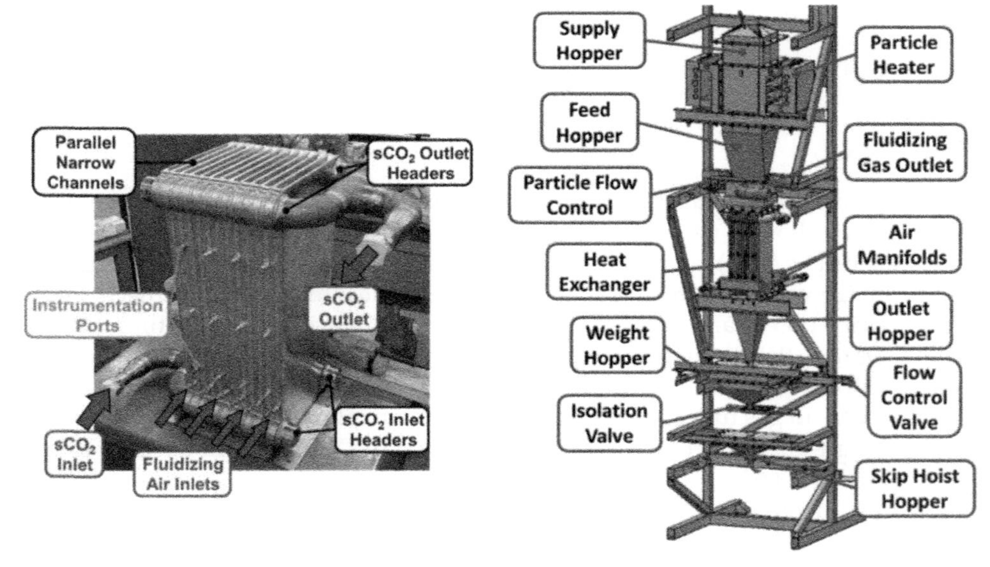

FIG. 12.9 Prototype ca. 40-kWth counterflow particle-sCO_2 narrow-channel fluidized bed heat exchanger and solid model of prototype heat exchanger integrated into the test facility at the NSTTF in Sandia National Laboratories [12].

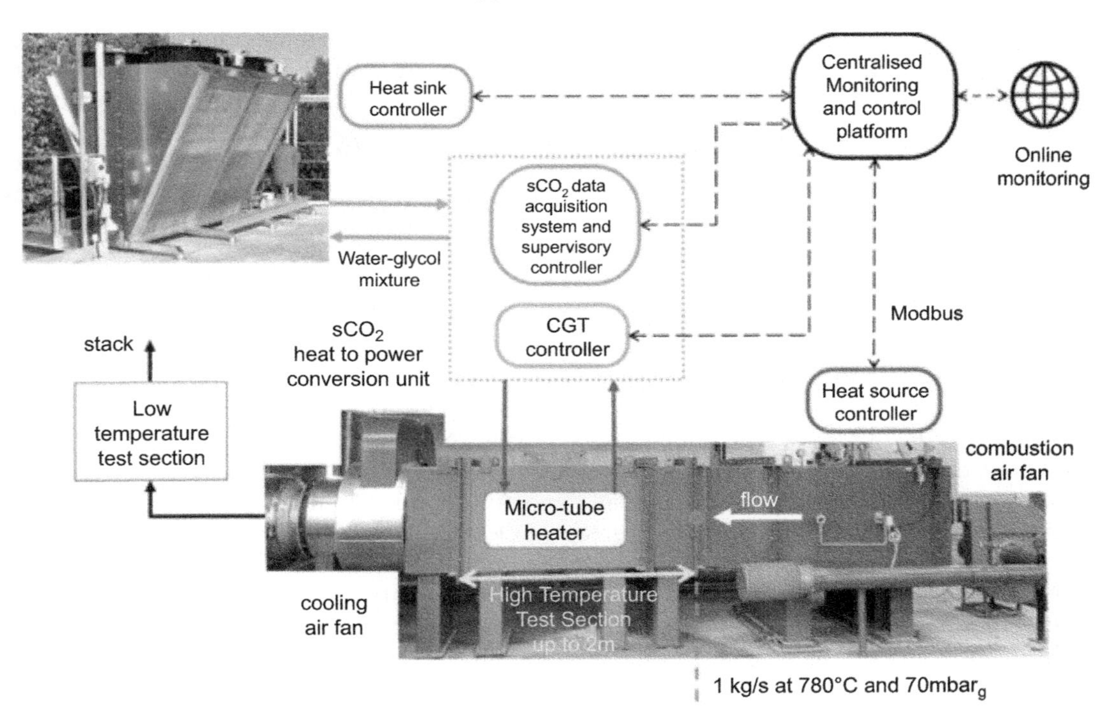

FIG. 12.10 The high temperature heat to power conversion facility at Brunel University London [13].

presents the facility, which includes a particle preheater, a skip hoist delivery system for particle recirculation, and a sCO_2 flow loop designed for controlled flow rates and temperatures.

The high temperature heat power conversion facility at Brunel University London is versatile, catering to various high-temperature source applications like steel, nuclear, and concentrated solar power industries. Fig. 12.10 details the configuration used for sCO_2 heat to power conversion experiments. The facility's core components include:

- Heat source: The e facility features an 830 kW gas-fired process air heater, noted for its operational flexibility.
- Control systems: Employs primary fan speed and flue gas flow rate as control inputs.
- High-temperature test section: This section can extend up to 2000 mm in length and has a cross-section of 500×500 mm, insulated with 200 mm alkaline earth silicate wool.
- Heat Sink: The e HT2C has a 500 kW dry cooler system, incorporating variable speed drives for the pump and fans.
- Auxiliary features: An electric heater is used to warm up the extra fluid in the cooling loop during the startup phase of the sCO_2 tests.
- sCO_2 Unit: A 50kWe sCO_2 unit operates on a simple recuperated Joule-Brayton cycle. The core of this unit is a compressor-generator-turbine (CGT) unit and a single shaft turbomachinery.
- Packaging and compliance: The experimental rig is housed in a 20 ft. container and complies with the Pressure Equipment Directive (PED) (2014/68/EU).

Safety research facilities

Transporting CO_2 also has inherent risks, in addition to the risks associated with traditional hydrocarbon fluids in pipelines. These include rapid decompression behavior and the asphyxiation properties associated with a rupture event. Published research has shown that CO_2 pipeline ruptures can be more catastrophic because the decompression rate of CO_2 within the pipeline is significantly slower when compared to natural gas [14]. Additionally, impurities in the CO_2 can increase the decompression time. Careful attention to pipe wall thickness and crack arrestor design must be executed to prevent long fractures in CO_2 pipelines, and more research is needed to understand the decompression behavior of pipelines that transport CO_2 with impurities. This includes large-scale rapid decompression tests to verify the pressure decay and temperature response of the CO_2 mixture that are used as inputs or validation for pipeline toughness and thermodynamic models.

In 2020, the only known CO_2 pipeline rupture event occurred in Satartia, Mississippi, for a pipeline operated by Denbury and was investigated by the Pipeline and Hazardous Materials Safety Administration [15]. This rupture was caused by a landslide below the pipeline support structure due to heavy rainfall in the area. Residents in close proximity to the rupture event were evacuated due to the high concentration of CO_2 in the atmosphere near the ground due to the higher density associated with CO_2 gas as compared to air vs the response when a hydrocarbon gas pipeline ruptures. This event demonstrated the need to educate emergency response personnel about the unique attributes of a CO_2 rupture event as compared to other pipeline failures. Additionally, the atmospheric dispersion of CO_2 plumes should be validated against traditional dispersion models to determine the required pipeline easements in populated areas.

Underground storage of CO_2 in deep saline aquifers has been achieved for many years by oil and gas companies. These storage sites were selected given their ability to permanently store CO_2 due to low permeability caprock and isolation from freshwater aquifers. The United States Environmental Protection Agency (EPA) has defined the permitting process for permanent sequestration in Class VI wells [16] but additional national, state, or regional requirements may be present based on the location of the storage site. More storage sites will need to be identified that are in proximity to CO_2 sources to reduce the amount of transportation energy needed to sequester CO_2. The National Energy Technology Laboratory (NETL) manages an online database called NATCARB that tracks potential CO_2 storage sites. It is important to note, though, that the industry views each storage site as unique and requires a comprehensive evaluation to ensure well integrity, subsurface isolation, and geomechanical stability. Additionally, the Bureau of Safety and Environmental Enforcement, jointly with the Bureau of Ocean Energy Management, is working with industry stakeholders to identify and disseminate regulations for the permanent storage of CO_2 in offshore wells, as NETL has identified significant potential storage capacity in coastal sedimentary formations.

As CO_2 transport becomes more commonplace, the need to carefully understand CO_2 dispersal modeling and detection becomes increasingly important, as a large-scale pipeline rupture in a dense metropolitan area may actually be substantially more deadly than a comparable hydrocarbon pipeline rupture. It is unclear what these facilities would require in terms of features; however, the test bed for CO_2 safety will include accumulated hours of operation from typical pipe loops used to study compression, pumping, and separation; however, much of this research will occur in the field as accidents occur in operation.

Natural gas compression research

While the list of natural gas research facilities is extensive, the role of these facilities has been thoroughly discussed in prior chapters. A few of these facilities merit additional discussion regarding their possible role in emerging pipeline research.

Metering research facility

The transportation of fluids, whether it be oil, natural gas, or water, is a crucial aspect of modern infrastructure. The pipeline industry is the backbone of this transportation network, enabling the safe and efficient transfer of resources over vast distances. Flowmeters, instruments that measure the rate of fluid flow within pipelines, are critical in ensuring that gas is accurately measured as it transfers ownership and is also aids in enhancing pipeline safety and leak detection. The Metering Research Facility at SwRI is a unique facility and plays a pivotal role in ensuring that flowmeters for the pipeline industry are properly calibrated.

Flowmeters are precision instruments designed to measure the volume, velocity, or mass of a fluid as it flows through a pipeline. They provide real-time data on the fluid's characteristics, helping operators make informed decisions about transportation, distribution, and allocation. Accurate flow measurements are vital for several reasons:

1. Resource management: Flowmeters enable operators to accurately quantify the amount of fluid being transported, aiding in resource allocation and management.
2. Safety: Precise measurements are crucial for detecting leaks or anomalies in the pipeline, reducing the risk of accidents and environmental damage.
3. Billing and revenue: Properly calibrated flowmeters ensure that customers are billed accurately for the resources they receive, preventing disputes and financial losses.
4. Environmental responsibility: Accurate flow measurements help minimize resource wastage, contributing to sustainable and responsible resource management.

The metering research facility performs calibration services by comparing industry equipment through a known standard. This facility is designed to allow for an array of real world conditions upstream and downstream of the flowmeter that may cause inaccuracies in the flow including multiphase behavior, elbows, restrictions, and temperature and pressure effects. This capability is essential for ensuring the accuracy of flowmeters used in diverse environments. Accurate flow measurements optimize the operation of pipelines. As the requirements of flowmeters in the industry are adapted to measure CO_2 in multiphase conditions or hydrogen, the importance of validating flow-meter technologies in relevant environments will continue to remain important.

Hydrogen pipeline R&D facilities

Hydrogen, often touted as the "fuel of the future," holds the promise of revolutionizing the energy landscape by providing a clean and efficient energy source. One of the critical challenges in harnessing hydrogen's potential is its compression. Hydrogen, in its gaseous state, possesses a low energy density, necessitating advanced compression techniques for storage and transportation. Several organizations and companies are actively engaged in hydrogen compression research to develop innovative technologies that will enable the widespread adoption of hydrogen as a clean energy carrier.

The significance of hydrogen compression

Hydrogen's unique advantages, including zero emissions when used in fuel cells, versatility, and compatibility with existing infrastructure, make it a key player in the transition to sustainable energy. However, its low volumetric energy density requires efficient compression to reduce storage and transportation costs. Hydrogen compression research facilities are instrumental in advancing compression technologies to address these challenges and make hydrogen economically viable and accessible.

Leading companies in hydrogen compression research

Numerous companies worldwide are contributing to hydrogen compression research. Some of the prominent ones include:

a. Siemens energy: Siemens Energy is a global leader in providing hydrogen solutions, including advanced compression technologies. Their expertise in mechanical and cryogenic compression systems is vital for enhancing hydrogen infrastructure.
b. Linde: Linde is known for its expertise in gas technology and engineering. They are actively involved in the development of hydrogen compression solutions, especially in the area of cryogenic hydrogen compression.
c. Hydrogenics (a subsidiary of Cummins Inc.): Hydrogenics specializes in electrolysis and hydrogen generation technologies. They are engaged in research related to hydrogen compression as part of their efforts to create end-to-end hydrogen solutions.
d. McPhy: McPhy is a company focused on solid-state hydrogen storage and purification technologies. Their work includes developing advanced materials for efficient hydrogen compression and storage.
e. Nel hydrogen: Nel Hydrogen is a global hydrogen technology company with expertise in electrochemical compression technologies. They are at the forefront of research in proton-exchange membrane (PEM) electrolysis and compression systems.
f. Air liquide: Air Liquide is actively involved in the development of hydrogen infrastructure, including compression solutions. Their research contributes to enhancing the efficiency and reliability of hydrogen compression.
g. Haskel: Haskel makes a linear hydrogen compressor that is capable of compressing hydrogen from 6 bar up to 1050 bar. The unit is either a hydraulic or electric driven piston style compressor.

Collaborative research initiatives

In addition to individual efforts, many of these companies collaborate with government agencies, research institutions, and other industry partners. Collaborations foster innovation and enable the sharing of resources, ultimately accelerating the development of technologies essential for a hydrogen-based economy. Hydrogen compression research undertaken by these companies aligns with the global energy transition toward renewable sources. As hydrogen production increasingly relies on clean methods such as electrolysis powered by renewable energy, efficient compression technologies become paramount for storing and transporting green hydrogen, reducing greenhouse gas emissions, and promoting sustainability. Any facility that is designed to study natural compression or other hydrocarbon compression can likely be adapted to study the compression of hydrogen; however, there may be some key differences that exist in the pressures and temperatures surrounding the compression process. As there are very few proven solutions for hydrogen compression at pipeline scale, these emerging technologies have not resulted in many R&D facilities for compression.

Hydrogen blending for engines

One area where hydrogen production and utilization have already been proven on an R&D/pilot scale includes the use of hydrogen in gas turbines and engines. This stems from a general trend requiring manufacturers of these pieces of equipment to show that their equipment supports the increased utilization of hydrogen to support market penetration.

Hydrogen blending has been investigated by multiple pipeline companies and equipment manufacturers to study the effects of introducing hydrogen into the natural gas fuel stream of gas engines. The importance of these studies stems from the general industry investigation into the possibility of blending hydrogen into natural gas pipelines. Many engines along the pipelines are fueled by the gas in the pipeline. If hydrogen is blended into the pipelines, there's a need to determine how that engine's new fuel will alter the engine's performance and reliability.

At the 2022 Gas Machinery Conference, there were presentations describing planned and completed testing of reciprocating engines with various percentages of hydrogen in the natural gas fuel lines. Waukesha (an INNIO Group company) and National Fuel Gas presented their plans to test hydrogen blending in the fuel gas of a Waukesha 7044GSI S5 engine. While discussing the testing plans, it was noted that Jenbacher (also an INNIO Group company) lean burn engines are capable of 25% hydrogen blends, and Jenbacher makes an engine capable of operating with a 100% hydrogen fuel source. Cooper Machinery Services also has published work with Southern Star and Williams on their testing of hydrogen blending in the fuel gas of two different Cooper legacy engines that have been installed and operating with a natural gas fuel source for many years [1].

Cooper Machinery Services worked with Southern Star and Williams to test the performance of two two-stroke lean-burn reciprocating engines when blending hydrogen into the fuel gas stream, which is traditionally fueled by natural gas. One engine is a GMVH-12 Cooper Bessemer engine (2400 hp), and the other is a Cooper 6V-250 (2000 hp). A hydrogen source was brought to the site of the compressors, which were Hugoton, KS, for the GMVH-12 and Kemmerer, WY, for the 6V-250, and a blending skid was necessary to accurately blend the hydrogen into the natural gas. The test setup for the Hugoton, KS, testing is shown in Fig. 12.11. Multiple emissions trends were analyzed during the successful testing of a fuel blend of up to 30% hydrogen. It was concluded from the testing that two-stroke lean-burn engines require minor modification to be able to handle operating on hydrogen and natural gas blends of up to 30% hydrogen. As plotted in Fig. 12.12, all measured emissions were found to decrease, while only the engine emissions of oxides of nitrogen (NOx) increased. It was found that NOx emissions can be controlled by either leaning the combustion mixture or delaying the ignition timing. Combustion performance criteria were also measured,

FIG. 12.11 Field Site Test Setup for Hydrogen Blending into Two-Stroke Lean-Burn Engine [1].

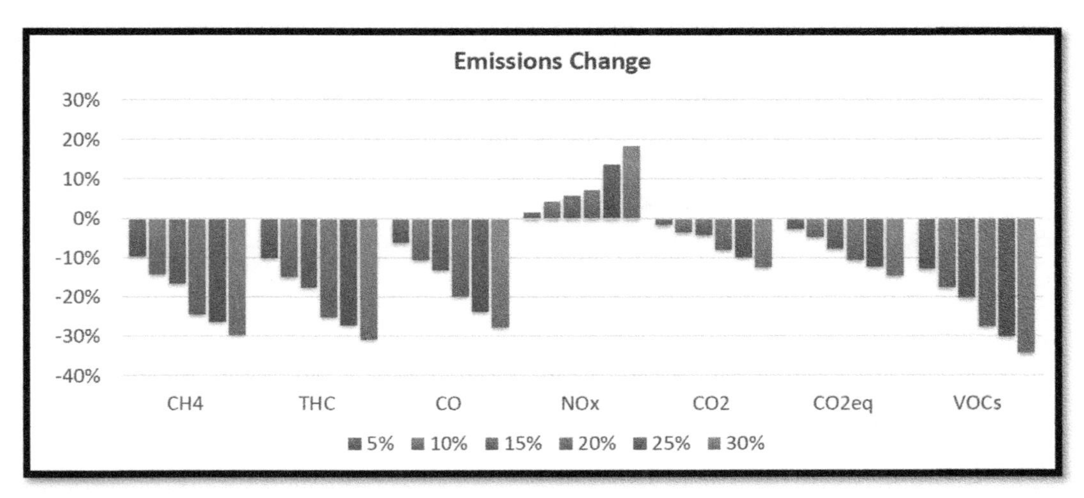

FIG. 12.12 Field site test results—emissions measurements [1].

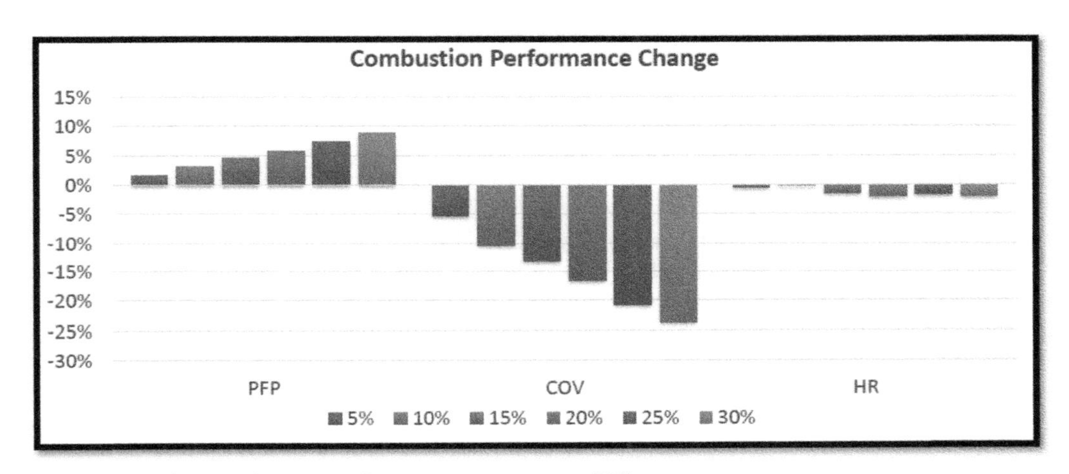

FIG. 12.13 Field site test results—combustion performance measurements [17].

which is shown in Fig. 12.13. Combustion stability was observed to improve as the percentage of hydrogen increased. Overall, combustion and emissions characteristics were found to be favorable to the addition of hydrogen. As noted previously, some minor modifications were necessary and/or recommended. As the percentage of hydrogen increased, typical orifice flow meters had an error of up to 13% compared to Coriolis meters, which would require orifice flow meters to be replaced with a more accurate flow meter when introducing hydrogen into the fuel gas of an engine. A turbocharger upgrade and the addition of an auto balance are recommended to handle the air flow increase that would be needed to maintain the baseline NOx levels. With the implementation of relatively minor changes, it was concluded that the addition of hydrogen up to 30% generally improves the engine emissions and performance [1].

Siemens has also noted testing of hydrogen and natural gas blended fuels on a two-stroke, precombustion chamber (PCC) test engine. The goals of the two test phases were to investigate the impact that adding hydrogen to the fuel line has on peak cylinder pressure standard deviation, peak cylinder pressure coefficient of variation (PP COV), brake specific fuel consumption, NOx emissions, carbon dioxide (CO_2), carbon monoxide (CO), and total hydrocarbon (THC) emissions. Phase one accomplished the demonstration that the PCCs could be operated with hydrogen/natural gas blends and pure hydrogen. When utilizing hydrogen as a prechamber fuel, improvements were observed in overall engine performance with a decrease in brake specific fuel consumption (BSFC), and the authors concluded that a leaner air-fuel ratio may be required to avoid preignition. Phase two evaluated increasing percentages of hydrogen blended with natural gas to fuel the main combustion chamber. Siemens observed no stall until the volumetric concentration of hydrogen reached approximately 93% when making no changes to the ignition timing throughout the test. As

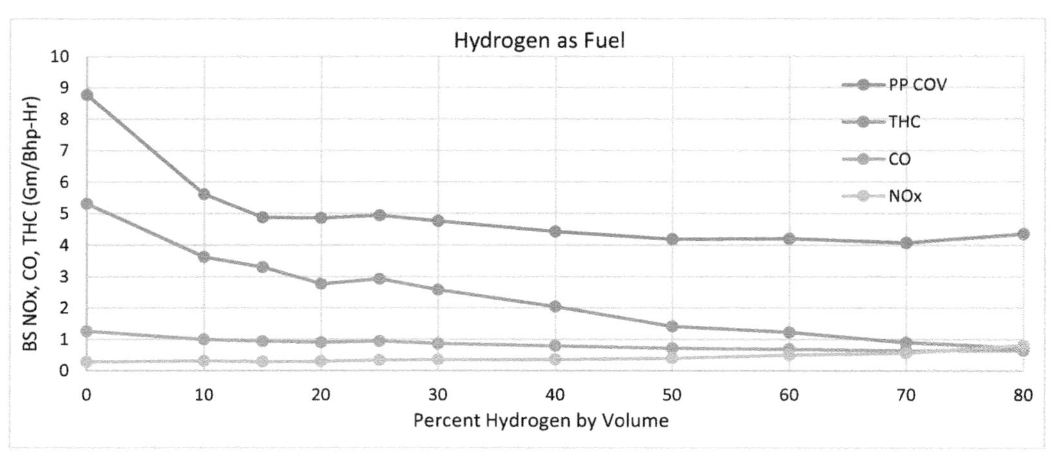

FIG. 12.14 Hydrogen as main combustion chamber fuel feasibility study [2].

hydrogen percentages increased, ignition timing retardation was necessary due to the higher burning velocity of hydrogen relative to natural gas. A 34% reduction in BSFC was measured after the percent of hydrogen was increased to 80%. As hydrogen percentages increased, emissions measurements summarized in Fig. 12.14 showed the following general trends: PP COV decreased, THCs decreased, CO levels decreased, and NOx emissions increased [18].

Wärtsilä, a manufacturer of large engines for the natural gas pipeline industry, is using their engine laboratory in Vaasa, Finland, to test their engines with up to 100% hydrogen fuel. "Company leaders have expressed hopes for an engine and power plant operating on pure hydrogen by 2025" [19].

As the potential for hydrogen blending into natural gas pipelines seems to become more probable, the need to understand hydrogen blended with natural gas as an engine fuel source becomes more important. It was found during testing performed by multiple engine manufacturers and pipeline operator companies that the blending of hydrogen and natural gas as a fuel gas for large, industrial engines generally results in improved engine performance and emissions. The need for system modifications to accommodate the addition of hydrogen was typically found to be relatively minor.

Fluid property testing

One area of research that impacts all areas of compression, transport, and operation are fluid properties. Though several facilities exist, these facilities are instrumental in ensuring that combinations of fluids and their behavior can be accurately modeled. Facilities used to characterize fundamental fluid property measurements for parameters like speed of sound, viscosity, density, and more are often operated by various organizations and institutions, including:

a. Research Institutions and Universities: Many universities and research institutions have dedicated laboratories and facilities for fluid property measurements. These facilities are often used for both academic research and collaborative projects with industry partners.
b. National Laboratories: National laboratories, such as the Lawrence Berkeley National Laboratory or the Oak Ridge National Laboratory in the United States, often house specialized research facilities for fluid property measurements. These laboratories conduct cutting-edge research and offer services to the scientific community and industry.
c. Private Research Organizations: Private research organizations, particularly those specializing in fluid dynamics, thermodynamics, and related fields, operate facilities for fluid property measurements. These organizations may offer testing services to companies and government agencies.
d. Industrial Research and Development Centers: Many large corporations, especially those in industries where precise fluid properties are crucial (e.g., petrochemical, aerospace, automotive), operate their research and development centers with specialized facilities for fluid property characterization.
e. Government Agencies: Government agencies involved in research and regulation, such as the National Institute of Standards and Technology (NIST) in the United States, often maintain laboratories for fluid property measurements. These facilities play a critical role in setting and enforcing industry standards.

f. Collaborative Research Facilities: Some facilities are operated jointly by multiple organizations, including government agencies, universities, and private companies. These collaborative research centers focus on specific areas of fluid dynamics and property measurement.

The specific organization operating a facility for fluid property measurements can vary depending on the nature of the research, the location, and the available resources. Researchers and engineers often collaborate with these facilities to conduct experiments, validate models, and gain a deeper understanding of fluid behavior, which is essential in various industries whether designing a pipeline, a compressor, or valve. Ridens et al. [20] notes the following: The primary objective of the fluid property research is to characterize representative gas mixtures, conduct testing, perform uncertainty analysis, and compare equation of state (EOS) predictions in the context of supercritical carbon dioxide (CO_2) compression applications. These equations of state are then used to design equipment such as compressors and gas turbines and this hold true for mixtures containing CO_2 with impurities, hydrocarbon and hydrogen mixtures, and the like. Ridens et al. emphasize the increasing interest in characterizing fluid property mixtures of supercritical CO_2 due to escalating greenhouse gas emission regulations and its relevance to various power plant applications. Furthermore, it highlighted the scarcity of publicly available data for verifying EOS calculations, particularly in the range of pressures, temperatures, and multi-species gas compositions relevant to compression and pipeline operations.

Conclusion

Research and development facilities are integral to supporting pipeline research for CO_2, methane, and hydrogen. Through driving technological advancements, addressing safety and environmental concerns, developing best practices, and enabling knowledge exchange, these facilities pave the way for efficient, safe, and sustainable pipeline systems that are essential for achieving global energy and climate goals. As we look to a future reliant on clean energy carriers, R&D facilities are indispensable in shaping the trajectory of pipeline transport and ushering in a more sustainable energy landscape.

References

[1] L. Rapp, Experimental Testing of a 1MW sCO2 Turbocompressor, in: Presented at the 7th International Supercritical CO2 Power Cycles Symposium, San Antonio,Texas, 2019. Apr.

[2] M.D. Carlson, Guidelines for the design and operation of supercritical carbon dioxide R&D systems, in: Presented at the SOLARPACES 2019: International Conference on Concentrating Solar Power and Chemical Energy Systems, Daegu, South Korea, 2020, p. 130003, https://doi.org/10.1063/5.0033262].

[3] E.M. Clementoni, T.L. Cox, Practical aspects of supercritical carbon dioxide Brayton system testing, in: Presented at the 4th International Symposium - Supercritical CO2 Power Cycles, Pittsburgh, Pennsylvania, 2014. Sep.

[4] A. Yu, W. Su, X. Lin, N. Zhou, Recent trends of supercritical CO2 Brayton cycle: bibliometric analysis and research review, Nucl. Eng. Technol. 53 (3) (2021) 699–714, https://doi.org/10.1016/j.net.2020.08.005.

[5] Wu, et al., Thermo-solar-driven sCO2 Brayton cycle [CSIRO], in: Solar-driven Supercritical CO2 Brayton Cycle (1-UFA004) Project results and lessons learnt, 2017. May 30. [Online]. Available http://www.csiro.au/energy.

[6] V. Illyes, S. Thaheise, P. Schwarzmayr, P.-L. David, X. Guerif, A. Werner, M. Haider, SCO2 test facility at TU Wien: design, operation and resutls, in: 5th European sCO2 Conference for Energy Systems, Prague, Czech Republic, 2023, https://doi.org/10.17185/duepublico/77261. March 14-16.

[7] https://doi.org/10.1016/j.ijmultiphaseflow.2021.103590.

[8] https://doi.org/10.1016/j.ijmultiphaseflow.2019.103139.

[9] https://www.sintef.no/en/latest-news/2023/sintef-to-use-its-world-class-co2-flow-facility-and-multiphase-modelling-expertise-to-develop-a-unique-simulator-for-co2-transport-and-injection/.

[10] J. Moore, T. Allison, N. Evans, J. Kerth, J. Pacheco, Development and testing of multi-stage internally cooled centrifugal compressor, in: Proceedings 44th Turbomachinery and 31st Pump Symposium, Houston, TX, Sept. 14–17, 2015.

[11] J. Richardson, R. Wardell, E. Fernandez, J. Kapat, Experimental and computational heat transfer study of sCO2 single jet impingement, ASME TurboExpo Paper (2023) (GT2023-102544).

[12] J. Fosheim, X. Hernandez, W. Arthur-Arhin, A. Thompson, C. Bowen, K. Albrecht, G. Jackson, Design of a 40-kWth Counterflow particle-supercritical carbon dioxide narrow-channel fluidized bed heat exchanger, AIP Conference Proceedings 2815 (2023) 160005. https://doi.org/10.1063/5.0165673.

[13] G. Bianchi, S.S. Saravi, R. Loeb, K.M. Tsamos, M. Marchionni, A. Leroux, S.A. Tassou, Design of a high-temperature heat to power conversion facility for testing supercritical CO2 equipment and packaged power units, Energy Procedia 161 (2019) 421–428, https://doi.org/10.1016/j.egypro.2019.02.109.].

[14] https://doi.org/10.1115/1.4034016.

[15] https://www.phmsa.dot.gov/sites/phmsa.dot.gov/files/2022-05/Failure%20Investigation%20Report%20-%20Denbury%20Gulf%20Coast%20Pipeline.pdf.

[16] https://www.epa.gov/uic/class-vi-wells-used-geologic-sequestration-carbon-dioxide#well_def.
[17] N. Alkadi, M. Toema, G. Blandford, H. Mathews, J. Saletsky, T. Kreuz, J. Penepent, Hydrogen testing in large bore slow speed integral engine, pushing the limits, in: Presented at the 2022 Gas Machinery Research Council (GMRC) Gas Machinery Conference, Fort Worth, TX, 2022. October 3.
[18] E. Glover, C. Adams, B. Castor, Hydrogen as Main Combustion Chamber Fuel Feasibility Study, Siemens Energy Presentation at the Engine Analyzer Reliability Workshop at SwRI, San Antonio, TX, 2022.
[19] https://www.energytech.com/energy-efficiency/article/21254445/important-hydrogen-blending-test-completed-at-michigan-gasfired-power-plant-using-wrtsil-engine.
[20] B. Ridens, K. Brun, High-pressure thermophysical gas property testing, uncertainty analysis, and equation of state comparison for supercritical CO_2 compression applications, in: Proc. Supercritical CO_2 Power Cycle Symposium, 2014. Sept. 9, San Antonio, TX.

Index

Note: Page numbers followed by *f* indicate figures and *t* indicate tables.

D

E

Printed and bound by CPI Group (UK) Ltd, Croydon, CR0 4YY
02/12/2024
01798486-0001